NATO ASI Series

Advanced Science Institutes Series

A series presenting the results of activities sponsored by the NATO Science Committee, which aims at the dissemination of advanced scientific and technological knowledge, with a view to strengthening links between scientific communities.

The Series is published by an international board of publishers in conjunction with the NATO Scientific Affairs Division

A Life Sciences B Physics	Plenum Publishing Corporation London and New York
C Mathematical and Physical Sciences D Behavioural and Social Sciences E Applied Sciences	Kluwer Academic Publishers Dordrecht, Boston and London
F Computer and Systems Sciences G Ecological Sciences H Cell Biology	Springer-Verlag Berlin Heidelberg New York London Paris Tokyo Hong Kong Barcelona

Series H: Cell Biology Vol. 45

The ASI Series Books Published as a Result of Activities of the Special Programme on CELL TO CELL SIGNALS IN PLANTS AND ANIMALS

This book contains the proceedings of a NATO Advanched Research Workshop held within the activities of the NATO Special Programme on Cell to Cell Signals in Plants and Animals, running from 1984 to 1989 under the auspices of the NATO Science Committee.

The books published as a result of the activities of the Special Programme are:

Vol. 1: **Biology and Molecular Biology of Plant-Pathogen Interactions.** Edited by J.A. Bailey. 1986.
Vol. 2: **Glial-Neuronal Communication in Development and Regeneration.** Edited by H.H. Althaus and W. Seifert. 1987.
Vol. 3: **Nicotinic Acetylcholine Receptor: Structure and Function.** Edited by A. Maelicke. 1986.
Vol. 4: **Recognition in Microbe-Plant Symbiotic and Pathogenic Interactions.** Edited by B. Lugtenberg. 1986.
Vol. 5: **Mesenchymal-Epithelial Interactions in Neural Development.** Edited by J.R. Wolff, J. Sievers, and M. Berry. 1987.
Vol. 6: **Molecular Mechanisms of Desensitization to Signal Molecules.** Edited by T.M. Konjin, P.J.M. Van Haastert, H. Van der Starre, H. Van der Wel, and M.D. Houslay. 1987.
Vol. 7: **Gangliosides and Modulation of Neuronal Functions.** Edited by H. Rahmann. 1987.
Vol. 9: **Modification of Cell to Cell Signals During Normal and Pathological Aging.** Edited By S. Govoni and F. Battaini. 1987.
Vol. 10: **Plant Hormone Receptors.** Edited by D. Klämbt. 1987.
Vol. 11: **Host-Parasite Cellular and Molecular Interactions in Protozoal Infections.** Edited by K.-P. Chang and D. Snary. 1987.
Vol. 12: **The Cell Surface in Signal Transduction.** Edited by E. Wagner, H. Greppin, and B. Millet. 1987.
Vol. 19: **Modulation of Synaptic Transmission and Plasticity in Nervous Systems.** Edited by G. Hertting and H.-C. Spatz. 1988.
Vol. 20: **Amino Acid Availability and Brain Function in Health and Disease.** Edited by G. Huether. 1988.
Vol. 21: **Cellular and Molecular Basis of Synaptic Transmission.** Edited by H. Zimmermann. 1988.
Vol. 23: **The Semiotics of Cellular Communication in the Immune System.** Edited by E.E. Sercarz, F. Celada, N.A. Mitchison, and T. Tada. 1988.
Vol. 24: **Bacteria, Complement and the Phagocytic Cell.** Edited by F.C. Cabello and C. Pruzzo. 1988.
Vol. 25: **Nicotinic Acetylcholine Receptors in the Nervous System.** Edited by F. Celementi, C. Gotti, and E. Sher. 1988.
Vol. 26: **Cell to Cell Signals in Mammalian Development.** Edited by S.W. de Laat, J.G. Bluemink, and C.L. Mummery. 1989.
Vol. 27: **Phytotoxins and Plant Pathogenesis.** Edited by A. Graniti, R.D. Durbin, and A. Ballio. 1989.
Vol. 31: **Neurobiology of the Inner Retina.** Edited by R. Weiler and N.N. Osborne. 1989.
Vol. 32: **Molecular Biology of Neuroreceptors and Ion Channels.** Edited by A. Maelicke. 1989.
Vol. 33: **Regulatory Mechanisms of Neuron to Vessel Communication in the Brain.** Edited by F. Battaini, S. Govoni, M.S. Magnoni, and M. Trabucchi. 1989.
Vol. 35: **Cell Separation in Plants: Physiology, Biochemistry and Molecular Biology.** Edited by D.J. Osborne and M.B. Jackson. 1989.
Vol. 36: **Signal Molecules in Plants and Plant-Microbe Interactions.** Edited by B.J.J. Lugtenberg. 1989.
Vol. 39: **Chemosensory Information Processing.** Edited by D. Schild. 1990.
Vol. 41: **Recognition and Response in Plant-Virus Interactions.** Edited by R.S.S. Fraser. 1990.
Vol. 43: **Cellular and Molecular Biology of Myelination.** Edited by G. Jeserich, H. H. Althaus, and T. V. Waehneldt. 1990.
Vol. 44: **Activation and Desensitization of Transducing Pathways.** Edited by T. M. Konijn, M. D. Houslay, and P. J. M. Van Haastert. 1990.
Vol. 45: **Mechanism of Fertilization: Plants to Humans.** Edited by B. Dale. 1990.

Mechanism of Fertilization: Plants to Humans

Edited by

Brian Dale

Stazione Zoologica, Villa Comunale, 80121 Naples, Italy

Springer-Verlag Berlin Heidelberg New York
London Paris Tokyo Hong Kong Barcelona
Published in cooperation with NATO Scientific Affairs Division

Proceedings of the NATO Advanced Research Workshop on Mechanism of Fertilization: Plants to Humans held at Sorrento, Italy, 1–5 October, 1989

ISBN-13: 978-3-642-83967-2 e-ISBN-13: 978-3-642-83965-8
DOI: 10.1007/978-3-642-83965-8

Library of Congress Cataloging-in-Publication Data
NATO Advanced Research Workshop on Mechanism of Fertilization: Plants to Humans (1989: Sorrento, Italy)
Mechanism of fertilization—plants to humans/edited by Brian Dale. p. cm.—(NATO ASI series. Series H, Cell biology; vol. 45)
"Published in cooperation with NATO Scientific Affairs Division." Includes bibliographical references.

1. Fertilization (Biology)—Congresses. I. Dale, Brian. II. North Atlantic Organization. Scientific Affairs Division. III. Title. IV. Series.
QP273.N33 1989 574.1'66—dc20 90-9873

Softcover reprint of the hardcover 1st edition 1990

2131/3140-543210 – Printed on acid-free-paper

Dedicated to the memory of Alberto Monroy

born, 26th July 1913, Palermo, Italy.

ORGANIZING COMMITTEE

Brian Dale
Stazione Zoologica, Naples, Italy

Colin Brownlee
Marine Biological Association, Plymouth, UK

Jacques Cohen
Emory University, Atlanta, GA, USA

Louis J. De Felice
Emory University, Atlanta, GA, USA

David Epel
Hopkins Marine Station, Pacific Grove, CA, USA

ACKNOWLEDGEMENTS

The Director and Organizing Committee wish to thank the following for their collaboration and financial support.

Prof. Gaetano Salvatore, President of the Stazione Zoologica, Naples, Italy

The Ares Serono Group

Commission of the European Communities

L'Istituto Italiano per gli Studi Filosofici

International Society of Developmental Biologists

European Developmental Biology Organization

The British Council

Carl Zeiss, SpA

Reproductive Biology Associates, Atlanta, GA, USA

Anatomy and Cell Biology Department, Emory University

Fertility Center srl, Naples, Italy

Stazione Zoologica di Napoli

Regione Campania

Consiglio Nazionale delle Ricerche

Contributors

Dr John AITKEN - MRC Reproductive Biology, 37 Chalmers Street, Edinburgh EH3 9EW /U.K.

Dr Anthony BELLVE' - Dept Anatomy and Cell Biology, College of Physicians and Surgeons, Columbia University, 650 West 168th St., New York 10032 / U.S.A.

Dr S.H. BRAWLEY - Dept of General Biology, Vanderbilt University, Nashville, Tennessee 37235 / U.S.A.

Dr C. BROWNLEE - Marine Biological Association, The Laboratory, Citadel Hill, Plymouth PL1 2PB / U.K.

Prof. J. CALLOW - School of Biological Sciences, University of Birmingham, P.O. Box 363, Birmingham B15 2TT /U.K.

Prof. Chiara CAMPANELLA - Dept of Evolutionary Biology, University of Naples, Via Mezzocannone 9, Napoli / Italy

Dr Michel CHARBONNEAU - Laboratoire de Biologie et Génétique du Développement, Université de Rennes 1, Campus de Beaulieu, 35042 Rennes Cédex / France

Dr Jacques COHEN - The Center for Reproductive Medicine and Infertility, Cornell University Medical Center, 505 Est 70 Street, New York, 10021, U.S.A.

Dr Pierre COLAS - Ecole Normale Superieure de Lyon, Laboratoire de Biologie Moleculaire et Cellulaire, 69634 Lyon Cedex 07 / France

Dr D.G. CRAN - Institute of Animal Physiology and Genetics Research, Babraham Hall, Cambridge CB2 4AT / U.K.

Prof. M. CRESTI - Dipartimento di Biologia Ambientale, Universitá di Siena, Via Mattioli 4, Siena / Italy

Dr Brian DALE - Stazione Zoologica di Napoli, Villa Comunale, 80121 Napoli / Italy

Prof. Eric H. DAVIDSON - Division of Biology 156-29, California Institute of Technology, Pasadena, California 91125 / U.S.A.

Prof. L.J. DE FELICE - Dept Anatomy and Cell Biology, Emory University, Atlanta, Ga 30322 / U.S.A.

Dr Rosaria DE SANTIS - Stazione Zoologica, Villa Comunale, 80121 Napoli / Italy

Prof. H. G. DICKINSON - Dept of Botany, Plant Science Laboratories, University of Reading, Whiteknights, P.O.Box 221, Reading RG6 2AS / U.K.

Prof. C. DUMAS - Université Claude Bernard, Lyons 1, RCAP Bal. 741 43, Bd du 11 Nov. 1918 / France

Prof. Richard ELINSON - Dept of Zoology, Ramsay Wright, Zoological Laboratoire University of Toronto, 25 Harbord Street, Toronto / Canada M5s 1A1

Prof. David EPEL - Hopkins Marine Station, Stanford University, Dept of Biological Sciences, Pacific Grove, California 93950-3094 / U.S.A.

Dr Tom FLEMING - Dept of Biology Medical and Biological Sciences Building, Basset Crescent East, Southampton SO9 3TU / U.K.

Prof. Raffaele GEREMIA - Dipartimento di Sanitá Pubblica e Biologia Cellulare, II Universitá di Roma "Torvergata", Via Orazio Raimondo 8, 00173 Roma / Italy

Prof. Giovanni GIUDICE - Istituto di Biologia dello Sviluppo, CNR, Via Archirafi 20, 90123 Palermo / Italy

Dr Roger GOSDEN - Dept of Physiology, University Medical School, Teviot Place, Edinburgh EH8 9AG / U.K.

Dr. G. GREEN - School of Biological Science, University of Birmingham, P.O. Box 363, Birmingham B15 2TT / U.K.

Prof. Pierre GUERRIER - Ecole Normale Superieure de Lyon, Laboratoire de Biologie Moleculaire et Cellulaire, 46 Allèe d'Italie, 69364 Lyon Cedex 07 / France

Prof. Ralph B.L. GWATKIN - Reproductive and Developmental Biology, The Cleveland Clinic Foundation, 9500 Euclid Avenue, Cleveland, Ohio 44106 / U.S.A.

Dr A.M. HETHERINGTON - Dept of Biological Sciences, The University, Lancaster, LA1 4A1 / U.K.

Prof. Motonori HOSHI, Dept of Life Science, Faculty of Science, Tokyo Institute of Technology / Japan

Dr L.F. JAFFE - Marine Biological Laboratory, Woods Hole, Ma 02643 / U.S.A.

Prof. W. JEFFEREY - Dept of Zoology, University of Texas, Austin, Texas 78712-1064 / U.S.A.

Dr Martin JOHNSON - Dept of Anatomy, University of Cambridge, Downing Street, Cambridge CB2 3DY / U.K.

Dr Douglas KLINE - Dept of Physiology, Ponce School of Medicine, P.O.Box 7004, Ponce, Puerto Rico 00732 /U.S.A.

Dr D. LAURIE - Institute for Plant Science Research, Maris Lane, Trumpington Cambridge CB2 2JB / U.K.

Prof. Frank J. LONGO - Dept of Anatomy, University of Iowa, Iowa City, Iowa 52242 / U.S.A.

Prof. Yoshio MASUI - Dept of Zoology, Ramsay Wright Zoological Labs, University of Toronto, 25 Harbord Street, Toronto, Ontario / Canada

Dr Masaaki MORISAWA - Misaki Marine Biological Station, University of Tokyo, 1024 Koajiro, Misaki cho, Miura / Japan

Dr J. MURNAME - Dept Anatomy and Cell Biology, Emory University, Atlanta, Ga 30322 / U.S.A.

Prof. Eizo NAKANO - Associazione Biologica Italo Giapponese, Nijigakoa 1-7, Meito, Nagoya 465 / Japan

Dr Isabelle NEANT - Ecole Normale Superieure de Lyon, Laboratoire de Biologie Moleculaire et Cellulaire, 69364 Lyon Cedex 07 / France

Dr S.C. NG - Centre for Reproductive Endocrinology, Dept of Obstetrics and Gynaecology, National University Hospital, Lower Kent Ridge Road, Singapore 0511

Dr G. PEAUCELLIER - INSERM U-249, CNRS-CRBM, Faculté de Pharmacie, BP-5015 34033, Montpellier Cedex / France

Dr Rosaria PINTO - Stazione Zoologica, Villa Comunale, 80121 Napoli / Italy

Prof. Floriana ROSATI - Dipartimento di Biologia Evolutiva, Universitá di Siena, Via Mattioli 4, 53100 Siena / Italy

Dr S.D. RUSSELL - Dept of Botany and Microbiology, University of Oklahoma, Norman, Oklahoma 73019 / U.S.A.

Prof. Dr Klaus SANDER - Institut fur Biologie 1 (Zoologie), Albert-Ludwigs Universitat, Albertstrasse 21a, D-7800 Freiburg / Germany

Dr L. SANTELLA - Stazione Zoologica di Napoli, Villa Comunale, 80121 Napoli / Italy

Dr Christian SARDET - Biologie Cellulaire Marine, URA 5 CNRS, Station Zoologique, Villefranche sur Mer / France 06230

Dr Henry SATHANANTHAN - Centre for Early Human Development, Monash Medical Centre, 246 Clayton Road, Clayton, Melbourne Victoria / 3168 Australia

Prof. Allen SCHUETZ - Dept Population Dynamics, John Hopkins School Hygiene and Public Health, 615 North Wolfe St. Baltimore, Md 21205 / U.S.A.

Dr Mario SOUZA - Departamento de Biologia Cellular, Instituto de Ciencias Biomedicas de Abel Salazar, Universidade do Porto, 4000 Porto / Portugal

Dr Annalisa SPEKSNIJDER - Zoologisch Laboratorium Rijksuniversiteit, Padualaan 8, 3584 CH Utrecht / The Netherlands

Prof. Norio SUZUKI - Noto Marine Laboratory Ogi Uchiura, Ishikava-Ken 927-05 / Japan

Dr Riccardo TALEVI - Dept of Evolutionary Biology, University of Naples, Via Mezzocannone), Napoli / Italy

Prof. Paul WASSARMAN - Dept of Cell and Developmental Biology, Roche Institute of Molecular Biology, Kingsland Street, Nutley, New Jersey 07110-1199 / U.S.A.

Dr Michael WHITAKER - Dept of Physiology, University College, Gower St. London WC1E 6BT / U.K.

Dr R. YANAGIMACHI - Dept of Anatomy and Reproductive Biology, University of Hawaii, Honolulu / Hawaii 96822

CONTENTS

Dedication to Alberto Monroy
Organizing Committee
Acknowledgements
Contributors

General Introduction 1
Brian Dale

LATE STAGES IN GAMETOGENESIS

Chapter 1
Isolation and characterization of the angiosperm gamete 3
Scott D. Russell

Chapter 2
The Cytoskeleton during pollen tube growth and sperm cell formation 17
A.Tiezzi and M.Cresti

Chapter 3
The Cytostatic Factor (CSF) that causes metaphase arrest in amphibian eggs 35
Y.Masui

Chapter 4
Intrafollicular mechanisms mediating oocyte maturation 45
A.Schuetz

Chapter 5
Tyrosine phosphorylation of MPF and membrane proteins during meiotic maturation of starfish oocytes 65
G.Peaucellier, A.Anderson, W.Kinsey and M.Dorée

Chapter 6
Protein synthesis and protein phosphorylation as regulators of MPF activity 79
P.Guerrier, I.Néant, P.Colas, L.Dufresne, J.Saint Pierre, F.Dubé

Chapter 7
Control of recruitment of preantral follicles in mammalian ovaries 101
R.Gosden

Chapter 8
Ultrastructure and AG staining of echinoderm spermatogenesis 115
M.Sousa and C.Azevedo

Chapter 9
Cell specific gene expression in mouse testis 131
R.Geremia, V.Sorrentino, M.Giorgi, P.Grimaldi, I. Sammarco and P.Rossi

Chapter 10
Regulation of sperm motility by osmotic pressure 143
M.Morisawa, T.Inoda and S.Oda

Chapter 11
Dynamic changes in the perinuclear matrix during spermiogenesis and sperm maturation in the mouse 155
A.Bellvé, R.Chandrika and A.Barth

Chapter 12
Reactive oxygen species and human sperm function 179
R.Aitken

SPERM EGG INTERACTION

Chapter 13
Fertilization in Fucus: exploring the gamete cell surfaces with monoclonal antibodies 189
J.Green, J.Jones, C.Stafford and J.Callow

Chapter 14
Gamete interaction and the initiation of egg activation in sea urchins 203
F.Longo, S.Cook, D.McCulloh, P.Ivonnet and E.Chambers

Chapter 15
Self-incompatibility in Brassica: The nature and role of female glycoproteins 219
H.Dickinson and M.Crabbe

Chapter 16
Induction of the acrosome reaction in starfish 239
M.Hoshi, T.Amano, Y.Okita, T.Okinaga and T.Matsui

Chapter 17
New data and concepts in angiosperm fertilization 253
C.Guilluy, T.Gaude, C.Digonnet-Kerhoas, A.Chaboud, P.Heizmann and C.Dumas

Chapter 18
Structure and function of egg-associated peptides of sea urchins 271
N.Suzuki

Chapter 19
Sperm-egg interaction in bivalves 287
F.Rosati and R.Focarelli

Chapter 20
Gamete interaction in ascidians: sperm binding and penetration through the vitelline coat 297
R.DeSantis and R.Pinto

Chapter 21
Cellular and molecular elements of mammalian fertilization 305
P.Wassarman

Chapter 22
Micromanipulation in the study of sperm-egg interactions 315
R.Gwatkin

Chapter 23
Human sperm-oocyte fusion 329
A.Sathananthan, S.Ng, A.Trounson, S.Ratnam and T.Bongso

Chapter 24
Micro-insemination sperm transfer (M.I.S.T.) into human oocytes and embryos 351
S.Ng, A.Bongso, A.Sathananthan and S.Ratnam

Chapter 25
Abnormalities of the human embryonic zona pellucida 377
J.Cohen

EGG ACTIVATION

Chapter 26
The roles of intermembrane calcium in polarizing and activating eggs 389
L.Jaffe

Chapter 27
Polyspermy blocks in fucoid algae and the occurence of polyspermy in Nature 419
S.Brawley

Chapter 28
How does a sperm activate a sea urchin egg? 433
M.Whitaker and I.Crossley

Chapter 29
The behaviour of sperm cells in cereal wide hybrids 445
D.Laurie and M.Bennett

Chapter 30
Calcium signals during fertilization and ooplasmic segregation in the ascidian egg 455
J.Speksnijder, C.Sardet and L.Jaffe

Chapter 31
Soluble sperm factors, electrical events and egg activation 475
B.Dale and L.DeFelice

Chapter 32
Can the handling of oocytes influence the success of therapeutic in vitro fertilization? 489
M.Johnson, P.Braude, C.Vincent and S.Pickering

Chapter 33
Is the egg activation-induced intracellular pH increase necessary for the embryonic development of Xenopus laevis (Anuran Amphibian)? 503
N.Grandin and M.Charbonneau

Chapter 34
The distribution and exocytosis of cortical granules in the mammal 519
D.Cran

Chapter 35
Receptors, G-proteins and activation of the amphibian egg 529
D.Kline, R.Kado, G.Kopf and L.Jaffe

Chapter 36
Electrical maturation of the mouse oocyte 543
J.Murnane and L.DeFelice

Chapter 37
The contribution of _Discoglossus pictus_ fertilization in the study of amphibian sperm-egg interaction 557
C.Campanella, R.Talevi, R.Gualtieri and P.Andreuccetti

CYTOPLASMIC SEGREGATION IN THE EARLY EMBRYO

Chapter 38
Intracellular signalling during fertilization and polarization in Fucoid algae 579
C.Brownlee

Chapter 39
Mechanism of dorsoventral axis determination in the ascidian embryo 591
W.Jeffery, B.Swalla, and J.Venuti

Chapter 40
The insect oocyte: fertilization, activation and cytoplasmic dynamics 605
K.Sander

Chapter 41
Fertilization in Ctenophores 625
D.Carre, C.Sardet and C.Rouviere

Chapter 42
The sea urchin egg: an old and new model 637
G.Giudice

Chapter 43
Collagen-binding proteins in sea urchin eggs and embryos 645
E.Nakano, M.Iwata and V.Matranga

Chapter 44
Evidence which supports the presence of stored messenger ribonucleoprotein (mRNP) in the unfertilized eggs of _Fucus serratus_ 653
A.Hetherington, J.Sommerville, A.Masters and A.Mitchell

Chapter 45
Dorsal specification in the fertilized frog egg 663
R.Elinson

Chapter 46
The cytoskeleton and polyspermy in sea urchin eggs 673
L.Santella

Chapter 47
Cell differentiation in the mouse preimplantation embryo 679
T.Fleming

Chapter 48
Fertilization and sex determination in the Rhizocephala (Cirripedia, Crustacea) 699
R.Yanagimachi

INTRODUCTION

The majority of scientists interested in fertilization and early developmental processes will undoubtably have encountered the works of Alberto Monroy at some time in their careers. Alberto's contribution to this field spans oogenesis to embryogenesis, where he used physiological, biochemical and morphological tools to answer a number of basic problems in cell biology. This multi-disciplinary approach, together with his remarkable intellectual flexibility and humour has had an enormous impact on this field and all those fortunate enough to have worked with him.

The chapters in this book have been divided into four sections. The initial presentations revolve around late events of gameteogenesis, that lead to a physiologically mature gamete. Probably the most exciting area for research at the moment is the identification of the cytoplasmic mechanisms responsible for the meiotic arrest of oocytes and the factors responsible for initiating their maturation (Chapters 3 and 4). Less is known about the physiological changes in the male gamete in preparation for fertilization and this may be identified as a major area for future research. Although comparable data for the plant kingdom is presently restricted to studies on marine algae, new techniques for isolating angiosperm gametes (Chapters 1 and 17) promise rapid advances in this field. The second section looks at the events and molecules involved in gamete recognition, binding and fusion. One of the most controversial topics is when does sperm-egg fusion actually occur (Chapter 14). Surface receptors seem to be comparable in gametes of algae, molluscs, ascidians and mammals, and their characterization is well under way.

There are two current ideas as to how a spermatozoon triggers the egg into metabolic activity, and these form the central theme for the third section. The first favours an externally located receptor and G-protein trans-membrane transduction mechanism (Chapter 35), and a second school points to a trigger factor in the spermatozoon that is released into the egg cytoplasm following gamete fusion (Chapters 26,28 and 31). Subsequent presentations look at the second messenger systems in the egg that translate the primary signal into a global cellular response. The last section of the book is dedicated to the changes in the egg following sperm penetration that lead to early pattern formation (Chapters 39,40, and 45). I would like to thank all participants for contributing to the success of this meeting and in particular Prof. R. Yanagimachi, who although unable to attend, wished to contribute to this book dedicated to Alberto.

Naples, 17th March,1990

Brian Dale

Isolation and Characterization of the Angiosperm Gamete

SCOTT D. RUSSELL
Department of Botany and Microbiology, University of Oklahoma, Norman, OK 73019-0245 U.S.A.

Abstract Isolation and characterization of angiosperm gametes has emerged as an area of special interest during the last five years. Modern attempts to isolate sperm cells, embryo sacs and eggs have yielded cells that appear intact and viable using the fluorescein diacetate reaction. Male gamete isolation has provided enrichments of up to 10 million cells per milliliter suitable for biochemical characterization and monoclonal antibody production. Female gamete isolation is still in its earliest stages, providing excellent microscopic preparations retaining fluorescein diacetate activity, but insufficient numbers for practical biochemical characterizations. Improvement of these procedures will promote the completion of knowledge about the unique features that the angiosperm gamete possess and may eventually allow in vitro fertilization ex ovulo.

Introduction

Although fertilization in angiosperms was first described 105 years ago (Strasburger, 1884), a true appreciation of the organization and uniqueness of the flowering plant gamete has only recently emerged, and as this area develops, the possibility of developing an in vitro model system for fertilization in higher plants increases. The complexities of the inter-relationship between the gametes, the gamete-producing gametophytes, and the female host sporophyte has its precedent in the early work of Wilhelm Hofmeister (1862), who first recognized that gametophytes and sporophytes could be homologized with the independent plant generations of lower plant groups. Thus, the retention of a degree of autonomy should not be unexpected in the developmental program of these plant generations during the course of their evolution.

The degree of autonomy of the gametophyte is confirmed by modern molecular studies (Stinson, et al., 1987; Hanson, et al., 1989; Mascarenhas, 1989), and is further expressed in physiological evidence that gametes express unique polypeptides not evident in surrounding vegetative tissues. This is indicative of a separate developmental program for gametophytes and gametes distinct from that of their surrounding tissues (Geltz and Russell, 1988). During the last five years, gamete isolation in angiosperms has emerged as an area of significant scientific progress that has potential for changing basic concepts of fertilization. This paper describes some of the early results obtained and fundamental methods by which the angiosperm gametes are currently being examined.

Male Germ Unit Concept and Male Gamete Organization

Ultrastructural studies of the male gamete in vivo have led to a concept known as the "male germ unit", which holds that the male gametes are physically associated with the vegetative nucleus forming together the functional unit of male DNA

NATO ASI Series, Vol. H 45
Mechanism of Fertilization
Edited by B. Dale

transmission (Russell and Cass, 1981; Dumas, et al., 1984). Although vegetative nucleus in the grasses is often unassociated with sperm in the pollen grain, the male germ unit may occurs later in development and is perhaps universal in angiosperms if all stages of development prior to fertilization are considered (Mogensen and Wagner, 1987).

Characterizations of microscopically isolated germ line cells using fluorescence microscopy have included localizations of plastid DNA (Corriveau and Coleman, 1988), nuclear materials (Coleman and Goff, 1985; Hough, et al., 1985) and cytoskeletal elements (Palevitz and Cresti, 1988, 1989). Visualization of mitochondria (Theunis and Went, personal communication) and selected membrane systems are also possible using the membrane potential-sensitive dye $DiOC_6$ (Matske and Matske, 1986), which allow surveys of cell organization to be conducted with great facility. These probes could be indispensible in the future for fluorescence-activated cell sorting strategies using isolated cells of the male germ lineage.

Numerous modern studies using carefully-applied cytological methods have provided significant new information concerning the organization of the male gamete in vivo. The techniques involved have included classical transmission electron microscopy (Cass, 1973; Zhu, et al., 1980; Russell and Cass, 1981; Mogensen and Wagner, 1987; Yu, et al., 1989), quantitative cytology (Russell, 1984; Mogensen and Rusche, 1985; Wilms, 1986; McConchie and Knox, 1986; McConchie, et al., 1987a, 1987b; Rusche, 1988; Rusche and Mogensen, 1988), three-dimensional reconstruction (Russell, 1984; Mogensen and Rusche, 1985; McConchie and Knox, 1986; McConchie, et al., 1987a, 1987b; Wagner and Mogensen, 1987; Rusche, 1988; Rusche and Mogensen, 1988) and stereology (Wagner, et al., 1989a). Additional studies have delved into the origin of the male germ unit from mitotic division (Palevitz and Cresti, 1988, 1989), to sperm cell inter-association (Charzynska, et al., 1988, 1989) and the physical association between male germ cells and the vegetative nucleus characteristic of the male germ unit concept (Russell and Cass, 1983; Kaul, et al., 1987; Mogensen and Wagner, 1987; Hu and Yu, 1988).

Among the more important findings about the male germ unit from classical transmission electron microscopic methods are that the sperm cells within a pollen grain may be dimorphic, either expressing cytoplasmic heterospermy (Russell, 1984; Wilms, 1986; McConchie, et al., 1987b) or nuclear heterospermy (Roman, 1948); in two of these cases sperm display a traceable preferential pattern of fusion during fertilization that may be shown using genetic (Roman, 1948) and ultrastructural methods (Russell, 1985). The concept of preferential fertilization is an important one in understanding the process of double fertilization, in general, because it provides specific markers in the germ line that predict this behavior prior to fusion.

Male Gamete Isolation

Although isolations of the male gamete have been conducted at a small scale before (Cass, 1973; Russell and Cass, 1981), en masse collection of isolated male germ cells is a relatively recent development (Russell, 1986a, Dupuis et al., 1987; Southworth and Knox, 1989). Viability testing (Vergne, et al., 1987), transmission electron (Cass and Fabi, 1988), biochemical characterization (Geltz and Russell, 1988, Knox, et al., 1988) and antibody production (Hough, et al., 1986; Pennell, et al., 1987) have emerged even more recently. This has clearly extended the early work in the field conducted using electron microscopy (Jensen and Fisher, 1968b; Larson, 1965), freeze-fracture of nuclei (LaFountain and LaFountain, 1973), differential interference microscopy (Cass, 1973) and biochemistry of isolated generative nuclei and pollen tubes (Mascarenhas, 1975; Pipkin and Larson, 1973).

The general protocols for isolation of sperm cells en masse involve osmotic pollen bursting (Russell, 1986a), pH-change-mediated pollen bursting (Dupuis, et al., 1987) or grinding (eg., Southworth and Knox, 1989) in order to liberate the sperm cells. Each of these procedures provides the potential of collecting from 30,000 to 10,000,000 sperm cells per milliliter for further experimentation. A protocol for the procedure used in isolating sperm cells of Plumbago is shown in Figure 1.

Early work on the physiology of the male germ cells focused on protein, RNA synthesis and nucleic acid compositional changes in pollen and generative nuclei (Mascarenhas, 1965; Pipkin and Larson, 1973). These studies used isolated nuclei, rather than cells and concentrated on bicellular pollen species. Their results showed dramatic developmental changes occur in the nucleus at the biochemical level. The early research approach, however, centered on the assumption that the pollen and embryo sac directed the important events of sexual reproduction (Jensen, 1974), without the direct, active participation of sperm cells in these processes.

Recent evidence, however, indicates that sperm cells possess the determinants that are involved in recognizing the egg (Russell 1985; Knox, et al., 1988) and that the cell cytoplasm appears to convey organelles in specific proportions (Russell, 1984, 1985), transmitting heritable organelles in specific ratios into the egg and central cell (Russell, 1987). Since recognition phenomena and the occurrence of dimorphic sperm cells are cytoplasmic in nature, studying the nuclear determinants would be expected to be initially more difficult than their expressional correlates, which are evident in the cytoplasm and on the surface (Knox, et al., 1988). In the matter of cytoplasmic inheritance, apparently both active and passive means exist for determining the content of cytoplasmic organelles during sperm maturation (Mogensen and Rusche, 1985), and for eliminating them during maturation and fertilization (Mogensen, 1988; Russell, et al., submitted). The more recent biochemical research committed to studying the intact male germ cells seeks to address these remaining questions.

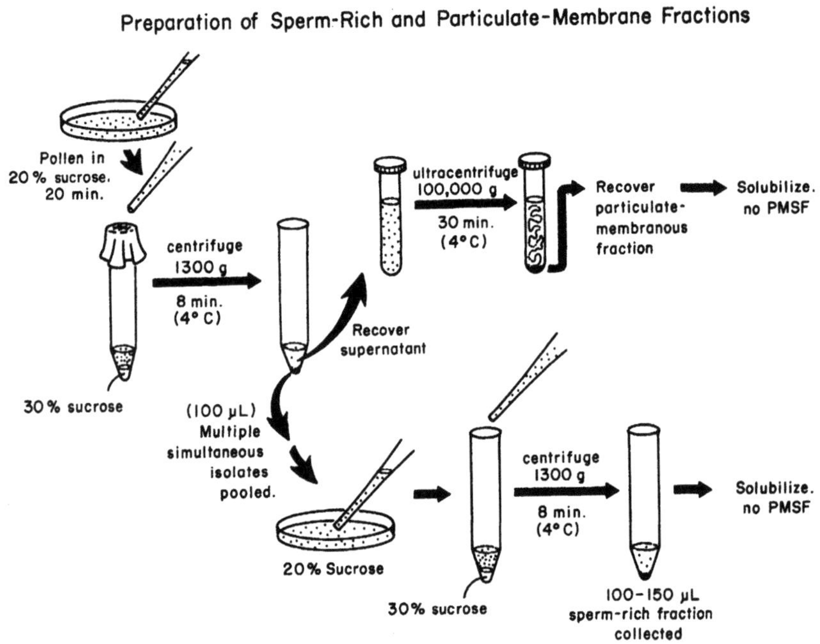

Figure 1. Isolation protocol used for sperm cells of *Plumbago zeylanica* (modified from protocol in Russell, 1986a) for biochemical characterization and antibody production.

Male Gamete Biochemical Characterization

Biochemical characterization of male gametes has been conducted on a small number of species, including *Brassica*, *Gerbera* (Knox, et al., 1988), *Plumbago* (Geltz and Russell, 1988) and *Zea* (Roeckel and Dumas, personal communication). The preliminary one-dimensional polyacrylamide gel electrophoretic results on these plants indicate the presence of polypeptide bands unique to the sperm-rich fraction. The possible effect that contamination from the vegetative nucleus, which is physically associated with the sperm cells in mature pollen of the former three species, may have on these results is not entirely known at present; however, in *Plumbago*, it is known that the vegetative nucleus is not intact after sperm isolation. Therefore, any contaminants that it did generate should be systematically eliminated by matching the sperm-rich fraction with comparable contaminant fractions. The material from the burst vegetative nucleus would be expected to cross-contaminate all of the fractions during protein isolation, and therefore, the effect of this material, if any, should be cancelled by presence/absence scoring of polypeptides between the various fractions (Geltz and Russell, 1988).

The apparent heterogeneity in one-dimensional polyacrylamide gel electrophoretic bands, however, does not reflect single polypeptide species, but groups of heterogeneous and likely unrelated polypeptides co-migrating to the same position in the gel. Therefore, two-dimensional electrophoretic gels were also conducted on Plumbago using isoelectric focusing in the first dimension and SDS separation in the second dimension. In addition to the sperm-enriched fraction, two other fractions enriched in contaminants of the sperm fraction were run under similar conditions (Geltz and Russell, 1988). The two contaminant fractions consisted of pelleted cytoplasmic-particulates (organelles, membranes and particles that sediment at 100,000 X g centrifugation; Figure 1) and water-soluble contaminants (excluding congealed lipids) that remain suspended at 100,000 X g; the latter fraction was mixed with PMSF to retard proteolysis at all stages in preparation. In this study, considerable polypeptide heterogeneity was evident in the sperm-enriched fraction (515 polypeptide spots observed from M_r 33,000 to 200,000). The cytoplasmic-particulate contaminant fraction contained fewer polypeptides (427 in the same M_r range) and the water-soluble fraction contained the fewest (285 polypeptides in the same M_r range).

Comparing selected polypeptide spots in this study revealed that 24.8% of them were found in all fractions. A remaining 48.9% occurred in two of the fractions, and 26.3% were unique to a single fraction. Of the unique spots, 13.5% of the total were found only in the sperm-rich fraction, 10.5% were found only in the water-soluble fraction, and 2% were found only in the cytoplasmic-particulate fraction (Geltz and Russell, 1988). Some of the male germ unit-specific proteins were of very high molecular weight (up to 205,000 daltons) and of a weight clearly sufficient for the complexity required to serve as a recognition polypeptide (Geltz and Russell, 1988). The highest molecular weight compounds appeared only in the male germ unit-rich fraction, and were entirely lacking from surrounding pollen material even in over-loaded gels.

Male Gamete Hybridoma Antibody Production

Antibody production in mouse hybridoma cells has been elicited using intact sperm cell inocula of Brassica (Hough, et al., 1986) and Plumbago (Pennell, et al., 1987). The antibodies elicited to Brassica and Plumbago sperm reveal that sperm antigens are effective in producing specific antibody-secreting lines (Hough, et al., 1986; Pennell, et al., 1987). In Plumbago, substantial heterogeneity in binding sites was evident, and pollen wall proteins did not appear to be immunodominant. Positive lines were found that were specific for the male germ unit (23%), vegetative cell organelles (4%) and soluble proteins (7%). These percentages strongly reflect the distribution of unique polypeptides described from two-dimensional electrophoretic studies in the previous section (Geltz and Russell, 1988). The antibody-secreting lines were predominantly IgM producers (61%), reflecting the high glycoprotein content of the inoculum and the pollen environment, in general (Pennell, et al., 1987). Data concerning the Brassica antibody lines are currently fragmentary but seem to suggest the same general trends with regard to IgM content and absence of immunodominant pollen wall polypeptides (Hough, et al., 1986; Knox, et al., 1988; Knox, personal communication). More sensitive screening for cell surface antibodies will be required to accomplish the more ambitious goal of producing a useful Mab library providing the keys to sperm cell determinants.

Developmental Changes in Male Expression

Part of the distinction between the gametic and somatic cell proteins will undoubtedly prove to be the result of expressional differences between genes activated in the sporophyte versus gametophyte generations (Stinson, et al., 1987; Hanson, et al., 1989), as for example, pollen wall proteins, which are often synthesized by the sporophyte. Other expressional differences, however, particularly between disparate cells of gametophytic origin may relate to proteins with unique functions in gametic cell interactions and physiology; these are likely to be expressed as part of the cellular program at specific stages of development.

One-dimensional polyacrylamide gel electrophoretic information on polypeptide changes in developing whole pollen and microspores are known for a very small group of plants. In Zea (Frova, et al., 1987), Triticum (Vergne and Dumas, 1988) and Brassica (Detchepare, et al., 1989), developmental phases appear to be reflected by a relatively small number of stage-specific changes. In Triticum, 11 new bands were detected in SDS-PAGE and IEF gels and only four bands were lost during pollen maturation. Changes reported from mixed anther-gametophyte collections during development of Zea pollen (Delvallee and Dumas, 1987), reveal similar stage-specific changes in both proteins and also enzyme activity. One presumes that during the next several years, enormous progress will be made in identifying specific polypeptides controlling stage-specific sperm cell behavior, particularly as radioisotope tracers are used to determine active patterns of protein translation.

Female Germ Unit Concept and Female Gamete Organization

The functional unit of female reproduction, termed the "female germ unit" (Dumas, et al., 1984), consists of the egg apparatus, composed of the egg and two synergids (sensu Jensen, 1965a, 1974), along with the central cell. These four cells together are essential for double fertilization, since the synergids form the delivery mechanism, and the egg and proendospermatic central cell constitute the essential participants in double fertilization. Support for the female germ unit as the functional entity comes from the fact that both the embryo and endosperm must be formed in order to produce viable seed.

The female gamete is a polarized cell in all of the flowering plants described to date (Jensen, 1974; Russell, 1986b). The micropylar end, that sited nearest the typical location of pollen tube entry, is typically in contact with the central cell, with both synergids and, over a short area, with the embryo sac wall. The micropylar cell wall is simple and without elaboration. The cytoplasm is predominated by a large vacuole located toward the micropylar end of the cell. The chalazal end contains the nucleus, which is commonly surrounded by many of the heritable organelles. The cell wall at the chalazal end is frequently interrupted near synergid and central cell boundaries. This wall-lacking region apparently provides the point of entry for sperm cells upon pollen tube discharge into one synergid (Jensen, 1974; Went and Willemse, 1984), providing a portal through which fertilization may occur.

The synergids are polarized largely in reverse to that of the egg, with a micropylar to mid-lateral nucleus and a chalazal vacuole, frequently composed of numerous vesicles (Jensen, 1965b; Went and Willemse, 1984). The micropylar end has surfaces in common with the egg, central cell, and embryo sac. The cell walls located at the micropylar end of the embryo sac are often elaborately thickened with internal projections, forming the classical "filiform apparatus," characteristic of synergids.

One of the two flanking synergids appears to provide the normal point of entry of pollen tube and gamete deposition in vivo (Pluijm, 1964; Jensen, 1974; Went and Willemse, 1984). In response to pollination, typically one synergid degenerates prior to fertilization (Jensen and Fisher, 1968a); however, instances have been reported where neither synergid appears to degenerate prior to normal fertilization (Went, 1970) or where both degenerate (Cass and Jensen, 1970; Went and Cresti, 1988). That the synergids are claimed to retain embryogenetic potential (Maheshwari and Sachar, 1963) would seem to indicate that synergids share a significant portion of their developmental program with the egg cell lineage. The occurrence of supernumerary eggs, evident in transmission electron microscopic studies of Linum (Secor and Russell, 1988), casts doubt on other reported cases of synergid embryogenesis and suggests that the programmatic divergence of synergids in some species may be at a considerably earlier stage.

The central cell is a voluminous cell occupying much of the embryo sac and possessing common boundaries with most of the cells of the embryo sac, including antipodals (Jensen, 1965a). This cell forms much of the boundary of the embryo sac, frequently possessing outer walls with elaborate cell wall ingrowths (Went and Willemse, 1984). The central cell therefore appears to be nutritionally essential to the function of the female gametophyte throughout its development. After fertilization, the polar nuclei, whether fused prior to fertilization, during it, or later in mitosis, merge and the central cell becomes the endosperm, which provides a nutritional source required for normal embyro development (Went and Willemse, 1984).

Although these cells are originally formed as free nuclei, the cells of the female germ unit appear to arise from a complex, synchronous mechanism of cytokinesis based on prior nuclear position (Cass, et al., 1985, 1986). The method by which these nuclei are positioned and how they sequester their characteristic cytoplasmic structure, however, is not presently described, nor is the mechanism for sperm cell migration within the female gametophyte (Russell, 1986b).

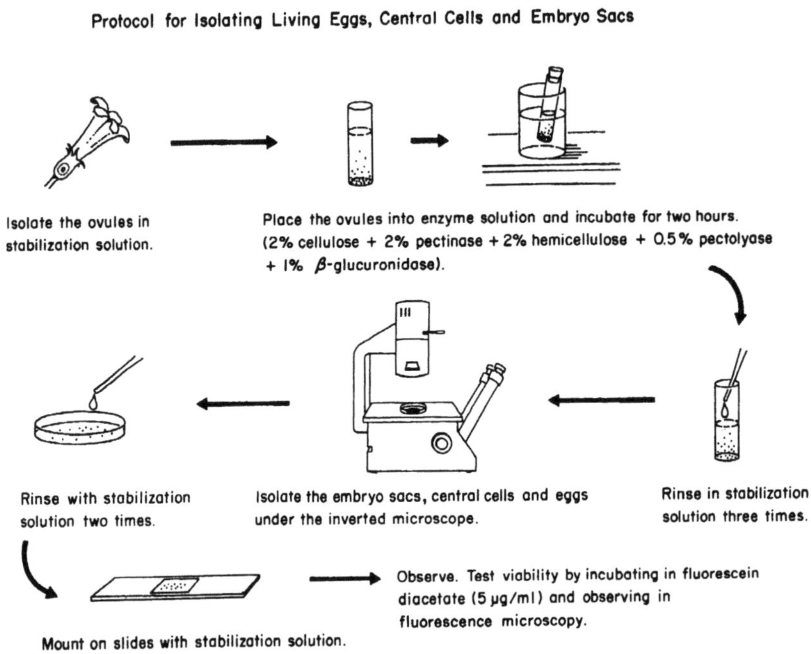

Figure 2. Isolation protocol used for viable eggs, central cells, lateral cells and embryo sacs of Plumbago zeylanica (based on protocol described in Huang and Russell, 1989).

Female Gametophyte and Gamete Isolation

Since the female germ unit is an integral part of the ovule, the isolation of a physiologically intact female gametophyte or female gamete has been an elusive goal that has only recently been attained (see Huang and Russell, 1989). Methods of obtaining female gametophytes can be separated into two groups: micromanipulation (Allington, 1985) and enzymatic maceration (Zhou and Yang, 1985, 1986; Hu, et al.,

1985; Mol, 1986. These methods have since been refined to produce viable female gametophytes (Hu, et al., 1985; Wagner, et al., 1989b; Huang and Russell, 1989) and viable gametophytic cells (Huang and Russell, 1989). The general procedure is outlined in Figure 2.

The current status of the area is one of promise, as the first biochemical and immunological characterizations are in progress. As these results become available, the behavior of the egg and central cells may be more specifically described in the framework of the polypeptide and genetic expression in which they occur.

Conclusions and Prospects

Improvement of long-term viability of isolated gametes will be needed before practical approaches to culturing flowering plant gametes can be successful. Methods of culturing cells in low concentration will also require further development. Obtaining massive numbers of living gametic cells will be needed to perform radiolabelling studies to trace gene expression, obtain specific monocolonal antibody lineages and to solve the molecular biological problems of the future.

Cytoplasmic organelle inheritance, previously examined using transmission electron microscopy, must be reexamined using probes sensitive to plastid and mitochondrial organellar DNA and also probes sensitive to parental origin. The use of molecular probes to allelic populations of cytoplasmic organelles would permit examination of both the origin and stability of chimeric cells.

Continuing questions involving the mechanisms of egg activation, initiation of embryogenesis, and triggering of apomictic development, for example, could be addressed by a well-characterized in vitro gametic culture system. The specific mechanisms by which these processes interact or whether particular polypeptide initiators, hormones or other factors function is unknown.

Ultimately, as these problems are resolved, sperm cells may become the optimum source of cells for zygotic formation of transformed gametes, vectors for introduction of stable chimeric products, and cells possibly capable of interspecific fusion. As these problems become resolved, fusion of isolated angiosperm male and female gametes will become a practical system for examining fertilization and obtaining recombinant progeny.

Acknowledgements

I thank Susan Gray and Huang Bing-Quan for preparation of the figures and US Dept of Agriculture grant 88-37261-3761 for support of parts of this research.

Allington PM (1985) Micromanipulation of the unfixed cereal embryo sac. In: Chapman GP, Mantell SH, Daniels RW (eds) Experimental manipulation of ovule tissues. Longman, Essex, pp 39-51

Cass DD (1973) An ultrastructural and Nomarski-interference study of the sperms of barley. Can J Bot 51: 601-605

Cass DD, Jensen WA (1970) Fertilization in barley. Amer J Bot 57: 62-70

Cass DD, Fabi GC (1988) Structure and properties of sperm cells isolated from the pollen of Zea mays. Can J Bot 66: 819-825

Cass DD, Peteya DJ, Robertson BL (1985) Megagametophyte development in Hordeum vulgare. 1. Early megagametogenesis and the nature of cell wall formation. Can J Bot 63: 2164-2171

Cass DD, Peteya DJ, Robertson BL (1986) Megagametophyte development in Hordeum vulgare. 2. Later stages of wall development and morphological aspects of megagametophyte cell differentiation. Can J Bot 64: 2327-2336

Charzynska M, Ciampolini F, Cresti M (1988) Generative cell division and sperm cell formation in barley. Sex Plant Reprod 1: 240-247

Charzynska M, Murgia M, Milanesi C, Cresti M (1989) Origin of sperm cell association in the "male germ unit" of Brassica pollen. Protoplasma 149: 1-4

Coleman AW, Goff LJ (1985) Applications of fluorochromes to pollen biology - 1 mithramycin and 4-6-diamidino-2-phenylindole (DAPI) as vital stains and for quantitation of nuclear DNA. Stain Technol 60: 145-154

Corriveau JL, Coleman AW (1988) Rapid screening method to detect potential biparental inheritance of plastid DNA and results for over 200 angiosperm species. Amer J Bot 75: 1443-1458

Delvallee I, Dumas C (1987) Anther development in Zea mays: changes in protein, peroxydase, and esterase patterns. J Plant Physiol 132: 210-217

Detchepare S, Heizmann P, Dumas C (1989) Changes in protein patterns and protein synthesis during anther development in Brassica oleracea. J Plant Physiol (in press)

Dumas C, Knox RB, McConchie CA, Russell SD (1984) Emerging physiological concepts in fertilization. What's New Plant Physiol 15: 17-20

Dupuis I, Roeckel P, Matthys-Rochon E, Dumas C (1987) Procedure to isolate viable sperm cells from corn (Zea mays L.) pollen grains. Plant Physiol 85: 876-878

Frova C, Binelli G, Ottaviano E (1987) Isozyme and HSP gene expression during male gametophyte development in maize. In: Isozymes: current topics in biological and medical research, vol 15, pp 97-120, Alan R Liss, New York

Geltz NR, Russell SD (1988) Two-dimensional electrophoretic studies of the proteins and polypeptides in mature pollen grains and the male germ unit of Plumbago zeylanica. Plant Physiol 88: 764-769

Hanson DD, Hamilton DA, Travis JL, Bashe DM, Mascarenhas JP (1989) Characterization of a pollen-specific cDNA clone from Zea mays and its expression. Plant Cell 1: 173-179

Hofmeister W (1862) On the germination, development and fructification of the higher cryptogamia and on the fructification of the Coniferae. Published for the Ray Society by Robert Hardwicke, London.

Hough T, Bernhardt P, Knox RB, Williams EG (1985) Applications of fluorochromes to pollen biology - 2 the DNA probes ethidium bromide and Hoechst 33258 in conjunction with the callose-specific aniline blue fluorochrome. Stain Technol 60: 155-162

Hough T, Singh MB, Smart IJ, Knox RB (1986) Immunofluorescent screening of monoclonal antibodies to surface antigens of animal and plant cells bound to polycarbonate membranes. J Immunol Meth 92: 103-108

Hu SY, Lg LI, C Zhu (1985) Isolation of viable embryos sacs and their protoplasts in Nicotiana tabacum. Acta Bot Sin 27: 337-344

Hu SY, Yu HS (1988) Preliminary observations on the formation of the male germ unit in pollen tubes of Cyphomandra betacea Sendt. Protoplasma 147: 55-63

Huang BQ, Russell SD (1989) Isolation of fixed and viable eggs, central cells, and embryo sacs from ovules of Plumbago zeylanica. Plant Physiol 90: 9-12

Jensen WA (1965a) The ultrastructure and composition of the egg and central cell of cotton. Amer J Bot 52: 781-797

Jensen WA (1965b) The ultrastructure and histochemistry of the synergids of cotton. Amer J Bot 52: 238-256

Jensen WA (1974) Reproduction in flowering plants. In: Robards AW (ed) Dynamic aspects of plant ultrastructure. McGraw-Hill, New York, pp 481-503

Jensen WA, Fisher DB (1968a) Cotton embryogenesis: the entrance and discharge of the pollen tube in the embryo sac. Planta 78: 158-183

Jensen WA, Fisher DB (1968b) Cotton embryogenesis: the sperm. Protoplasma 65: 277-286

Kaul V, Theunis BF, Palser BF, Knox RB, Williams EG (1987) Association of the the generative cell and vegetative nucleus in pollen tubes of Rhododendron. Ann Bot 59: 227-235

Knox RB, Southworth DA, Singh MB (1988) Sperm cell determinants and control of fertilization in plants. In: Chapman GP, Ainsworth CC, Chatham CJ (eds) Eukaryotic cell recognition. Cambridge University Press, Cambridge, pp 175-193

LaFountain JR, LaFountain KL (1973) Comparison of density of nuclear pores on vegetative and generative nuclei in pollen of Tradescantia. Exp Cell Res 78: 472-476

Larson DA (1965) Fine structural changes in the cytoplasm of germinating pollen. Amer J Bot 52: 139-154

Maheshwari P, Sachar RC (1963) Polyembryony. In: Maheshwari P (ed) Recent advances in the embryology of angiosperms. University of Delhi, Delhi, pp 265-296

Mascarenhas JP (1965) Pollen tube growth and ribonucleic acid synthesis by vegetative and generative nuclei of Tradescantia. Amer J Bot 52: 605-610

Mascarenhas JP (1975) The biochemistry of angiosperm pollen development. Bot Rev 41: 259-314

Mascarenhas JP (1989) The male gametophyte of flowering plants. Plant Cell 1: 657-664

Matske MA, Matske AJM (1986) Visualization of mitochondria and nuclei in living plant cells by the use of a potential-sensitive fluorescent dye. Plant Cell Environ 9: 75-77

McConchie CA, Hough T, Knox RB (1987a) Ultrastructural analysis of the sperm cells of maize, Zea mays. Protoplasma 139: 9-19

McConchie CA, Knox RB (1986) The male germ unit and prospects for biotechnology. In: Mulcahy DL, Mulcahy GB, Ottaviano E (eds) Pollen biotechnology and ecology. Elsevier Biomedical Press, New York, pp 289-296

McConchie CA, Russell SD, Dumas C, Tuohy M, Knox RB (1987b) Quantitative cytology of the egg and central cell of Brassica campestris and B. oleracea. Planta 170: 446-452

Mogensen HL (1988) Exclusion of male mitochondria and plastids during syngamy as a basis for maternal inheritance. Proc Natl Acad Sci (USA) 85: 2594-2597

Mogensen HL, Rusche ML (1985) Quantitative analysis of barley sperm: occurrence and mechanism of cytoplasm and organelle reduction and the question of sperm dimorphism. Protoplasma 128: 1-13

Mogensen HL, Wagner VT (1987) Associations among components of the male germ unit following in vivo pollination in barley. Protoplasma 138 161-172

Mol R (1986) Isolation of protoplasts from female gametophytes of Torenia fournieri. Plant Cell Rep 3: 202-206

Palevitz BA, Cresti M (1988) Microtubule organization in the sperm of Tradescantia virginiana. Protoplasma 146: 28-34

Palevitz BA, Cresti M (1989) Cytoskeletal changes during generative cell division and sperm formation in Tradescantia virginiana. Protoplasma 150:54-71
Pennell RI, Geltz NR, Koren E, Russell SD (1987) Production and partial characterization of hybridoma antibodies elicited to the sperm of Plumbago zeylanica. Bot Gaz 148: 401-406
Pipkin JL, Larson DA (1973) Changing patterns of nuclei acids, basic and acidic proteins in generative and vegetative nuclei during pollen germination and pollen tube growth in Hippeastrum belladona. Exp Cell Res 79: 78-
Pluijm JE van der (1964) An electron microscopic investigation of the filiform apparatus in the embryo sac of Torenia fournieri. In: Linskens HF (ed) Pollen physiology and fertilization. North Holland, Amsterdam, pp 8-16
Roman, H (1948) Directed fertilization in maize. Proc Nat Acad Sci (USA) 34: 36-42
Rusche ML (1988) The male germ unit of Zea mays in the mature pollen grain. In: Wilms HJ, Keijzer CJ (eds) Plant sperm cells as tools for biotechnology. PUDOC, Wageningen, pp 61-68
Rusche ML, Mogensen HL (1988) The male germ unit of Zea mays: quantitative ultrastructure and three-dimensional analysis. In: Cresti M, Gori P, Pacini E (eds) Sexual reproduction in higher plants. Springer-Verlag, New York, pp 221-226
Russell SD (1984) Ultrastructure of the sperm of Plumbago zeylanica: 2. Quantitative cytology and three-dimensional reconstruction. Planta 162: 385-391
Russell SD (1985) Preferential fertilization in Plumbago: ultrastructural evidence for gamete-level recognition in an angiosperm. Proc Natl Acad Sci (USA) 82: 6129-6132
Russell SD (1986a) A method for the isolation of sperm cells in Plumbago zeylanica. Plant Physiol 81: 317-319
Russell SD (1986b) Dimorphic sperm cells, cytoplasmic transmission, and preferential fertilization in the synergid-less angiosperm, Plumbago zeylanica. In: Mantell SH, Chapman GP, Street PFS (eds) The chondriome. Longman, Essex, pp 69-116
Russell SD (1987) Quantitative cytology of the egg and central cell of Plumbago zeylanica and its impact on cytoplasmic inheritance patterns. Theoret Appl Genet 74: 693-699
Russell SD, Cass DD (1981) Ultrastructure of the sperm of Plumbago zeylanica: 1. Cytology and association with the vegetative nucleus. Protoplasma 107: 85-107
Russell SD, Cass DD (1983) Unequal distribution of plastids and mitochondria during sperm cell formation in Plumbago zeylanica. In: Mulcahy DL, Ottaviano E (eds) Pollen: biology and implications for plant breeding, Elsevier, New York, pp 135-140
Russell SD, Rougier M, Dumas C (Submitted) Organization of the early post-fertilization megagametophyte of Populus deltoides: ultrastructure and implications for male cytoplasmic transmission. Protoplasma
Secor DL, Russell SD (1988) Organization of the megagametophyte in a polyembryonic line of Linum usitatissimum. Amer J Bot 75: 114-122
Southworth D, Knox RB (1989) Flowering plant sperm cell: isolation from pollen of Gerbera jamesonii (Asteraceae). Plant Sci 60: 273-277
Stinson JR, Eisenberg AJ, Willing RP, Pe ME, Hanson DD, Mascarenhas JP (1987) Genes expressed in the male gametophyte of flowering plants and their isolation. Plant Physiol 83: 442-447
Strasburger E (1884) Neue Untersuchen uber den Befruchtungsvorgang bei den Phanerogamen als Grundlage fur eine Theorie der Zeugung. Gustav Fischer, Jena
Vergne P, Dumas C (1988) Isolation of viable wheat gametophytes of different stages of development and variations in their protein patterns. Plant Physiol 88: 969-972

Vergne P, Delvalee I, Dumas C (1987) Rapid assessment of microspore and pollen development stage in wheat and maize using DAPI and membrane permeabilization. Stain Tech 62: 299-304
Wagner VT, Mogensen HL (1987) The male germ unit in the pollen and pollen tubes of Petunia hybrida: ultrastructural quantitative and three dimensional features. Protoplasma 143: 93-100
Wagner VT, Dumas C, Mogensen HL (1989a) Morphometric analysis of the sperm cells of isolated sperm cells of Zea mays L. J Cell Sci 93: 179-184
Wagner VT, Song YC, Matthys-Rochon E, Dumas C (1989b) Observations on the isolated embryo sac of Zea mays L. Plant Sci 59: 127-132
Went JL van (1970) The ultrastructure of the fertilized embryo sac of petunia. Acta Bot Neerl 19: 468-480
Went JL van, Cresti M (1988) Pre-fertilization degeneration of both synergids in Brassica campestris ovules. Sex Plant Reprod 1: 208-216
Went JL van, Willemse MTM (1984) Fertilization. In: Johri BM (ed) Embryology of Angiosperms. Springer-Verlag, New York, pp 273-317
Wilms HJ (1986) Dimorphic sperm cells in the pollen grain of Spinacia. In: Cresti M, Dallai R (eds) Biology of reproduction and cell motility in plants and animals. University of Siena, Siena, pp 193-198
Yu HS, Hu SY, Zhu C (1989) Ultrastructure of sperm cells and the male germ unit of Nicotiana tabacum. Protoplasma (in press)
Zhou C, Yang HY (1985) Observations on enzymatically isolated, living and fixed embryo sacs in several angiosperm species. Planta 165: 225-231
Zhou C, Yang HY (1986) Isolation of embryo sacs by enzymatic maceration and its potential in haploid study. In: Hu H, Yang HY (eds) Haploids of higher plants in vitro. Springer-Verlag, Berlin, pp 192-203
Zhu C, Hu SY, Xu LY, Li XR, Shen JH (1980) Ultrastructure of sperm cell in mature pollen grain of wheat. Sci Sinica 23: 371-379

THE CYTOSKELETON DURING POLLEN TUBE GROWTH AND SPERM CELL FORMATION

A. Tiezzi and M. Cresti
Dipartimento di Biologia Ambientale
Università degli Studi di Siena
Via P.A. Mattioli n. 4
53100 Siena
Italy

Introduction and Summary of Pollen Development

Angiosperm plants produce flowers in which their reproductive development occurs. In flowering plants and other groups of plants a diploid generation (sporophyte) alternates with a haploid generation (gametophyte). In angiosperms the male and female gametophytes are reduced to microscopic structures that are dependent, for their development, on the tissue of the sporophyte. The flower contains specialized structures, the anthers and pistil or gyneaceum, in which the male and female gametophytes are formed. The functions of the gametophytes are the production of the male sperm cells and the female egg cell, and their fusion in the fertilization. The male gametophyte completes its development in the anthers. In flowering plants, the pollen grain is the male gametophyte and the embryo sac is the female one.

For a short while following the dehiscence (rupture of the anther and release of pollen), the mature pollen grain exists as a free organism until it is transported by wind, animals or other agents to the stigma of an appropriate pistil. After recognition the pollen begins another phase of its life and development. After hydration, each pollen grain germinates by the emission of a tube through a germ pore. The tube then grows into the style whereas the

NATO ASI Series, Vol. H 45
Mechanism of Fertilization
Edited by B. Dale

vegetative nucleus and, depending on the species, the generative cell (bicellulate grains) or sperm cells (tricellulate grains) move out of the pollen grain and penetrate into the tube. Germination and pollen tube growth are relatively rapid events. In most plants, the period from pollination to fertilization ranges from 1 hr to 48 hrs. The rate of tube growth depends upon the species; in most species it is extremely rapid (35 mm per hr). The tube grows through the transmitting tissue or canal cavity of the style, penetrates the micropylar part of the ovule and finally reaches the embryo sac. The contents of the tube are discharged in one of the two synergids (normally the one that has begun to degenerate). Inside the latter, the two sperm cells move by a presently unknown mechanism, one fusing with the egg cell to form the zygote and the other one fuses with the generally diploid central cell, giving rise to the primary endosperm nucleus. Recent research (Russell, 1985; Mc Conchie et al., 1987) showed the presence of two morphologically different sperm cells in the pollen grain, perhaps functionally specialized to fuse with the egg and the central cell respectively. In higher plants the complete process is termed "double fertilization" (for a complete review see Mascarenhas, 1989).

Recently ultrastructural and biochemical research has put in evidence and charaterized a complex cytoskeleton inside pollen and pollen tube; in addition the presence of a high concentration of Ca^{2+}at the tip region allowed to move to new investigations on physiological roles of specific ions in the process of pollen tube growth. These findings add significant details to the modalities of pollen tube growth, generative cell locomotion and sperm cell formation and movement, contributing to a better understanding of the male gamete role in the fertilization process of the higher plants.

Cytoskeletal Apparatus

Until some years ago very few was reported on the presence of a cytoskeletal apparatus in pollen tubes. Histological and ultrastructural observations better characterized the organelles and defined their distribution in the cytoplasm, whereas little research was focussed on microtubules and/or other fibrillar systems present inside either the pollen tube cytoplasm or the generative cell.

The term "cytoskeleton" was developed at the end of the '70, when this new biological concept allowed to consider all together the new findings that biochemical, ultrastructural and histological investigations gave about fibrillar structures in the eukaryotic cells. Particular attention was focussed on cells, especially animal cells, that showed a high content of such fibrillar structures, mainly microtubules and microfilaments. Detailed studies were started on particular cells that allowed to relate the presence of such polypeptides to some cellular functions and that were termed and used as "model cells".

In plant biology the protoplasts of meristematic tissues were first intensely investigated as they significantly can contribute to a detailed structural definition of the mitotic process.

Another plant cell model investigated for its biological particularities was the pollen grain, a specialized cell from flowering plants and the pollen tube, a structure which is growing from the pollen grain to fertilize the ovule. Both these structures can be easily cultured "in vitro" and showed a rich content of cytoskeletal structures in the first ultrastructural an biochemical investigations (Van der Woude and Morre', 1968; Franke et al., 1972; Condeelis, 1974). These reports contributed very much to improve research on the pollen tube cytoskeleton, especially when the intense phenomenon of cytoplasmic streaming was related to the cytoskeletal structures. In this view recent investigations (Miki-Hirosige and Nakamura, 1982; Cresti et al., 1984; Derksen et al., 1985; Parthasaraty et al., 1985; Perdue and Parthasaraty, 1985; Van Lammeren et al., 1985; Cresti et al., 1986b; Heslop-Harrison et al., 1986; Pierson et al., 1986; Tiezzi et al., 1986; Cresti et al., 1987; Heslop-Harrison and Heslop-Harrison, 1987; Lancelle et al., 1987; Raudaskosky et al., 1987; Tiezzi et al., 1987; Heslop-Harrison and Heslop-Harrison, 1988a, 1988b; Heslop-Harrison et al., 1988; Palevitz and Cresti, 1988; Pierson, 1988; Tiezzi et al., 1988a, 1988b; Tiwari and Polito, 1988a, 1988b; Cresti and Tiezzi, 1989; Heslop-Harrison and Heslop Harrison, 1989; Palevitz and Cresti, 1989; Steer and Steer, 1989; Tang et al., 1989; Tiezzi et al., 1989) contributed to a detailed characterization of the cytoskeletal architecture opening even exciting new perspectives in the reproductive process of higher plants.

It is now evident that in every plant the process of pollen tube growth is related to the functionality of its cytoskeletal apparatus and the modalities of pollen tube growth, i.e. the growth rate, seem to differ in different plants. The results reported on the cited literature have shown the presence of microfilaments and microtubules as major components of the pollen tube cytoskeleton, suggesting their involvement in the processes of intracytoplasmic movements observed inside the growing pollen tube. Furthermore myosin and some not well characterized microtubule interacting proteins contribute to the cytoskeletal architecture and its functionality. In addition, the cited literature significantly extends our knowledge on the system termed "pollen and pollen tube", where a cell, the vegetative cell and properly the grain and tube, contains other cells, the generative cell and/or the sperm cells, assuring the conditions for their formation, viability, transport and structural integrity for the fertilization process. As it has been shown that all of the cell types mentioned above show an own cytoskeletal apparatus, below we describe these different cytoskeletons separately in order to contribute and provide defined directions for future research in the field.

The Vegetative Cell Cytoskeletal Apparatus

Although each pollen species shows own modalities of germination and growth rate, cytoskeletal elements, mostly microfilaments and microtubules, are always involved in the process of pollen tube growth, suggesting that the growth process is generally depending upon the correct functionality of the cytoskeleton. As microfilaments and microtubules represent the most prominent part of the cytoskeletal structure, it is obvious that they are the most studied components of the pollen tube cytoskeleton and are the main source of our present knowledge.

In some pollen grains microfilaments can be organized as paracrystalline structures, apparently showing random distribution in the cytoplasm (Cresti et al., 1986a; Heslop-Harrison et al., 1986). After pollen grain activation and pollen tube emission the paracrystalline structures disappear and microfilaments become evident and specifically locate in the growing pollen tube. Different systems of microfilaments with cortical and cytoplasmic distribution are longitudinally oriented from the grain to the tube tip. The cytoplasmic microfilaments are predominantly axially oriented. Detailed ultrastructural observations carried out in Nicotiana tabacum showed cytoplasmic microfilaments essentially free in the cytoplasm and microfilaments running parallel to microtubules in the cortical zone of the tube (Lancelle et al., 1987). Further immunofluorescence investigations colocalized microfilaments and microtubules in the cortical region (Pierson et al., 1989). The interpretation of the diffuse fluorescing zone in the tube apex is still unclear: the rhodamine-phalloidine staining might suggest the presence of a complex system of short microfilaments (Tiwari and Polito, 1988a), perhaps in association with the secretory bodies abundantly present in the tip zone.

A detailed study on the motility of the pollen tube cytoplasm was effectuated in Lilium, where the cytoplasmic streaming was shown to be regulated by Ca^{2+}, leading to the conclusion that the pollen tube streaming accordingly linked to both actin and myosin (Khono and Shimmen 1987, 1988). The presence of myosin has recently been demonstrated in the pollen tube of Nicotiana tabacum (Tang et al., 1989) by both western blot and immunofluorescence techniques. Myosin is specifically localized on the organelles, generative cell, vegetative nucleus and vesicle surfaces distributed all along the pollen tube. Using another antibody to myosin a specific staining of spherical particles was observed in the proximity of and at the wall of the pollen tube apex, perhaps suggesting the presence of different isoforms of myosin. The characterization of the actin-myosin system now allows to rethink on the process of cytoplasmic streaming and pollen tube growth.

In conclusion the pollen tube system now can really be compared with other plant and animal cells whose mechanisms of cytoplasmic streaming are subject of similar investigations.

In the pollen grain before activation and pollen tube emission, microtubules are not organized like microfilants and are not evident. In Nicotiana only after 30-40 minutes after germination they become clearly visible. As centrioles are absent in plant cells, there is no clarity as to the origin sites of microtubules, although there are some indications to believe that the microtubule system could begin behind the tube tip region (Heslop-Harrison and

Heslop-Harrison, 1988b). In Nicotiana microtubules are organized as singles, locate in the cortical region and their orientation is mainly axial; microfilaments run parallel to cortical microtubules (Lancelle et al., 1987). Immunolabelling using specific antibodies showed a diffuse staining at the tube apex (Derksen et al., 1985; Pierson et al., 1989) perhaps suggesting the presence of unpolymerized tubulin. Also ultrastructural investigations showed the absence of microtubules at the tube tip (Franke et al., 1972; Lancelle et al., 1987).

Whereas the discovery of myosin allowed to better understand the function of the microfilament system, the investigations carried out till now do not help very much to define the functions of microtubules. For example nocodazole treatment and subsequent microtubule depolymerization does not seem to affect the pollen tube growth process (Heslop-Harrison et al., 1988). On the other hand in animal cells microtubules cooperate with some specific ATPases to assure specialized intracytoplasmic movements. In addition processes of positioning and movement of vesicles are dependent from kinesin, specifically located on the vesicle surface. This well characterized ATPase interacts with microtubules by a specialized region of its molecule extruded from the vesicle membrane to form an arm-like structure (Vale, 1987).

In the pollen tube an active bidirectional cytoplasmic streaming occurs parallel to the longitudinal axis and a large number of vesicles, probably formed by Golgi, move thereby to the apical zone, fusing with the plasma membrane at the tube tip thus allowing the growth of the tube (see Heslop-Harrison and Heslop-Harrison, 1987). Using a monoclonal antibody to mammalian kinesin a specific punctate staining was shown at the apex of Nicotiana pollen tubes (Moscatelli et al., 1988), opening the possibility for the presence of another specialized motile system consisting of kinesin or a kinesin-like protein located on the surface of the vesicles and related to microtubules. This possibility can be supported by ultrastructural observations in freeze-substituted material where arm-like structures extruding from the vesicles at the tube apex are present (Cresti et al., unpublished data); in addition, as observed for the purification of kinesin in animal tissues (Vale et al., 1985), also in Nicotiana pollen tube extracts some polypeptides specifically bind to microtubules in an AMP-PNP sensitive manner (Tiezzi et al., in preparation). On the other hand there are evidences that disagree completely with the possible presence of a motile system related to microtubules: (i) nothing has been reported on the polarity of microtubules; (ii) the punctate zone at the tube tip (Moscatelli et al., 1988), where the stained material might consist of vesicles, does not contain polymerized microtubules (iii) in the tube apex a similar staining is observed after anti-myosin treatment. However, it must be considered that the findings of closely association of microfilaments and microtubules in the pollen tube of Nicotiana (Lancelle et al., 1987; Pierson et al., 1989) could suggest a sort of functional cooperation between the two different cytoskeletal systems. Perhaps experiments of double-immunofluorescence and/or immunogold staining could help to understand whether myosin and kinesin or and kinesin-like protein colocalize on the same structures or that they specifically locate on different vesicles, indicating a kind of vesicle specialization at the tube apex. Nevertheless one should remember that the vesicle movement process seems to differ among different plants. In the grasses and other groups where vesicles-containing wall and membrane precursor materials are stored in the pollen grain, vesicles apparently move out with cyclotic current and the process of cytoplasmic streaming and vesicle transport is probably mediated by an actin-myosin system, since it has been established that both processes are inhibited by cytochalasin-B (Mascarenhas and Lafountain, 1972).

The Generative Cell Cytoskeletal Apparatus

Several papers have reported about the presence of a cytoskeletal apparatus in the generative cell of many angiosperms consisting mostly of microtubules and some minor microtubule associated structures essentialy described at electron microscopical level (Franke et al., 1972; Lancelle et al., 1987; Tiezzi et al., 1988b). The generative cell is spindle-shaped as is the distribution of the microtubules. Microtubules are organized as bundles and specifically locate in the peripheral part of the generative cell cytoplasm, surrounding the cell and forming a basket-like structure. The associated structures cross-link microtubules contributing to their organisation in bundles and moduling them during the generative cell reshaping inside the pollen tube.

In binucleate pollen the generative cytoskeletal apparatus has to mediate the generative cell division inside the pollen tube and the subsequent sperm cell formation. As has been already reported in this article, inside the pollen tube other movement processes are driven by a biochemical cooperation of microfilaments and myosin molecules (Tang et al., 1989) but since ultrastructural and fluorescence studies have demonstrated the absence of actin inside the generative cell (Pierson et al., 1986; Lancelle et al., 1987; Pierson et al., 1989), it is obvious that microfilaments cannot have roles in preparing the generative cell to the karyokinesis process.

The presence of a large number of microtubules organized as bundles could instead suggest the possibility of some specific interactions of microtubules with other specialized polypeptides and their consequent involvement in the division process. In fact both "in vitro and "in vivo" biochemical studies have shown that processes of movement related to the microtubules are depending on other polypeptides, mainly showing ATPases activity (Kuznetsov and Gelfald, 1986; Vallee et al., 1988) and it is possible that similar polypeptides are cooperating with microtubules inside the generative cell. It has to be investigated if they can be identified as the microtubule associated structures: a positive response could perhaps mean that the microtubule associated structures play a role similar to the dynein role in the axoneme (Cresti et al., 1989). Nevertheless it must be pointed out that an antibody anti-kinesin (Moscatelli et al., 1988) stained no structures inside the generative cell, suggesting either the absence of such polypeptide or the presence of a kinesin isoform, molecularly differing from the kinesin of the vegetative cell.Other than the biomechanical activity related to the reshaping process, microtubules appear to have specific interactions with vesicles apparently dispersed in the cytoplasm (Wagner et al., 1989) and with chromosomes during the generative cell division and sperm cell formation. Nothing has been reported on the molecular motor(s) driving these processes.

The Sperm Cell Formation and the Cytoskeletal Apparatus

As considerable interest has been recently focussed on the organization of the sperm cells (Kejzer et al., 1988), detailed studies on the role of cytoskeletal structures in generative cell division and sperm cell formation have been carried out.

The mechanism of cell division does not seem to be always the same, but as other biological processes it can be different in different plants. In fact some studies indicate that sperm cells are formed through the action of a typical spindle and phragmoplast (Johnston, 1941; Ota, 1957; Karas and Cass, 1976; Raudaskoski et al., 1987), whereas other reports refer to tandemly aligned kinetochores and cytokinesis mediated by a furrowing process (O'Mara, 1933; Sax and O'Mara, 1941; Stanley and Linskens, 1974; Lewandowska and Charzynska, 1977). In the latter papers detailed observations on cytoskeletal structures were substantially absent and only recently the process has been investigated with particular regard to the cytoskeleton.

Detailed research has been carried out on Tradescantia virginiana (Palevitz and Cresti, 1989). The generative cell microtubular apparatus continuously reorganizes during the process of cell division and the mitotic spindle is directly formed by the microtubules in interphase array. Cytokinesis appears to utilize a furrowing process and no phragmoplast or cell plate were observed. When the generative cell division process is finishing, the microtubules interconnecting the two sperm cells have a fragmentation in the zone where a new formed wall divides the sperm cells (Palevitz and Cresti, 1989). It is intersting to note that in Nicotiana after sperm cell formation, a tail-like structure reorganizes in both the sperm cells and that microtubules elongate from the sperm cell body to the tail, showing a microtubular array similar to that observed in generative cell (Tiezzi et al., in preparation).

Cytoskeleton and Sperm Cell Motility in Animals and Higher Plants

The fertilization process is depending on specialized cells named egg cells and sperm in both animals and higher plants. Whereas the egg belong and represent the female contribution to the zygote formation and are essentially blocked in a specialized apparatus, the sperm represent the male contribution to the zygote formation and are equipped with specialized facilities to allow processes of movement in order to favorite the interaction with the egg.
In animals the locomotor apparatus of the sperm is constituted by the flagellum. Inside the flagellum is present the functional apparatus generating movement, the axoneme, a specialized structure well characterized both at ultrastructural and biochemical level. The movement process is due to the specific interaction between microtubules and dynein, a protein located on the microtubule surface and having ATPase activity. Sperm move in a water rich environment with modalities resembling the movement of some algae in which the locomotor system can be formed by flagellae or ciliae but whose functional apparatus generating movement is mostly the axoneme.
In higher plants the sperm cells move inside the pollen tube in environmental conditions surely different with respect to animal sperm. As the movement takes place in the cytoplasm, the water content is lower than in animals and the modalities of movement are consequently related to the structure of the pollen tube.
In a large part of the studied plants the sperm cells do not seem to be equipped with an autonomous system of locomotion; nevertheless, it should be noted that in some plants the sperm cells show the presence of a specialized structure specifically located on the tail and resembling the typical 9+2 organization of the axoneme. These findings, first reported on the sperm cells of Pteridium aquilinum (Duckett and Bell, 1971) and Zamia integrifolia (Norstog, 1975), were recently showed even in Ginkgo biloba (Li et al., 1989).
From this point of view more general considerations can be added evaluating the sperm cell cytoskeleton of the plants where microtubules are tightly cross linked into bundles by characteristic microtubule side arms (Cresti et al., 1989). In these cases the microtubular organization is not characteristically ordered like in the axoneme, nevertheless some

functional relations between the mentioned microtubular bundles and the axoneme may be present. Remembering that in animals and lower plants the sperm motility is essentially based on the ciliary-flagellar axoneme, it has to be confirmed if during evolution several higher plant sperm cells, although lacking the 9+2 axonemal organization, may have retained a part of the motility apparatus in the form of microtubular bundles (Cresti et al., 1989) partially supporting their movement inside the tube by a process of intermicrotubular sliding.

Role of Ca^{2+} in the Pollen Tube Growth

Pollen grains of many species germinate, when sown in an artificial medium, forming a pollen tube which has a peculiar tip growth. According to the species several growth media can be used. Apart from sucrose, polyethylene glycol etc. the right medium needs only to contain potassium ions, calcium and borate with appropriate anions such as phosphate, nitrate or chlorine. Ca^{2+} is essential for tube growth, but at high concentration (above 10^{-2} M) the effect on the growth is negative (for reviews see Hepler and Wayne, 1985; Steer, 1988).
One of the most significant indications that Ca^{2+} is involved in the tip growth system was studies in Lilium (Jaffe et al., 1975), where the Ca^{2+} accumulates in the tip region and the amount decreases with the distance from the apex. In fact Ca^{2+} enters at the tip region, with other positive iones, and it is absent at the basal region of pollen tube (Weisenseel et al., 1975). The Ca^{2+} gradient in the pollen tube was also recently demostrated by chlorotetracycline (Reiss and Hert, 1978; Polito, 1983) and by ionophore A 23187 (Herth, 1978; Reiss et al., 1985). The data reported in Lilium are not confirmed for two Poaceae (Zea mays and Pennisetum americana) where the Ca^{2+} gradient is absent this is probably due to the different pollen and pollen tube organization in these species (Heslop-Harrison et al., 1985).
The Ca^{2+} gradient seems to be optimal for cytoskeleton activity, which gives an important contribution to the movement of the cells and organelles inside the tube. The secretory

vesicles maintain the supply of new plasma membrane and wall extension throught calcium ion channels (Steer, 1988). Probably also the mitochondrial activity and the organelles transport system are connected to the Ca^{2+} gradient (Steer and Steer, 1989).

In order to elucidate the role of Ca^{2+} in the pollen tube growth, cell motility and transport of organelles, several experiments are in progress in our laboratory in collaboration with Dr. Hepler (University of Massachusetts) using BAPTA buffer that in an appropriate concentration can destroid the Ca^{2+} gradient in the living pollen tube.

REFERENCES

Condeelis J.S. (1974) The identification of F-actin in the pollen tube and protoplast of Amaryllis belladonna. Exp Cell Res 88:435-438

Cresti m, Ciampolini F, Kapil RN (1984) Generative cells of some angiosperms with particular emphasis on their microtubules. J Submicrosc Cytol 16:317-326

Cresti M, Hepler PK, Tiezzi A, Ciampolini F (1986a) Fibrillar structures in Nicotiana pollen: changes in ultrastructure during pollen activation and tube emission. In: Biotechnology and ecology of pollen (Mulcahy DL, Mulcahy GB, Ottaviano E eds) pp 283-288, Springer-Verlag, Berlin, Heidelberg, N.Y.

Cresti M, Tiezzi A, Moscatelli A (1986b) Pollen and pollen tube cytoskeleton. In: Genetic and cellular engineering of plants and microorganisms important for agriculture (Magnien E ed) pp 86-88, Commission of the European Communities, Louvain-La-Neuve

Cresti M, Lancelle SA, Hepler PK (1987) Structure of the generative cell wall complex after freeze-substitution in pollen tubes of Nicotiana and Impatiens. J Cell Sci 88:373-378

Cresti M, Tiezzi A (1989) Germination and pollen tube formation. In: Microsporogenesis: ontogeny and evolution (Knox RB, Blackmore S eds) Academic Press, In press

Cresti M, Murgia M, Theunis CH (1989) Microtubules organization in sperm cells in the pollen tubes of Brassica oleracea L. Protoplasma, In press

Derksen J, Pierson ES, Trass JA (1985) Microtubules in vegetative and generative cells of pollen tubes. Eur J Cell Biol 38:142-148

Ducket JG, Bell PR (1971) Studies on fertilization in archegoniate plants. Changes in the structure of the spermatozoids Pteridium aquilinum L. Kuhn during Cytobiology 4:421-426

Franke WW, Herth W, Van der Woude JW, Morrè JD (1972) Tubular and filamentous structures in pollen tube: possilbe involvement as guide elements in protoplasmic streaming and vectorial migration of secretory vesicles. Planta 105:317-341

Hepler PK, Wayne RO (1985) Calcium and plant development. Ann Rev Plant Physiol 36:397-439

Herth W (1978) Ionophore A 23187 stops tip growth, but not cytoplasmic streaming in the pollen tubes of Lilium longiflorum. Protoplasma 96:275

Heslop-Harrison JS, Heslop-Harrison J, Heslop-Harrison Y, Reger BJ (1985) The distribution of calcium in the grass pollen tube. Proc R Soc Lond B 225:315-327

Heslop-Harrison J, Heslop-Harrison Y, Cresti M, Tiezzi A, Ciampolini F (1986) Actin during pollen germination. J Cell Sci 86:1-8

Heslop-Harrison J, Heslop-Harrison Y (1987) An analysis of gamete and organelle movement in the pollen tube of Secale cereale L. Plant Science 51:203-213

Heslop-Harrison J, Heslop-Harrison Y (1988a) Organelle movement and fibrillar elements of the cytoskeleton in the angiosperm pollen tube. Sex Plant Reprod 1:16-24

Heslop-Harrison J, Heslop-Harrison Y (1988b) Sites of origin of the peripheral microtubule system of the vegetative cell of the angiosperm pollen tube. Ann Bot 62:455-461

Heslop-Harrison J, Heslop-Harrison Y, Cresti M, Tiezzi A, Moscatelli A (1988) Cytoskeletal elements, cell shaping and movement in the angiosperm pollen tube. J Cell Sci 91:49-60

Heslop-Harrison J, Heslop-Harrison Y (1989) Actomyosin and movement in the angiosperm pollen tube: and interpretatioon of some recent results. Sex Plant Reprod, in press

Jaffe LA, Weisenseel MH, Jaffe LF (1975) Calcium accumulations within the growing tips of pollen tubes. J Cell Biol 67:488-492

Johnston GW (1941) Cytological studies on male gamete formation in certain angiosperms. Amer J Bot 28:306-319

Karas I, Cass DD (1976) Ultrastructural aspects of sperm formation in rye: evidence for cell plate involvement in generative cell division. Phytomorphology 26:36-45

Kohno T, Shimmen T (1987) Ca^{2+} induced F-actin fragmentation in pollen tubes. Protoplasma 141:177-179

Kohno T, Shimmen T (1988) Mechanism of Ca^{2+} inhibition of cytoplasmic streaming in Lily pollen tubes. J Cell Sci 91:501-509

Keijzer CJ, Wilms HJ, Mogensen HL (1988) Sperm cell research: the current status and application for plant breeding. In: Plant sperm cells as tools for biotechnology (Wilms HJ, Keijzer CJ eds) Pudoc Wageningen pp 3-8

Lancelle SA, Cresti M, Hepler PK (1987) Ultrastructure of cytoskeleton in freeze-substituted pollen tubes of Nicotiana alata. Protoplasma 140:141-150

Kuznetsov SA, Gelfand VI (1986) Bovine brain kinesine is a microtubule-activated ATPase. Proc Natl Acad Sci USA 83:8530-8534

Lewandowska E, Charzynska M (1977) Tradescantia bracteata pollen in vitro: pollen tube development and mitosis. Acta Soc Bot Polon 46:587-597

Li Y, Wang FH, Knox RB (1989) Ultrastructural analysis of the flagellar apparatus in sperm cells of Ginkgo biloba. Protoplasma 149:57-63

Mascarenhas JP (1989) The male gametophyte of flowering plants. The Plant Cell 1:657-664

Mascarenhas JP, Lafountain J (1972) Protoplasmic streaming, cytochalasin B and the growth of the pollen tube. Tissue Cell 4:11-14

McConchie CA, Russel SD, Dumas C, Tuohy M, Knox RB (1987) Quantitative cytology of the sperm cells of Brassica campestris and B. oleracea. Planta 170:446-452

Miki-Hirosige H, Nakamura S (1982) Process of metabolism during pollen tube wall formation. J Elect Microsc 31:51-62

Moscatelli A, Tiezzi A, Vignani R, Cai G, Bartalesi A, Cresti M (1988) Presence of kinesin in Tobacco pollen tube. In: Sexual reproduction in higher plants (Cresti M, Pacini E, Gori P eds) pp 205-209, Springer-Verlag, Berlin, Heidelberg, N.Y., London, Paris

Norstog K (1975) The motility of cycad spermatozoids in relation to structure and function. In: The biology of the male gamete (Duckett JG, Racey PA eds) pp 135-142 Academic Press, London

O'Mara J (1933) Mechanism of mitosis in pollen tubes. Bot Gaz 102:629-636

Ota T (1957) Division of the generative cell in the pollen tube. Cytologia 22:15-27

Palevitz BA, Cresti M (1988) Microtubule organization in the sperm of Tradescantia virginiana. Protoplasma 146:28-34

Palevitz BA, Cresti M (1989) Cytoskeletal changes during generative cell division and sperm formation in Tradescantia virginiana. Protoplasma 150:54-71

Parthasarathy MV, Perdue TD, Witztum A, Alvernaz J (1985) Actin network as a normal component of the cytoskeleton in many vascular plant cells. Amer J Bot 72:1318-1323

Perdue TD, Parthasarathy MV (1985) In situ localization of F-actin in pollen tube. Europ J Cell Biol 39:13-20

Pierson ES, Derksen J, Traas JA (1986) Organization of microfilaments and microtubules in pollen tubes grown in vitro or in vivo in various angiosperms. Eur J Cell Biol 41:14-18

Pierson ES (1988) Rhodamine-phalloidin staining of F-actin in pollen after dimethylsulphoxide permeabilization. A comparison with the conventional formaldehyde preparation. Sex Plant Reprod 1:83-87

Pierson ES, Kengen HMP, Derksen J (1989) Microtubules and actin filaments co-localize in pollen tubes of Nicotiana tabacum L and Lilium longiflorum Thunb. Protoplasma 150:75-77

Polito V (1983) Membrane-associated calcium during pollen grain germination: a microfluorometric analysis. Protoplasma 117:226-232

Raudaskoski M, Astrom H, Perttila K, Virtanen I, Louhelainen J (1987) Role of the microtubule cytoskeleton in pollen tubes: an immunocytochemical and ultrastructural approach. Biol Cell 61:177-188

Reiss HD, Herth W (1978) Visualization of Ca^{2+} gradient in growing pollen tubes of Lilium longiflorum with chlorotetracycline fluorescence. Protoplasma 97:373-377

Reiss HD, Herth W, Nobling R (1985) Development of membrane and calcium-gradients during pollen germination of Lilium longiflorum. Planta 163:84-90

Russell SD (1985) Preferential fertilization in Plumbago: ultrastructural evidence for gamete-level recognition in an angiosperm. Proc Natl Acad Sci USA 82:6129-6132

Sax K, O'Mara J (1941) Mechanism of mitosis in pollen tubes. Bot Gaz 102:629-636

Stanley RG, Linskens HF (1974) Pollen: biology, biochemistry, management. Springer-Verlag, NY

Steer MW (1988) Calcium control of pollen tube tip growth. Biol Bull 176S:18-20

Steer MW, Steer JM (1989) Pollen tube tip growth. New Phytol 111:323-358

Tang X, Hepler PK, Scordilis SP (1989) Immunochemical and immunocytochemical identification of a myosin heavy chain polypeptide in Nicotiana pollen tubes. J Cell Sci 92:569-574

Tiezzi A, Cresti M, Ciampolini F (1986) Microtubules in Nicotiana pollen tubes: ultrastructure, immunofluorescence and biochemical data. In: Biology of reproduction and cell motility in plants and animals (Cresti M, Dallai R eds) pp 87-94, University of Siena, Siena

Tiezzi A, Moscatelli A, Milanesi C, Ciampolini F, Cresti M (1987) Taxol-induced structures derived from cytoskeletal elements of Nicotiana pollen tube. J Cell Sci 88:657-661

Tiezzi A, Moscatelli A, Cresti M (1988a) Taxol-induced microtubules from different sources: an ultrastructural comparison. J Submicrosc Cytol Pathol 20:613-617

Tiezzi A, Moscatelli A, Ciampolini F, Milanesi C, Murgia M, Cresti M (1988b) The cytoskeletal apparatus of the generative cell in several angiosperm species. In: Sexual reproduction in higher plants (Cresti M, Pacini E, Gori P eds) pp 215-220, Springer-Verlag, Berlin, Heidelberg, NY, London, Paris

Tiezzi A, Moscatelli A, Murgia M, Russell SD, Del Casino C, Bartalesi A, Cresti M (1989) Immunofluorescence studies on microtubules in the male gamete of Hyacinthus orientalis and Nicotiana tabacum using confocal scanning laser microscopy. Sperm Cell Club Meeting, 8-12 June, Budapest, Hungary (in press)

Tiwari SC, Polito VS (1988a) Spatial and temporal organization of actin during hydration, activation and germination of pollen in Pryus communis L.: a population study. Protoplasma 147:5-15

Tiwari SC, Polito VS (1988b) Organization of the cytoskeleton in pollen tubes of Pyrus communis: a study employing conventional and freeze-substitution electron microscopy, immunofluorescence and Rhodamine-Phalloidin. Protoplasma 147:100-112

Vale RD, Reese TS, Sheetz MP (1985) Identification of a novel force-generating protein, kinesin, involved in microtubule-based motility. Cell 42:39-50

Vale RD (1987) Intracellular transport using microtubules-based motors. Ann Rev Cell Biol 3:347-378

Vallee RB, Wall JS, Paschal BR, Shpetner HS (1988) Microtubules associated protein 1C from Brain is a two-headed cytosolic dynein. Nature 332:561-563

Van der Woude WJ, Morrè DJ (1968) Endoplasmic reticulum-dictyosome-secretory vesicle associations in pollen tubes of Lilium longiflorum Thunb. Proc Indian Acad Sci 77:164-170

Van Lammeren AAM, Keijzer CJ, Willemse MTM, Kieft H (1985). Structure and function of the microtubular cytoskeleton during pollen development in Gasteria verucosa (Mill) H Duval. Planta 165:1-11

Wagner VT, Murgia M, Ciampolini F, Milanesi C, Tiezzi A, Cresti M (1989) Generative cell cytoskeleton of Hyacinthus oritentalis. Submitted for pubblication

Weisenseel MHR, Nuccitelli R, Jaffe LF (1975) Large electrical currents traverse growing pollen tubes. J Cell Biol 66:556-567

THE CYTOSTATIC FACTOR (CSF) THAT CAUSES METAPHASE ARREST IN AMPHIBIAN EGGS

Yoshio Masui
Department of Zoology
University of Toronto
Toronto, Ontario M5S 1A1
Canada

Meiotic Arrest of Animal Oocytes

Shortly after gamete fusion, chromosome cycles of female and male pronuclei are synchronized under the influence of egg cytoplasm before they form a zygote nucleus. However, if fertilization took place between gametes whose chromosome cycles were continuing, it would be difficult for the zygote to synchronize pronuclear chromosome cycles. If this were the case, perhaps, gametes with chromosomes which were arriving at various phases of the cycle would conjugate randomly, thus more frequently bringing about asynchrony between pronuclear chromosome cycles, and also chromosome aberration in the zygote nucleus. Therefore, it seems reasonable that in nature fertilization takes place only between gametes whose chromosome cycles are arrested at definite phases.

While chromosome cycles in male gametes of all species are arrested after completing meiosis, those in female gametes are arrested at various phases during meiosis in a species-specific fashion. In some species meiotic arrest occurs at the germinal vesicle stage, and in others at the pronuclear stage. However, in a majority, it occurs at metaphase of the first or second meiosis (Masui 1985 for review). In amphibians the arrest occurs at the second metaphase (M II) as in other vertebrates.

In fact, metaphase arrest of chromosome cycles is a rather singular event in nature. In cells other than eggs, chromosome cycles are arrested usually at interphase except for the zygote of Polychoerus carmelensis whose first mitosis is arrested at metaphase (Costello 1961). The reason for this species-specific pattern of meiotic arrest of animal oocytes has remained a mystery, although its evolutionary significance is considered an adaptation of reproductive strategy to optimize fertilization in a given environment (Masui 1985).

NATO ASI Series, Vol. H 45
Mechanism of Fertilization
Edited by B. Dale

As well, the mechanism that causes meiotic inhibition in oocytes has not been clarified. The late professor Alberto Monroy suggested that "inhibition may, in fact, depend on the formation of specific enzyme inhibitors, or it may be due to the upsetting of the equilibrium of key metabolic reactions, either as a result of an enzymatic intermediate step or of the excess production of a metabolite. On the other hand, it may be brought about by the formation of allosteric effectors; or finally, it may depend on the alteration of the steric configuration of some proteins, whether endowed with enzymatic activity or not"(p 377 Monroy & Tyler 1967). In fact, his final suggestion has been generally supported by experiments which will be described in the following chapters.

Cytostatic Factor in Frog Oocytes

Oocytes of the frog, *Rana pipiens*, produce two cytoplasmic factors during maturation which control their nuclear activities : "Maturation Promoting Factor(MPF)" which,when injected into oocytes at the germinal vesicle stage, induces the nucleus to change its state from interphase to metaphase, and "Cytostatic Factor (CSF)" that causes the arrest of the chromosome cycle at metaphase when it is injected into zygotes (Masui & Markert 1971). These factors disappear from egg cytoplasm shortly after eggs have been activated. This suggests that they are responsible for the maintenance of the metaphase arrest in unfertilized eggs.

MPF is now known to appear at metaphase not only during meiosis, but also during mitosis, in a variety of cells, but always disappears during interphase,suggesting it controls chromosome cycles (Lohka 1989 for review). Therefore, it seems that the maintenance of chromosomal state at metaphase requires persisting MPF activity. The fact that blastomeres of 2-cell embryos injected with unfertilized egg cytoplasm remain at metaphase for a long time suggests that CSF helps the cell maintain MPF activity in the cytoplasm.

In an early experiment to determine localization of CSF in *Rana pipiens* eggs, egg cytoplasm was stratified by a weak centrifugal force, and CSF activities in different layers of cytoplasm were examined (Masui 1974). The results showed that most CSF activity had accumulated in the cytosol. However,cytosols obtained from homogenized eggs never exhibited CSF activity. Instead, the cytosol containing CSF activity could be extracted together with MPF only from eggs crushed by a centrifugal force (Wasserman & Masui

1976, Meyerhof and Masui 1977, Shibuya & Masui 1988). Also, in order to maintain CSF activity in cytosols, EGTA and Mg must be present in extraction medium to prevent Ca action (Meyerhof and Masui 1977). Notwithstanding these precautions, CSF activity was unstable.

However, it was intriguing that while CSF extracted with Ca-free medium was short-lived,CSF extracted with medium containing Ca could persist longer (Masui 1974).This paradox was resolved by the finding that while Ca addition to fresh cytosols could rapidly inactivate CSF, it also caused CSF activity to reappear during storage (Meyerhof & Masui 1977, Shibuya & Masui 1989b). Hence, Ca-sensitive CSF in fresh cytosols, which is unstable, has been designated primary CSF (CSF-1), and Ca-induced CSF in stored cytosols, which is very stable, secondary CSF (CSF-2).

Xenopus laevis eggs also have CSF. However, when egg cytoplasm is injected into blastomeres of 2-cell embryos, recipient cells frequently degenerate. This has made it difficult to demonstrate the effect of CSF. It was found that the degeneration of recipient cells could be prevented when cytoplasm was transferred from donor eggs which had been injected with EGTA, and then the CSF effect was observed(Meyerhof & Masui 1979a). Further, when the K level of the external medium in which recipient blastomeres were injected was increased, degeneration of recipient blastomeres was prevented, but this time without prior injection of EGTA into donor eggs. Thus, when cytoplasm of unfertilized eggs was injected into recipient embryos in medium containing 50 mM NaCl, 30 mM KCl, 2 mM $MgSO_4$ and 10% polyvinylpyrolidone (pH 6.3),recipient blastomeres were arrested at metaphase, while blastomeres injected with cytoplasm from 2-cell embryos under the same condition continued cleavage (Moses & Masui in press).

Extraction of CSF from Xenopus eggs also requires special precautions, since Xenopus eggs are usually activated if they are immersed in extraction medium colder than 6°C for 10 to 20 min (Masui unpublished). This is not the case for Rana eggs. However, this cold shock activation can be avoided by treatment of eggs with CO_2. Therefore, cytosols containing CSF activity should be extracted from eggs which have been exposed to CO_2-saturated 50 mM Na_2CO_3 for 10 min before they are crushed by centrifugation. Furthermore, to preserve CSF activity in extracted cytosols, it is necessary to add β-glycerophosphate to the extraction medium. It was noted that although fresh cytosols extracted without β-glycerophosphate showed no CSF activity immediately, they exhibited strong CSF activity a week later if Ca had been added to the fresh extracts (Moses & Masui in press). This observation suggests that Xenopus egg cytosols also develop CSF-2 activity during storage in the presence of Ca.

Molecular Characteristics of CSF

The activity of CSF-1 appears to be dependent on protein phosphorylation in fresh cytosols. This is suggested by the fact that CSF-1 activity in Rana cytosol can be stabilized and even increased when ATP, γ-S-ATP, or NaF (phosphatase inhibitor) is added to the cytosols (Shibuya & Masui 1988). Thus, only in the presence of ATP and NaF, CSF-1 can be precipiated with ammonium sulphate (AmS) solution between 20 and 40% (Shibuya & Masui 1989b). Similarly, Xenopus egg cytosols exhibited no CSF activity when β-glycerophosphate, another phosphatase inhibitor, was removed from extraction medium (Moses & Masui in press). In Rana pipiens egg extracts, however, it was found that once its initial CSF-1 activity was lost, it could not be reactivated by addition of ATP, and that after precipiated with AmS, ATP and NaF had no effect on its activity (Shibuya & Masui 1988). These observations suggest that the maintenance of CSF-1 activity requires continual protein phosphorylation, which must be dependent on its own activity and also on some other factor(s) which may be inactivated or separated from CSF-1 by AmS fractionation.

In contrast, CSF-2 activity is very stable. Once it appears in stored cytosol, under the influence of Ca, it is no longer dependent on biochemical activities of cytosols, and resistant to treatments with 4M LiCl, 8M urea and alkaline pH (11.3) (Shibuya & Masui 1989a), but not to heat treatment above 55° C (Masui 1974). Once CSF-2 developed in stored cytosols, it could easily be precipitated with AmS solution between 20 and 40% saturation. However, CSF-2 does not develop from the fractions of fresh cytosols which can be precipitated with AmS solution between 20 and 40% saturation and contains CSF-1. Clearly, CSF-1 is not the precursor of CSF-2. Instead, CSF-2 develops from fractions which can be precipitated in AmS solutions above 50% saturation and have no CSF activity (Shibuya & Masui 1989b).

CSF-1 activity appears to be associated with rather small molecules having a low sedimentation coeficient(3s),whereas CSF-2 is a molecule larger than $2x10^{6}$ daltons. After purification by column chromatography, none of proteins contained in CSF-2 samples can enter electrophoresis gels. However, partially purified samples of CSF-2 as well as CSF-1 show susceptibility to proteolysis. They are inactivated by proteolytic enzymes, and at the same time degraded into small molecules which can be electrophoresed, but rather resistant to RNAse treatment (Shibuya & Masui 1989a). It may be concluded that CSF-1 and CSF-2 are both protein molecules.

Relative Roles of CSF, MPF and Protein Synthesis in the Control of Chromosome Cycles

The cytoplasm of Rana pipiens eggs which have been arrested at MII has the ability to convert injected nuclei into metaphase chromosomes, called "Chromosome Condensation Activity (CCA)" (Ziegler & Masui 1973, 1976). CCA disappears, together with MPF and CSF activities, when eggs are activated. The cytoplasm of CSF-arrested blastomeres also exhibits CCA when brain nuclei or demembranated sperm nuclei are injected(Meyerhof and Masui 1979b, Shibuya & Masui 1982). Conversely, when fresh cytosols extracted from unfertilized eggs of Rana pipiens are injected into fertilized eggs within 30 min of insemination, these extracts show no CSF activity while they are able to arrest fertilized eggs at metaphase of the first mitosis if injected at later times. This result implies that Rana pipiens zygotes have the ability to inactivate CSF-1 at the very early stage of fertilization, the time of egg activation, whereas those at later stages lose this ability.

On the other hand,CSF as well as MPF in fresh cytosolic extracts from unfertilized Rana pipiens eggs quickly lose their activities when Ca is added at concentration as low as 10 M (Wasserman & Masui 1976, Meyerhof & Masui 1977). Therefore it seems reasonable to assume that injected CSF-1 must be inactivated in the cytoplsm of activated eggs due to elevation of the cytosolic Ca level shortly after fertilization (Busa & Nuccitelli 1985).

Furthermore, it was found that CSF-arrested blastomeres could also be released from metaphase block when a large number of sperm nuclei (Shibuya & Masui 1982) or high doses of Ca were injected (Newport & Kirschner 1984). In both cases MPF as well as CCA disappear from cytoplasm of the recipient cells, and the cells resume cell cycle acativities. These observations suggest that sperm nuclear factor as well as Ca have a similar effect that deprives the egg cytoplasm of MPF and CCA to destabilize its metaphase state maintained by CSF.

Although the previous study showed that MPF could be inactivated quickly in the presence of Ca in cytosols (Wasserman & Masui 1976), recent studies with partially purified MPF have indicated that MPF itself is rather insensitive to Ca (Kishimoto 1988 for review). Therefore, it is highly probable that a loss of MPF avativity, which leads to destabilization of the metaphase state, is due to the Ca action on cytoplasmic factors other than MPF. One of the possible targets for Ca action may be CSF.

Destabilization of the metaphase state by sperm nuclei injected in a large number might be explained if some components of the sperm nuclei neutralize cytoplasmic factors necessary for the maintenance of the metaphase state. In the mouse it was found that the capacity of the metaphase oocyte cytoplasm to transform sperm nuclei to metaphase chromosome, i.e. CCA is proportional to its volume (3 nuclei per oocyte), and therefore, if the number of sperm nuclei exceeds this capacity, all nuclei are transformed into interphase nuclei (Clarke & Masui 1987). This suggest that metaphase factors required for chromsome condensation are consumed by sperm nuclei in a dose-dependent manner.

In order for amphibian oocytes at the germinal vesicle stage to develop MPF protein synthesis is required (Wasserman & Masui 1975), and even after MPF has appeared inhibition of protein synthesis at M I causes a loss of CCA (Ziegler & Masui 1976). In fertilized eggs MPF reappears just before metaphase (Wasserman & Smith 1978), but the inhibition of protein synthesis suppresses the reappearance of MPF (Miyake-Lye & Kirschner 1983). In sea urchin zygotes, it was found that protein synthesis was required for cells to progress from interphase to metaphase,and inhibition of protein synthesis causes the arrest of the cells at G_2 phase (Wagenaar 1983). Also, in zygotes of marine invertebrates it was observed that a group of a few proteins called "cyclins", were continuously synthesized and accumulated during interphase, but they were rapidly degraded at the end of metaphase, exhibiting cyclic changes in concentration during the chromosome cycle (Evans et al. 1983). In Xenopus, it was shown that translation of cyclin mRNA injected oocytes could induce the cells to enter metaphase (Swenson et al. 1986). These observations taken together strongly suggest that continuous synthesis of a special kind of proteins is required for interphase cells to enter metaphase.

However, in frog eggs at M II, after CSF has appeared in the cytoplasm, protein synthesis inhibition has no effect on CCA suggesting that there is no requirement for protein synthesis to maintain a metaphase state (Ziegler & Masui 1976, Masui et al. 1979). Therefore, CSF may play an important role in maintaining metaphase conditions of the cytoplasm in the absence of protein synthesis. In fact, it was shown that while MPF injected into zygotes whose protein synthesis had been inhibited was rather quickly inactivated, it remained active for a long period of time if MPF and CSF were both together injected (Newport & Kirschner 1984). Evidently, CSF has the ability to stabilize MPF activity in living cells lacking protein synthesis activity.

Recently, the role of CSF in the maintenance of cytoplasmic state at metaphase has been analyzed using cell-free systems. When demembranated sperm nuclei of Xenopus, having highly condensed chromatin, were incubated with extracts prepared from activated Rana eggs which had been crushed by centrifugation at 10,000 X G for 15 min, they were transformed into interphase nuclei (Lohka & Masui 1983). However, when sperm nuclei were incubated with extracts prepared from unactivated eggs crushed in the presence of EGTA, the nuclei were transformed into metaphase chromosomes (Lohka & Masui 1984). Similar results were obtained using Xenopus eggs (Lohka & Maller 1985). Further, it was shown that interphase nuclei formed in extracts of activated eggs could be transformed into metaphase chromosomes if cyclin mRNA was translated in the extracts (Minshull et al. 1989, Murray & Kirschner 1989). In fact, the presence of cyclin is prerequisite for the activation of MPF and maintenance of its activity in extracts of activated eggs; inhibition of protein synthesis quickly leads to disappearance of cyclin, and a loss of MPF activity. On the other hand, cyclin and MPF remain stable in CSF-arrested extracts prepared from unactivated eggs using medium containing EGTA even long after protein synthesis has been inhibited (Murray et al. 1989). Apparently, CSF maintains eggs at metaphase by preventing cyclin degradation.

Two Mechanisms of Meiotic Arrest

The conclusion reached above agrees well with the previous observation that when protein synthesis of Rana pipiens oocytes was inhibited, those at M I, whose cytoplasm showed no CSF activity in the cytoplasm, lost CCA, whereas those at M II with CSF acativity in the cytoplasm did not (Ziegler & Masui 1976, Masui et al. 1979). In all probability, this implies that eggs which have developed CSF in the cytoplasm cannot be activated by a protein synthesis inhibitor alone. If so, conversely, eggs arrested at metaphase that can be activated by protein synthesis inhibition do not have CSF in the cytoplasm. In these eggs the maintenance of metaphase before fertilization require continual protein synthesis. Examples of such eggs include those marine invertebrates arrested at M I, and mouse eggs at M II (Masui 1985 for review).

Therefore, there appear to be two different types of metaphase arrest; one dependent on CSF,and the other on continual protein synthesis for their meiotic arrest. However, in both types of eggs, active protein phosphorylation plays an essential role in the maintenance of the metaphase state. As seen in Rana pipiens eggs, CSF-1 activity in the cytosol requires the presence of ATP for its maintenance, suggesting dependency of CSF activity on protein phosphorylation (Shibuya & Masui 1988). In Patella vulgata, eggs are arrested at M I before fertilization, but they can be activated by emetine (protein synthesis inhibitor). However,it was found that treatment with 6-DMAP (protein phosphorylation inhibitor) could cause a loss of CCA without protein synthèsis inhibition(Neant & Guerrier 1988). These examples suggest that meiotic arrest of unfertilized eggs at metaphase requires incessant protein phosphorylation whether it is dependent on the presence of CSF or on continual protein synthesis.

Acknowledgement

The author thanks Mr. N. Duesbery for his proof-reading and helpful suggestions.

References

Busa WB, Nuccitelli, R (1985) An elevated free cytosolic C^{2+} wave follows fertilization in eggs of the frog, Xenopus laevis. J Cell Biol 100: 1325-1329

Clarke HJ, Masui Y (1987) Dose-dependent relationship between oocyte cytoplasmic volume and transformation of sperm nuclei to metaphase chromosome. J Cell Biol 104:831-840

Costello DP (1961) On the orientation of centrioles in dividing cells, and its significance: A new contribution to spindle mechanics. Biol Bull (Woods Hole, Mass) 120:285-321

Evans T, Rosenthal ET, Youngblom J, Disted D, Hunnt T (1983) Cyclin: a protein specified by maternal mRNA in sea urchin eggs that is destroyed at each cleavage division. Cell 33:389-396

Kishimoto T (1988) Regulation of metaphase by a maturation promoting factor Develop Growth & Differ 30:105-115

Lohka MJ (1989) Mitotic control by metaphase-promoting factor and cdc proteins. J Cell Sci 92:131-135

Lohka MJ, Masui Y (1983) Formation in vitro of sperm pronuclei and mitotic chromosomes induced by amphibian ooplasmic components. Science 220: 719-721

Lohka MJ, Masui Y (1984) Effects of Ca^{2+} ion on the formation of metaphase chromosomes and sperm pronuclei in cell-free preparations from unactivated Rana pipiens eggs. Dev Biol 103:434-442

Lohka MJ, Maller JL (1985) Induction of nuclear envelope breakdown, chromosome condensation, and spindle formation in cell-free extracts. J Cell Biol 101:518-523

Masui Y (1974) A cytostatic factor in amphibian oocytes: Its extraction and partial characaterization. J Exp Zool 187:141-147

Masui Y (1985) Meiotic arrest in animal oocytes. In "Bioilogy of Fertilization" (Metz CB, Monroy A eds.) Vol. 1 pp189-219 Academic Press, New York

Masui Y, Markert CL (1971) Cytoplasmic control of nuclear behavior during meiotic maturation of frog oocytes. J Exp Zool 177:129-145

Masui Y, Meyerhof PG, Ziegler DH (1979) Control of chromosome behavior during progesterone-induced maturation of amphibian oocytes. J Ster Bioche 11:715-722

Myerhof PG, Masui Y(1977) Ca and Mg control of cytostatic factors from Rana pipiens oocytes which cause metaphase and cleavage arrest. Dev Biol 61:214-229

Meyerhof PG, Masui Y (1979a) Properties of a cytostatic factor from Xenopus laevis eggs. Dev Biol 72:182-187

Minshull J, Blow JJ, Hunt T (1989) Translation of cyclin mRNA is necessary for extracts of activated Xenopus eggs to enter mitosis.Cell 56:947-956

Miyake-Lye R,Newport J,Kirschner M (1983) Maturation-promoting factor induces nuclear envelope breakdown in cycloheximide-arrested embryos of Xenopus laevis. J Cell Bio 97:81-89

Monroy A, Tyler A (1967) The activation of the egg. In "Fertilization" (Metz CB, Monroy A eds.) Vol. 1 pp369-412, Academic Press, New York

Moses RM, Masui Y (in press) Cytostatic factor (CSF) in the eggs of Xenopus laevis. Exp Cell Res

Moses RM, Masui Y (in press) Cytostatic factor (CSF) activity in cytosols extracted from Xenopus laevis eggs. Exp Cell Res

Murray AW, Kirschner MW (1989) Cyclin synthesis drives the early embryonic cell cycles. Nature 399:275-280

Murray AW, Solomon MJ, Kirschner MW (1989) The role of cyclin synthesis and degradation in the control of maturation promoting factor activity. Nature 339:280-286

Neant I,Guerrier P (1988) Meiosis regulation in the mollusc Patella vulgata. Regulation of MPF, CSF and chromosome condensation activity by intracellular pH, protein synthesis and phosphorylation. Development 102: 505-516

Newport JW, Kirschner MW (1984) Regulation of the cell cycle during early Xenopus development. Cell 37:731-742

Shibuya EK, Masui Y (1982) Sperm-induced cell cycle activities in blastomeres arrested by the cytostatic factor of unfertilized eggs in Rana pipiens. J Exp Zool 220:381-385

Shibuya EK, Masui Y (1988) Stabilization and enhancement of primary cytostatic factor (CSF) by ATP and NaF in amphibian egg cytosols. Dev Biol 129:253-264

Shibuya EK, Masui Y (1989a) Molecular characateristics of cytostatic factors in amphibian egg cytosols. Development 106: 799-808

Shibuya EK,Masui Y (1989b) Fractionation of cytostatic factors from cytosols of amphibian eggs. Dev Biol 135:212-219

Swenson KI, Farrell KM, Ruderman JV (1986) The clam embryo protein cyclin A induces entry into M phase and the resumption of meiosis in Xenopus oocytes. Cell 47:861-870

Wagenaar EB (1983) The timing of synthesis of proteins required for mitosis in the cell cycle of the sea urchin embryos. Exp Cell Res 144:393-403

Wasserman WJ,Masui Y (1975) Effects of cycloheximide on a cytoplasmic factor initiating meiotic maturation in Xenopus oocytes. Exp Cell Res 91:381-388

Wasserman WJ, Masui Y(1976) A cytoplasmic factor promoting oocyte maturation: Its extraction and preliminary characterization. Science 191:1266-1268

Wasserman WJ, Smith LD (1978) The cyclic behavior of a cytoplasmic factor controlling nuclear membrane breakdown. J Cell Biol 78:R15-20

Ziegler DH,Masui Y (1973) Control of chromosome behavior in amphbian oocytes I. The acativity of maturing oocytes inducing chromosome condensation in transplanted brain nuclei. Dev Biol 35:283-292

Ziegler DH, Masui Y (1976) Control of chromosome behavior in amphibian oocytes II. the effect of inhibitors on RNA and protein synthesis on the induction of chromosome condensation in transplanted brain nuclei by oocyte cytoplasm. J Cell Biol 68:620-628

INTRAFOLLICULAR MECHANISMS REGULATING OOCYTE MATURATION

Allen W. Schuetz, Ph.D.
Dept. Population Dynamics
Johns Hopkins University
615 N. Wolfe Street
Baltimore, MD 21205

Summary. Arrest and reinitiation of oocyte maturation occur in association with the processes of folliculogenesis and ovulation. Somatic cells in the ovarian follicle play important and variable roles in regulating these events. This review considers 2 aspects of somatic cell physiology in relation to maturational events in the ovarian follicle: 1) their endocrine metabolic activity (amphibians) and 2) cell cycle characteristics (mammals). In amphibians reinitation of meiosis following gonadotropin stimulation is linked to intrafollicular production and action of progesterone. Endogenous intrafollicular levels of progesterone fluctuate in response to physiological and experimental treatments. Amphibian gonadotropins (pituitary extracts) stimulate a rapid and progressive increase in intrafollicular levels of progesterone which precedes GVBD during in vitro culture. Subsequently, follicular and oocyte levels of the steroid decline, presumably as a result of its metabolism. Intrafollicular progesterone levels respond to manipulation of endogenous cAMP levels. Coordinated stimulation of adenylate cyclase and inhibition of phosphodiesterase enzymes maximize intrafollicular progesterone levels and replicate the steroidogenic effects of gonadotropins. Less dramatic changes in intrafollicular progesterone levels occur in response to alterations in endogenous and exogenous calcium concentrations. The magnitude of the steroidogenic response of follicles varies depending on the nature of the stimulation. Oocyte maturation is not assured in the presence of elevated intrafollicular levels of steroid and cAMP. Evidence suggests that the relative amounts of cAMP and steroid play key roles in determining whether and when oocyte maturation occurs within ovarian follicles. The pathway for gonadotropin stimulation of progesterone synthesis appears to proceed via cAMP synthesis, cholesterol mobilization and pregnenolone synthesis. Continuous protein synthesis is required for progesterone production and involves early events in the steroidogenic cascade prior to enzymatic conversion of pregnenolone to progesterone. Maturation of oocytes, without evident hormone stimulation, can be achieved in the presence or absence of the somatic cells, following activation of protein kinase C or transient exposure to a synthetic protease inhibitor (TPCK) of chymotrypsin. Results indicate that several pathways exist for regulating arrest and reinitation of meiosis in the germ cell and all are sensitive to inhibition by cAMP. In addition to endocrine changes, somatic granulosa cells (mammals) undergo characteristic changes in their cell cycle profiles in association with follicle growth and ovulation of mature oocytes. The proportion of granulosa cells engaged in DNA synthesis (S-phase of the cell cycle) decreases markedly during the periovulatory period. Results suggest that endocrine and cell cycle changes in somatic cells which occur during follicle and oocyte maturation may be inter-dependent processes and an important area for future research.

NATO ASI Series, Vol. H 45
Mechanism of Fertilization
Edited by B. Dale

Introduction

Somatic elements of ovarian follicles play key and variable roles in controlling growth and maturation of oocytes throughout folliculogenesis and oogenesis. During a major portion of their growth oocytes are maintained in meiotic arrest (germinal vesicle stage i.e. G_2 of the cell cycle). A key, early necessary and readily visible event in the meiotic reinitiation process is breakdown of the nucleus or germinal vesicle (GVBD). Following appropriate stimulation by gonadotropins meiosis is reinitiated and nuclear and cytoplasmic differentiation proceed in a characteristic and orderly manner with oocytes becoming developmentally competent. Oocytes, in response to an appropriate maturational stimulus, undergo complex structural, metabolic, physiological and biochemical changes, many of which appear to be mediated or influenced by ions, hormones and endogenous cytoplasmic factors, such as maturation promoting factor (MPF), (Schuetz, 1985). Experimental evidence, obtained more than 20 years ago, linked interruption of meiotic arrest to the local intragonadal production and action of small molecular weight intermediary factors in starfish (Schuetz and Biggers 1967, Schuetz 1969) and amphibians (Schuetz 1967, a b). Accumulating evidence increasingly indicates that this basic regulatory mechanism exists and operates to varying degrees throughout the animal kingdom including mammalian species. Extensive observations and indirect analyses further attest to the functional importance of intrafollicular somatic cell interactions in regulating the progressive growth and differentiation of the oocyte, mediating meiotic maturation and establishing conditions for fertilization and embryogenesis. Striking evidence of this functional and symbiotic relationship is that oocyte development within the ovary does not occur in the absence of or following the loss of viable somatic cells. Furthermore, breakdown of this intimate somatic-germ cell relationship has profound and dire consequences for numerous reproductive processes: follicular and oocyte atresia, failure of ovulation and fertilization, cessation of follicle and oocyte growth, sterility, menopause, parthenogenesis, teratogenesis, gamete aging and embryonic loss (Schuetz 1979).

Interestingly, in contrast to the considerable progress made in deciphering the intra-oocyte (nuclear, cytoplasmic) maturational processes, our understanding of molecular mechanisms involved in intrafollicular transduction of the gonadotropin stimulus (ie. formation, transfer and action of meiotic stimulators) remains fragmentary and incomplete. Several factors have contributed to this situation and include: (1) lack of sensitive and quantitative methods for monitoring changes in specific molecules, (2) the realization that multiple processes may be happening simultaneously, change over time and occur in both the somatic and germ cell compartments of the follicle, (3) that maturational events in follicles and oocytes occur rapidly, and (4) the use of artificial systems or isolated components of the gonads or follicles to study various aspects of the problem. Additionally, although mammalian granulosa cells are one of the most extensively analyzed and studied cell types, the physiological relevance of information obtained as it applies to the oocyte remains obscure.

In recent years we have studied intrafollicular processes in amphibian and mammalian species. Results obtained provide clues, concerning somatic cell steroidogenic and cell cycle changes during folliculogenesis in relation to oocyte maturation. In view of extensive evidence linking progestational steroids to the maturation stimulus in amphibians, we undertook some years ago the development of steroid radioimmunological assay procedures in order to monitor the intrafollicular endocrine environment and study the physiological factors and mechanisms which regulate and mediate such changes (Fig 1).

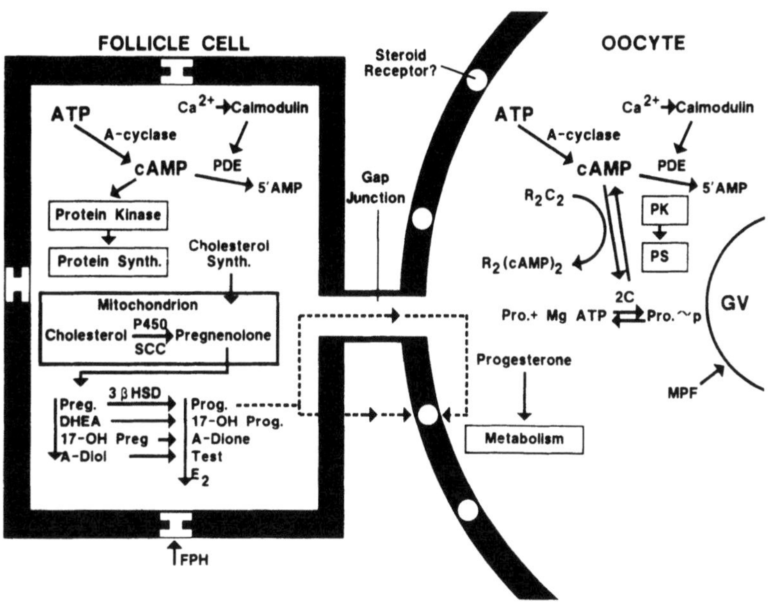

Figure 1: Diagram of Intrafollicular (somatic cell-germ cell) mechanisms which are thought to regulate oocyte meiotic events in amphibians.

Essentially all of our amphibian studies have been performed on isolated Rana pipiens ovaries, follicles and oocytes using standardized culture procedures. A few experiments have been carried out using Xenopus follicles and a number of interesting differences were noted between the species with respect to in vitro steroidogenic responses, (unpublished data).

INTRAFOLLICULAR STEROIDOGENESIS

Early studies showed that the nature and time courses for steroid and pituitary induced oocyte maturation were quite similar (Schuetz 1967, 1971). Thus it seemed feasible that gonadotropins may act simply by transferring stored preformed steroid from somatic cells to the oocyte. Consequently initial experiments focused on determining how the endogenous intrafollicular environment changed during the course of gonadotropin (frog pituitary homogenate-FPH) induced maturation. Time course studies showed repeatedly that a rapid and significant intrafollicular accumulation of progesterone occurred within minutes of gonadotropin stimulation and continued for some hours before peaking (Lin and Schuetz 1984, Petrino and Schuetz 1986a). Subsequently this peak was followed by a rapid decline in intrafollicular steroid levels without a reciprocal increase in medium levels of hormone. Presumably this occured as a result of intrafollicular progesterone catabolism. (See Fig. 2) Invariably the accumulation of steroid following gonadotropin stimulation preceded onset of GVBD by many hours. Typically, the levels of progesterone secreted into the culture medium were a small and variable proportion of the total amount of hormone secreted by the follicles. The amount of steroid synthesized and accumulated within ovarian follicles varied considerably between animals and different physiological or culture conditions. High intrafollicular concentrations of progesterone in excess of ng/follicle, were obtained routinely. If one assumes that the volume of a single follicle is 1 μl (1 mm diameter follicle) 1 ng of progesterone/follicle is equivalent to 1 μg of steroid added to 1 ml of culture medium. Clearly high intrafollicular concentrations of steroid are achievable under physiological conditions which replicate exogenous steroid doses required to induce oocyte maturation under in vitro conditions. A less dramatic elevation of intrafollicular progesterone levels was observed in follicles in which spontaneous oocyte maturation occurred during in vitro culture (Lin and Schuetz 1985). Interestingly, purified mammalian gonadotropic hormones (pituitary, placental) were relatively ineffective in increasing intrafollicular levels of progesterone (unpublished data). Ovarian follicles were also responsive to other polypeptide hormones. Insulin was shown to trigger maturation of Rana oocytes (Lessman and Schuetz 1981, 1982) and results obtained suggest that insulin's effects were partially mediated via actions on the follicle wall. Insulin, compared to FPH, was weakly steroidogenic with respect to progesterone accumulation and augmented effects of progesterone on oocyte maturation. A steroidogenic effect of insulin on mammalian granulosa cells has been observed in a number of laboratories and demonstrates that the action of insulin is not unique to amphibian granulosa cells.

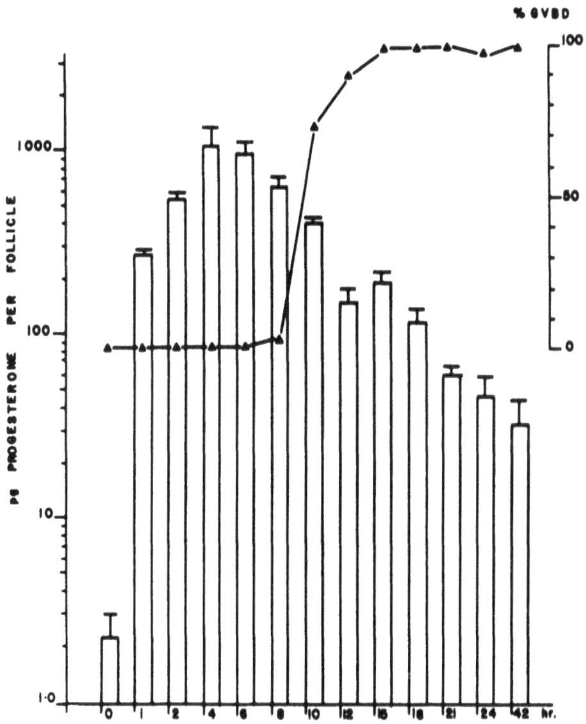

Figure 2: Time course of FPH-induced GVBD and progesterone production. Pooled follicles were randomized and distributed into culture wells (20 follicles/2 ml AR/well). Frog pituitary homogenate (FPH, 0.01 pit. equiv. per well) was added to each well at time zero. At each time point thereafter, follicles were assessed for progesterone and checked for GVBD. Follicles were then extracted with methanol and the extract analyzed for its progesterone content by RIA. Concomitantly, the fixed follicles were cracked open and checked for oocyte maturation (GVBD). Bar graph represents pg progesterone/follicle (mean $\pm$ SEM; n=4). Line graph shows oocyte maturation (%GVBD) at the same time points. (From Lin and Schuetz, 1985).

CAMP STEROIDOGENESIS - OOCYTE MATURATION

Extensive literature exists which documents or implicates cAMP in multiple aspects of follicle differentiation and oocyte maturation in mammals and amphibians. The cyclic nucleotide consistently has been shown or used to inhibit hormone induced or spontaneous in vitro maturation of oocytes. In contrast, cAMP has been identified, particularly in mammals, as a common mediator of steroidogenesis in numerous gonadal and nongonadal cells and tissues. Thus we were particularly interested in elucidating its role in controlling steroidogenesis and oocyte maturation within cultured amphibian ovarian follicles.

Forskolin and IBMX (isobutyl-methyl xanthine) were utilized to manipulate intrafollicular cAMP levels via stimulation of cAMP production (adenylate cyclase) or inhibition of its metabolism by phosphodiesterase. Initial studies demonstrated that appropriate doses of forskolin added to intact follicles, defolliculated oocytes or oocytes denuded of all somatic cells, inhibited progesterone induced oocyte maturation (Kwon and Schuetz 1985). IBMX likewise inhibited FPH and steroid induced maturation of follicular oocytes (Kwon and Schuetz 1986). Results suggested that oocyte cAMP levels were sufficiently responsive to manipulation by either mechanism in the presence or absence of the somatic cells to prevent progesterone from triggering oocyte maturation. Subsequent analyses demonstrated that both Forskolin and IBMX effectively stimulated limited progesterone production in cultured follicles, with Forskolin being somewhat more potent in this regard (Kwon and Schuetz 1986). Neither agent produced progesterone levels equal to those obtained following FPH treatment of cultured ovarian follicles. However, intrafollicular levels of progesterone essentially duplicated those seen in FPH stimulated follicles when follicles were simultaneously exposed to Forskolin or cAMP and IBMX (Figs 3,4 5). These results suggest strongly that maximal progesterone levels are achieved via cAMP when synthesis of the nucleotide is maximized and its metabolism minimized. This hypotheses also implies that the maximum steroidogenic response results from effects of the gonadotropin(s) on both enzyme systems in the follicles. It remains to be determined whether these 2 cAMP regulatory enzymes are responsive to the same or different gonadotropic hormones. Steroidogenic effects of FPH as well as those of Forskolin and IBMX required the presence of the follicle wall somatic components. Based on this information, it was anticipated that oocyte maturation could be achieved via elevation of endogenous cAMP levels without the use of FPH. However, despite intrafollicular accumulation of large amounts of steroid via cAMP manipulation, oocyte maturation was not observed. Furthermore, short term exposure to such treatment regimens followed by extended culture in control medium, produced no oocyte maturation in Rana pipiens oocytes. This result contrasts with the reversible inhibition of oocyte maturation observed following cycloheximide and steroid treatment of follicles (Samson and Schuetz, 1979). More recently oocyte maturation was obtained in another amphibian species (Rana Dybowskii), following transient in vitro stimulation of endogenous cAMP in ovarian follicles (Kwon, Park and Schuetz 1990). It seems likely that failure to induce maturation in Rana pipiens oocytes was due to an inappropriate balance between cAMP and steroid levels. On the basis of all the data it is reasonable to conclude that fluctuations in endogenous intrafollicular cAMP levels are responsible for progesterone synthesis and accumulation in cultured follicles.

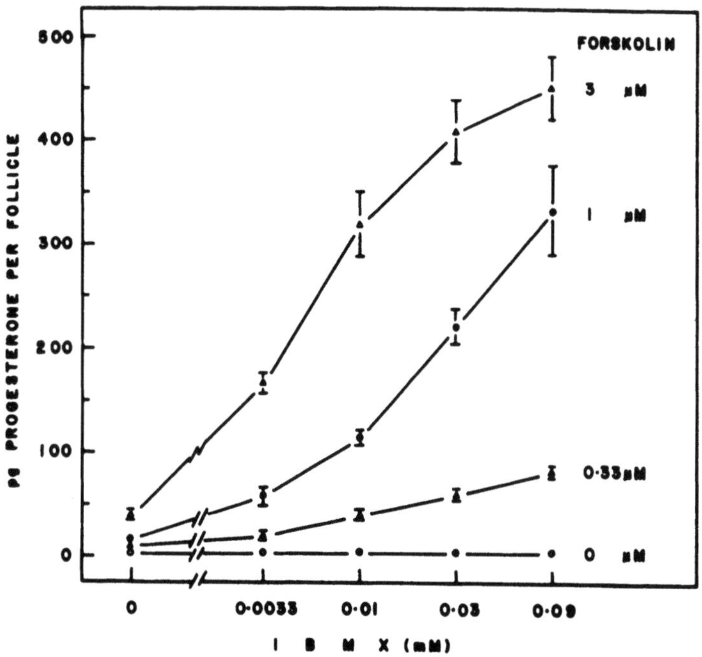

Figure 3: Potentiating effect of IBMX on forskolin-induced follicular progesterone levels. Follicles were cultured for 6 hr in the various concentrations of forskolin (03 μM) and IBMX (0-0.09 mM) and extracted for steroid RIA. Each value represents the mean of 8 determinations (4 well replicate culture with 2 animals, 2 assay duplicates). (From Kwon and Schuetz, 1986).

PROTEIN SYNTHESIS INVOLVEMENT IN FOLLICULAR MATURATION

Early studies demonstrated that metabolic inhibitors of protein synthesis (cycloheximide, puromycin) prevented steroid induced oocyte maturation (Schuetz 1967). Similarly oocyte maturational effects of FPH were inhibited by these same compounds. Based on this information experiments were undertaken to determine whether protein synthesis was involved in production or action of steroids produced following FPH stimulation (Petrino & Schuetz 1986, a,b). Cycloheximide suppressed intrafollicular progesterone production and accumulation when added before or with FPH. Increasing amounts of progesterone were produced with delayed addition of cycloheximide to FPH treated follicles (Fig 6) and suggest that continuous protein synthesis was required for the steroidogenic response.

Cycloheximide inhibition of progesterone production was not observed when follicles were cultured with pregnenolone, the immediate precursor of progesterone (Petrino and Schuetz 1986b). This result suggests that the enzyme (3ß Hydroxysteroid dehydrogenase (3ß HSD)), responsible for pregnenolone conversion to progesterone was already present and functional in the follicles. Thus, the protein(s) necessary for steroid production appear to be synthesized early in the steroidogenic cascade which apparently proceed via cAMP synthesis, cholesterol synthesis mobilization and conversion to pregnenolone (Petrino and Schuetz 1987). Additionally intrafollicular progesterone synthesis and accumulation were shown to be sensitive to inhibition by estrogenic steroids, apparently acting by somewhat different mechanisms (Lin and Schuetz 1983, 1984, Lin et al 1988).

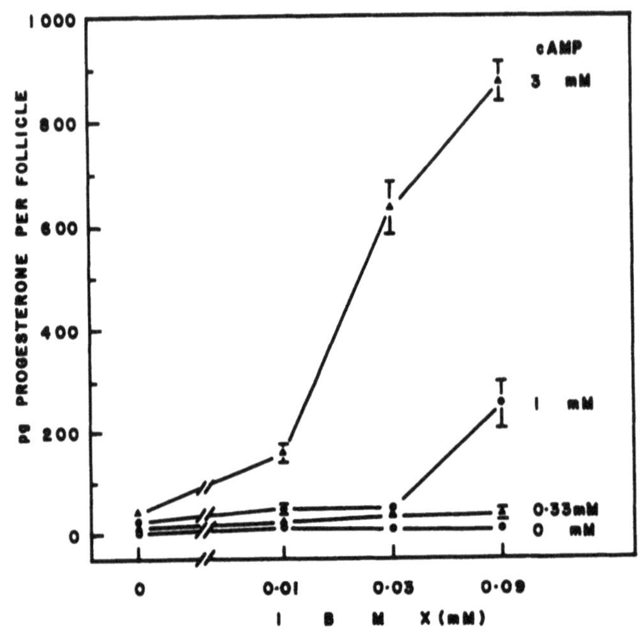

Figure 4: Potentiating effect of IBMX on cAMP-induced follicular progesterone levels. Follicles were cultured for 6 hr in the presence of various concentrations of cAMP (0-3 mM) and IBMX (0-0.09 mM) and extracted for RIA. Each value represents the mean ± SEM of 8 determinations (4 well replicate culture with 2 animals, 2 assay duplicates). (From Kwon and Schuetz, 1986).

ROLE OF CALCIUM IONS STERIODOGENESIS AND OOCYTE MATURATION

Calcium ions have been implicated as an intracellular regulator of numerous oocyte and somatic cell responses including steroidogenesis. Typically, isolated ovarian follicles can be maintained for extended periods in culture medium containing physiological levels of calcium. Under these conditions meiotic arrest is maintained and only basal levels of progesterone are detectable. Simply increasing the extracellular concentration of calcium ions (2-10 fold) bathing non-hormonally treated follicles significantly increased follicular levels of progesterone and in some cases induced oocyte maturation. The elevated progesterone levels were invariably a fraction of that accumulated in response to FPH, although the time course of accumulation was similar in both cases (Fig. 7). Addition of the calcium ionophore A-23187 had dichotomous (stimulatory, inhibitory) effects on accumulated follicular progesterone depending upon the level of calcium present in the culture medium (Kleis-San Francisco and Schuetz 1986). A-23187 increased progesterone accumulation in a dose dependent manner when added to standard culture medium whereas it inhibited progesterone accumulation in the presence of elevated calcium concentrations or following FPH stimulation. In contrast depletion of calcium from the culture medium decreased FPH induced progesterone accumulation and oocyte maturation. Addition of the calcium channel blocker (verapamil) significantly inhibited FPH induced progesterone accumulation and oocyte maturation (Kleis-San Francisco and Schuetz 1987). Likewise progesterone accumulation in response to FPH or exogenous cAMP + IBMX was inhibited by decreasing the level of calcium available to the follicles, whereas pregnenolone conversion to progesterone was unaffected by the same treatments. Furthermore, addition of calmodulin antagonists to FPH treated follicles furthermore inhibited follicular progesterone accumulation as well as exogenous progesterone induced GVBD. Results suggested that calcium acting through calmodulin, plays a role in mediating progesterone accumulation within the ovarian follicle. Both extracellular and intracellular calcium appear to have common as well as different roles to play in mediating gonadotropin induced intrafollicular progesterone accumulation and secondarily, oocyte maturation. Calcium ions may also play a role in regulating spontaneaus oocyte maturation which is occasionaly observed in cultured Rana follicles (Len & Schuetz 1985). Whether effects of Calcium on progesterone production are mediated by one or both cAMP regulating enzymes remains to be determined and of considerable interest.

IN VITRO AND IN VIVO STEROIDOGENESIS

Our data show that synthesis and intrafollicular accumulation of progesterone occured rapidly and progressively following FPH stimulation *in vitro*. However, we were concerned whether this response was unique to cultured follicles or also occured under normal physiological conditions.

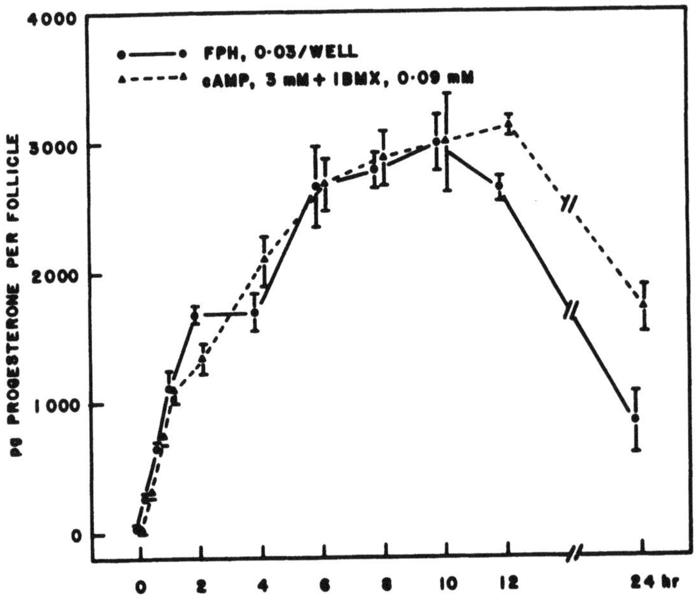

Figure 5: Time course of changes of follicular progesterone levels following FPH or cAMP plus IBMX treatment of follicles. Follicles from one animal were cultured for up to 24 hr in the presence of FPH (0.03 pit. eq./well, 2 ml) or cAMP (3 mM) plus IBMX (0.09 mM) extracted with methanol for steroid RIA at designated time points. Each value represents the mean $\pm$ SEM of 4 determination 2 well (replicates, 2 assay duplicates).(From Kwon and Schuetz, 1986).

To address this question animals were injected with FPH and follicles were subsequently collected and analyzed for their progesterone content before and after different periods of in vitro culture. Results of one such experiment are presented in Table 1 and show that high intrafollicular progesterone levels were present some 8.5 hours after FPH injection and prior to ovulation. (Schuetz and San Francesco, unpublished data). Progesterone levels decreased rapidly during subsequent in vitro culture and similar amounts of steroid were detected in ovulated oocytes and unovulated follicles. Also, elevated levels of steroid were measured in recently ovulated body cavity oocytes (unpublished data) devoid of all somatic cells; presumably this is a result of the intrafollicular transfer of steroid from somatic cells to the oocyte. Both results suggest strongly that the oocyte, rather than the follicle wall, is the primary repository for newly synthesized steroid.

Various attempts have been made to discern (by interrupting granulosa cell-oocyte connections) whether the gap junctions play a role in intrafollicular steroid transfer; however, our results to date have been equivocal (unpublished data). Significantly follicle (granulosa) cells uncoupled from immature oocytes retain responsiveness to FPH and exhibit steroid secretory function and oocyte maturation inducing activity during in vitro culture (Schuetz & Glad, 1985).

PROGESTERONE CONTENT OF FOLLICLES AND OOCYTES FOLLOWING IN VIVO FPH* INJECTION AND IN VITRO CULTURE

Time (Hrs) Collected after FPH Inj.	Duration (Hrs) Culture	(n)	Prog (pg/follicle)
8.5	0	4	1230.5 ± 40.7
23.5**	15.0	4	325.1 ± 23.9
30**	21.5	3	312.0 ± 29.6
In Vitro ovulated oocytes			
24	15.5	4	278.1 + 16.3
29.5	21.0	4	309.8 ± 6.0
31.5	23.0	4	240.0 ± 7.0

* Frog Pituitary Homogenate

** Ovulated follicles following In Vitro culture

From Schuetz and San Francisco unpublished data

The preponderance of data support the idea that steroids produced by the somatic cells move rapidly to the oocyte by simple diffusion. Other studies suggest that the somatic component of the follicle may also carry out other functions relative to steroidogenesis. Endogenous levels of progesterone were monitored following exogenous addition of graded amounts of progesterone to intact follicles or denuded oocytes obtained from the same animal. Results from such experiments showed that oocytes consistently contained higher progesterone levels and also exhibited a higher incidence of maturation than intact follicles (Schuetz & Kwon unpublished data). Thus it appears that in the presence of the follicle wall progesterone accumulation from the culture medium was hindered by some mechanism.

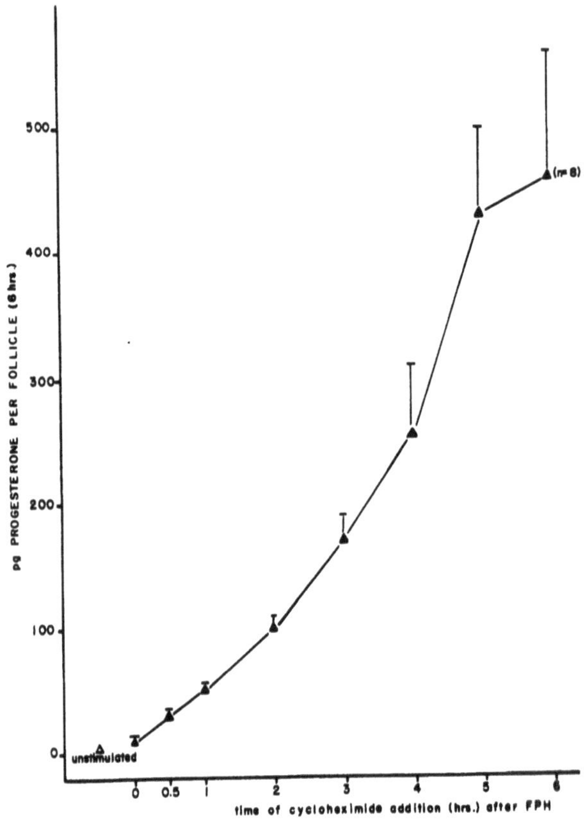

Figure 6: Time of cycloheximide addition relative to FPH stimulation: effect on progesterone production. Treatment groups consisted of follicles (20 follicles per well) stimulated with FPH (0.03 pit. equiv.) at time zero. At subsequent intervals (depicted on x-axis) cycloheximide was added to FPH-stimulated follicles. Follicles in all treatment groups, regardless of the time of cycloheximide addition, were extracted for progesterone at 6 hours after FPH addition. Each point represents pg (mean ± SEM) progesterone per follicle. (From Petrino and Schuetz, 1986a).

SOMATIC CELL-HORMONE INDEPENDENT INDUCTION OF OOCYTE MATURATION

Although follicular somatic cells play crucial role(s) in mediating arrest and reinitiation of meiosis, neither gonadotropins, steroids nor the follicle wall are absolutely required for induction of oocyte maturation (GVBD). A wide range of non-physiological treatments or pharmacological agents have been shown to trigger maturation of follicle enclosed oocytes and, in some instances, denuded oocytes. In recent studies it was demonstrated that activation of protein kinase C by phorbol ester (TPA) induced maturation of follicular and denuded oocytes (Kleis-San Francisco and Schuetz 1988). Similarly transient exposure of non-hormone treated follicles or oocytes to the synthetic protease inhibitor of chymotrypsin (TPCK) triggered reinitation of meiosis (Ishikawa, Schuetz and San Francisco 1989). Induction of follicular oocyte maturation occurred without evident stimulation of progesterone synthesis. Furthermore, maturational effects of TPCK and TPA were readily inhibited by either cycloheximide or elevation of endogenous cAMP levels. Based on such data it is concluded that basic structural and biochemical machinery necessary for mediating oocyte maturation had already differentiated when the follicles were collected for experimentation. Clearly several pathways, or mechanisms exist within the follicle and oocyte for regulating meiotic arrest and for reinsteating meiotic maturation. Interestingly, all of these pathways are sensitive to inhibition by cAMP.

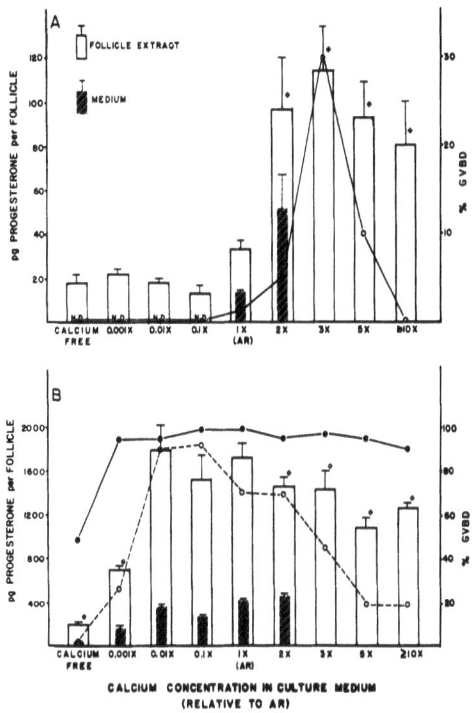

Figure 7: Effects of extracellular Ca^{2+} on endogenous and hormonally stimulated progesterone accumulation and germinal vesicle breakdown (GVBD). Data represent progesterone accumulation over a range of Ca^{2+} obtained by varying the amount of $CaCl_2$ added to amphibian Ringer's (AR), with or without graded doses of EGTA. Data are pooled from ten animals. For each animal intact follicles were cultured in groups of 20 per well in triplicate for each culture condition. No exogenous hormone was added (A) or 0.02 pituitary equivalents of frog pituitary homogenate (FPH; B) were added to cultures. After 4 hours of culture, follicles were extracted with MeOH for RIA determination of progesterone content (open bars). For some animals, incubation media was also collected for direct RIA of progesterone accumulated in the medium (dark bars). Data are expressed as the mean ± S.E.M. for each range of calcium concentration tested. N.D. = not detectable. A diamond indicates significant differences from AR as determined by ANOVA followed by Duncan's multiple range test ($P < .05$). GVBD was assessed in cocultures under the same conditions at 24 hours; % maturation with no exogenous hormone (o - o), 0.02 FPH (o---o), or exogenous progesterone (●-●). (From San Francisco and Schuetz, 1986).

THE CELL CYCLE AND SOMATIC CELL DIFFERENTIATION

Preceding data demonstrate that the somatic cells in fully grown amphibian ovarian follicles, are poised to respond dramatically and rapidly to a gonadotropin stimulus with a cascade of metabolic changes linking cAMP to steroid production and oocyte maturation. Significantly, somatic granulosa cells also undergo other fundamental changes in association with synchronization of follicle growth, oocyte maturation and ovulation and endocrine changes (Readhead et al 1979,Schuetz and Swartz 1979, Schuetz and Dubin 1981). In recent studies using a mammalian animal model, the prepubertal rat, we monitored the cell cycle characteristics of granulosa cells collected during the periovulatory period following gonadotrophin synchronization of follicle growth and ovulation (Schuetz, Whittingham and Legg, 1989).

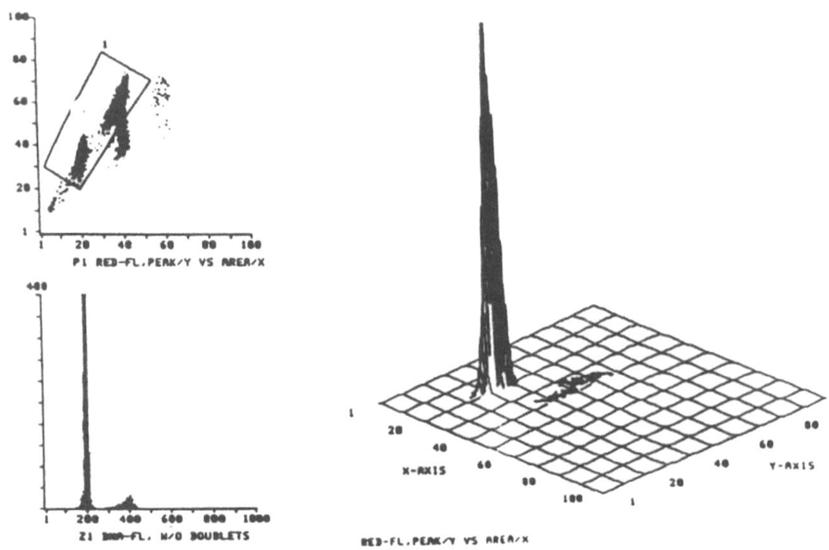

Figure 8: DNA content of nuclei from cumulus granulosa cells separated from ovulated rat oocytes. Approximately 20,000 nuclei were examined in each test using the Orthocomputer program DNADISC. DNA fluorescence was excited at 488 nm and emission was measured above 630 nm. The upper left panel depicts a two dimensional scatter plot, each dot representing one nucleus. The instrument computes the data in histogram form (lower left panel) which permits percentage phase values to be calculated. The number of nuclei are plotted on the Y axis and the amount of DNA on the X axis. The program discriminates against doublets in the population of nuclei examined and they are excluded from data in lower panel. The same data are presented in three dimensional form in the right panel. Coefficent of variation (C.V.) = 5.0% (G_1); 6.0% (G_2). (From Schuetz, Whittingham and Legg, 1989).

The DNA content of individual granulosa cell nuclei was assessed by flow cytometry and the data were utilized to define the proportions of cells at 3 cell cycle phases (G_0+G_1,S,G_2+M_2) on the basis of their relative DNA (2C, intermediate and 4C) content. Examination of the cell cycle profile of cumulus granulosa cells recovered from ovulated oocytes revealed that the majority of cells existed at the G_0+G_1 (2C DNA content) phase of the cell cycle with a small percentage of cells in G_2+M_2 (4C DNA content).

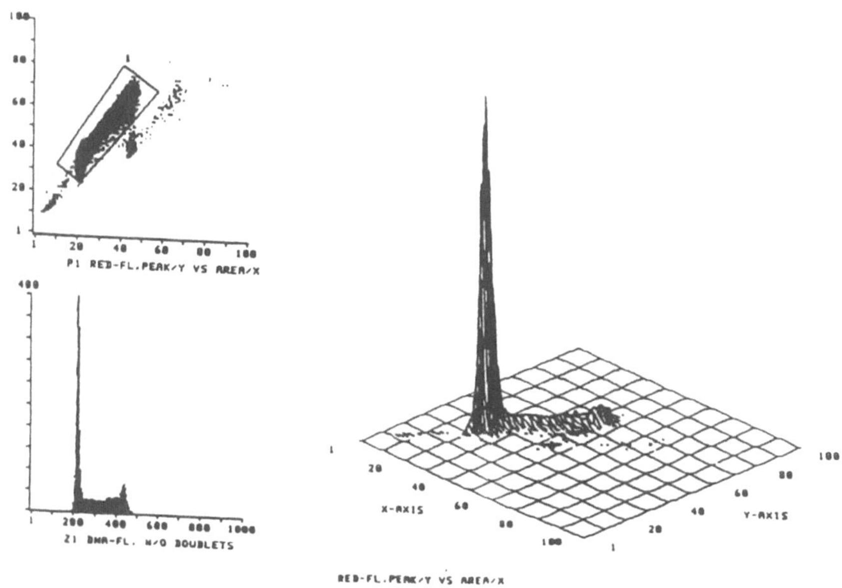

Figure 9: DNA content of nuclei from intrafollicular (cumulus + mural) granulosa cells. Data analyzed as described in Figure 1. (From Schuetz, Whittingham and Legg, 1989).

As indicated by the virtual absence of an S phase (Fig. 8) essentially no cumulus granulosa cells were engaged in DNA synthesis. In contrast, follicular granulosa cells (consisting of both mural and cumulus granulosa cells) collected from growing follicles contained a significant percentage of cells undergoing DNA synthesis (Fig. 9). The proportion and number of granulosa cells at different cell cycle stages changed in a characteristic manner over the course of preovulatory growth of the follicle (Fig. 10). Less dramatic changes were noted in these proportions in cumulus cells collected at different times following ovulation. It is unclear whether these changes reflect the continuation of processes initiated within the ovarian follicle or are the result of oviductal influences. A significant proportion of the cells were engaged in DNA synthesis prior to the ovulatory stimulus (endogenous LH or exogenous HCG). The close association of suppression of DNA synthesis with maturation of the oocyte, expansion of the cumulus cells and ovulation suggests important functional relationships. Clearly, the oocyte and its surrounding somatic cells both exhibit a type of cell cycle arrest in preovulatory follicles; however, this arrest occurs at different stages of the cell cycle in the 2 types of cells.

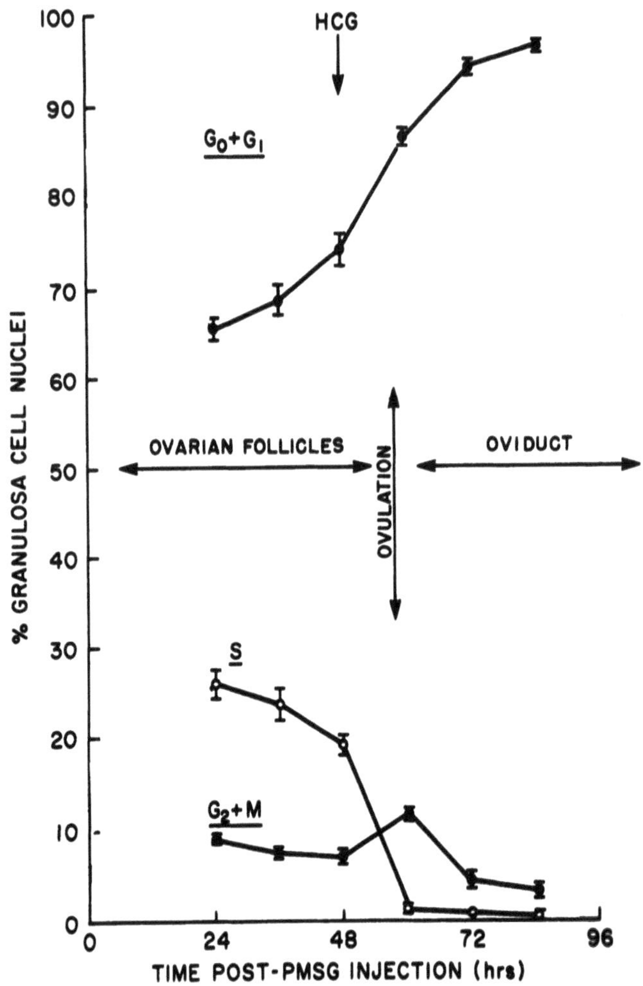

Figure 10: Changes in cell cycle parameters of intraovarian and oviductal granulosa cells collected during the periovulatory period in PMSG injected prepubertal rats. (data from Table 1, Schuetz, Whittingham and Legg, 1989).

Additional studies will be required to determine whether cell cycle changes play a significant functional role in mediating oocyte and granulosa cell differentiation or occur as a consequence of such changes. Further experiments will also be required to define the intrafollicular and intracellular factors, which mediate cell cycle changes within ovarian follicles during the periovulatory periods.

Acknowledgements

I wish to acknowledge the invaluable support for these studies by the NIH and NSF. The secretarial services of Bettie Martin and Janice C. Staton are likewise recognized and appreciated.

Data discussed and reviewed in this article are primarily the results of studies performed in collaboration with students, fellows and international colleagues during recent years. Invaluable contributions of these individuals (P. Lin, T. Petrino, S. Kleis San Francisco, C. Lessman, H.B. Kwon, R. Glad, D.G. Whitingham, R. Legg) in deciphering the complex mechanisms regulating follicular and oocyte maturation are gratefully acknowledged. Furthermore, it is the author's privilege to have known and interacted with Alberto Monroy for many years following our initial meeting as participants (fellow and faculty) on The Fertilization and Gamete Physiology Training Program at The Marine Biological Laboratory at Woods Hole, Massachusetts in 1965. His dynamics personality, interest and encouragement in pursuing the study of oocyte maturation mechanisms are a cherished component of the author's scientific and personal heritage. It is a pleasure to dedicate this article to his memory.

REFERENCES

Ishikawa K, Schuetz AW and San Francisco SK (1989) Induction and inhibition of amphibian (Rana pipiens) oocyte maturation by protease inhibitior (TPCK). Gamete Research 22:339-354.

Kleis-San Francisco S and Schuetz AW (1986) Calcium effects on progesterone accumulation and oocyte maturation in cultured follicles of Rana pipiens. J. Exp. Zool. 240:265-273.

Kleis-San Francisco S and Schuetz AW (1987) Sources of calcium and the involvement of calmodulin during steroidogenesis and oocyte maturation in follicles of Rana pipiens. J. Exp. Zool. 244:133-143.

Kleis-San Francisco S and Schuetz AW (1988) Role of protein kinase C activation in oocyte maturation and steroidogenesis in ovarian follicles of Rana pipiens: Studies with phorbol 12-myristate 13-acetate. Gamete Research 21:323-334.

Kwon HB and Schuetz AW (1985) Dichotomous effects of forskolin on somatic and germ cell components of the ovarian follicle: Evidence of cAMP involvement in steroid production and action. J. Exp. Zool. 236: 219-228.

Kwon HB and Schuetz AW (1986) Role of cAMP in modulating intrafollicular progesterone levels and oocyte maturation in amphibians (Rana pipiens). Developmental Biology 117: 354-364.

Kwon HB, Park HJ and Schuetz AW (1989) Induction and Inhibition of Meiotic Maturation of Amphibian. (Rana Dybowskii) follicular oocytes by forskolin and cAMP in vitro. Gamete Res. In Press.

Lessman CA and Schuetz AW (1981) Role of the follicle wall in meiosis reinitiation induced by insulin in Rana pipiens oocytes. Am. J. Physiol. (Endocrinol. Metab. 4). 241 E51-E56.

Lessman CA and Schuetz AW (1982) Insulin induction of meiosis in Rana pipiens oocytes: Relation to endogenous progesterone. Gamete Research 6:95.

Lin YP and Schuetz AW (1983) In vitro estrogen modulation of pituitary and progesterone-induced oocyte maturation in Rana pipiens. J. Exp. Zool. 116: 281-291.

Lin YP and Schuetz AW (1984) Intrafollicular action of estrogen in regulating pituitary-induced ovarian progesterone synthesis and oocyte maturation in Rana pipiens: Temporal relationship and locus of action. Gen. Comp. Endocrinol. 58: 521-435.

Lin YP and Schuetz AW (1985) Spontaneous oocyte maturation in Rana pipiens: Estrogen and follicle wall involvent. Gamete Research 12: 11-28.

Lin Yp, Kwon HB, Petrino TR and Schuetz AW (1988) Studies on the mechanism of action of estradiol in regulating follicular progesterone levels: Effects of cAMP mediated events and 3ß-Hydroxysteroid dehydrogenase. Develop. Growth and Differ 30(6) 611-618.

Petrino TR and Schuetz AW (1986) Protein synthesis and steroidogenesis in amphibian (Rana pipiens) ovarian follicles. Studies on the conversion of pregnenolone to progesterone. Gen. and Comp. Endocrinol. 63: 441-450.

Petrino TR and Schuetz AW (1986) Protein involvement in regulating pituitary induced progesterone levels in ovarian follicles of Rana pipiens J. Exp. Zool. 239:411-421.

Petrino TR and Schuetz AW (1987) Cholesterol mediation of progesterone production and oocyte maturation in cultured amphibian (Rana pipiens) ovarian follicles. Biol. Reprod. 36: 1219-1228.

Readhead CA, Kaufman MH, Schuetz AW and Abraham GE (1979) Relationship between steroidogenesis and oocyte maturation in rat graafian follicles cultured in vitro. In Ovarian Follicular and Corpus Luteum Function (Channing CP, Marsh JM and Sadler WA eds) Plenum Press, New York 293-300.

Samson GA and Schuetz AW (1979) Progesterone induction of oocyte maturation in Rana pipiens: Reversibility of cycloheximide inhibition. J. Exp Zool 208:213-220.

Schuetz AW (1967a) Effect of steroids on the germinal vesicle of oocytes of the frog (Rana pipiens) In vitro Proc. Soc. Exp. Biol. Med. 124:1307-1310.

Schuetz AW (1967b) Action of hormones in germinal vesicle breakdown in frog (Rana pipiens) oocytes. J. Exp. Zool. 166: 347-354.

Schuetz AW and Biggers JD (1967) Regulation of germinal vesicle breakdown in starfish oocytes. Exp. Cell Res. 46: 624-627.

Schuetz AW (1969) Chemical properties and physiological actions of a starfish radial nerve factor and ovarian factor. Gen. and Comp. Endocrinol. 12: 209-221.

Schuetz AW and Swartz WJ (1979) Intrafollicular cumulus cell transformations associated with oocyte maturation following gonadotrophic hormone stimulation of adult mice. J. Exp. Zool. 207: 399-496.

Schuetz AW and Dubin NH (1981) Progesterone and prostaglandin secretion by ovulated rat cumulus cell-oocyte complexes. Endocrinology 108: 457-463.

Schuetz AW (1985) Local Control Mechanisms during Oogenesis and Folliculogenesis, In Developmental Biology: A Comprehensive Synthesis Vol. 1: Oogenesis (Browder, L. ed.) Plenum Press. New York, pp. 3-83.

Schuetz AW and Glad, R (1985) In vitro production of meiosis inducing substance (MIS) by isolated amphibian (Rana pipiens) follicle cells. Develop Growth and Differ 27(3) 201-211.

Schuetz AW, Whittingham DG and Legg RF (1989) Alterations in the cell cycle characteristics of granulosa cells during the periovulatory period: Evidence of ovarian and oviductal influences. J. Exp. Zool. 249: 105-110.

TYROSINE PHOSPHORYLATION OF MPF AND MEMBRANE PROTEINS DURING MEIOTIC MATURATION OF STARFISH OOCYTES

G. Peaucellier, A.C. Andersen, W.H. Kinsey[1] and M. Dorée[2]
Station Biologique
F29211 Roscoff
France

SUMMMARY

It has been reported that the starfish homolog of the fission yeast *cdc2* protein is a component of maturation promoting factor (MPF) that controls entry of eukaryotic cells into M-phase. The p34*cdc2* protein is phosphorylated during interphase and dephosphorylated during M-phase. We show that starfish p34*cdc2* is phosphorylated *in vivo* on tyrosine, threonine and serine in G2 (prophase I) and G1 (after completion of meiotic divisions). Rephosphorylation of p34*cdc2* is not prevented or modified by inhibition of protein synthesis during M-phase.

Tyrosine phosphorylation was found to occur in proteins from oocytes cortices, by immunoprecipitations with an antibody specific for phosphotyrosine. In ^{32}P-preloaded oocytes, labeled phosphotyrosine containing proteins were noticeable only after hormonal induction of meiotic divisions. Labeling increased in five major phosphoproteins of 195, 155, 100, 45 and 35 kDa until first polar body emission, then decreased upon completion of meiosis. Endogenous *in vitro* phosphorylation of cortices showed a high tyrosine kinase activity towards a 68-kDa protein but no difference between cortices from oocytes treated or not with the hormone.

These findings suggest that tyrosine phosphorylation contributes both to the transduction of the hormonal signal for meiosis resumption and to several steps of the ensuing divisions.

---

1 Department of Anatomy and Cell Biology, University of Miami School of Medicine, Miami, Florida 33101, USA.
2 CNRS and INSERM, BP 5051, 34033 Montpellier Cedex, France.

NATO ASI Series, Vol. H 45
Mechanism of Fertilization
Edited by B. Dale

INTRODUCTION

Protein phosphorylation is a major postranslational regulatory mechanism which plays an important role in the control of cell division. During meiotic divisions of starfish oocytes, for instance, many proteins have been shown to undergo cyclic phosphorylations (Dorée *et al.,* 1983). Most of these changes are correlated with the activity of a major M phase-specific protein kinase, independent of calcium and cyclic nucleotides (Capony *et al.,* 1986). This kinase has been recently identified as the starfish homolog of the p34*cdc2* protein kinase from fission yeast (Labbé *et al.,* 1988; Arion *et al.,* 1988). The *cdc2* protein kinase is an active component of the Maturation Promoting Factor (MPF), which is found in the cytoplasm of a wide range of eukaryotic cells at M phase and is active across phyletic boundaries (Kishimoto, 1982). This has led to the proposal that *cdc2* acts as an universal regulator of M phase in eukaryotic cells (Lee and Nurse, 1988; Dunphy and Newport, 1988).

The regulation of *cdc2* activity is still unclear but there are evidences that it involves both association with polypeptides such as p13, encoded by the Sucl gene in fission yeast (Brizuela *et al.,* 1987), and cyclins (Brizuela *et al.,* 1989; Draetta *et al.*, 1989; Labbé *et al.*, 1989; Meijer *et al.*, 1989) and phosphorylation of p34 itself. Dephosphorylation of *cdc2* has been correlated with its activation at M phase in starfish (Labbé *et al.*, 1989), Xenopus (Gautier *et al.*, 1989; Dunphy and Newport, 1989) and mouse (Morla *et al.*, 1989). In the latter two species, tyrosines were among the main phosphorylated residues. This point needed to be investigated in starfish, which is part of the present work.

Before this recent discovery, tyrosine phosphorylation was known to play an important role in cellular growth control and in the malignant transformation of vertebrate cells (Hunter and Cooper, 1985). Insulin and growth factor receptors possess a tyrosine kinase activity which is necessary for signal transduction (Chen *et al.*, 1987; Chou *et al.*, 1987). This activity is also preserved in oncogenes derived from such receptors (Heldin and Westermark, 1985; Yarden and Ulrich, 1988). Protein tyrosine kinase activity has also been demonstrated in sea urchin egg membranes and shown to increase following fertilization (Dasgrupta and Garbers, 1983; Ribot *et al.*, 1984; Kinsey, 1984; Satoh and Garbers, 1985), whith the exception of a *src*-

related tyrosine kinase whose activity decreased (Kamel *et al.*, 1986). A high molecular weight membrane protein was also found to be transiently phoshorylated on tyrosine *in vivo* (Peaucellier *et al.*, 1988)

In the present study we have used starfish oocytes to study tyrosine phosphorylation changes associated with mitogenic signalling. In this species the G2 to M-phase transition of first meiotic division is triggered by the binding of 1-methyladenine (1-MeAde), an hormone produced by follicular cells under neurosecretory control, to its still unisolated receptor (Yoshikuni, 1988). This study was made feasible by the high rate of phosphate uptake in prophase-blocked oocytes, which allowed ^{32}P-phosphate preloading at high specific activities, and the availability of an antibody specific for phosphotyrosine (Peaucellier *et al.*, 1988)

METHODS

Handling of oocytes - Ripe females of *Marthasterias glacialis* were obtained from the Biological Station of Roscoff. Fully grown prophase-blocked oocytes were prepared free of follicle cells by washing them several times in artificial Ca^{2+}-free seawater (Dorée and Guerrier 75).

[^{32}P]Phosphate in vivo labeling - A 12% suspension of prophase-blocked oocytes was incubated with gentle stirring in sea water containing 0.5 mCi/ml carrier-free [^{32}P]orthophosphate (Amersham). After 3 hours incubation, oocytes were washed three times with Ca^{2+}-free seawater. Meiosis reinitiation was initiated by addition of 1 mM 1-MeAde to a final concentration of 2 µM. In some experiments, proteins synthesis was inhibited by addition of 0.2 mM emetine, 15 min after 1 MeAde (Guerrier and Dorée 75).

Isolation of p34^{cdc2} - Packed oocytes (0.8 ml) were suspended in 9 ml of ice-cold extraction buffer (50 mM Tris pH 7.5; 0.15 M NaCl; 0.1% Triton X100; 0.1mg/ml SBTI; 1 mM PMSF; 50 mM NaF; 10 mM Na pyrophosphate; 10 mM α-naphthyl phosphate; 10 mM phenyl phosphate; 0.1 mM $ZnCl_2$; 0.1 mM Na orthovanadate), homogenized by hand with a teflon peste, frozen in liquid nitrogen and stored at -30 ^{o}C. After thawing in a cooled water bath at 4 ^{o}C, and sonication for 15 sec at 120 V, samples were centrifuged for 30 min at 10^5 g. Each supernatant was incubated for 1 h at 4° with 0.25 ml Sepharose-coupled suc1 (3.5 mg suc1/ml gel) prepared as previously described (Labbé *et al.*, 1989). The gel was transferred in a small column washed with 12.5 ml of buffer (50 mM Tris pH 7.5; 0.15 M NaCl; 0.1% Triton X100; 50 mM NaF; 10 mM Na pyrophosphate; 0.1 mM Na orthovanadate), 1 ml of 50 mM Tris pH 7.5, mixed with 0.25 ml gel sample buffer and boiled for 3 min. Samples were run on a 10% SDS-polyacrylamide gel and transferred electrophoretically to an Immobilon membrane (Millipore) for 2 h at 20 v in 25 mM Tris pH 8.3, 192 mM glycine, 20 % Methanol, 0.1 % SDS. The membrane was washed 3 times with distilled water (10 min each) and air dryed. The p34^{cdc2} band was located by overnight autoradiography, cut out and briefly wetted with methanol and distilled water before HCl hydrolysis and phosphoamino acid analysis.

Analysis of labeled phosphoamino acids - Labeled phosphoamino acid analysis was carried out as described by Cooper *et al.*, (1983). Samples were hydrolyzed in 6 N HCl for 1 at 110 °C. After addition of unlabeled phosphoserine, phosphothreonine and phosphotyrosine, they were analyzed by two-dimensional thin-layer electrophoresis on 100 µm-thick cellulose plates (Sigma) with migration at pH 1.9 (1500 V for 25 min) in the first dimension and at pH 3.5 (1500 V for 25 min) in the second dimension. The position of standards was determined by ninhydrin staining and radioactivity was scored by autoradiography or by scintillation counting of cellulose scraped from the spots.

Antibody production - The polyclonal anti-phosphotyrosine antibody was prepared as previously described (Peaucellier *et al.*, 1988), according to the method of Ross *et al.* (1981). Briefly, New Zealand White rabbits were immunized with keyhole limpet hemocyanin which had been derivatized with the synthetic phosphotyrosine analog *p*-azobenzyl phosphonate. Antibodies were purified by affinity chromatography on cyanogen bromide-activated Sepharose (Pharmacia LKB biotechnologie) derivatized with L-phosphotyrosine. They were eluted with 40 mM phenyl phosphate at 40° C.

Preparation of cortices - Cortices were prepared according to the method of Detering *et al.* (1977). Briefly, 0.6 ml aliquots of oocytes were washed in ice-cold seawater C (0.5 M NaCl; 10 mM KCl; 2 mM $NaHCO_3$; 25 mM EGTA; pH 8.0) and homogenized by hand with a teflon pestle in 10 ml of seawater C containing protease and phosphatase inhibitors (1mg/ml SBTI; 1 mM PMSF; 50 mM NaF; 10 mM Na pyrophosphate; 10 mM α-naphthyl phosphate; 10 mM phenyl phosphate; 0.5 mM $ZnCl_2$; 0.1 mM Na orthovanadate). Cortices were pelleted by low speed centrifugation (1 min at 1000 g) and washed three times with homogenization medium. Phenyl phosphate and o-naphthyl phosphate, which may inhibit antibody binding, were omitted in the last washing. In some experiments, cortical granules were removed according to Kinsey *et al.* (1980), by three washes in isotonic sucrose (1 M sucrose; 0.1 mg/ml SBTI) separated by centrifugation at 5000g for 20 min. When not used immediately (*in vitro* experiments only), cortices and membranes were frozen in liquid nitrogen and stored at -30 °C.

Immunoprecipitations - All operations were performed at 4°C. Each cortex sample was suspended in 3ml of immunoprecipitation buffer (50 mM Tris pH 7.5; 0.15 M NaCl; 1% Triton X100; 0.1% SDS; 0.1mg/ml SBTI; 1 mM PMSF; 50 mM NaF; 10 mM Na pyrophosphate; 0.5 mM $ZnCl_2$; 0.1 mM Na orthovanadate), frozen in liquid nitrogen and stored at -30 °C. After thawing in a cooled water bath at 4 °C, and sonication for 5 sec at 100 V, samples were centrifuged for 30 min at 10^5 g. Two 1.3 ml aliquots were taken from each supernatant and incubated with anti-phosphotyrosine antibody (10 ug/ml) for 2 h, one aliquot used as control also contained 5 mM L-phosphotyrosine. Protein A-Sepharose (Pharmacia), was added (25 ul/sample) and, after 1 h, the immune complexes were collected by centrifugation, washed four times with immunoprecipitation buffer, twice with 50 mM Tris pH 7.5, solubilized in SDS sample buffer and heated at 100 °C for 4 min.

in vitro phosphorylations - They were performed as described previously (Peaucellier *et al.*, 1988). Briefly, cortices were incubated in a phosphorylation buffer containing 10 mM HEPES pH 7.5, 10 mM $MnCl_2$, 10 µM Na_3VO_4, 0.1 mg/ml Leupeptin (Bachem), 0.15 % Nonidet P-40 (BDH). The reaction was started by addition of [γ-^{32}P]ATP (Amersham) to a final concentration of 425 uCi/ml (about 0.1 µM ATP), after a 5 min incubation at

20 ^{o}C, unlabeled ATP was added to a concentration of 25 μM and samples were incubated for another 5 min period. Ten volumes of ice cold immunoprecipitation buffer were added and samples were frozen in liquid nitrogen. Immunoprecipitation was performed as described for *in vivo* experiments.

Gel electrophoresis and autoradiography - For SDS-polyacrylamide slab gel electrophoresis, linear gradients (7.5-20 % acrylamide, 0.28-0.05 % bisacrylamide) were used with the discontinuous buffer system of Laemmli (1970). Molecular weight markers were purchased from Pharmacia. Gels were stained with Coomassie blue R-250, dried under vacuum and processed for autoradiography using X-Omat AR films (Kodak). Some gels were incubated with alkali (2 h in 1 N KOH at 56 ^{o}C) after glutaraldehyde treatment (30 min in 10 % glutaraldehyde) according to Bourassa *et al.* (1988).

RESULTS AND DISCUSSION

Tyrosine phosphorylation of cdc2.

As previously reported, the starfish p34^{cdc2} kinase binds strongly to the p13 protein encoded by the suc1 gene from fission yeast (Labbé *et al.*, 1989). This allowed its extraction, with a high recovery rate, from whole cell homogenates by Sepharose-coupled suc1.

One batch of prophase-arrested oocytes of *M. glacialis*, preloaded with ^{32}P, was divided into three aliquots. One was left untreated, 1 MeAde was added to the second to induce resumption of meiotic divisions, the third one was also treated with 1 MeAde but protein synthesis was inhibited by addition of emetine 15 min later. In oocytes treated with 1 MeAde alone, both polar bodies were emitted and development stopped in G1 after formation of the female pronucleus. In emetine treated eggs, the second meiotic division was prevented and a pronucleus reformed after first polar body emission, as previously reported (Picard *et al.*, 1985). Arrested oocytes from these three batches were homogenized in the presence of phosphatase inhibitors and centrifugated as described in methods. p34^{cdc2} was extracted from the supernatant by incubation with Sepharose-coupled suc1. The proteins retained on the gel were submitted to SDS-polyacrylamide gel electrophoresis and transferred electrophoretically to Immobilon membranes. The p34^{cdc2} band was located by autoradiography of the blot, cut out and hydrolyzed in 6 N HCl at 110° for 1 h, before separation of phosphoamino acids by thin layer electrophoresis. The autoradiograms in Fig. 1 show ^{32}P-incorporation in phosphotyrosine, phosphothreonine and phosphoserine in all three samples. No significant difference was found between the samples even scintillation counting of the radioactives spots: phosphotyrosine, phosphothreonine and phosphoserine accounted for about 15%, 30% and 50% of total incorporation respectively (Table 1).

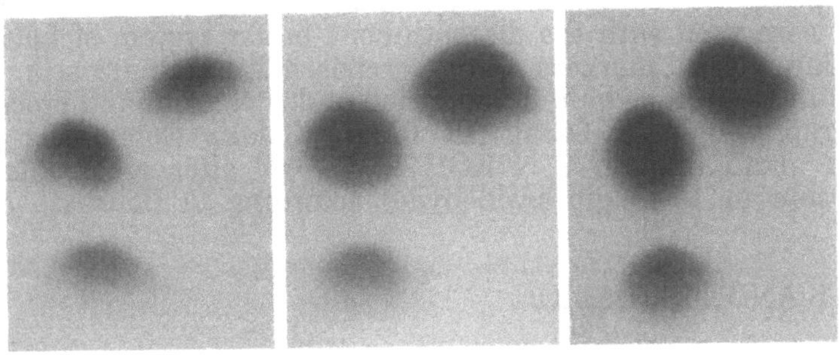

Fig. 1. Phosphoamino acid analysis of *in vivo* labeled p34*cdc2*.
p34*cdc2* was prepared, as described in methods, from 32P-preloaded oocytes blocked in G2 (*Prophase I*) or in G1 after treatment with 1 MeAde in the presence (*G1/emetine*) or absence (*G1/no emetine*) of emetine (added 15 min pha). After HCl hydrolysis for 1 h at 110°C, phosphoamino acids were separated by two-dimensional thin layer electrophoresis at pH 1.9 (*left to right*) and pH 3.5 (*bottom to top*). *P-Ser*, phosphoserine; *P-Thr*, phosphothreonine; *P-Tyr*, phosphotyrosine.

Table 1. 32P-incorporation in p34*cdc2* from arrested oocytes.

Samples	Prophase I	G1/emetine	G1/no emetine
P-Tyr	18 %	12 %	12 %
P-Thr	39 %	34 %	38 %
P-Ser	43 %	54 %	50 %
Total cpm	414	1203	1319

The phosphoamino acid spots were scraped from the plates after autoradiography (cf Fig. 1) and radioactivity was measured by liquid scintillation.

These results are in agreement with those obtained with mammalian cells and *Xenopus* oocytes. They support the assumed correlation between *cdc2* phosphorylation and lack of kinase activity at the G2 (prophase I) as well as the G1 stage (completion of meiosis). However, they do not imply by themselves a special role for phosphorylation on tyrosine versus serine end threonine since total dephosphorylation was observed upon entry in M phase (Labbé *et al.*, 1989). At difference with results obtained in non mitotic 3T3 mouse fibroblasts (Morla *et al.*, 1989) serine phosphorylation seems significantly higher in starfish oocytes. A special feature of the present experiment is that *cdc2* rephosphorylation is not affected by inhibition of protein synthesis, which argues against the hypothesis advanced by Morla *et al.* that *cdc2* tyrosine phosphorylation may be related to its synthesis.

In contrast to results obtained in other species (Draetta *et al.*, 1988; Morla *et al.*, 1989; Dunphy and Newport, 1989), no phosphotyrosine containing protein was detectable with our anti-phosphotyrosine antibodies in western blots from whole starfish oocytes at any stage of meiosis. Negative results were also obtained in immunoprecipitation experiments with anti-phosphotyrosine antibodies on homogenates from ^{32}P-preloaded oocytes. This is probably due to the use of different anti-phosphotyrosine antibodies since significant differences in affinity have been reported, in relation with the antigens used for immunization (Wang, 1988; Kamps and Sefton, 1989)

In vivo tyrosine phosphorylation of membrane proteins

In previous experiments with sea urchin eggs, most tyrosine kinase activity was found in membrane fractions (Dasgrupta and Garbers, 1983; Ribot *et al.*, 1984). Thus, cortices were prepared from ^{32}P-preloaded oocytes, at various times after 1 MeAde addition, in the presence of phosphatase and protease inhibitors as described in methods. These cortices were extracted with detergents (1 % Triton X100 plus 0.1 % SDS) and the high speed supernatant used for immunoprecipitation with anti-phosphotyrosine antibodies. Immunoprecipitates were analyzed by SDS gel electrophoresis and autoradiography.

As shown in Fig. 2A, no significant label was found in immunoprecipitates from prophase I blocked oocytes but several bands were seen after induction of meiosis reinitiation by 1 MeAde. No radioactive band was seen in controls (not shown) : equal aliquots of the same samples incubated with the antibody in presence of 5 mM phosphotyrosine. The first band to appear was a 155-kDa protein, the labeling of which was already significant 2 min after 1 MeAde addition and increased until germinal vesicle breakdown (GVBD) at 20 min posthormone addition (pha). Other phosphoproteins of 195, 100, and 45 kDa were labeled somewhat later, together with some minor ones, as a 85-kDa. Incorporation increased in these bands up to 40 min pha, at which time oocytes approach metaphase I. A 35-

kDa phosphoprotein was the latest to be labeled, becoming apparent at GVBD (20 min pha) and heavily labeled at 40 min pha.

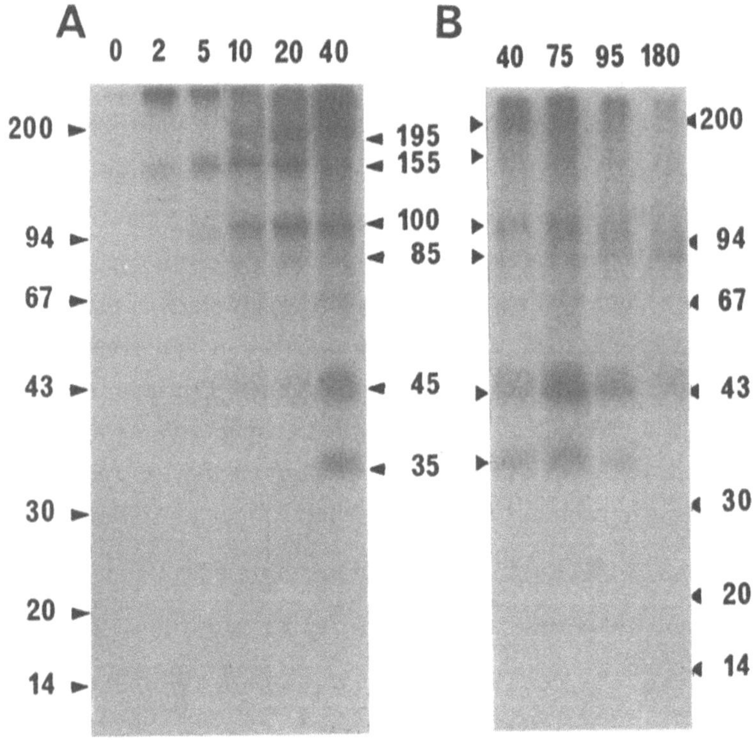

Fig. 2. Imunoprecipitation with anti-phosphotyrosine antibodies of *in vivo* labeled cortices.

Cortices were prepared from ^{32}P-preloaded oocytes at various times after hormone addition and extracted with detergents. The high speed supernatant were incubated with anti-phosphotyrosine antibody and the immunoprecipitates analyzed by SDS gel electrophoresis and autoradiography. Autoradiographs from two experiments on differents batches of oocytes are shown. The times (in minutes post-hormone addition) at which the samples were taken is indicated at the top of each lane. The migrations and molecular weights ($x10^{-3}$) of markers proteins are indicated on the left.

Upon completion of meiotic divisions, as shown in Fig. 2B, the label was seen to decrease in most bands after first polar body emission (occuring 85 min pha), and disappear almost completely after second polar body emission (occuring 120 min pha). The 85-kDa phosophoprotein was the most noticeable exception, since its labeling seemed even slightly increased.

This kind of experiment gives no information on what actually happens to individual phosphoproteins. Increased 32P-incorporation may result from global increase in phosphate content of the protein, protein synthesis, or increased phosphate turnover by activation of both protein kinases and phosphatases activities. The first hypothesis seems the more plausible since meiotic maturation, up to first polar body emission does not require protein synthesis, as previously indicated, and no change in protein phosphatase activity was observed under similar conditions (Pondaven *et al.*, 1987). Moreover, several bands showed a slightly reduced mobility, correlated with increased labeling, which may indicate hyperphosphorylation. Dephosphorylation is the more likely explanation for the observed decreased incorporation upon completion of meiosis but proteolysis cannot be excluded, despite the absence of labeling at lower molecular weights. A third explanation would be a decrease in the specific activity of intracellular ATP but this is unlikely since no change was reported in a previous study under similar conditions (Dorée *et al.*, 1983).

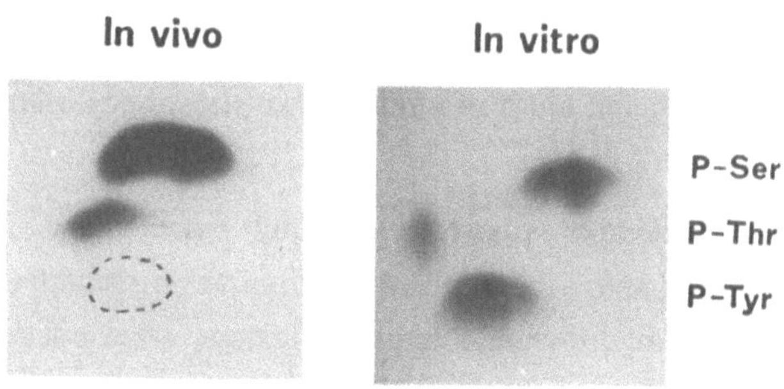

Fig. 3. Phosphoamino acid analysis of proteins from *in vivo* and *in vitro* labeled cortices.

Cortices were prepared from 32P-preloaded (*in vivo*) or unlabeled oocytes (*in vitro*) 20 min after hormone addition. Unlabeled cortices were incubated *in vitro* with 32P-ATP as described in methods. Detergents extracts were precipitated with 10% TCA, washed with acetone and hydrolyzed with HCl for 2 h. Phosphoamino acids were separated by two-dimensional thin layer electrophoresis at pH 1.9 (*left to right*) and pH 3.5 (*bottom to top*). *P-Ser*, phosphoserine; *P-Thr*, phosphothreonine; *P-Tyr*, phosphotyrosine.

Tyrosine phosphorylation is a very rare event in normal cells and it was under the detection limit in phosphoamino acid analysis of whole *in vivo* labeled cortices (Fig. 3 A). The specificity of the antibody for phosphotyrosine was indicated by the observed competition with excess phosphotyrosine and previous experiments with the same batch of antibody which showed no inhibition with phosphoserine or phosphothreonine (Peaucellier *et al.,* 1988). Specific activity in immunoprecipitates was too low to allow phosphoamino acid analysis but the same pattern of incorporation was observed after alkali treatment of the gel, which selectively removes phosphoserine and, to a lesser extent, phosphothreonine (Hunter and Cooper, 1985) (results not shown). There is thus good presumption that all major phosphoproteins detected in immunoprecipitates were significantly labeled on phosphotyrosine.

In vitro tyrosine phosphorylation of membrane proteins

Previous experiments on sea urchin eggs had shown that membranes contained a high endogenous tyrosine kinase activity (Peaucellier *et al.,* 1988). A similar result was obtained in starfish, as shown in Fig. 3B, when phosphoamino acid analysis was performed on *in vitro* phosphorylated cortices. Incorporated radioactivity was approximately equal in phosphotyrosine and phosphoserine, while phosphothreonine accounted for only 15% of total. This imply that tyrosine kinase activity is a major endogenous kinase activity in cortices and that diffusible serine/threonine kinases are responsible for most of the *in vivo* phosphorylations of cortices proteins.

Endogenous tyrosine phosphorylations were also analyzed by immunoprecipitation with anti-phosphotyrosine antibodies. Cortices, prepared from oocytes at different stages of meiosis, were phosphorylated *in vitro* and processed as for *in vivo* experiments. As shown in Fig. 4, progression through meiosis was not correlated with major changes in immunoprecipitated proteins, a 68-kDa phosphoprotein being the most heavily labeled at any stage. Only minor bands showed increased incorporation in cortices from oocytes at first meiotic division (40 min pha) and after completion of meiosis (180 min pha), at the 85 and 80 kDa level for instance. These results indicate that one or several functional tyrosine kinases are already present in cortices of prophase-blocked oocytes. The differences between *in vivo* and *in vitro* experiments can be explained in many ways, such as the existence of an inhibitor which would dissociate under *in vitro* conditions or the inaccessibility of the substrates under *in vivo* conditions. The latter hypothesis would explain why the 68-kDa protein substrate was phosphorylated *in vitro* but not *in vivo*. The cortices used in these experiments were rather crude preparations, obtained by low speed centrifugations, but a similar result was obtained after further purification with isotonic sucrose (Kinsey *et al.*, 1980) (result not shown). This treatment removes attached cortical granules, which account for nearly 3/4 of total cortices proteins. This indicates that both the tyrosine kinase activity and the 68-kDa protein substrate are not associated with cortical granules, but rather with the plasma membrane or associated peripheral proteins.

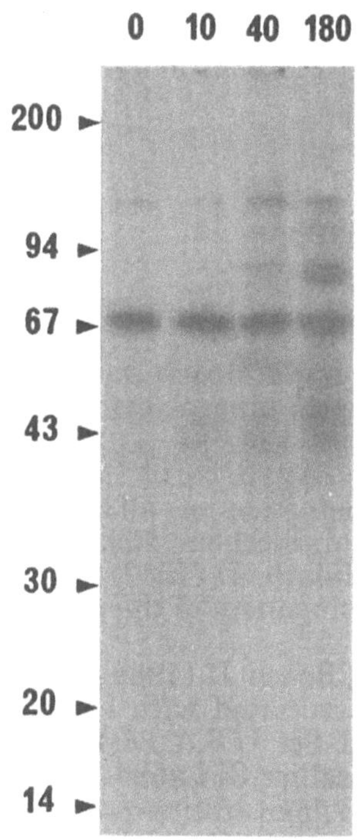

Fig. 4. Imunoprecipitation with anti-phosphotyrosine antibodies of *in vitro* labeled cortices.

Cortices were prepared from unlabeled oocytes at different times after hormone addition (in minutes pha, across the *top*). They were incubated with ^{32}P-ATP and detergents extracts were used for immunoprecipitation as in *in vivo* experiments. An autoradiograph of the gel is shown, with the migrations and molecular weights (x10^{-3}) of markers proteins indicated on the left.

These results indicate that tyrosine phosphorylation in membranes is involved in meiotic divisions and no only in the G0/G1 transition, as in most other previous studies. However they are still purely descriptive since homologies of the protein substrates with those identified in other materials cannot be established from the sole molecular weights. Indications from temporal correlations with the meiotic process are also highly speculatives but give some hints, such as the early phosphorylation of the 155-kDa protein, that tyrosine phosphorylation may be involved in the early phase of hormonal signal transduction. If they do not imply a key role for tyrosine phosphorylation in meiotic divisions, these findings agree with other results obtained in the homolog system of *Xenopus* oocytes, such as the effect of $pp60^{v\text{-}src}$ on meiosis reinitiation (Spivack *et al.*, 1984) or the presence in the plasma membrane of a tyrosine phosphorylated serine kinase (Sakanoue *et al.*, 1988).

REFERENCES

Arion D, Meijer L, Brizuela L and Beach D (1988) *cdc2* is a component of the M phase-specific histone H1 kinase : evidence for identity with MPF. Cell 55:371-378

Bourassa C, Chapdelaine A, Roberts K D and Chevalier S (1988) Enhancement of the detection of alkali-resistant phosphoproteins in polyacrylamide gels. Anal. Biochem. 169:356-362

Brizuela L, Draetta G and Beach D (1987) $p13^{suc1}$ acts in the fission yeast cell division cycle as a component of the $p34^{cdc2}$ protein kinase. EMBO J. 6:3507-3514

Brizuela L, Draetta G and Beach D (1989) Activation of *cdc2* protein as a histone H1 kinase is associated with complex formation with the p62 subunit.Proc. Natl. Acad. Sci. U.S.A. 86:4362-4366

Capony J-P, Picard A, Peaucellier G, Labbé J-C and Dorée M (1986) Changes in the activity of the maturation-promoting factor during meiotic maturation and following activation of amphibian and starfish oocytes : their correlation with protein phosphorylation. Dev. Biol. 117:1-12

Chen WS, Lazar CS, Poenie M, Tsien RY, Gill GN and Rosenfeld MG (1987) Requirement for intrinsic protein tyrosine kinase in the immediate and late actions of the EGF receptor. Nature 328:820-823

Chou CK, Dull TJ, Russell DS, Gherzi R, Lebwohl D, Ullrich U and Rosen OM (1987) Human insulin receptors mutated at the ATP-binding site lack protein tyrosine kinase activity and fail to mediate postreceptor effects of insulin. J. Biol. Chem. 262:1842-1847

Cooper JA, Sefton BM and Hunter T (1983) Detection and quantification of phosphotyrosine in proteins. Methods Enzymol. 99:387-401

Dasgrupta JD and Garbers DL (1983) Tyrosine protein kinase activity during embryogenesis. J. Biol. Chem. 258:6174-6178

Detering NK, Decker GL, Schmell ED and Lennarz WJ (1977) Isolation and characterization of plasma membrane-associated cortical granules from sea urchin eggs. J. Cell Biol. 75:899-914

Dorée M and Guerrier P (1975) Site of action of 1-methyladenine in inducing oocyte maturation in starfishes : Kinetical evidences for receptors localized on the cell surface. Exp. Cell Res. 91:296-300

Dorée M, Peaucellier G and Picard A (1983) Activity of the maturation-promoting factor and the extent of protein phosphorylation oscillate simultaneously during meiotic maturation of starfish oocytes. Dev. Biol. 99:489-501

Dunphy WG and Newport JW (1988) Unraveling of mitotic control mechanisms. Cell 55:925-928

Dunphy WG and Newport JW (1989) Fission yeast p13 blocks mitotic activation and tyrosine dephosphorylation of the Xenopus *cdc2* protein kinase. Cell 58:181-191

Draetta G, Piwnica-Worms H, Morrison D, Druker B, Roberts T and Beach D (1988) Human *cdc2* protein kinase is a major cell-cycle regulated tyrosine kinase substrate. Nature 336:738-743

Gautier J, Matsukawa T, Nurse P and Maller J (1989) Dephosphorylation and activation of "Xenopus" p34*cdc2* protein kinase during the cell cycle. Nature 339:626-628

Guerrier P and Dorée M (1975) Hormonal control of reinitiation of meiosis in starfish. The requirement of 1-methyladenine during nuclear maturation. Dev. Biol. 47:341-348

Heldin CH and Westermark B (1984) Growth factors : mechanism of action and relation to oncogenes. Cell 37:9-20

Hunter T and Cooper J.A. (1985) Protein-tyrosine kinases. Annu. Rev. Biochem. 54:897-930

Kamel C, Veno PA and Kinsey WH (1986) Quantitation of a *src*-like tyrosine protein kinase during fertilization of the sea urchin egg. Biochem. Biophys. Res. Commun. 138:349-355

Kamps MP and Sefton BM (1988) Identification of multiple novel polypeptide substrates of the v-*src*, v-*yes*, v-*fps*, v-*ros*, and v-*erb*-B oncogenic protein kinases utilizing antisera against phosphotyrosine. Oncogene 2:305-315

Kinsey WH (1984) Regulation of tyrosine-specific kinase activity at fertilization. Dev. Biol. 105:137-143

Kinsey WH, Decker GL and Lennarz WJ (1980) Isolation and partial characterization of the plasma membrane of the sea urchin egg. J. Cell Biol. 87:248-254

Kishimoto T, Kuriyama R, Kondo H and Kanatani H (1982) Generality of the action of various maturation-promoting factors. Exptl. Cell Res. 137: 121-126

Labbé J-C, Capony J-P, Caput D, Cavadore J-C, Derancourt J, Kaghad M, Lelias J-M, Picard A and Dorée M (1989) MPF from starfish oocytes at first meiotic metaphase is an heterodimer containing one molecule of *cdc2* and one molecule of cyclin B. EMBO J. 8:3053-3058

Labbé J-C, Lee MG, Nurse P, Picard A and Dorée M (1988) Activation at M-phase of a protein kinase encoded by a starfish homologue of the cell cycle control gene *cdc2+*. Nature 335:251-253

Labbé JC, Picard A, Peaucellier G, Lee M, Nurse P and Dorée M (1989) Purification of MPF from starfish : identification as the H_1 histone kinase p34*cdc2* and a possible mechanism for its periodic activation. Cell 57:253-263

Laemmli UK (1970) Cleavage of structural proteins during the assembly of the head of bacteriophage T4. Nature 227:680-685

Lee M and Nurse P (1988) Cell cycle control genes in fission yeast and mammalian cells. Trends Genet. 4:287

Meijer L, Arion D, Goldstein R, Pines J, Brizuela L, Hunt T and Beach D (1989) Cyclin is a component of the sea urchin egg M-phase specific histone H1 kinase. EMBO J. 8:2275-2282

Morla AO, Draetta G, Beach D and Wang JY (1989) Reversible tyrosine phosphorylation of *cdc2*: dephosphorylation accompanies activation during entry into mitosis. Cell 58:193-203)

Peaucellier G, Veno PA and Kinsey WH (1988) Protein tyrosine phosphorylation in response to fertilization. J. Biol. Chem. 263:3806-13811

Picard A, Peaucellier G, Le Bouffant F, Le Peuch C and Dorée M (1985) Role of protein synthesis and proteases in production and inactivation of maturation-promoting activity during meiotic maturation of starfish oocytes. Dev. Biol. 109:311-320

Pondaven P, Meijer L and Pelech SL (1987) Protein phosphorylation in starfish oocyte meiotic divisions and sea urchin egg mitotic divisions. Adv. Protein Phosphatases 4:229-251

Ribot HD, Eisenman EA and Kinsey WH (1984) Fertilization results in increased tyrosine phosphorylation of egg proteins. J. Biol.Chem. 259:5333-5338

Ross AH, Baltimore D and Eisen HN (1981) Phosphotyrosine-containing proteins isolated by affinity chromatography with antibodies to a synthetic hapten. Nature 294:654-656

Sakanoue Y, Hashimoto E, Nakamura S-I and Yamamura H (1988) Insulin-stimulated serine kinase in *Xenopus* oocyte plasma membrane. Biochem. Biophys. Res. Commun. 150:1176-1184

Satoh N and Garbers DL (1985) Protein tyrosine kinase activity of eggs of the sea urchin *Strongylocentrotus purpuratus* : The regulation of its increase after fertilization. Dev. Biol. 111:515-519

Spivack JG, Erikson RL and Maller JL (1984) Microinjection of pp60v-src into *Xenopus* oocytes increases phosphorylation of ribosomal protein S6 and accelerates the rate of progesterone-induced meiotic maturation. Mol. Cell Biol. 4:1631-1634

Wang JYJ (1988) Antibodies for phosphotyrosine : analytical and preparative tool for tyrosyl-phosphorylated proteins. Anal. Biochem. 172:1-7

Yarden Y and Ullrich A (1988) Growth factor receptor tyrosine kinases. Ann. Rev. Biochem. 57:443-478

Yoshikuni M, Ishikawa K, Isobe M, Goto T and Nagahama Y (1988b) Characterization of 1-methyladenine binding in starfish oocyte cortices. Proc. Natl. Acad. Sci. U.S.A. 85:1874-1877

PROTEIN SYNTHESIS AND PROTEIN PHOSPHORYLATION AS REGULATORS OF MPF ACTIVITY

P. Guerrier, I. Néant , P. Colas (1)
L. Dufresne, J. Saint Pierre, F. Dubé (2)

(1) Laboratoire de Biologie Moléculaire et Cellulaire
Ecole Normale Supérieure de LYON
46 Allée d'Italie, 69364 Lyon Cedex 07
FRANCE

(2) Université du Québec
Rimouski, G5L3A1
CANADA

ABSTRACT

Oocytes stimulated to reinitiate meiosis produce an intracytoplasmic maturation promoting factor (MPF). MPF can trigger maturation of recipient prophase-arrested oocytes even after a number of serial microinjection transfers performed in the presence of protein synthesis inhibitors. This indicates that arrested oocytes contain a stockpile of MPF precursor molecules which are posttranslationally activated following MPF seed microinjection and accounts for MPF amplification.

In *Patella* and the starfish, MPF activity has been shown to be associated with protein phosphorylation. Emetine treatment does not preclude the initiation of protein phosphorylation and MPF formation, which results in nuclear breakdown, chromosome condensation and first maturation spindle formation. Under these conditions, however, dephosphorylation soon occurs, which induces spindle destruction and the precocious formation of interphase-type resting nuclei. The same cytological effects were promoted by the drug 6- dimethyladenine which directly inhibited protein kinase

NATO ASI Series, Vol. H 45
Mechanism of Fertilization
Edited by B. Dale

without affecting protein synthesis. Similar results were obtained with sea urchins, amphibians and mammals.

We conclude that (1) newly formed short lived proteins are required for maintaining metaphase 1 conditions in *Patella* or for driving the next cycle ; (2) these proteins can only act if they are further activated directly or indirectly via a phosphorylation dependent process.

Presently, even when major advances have been made in our understanding of the molecular and genetic factors which drive the cell cycle, a number of questions remain to be answered. These will be soon resolved provided that one can choose, for each one, the appropriate key and the best available model.

INTRODUCTION

It is well known that oocytes must reach a certain stage of physiological maturity before they can be fertilized. This stage does not necessarily coincide with completion of the maturation divisions as is the case for sea urchins. In fact, female gametes of most vertebrates and invertebrates are usually blocked during the prophase of the first maturation division (diplotene stage), after they have duplicated their DNA.

They present a huge nucleus, the germinal vesicle (GV). External signals such as the spermatozoon as found in the bivalves *Spisula*, *Pholas* or *Barnea* or, more often, an hormone will stimulate them to reinitiate meiosis. While, in the bivalves or the starfish, meiosis goes to completion, a second block may occur, either in metaphase 1 as observed in the annelids *Sabellaria* and *Arenicola*, the molluscs *Mytilus*, *Dentalium* and *Patella* or in metaphase 2, after extrusion of the second polar body, as observed in amphibian or mammalian oocytes. These last blocks are only released upon further fertilization or activation.

HORMONE - INDUCED MEIOSIS REINITIATION

The general scheme leading to meiosis reinitiation is not very different in starfish and amphibian oocytes which represent so far the most extensively studied systems. In both cases, the follicle cells have been found to respond to neurohormones (MASUI & CLARKE, 1970 ; KANATANI, 1964) by synthesizing and releasing a relay hormone which is directly active on bare defolliculated oocytes. These relay hormones are respectively 1- methyladenine (1- MeAde) for the starfish (KANATANI et al. , 1969) and progesterone for the amphibians (FORTUNE et al., 1975). A similar system seems to be involved in triggering maturation of the prosobranch mollusc *Patella vulgata*. However, in this last case, the nature of the relevant hormones have not been yet characterized (SHIRAI et al., 1987).

Besides ovulation (KANATANI & SHIRAI, 1972 ; MASUI & CLARKE ; GUERRIER et al., 1986 a), these hormones appear to trigger both nuclear and cytoplasmic maturation, this last process being largely independent on the presence of the germinal vesicle.Thus, enucleated starfish or amphibian oocytes may be activated or fertilized provided that they are first stimulated with the hormone (HIRAI & KANATANI., 1971 ; MASUI & MARKERT, 1971). They exhibit the cortical reaction but, usually, remain unable to condense foreign chromosomes and to cleave unless some germinal vesicle nucleoplasm is added (MASUI et al., 1979 ; REYNHOUT & SMITH, 1974 ; SMITH & ECKER, 1969). The acquisition of cytoplasmic maturity by *Patella* oocytes also requires the presence of GV nucleoplasm in addition to a change in the intracellular pH (GUERRIER et al., 1986 b).

In the starfish and the amphibian, it has been clearly shown that the process of nuclear and cytoplasmic maturation could not be directly mediated by the hormone itself or by any of its metabolites since :

1) There is no obvious relation between uptake or metabolism of the hormone and meiosis reinitiation (DOREE & GUERRIER, 1975 ; KANATANI & HIRAMOTO, 1970 ; SMITH & ECKER, 1971).

2) Meiosis cannot be triggered by direct intracellular hormone injection (MASUI & MARKERT, 1971 ; REYNHOUT & SMITH, 1974 ; SMITH & ECKER, 1969).

3) Meiosis reinitiation can be obtained using a variety of mimetics unrelated to the hormone (GUERRIER et al. 1982 ; MEIJER & GUERRIER, 1984). Quite recently, it has been shown, for example, that okadaic acid, a phosphoprotein phosphatase inhibitor, promoted meiosis reinitiation in *Xenopus* within a very short time, i. e. 1 hr versus 6 hr in the case of progesterone (GORIS et al., 1989).

All these signals which release oocytes from the prophase block were found to produce an intracytoplasmic maturation promoting factor or MPF which is responsible for germinal vesicle breakdown (GVBD) and chromosome condensation. Main properties of this factor have been revealed through microinjection and fusion experiments.

THE MATURATION PROMOTING FACTOR

Activation and amplification

One of the first recognized characteristics of MPF was its ability to be amplified after its transfer from a maturing to a GV-arrested recipient oocyte. This amplification property is particularly well evidenced in serial transfer experiments which involve, at each step, a 1/10th dilution of the originally taken cytoplasmic sample. (Fig. 1). This amplification occurs normally even in the presence of protein synthesis inhibitors. It thus appears that such an amplification requires the recruitment and the posttranslational activation of preexisting latent molecules which constitute the pre MPF. In the starfish, it is worth to note that these molecules are no longer activable at the end of the maturation process since the ootid (i. e. the oocyte which has extruded both polar bodies and presents an haploïd female pronucleus) cannot amplify microinjected MPF and preserves

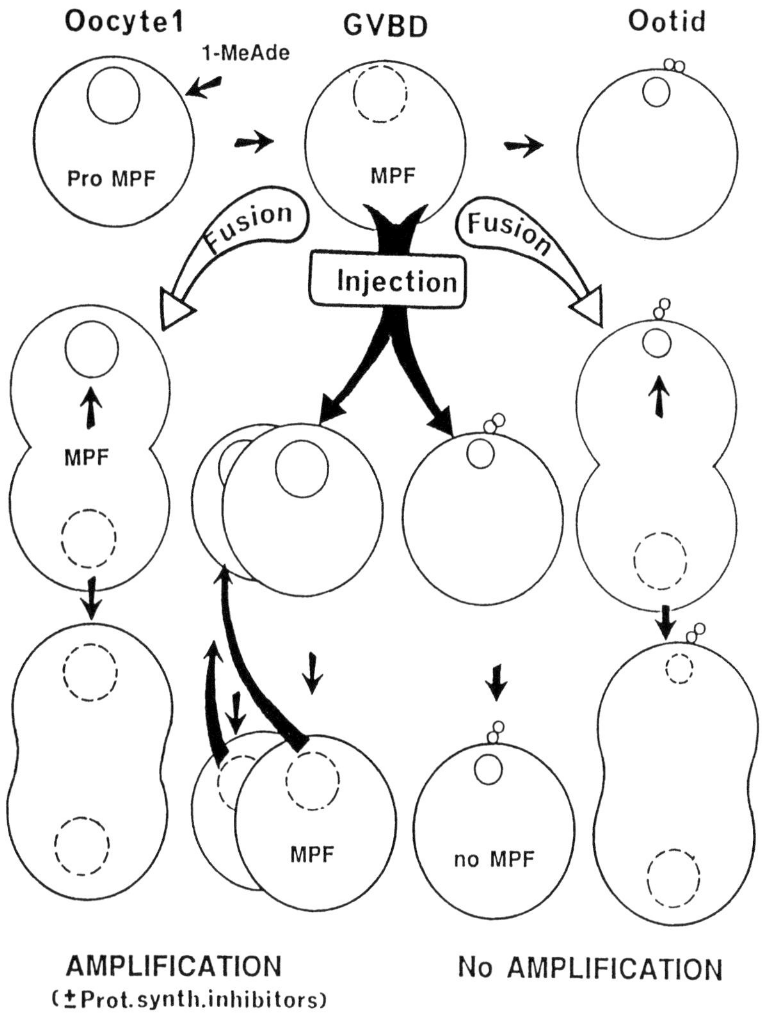

Fig.1 : MPF activity as revealed by microinjection and by fusion in the starfish oocyte.
The microinjection of MPF taken from the oocyte undergoing GVBD does not trigger female pronucleus breakdown in the ootid but induces GVBD in the oocyte 1. Response of the oocyte 1 is maintained through serial transfers and occurs even in the presence of inhibitors of the protein synthesis. Cell fusion shows that the female pronucleus is sensitive to MPF.

In fact, it is now agreed that the formation of active MPF corresponds to the activation of a M- phase specific histone H1 kinase. In the starfish, the two foundamental elements of this complex, i. e. a 34 kd protein related to the yeast cdc2/CDC 28 gene product (ARION et al., 1988 ; LABBE et al., 1988) and a p 54 kd cyclin protein related to the fission yeast cdc 13+ gene product are present both in the GV-arrested oocyte and the female pronucleus ootid (STANDART et al. 1987 ; PONDAVEN et al., unpublished). Since it has been shown, in various models, that activation of this complex requires tyrosine dephosphorylation of the p 34 cdc2 protein (amphibians : DUMPHY & NEWPORT, 1989 ; GAUTIER et al. , 1989; mammalian cells : DRAETTA et al., 1988 ; MORLA et al., 1989 ; starfish : LABBE et al., 1989 ; PONDAVEN et al., unpublished), it is possible that protein synthesis was required, in our fusion experiments, to produce either an activator of a p 34 cdc2 specific phosphotyrosine phosphatase or an inhibitor of the corresponding phosphotyrosine kinase (Fig 2).

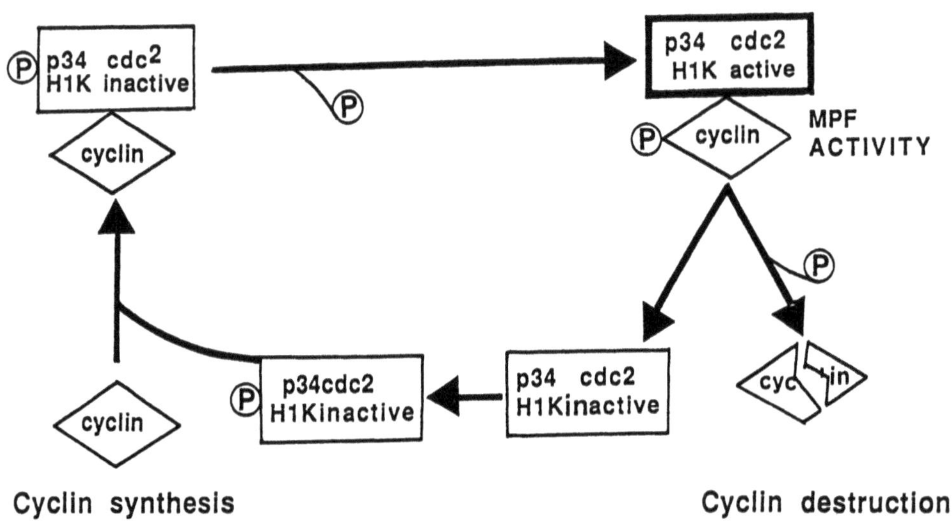

Fig.2 : A simplified model of the main regulations controlling MPF activity. Another protein p13, produced by the gene suc 1, may be involved in the complex.

an intact female pronucleus. Simultaneously, however, one can show, by cell fusion, that the female pronucleus remains perfectly sensitive to active MPF (Fig 1, right column).

MPF amplification was also demonstrated to occur in *Xenopus* cell free extracts containing demembranated sperm nuclei which entered mitosis following MPF seed addition (CYERT & KIRSCHNER, 1988). DUMPHY and NEWPORT (1988) also demonstrated that interphase *Xenopus* cell free extracts did contain a latent MPF which could express itself following ammonium sulfate precipitation, a step which might have removed some kind of inhibitor from the crude preparation.

"In vivo", we also obtained the same restoration or resuscitation of MPF activity by fusing a GV arrested oocyte which only contained pre MPF and a female pronucleus - arrested oocyte which has been shown unable to amplify microinjected seed MPF (GUERRIER & NEANT, 1986). This process, which now requires 45 min instead of the 20 min needed after hormone stimulation or MPF transfer does not correspond to the amplification of an eventual residual MPF present in the ootid. Indeed, while MPF amplification is possible in the presence of protein synthesis inhibitors or inhibitors of protease activity, these two groups of inhibitors block the reappearance of MPF resulting from the fusion between the oocyte 1 and the ootid. Inhibitors of protein phosphorylation such as diamide (GUERRIER & NEANT, 1986) or 6- dimetylaminopurine (NEANT et al., 1989) also block this process. One must conclude that each partner of the heterologous pair contains complementary elements which may cooperate when brought in contact. If some proteases present in the ootid were only required to activate the MPF molecules present in the oocyte 1, this process should have taken place in the absence of protein synthesis as is the case during hormone-induced maturation. In the starfish, hormone-induced GVBD is indeed blocked by antiproteases, whereas it can occur in the absence of protein synthesis (GUERRIER & DOREE, 1975 ; SANO & KANATANI, 1983 ; SAWADA et al. 1989). The fact that protein synthesis is required points to a more complex situation.

Species specificity

A second property of MPF is that this factor is devoid of any species specificity. It acts efficiently in any kind of cross transfer one can imagine to undertake between somatic cells in mitosis and germ cells in meiosis even when quite different phylla are considered (KISHIMOTO et al., 1984 ; SUNKARA et al., 1979 ; WEINTRAUB et al., 1982 ; NEANT & GUERRIER, 1988 a). This is not surprising now in view of the high conservation of the proteins which constitute the core of the M- phase histone H1 kinase.

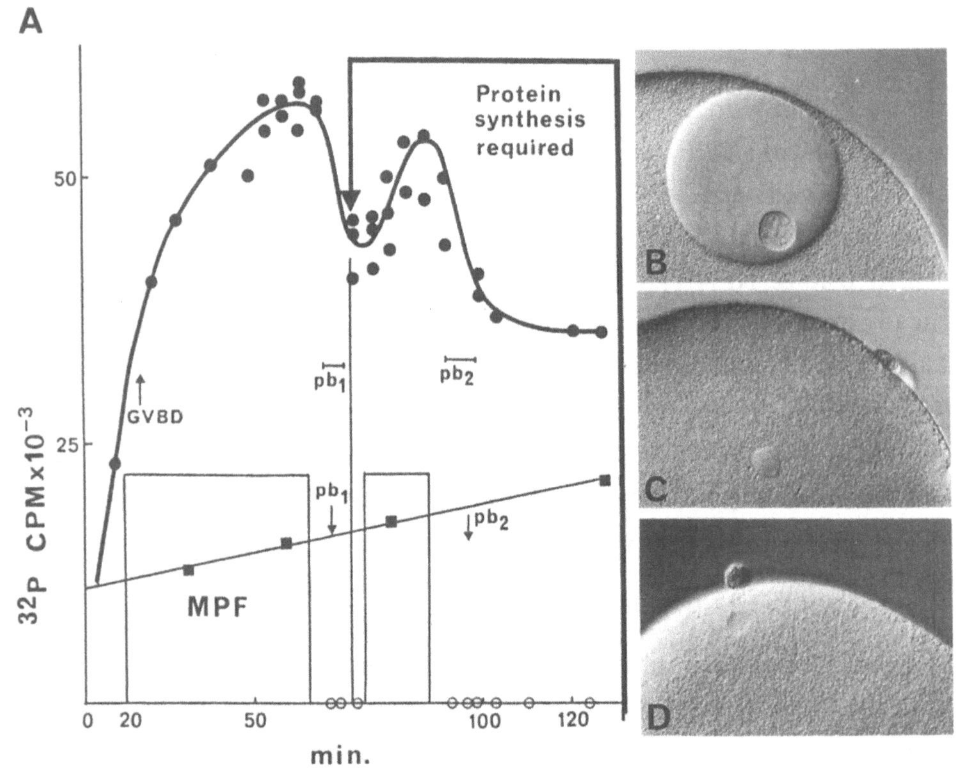

Fig.3 : Simultaneous oscillations of protein phosphorylation and MPF activity in the starfish oocyte.
A. Synthetic graph modified from the data of Dorée et al., 1983. B. Germinal vesicle stage. C. Female pronucleus stage after extrusion of the first and second polar bodies (140 min). D. Precocious formation of a resting nucleus after extrusion of the first polar body in oocytes treated with emetine and stimulated with 1- MeAde (90 min).

Phosphorylation and protein synthesis

Finally, a last important characteristic of MPF expresses itself in the fact that the level of MPF activity is directly related to the extent of phosphorylation of the oocyte endogenous proteins. Indeed, protein phosphorylation and MPF activity have been shown to oscillate in agreement with the cell cycle. Thus, in the starfish (Fig 3), there are two cycles of phosphorylation and MPF activity which peak in metaphase and fall during first and second polar body extrusion (Dorée et al., 1983).

PROTEIN SYNTHESIS AS REGULATOR OF MPF ACTIVITY

In the starfish, the first cycle of phosphorylation and MPF activity occurs normally in the absence of protein synthesis (GUERRIER & DOREE, 1975 ; DOREE et al., 1983). In contrast, the second cycle, as well as all the other cycles that will follow activation or fertilization, require protein synthesis (DOREE et al., 1983 ; PICARD et al., 1987 ; GUERRIER & NEANT, 1986).
Thus, in the presence of emetine, there is no second phosphorylation burst and no second peak of MPF activity. As a direct effect, resting nuclei form quite precociously, right after first polar body extrusion and before than the second polar body and the female pronucleus are formed in control hormone stimulated oocytes (Fig 3, C, D). Since it has been shown that cyclin is destroyed at each cytokinesis (STANDART et al., 1987), one can suspect that protein synthesis is at least required to reconstitute this essential element of the M- phase kinase complex.

In the prosobranch mollusc *Patella* , there is only one burst of phosphorylation which gives rise to a plateau characteristic for the metaphase 1 stage (Fig 4). In the presence of emetine, the signal used to reinitiate meoisis (10mM ammonia) does stimulate protein phosphorylation, GVBD, chromosome condensation and formation of the metaphase spindle. This indicates that, in *Patella,* alike in the starfish, MPF formation does not require the synthesis of a protein initiator. However, under these conditions, dephosphorylation soon

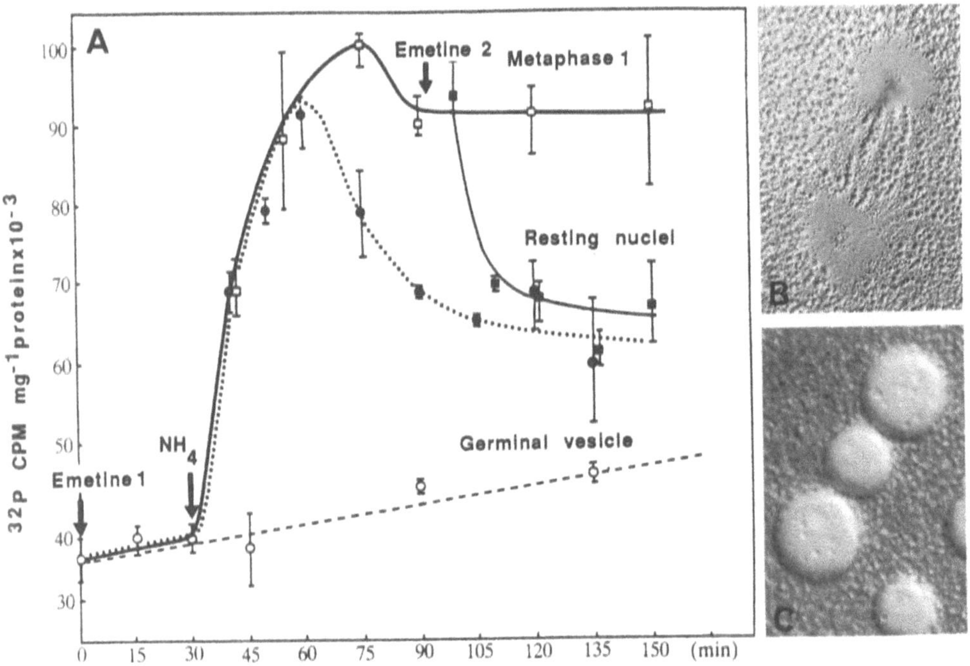

Fig.4 : Effect of emetine on ammonia-induced meiosis reinitiation and protein phosphorylation in *Patella vulgata* . A, Emetine (90 µM) was added 30 min before the addition of 10 mM NH4CL or 30 min later, at the metaphase-1 stage. O, control oocytes ; □ , stimulated oocytes ; ■ , metaphase 1 oocytes treated with 6-DMAP. B, metaphase 1 spindle 60 min after ammonia stimulation (Nomarsky). C, resting nuclei 90 min after emetine addition (Nomarsky).

occurs, which promotes spindle desorganisation, chromosome decondensation and the formation of resting nuclei. Identical results are obtained when emetine is directly added to the metaphase 1-arrested oocyte (NEANT & GUERRIER, 1988 a).

In the bivalve *Mytilus*, where freshly shed oocytes are blocked in first metaphase, the inhibition of protein synthesis by puromycin lead, within 30 min, to the extrusion of the first polar body, while cyclin disappears in the mean time (DUBE et al., unpublished). The second polar body is never formed under these conditions and a huge nucleus reconstitutes in the cytoplasm (Fig 5).

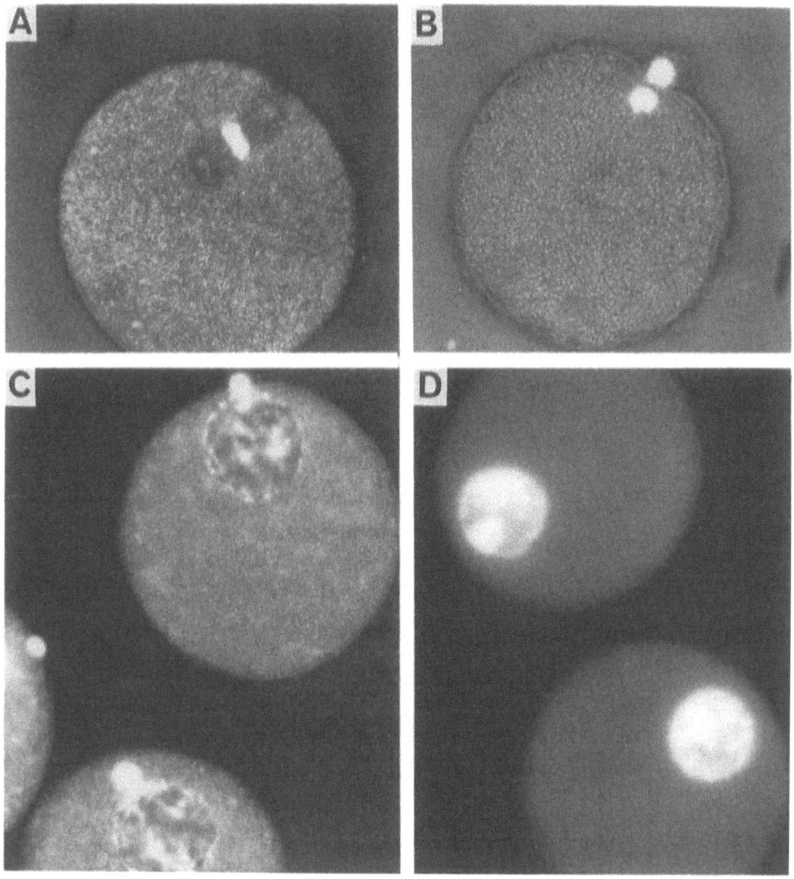

Fig. 5 : Response of the metaphase 1-arrested oocyte of *Mytilus* to inhibitors of protein synthesis or protein phosphorylation A, the metaphase 1 resting oocyte ; B, the oocyte extrudes its first polar body within a 40 min treatment with 150 µM emetine ; C, oocytes of the same batch1.5 hr after treatment ; D, oocytes treated with 300 µM 6-DMAP directly decondense their chromosomes to produce a big resting nucleus.

From all these data, it appears clearly that a continuous synthesis of short lived proteins such as the cyclins is required for maintaining that high phosphorylation level which looks necessary for insuring condensation of the metaphase chromosomes and stabilization of the MPF activity.

However, it remains that the inhibition of protein synthesis is always accompanied by a decrease in the level of phosphorylation of endogenous proteins. One can thus seriously question whether the cytological consequences of such a treatment are mainly related to the absence of a particular protein or to the fact that phosphorylation was indirectly affected. In order to answer this point, we tryed to modulate protein phosphorylation without affecting protein sythesis. To do that, we used the puromycin analogue 6-dimethylaminopurine (6-DMAP) which was known to block the cell cycle in *Spisula* and the sea urchin without affecting protein synthesis (REBHUN et al., 1973).

PROTEIN PHOSPHORYLATION AND THE CELL CYCLE

Our results confirm that 6-DMAP did not affect protein synthesis both in *Patella* (NEANT & GUERRIER, 1988 a), the starfish (NEANT & GUERRIER, 1988 b), the sea urchin (NEANT et al., 1989), the mouse (RIME et al., 1989) and Mytilus (DUBE et al., unpublished). In the sea urchin and the starfish, we even observed that 6-DMAP did not block the cyclic evolution of the 57 kd cyclin which was transiently destroyed at the metaphase - anaphase transition and disappeared completely after treatment with emetine. Thus, even when these proteins are present continuously and behave as usual, 6-DMAP promotes the same cytological effects as those observed with emetine, i. e. decondensation of the chromosomes, destruction of the spindle structure and formation of resting nuclei. This holds true for *Patella* (NEANT & GUERRIER, 1988 a), the starfish (NEANT & GUERRIER, 1988 b), the sea urchin, (NEANT et al., 1989), the mouse (RIME et al., 1989) and *Mytilus* (Fig 5).

It is thus clear that both newly formed proteins such as the cyclins or other regulatory proteins which may preexist in the oocyte must be modified through a process which is inhibited by 6- DMAP.

It appears likely that such a posttranslational modification occurs via a phosphorylation process. Indeed, in the sea urchin (NEANT et al., 1989), we observed that 6-DMAP did suppress the phosphorylation bursts which proceed before each cytokinesis. In *Patella* (NEANT & GUERRIER, 1988 a) and the starfish (NEANT & GUERRIER, 1988 b), we showed moreover that 6-DMAP instantaneously induced an important protein dephosphorylation. We also found that this effect was reversible as was the cytological modifications induced by the drug. Indeed, washing out the inhibitor restored both phosphorylation, chromosome condensation and tubulin reorganization. The same cytological modifications were observed with the mouse oocyte (RIME et al., 1989). Here again, like in *Patella*, the starfish and *Xenopus* (unpublished), 6-DMAP inhibits meiosis reinitiation and this inhibition is reversible. It is very clear that this effect only depends on the inhibition of protein phosphorylation induced by 6-DMAP since protein synthesis inhibitors do not block meiosis reinitiation up to GVBD in the mouse. Later on, when puromycin is given to Metaphase 1 or Metaphase 2 oocytes, supression of the protein synthesis gives rise to resting nuclei (CLARKE & MASUI, 1983). However, the proteins which are synthesized at these two stages must share different properties. Indeed, when 6-DMAP is given to metaphase 1 oocytes, phosphorylation is inhibited and resting nuclei are produced (Fig 6).On the contrary, when metaphase 2 oocytes are treated with 6-DMAP, no significant dephosphorylation is observed and the chromosomes remain condensed and do not produce resting nuclei (RIME et al., 1989).

It is important to note that 6-DMAP does not affect directly tubulin molecules since this drug has no effect on the cortical microtubules array, while it efficiently destroys aster and spindle microtubules (Fig 7), by which pronuclear migration and chromosome separation are dramatically affected (NEANT et al., 1989 ; SAINT PIERRE et al., unpublished). It is likely that 6-DMAP acts on spindle microtubules by modulating the phosphorylation of some regulatory microtubule-associated proteins which are not present in the cortex.

SITE AND MODE OF ACTION OF 6-DMAP

This constitutes a still unresolved question. Presently, we can only say that 6-DMAP does not stimulate protease or

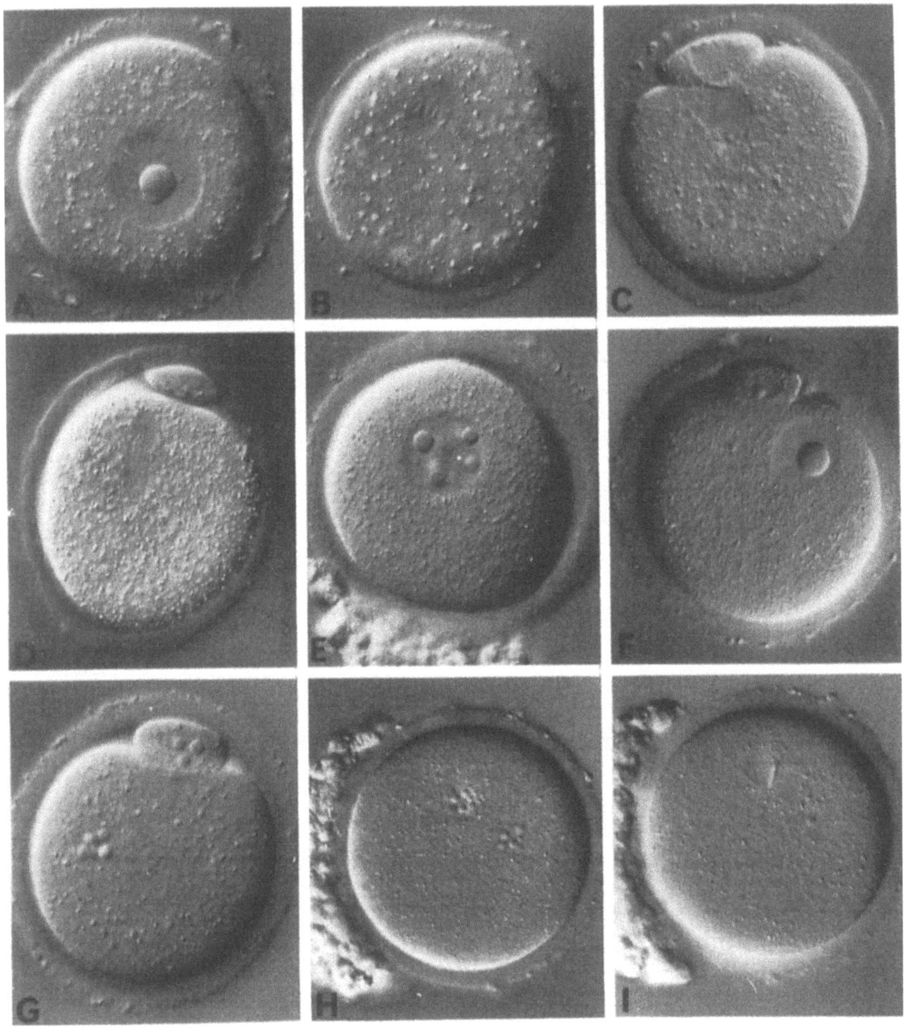

Fig.6 : Effects of 6-Dimethylaminopurine (6-DMAP) on spindle structure and chromosome condensation during spontaneous maturation of the mouse oocytes. Nomarsky contrast.

A-D, control oocyte : A, germinal vesicle stage after removal of the corona radiata follicle cells; B, Metaphase 1 spindle observed 4 hr later ; C, Extrusion of the first polar body at 9 hr. D, Unfertilized oocyte arrested in metaphase 2, 14 hr after isolation.

E-I, oocytes treated with 2.5 mM 6-DMAP in metaphase 1 and observed at 14hr when controls are as in D. E, first polar body did not form and a single central nucleus is present ; F, G, two nuclei are present, one in the oocyte, the other in the polar body ; H, I, two different focus show the presence of two nuclei (H) and a persisting mid-body : the 1st polar body fused to the oocyte.

phosphoproteine phosphatase activities but that it directly affects protein kinase activities both in vivo and in vitro (NEANT & GUERRIER, 1988 a, b). RIME et al. (unpublished) have also shown that 6-DMAP up to a concentration of 2.5 mM has no effect in vitro on isolated type 1 or type 2 A phosphatases or on phosphotyrosyl phosphatases. This suggests that cyclin phosphorylation, which is blocked by 6-DMAP, might play a significant role in the activation of the cdc2-cyclin kinase complex. Anyway, it would be interesting to verify if the p34 cdc2 tyrosine is normally dephosphorylated when starfish oocytes or sea urchin eggs are stimulated in the presence of 6-DMAP.

Another possibility would be that 6- DMAP acts only indirectly on the phosphorylation processes required during the meiotic and mitotic cell cycles. One may conceive that it will physically inhibit or destroy the association of those foundamental subunits which constitute the active MPF, i. e. the M-phase histone H1 kinase complex. This hypothesis is now open to verification through immunoprecipitation since p13, p34 and cyclin antibodies are becoming available.

CONCLUDING REMARKS

In this paper, we were still examining some experimental data which are not yet fully understood. However, during the two current years, very important advances have been made which point to those specific genes and products that are surely involved in controlling major transitions in the cell cycle. Most of the basic components of the system have been identified and specific tools have been produced to label and follow them in their interactions. Despite this, hopefully, the wine still clings to the mouldering wall and a number of problems remain to be solved. Among them :

1) How proteases are mobilized to activate pre MPF on one hand and to inactivate cyclin on the other hand ?

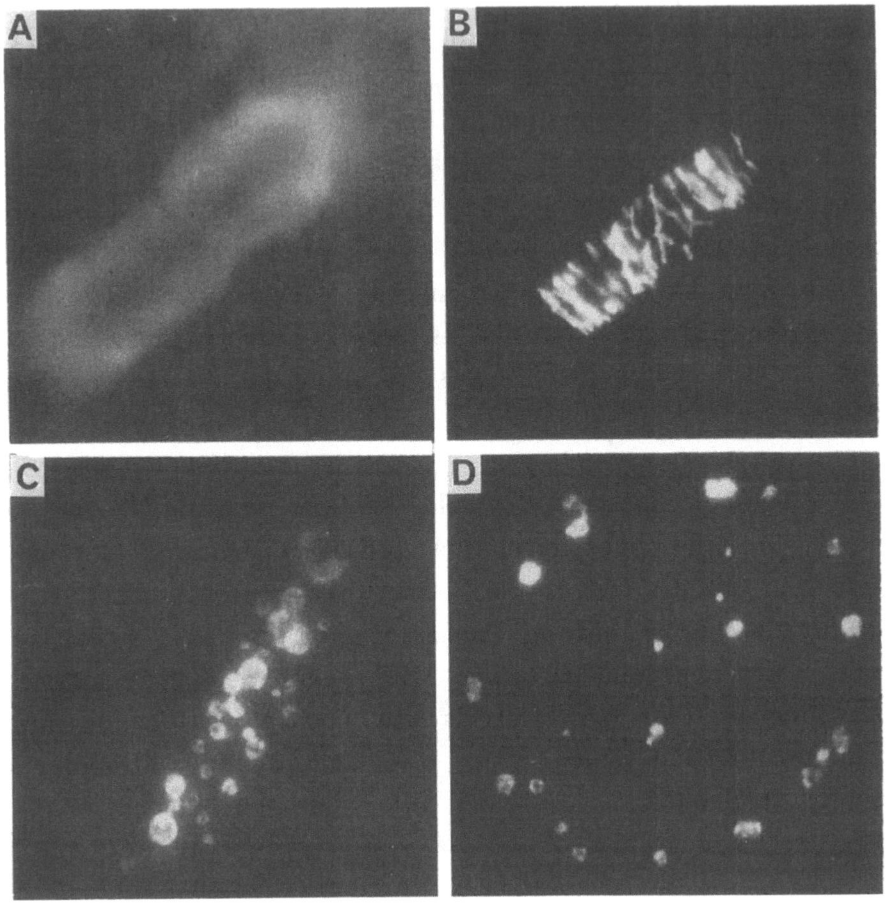

Fig.7 : Effects of 6-Dimethylaminopurine (6-DMAP) on spindle structure in the sea urchin *Strongylocentrotus droebachiensis*.
The eggs received 600 µM 6-DMAP when they reached metaphase of the first cleavage, 110 min after insemination. 10 min later, asters have disappeared and the spindle has flattened around the equatorial plane as shown in A (tubulin immunostaining) and in B (corresponding view with Hoechst fluorochrome). Chromosomes are in early anaphase configuration and begin to decondense. 20 min after 6-DMAP addition, chromosomes give rise to telophase caryomeres which remain in the original spindle equatorial plane as shown in C (view perpendicular to the original spindle axis, as in A) and in D (view along the original spindle axis).

2) How phosphoprotein phosphases, their activators or inhibitors may play a role in driving the cell cycle ?

3) What are the in vivo substrates of the MPF and how can they be involved in controlling the highly integrated spatial and temporal patterns exhibited by the mitotic apparatus components?

4) What are the intimate mechanisms which are responsible for blocking the cell in metaphase or in interphase ? The question looks especially interesting in the case of the starfish since cyclin and p34 cdc2 are known to be simultaneously present at the female pronucleus stage. These proteins are also present in those oocytes which are arrested in metaphase.

LITERATURE CITED

Arion D, Meijer L, Brizuela L, Beach D (1988)
cdc2 is a component fo the M-phase specific histone H1-kinase : evidence for identity with MPF. Cell 55 : 371-378

Clarke HJ, Masui Y (1983)
The induction of reversible and irreversible chromosome condensation by protein synthesis inhibition during meiotic maturation of mouse oocytes. Dev Biol 97 : 291-301

Cyert MS, Kirschner MW (1988)
Regulation of MPF activity in vitro. Cell 53 : 185-195

Dorée M, Guerrier P (1975)
Site of action of 1-methylademine in inducing oocyte maturation in starfish : kinetic evidence for receptors localized on the cell membrane. Exp Cell Res 96 : 296-300

Dorée M, Peaucellier G, Picard A (1983)
Activity of the maturation promoting factor and the extent of the protein phosphorylation oscillate simultaneously during meiotic maturation of starfish oocytes. Dev Biol 99 : 489-501

Draetta G, Piwnica-Worms H, Morrison D, Druker B, Roberts T, Beach D (1988)
Human cdc2 protein kinase is a major cell-cycle regulated tyrosine kinase substrate. Nature 336 : 738-744

Dumphy WG, Newport JW (1988)
Mitosis-inducing factors are present in a latent form during interphase in the *Xenopus* embryo. J Cell Biol 106 : 2047-2056

Dumphy WG, Newport JW (1989)
Fission yeast p13 blocks mitotic activation and tyrosine dephosphorylation of the *Xenopus* cdc 2 protein kinase. Cell 58 : 181-191

Fortune JE, Concannon PW, Hansel W (1975)
Ovarian progesterone level during in vitro oocyte maturation and ovulation in *Xenopus laevis* . Biol Reprod 13 : 561-567

Gautier J, Matsukanva T, Nurse P, Maller J (1989)
Dephosphorylation and activation of *Xenopus* p34 cdc2 protein kinase during the cell cycle. Nature 339 : 626-629

Goris J, Hermann J, Hendrix P, Ozon R, Merlevede W (1989)
Okadaic acid, a non-TPA tumor promotor, inhibits specifically protein phophatases, induces maturation and MPF formation in *Xenopus laevis* oocytes. Adv Prot Phosphatases 5 : 579-592

Guerrier P, Dorée M (1975)
Hormonal control of reinitiation of meiosis in starfish. The requirement of 1- methyladenine during nuclear maturation. Dev Biol 47 : 341-348

Guerrier P, Moreau M, Meijer L, Mazzei G, Vilain JP, Dubé F (1982)
The role of calcium in meiosis reinitiation. Progress in clinical and Biological Research 91 : 139-155

Guerrier P, Brassart M, David C, Moreau M (1986 a)
Sequential control of meiosis reinitiation by pH and Ca2+ in the prosobranch mollusk *Patella vulgata*. Dev Biol 114 : 315-324

Guerrier P, Guerrier C, Néant I, Moreau M (1986 b)
Germinal vesicle nucleoplasm and intracellular pH requirements for cytoplasmic maturity in oocytes of the prosobranch mollusk Patella vulgata. Dev Biol 116 : 92-99

Guerrier P, Néant I (1986)
Metabolic cooperation following fusion of starfish ootid and primary oocyte restores meiotic phase-promoting activity. Proc Natl Acad Sci USA 83 : 4814-4818

Hirai S, Kanatani H (1971)
Site of production of meiosis-inducing substance in ovary of starfish. Exp Cell Res 67 : 224-227

Kanatani H (1964)
Spawning of starfish : action of gamete-shedding substances obtained from radial nerves. Science 146 : 1177-1179

Kanatani H, Shirai H, Nakanishi K, Kurokawa T (1969)
Isolation and identification of meiosis-inducing substance in starfish *Asterias amurensis* . Nature 221 : 273-274

Kanatani H, Shirai H (1972)
On the maturation-inducing substance produced in starfish gonad by neural substance. Gen Comp Endocrinol Suppl 3 : 571-579

Kanatani H, Hiramoto Y (1970)
Site of action of 1-methyladenine in inducing oocyte maturation in starfish. Exp Cell Res 61 : 280-284

Kishimoto T, Yamazaki K, Kato Y, Koide S, Kanatani H (1984)
Induction of starfish oocyte maturation by maturation promoting factor of mouse and surf clam oocytes. J Exp Zool 231 : 293-295

Labbé JC, Lee MG, Nurse P, Picard A, Dorée M (1988)
Activation at M- phase of a protein kinase encoded by à starfish hormologue of the cell cycle controgene cdc2+. Nature 335 : 251-254

Labbé JC, Picard A, Peaucellier G, Cavadore JC, Nurse P, Dorée M (1989)
Purification of MPF from starfish : identification as the H1 histone kinase p34 cdc2 and a possible mechanism for its periodic activation. Cell 57 : 253-263

Masui Y, Clarke HJ (1979)
Oocyte maturation. Int Rev Cytol 57 : 185-282

Masui Y, Markert CL (1971)
Cytoplasmic control of nuclear behavior during meiotic maturation of frog oocytes. J Exp Zool 177 : 129-146

Masui Y, Meyerhof PG, Ziegler DH (1979)
Control of chromosome behavior during progesterone induced maturation in amphibian oocytes. J Steroid Biochem 11 : 715-722

Meijer L, Guerrier P (1984)
Maturation and fertilization in starfish oocytes. Int Rev Cytol 86 : 129-196

Morla AO, Draetta G, Beach D, Wang JYJ (1989)
Reversible tyrosine phosphorilation of cdc2 : dephosphorylation accompanies activation during entry into mitosis. Cell 58 : 193-203

Néant I, Guerrier P (1988 a)
Meiosis réinitiation in the mollusc *Patella vulgata*. Regulation of MPF, CSF and chromosome condensation activity by intracellular pH, protein synthesis and phosphorylation. Development 102 : 505-516

Néant I, Guerrier P (1988 b)
6- dimethylaminopurine blocks starfish oocyte maturation by inhibiting a relevant protein kinase. Exp Cell Res 176 : 68-79

Néant I, Charbonneau M, Guerrier P (1989)
A requirement for protein phosphorylation in regulating the meiotic and mitotic. cell cycles in echinoderms. Dev Biol 132 : 304-314

Picard A, Labbé JC, Peaucellier G, Le Bouffant F, Le Peuch C, Dorée M (1987)
Changes in the activity of the maturation promoting factor are correlated with those of a major cyclin AMP and calcium-independent protein kinase during the first meiotic cell cycles in the early starfish embryo. Develop Growth Differ 29 : 93-103

Rebhun LL, White D, Sander G, Ivy N (1973)
Cleavage inhibition in marine eggs by puromycin and 6-dimethylaminopurine. Exp Cell Res 77 : 312-318

Reynhout JK, Smith LD (1974)
Studies of the appearance and nature of a maturation inducing factor in the cytoplasm of amphibian oocytes exposed to progesterone. Dev Biol 38 : 394-400

Rime H, Néant I, Guerrier P, Ozon R (1989)
6- Dimethylaminopurine (6 DMAP), a reversible inhibitor of the transition to metaphase during the first meiotic cell division of the mouse oocyte. Dev Biol 133 : 164-179

Sano K, Kanatani H (1983)
Effect of various protease inhibitors on starfish oocyte maturation. Biomed Res 4 : 139-146

Sawada MT, Someno T, Hoshi M, Sawada H (1989)
Inhibition of starfish oocyte maturation by leupeptin analogs, potent trypsin inhibitors. Dev Biol 133 : 609-612

Shirai H, Néant I, Guerrier P (1987)
Induction of oocyte maturation by a ganglion extract in *Patella vulgata* . Develop Growth Differ 29 : 398

Smith LD, Ecker RE (1969)
Role of the oocyte nucleus in physiological maturation in *Rana pipiens* . Dev Biol 19 : 281-309

Smith LD, Ecker RE (1971)
The interaction of steroids with *Rana pipieus* oocytes in the induction of maturation. Dev Biol 25 : 232-247

Standart N, Minshull J, Pines J, Hunt T (1987)
Cyclin synthesis, modification and destruction during meiotic maturation of the starfish oocyte. Dev Biol 124 : 248-258

Sunkara PS, Wright DA, Rao PN (1979)
Mitotic factors from mammalian cells induce germinal vesicle breakdown and chromosome condensation in amphibian oocytes. Proc natl Acad Sci USA 76 : 2799-2802

Weintraub H, Buscaglia M, Ferrez M, Weiller S, Boulet A, Fabre F, Baulieu EE (1982)
Mise en évidence d'une activité MPF chez Saccharomyces cerevisiae. CR hebd Seanc Acad Sci Paris : 295 : 787-790

CONTROL OF RECRUITMENT OF PREANTRAL FOLLICLES IN MAMMALIAN OVARIES

ROGER G GOSDEN
Department of Physiology
University Medical School
Edinburgh EH8 9AG
United Kingdom

Mature oocytes in Graafian follicles are at the end of a developmental continuum which began when they entered meiotic prophase months or even years earlier. The greater part of the lifespan of those in adult ovaries is represented by the population of primordial follicles. These follicles provide an irreplaceable store for recruitment towards ovulatory ripeness, though earlier termination of development by atresia is the fate of the majority. Considering their significance in the life history of the species, it is remarkable that so little is known about the biology of small follicles compared with the extensive knowledge about Graafian stages evolving from them. New experimental opportunities that are now at hand promise to bring greater research attention and a better understanding to primordial follicles. In the following sections their salient characteristics will be described in terms of population dynamics and developmental biology of individual units.

1) Follicle numbers and dynamics

The numbers of primordial follicles in ovaries differ by up to several orders of magnitude inter-specifically and between members of the same species at different ages. A survey of 19 mammalian species of widely separated taxonomic groupings showed that at the beginning of reproductive life the numbers of follicles (N) vary allometrically with body mass (M) as:
$\log N = \log 27700 + 0.47 \log M$ ......(i) (Gosden & Telfer, 1987b). This is not surprising since large animals have greater longevities than small ones and the germ cell stores of all species would be expected to be sufficient for the maximum lifespan, human menopause being a notable exception (Gosden, 1985).

NATO ASI Series, Vol. H 45
Mechanism of Fertilization
Edited by B. Dale

Follicle numbers vary across the lifespan because recruitment and wastage by atresia occur continuously while folliculogenesis is restricted to prenatal or early postnatal life. The numbers of primordial follicles present at a given age are determined by (i) the rate of depletion by recruitment/death and (ii) the initial population size. Since the age distribution of small follicle numbers is approximately exponential (Faddy *et al.*, 1983, 1987), the rates of follicle attrition can be described by the population half-lives. However, few sets of data, apart from the following, are sufficiently extensive for reliable estimates to be made.

Table 1. Half-lives of primordial follicle populations (T) in 5 species estimated by fitting exponential functions to the age distribution for follicle numbers

Species	Adult wt (kg)	T (days)	Source of data
Mouse (strain A)	0.025	100	Jones & Krohn (1961)
Mongolian gerbil	0.085	200	Norris & Adams (1982)
Norway rat	0.25	274	Mandl & Shelton (1959)
Tamarin monkey	0.45	876	Tardiff (1985)
Human	65.0	2550	Block (1952)

From this admittedly limited group it is evident that the pace of follicle disappearance is much greater in small, short-lived species than in others. When these data are expressed on logarithmic coordinates the relationship between follicle population half-life (T) and body mass (M) can be expressed by the following equation, the fit to this model having a coefficient of determination (r^2) of 89.7:

$log T = log 1.55 + 0.404 \, log M$ ......(ii)

From these two allometric expressions (i) and (ii) we obtain:

$log N = log 27700 + 0.47(log T - log 1.55)/0.44$ ......(iii) Hence, $log N = 4.2 + 1.2 \, log T$ ...... (iv)

In conclusion, therefore, the life history strategy adopted by longer-lived animals involves formation of a large follicular store *and* a sparing rate of utilization. The balance of factors contributing to this phenomenon by modulating follicular death and growth rates (including oogonial mitoses) is unknown.

The dynamics of primordial follicle growth initiation and death can be estimated from differential follicle counts using compartmental mathematical models (Faddy *et al.*,

1976). The rates at which murine follicles enter the growing population are approximately constant throughout life. In infancy many follicles die at primordial stages, while others reaching antral sizes undergo atresia as a result of an unfavourable endocrine environment before puberty. Death rates vary more between inbred strains than do growth rates (Table 2). In CBA strain wastage among the primordial follicle population continues at a relatively high level throughout adult life and is responsible for premature sterility at about 1 year of age (Jones & Krohn, 1961). In human ovaries there is tentative evidence that, following an exponential phase, follicle depletion is accelerated in the years immediately preceding menopause (Richardson *et al.*, 1987; Gosden, 1987). Additional studies are required to determine whether this is due to an increase of either (or both) cell death or follicular recruitment as the population of growing stages wanes.

Table 2. Daily rates of primordial follicle growth initiation and death in 3 strains of mice and an F1 hybrid at two age phases (P1 & P2). (Adapted from Faddy *et al.*, 1983).

Strain	Age at P1/P2 transition(days)	Growth initiation rate		Death rate	
		P1	P2	P1	P2
A	30	.0036	.0047	.023	.001
CBA	20	.0033	.0039	.035	.011
A x CBA	25	.0025	.0033	.028	.001
RIII	35	.0070	.0039	.028	0

Primordial follicle numbers continue to dwindle throughout life irrespective of major physiological changes such as puberty, seasonal anoestrus, pregnancy and lactation. Nevertheless, this process is retarded in laboratory rodents by hypophysectomy and through a reduced dietary intake, which cause significant conservation of follicle numbers after several weeks (Lintern-Moore & Everitt, 1978; Faddy *et al.*, 1983; Nelson *et al.*, 1985). The mechanism(s) responsible for retarding ovarian ageing have not been identified, but could involve fundamental effects on cellular metabolism rather than changes in circulating levels of gonadotrophic hormones, FSH and LH, which differ in the two conditions (see **3**).

2) Cytological profile of small follicles

Primordial follicles vary from 15-25 microns in diameter in myomorphic rodents and some insectivores and bats to about 90 microns in cats and rabbits (Gosden & Telfer, 1987a). This variation is due to differences in either or both the volumes of ooplasm and the shape/size of the associated somatic cells. Oocytes are arrested at diplotene and enter the dictyate stage although the degree of chromatin condensation varies between species. Most mouse oocytes are incapable of resuming meiotic maturation until 60 microns when they are almost full-grown (Canipari *et al.*, 1984). The chromosomes of oocytes in primordial and growing follicles are not strictly comparable to the lampbrush chromosomes of amphibia, though relatively stable RNA is accumulated up to the stage when the germinal vesicle nucleus breaks down and meiosis resumes (Bachvarova, 1981; Fourcroy, 1982; De Leon *et al.*, 1983).

While primordial follicles are obviously store-houses for germ cells, it is misleading to portray them as "resting" stages. The oocyte possesses RNA polymerase activity and all the organelles required for protein synthesis (Baker, 1973; Lintern-Moore & Moore, 1979). The ultrastructural appearance of ooplasm is unremarkable apart from the possession in most species of conspicuous paranuclear material, which includes RNA (Ullmann, 1978). Interestingly, the genes responsible for the zona pellucida are not expressed until oocytes begin growing (Philpott *et al.*, 1987; Wassarman, 1988), but the pregranulosa cells rest on a delicate basement membrane from the earliest stages (Bagavandos *et al.*, 1983).

The somatic cells which envelope (sometimes incompletely) primordial oocytes are presumed to be sole progenitors of the follicular epithelium, which numbers about 50000 cells in mouse Graafian follicles. The epithelium consists of about five clones, corresponding closely to the numbers of squamous pregranulosa cells in intimate contact with the oocyte (Lintern-Moore & Moore, 1979; Telfer *et al.*, 1988). These stem cells make approximately equal contributions to the differentiated sub-populations of cumulus and mural granulosa cells. In ultrastructural appearance they appear to be undifferentiated compared with the actively secreting granulosa cells in larger follicles. Pregranulosa cells are mitotically inactive and the following Figure shows that they are suspended at the $G_1(G_0)$ phase of the cell cycle.
Having entered the growth cycle, granulosa cells divide continuously and asynchronously through approximately 10 cell cycles to produce a large multilaminar follicle numbering

2-3000 cells over a period of about 2 weeks, depending on age (Pedersen, 1970; Faddy *et al.* 1987).

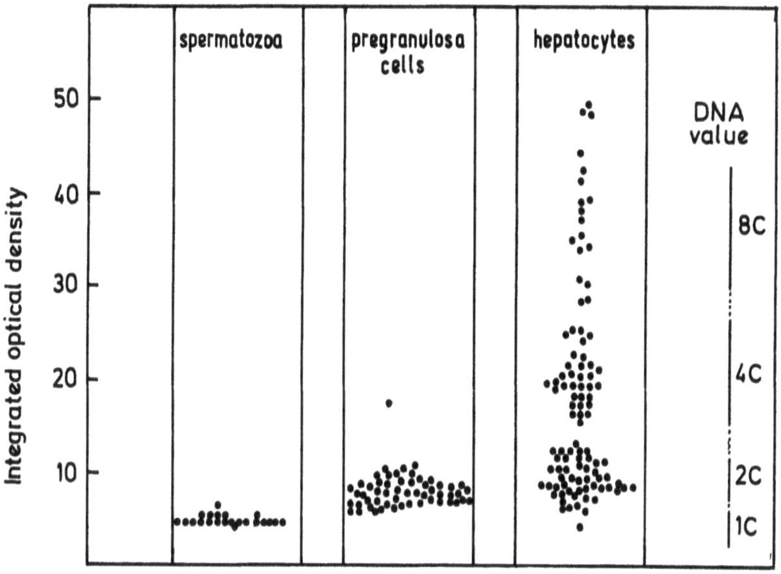

Fig 1. Comparison of DNA content in air-dried preparations of murine pregranulosa cells, spermatozoa and hepatocytes. Preparations, which were stained with the Feulgen reagent, were measured at 560 nm with a Vickers M86 scanning and integrating microdensitometer. (RG Gosden, unpublished data).

Subsequent growth becomes increasingly restricted to the central cells which will form the cumulus oophorus and the inner border of mural cells (Gosden *et al.*, 1983).

The nature of the physical and metabolic relationships between oocytes and somatic cells in primordial follicles has not been well-characterized. Although gap junctions can be observed between all follicle cells at all stages (Anderson & Albertini, 1976; Mitchell *et al.*, 1986) functional tests of cell-cell coupling and of metabolic cooperativity have been obtained hitherto only for growing follicles (Moor *et al.*, 1980; Heller *et al.*, 1981; Brower & Schultz, 1982; Colonna *et al.*, 1989). Pregranulosa cells would appear to provide a supporting function for oocytes since few naked oocytes persist in the mouse ovary beyond the juvenile phase and initiation of growth requires cell contact (see 3).

3) Initiation of follicle growth

Despite much speculation about the mechanism driving the recruitment of primordial follicles this remains one of the least understood and most important areas of ovarian biology. A number of hypotheses have been offered but there remains a lack of critical experimental evidence. According to one hypothesis, which is based on cytogenetic evidence only, the age at which follicles start to grow is predetermined by the order in which they are formed: the first formed are the first to grow, etc (Henderson & Edwards, 1968). While folliculogenesis extends over a period of months in human fetuses the equivalent phase lasts only 2-4 days in hamsters, mice and rats; such close synchrony would seem to vitiate developmental differences within the germ cell population which the production line hypothesis depends on.

Some authors have proposed a role for gonadotrophic hormones, particularly FSH (Mulheron *et al.*, 1987). The experimental evidence is frequently conflicting and so the involvement of the gonadotrophins remains doubtful. While some studies indicate that gonadotrophins are required for initiating the first wave of follicle growth in the neonatal mouse (Baker & Neal, 1973; Lintern-Moore, 1978) others deny this (Peters *et al.*, 1973) and the results of hypophysectomy in adult mice demonstrate that pituitary hormones are not obligatory for follicles to begin growing (Nakano *et al.*, 1975; Faddy *et al.*, 1983). Nevertheless, data obtained by modelling follicle populations in *hypogonadal* mice and in normal animals after ovariectomy indicate that the pituitary gland may exert subtle influences on the process of recruitment (Halpin *et al.*, 1986; Gosden *et al.*, 1989). This conclusion is compatible with evidence of increasing dependency on FSH from early stages of follicular growth (Arendsen de Wolff-Exalto, 1982; Roy & Greenwald, 1986, 1989).

This author favours the hypothesis that the step of commitment to growth involves a stochastic event originating in the oocyte. A similar view was expressed years ago by Baker (1973), although others have suggested the primacy of granulosa cell activity (Lintern-Moore & Moore, 1979). Conjecturally, oocyte growth might be triggered by the building up or breaking down by protease of signal molecules, perhaps alleviating inhibition of the cell cycle of pregranulosa cells. On *a priori* grounds one might expect that fundamental processes in cellular metabolism are involved rather than, for example, a more highly evolved ligand-receptor interaction in such a crucial step in the life history of every species. This hypothesis is consistent with aforementioned data showing that a constant proportion of follicles is recruited daily, implying that signals are initiated independently throughout the oocyte population. Furthermore, we do not know as yet of any physiological intervention that can completely halt the recruitment process, and although genetic loci affecting later stages of development have been identified (Kuroda *et al.*, 1988), no mutations have been described in mice for this step. The "resistant ovary syndrome" in humans bears some resemblance to such state, albeit with incomplete genetic penetration (Jones & de Moraes-Ruehsen, 1969).

The assumption that the stimulus for growth originates in the oocyte rather than in the somatic cells subserving it finds tentative support from three independent sets of experimental observations. Firstly, when intact primordial follicles are cultured in plasma clots a large majority of the oocytes begin to grow spontaneously while pregranulosa cells do not respond unless the clots are transplanted to host animals (see section 4). While this behaviour is still unaccounted for, it is evident that oocyte growth can begin independently of growth and differentiation of the pregranulosa cells, although requiring contact with them (Eppig, 1977; Bachvarova *et al.*, 1980).

Secondly, when follicles are plated out in tissue culture vessels in medium containing serum the 3-D structure breaks down within a few hours as pregranulosa cells attach, spread and multiply to form monolayers. The oocyte is released or remains only delicately attached (Fig 2).

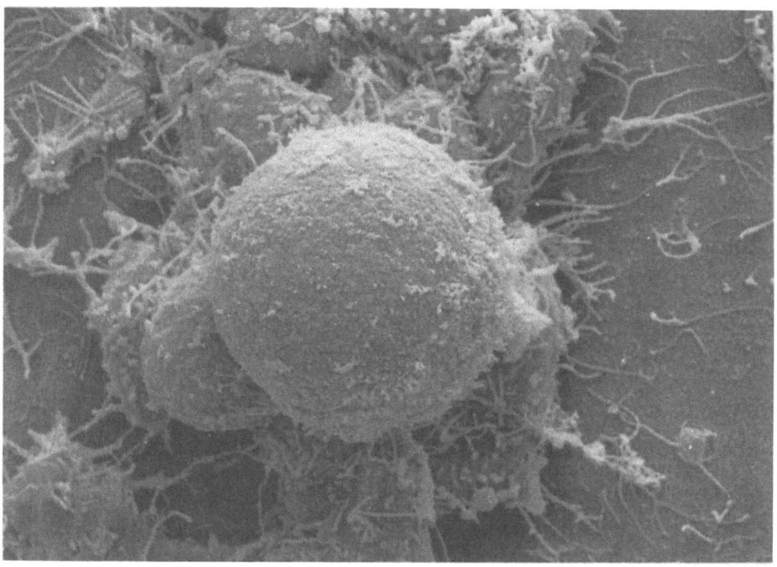

Fig 2. Scanning electron micrograph of a murine primordial follicle after 4 h in culture. The pregranulosa cells have attached to the substratum exposing the oocyte which may subsequently be released. x2180 (N Brown & RG Gosden, unpublished).

That the freed oocytes remain viable for several days was indicated by exclusion of trypan blue dye, maintenance of a steady resting potential (c.-10mv) and growth to form morphologically normal follicles after recombining with granulosa cells followed by transplantation to a host animal. These findings are compatible with a hypothetical inhibitory action exerted by oocytes on the cell cycle of their pregranulosa cells. Finally, a primary controlling influence of the oocyte is consistent with findings from the natural occurrence of polyovular follicles, which are abundant in ovaries of domestic bitches. If growth signals arise spontaneously and independently in all members of the oocyte population one might expect that the probability of polyovular follicles being recruited early in life would be greater than that of normal follicles, assuming that a signal in one cell would be passed to others in the same follicle unit. Polyovular follicles would then disappear during the process of ageing, which is the case (Telfer & Gosden, 1987). While such circumstantial evidence is helpful in formulating hypotheses, further progress on this question is likely to be delayed until better experimental models become available.

4) Experimental transplantation of primordial follicles

During the course of ageing, ovaries move towards a final state of sterility, although death normally intervenes (see section 1). Follicular deficiency can be overcome by transplanting animal ovaries when inbred strains are available (Krohn, 1977), but an alternative technique has been devised recently for transferring primordial follicles to create chimaeric organs (Gosden, 1989). Follicles were isolated by disaggregating neonatal mouse ovaries with collagenase to allow manipulation and control *in vitro* (Torrance *et al.*, 1989). Since it is impracticable to inject follicles into diminutive mouse ovaries they were incorporated into a droplet of plasma which served as a vehicle. Over a period of 1 or 2 days in culture the clots contracted to form compact cell masses. Preliminary results suggested fibrinolysis in these cultures was involving serum protease and possibly secretion of plasminogen activator(s), which are known products of mature ovaries (Canipari *et al.*, 1987). The grafts were inserted into resected ovaries of mice sterilized by abdominal X-irradiation, although subsequent studies showed that transfer to the evacuated bursa of ovariectomized animals could be equally effective. Follicles grew within the graft and reached Graafian sizes within three weeks (Fig 3). Many specimens were morphologically and functionally indistinguishable from normal organs, apart from wide variation in follicle numbers and occasional binovular follicles. After a latent period of about 2 weeks post-transplantation, most animals presented oestrous behaviour and, in some cases, pregnancies were established with delivery at term of normal pups.

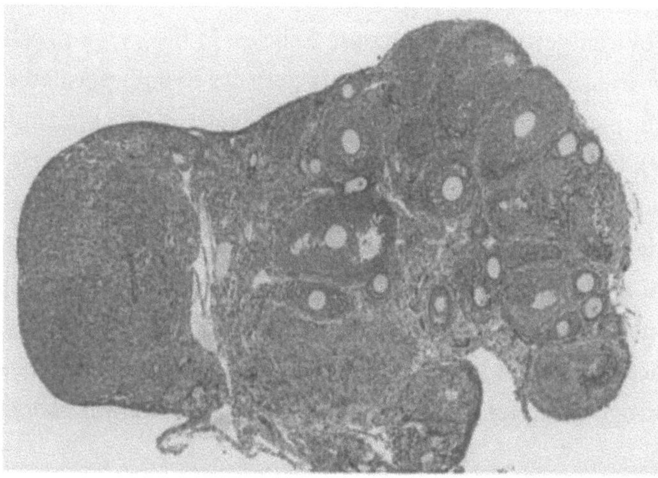

Fig 3. A chimaeric mouse ovary consisting of the X-irradiated remnant of the host with a protruding segment of grafted tissue in which primordial follicles have grown to multilaminar stages during a three-week period. H & E x50

5) Scope for future work

Elucidation of the signal committing primordial follicles to growth is a major goal for ovarian biology. The character of this signal will indicate the limits to which recruitment can be manipulated artificially. If determined by stochastic events operating at the level of the cell membrane or fundamental metabolic pathway rather than by extrinsic hormonal control by, for example, gonadotrophic hormones it may prove impossible to manipulate the process without disturbing cellular physiology more generally. The biology of primordial follicle death is of almost equal significance because wastage of germ cells early in life reduces fecundity later. It remains unclear whether these deaths affect abnormal oocytes selectively or have a more random distribution, perhaps having evolved as a mechanism for regulating excess follicle numbers and, hence, of ovarian size.

Although progress on these problems has been hampered by the small sizes of the follicles, it is now plausible for cellular and molecular biology to be applied to the study of cell membrane properties and gene expression in single follicle cells. Furthermore, the ability to reconstruct functional ovaries from disaggregated cells provides new possibilities for looking at the developmental biology of specific cell lineages.

It is too early to predict what practical implications, if any, will follow from the demonstration of transgermination of mouse follicles. In theory, germ cell transfer might prove useful for transmitting favourable characteristics in domesticated animals and for treating human patients presenting gonadal dysgenesis or premature menopause (Edwards, 1989). Frozen storage of small follicles, which has been developed recently (J Carroll, pers commun), could further these aims because one immature ovary contains sufficient follicles for many adult recipients. However, extension of germ cell transfer techniques from rodents to other species will require overcoming the potentially serious obstacles of transplant rejection and availability of suitable donor organs. However, even if these factors turn out to bar applications to reproductive medicine there can be little doubt that opportunities for manipulating small follicles will benefit experimental science.

Acknowledgements. I thank the Wellcome Trust, Medical Research Council and Galton Institute (London) for financing my research and Noel Brown and Kay Grant for assistance with microscope preparations depicted in this paper.

References

Anderson E, Albertini DF (1976) Gap junctions between the oocyte and companion follicle cells in the mammalian ovary. J Cell Biol 71:680-686

Arendsen de Wolff-Exalto E (1982) Influence of gonadotrophins on early follicle cell development and early oocyte growth in the immature rat. J Reprod Fert 66:537-542

Bachvarova R (1981) Synthesis, turnover and stability of heterogeneous RNA in growing mouse oocytes. Dev Biol 86:384-392

Bachvarova R, Baran MM, Tejblum A (1980) Development of naked growing mouse oocytes in vitro. J Exp Zool 211:159-169

Bagavandoss P, Midgley Jr AR, Wicha M (1983) Developmental changes in the ovarian follicular basal lamina detected by immunofluorescence and electron microscopy. J Histochem Cytochem 31:633-640

Baker TG (1973) Gametogenesis. In: Symposium on the use of non-human primates in research on problems of human reproduction (WHO, Sukhumi USSR). Acta Endocr Suppl 166:18-45

Baker TG, Neal P (1973) Initiation and control of meiosis and follicular growth in ovaries of the mouse. Ann Biol Anim Bioch Biophys 13:137-144

Block E (1952) Quantitative morphological investigations of the follicular system in women. Variations at different ages. Acta Anat 14:108-123

Brower PT, Schultz RM (1982) Intercellular communication between granulosa cells and mouse oocytes: existence and possible nutritional role during oocyte growth. Dev Biol 90: 144-153

Canipari R, O'Connell ML, Meyer G, Strickland S (1987) Mouse ovarian granulosa cells produce urokinase-type plasminogen activator, whereas the corresponding rat cells produce tissue-type plasminogen activator. J Cell Biol 105:977-981

Canipari R, Palombi F, Riminucci M, Mangia F (1984) Early programming of maturation competence in mouse oogenesis. Dev Biol 102:519-524

Colonna R, Cecconi S, Tatone C, Mangia F, Buccione R (1989) Somatic cell - oocyte interactions in mouse oogenesis: stage - specific regulation of mouse oocyte protein phosphorylation by granulosa cells. Dev Biol 133:305-308

Edwards RG (1989) Life before birth - reflections on the embryo debate, Hutchinson, London, p 81

Eppig JJ (1977) Mouse oocyte development *in vitro* with various culture systems. Dev Biol 60:371-388

Faddy MJ, Gosden RG, Edwards RG (1983) Ovarian follicle dynamics in mice: a comparative study of three inbred strains and an F1 hybrid. J Endocr 96:23-33

Faddy MJ, Jones EC, Edwards RG (1976) An analytical model for ovarian follicle dynamics. J Exp Zool 197:173-185

Faddy MJ, Telfer E, Gosden RG (1987) The kinetics of preantral follicle development in ovaries of CBA/Ca mice during the first 14 weeks of life. Cell Tissue Kinet 20:551-560

Fourcroy JL (1982) RNA synthesis in immature mouse oocyte development. J Exp Zool 219:257-266

Gosden RG (1985) Biology of menopause - the causes and consequences of ovarian ageing, Academic Press, London

Gosden RG (1987) Follicular status at the menopause. Hum Reprod 2:617-621

Gosden RG (1989) Restoration of fecundity to sterilized mouse ovaries by transferring immature follicles. Proceedings of the Physiological Society, D10 (Edinburgh Meeting) J Physiol (in press)

Gosden RG, Laing SC, Flurkey K, Finch CE (1983) Graafian follicle growth and replacement in anovulatory ovaries of ageing C57BL/6J mice. J Reprod Fert 69:453-462

Gosden RG, Telfer E (1987a) Scaling of follicular sizes in mammalian ovaries. J Zool 211:157-168

Gosden RG, Telfer E (1987b) Numbers of follicles and oocytes in mammalian ovaries and their allometric relationships. J Zool 211:169-175

Gosden RG, Telfer E, Faddy MJ, Brook JD (1989) Ovarian cyclicity and follicular recruitment in unilaterally ovariectomized mice. J Reprod Fert 87:257-264

Halpin DMG, Charton HM, Faddy MJ (1986) Effects of gonadotrophin deficiency on follicular development in *hypogonadal* (*hpg*) mice. J Reprod Fert 78:119-125

Heller DT, Cahill DM, Schultz RM (1981) Biochemical studies of mammalian oogenesis: metabolic cooperativity between granulosa cells and growing mouse oocytes. Dev Biol 84:455-464

Henderson SA, Edwards RG (1968) Chiasma frequency and maternal age in mammals. Nature 218:22-28

Jones EC, Krohn PL (1961) The relationships between age, numbers of oocytes and fertility in virgin and multiparous mice. J Endocr 21:469-495

Jones GS, de Moraes-Ruehsen M (1969) A new syndrome of amenorrhea in association with hypergonadotropism and apparently normal follicular apparatus. Am J Obstet Gynec 104:597-600

Krohn PL (1977) Transplantation of the ovary. In: Zuckerman Lord, Weir BJ (eds) The Ovary, vol II. Academic Press, New York, p 101

Kuroda H, Terada N, Nakayama H, Matsumoto K, Kitamura Y (1988) Infertility due to growth arrest of ovarian follicles in Sl/Sl^t mice. Dev Biol 126:71-79

De Leon V, Johnson A, Bachvarova R (1983) Half-lives and relative amounts of stored and polysomal ribosomes and Poly(A)+ RNA in mouse oocytes. Dev Biol 98:400-408

Lintern-Moore S (1978) Initiation of follicle growth in the infant mouse ovary by exogenous gonadotrophin. Biol Reprod 17:635-639

Lintern-Moore S, Everitt AV (1978) The effect of restricted food intake on the size and composition of the ovarian follicle population in the Wistar rat. Biol Reprod 19:688-691

Lintern-Moore S, Moore GPM (1979) The initiation of follicle and oocyte growth in the mouse ovary. Biol Reprod 20:773-778

Mandl AM, Shelton M (1959) A quantitative study of oocytes in young and old nulliparous laboratory rats. J Endocr 18:444-450

Mitchell PA, Burghardt RC (1986) The ontogeny of nexuses (gap junctions) in the ovary of the fetal mouse. Anat Rec 214:283-288

Moor RM, Smith MW, Dawson RMC (1980) Measurement of intercellular coupling between oocytes and cumulus cells using intracellular markers. Exp Cell Res 126:15-29

Mulheron GW, Quattropani SL, Nolin JM (1987) On the intrinsic ovarian control of the developmental transition from primordial to primary follicle. Adv Exp Med Biol 219:737-742

Nakano R, Mizuno T, Katayama K, Tojo S (1975) Growth of ovarian follicles in rats in the absence of gonadotrophins. J Reprod Fert 45:545-546

Nelson JF, Gosden RG, Felicio LS (1985) Effect of dietary restriction on estrous cyclicity and follicular reserves in aging C57BL/6J mice. Biol Reprod 32:515-522

Norris ML, Adams CR (1982) Effect of unilateral ovariectomy on the population of ovarian follicles relative to age in the Mongolian gerbil (*Meriones unguiculatus*). J Reprod Fert 66:335-340

Pedersen T (1970) Determination of follicle growth rate in the ovary of the immature mouse. J Reprod Fert 21:81-93

Peters H, Byskov AG, Lintern-Moore S, Faber M, Andersen M (1973) The effect of gonadotrophin on follicle growth initiation in the neonatal mouse ovary. J Reprod Fert 35:139-141

Philpott CC, Ringuette MJ, Dean J (1987) Oocyte-specific expression and developmental regulation of ZP3, the sperm receptor of the mouse zona pellucida. Dev Biol 121:568-575

Richardson SJ, Senikas V, Nelson JF (1987) Follicular depletion during the menopausal transition: evidence for accelerated loss and ultimate exhaustion. J Clin Endocr Metab 65:1231-1237.

Roy SK, Greenwald GS (1986) Effects of FSH and LH on incorporation of [^{3}H]-thymidine into follicular DNA. J Reprod Fert 78:201-209

Roy SK, Greenwald GS (1989) Hormonal requirements for the growth and differentiation of hamster preantral follicles in long-term culture. J Reprod Fert 87:103-114

Tardiff SD (1985) Histologic evidence for age-related differences in ovarian function in Tamarins (*Saguinas* sp., Primates) Biol Reprod 33:993-1000

Telfer E, Gosden RG (1987) A quantitative cytological study of polyovular follicles in mammalian ovaries with particular reference to the domestic bitch (*Canis familiaris*). J Reprod Fert 81:137-147

Telfer E, Ansell JD, Taylor H, Gosden RG (1988) The number of clonal precursors of the follicular epithelium in the mouse ovary. J Reprod Fert 84:105-110

Torrance C, Telfer E, Gosden RG (1989) Quantitative study of the development of isolated mouse pre-antral follicles in collagen gel culture. J Reprod Fert 87:367-374

Ullmann SL (1978) Observations on the primordial oocyte of the bandicoot *Isoodon macrourus* (Peramelidae, Marsupialia), J Anat 128:619-631

Wassarman PM (1988) Zona pellucida glycoproteins. Ann Rev Biochem 57:415-442

ULTRASTRUCTURE AND AG STAINING OF ECHINODERM SPERMATOGENESIS

M. Sousa and C. Azevedo
Department of Cell Biology, Institute of Biomedical Sciences
University of Porto
4000 Porto
Portugal

During the past 20 years, only a few fine structural observations have been made on the entire spermatogenetic process of echinoderms. In this paper we review the process of acrosome formation in the Echinodermata with emphasis on those aspects which still remain obscure. Silver nitrate staining at the ultrastructural level is also discussed and its application to the study of echinoderm spermatogenesis is presented.

1. PROACROSOMAL VESICLE FORMATION

As to the beginning of the acrosome formation with an appearance of the proacrosomal vesicles, variation is found in the classes of the Echinodermata. In the Asteroidea, the proacrosomal vesicles are initially recognized in the spermatogonia (Sousa and Azevedo, 1988a) (Figs. 1,2,5), in the primary spermatocytes (Yamagata, 1988), or in the spermatids (Dan and Sirakami, 1971; Janssen, 1984); in the Echinoidea it is in the spermatids (Longo and Anderson, 1969); in the Ophiuroidea it is in the spermatogonia (Yamashita, 1983; Yamashita and Iwata, 1983) or in the spermatids (Buckland-Nicks et al., 1984); in the Holothuroidea it is in the spermatogonia (Atwood, 1974; Pladellorens and Subirana, 1975) or in the spermatids (Tilney, 1976a); and in the Crinoidea it is in the primary spermatocytes (Bickell et al., 1980) or in the spermatids (Afzelius, 1977). Therefore, acrosome formation in the echinoderms is initiated at the latest in the spermatogonia, with the exception of the echinoids. The formation of the proacrosomal vesicles, wherever it begins, is always associated with the presence of a well developed Golgi apparatus. Yamashita and Iwata (1983) suggested that the presence of proacrosomal vesicles in those stages less mature than spermatids could express the failure to evolve the necessary machinery for the rapid production of acrosomal materials at a specific later period of spermatogenesis or the

NATO ASI Series, Vol. H 45
Mechanism of Fertilization
Edited by B. Dale

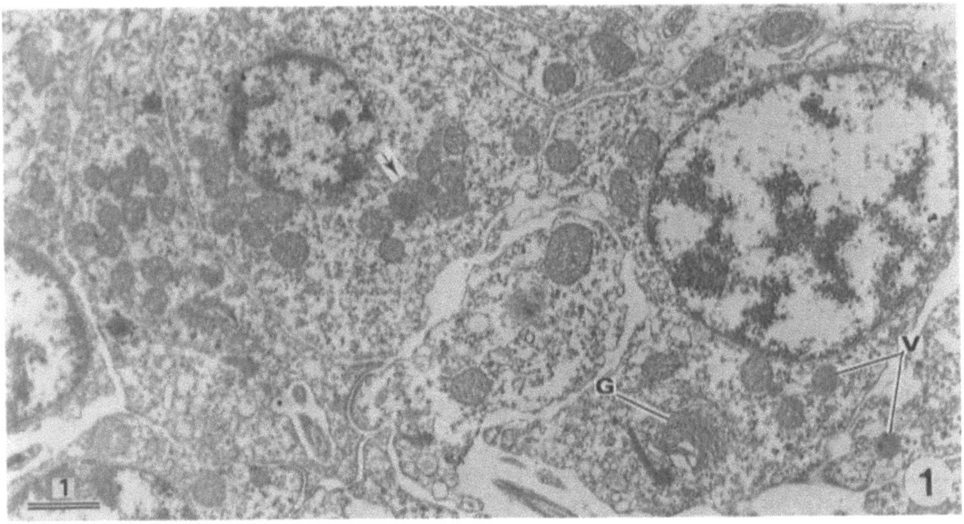

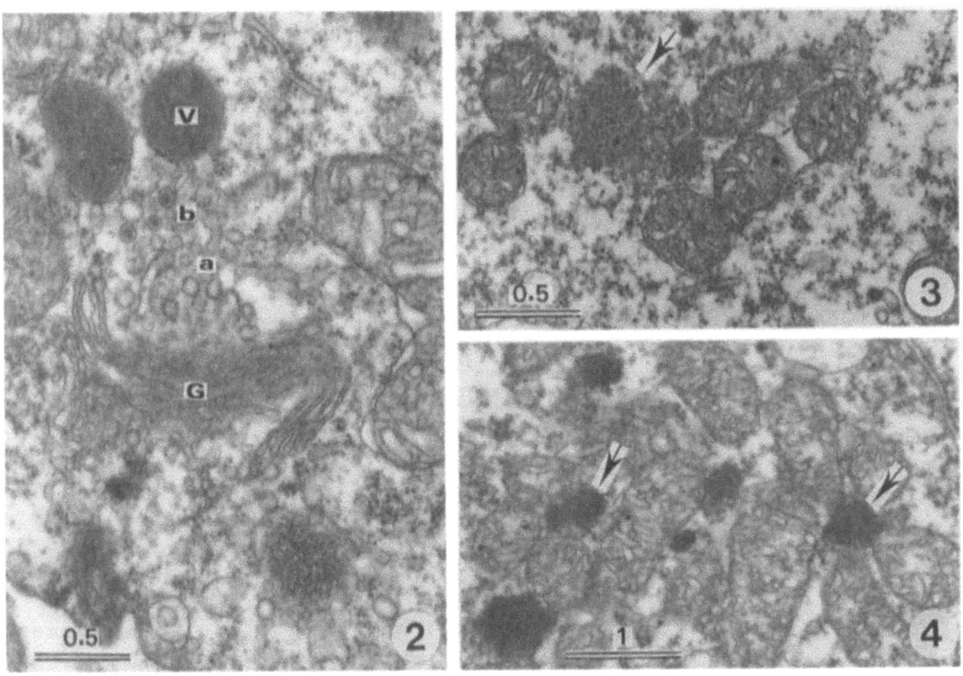

Figs. 1-13. Spermatogenesis of M. glacialis (Asteroidea) (From Sousa and Azevedo, 1988a). Bar = μm.
Figs. 1,3. Early spermatogonium with a large granulo-fibrillar mass (arrow) and late spermatogonium with a well-developed Golgi apparatus (G) and large proacrosomal vesicles (V).
Figs. 2,4. Late spermatogonia. Small coated vesicles with light contents (a) are originated in the Golgi apparatus (G), then acquire a dense core (b), and finally give origin to large proacrosomal vesicles (V). Small dense bodies (arrows) originate from large granulo-fibrillar masses (Fig.4).

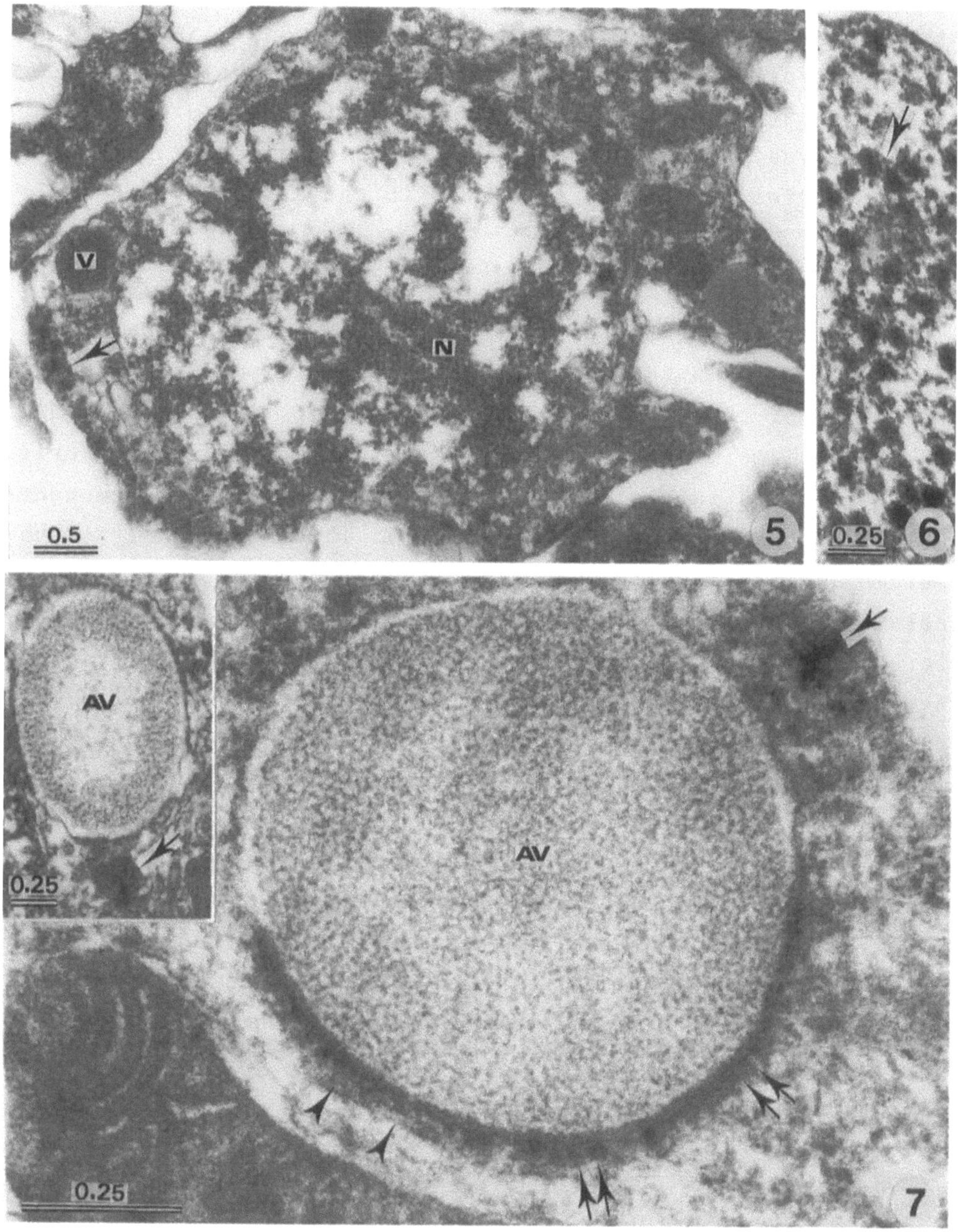

Figs. 5,6. Primary spermatocytes accumulate large proacrosomal vesicles (V) and small dense bodies (arrows). N, nucleus.
Fig.7. Mid spermatid. The basal half of the acrosomal vesicle (AV) is coated by dense material (double arrows) which is similar to the associated dense body (arrow). Arrowheads point to the developing actomere.

short duration of spermiogenetic stage insufficient to produce enough amount of the acrosomal materials. To solve this question, proacrosomal vesicles should be searched since the spermatogonia stage, as in many reports only spermiogenesis has been observed. If the above variation is confirmed, autoradiographic studies could help to answer to the second hypothesis.

2. ACROSOMAL VESICLE DIFFERENTIATION

The antero-posterior differentiation of the acrosomal vesicle is completed in the early spermatid stage immediately after the fusion of the proacrosomal vesicles (Fig.7). The basal half of the vesicle membrane develops more osmiophilic characteristics due to its interaction with the acrosomal vesicle contents and periacrosomal materials (Figs.7-8). This denser half of the vesicle membrane will come to lie on the nuclear side, it is resistant to the fusion process that occurs during the exocytotic step of the acrosomal reaction, and will cover the tip of the acrosomal process. The anterior half of the vesicle membrane, on the contrary, will stay in contact with the plasma membrane and has the capacity to fuse, not only with the plasma membrane during the acrosomal reaction, but also with smaller proacrosomal vesicles during the maturation process (Sousa and Azevedo, 1985,1988b). This differentiation of the acrosomal vesicle may be primarily determined by its own molecular composition, as different molecules could become differently segregated accordingly to their own affinities and/or due to the interaction with other intra and/or extravesicular molecules (Sousa and Azevedo, 1986; Yanagimachi, 1986; Garbers, 1989; Longo et al., 1989). In those cases where the proacrosomal vesicles are precociously formed , if they already appear differentiated (Sousa and Azevedo,1988a; Yamagata, 1988), cells just need that some factor inhibits further vesicle fusion until the spermatid stage; but if no precocious differentiation occurs, some later formed factor must also interact with the vesicle to induce differentiation. Therefore, it appears of immediate importance (1) to search for a precocious differentiation in those cases where the proacrosomal vesicles are earlier formed, and (2) to isolate the acrosomal vesicles and

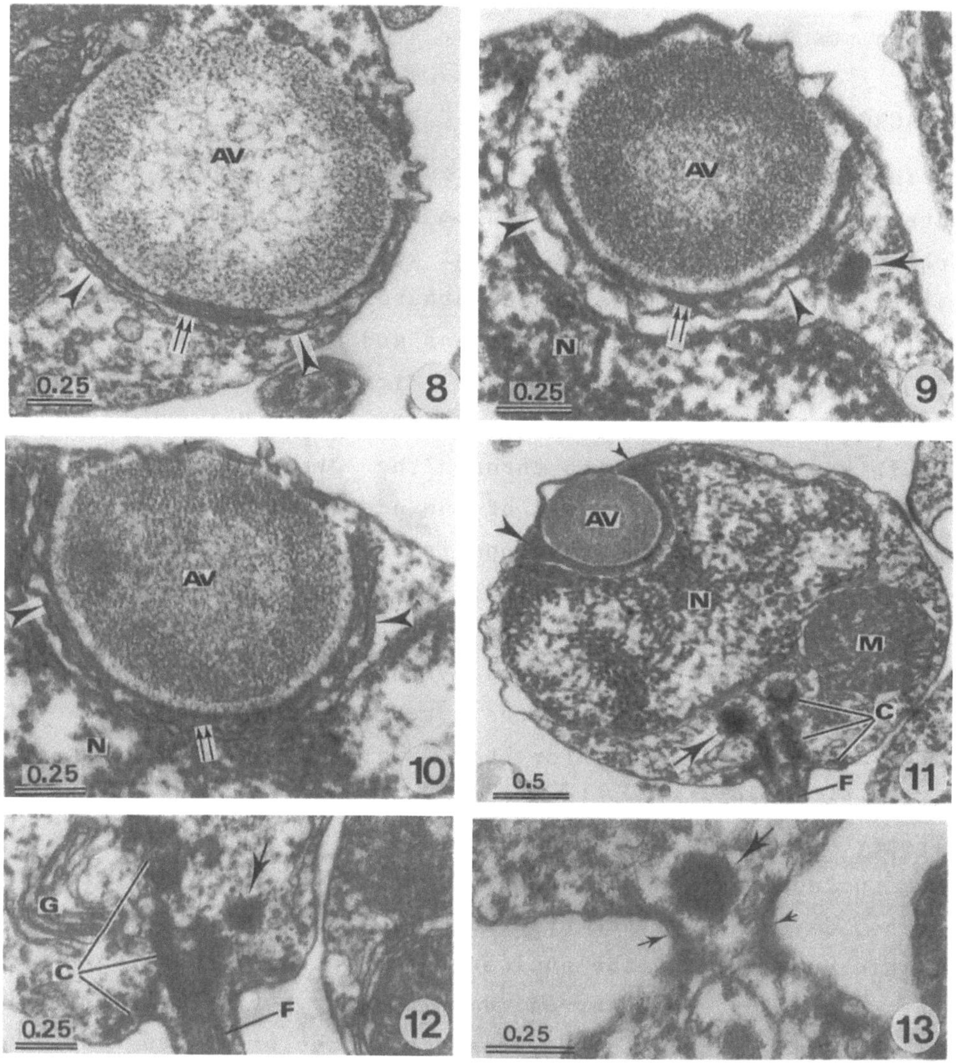

Figs. 8-10. Mid spermatids. Endoplasmic reticulum cisternae (arrowheads) surround the coated basal half (double arrows) of the acrosomal vesicle (AV) and are incorporated in the acrosomal fossa. N, nucleus; dense body (arrow).

Figs. 11-13. Late spermatids. Dense bodies (arrow) are associated with centrioles (C) and intercellular bridges (small arrows). Periacrosomal material (arrowheads); M, mitochondria; G, Golgi apparatus; F, flagellum.

study by immunocytochemistry the distribution and fate of their components during spermatogenesis.

3. ACROSOMAL VESICLE MIGRATION

The transportation of the acrosomal vesicle from caudal to anterior pole of the spermatid is differently achieved in the Echinodermata : (1) The acrosomal vesicle migrates through the cytoplasm in asteroids and holothuroids (Dan and Sirakami, 1971; Atwood, 1974; Pladellorens and Subirana, 1975; Tilney, 1976a; Sousa and Azevedo, 1988a). Microtubules have been noted in close association with the vesicle in the Holothuroidea and possibly aid in migration through the cytoplasm. (2) During migration, the basal half of the acrosomal vesicle membrane attaches to the nuclear envelope in echinoids, ophiuroids and crinoids (Bickell et al., 1980; Yamashita, 1983; Yamashita and Iwata, 1983; Sousa and Azevedo, 1988b). This acrosomal vesicle displacement could be explained by the rotation of the nucleus-vesicle complex relative to the cytoplasm. This model, however, seems to be inadequate for the sea urchin, because in this case the posterior region of the nucleus and of the cell is already determined by the attachement of the two centrioles to the nucleus and the axoneme (Fig. 22) (Sousa and Azevedo, 1988b). Severance of the hypothesized binding sites between the vesicle membrane and the nuclear envelope would then be necessary to allow the observed displacement. Finally, (3) the vesicle may attach to the plasma membrane with its anterior portion, as described in a starfish (Yamagata, 1988).

Whatever the mechanism of transportation, there must be some positional stored molecules on the involved membranes that specifically bind to each other, either directly or via some exogenous material. Again, biochemical isolation of these membranes and immunocytochemical localization of their molecules are necessary to further understand the mechanisms of these organelle's interaction.

4. ACROSOMAL VESICLE COMPOSITION

To date, only the acrosomal vesicle contents have been studied, while the vesicle membrane remains uninvestigated. In echinoids the vesicle contains the enzymes acid and alkaline phosphatase, phospholipase, protease, arylsulfatase, catalase and peroxidase (Anderson, 1968; Conway and Metz, 1976; Levine and Walsh, 1979; Green and Summers, 1980; Hoshi and Moriya, 1980; Yamada and Aketa, 1981; Boldt et al., 1984), which are involved in vesicle exocytosis and help the sperm to penetrate the oocyte investments; acid mucopolysaccharides (Summers and Hylander, 1974); and the protein bindin (Vacquier and Moy, 1977; Moy and Vacquier, 1979) that is responsible for adhesion of sperm to the oocyte. In asteroids, the vesicle is morphologically and functionally compartmentalized, as it contains the enzyme Ca-Mg-ATPase and calcium in the peripheral component, that binds the sperm to the oocyte jelly, and acid and alkaline phosphatases in the central and apical components, which are involved in vesicle exocytosis and help the acrosomal process to penetrate the oocyte coats (Mabuchi and Mabuchi, 1973; Sousa and Azevedo, 1985, 1986, 1988d,e, 1989a). In the ophiuroids, the vesicle contains acid mucopolysaccharides (Hylander and Summers, 1975).

5. FORMATION AND COMPOSITION OF THE PERIACROSOMAL MATERIAL

After the acrosomal vesicle has reached its definitive location, the nucleus invaginates around the basolateral circunference of the acrosomal vesicle to form the acrosomal fossa, which becomes filled with the periacrosomal material (Figs. 9-11). Except in the Holothuroidea, all other echinoderms develop a second nuclear invagination beneath the vesicle, the subacrosomal fossa, which contains the actomere (Tilney, 1978). It is supposed that the nucleus invaginates due to the interaction between the two membranes and the periacrosomal molecules.

The periacrosomal material has been studied in holothuroids and asteroids, and consists of G actin associated with spectrin, fascin and profilin-like molecules (Tilney, 1976b; Maekawa and Sakai, 1982; Maekawa et al., 1982), as well as membrane precursors. During the acrosomal reaction and after vesicle exocytosis, the actomere initiates the assembly of actin filaments by making use of the stored G actin molecules. Spectrin and profilin regulate G actin polymerization, while fascin

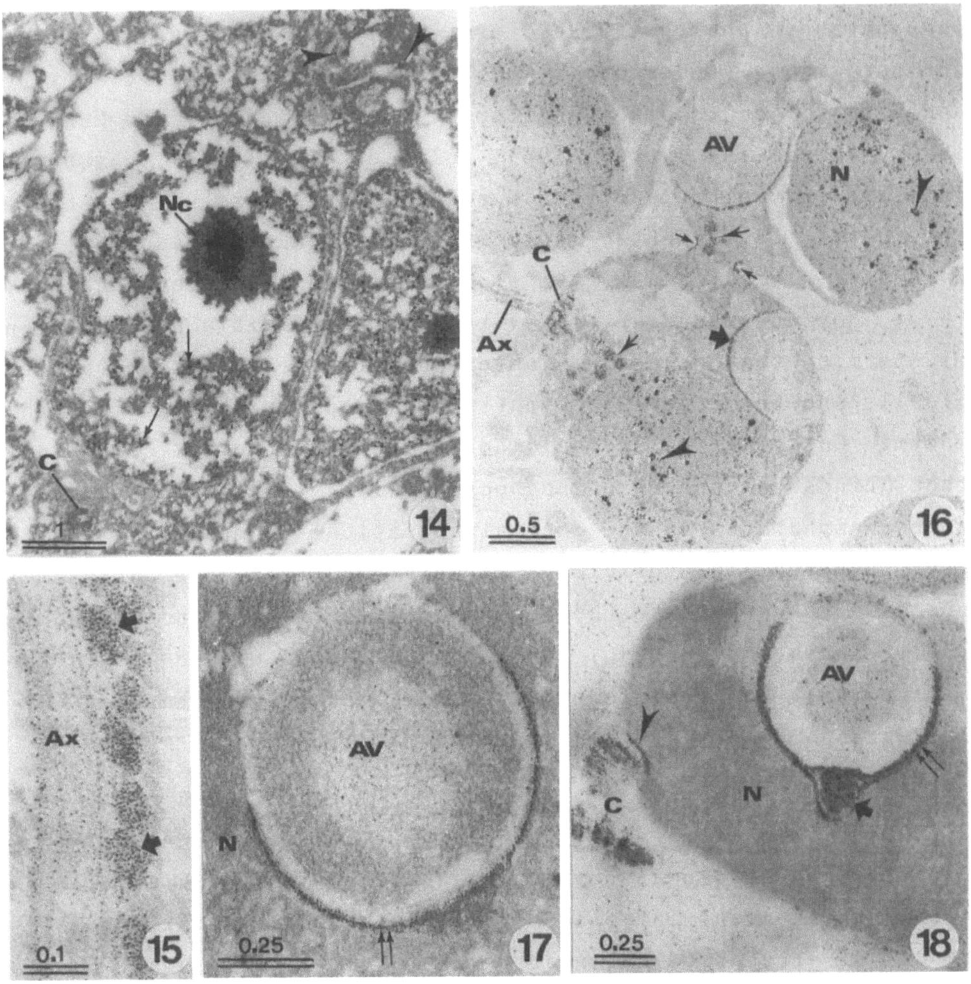

Figs. 14-18. Silver staining in the starfishes M. glacialis (Figs. 14-17) and A. rubens (Fig. 18) (From Sousa and Azevedo, 1988c, 1989b). Bar = µm.

Fig. 14. Spermatogonium. Silver stains the nucleolus (Nc), specific sites at the chromatin periphery (arrows), dense bodies (arrowheads) and centrioles (C).

Fig. 15. Primary spermatocyte. Argyrophilic bodies (arrows) interact with the axonemal microtubules (Ax).

Fig. 16. Mid spermatids. Silver granules accumulate in those nuclear spaces not yet condensed (arrowheads). In the cytoplasm silver stains the outer face (large arrow) of the basal acrosomal vesicle (AV) membrane, centrioles and pericentriolar complex (C), axoneme (Ax), intercellular bridges (small arrows) and dense bodies (arrows). N, nucleus.

Figs. 17,18. Spermatozoa. Silver stains the centriolar fossa (arrowhead), centrioles and pericentriolar complex (C), the outer face of the inner acrosomal vesicle membrane (double arrows) (Figs. 17, 18) and the subacrosomal material (large arrow) (Fig. 18).

causes actin filaments to bundle. The F actin bundle then everts the remaining basal half of the vesicle membrane, forming a thin tubule filled with actin filaments, the acrosomal process, that interacts with the oolemma (Tilney *et al*., 1973).

In echinoids, the acrosomal process membrane is formed by the simple eversion of the basal portion of acrosomal vesicle membrane, but in non-echinoids neither the eversion of vesicle membrane nor a forward slippage of plasma membrane can account for the vast amount of tubule membrane present. A source of new membrane must then exist and it is supposed to derive from the periacrosomal material. Several evidences support this view : (1) membranous vesicles suddenly appear within the ophiuroid and holothuroid periacrosomal material during the acrosome reaction and fuse at the base of the acrosomal process (Hylander and Summers, 1975; Tilney and Inoué, 1982); (2) lipids are present in the starfish periacrosomal material (Sardet and Tilney, 1977); and (3) cisternae of endoplasmic reticulum are incorporated in the periacrosomal region during the starfish acrosome maturation process (Figs. 8-10) (Sousa and Azevedo, 1988a).

6. ORIGIN AND FATE OF THE PERIACROSOMAL MATERIAL

Concerning the origin of the periacrosomal material, Tilney (1976a) has suggested that the periacrosomal molecules associate and bind to the specialized regions of the acrosomal vesicle membrane and of the nuclear envelope. In fact, in all cases, the periacrosomal material seems to accumulate between two membranes, namely, the basal vesicle membrane and the nuclear envelope in non-asteroids (Atwood, 1974; Tilney, 1976a; Afzelius, 1977; Yamashita, 1983; Buckland-Nicks *et al*., 1984; Sousa and Azevedo, 1988b), or the basal vesicle membrane and cisternae of endoplasmic reticulum in asteroids (Dan and Sirakami, 1971; Sousa and Azevedo, 1988a), wherever this occurs at the end of acrosomal vesicle migration (in holothuroids) or during acrosomal vesicle displacement (in non-holothuroids) (Figs. 8-11).

Alternatively, the periacrosomal material could be synthesized by the nuclear envelope or the endoplasmic reticulum. The

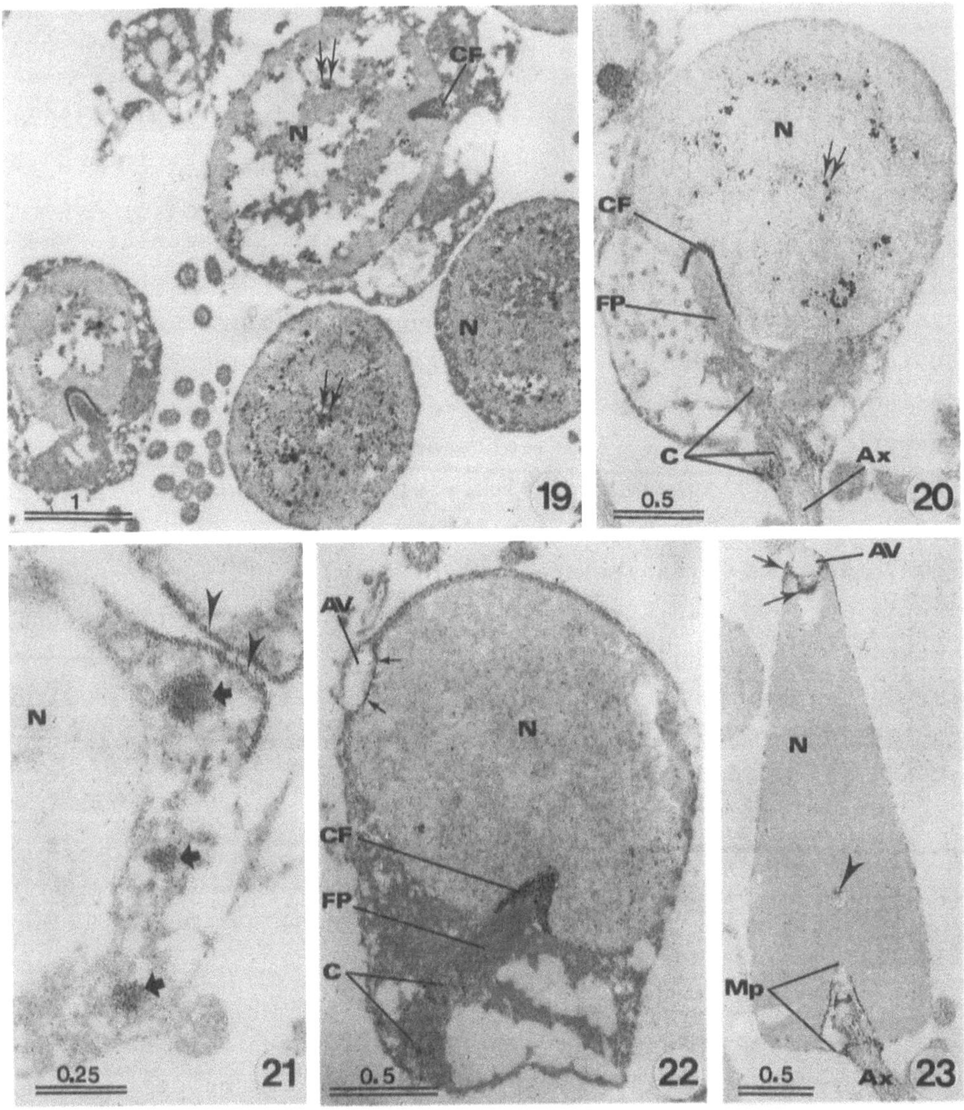

Figs.19-23. Silver staining in the echinoid P. lividus (From Sousa and Azevedo, 1988b). Bar = μm.
Figs.19-22. During spermiogenesis, silver granules accumulate in those nuclear spaces not yet condensed (double arrows) and finally disappear (Fig.22). Silver also stains the centriolar fossa (CF), centrioles and pericentriolar complex (C), intercellular bridges (arrowheads), dense bodies (arrows), and the material (small arrows) that connects the acrosomal vesicle (AV) to the nucleus (N). FP, fibrous process; Ax, axoneme.
Fig.23. Spermatozoon. Silver stains the periacrosomal material (arrows), nuclear vacuole (arrowhead), midpiece structures (Mp) and the axoneme (Ax).

presence of ribosomes, however, has only been documented in a starfish (Yamagata, 1988), and autoradiographic studies did not show any protein synthetic activity in the acrosomal region of sea urchin spermatids (Nicotra et al., 1984).

More recently, a new model for the origin of the periacrosomal material appeared (Sousa and Azevedo, 1988a,b, 1989b). We have described the presence of cytoplasmic dense bodies since the spermatogonia stage in echinoid and asteroid spermatogenesis, and showed that they seem to be involved in the formation of the periacrosomal material as well as other organelles (Figs. 3-13). This morphologic relationship was further confirmed by cytochemical methods, as dense bodies and those organelles selectively stained with silver (Figs. 14-23).

Cytoplasmic dense bodies have also been described in spermatogonia of echinoids and ophiuroids, in primary spermatocytes of holothuroids and in spermatids of asteroids, but their fate has not been studied (Longo and Anderson, 1969; Dan and Sirakami, 1971; Atwood, 1974; Houk and Hinegardner, 1981; Yamashita, 1983; Yamashita and Iwata, 1983).

Ag-NOR staining is due to the acidic, highly phosphorylated protein C23, which binds to histone H1 and induces chromatin decondensation (Erard et al., 1988). Our findings using silver staining show that besides nucleolar staining, some regions at the periphery of chromatin also appear silver stained (Figs. 14,15,19,20), and we suggested that these sites also contain decondensed portions of chromatin (Sousa and Azevedo, 1988b). Further substantiating this view, we also have found that as chromatin condenses during spermiogenesis, silver deposits become trapped in those nuclear regions not yet condensed and finally disappear in mature nuclei (Figs. 14-23) (Sousa and Azevedo, 1988b, 1989b).

The shared property of argyrophilia of the nucleolar, nuclear and cytoplasmic structures does not imply identity in the proteins associated with these structures. Our studies in the echinoids and asteroids, however, suggest that Ag-staining of cytoplasmatic structures may be due to a common set of structural argyrophilic proteins. The morphological association between argyrophilic bodies and other cytoplasmic structures in the acrosome and midpiece, as well as their common Ag-staining, suggest that dense bodies may be primordial storage depots containing the argyrophilic proteins that will be needed for the formation of other argyrophilic organelles. Argyrophilic bodies, as newly described organelles, must now be further investigated. As silver specifically stains dense bodies, the presence of these structures in other echinoderms should be searched using this cytochemical method. The presence of C23 and G actin in the argyrophilic bodies should be investigated by immunocytochemistry. The isolation of the sea urchin periacrosomal material would allow to know which proteins are silver stained and study their evolution during spermatogenesis.

Alternatively, dense bodies could be nuage-like material and contain the necessary protein synthetic machinery to give origin to some components of those cytoplasmic structures. Selective enzymatic digestions, nuclease-gold experiments and autoradiographic studies would help to solve this question.

ACKNOWLEDGEMENTS

This work was supported by CEM-INIC, JNICT, University of Porto (contract 4/86/87) and the Eng. A. Almeida Foundation. We thank Mr. J. Carvalheiro for preparation of the figures.

REFERENCES

Afzelius BA (1977) Spermatozoa and spermatids of the crinoid Antedon petasus, with note on primitive spermatozoa from Deuterostome animals. J Ultrastruct Res 59 : 272-281

Anderson WA (1968) Cytochemistry of sea urchin gametes. III. Acid and alkaline phosphatase activity of spermatozoa and fertilization. J Ultrastruct Res 25 : 1-14

Atwood DG (1974) Fine structure of spermatogonia, spermatocytes, and spermatids of the sea cucumbers Cucumaria lubrica and Leptosynapta clarki (Echinodermata:Holothuroidea). Can J Zool 52 : 1389-1396

Bickell LR, Chia F-S, Crawford BJ (1980) A fine structural study of the testicular wall and spermatogenesis in the crinoid, Florometra serratissima (Echinodermata). J Morphol 166 : 109-126

Boldt J, Alliegro MC, Schuel H (1984) A separate catalase and peroxidase in sea urchin sperm. Gamete Res 10 : 267-281

Buckland-Nicks J, Walker CW, Chia F-S (1984) Ultrastructure of the male reproductive system and of spermatogenesis in the viviparous brittle-star, Amphipholis squamata. J Morphol 179 : 243-262

Conway AF, Metz CB (1976) Phospholipase activity of sea urchin sperm: its possible involvement in membrane fusion. J Exp Zool 198 : 39-48

Dan JC, Sirakami A (1971) Studies on the acrosome. X. Differentiation of the starfish acrosome. Dev Growth Differ 13 : 37-52

Erard MS, Belenguer P, Caizergues-Ferrer M, Pantaloni A, Amalric F (1988) A major nucleolar protein, nucleolin, induces chromatin decondensation by binding to histone H1. Eur J Biochem 175 : 525-530

Garbers DL (1989) Molecular basis of signalling in the spermatozoon. J Androl 10 : 99-107

Green JD, Summers RG (1980) Ultrastructural demonstration of trypsin-like protease in acrosomes of sea urchin sperm. Science 209 : 398-400

Hoshi M, Moriya T (1980) Arylsulfatase of sea urchin sperm. 2. Arylsulfatase as a lysin of sea urchins. Dev Biol 74 : 343-350

Houk MS, Hinegardner RT (1981) Cytoplasmic inclusions specific to the sea urchin germ line. Dev Biol 86 : 94-99

Hylander BL, Summers RG (1975) An ultrastructural investigation of the spermatozoa of two ophiuroids, Ophiocoma echinata and Ophiocoma wendti: acrosomal morphology and reaction. Cell Tiss Res 158 : 151-168

Janssen HH (1984) Development and ultrastructure of spermatozoa of Archaster typicus Mull. and Trosch. (Echinodermata, Asteroidea). Int J Invert Reprod Dev 7 : 333-344

Levine AE, Walsh KA (1979) Involvement of an acrosin-like enzyme in the acrosome reaction of sea urchin sperm. Dev Biol 72 : 126-137

Longo FJ, Anderson E (1969) Sperm differentiation in the sea urchins Arbacia punctulata and Strongylocentrotus purpuratus. J Ultrastruct Res 27 : 486-509

Longo FJ, Canvin AT, Bailey JL (1989) Membrane specializations associated with the acrosome complex of sea urchin sperm as revealed by immunocytochemistry and freeze fracture replication. Gamete Res 23 : 429-440

Mabuchi Y, Mabuchi I (1973) Acrosomal ATPase in starfish and bivalve mollusk spermatozoa. Exp Cell Res 82 : 271-279

Maekawa S, Sakai H (1982) Inhibitor protein of actin polymerization from starfish sperm head: purification and localization. Biomed Res 3 : 46-53

Maekawa S, Endo S, Sakai H (1982) A protein in starfish sperm head which bundles actin filaments in vitro: purification and characterization. J Biochem 92 : 1959-1972

Moy GW, Vacquier VD (1979) Immunoperoxidase localization of bindin during the adhesion of sperm to sea urchin eggs. Curr Topics Dev Biol 13 : 31-44

Nicotra A, Arizzi M, Gallo PV (1984) Protein synthetic activities during spermiogenesis in the sea urchin: an high resolution autoradiographic study of H-leucine incorporation. Dev Growth Differ 26 : 273-280

Pladellorens M, Subirana JA (1975) Spermiogenesis in the sea cucumber Holothuria tubulosa. J Ultrastruct Res 52 : 235-242

Sardet C, Tilney LG (1977) Origin of the membrane for the acrosomal process: is actin complexed with membrane precursors ? Cell Biol Int Rep 1 : 193-200

Sousa M. Azevedo C (1985) Acrosomal reaction and early events at fertilization in Marthasterias glacialis (Echinodermata:Asteroidea). Gamete Res 11 : 157-167

Sousa M, Azevedo C (1986) Cytochemical study on the spermatozoon and at early fertilization in Marthasterias glacialis (Echinodermata, Asteroidea). Biol Cell 56 : 79-84

Sousa M, Azevedo C (1988a) Fine structural study of the acrosome formation in the starfish Marthasterias glacialis (Echinodermata, Asteroidea). Tiss Cell 20 : 621-628

Sousa M, Azevedo C (1988b) Ultrastructure and silver-staining analysis of spermatogenesis in the sea urchin Paracentrotus lividus (Echinodermata, Echinoidea). J Morphol 195 : 177-188

Sousa M, Azevedo C (1988c) Comparative silver staining analysis on spermatozoa of various invertebrate species. Int J Invert Reprod Dev 13 : 1-8

Sousa M, Azevedo C (1988d) Starfish acrosomal acid phosphatase: a cytochemical and biochemical study. Biol Cell 63 : 101-104

Sousa M, Azevedo C (1988e) Presence of ATPase and alkaline phosphatase activities in the starfish sperm acrosome. Cell Biol Int Rep 12 : 1049-1054

Sousa M, Azevedo C (1989a) Ultrastructural localization of calcium in the acrosome and jelly coat of starfish gametes. Dev Growth Differ 31 : 227-232

Sousa M, Azevedo C (1989b) Silver staining of spermatogenesis in the starfish Marthasterias glacialis. Invert Reprod Dev 15 : 105-108

Summers RG, Hylander BL (1974) An ultrastructural analysis of early fertilization in the sand dollar, Echinarachnius parma. Cell Tiss Res 150 : 343-368

Tilney LG, Hatano S, Ishikawa H, Mooseker MS (1973) The polymerization of actin: role in the generation of the acrosomal process of certain echinoderm sperm. J Cell Biol 59 :109-126

Tilney LG (1976a) The polymerization of actin. II. How non filamentous actin becomes nonrandomly distributed in sperm: evidence for the association of this actin with membranes. J Cell Biol 69 : 51-72

Tilney LG (1976b) The polymerization of actin. III. Aggregates of nonfilamentous actin and its associated proteins: a storage form of actin. J Cell Biol 69 : 73-89

Tilney LG (1978) Polymerization of actin. V. A new organelle, the actomere, that initiates the assembly of actin filaments in Thyone sperm. J Cell Biol 77 : 551-564

Tilney LG, Inoué S (1982) Acrosomal reaction of Thyone sperm. II. The kinetics and possible mechanism of acrosomal process elongation. J Cell Biol 93 : 820-827

Vacquier VD, Moy GW (1977) Isolation of bindin: the protein responsible for adhesion of sperm to sea urchin eggs. Proc Natl Acad Sci USA 74 : 2456-2460

Yamada Y, Aketa K (1981) Vitelline layer lytic activity in sperm extracts of sea urchin, Hemicentrotus pulcherrimus. Gamete Res 4 : 193-202

Yamagata A (1988) Ultrastructure of spermatogenesis in the sea star, Asterina minor. Gamete Res 19 : 215-225

Yamashita M (1983) A fine structural study of spermatogenesis in the brittle-star Ophiura sarsii (Echinodermata: Ophiuroidea), with a demonstration of the precocious formation of the acrosome. J Fac Sci Hokkaido Univ 23 : 254-265

Yamashita M, Iwata F (1983) Ultrastructural observations on the spermatogenesis of the brittle-star Amphipholis kochii Lutken (Echinodermata: Ophiuroidea). Seto Mar Biol Lab 5/6 : 403-415

Yanagimachi R (1986) The acrosome reaction: analysis of its mechanism using guinea pig spermatozoa. Dev Growth Differ 28 : 3-4

CELL SPECIFIC GENE EXPRESSION IN MOUSE TESTIS

R. Geremia, V. Sorrentino*, M. Giorgi, P. Grimaldi, I. Sammarco and P. Rossi

Dept. Public Health and Cell Biology
II University of Rome "Torvergata"
Via O. Raimondo 8,
00173 Roma,
Italy.

Spermatogenesis is a complex differentiative process, highly conserved in the course of evolution, represented by the ordered sequence of mitotic, meiotic and differentiative stages. The pioneering work of Oakberg (1956), Monesi (1965) and Clermont (1972) have clearly shown that these stages are characterized by a complex series of biosynthetic processes which make spermatogenesis an excellent model system for the study of molecular events related to cell differentiation. DNA, RNA and protein synthesis were initially studied by autoradiographic and histochemical techniques, and only later with biochemical analysis after the introduction of cell fractionation techniques (Lam et al., 1970; Meistrich, 1973) which made available germ cells at defined stage of differentiation. Further insight into the genetic events and the mechanisms which control germ cell differentiation seems now possible with the use of powerful tools like molecular biology and transgenic techniques.

In this chapter we will review studies from our and others' laboratories focused on testis-specific and cell stage-specific gene expression, in the attempt to select informations relevant to the understanding of the control mechanisms and molecular events which characterize spermatogenesis.

---

* European Molecular Biology Laboratory
Meyerhoffstrasse 1
6900 Heidelberg
Germany

NATO ASI Series, Vol. H 45
Mechanism of Fertilization
Edited by B. Dale

RNA and protein synthesis

DNA transcription in spermatogenesis was shown to occur both in meiosis and spermiogenesis, although with different rates (Monesi, 71), and both ribosomal and poly (A+) RNA were shown to be synthesised in the diploid and haploid stages (Geremia et al., 1978).

Amino acid incorporation was also autoradiografically demonstrated to occur in meiosis and after a drop in the early stages of the haploid phase, to be resumed in late spermiogenesis (Monesi 1971).It was later shown by using bidimensional elecrophoretic analysis that this stage-specific expression is characterized not only by quantitative but also by several qualitative differences in the proteins synthesized. (Boitani et al., 1980; Kramer and Erickson, 1982).

A growing body of studies is providing indications of tissue-specific and stage-specific expression of several known genes. Human testis-specific lactate-dehydrogenase, a form functionally similar but genetically distinct from that present in somatic tissues, constitutes a well characterized example of tissue-specific transcription in differentiating germ cells (Millan et al., 1987). Specific histones are synthesized during meiotic prophase of the mouse (Bhatnagar et al., 1985) and of the rat (Meistrich et al., 1985) spermatogenesis, and a germ cell-specific variant of H3 has been isolated from rat testis (Trostle-Weige et al., 1984). Three mouse t-complex proteins have been shown to be synthesized specifically in the haploid phase (Silver et al., 1987), and mouse heat shock proteins studied at mRNA level (Zakeri and Wolgemuth, 1987) also showed a stage-specific transcription in the haploid phase of a unique sized hsp 70 mRNA of 2.7 Kb. Distinct isotypes of actin and tubulin are synthesized in meiotic with respect to post-meiotic cells (Hecht et al., 1984). Two size classes of actin mRNA are differentially expressed during mouse spermatogenesis (Waters et al., 1985), and two RNA transcripts for alpha tubulin present in post meiotic cells of the testis have a 3' untranslated end different from that present in transcripts from meiotic cells and from somatic cells (Distel et al., 1984).

The modification of the protein synthetic pattern can depend either on the regulation of the transcriptional activity or on the control at post-transcriptional level. A major example of tissue-specific transcriptional regulation in germ cells is the adaptation that accompanies X chromosome inactivation in meiotic prophase. Essential genes located in the X chromosome, which cannot be expressed during spermatogenesis, are functionally replaced by autosomal genes specifically expressed in germ cells, whose protein products strictly resemble those expressed by X chromosome in somatic cells. This has been shown for the human testis-specific phosphoglycerokinase gene, an intronless gene which probably arose from retrovirus-mediated reverse transcription of the somatic message, which is codified by a gene localized on the X chromosome (McCarrey and Thomas, 1987).

A cell-specific transcriptional and translational regulation of gene expression has been demonstrated for the synthesis of protamine that occurres in elongating spermatids. In the trout the message is synthesized in meiosis and stored as inactive ribonucleoprotein particles up to spermiogenesis where translation occurs (Iatrou et al., 1978). In the mouse, poly (A+) RNAs for the genetically distinct protamines, mP1 and mP2 (Yelick et al., 1987; Hecht, 1989), first appear at detectable levels in round spermatids (Kleene et al., 1983), whereas translation occurs at later stages (Kleene et al., 1984).

This complex mechanism of regulation allowing gene expression while the genome is completely inactive, has stimulated studies directed to the identification of the gene elements responsible for tissue-specific regulation of expression, and of the message elements responsible for stage-specific regulation of translation. Recently, using transgenic mice, the cis-acting sequence required for cell-specific transcription in round spermatids of the cloned mP1 gene has been identified (Peschon et al., 1987). Moreover transgenic mice have been generated in which the 5' flanking region of mP2 drives the expression of c-myc and SV40 T-antigen mRNA (Stewart et al., 1988). Messenger RNAs are specifically transcribed in spermatids, however T-antigen synthesis was not detectable, thus suggesting that sequences in the protamine mRNA, presumably at the 3' untranslated region, are required for mRNA stability and/or the regulation of stage-specific translation in elongating spermatids.

The observation that RNA synthesis occurs in round spermatids would indicate that the spermatozoa phenotype depends at least in part on the haploid genotype. However this is probably not the case, since germ cells are connected up to spermiation in syncytium by cytoplasmic bridges (Dym and Fawcett, 1971) through which mRNA or protein molecules could diffuse freely, thus allowing diploid phenotype to be expressed even from haploid genomes. This hypothesis has been elegantly demonstrated observing that the product of a transgene which is expressed in post-meiotic cells in hemizygous transgenic mice is present in almost all the sperm cells instead of the predictable 50% (Braun et al., 1989).

Enzyme activities related to signal transduction mechanisms

Spermatogenesis is under Sertoli cell mediated hormonal control via signals and/or interactions (as a review see: Stefanini et al., 1984) whose nature is not known, but that would anyhow require receptors and signal transduction systems in germ cells for a communication to occur. We then became interested in the stage specific espression of components of systems transducing exogenous stimuli in germ cells.

The signal transduction mechanism that we have more extensively studied is the cyclic nucleotide system, in view of the implications of cAMP levels in oogenesis, in the maintainance of the meiotic block (Schultz et al., 1983), and in spermatogenesis and fertilization, in the control of sperm maturation and capacitation (Sanborn et al., 1980). Hormone sensitive membrane associated adenylate cyclase could not be found in mouse germ cells, while a soluble hormone independent form was identified, whose activity varied during spermatogenesis without any obvious correlation with cAMP levels (Adamo et al., 1980). Since cAMP level in the cells is the result of the equilibrium between adenylate cyclase synthetic activity and phosphodiesterase degradative activity, we reasoned that degradation might be the regulated activity in germ cells, via stage-specific quantitative variations or still unidentified regulation of phosphodiesterase activity.

Germ cell-specific isoforms of phosphodiesterase have been identified in both rat (Geremia et al., 1982) and mouse testis (Rossi et al., 1985). Mouse germ cells at all stage of differentiation were shown to possess a calcium-calmodulin dependent high affinity cAMP and cGMP isoform, while a calcium-independent high affinity cAMP isoform appears only in post-meiotic stages.

The calcium-dependent form has been characterized (Geremia et al., 1984) and purified to almost homogeneity (Rossi et al., 1988). This form possesses distinct physical and kinetic properties, however it is immunologically related to the somatic cell enzyme from brain. Genetic data are not yet available, it cannot therefore be established whether the somatic and the germ cell forms are products of different genes derived from a common ancestral gene, or products of the alternative processing of a common precursor mRNA.

Evidence for the existence of distinct genes codifying for germ cell-specific phosphodiesterases has been recently obtained for a calcium-independent form. Molecular cloning from a rat testis cDNA library (Swinnen et al., 1989) was performed using a cDNA probe for the Drosophila melanogaster cAMP phosphodiesterase codified by the "dunce" locus (Chen et al., 1987). We are therefore facing a tissue and cell stage-specific expression of phosphodiesterases which might be relevant to the control of gametogenesis. To this regard it is worth mentioning that mutations affecting the "dunce" locus produce, together with an altered expression of cAMP phosphodiesterase, a phenotype characterized by female fly sterility (Saltz et al., 1982).

It cannot be excluded that phosphodiesterase activity in germ cells might be functionally regulated by still unknown factors. To this regard, a line of evidence suggesting both a role for cAMP phosphodiesterase in the regulation of gametogenesis, and the possibility to control phosphodiesterase activity in gametes, has been provided by the observation that the proto-oncogene c-RAS, which is known to induce meiotic resumption in oocytes (Birchmeier et al., 1985), can activate the oocyte cAMP phosphodiesterase (Sadler and Maller, 1989).

Oncogene expression

The possible involvement of oncogene products in germ cell differentiation has recently attracted the attention of many investigators (as reviews see: Propst et al., 1988; Ponzetto, 1989). A growing number of proto-oncogenes has been, in fact, demonstrated to be involved in the signal transduction cascade which controls cell division and differentiation. Many oncogenes codify for growth factors (e.g.: c-sis), growth factor receptors and transducers (e.g.: c-erb and c-kit), GTP-binding proteins (e.g.: RAS family) and nuclear proteins (e.g.: c-fos, c-jun, c-myc).

Muller et al. (1982) demonstrated the presence of small size c-abl mRNA in mouse adult testis, which was later shown to be specific of haploid germ cells (Ponzetto and Wolgemuth, 1985). Transcripts of unusual size in early spermatids, that differ from the transcripts expressed in meiotic germ cells and somatic cells have also been shown by us and others for c-mos proto-oncogene (Propst et al., 1987; Mutter and Wolgemuth, 1987; Goldman et al., 1987; Sorrentino et al., 1988). Similar result was obtained for pim-1 which is espressed in post-meiotic germ cells as a unique 2.4 Kb transcript (Sorrentino et al., 1988). Nucleotide sequence of a cDNA clone for the testis-specific transcript of c-abl showed that it arises as a result of 3' truncation leaving an intact open reading frame for the c-abl tyrosine kinase (Oppi et al., 1987). For c-Mos and Pim-1 no such data are available, and it remains to be understood whether the short size reflects differential transcription or splicing, and whether it results in the formation of gene products different from those codified by the normal size messages.

We have also studied the expression of a larger group of proto-oncogenes in mouse male germ cells showing stage-specific regulation, while the size of the message is not modified (Sorrentino et al., 1988). Low levels of Ha-ras transcript are present in both meiotic and post-meiotic cells, whereas a substantial expression of Ki-ras can be observed at the pachytene stage but not in early spermatids. N-ras transcripts can be barely detected in pachytene cells, whereas they are very abundant in early spermatids. c-raf and c-fos are expressed only in meiotic stages, and finally no c-myc transcript could be detected in either meiotic or post-meiotic cells.

More recently, we have focused our studies on the expression of c-kit proto-oncogene. The cellular homolog of v-kit encodes for a transmembrane tyrosine-kinase receptor that is structurally similar to the receptors for CSF-1 and PDGF (Besmer et al., 1986), and the special interest to study its expression in spermatogenesis is that it maps in the W locus (Chabot et al., 1988; Geissler et al., 1988). Mice carrying mutations at this locus are characterized by anemia, lack of pigmentation and sterility (Russel, 79); the mutation appears to affect cell proliferation during embryogenesis and the ability of stem cells to respond to an unknown ligand. Northern blots of poly (A+) RNA and total RNA from mice of different age and from different cell components of the testis, isolated by the use of several fractionation procedures, showed that both somatic and germ cells express c-kit and that spermatogonia are by far the major site of transcription. Furthermore, the size of the transcript showed variations: spermatids showed two signals of 3.5 and 2.3 Kb different from the expected 5.5 Kb signal, which is the only present in the other cell types of the male gonad. Our data add a further example to the intriguing phenomenon of transcript size modification in haploid germ cells, and, more interestingly, indicate that spermatogonia are a major site of expression of c-kit protooncogene in mouse testis (Sorrentino et al., 1989). This is in agreement with genetic analysis which links W mutation to an intrinsic defect in stem cell populations (Harrison and Astle, 1976), and suggests that the receptor encoded by c-kit might be involved in the regulation of germ cell division and progression of spermatogenesis.

Homeobox expression

Homeobox containing genes, originally identified as regulators of segmentation in Drosophyla embryogenesis, have been demonstrated to be widely distributed from yeast to man (as a review see: Fienberg et al., 1987). Recently homeobox gene products have been shown to be sequence-specific transcription factors which control cell-specific expression of a number of genes (Levine and Hoey, 1988). Screening a mouse testicular cDNA library with the 180 bp homeobox domain from the antennapedia gene of Drosophyla, a homeobox containing gene, Hox-1.4 was isolated. Transcripts for Hox-1.4 were found to be specifically expressed at high abundance in the testis of the adult animal (Wolgemuth et al., 1986). A developmental study demonstrated that in adult tissues Hox-1.4 transcription is restricted to the germ cell line, whereas in the embryo, transcripts larger in size can be detected in varying abundance in somatic tissues (Wolgemuth et al., 1987).

Concluding remarks

The answers to questions concerning the regulation of gene expression in spermatogenesis and the identification of the signal transduction mechanisms which control gene expression and germ cell differentiation are still limited. However many efforts are devoted to these studies, and, thanks to recombinant DNA technology, many informations are becoming available on the genetic mechanisms of transcriptional and post-transcriptional regulation of tissue- and stage-specific gene expression in differentiating germ cells. More efforts should be addressed to improve in vitro systems which reproduce spermatogenesis. This would represent a major breakthrough to the understanding of the regulative mechanisms involved in spermatozoa differentiation.

Acknowledgements: Works from our laboratories were supported by MPI 40 and 60%, CNR special project "Biotecnologie e Biostrumentazione" n.89.00165.70, and CNR project n. 88.02100.04.

REFERENCES

Adamo S, Conti M, Geremia R, and Monesi V (1980) Particulate and soluble adenylate cyclase activities of mouse male germ cells. Biochem. Biophys. Res. Commun. 97:607-613.

Besmer P, Murphy PC, George PC, Qui F, Bergold PJ, Lederman L, Snyder HW, Brodeur D, Zuckerman EE, and Hardy WD (1986) A new acute transforming feline retrovirus and relationship on its oncogene v-kit with the protein kinase gene family. Nature 320:415-421

Bhatnagar YM, Romrell LJ, and Bellve AR (1985) Biosynthesis of specific histones during meiotic prophase of mouse spermatogenesis. Biol. Reprod. 32:599-609

Birchmeier C, Broek D, and Wigler M (1985) Ras proteins can induce meiosis in Xenopus oocytes. Cell 43:615-621

Boitani C, Geremia R, Rossi P, and Monesi V (1980) Electrophoretic pattern of polypeptide synthesis in spermatocytes and spermatids of the mouse. Cell Differentiation 9:41-49

Braun RE, Behringer RR, Peschon JJ, Brinster RL, Palmiter RD (1989) Genetically haploid spermatids are phenotypically diploid. Nature 337:373-376

Chabot B, Stephenson DA, Chapman VM, Besmer P, and Bernstein A (1988) The proto-oncogene c-kit encoding a transmembrane tyrosine kinase receptor maps to the mouse W locus. Nature 335:88-89

Chen C, Malone T, Beckendorf SK, and Davis RL (1987) At least two genes reside within a large intron of the dunce gene of Drosophila. Nature 329:721-724

Clermont Y (1972) Kinetics of spermatogenesis in mammals: seminiferous epithelium cycle and spermatogonial renewal. Physiol. Rev. 52:198-236

Distel RJ, Kleene KC, and Hecht NB (1984) Haploid expression of a mouse testis alpha-tubulin gene. Science 224:68-70

Dym M, and Fawcett DW (1971) Further observations on the numbers of spermatogonia, spermatocyte, and spermatids connected by intercellular bridges in the mammalian testis. Biol. Reprod. 4:195-215

Fienberg AA, Utset MF, Bogarad LD, Hart CP, Awgulewitsch A, Ferguson-Smith A, Fainsod A, Rabin M, and Ruddle FH (1987) Homeo box genes in murine development. In: Current Topics in Developmental Biology. Vol 23, pp. 233-256. Academic Press, New York.

Geissler EN, Ryan MA, and Housman DE (1988) The dominant-white spotting (W) locus of the mouse encodes the c-kit protooncogene. Cell 55:185-192

Geremia R, D'Agostino A, and Monesi V (1978) Biochemical evidence of haploid gene activity in spermatogenesis of the mouse. Exp. Cell Research 111:23-30

Geremia R, Rossi P, Pezzotti R, and Conti M (1982) Cyclic nucleotide phosphodiesterase in developing rat testis: identification of somatic and germ cell forms. Mol. Cell. Endocrinol. 28:37-53

Geremia R, Rossi P, Mocini D, Pezzotti R, and Conti M (1984) Characterization of a calmodulin-dependent high-affinity cyclic AMP and cyclic GMP phosphodiesterase from male mouse germ cells. Biochem. J. 217:693-700

Goldman DS, Kiessling AA, Millette CF, and Cooper GM (1987)

Expression of c-mos RNA in germ cells of male and female mice. Proc. Natl. Acad. Sci. USA 84:4509-4513
Harrison DE, and Astle CM (1976) Population of lymphoid tissues in cured W anemics by donor cells. Transplantation. 22:42-46
Hecht NB, Kleene KC, Distel RJ, and Silver LM (1984) The differential expression of the actins and tubulins during spermatogenesis in the mouse. Exp. Cell Res. 153:275-280
Hecht NB (1989) The molecular biology of spermatogenesis: regulation of mammalian protamines from gene to protein. In: Perspectives in Andrology. Serio M (ed). Serono Symposia. Vol. 53, pp. 25-35, Raven Press, New York
Iatrou K, Spira AW, and Dixon GH (1978) Protamine messenger RNA: evidence for early synthesis and accumulation during spermatogenesis in rainbow trout. Develop. Biol. 64:82-98
Kleene KC, Distel RJ, and Hecht NB (1983) cDNA clones encoding cytoplasmic poly(A)+ RNAs which first appear at detectable levels in haploid phases of spermatogenesis in the mouse. Develop. Biol. 98:455-464
Kleene KC, Distel RJ, and Hecht NB (1984) Translational regulation and deadenylation of a protamine mRNA during spermiogenesis in the mouse. Develop. Biol. 105:71-79
Kramer JM, and Erickson RP (1982) Analysis of stage-specific protein syntesis during spermatogenesis of the mouse by two-dimensional gel electrophoresis. J. Reprod. Fert. 64:139-144
Lam DMK, Furrer R and Bruce WR (1970) The separation, physical characterization, and differentiation kinetics of spermatogonial cells of the mouse. Proc. Nat. Acad. Sci. USA 65:192-199
Levine M, and Hoey T (1988) Homeobox proteins as sequence-specific transcription factors. Cell 55:537-540
McCarry JR, Thomas K (1987) Human testis-specific PGK gene lacks introns and possesses characteristics of a processed gene. Nature 326:501-505
Meistrich ML, Bruce WR, and Clermont Y (1973) Cellular composition of fractions of mouse testis cells following velocity sedimentation separation. Exp. Cell Res. 79:213-227.
Meistrich ML, Bucci LR, Trostle-Weige PK, and Brock WA (1985) Histone variants in rat spermatogonia and primary spermatocy- tes. Develop. Biol. 112:230-240
Millan JL, Driscoll CE, LeVan KM, and Goldberg E (1987) Epitopes of human testis-specific lactate dehydrogenase deduced from a cDNA sequence. Proc. Natl. Acad. Sci. USA 84:5311-5315
Monesi V (1965) Synthetic activities during spermatogenesis in the mouse: RNA and protein. Exp. Cell Res. 39:197-224
Monesi V (1971) Chromosome activities during meiosis and spermiogenesis. J. Reprod. Fert. 13:1-14
Muller R, Slamon DJ, Tremblay JM, Cline MJ, and Verma IM (1982) Differential expression of cellular oncogenes during pre- and postnatal development of the mouse. Nature 299:640-644
Mutter GL, and Wolgemuth DJ (1987) Distinct developmental patterns of c-mos protooncogene expression in female and male mouse germ cells. Proc. Natl. Acad. Sci. USA 84:5301-5305
Oakberg EF (1956) Duration of spermatogenesis in the mouse and timing of stages of cycle of the seminiferous epithelium. Am. J. Anat. 99:507.
Oppi C, Shore SK, and Reddy EP (1987) Nucleotide sequence of

testis-derived c-abl cDNAs: implications for testis-specific transcription and abl oncogene activation. Proc. Natl. Acad. Sci. USA 84:8200-8204
Peschon JJ, Behringer RR, Brinster RL, and Palmiter RD (1987) Spermatid-specific expression of protamine 1 in transgenic mice. Proc. Natl. Acad. Sci. USA 84:5316-5319
Ponzetto C, and Wolgemuth DJ (1985) Haploid expression of a unique c-abl transcript in the mouse male germ line. Mol. Cell. Biol. 5:1791-1794
Ponzetto C (1989) Proto-oncogene expression in the testis. In: Perspectives in Andrology. Serio M (ed). Serono Symposia Publications. Vol. 53, pp. 251-258, Raven Press, New York
Propst F, Rosenberg NP, Iyer A, Kaul K, and Vande Woude GF (1987) c-mos proto-oncogene RNA transcripts in mouse tissues: structural features, developmental regulation, and localization in specific cell types. Mol. Cell. Biol. 7:1629-1637
Propst F, Rosenberg MP, and Vande Woude JF (1988) Proto-oncogene expression in germ cell development. Trends in Genetics. 4:183-187
Rossi P, Pezzotti R, Conti M, and Geremia R (1985) Cyclic nucleotide phosphodiesterases in somatic and germ cells of mouse seminiferous tubules. J. Reprod. Fert. 74:317-327
Rossi P, Giorgi M, Kincaid RL, and Geremia R (1988) Testis-specific calmodulin-dependent phosphodiesterase: a distinct high affinity cAMP isoenzyme immunologically related to brain calmodulin-dependent cGMP phosphodiesterase. J. Biol. Chem. 263:15521-15527
Russel ES (1979) Hereditary anemias of the mouse; a review for geneticists. Adv. Genet. 20:357-459
Sadler SE, and Maller JL (1989) A similar pool of cyclic AMP phosphodiesterase in Xenopus oocytes is stimulated by insulin, insulin-like growth factor 1, and (Val12, Thr59) Ha-ras protein. J. Biol. Chem. 264:856-861
Saltz HK, Davis RL, and Kiger JA (1982) Genetic analysis of chromomere 3D4 in Drosophila Melanogaster: the Dunce and sperm-amotile genes. Genetics 100:587-596
Sanborn BM, Heindel JJ, and Robison GA (1980) The role of cyclic nucleotides in reproductive processes. Ann. Rev. Physiol. 42:37-57
Schultz RM, Montgomery RR, and Belanoff JR (1983) Regulation of mouse oocyte meiotic maturation: implication of a decrease in oocyte cAMP and protein dephosphorylation in commitment to resume meiosis. Develop. Biol. 97:264-273
Silver LM, Kleene KC, Distel RJ, and Hecht NB (1987) Synthesis of mouse t complex proteins during haploid stages of spermatogenesis. Develop. Biol. 119:605-608
Sorrentino V, McKinney MD, Giorgi M, Geremia R, and Fleissner E (1988) Expression of cellular proto-oncogenes in the mouse male germ line: a distinctive 2.4-kilobase pim-1 transcript is expressed in haploid postmeiotic cells. Proc. Natl. Acad. Sci. USA 85:2191.2195
Sorrentino V, Nocka K, Giorgi M, Geremia R, Besmer P, and Rossi P (1989) c-kit proto-oncogene expression in different mouse testicular cell populations: selective expression of testis-specific transcripts in postmeiotic cells. (in preparation).
Stefanini M, Conti M, Geremia R, and Ziparo E (1984) Regulatory

mechanisms of mammalian spermatogenesis. In: Biology of Fertilization. Monroy A, and Metz CB (eds) Vol. 2, pp. 3-45, Academic Press, New York

Stewart TA, Hecht NB, Hollingshead PG, Johnson PA, Leong JC, and Pitts SL (1988) Haploid-specific transcription of protamine-myc and protamine-T-antigen fusion genes in transgenic mice. Mol. Cell. Biol. 8:1748-1755

Swinnen J, Joseph DR, and Conti M (1989) Molecular cloning of rat homologues of the Drosophila Melanogaster dunce cAMP phosphodiesterase: evidence for a family of genes. Proc. Natl. Acad. Sci. USA 86:in press

Trostle-Weige PK, Meistrich ML, Brock WA, and Nishioka K (1984) Isolation and characterization of TH3, a germ cell-specific variant of histone 3 in rat testis. J. Biol. Chem. 259:8769-8776

Waters SH, Distel RJ, and Hecht NB (1985) Mouse testes contain two size classes of actin mRNA that are differentially expressed during spermatogenesis. Mol. Cell. Biol. 5:1649-1654

Wolgemuth DJ, Engelmyer E, Duggal R, Gizang-Ginsberg E, Mutter GM, Ponzetto C, Viviano CM, and Zakeri ZF (1986) Isolation of a mouse cDNA coding for a developmentally regulated, testis-specific transcript containig homeo box homology. EMBO J. 5:1229-1235

Wolgemuth DJ, Viviano CM, Gizang-Ginsberg E, Frohman MA, Joiner AL, and Martin GR (1987) Differential expression of the mouse homeobox-containing gene Hox-1.4 during male germ cell differentiation and embryonic development. Proc. Natl. Acad. Sci USA 84:5813-5817

Yelick PC, Balhorn R, Johnson PA, Corzett M, Mazrimas JA, Kleene KC, and Hecht NB (1987) Mouse protamine 2 is synthesized as a precursor whereas mouse protamine 1 is not. Mol. Cell. Biol. 7:2173-2179

Zakeri ZF, and Wolgemuth DJ (1987) Developmental-stage-specific expression of the hsp70 gene family during differentiation of the mammalian male germ line. Mol. Cell. Biol. 7:1791-1796

REGULATION OF SPERM MOTILITY BY OSMOTIC PRESSURE

M. Morisawa, *T.Inoda, and S. Oda
Misaki Marine Biological Station,
Faculty of Science,
University of Tokyo,
1024, Koajiro, Misaki-cho, Miura 238-02,
Japan

I. Introduction

The life of many higher organisms starts with the time when two gametes, an egg and a spermatozoon, fuse each other. At that moment, several processes, for example acrosome reaction, sperm-egg binding, exocytosis of cortical granules occur and the dramatic event, fertilization is completed. On the other hand, it is true that there must be several prologues; maturation and spawning of gametes are prerequisite processes for the accomplishment of fertilization. The former problem, that of oocyte maturation has been recognized to involve the problem of cell division and in recent years there is a growing body of data that should help to permit understanding of the mechanism underlying the critical events. The latter problem, that of the spawning of gametes, and especially the change of sperm behavior on spawning has been investigated for a long time.

Gray (1928) described that spermatozoa of sea urchin which are quiescent in the male reproductive organ burst into motility when they are released into the spawning ground, sea water, and suggested that initiation of sperm motility occurs by mechanical dilution whereby spawned sperm obtain open space for movement. A number of investigators then tried to find the factor(s) definitively affecting the mechanism of the initiation of sperm motility by studying mainly marine invertebrates (see review by Morisawa, 1985), while the

*Department of Biology,
Faculty of Science,
Toho University,
Funabashi, Chiba 274,
Japan

NATO ASI Series, Vol. H 45
Mechanism of Fertilization
Edited by B. Dale

problem has not been clearly explained despite more than half a century of effort since Gray.

Recently, we have focused again on this problem, i.e. the initiation of sperm motility at spawning since 1980 when we found actual physiological factors which can control the on-off switch of motility in the sperm of teleosts; potassium in salmonid fishes and osmolality in fresh water cyprinid and marine fishes (Morisawa, Suzuki 1980). In this review, we would like to summarize the K^+-dependent initiation system in salmonid fishes and then describe the process of the initiation of sperm motility which is triggered by osmolality.

II. Potassium-dependent initiation system

Based on the above discovery in 1980, details of the initiation mechanism triggered by dilution of potassium have been investigated in salmonid fishes for the past ten years. As summarized in reviews (Morisawa 1985; Morisawa 1987) and schematically illustrated in Figure 1, spermatozoa of salmonid

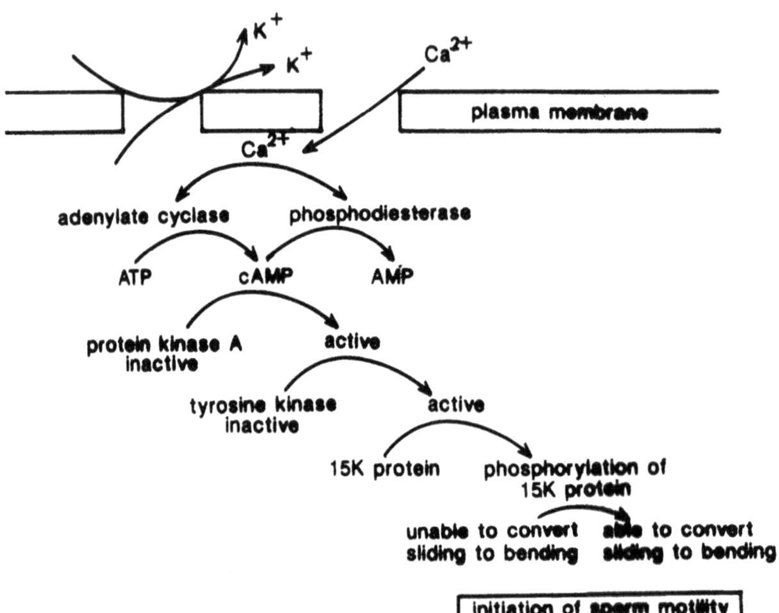

Fig. 1. Cascade process of the initiation of sperm motility in salmonid fishes.

fishes are forced to be completely quiescent in the sperm duct by the presence of high concentration of potassium contained in the seminal plasma, and reduction of potassium concentrations surrounding spermatozoa at spawning in potassium-poor fresh water proceeds the cascade process for the initiation of sperm motility which occurs in the sperm cells; The decrease in external potassium affects the plasma membrane of sperm flagellum, and induces potassium efflux and calcium influx through ion channels; The intruding calcium stimulates adenylate cyclase, resulting in transient increase in intracellular cyclic AMP; The synthesized cyclic AMP causes the activation of protein kinase A which subsequently phosphorylates and activates tyrosine kinase; The active tyrosine kinase phosphorylates the tyrosine residue of an axonemal protein whose molecular weight is 15,000 which in turn causes the initiation of flagellar motility.

III. Osmotic pressure-dependent initiation system

Osmolality has been discovered as another external signal for the initiation of sperm motility in freshwater cyprinid fishes (Morisawa, Suzuki 1980; Morisawa et al. 1983a). We recently found that calcium plays an important role as an intracellular factor which may regulate flagellar motility.

1. First signal: Osmotic pressure

The change of osmotic pressure which occurs around spermatozoa at spawning is the actual physiological initiator of motility in vertebrate species which release gametes in aquatic solution. As is generally known, the osmolality of body fluid, such as blood plasma, seminal plasma in vertebrates, is approximately 300 mOsm/kg which is extremely different from the osmolality of the spawning ground, for example river fresh water and sea water. Consequently, spermatozoa must be exposed respectively to hypotonic fresh water or hypertonic sea water at spawning.

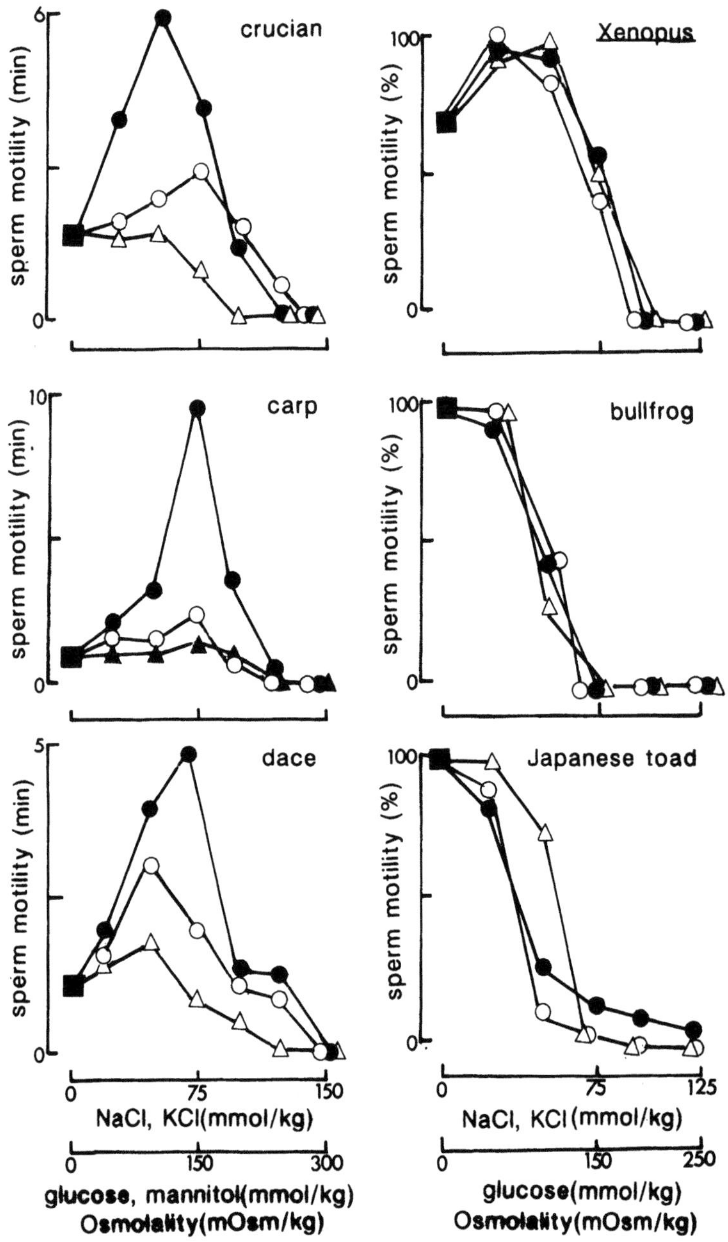

Fig. 2. Effect of osmolality on sperm motility of freshwater cyprinid fishes and amphibia. Semen was added to the media containing various concentrations of NaCl (○), KCl(●), mannitol(▲) or glucose(△). Motility was measured as described previously (Morisawa et al, 1983b; Inoda & Morisawa, 1987).

In the experimental conditions, spermatozoa of all freshwater cyprinid fishes examined, goldfish, crucian, dace, carp (Morisawa, Suzuki 1980; Morisawa et al. 1983a), were immotile when semen of these species were diluted in electrolyte (NaCl, KCl) at a concentration of 150 mmol/kg or non-electrolyte (mannitol, glucose) at a concentration of 300 mmol/kg which possess osmolality isotonic to the seminal plasma (300 mOsm/kg). Motility was observed by microscope when spermatozoa of these species were diluted in hypotonic solutions (Fig. 2).

A similar phenomenon is observed in other freshwater spawning vertebrate, amphibia. As is well known, the body fluid of toads and frogs is more dilute than that of other vertebrates, exhibiting osmolality around 250 mOsm/kg. In our study (Inoda, Morisawa 1987), sperm motility in Xenopus, bullfrog and Japanese toad did not occur in electrolyte and nonelectrolyte solutions isotonic to the seminal plasma (Fig. 2). Motility occurred only when the semen was diluted with a solution of lower osmolality, suggesting that the osmolality is the trigger for the initiation of sperm motility in amphibia as well as in freshwater cyprinid fishes.

The osmolality of seminal plasma in marine teleosts is also approximately 300 mOsm/kg. Thus, their gametes of them must be exposed to hypertonic sea water which has osmolality of approximately 1100 mOsm/kg. Spermatozoa of all marine fishes examined by us, puffer (Morisawa, Suzuki 1980), flounder, cod (Fig. 3) etc., were immotile in the solution isotonic to the seminal plasma and motility was initiated only when semen was mixed with the hypertonic solution, indicating that high osmolality is the only trigger for the initiation of sperm motility in marine teleosts.

The above experiments were designed to parallel the natural conditions occurring at spawning of spermatozoa in which motility was suppressed by the osmolality isotonic to the seminal plasma and initiated by a decrease or an increase in the osmolality in freshwater or seawater spawning species, respectively, suggesting that the change of osmolality surrounding sperm at natural spawning in hypo- or hypertonic environment is the first signal for the initiation of sperm motility in freshwater cyprinid fishes, amphibia and marine fishes.

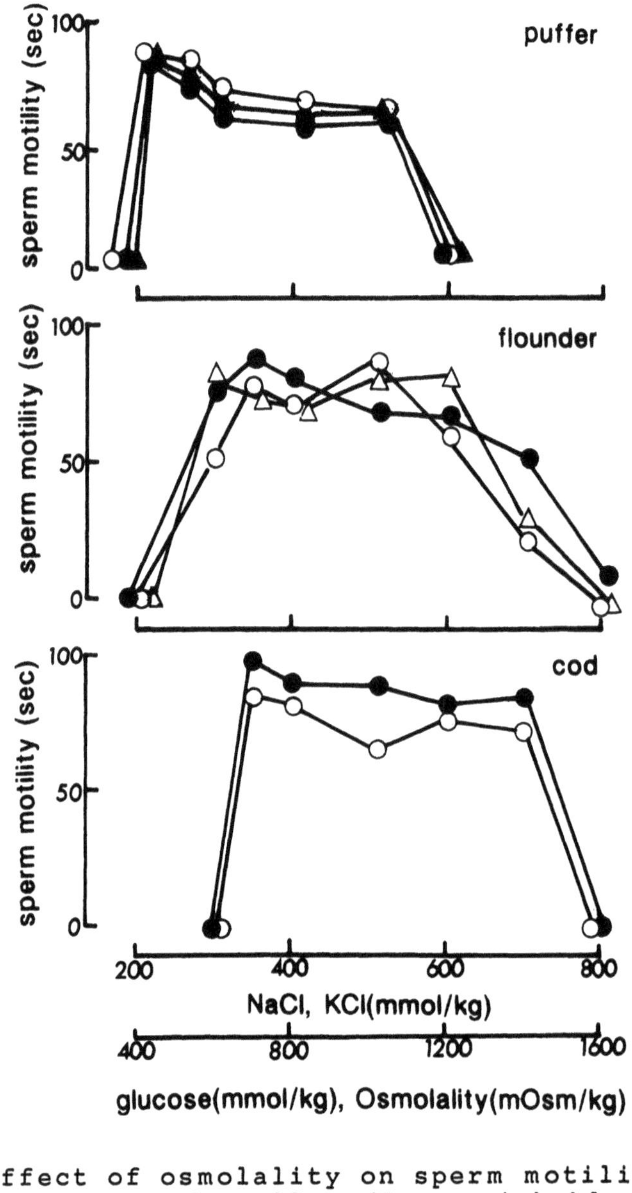

Fig. 3. Effect of osmolality on sperm motility in marine teleosts. Semen of puffer (Fugu niphobles), flounder (Microstomus achne) and cod (Gadus morrhua macrocephalus) was added to the media containing various concentrations of NaCl (○), KCl (●), mannitol (▲) or glucose (△). Sperm motility was measured as described (Morisawa, Suzuki 1980).

2. Osmolality as a convertor

The motility of spermatozoa can be manipulated by changing the external osmolality. In the case of freshwater cyprinid fish, goldfish, spermatozoa which were completely immotile in an isotonic solution (300 mOsm/kg) started to move if the osmolality were slightly decreased, for example to 275 mOsm/kg, by adding an appropriate volume of hypotonic solution. The motility continued for several minutes and then ceased. It is of interest that immotile sperm became motile sperm again when the osmolality of the sperm suspension was further decreased again by further addition of hypotonic solution. The reinitiation of sperm motility could be repeated several times. (Inoda, Morisawa unpublished data).

Amphibian sperm have a long motility duration time. In these animals, initiation and termination of sperm motility can be caused arbitrarily by a decrease or an increase in external osmolality. As described in a previous report (Inoda, Morisawa 1987), spermatozoa of Xenopus exhibit full motility in a hypotonic solution with a duration time of more than 20 minutes. When the osmolality of the sperm suspension was increased to the isotonic level by addition of hypertonic solution, spermatozoa became completely immotile. The conversion from a motile to immotile state, and vice versa, could be repeated at least 3 times by an increase or a decrease in environmental osmolality (Inoda, Morisawa unpublished data).

In some marine teleosts, the motility can be switched by changing the external osmolality. Because they have a long motility duration time, spermatozoa of some deep-sea fishes are convenient for analyzing the regulatory mechanism of sperm motility. Mature Cyclothone microdon with about 6 cm body length can easily be collected using an Isaacs-Kidd plankton trawl with a mouth area of 7.3 m^2 from 600-1000 m depth in Sagami Bay where the Misaki Marine Biological Station is located. Spermatozoa of this species can continue movement for several hours in artificial sea water. The motility completely halted if the external osmolality dropped to a level isotonic to the seminal plasma. Reinitiation of motility could be induced by increasing the osmolality of sperm suspension. The state of immotility or motility could be converted at least 3 times (Oda, Morisawa unpublished data). From these results, it is conceivable that the osmolality-dependent machinery which controls the mechanism underlying the initiation of sperm motility may have the potential to convert the on-off switch of flagellar motility.

3. Accelerator: Potassium

The osmolalities of the seminal plasma in all groups of teleosts and amphibia are almost identical; isotonic to the blood plasma. However, the ionic composition, especially the potassium concentration, is different. In salmonid fishes, cyprinid fishes and amphibia, the seminal plasma contains a higher concentration of potassium (rainbow trout, 37mM; chum salmon, 86 mM; goldfish, 70 mM; carp, 82 mM; Xenopus,73 mM) than does the blood plasma (1-14 mM) (Morisawa 1985; Inoda, Morisawa 1987). Conversely, seminal potassium in marine teleosts (puffer, 6mM; black sea bream, 2mM) is somewhat lower than that in the blood plasma (Morisawa 1985).

Despite the similar high concentration of seminal potassium in salmonid fishes, cyprinid fishes and amphibia, the effect of this ion on the sperm behavior is quite different. As described above, potassium is a potent physiological inhibitor of sperm motility in salmonid fishes; motility is completely suppressed in the presence of 3mM KCl (Morisawa et al. 1983b). In contrast, this ion increases both duration and velocity of sperm motility in cyprinid fishes. For example, duration of sperm motility in carp became 3.5 times longer in the presence of 10 mM KCl and maximum 5 times duration was obtained at a concentration of 30 mM KCl (Morisawa et al. 1983a). Acceleration of sperm motility also observed in other

Cyprinidae (Fig. 2). However, in another freshwater spawner, amphibia, potassium did not affect the sperm motility, although seminal plasma contains a high concentration of potassium. In marine teleosts, potassium, which is included at lower concentration in the seminal plasma, had no special effect on sperm motility.

It is still unknown why potassium has opposite effects on the motility of sperm in Salmonidae and Cyprinidae and how it acts as an accelerator. However, it is clear that spermatozoa possess the potassium-dependent system which suppresses or accelerates flagellar motility. The former system may be used in Salmonidae as the trigger of the initiation of sperm motility through the release from the suppression effect of potassium. In Cyprinidae on the other hand, spermatozoa spawned together with the seminal plasma seem to be able to maintain

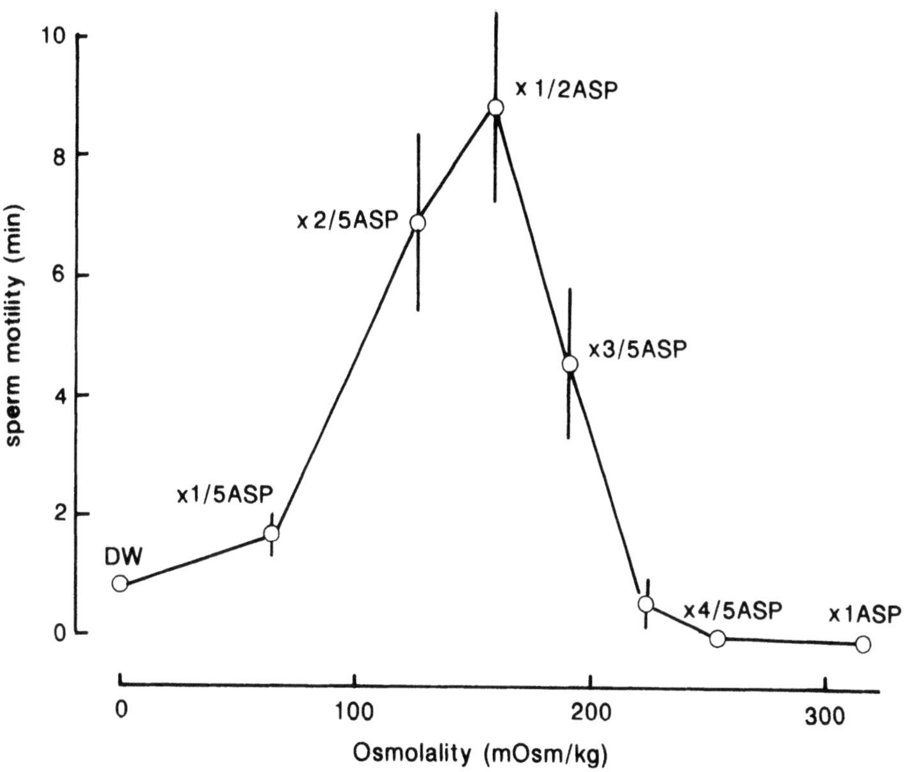

Fig. 4. Sperm motility of carp in the artificial seminal plasma and it's diluents. Semen was mixed with the artificial seminal plasma (ASP) whose ionic composition and pH is mimic to the seminal plasma of the carp (NaCl, 155 mM; KCl, 55 mM; $CaCl_2$, 2.5 mM; $MgCl_2$, 1.5 mM; pH 8.2) and diluents of ASP.

the long life span in the potassium-rich atmosphere formed by the mixing of the seminal plasma and fresh water.

It is well known that the male of freshwater cyprinid fishes approaches the female before spawning and releases spermatozoa immediately after oviposition, thus it is possible that spermatozoa can reach the eggs within short period. During the period, spermatozoa may swim with high velocity and long duration in the environment containing potassium. As shown in Figure 4, spermatozoa which are completely immotile in the ASP, become motile in the diluents of ASP and exhibit maximum duration of motility in the one to one correspondence mixture of ASP and fresh water, suggesting that the environment consisting of seminal plasma and river fresh water may be convenient for the complete of fertilization in freshwater cyprinid fishes. Decrease in motility at osmolality less than half of seminal plasma may be due to the disruption of sperm structure by the hypotonic shock.

4. Second messenger: Calcium

In the potassium-dependent initiation system of sperm motility in salmonid fishes, intrusion of calcium into sperm cells from the outside through ion channels is indispensable for the further advance of the intracellular cascade process of the initiation of sperm motility; cyclic AMP dependent phosphorylation of a 15K protein (Fig. 1). This fact was first demonstrated by the following experiment; spermatozoa which were completely immotile in medium containing potassium at a concentration of 5 mM exhibited full motility by the subsequent addition of several mM calcium (Tanimoto, Morisawa 1988). In contrast to this, addition of calcium to the sperm cells usually did not induce motility of spermatozoa in the cyprinid fishes, the motility being suppressed by osmolality isotonic to the seminal plasma. This was also true in amphibia and marine fishes. Furthermore, removal of calcium from the external environment of the cells with EGTA did not affect the motility, suggesting that sperm motility of these animals is independent of extracellular calcium.

On the other hand, we found that spermatozoa of Xenopus became immotile if calcium ionophore A 23187 was added to sperm whose motility has been initiated by a decrease in environmental osmolality. Subsequent addition of calcium to the ionophore-dependent immotile sperm induced the reinitiation of sperm motility. Neither cyclic nucleotides nor other ions such as magnesium could induce the initiation of motility of Xenopus sperm from which the plasma membrane was removed by Triton X-100. However, initiation of motility could be induced by calcium. From these results, it is possible that calcium is the intracellular factor indispensable for the initiation of flagellar motility in amphibia and possibly fresh water cyprinid fishes and marine fishes.

III. Epilogue

The molecular mechanism by which the potassium-cyclic AMP-dependent system triggers the initiation of flagellar motility in salmonid fishes is becoming clear. However, the other process defined by osmotic pressure and calcium remains obscure. How does the change of extracellular osmotic pressure cause the change of behavior of intracellular calcium? What is the target of calcium and how does calcium interact with motile machinery present in the sperm cell? These questions must be addressed by future studies.

References

Gray J (1928) The effect of dilution on the activity of spermatozoa. Brit J exp Biol 5: 337-344

Inoda T, Morisawa M (1987) Effect of osmolality on the initiation of sperm motility in *Xenopus laevis*. Comp Biochem Physiol 88A: 539-542

Morisawa M (1985) Initiation mechanism of sperm motility at spawning in teleosts. Zool Sci 2: 605-615

Morisawa M (1987) The process of initiation of sperm motility at spawning and ejaculation. In: Mohri H(ed) New horizons in sperm cell research, Japan Sci Soc Press, Tokyo/Gordon and Breach Sci Publ, New York, p 137-157

Morisawa M, Suzuki, K (1980) Osmolality and potassium: Their roles in initiation of sperm motility in teleosts. Science 210: 1145-1147

Morisawa M, Okuno M (1982) Cyclic AMP induces maturation of trout sperm axoneme to initiate motility. Nature 295:703-704

Morisawa M, Suzuki K, Shimizu H, Morisawa S, Yasuda K (1983a) Effects of osmolality and potassium on motility of spermatozoa from freshwater cyprinid fishes. J exp Biol 107:95-103

Morisawa M, Suzuki K, Morisawa S (1983b) Effects of potassium and osmolality on spermatozoan motility of salmonid fishes. J exp Biol 107: 105-113

Tanimoto S, Morisawa M (1988) Role for potassium and calcium channels on the initiation of sperm motility in rainbow trout. Develop Growth & Differ 30: 117-124

DYNAMIC CHANGES IN THE PERINUCLEAR MATRIX DURING SPERMIOGENESIS AND SPERM MATURATION IN THE MOUSE

ANTHONY R. BELLVÉ, R. CHANDRIKA AND ALINDA H. BARTH

Departments of Anatomy and Cell Biology, and Urology,
and Center for Reproductive Sciences
College of Physicians and Surgeons
Columbia University
West 168th Street
New York, NY 10032

NATO ASI Series, Vol. H 45
Mechanism of Fertilization
Edited by B. Dale

INTRODUCTION

During mammalian spermatogenesis the germ cell undergoes mitotic proliferation, meiosis, and spermiogenesis (Bellvé, 1979). In meiosis the nucleus passes through distinctive cytological events that reflect the molecular processes of genetic recombination, chromosome segregation and two reduction divisions. Later, during early spermiogenesis, the nucleus acquires polarity by forming a cap of heterochromatin, fusing the justaposed region of nuclear membranes and depositing a layer of perinuclear material. Then, as the nucleus condenses and becomes falciform in shape, the histones are replaced by protamines and quantitatively the bulk of the nonhistone proteins are lost. Of the many "non-histone" or "nonprotamine" proteins that remain associated with the nucleus after spermiation, nothing is known about their functions (Bellvé and O'Brien, 1983). Yet, these proteins may have structural or regulatory roles essential to sperm maturation, capacitation, acrosome reaction and/or fertilization.

In this review evidence is presented for a morphological and biochemical transition of germ cell-specific epitopes present in the perinuclear theca during mouse spermiogenesis and epididymal maturation. A family of proteins, M_rs 80,000, 77,000 and 75,000, each germ cell-specific and immunologically-related, are assembled during early spermiogenesis into a cap-like structure on the anterior pole of the nucleus. Then, during the transit of sperm from the testis to the caput epididymis, this high-M_r family is lost. Instead, the same epitope domain is detected on two proteins with M_rs of 50,000 and 48,000. At the same time the epitopes become distributed in a narrower region along the dorsal edge of the anterior pole and in a new ventral-caudad region on the posterior pole. The evidence is consistent with the high-M_r family undergoing endoproteolytic processing during sperm maturation to form two low-M_r peptides. The possibility also exists that some of the product then migrates from the anterior onto the posterior pole of the nucleus.

THE PERINUCLEAR THECA

The mouse sperm head contains a perinuclear theca applied against the external surface of the nuclear envelope, except for a small region around the implantation fossa. The theca consists of an apical perforatorium, a three-pronged structure enveloping the anterior pole of the nucleus, and the postacrosomal dense lamina, sheath or calyx that surrounds the posterior pole (Koehler, 1972; Lalli and Clermont, 1981; Olson et al., 1983; Longo et al., 1987). This complex entity is comprised of multiple proteins, some of which are localized in different domains on the anterior and posterior poles of the nucleus. Thus, a 14,000-16,000-M_r perforatorium protein is present on the anterior pole of rat sperm (Olson et al., 1976; Pruslin and Rodman, 1985; Oko and Clermont, 1988), and the 60,000-M_r calicin is a major component of the postacrosomal calyx of sperm from several mamalian species (Longo et al., 1987; Paranko et al., 1988). Many other as yet unidentified proteins also have been localized to different regions (Bellvé et al., 1990a). By contrast, the multiple band polypeptides (Longo et al., 1987) and the lamin B-like protein (Moss et al., 1990) are distributed around the entire periphery of the perinuclear theca or nucleus. The latter is considered, by analogy to somatic cells, to be a structural component of the nucleus. Thus, while some constituents associated with the sperm nucleus are uniformly distributed, others are organized into discrete regional domains.

THE PERINUCLEAR MATRIX

The mouse sperm nucleus, on displacement of the protamines, P1 and P2, and the hydrolysis of DNA, yields a structural element consisting of the peripheral theca and coarse intranuclear fibers. This entity is formed from the nucleus in two steps (Fig. 1). First, protamine cystinyl disulfide bonds are reduced with dithiothreitol (DTT), and second, nucleoprotamine is removed by treating with 25 μg DNase I/ml and $CaCl_2$, $MgCl_2$ (3:2; mol:mol), at concentrations increasing from 125, 175, 200 to 250 mM. The resulting structure, due to it's dual cyto- and karyoskeletal origins, is referred to as the perinuclear matrix (Bellvé et al., 1990a). After stringent extraction with 1% SDS, followed by high concentrations of $CaCl_2$, $MgCl_2$ (3:2: mol:mol), the perinuclear matrix retains the original configuration of the nucleus. This suggests the entity may have a structural role in determining and/or maintaining nuclear shape (Fig. 2). Moreover, by IEF and NEPHGE analyses, the perinuclear matrix is composed of ≤ 230 proteins (Bellvé et al., 1990a). Given this level of complexity and their intra-cellular location, it is anticipated many of these proteins will prove to have either structural or regulatory functions.

Novel proteins have been identified, characterized, and localized to different regions of the perinuclear theca and the nucleus by using

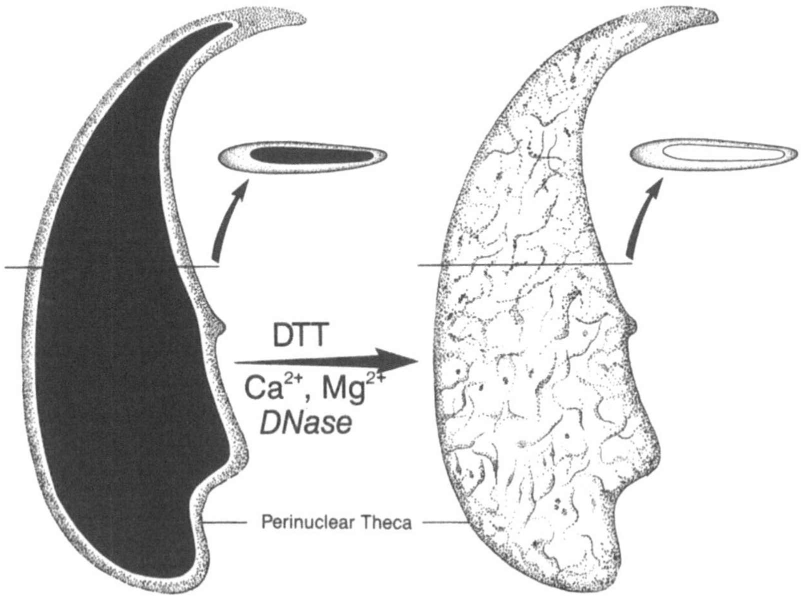

Fig. 1. Schematic of the procedure used to form perinuclear matrices from mouse sperm. Nuclei were isolated to ≥99% purity by a 2-min exposure to 1% SDS and centrifugation over 1.6 M sucrose (Bellvé et al., 1988). The samples were incubated in 50 mM DTT for 20 min, and at 10-min intervals with 25 µg DNase I/ml and $CaCl_2$, $MgCl_2$ (3:2) at concentrations of 125, 175, 200 and 250 mM, in the presence of 1 µM PMSF, 5 mM EDTA and 1 µg leupeptin/ml. The resulting perinuclear matrices were recovered by centrifugation over 1.0 M sucrose at 100,000g for 12 hr.

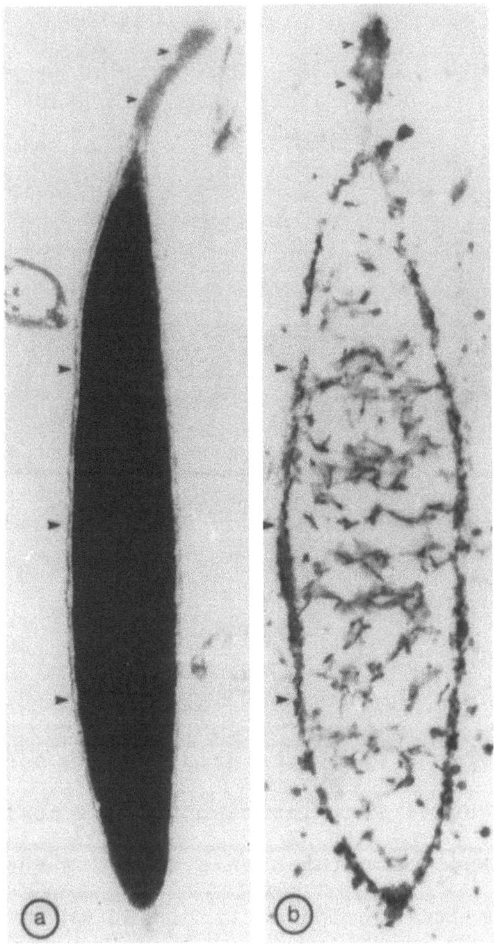

Fig. 2. Electron micrographs of a mouse sperm nucleus and the perinuclear matrix. [a] In this longitudinal section, the nucleus consisted of dense chromatin and was surrounded by the perinuclear theca (arrowheads). The perforatorium was present at the anterior pole (open triangles). [b] A corresponding section of a perinuclear matrix showing the peripheral theca and the intranuclear, coarse fibers. Reproduced from Bellvé et al., (1990a) by permission of the Editor

an array of monoclonal antibodies (mAbs) (Bellvé et al., 1990a). Xenogeneic mAbs were prepared by immunizing two adult rats with perinuclear matrices and fusing the activated spleen cells with P3-X63-Ag 8.653, mouse myeloma cells (Bellvé and Moss, 1983). Antibody-producing hybridomas were identified with an enzyme-linked, immunosorbent solid-phase assay (ELISA). From these hybridomas, 45 were selected for further study based on data from immunofluorescence, ELISA and SDS-PAGE immunoblot analyses of differentiating and mature germ cells. This report will focus on a subset of 10 mAbs, PNT-1 to -10, that recognize two families of immunologically-related proteins, denoted as the "thecins".

POLAR EXPRESSION OF THECINS DURING SPERMIOGENESIS AND EPIDIDYMAL MATURATION

The thecins were first detected in spermatids with mAbs PNT-1 (IgG_{2b}) and PNT-2 (IgG_{2a}), at ~step 5 of spermiogenesis as a cap-like structure on the apical pole of the haploid nucleus (Fig. 3). As the cell's progressed through spermiogenesis, the zone of immunofluorescence gradually extended caudad along with the developing acrosomic system, until in step 16 spermatids and testicular sperm, the antigens covered the entire anterior pole of the nucleus. In whole mounts of spermatids at steps 6 to 12, frequent circular patterns of reactivity around the periphery of the nucleus were interpreted as cross-sections through the mid region of the cap (Bellvé et al., 1990b).

Comparable patterns of immunofluorescence were present on nuclei after treating the cells with 1% Triton X-100, and on nuclear matrices

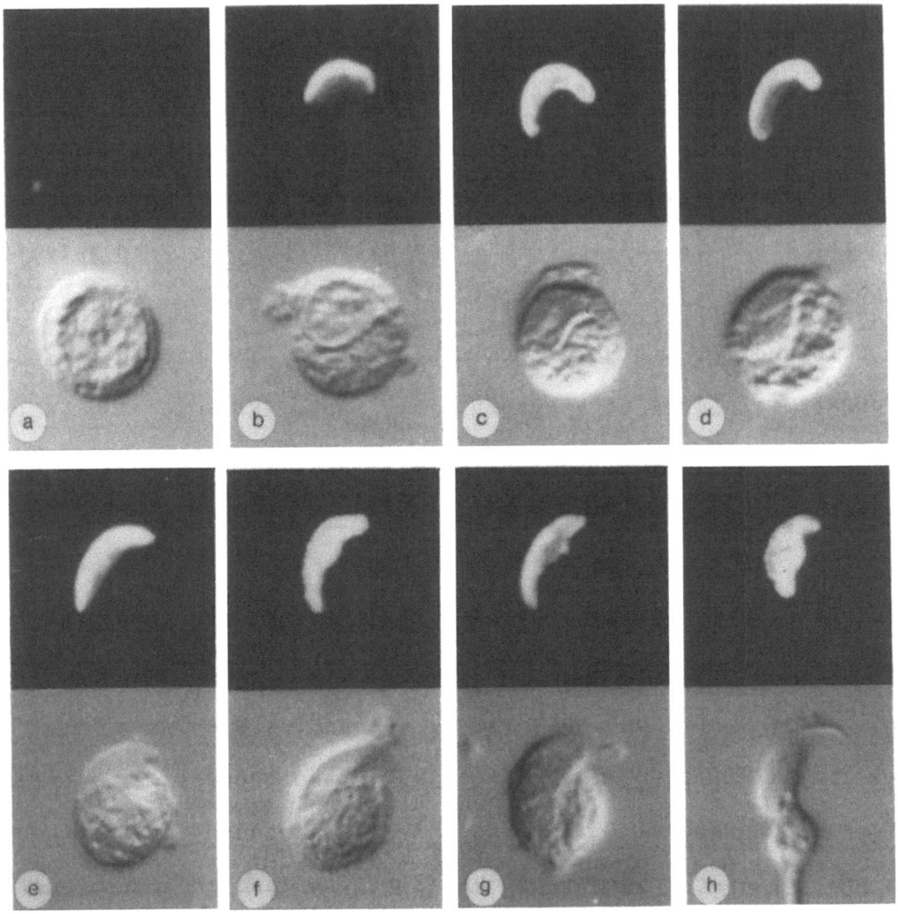

Fig. 3. Differential interference and corresponding immunofluorescence micrographs of haploid germ cells at different stages of spermiogenesis. The cells were isolated by sedimentation velocity at unit gravity (Bellvé et al., 1977), fixed in 4% paraformaldehyde, reacted with mAb PNT-1 or PNT-2, followed by an anti-rat IgG, rabbit IgG conjugated to fluorescein (Bellvé et al., 1990b). Spermatids: (a) step 2; (b) step 5, (c) step 8, (d) step 12, (e) step 14, (f) step 16, and (g) testicular sperm. x 4,000. [Reproduced from Bellve et al., (1990b) by permission of the Editor,].

after extracting nuclei with a sequence of 25, 50, 75 and 100 mM $CaCl_2$, $MgCl_2$, followed by 1 M NaCl. However, no reactivity was seen after exposing nuclei to 1% cetyltrimethylammonium bromide (CTAB), a cationic detergent that removes the theca from sperm (Balhorn et al., 1977). Also, no mAb binding was detected in other areas of either differentiating or mature germ cells, including all flagella structures. These observations are consistent with the thecins being located in the anterior region of the developing perinuclear theca.

The distribution of thecin epitopes changed substantially during the early phase of sperm maturation. Sperm from the caput epididymis, rather than expressing antigen just in the anterior theca, exhibited a bipolar distribution of epitopes. Reactivity was confined to a narrower zone at the anterio-dorsal margin of the perforatorium and in a new caudad-ventral region at the posterior pole, around the perifossal region of the sperm head (Fig. 4). The intermediate equatorial zone was negative. The same pattern of mAb binding was seen in all sperm from the caput, corpus and cauda epididymis and the vas deferens.

TRANSITION IN PEPTIDIC EPITOPES DURING SPERM MATURATION

Three proteins of 80,000, 77,000 and 75,000-M_rs present in seminiferous tubules at Day 24 and later ages of testicular development reacted specifically with MAbs PNT-1 and PNT-2 (Bellvé et al., 1990b). Comparable immunoreactivity was detected in isolated populations of spermatids, at steps 1-8 and 12-16 of spermiogenesis, but not in germ cells at earlier stages of differentiation (Fig. 5). On higher

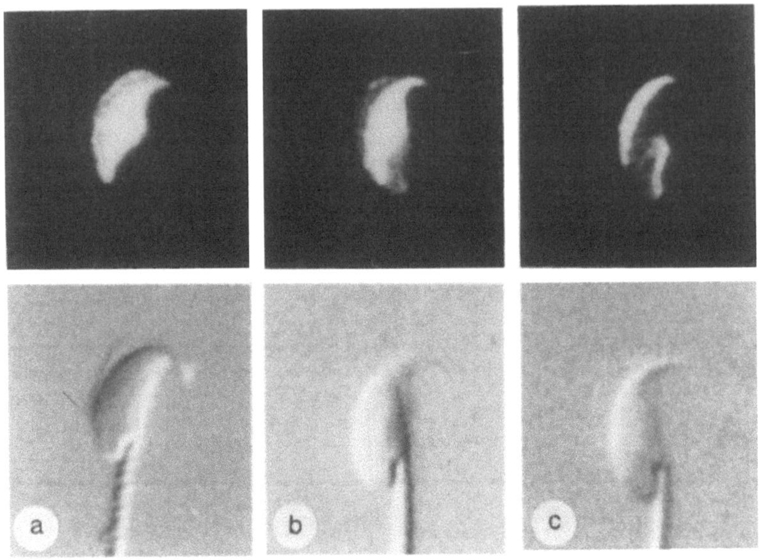

Fig. 4. Differential interference and corresponding immunofluorescence micrographs of sperm at different stages of sperm maturation. Sperm were recovered from the designated regions, fixed in 4% paraformaldehyde, and reacted with mAb PNT-1, followed by FITC-IgG. Sperm from: (a) testis; (b) caput epididymis; (c) corpus epididymis. x 4,000. [Reproduced from Bellvé et al., (1990b) by permission of the Editor,].

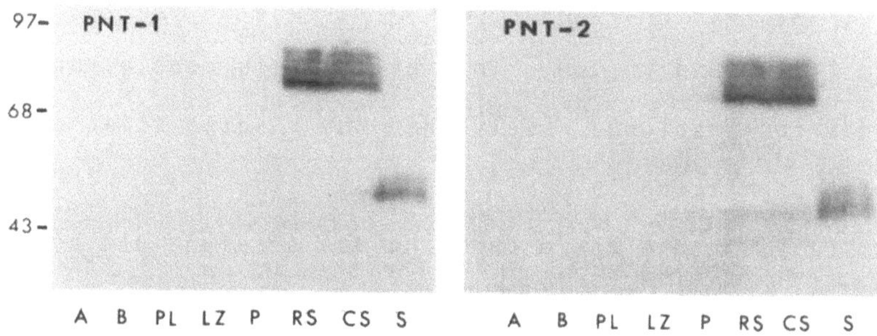

Fig. 5. Immunoblots of total protein of germ cells isolated at different stages of mouse spermatogenesis. Proteins were extracted in Sample Buffer containing 4% SDS and protease inhibitors, sonicated, subjected to 10% SDS-PAGE, and immunoblotting. Samples included: (A) type A spermatogonia; (B) type B spermatogonia; (PL) preleptotene spermatocytes; (LZ) leptotene spermatocytes; (P) pachytene spermatocytes; (RS) spermatids (steps 1-8); (CS) spermatids (steps 12-16); and (S) mature sperm. [Reproduced from Bellvé et al., (1990b) by permission of the Editor,].

resolution gels, four proteins were resolved, with M_rs of ~81,000 and 79,000, and 77,000 and 75,000. This evidence is consistent with the morphological observations, namely that the thecins first apppear during early spermiogenesis.

Interestingly, mature sperm from the vas deferens did not express the higher-M_r family; instead these cells contained two immunoreactive proteins of M_rs 50,000 and 48,000 (Fig. 5). Thus, a major change in proteins containing the epitopes occurred during sperm maturation. It had been noted that both thecin families, 80,000-75,000 and 50,000-48,000-M_r, co-localized on cellular fractionation with the nucleus, perinuclear matrix and, in particular, with the perinuclear theca (Bellvé et al., 1990b). Therefore, to determine when the transition occurred, nuclear proteins were prepared from germ cells at different stages of spermiogenesis and epididymal maturation and the binding of mAbs PNT-1 and PNT-2 was examined on immunoblots. Significantly, the transition of the thecin epitopes from the high- to the low- M_r forms was seen to occur abruptly and coordinately as sperm moved from the testis to the caput epididymis (Fig. 6). This transition occurred just prior to major changes, both quantitative and qualitative, in the complement of nuclear proteins present in sperm from the caput, corpus and cauda epididymis. These latter changes most likely were due to the acquisition of pre-existing proteins by de novo formation of inter-molecular disulfide bonds.

The thecins were not sensitive to glycosidases, either in terms of a shift in M_r or reactivity to mAbs PNT-1 and PNT-2 on immunoblots. Aminoethylation in dissociating conditions (6 M G.HCl) had no dis-

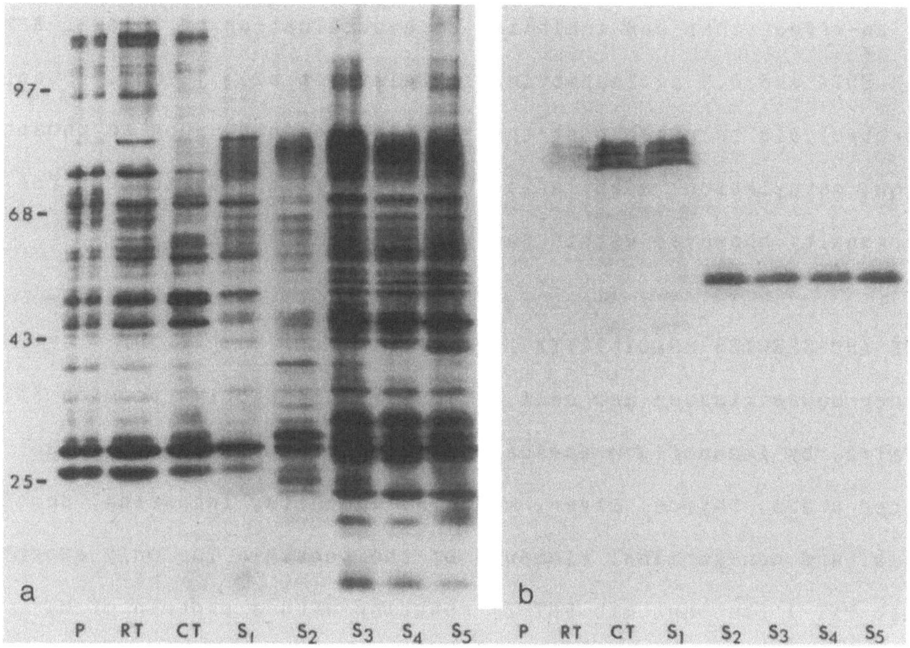

Fig. 6. SDS-PAGE and the corresponding immunoblots of nuclear proteins isolated from germ cells at different stages of spermiogenesis and epididymal maturation. Protein from an equivalent number of nuclei was applied to each lane. [A] Silver-stained gel. [B] Immunoblot. Samples included: (P) pachytene spermatocyte; (RS) spermatid, steps 1-8; (CS) spermatid, steps 12-16; and sperm from the (S_1) testis; (S_2) caput epididymis; (S_3) corpus epididymis; (S_4) cauda epididymis; (S_5) vas deferens. (Reproduced from Bellvé et al., (1990b) by permission of the Editor,].

cernible effect. The proteins were susceptible to proteolysis, however; an effect that was inhibited by a combination of 25 mM ρ-APMSF, 0.5 mM EDTA and 0.5 μg leupeptin/ml (Bellvé et al., 1990b). Specific endoproteolysis or other post-translational events, such as phosphorylation, methylation, acetylation or ribosylation may account for the heterogeneity observed within the two protein families.

TISSUE AND SPECIES SPECIFICITY

Other mouse tissues and cell types were nonreactive to mAbs PNT-1 and PNT-2, by immunofluorescence and immunoblot analyses. These included brain, thymus, liver, kidneys, adrenals, intestine, prostate, ovaries, and non-germinal elements of the testes. The only exception was mAb PNT-1 (but not mAb PNT-2) reacting with a protein of M_r 43,000 on immunoblots of skeletal muscle.

The cell lines, Balb/c 3T3 fibroblasts, TM_3 Leydig and TM_4 Sertoli cells, during interphase, colcemide-arrested metaphase, or 3-6 hr after induction of heat shock, were all negative. Mature sperm of Xenopus laevis, dove, cat, guinea pig, rabbit, dog, boar and baboon also were non-reactive.

CHARACTERIZATION OF THE EPITOPE DOMAIN

The binding affinities of mAbs PNT-1 (IgG_{2b}) and PNT-2 (IgG_{2a}) were determined by Scatchard analyses. Partially-purified sperm proteins were attached to microtiter plates and used as target antigens (Bellvé et al., 1990b). The ratio of bound over free [^{35}S]methionine-labelled mAb was determined as a function of total mAb bound to 20 pmoles of

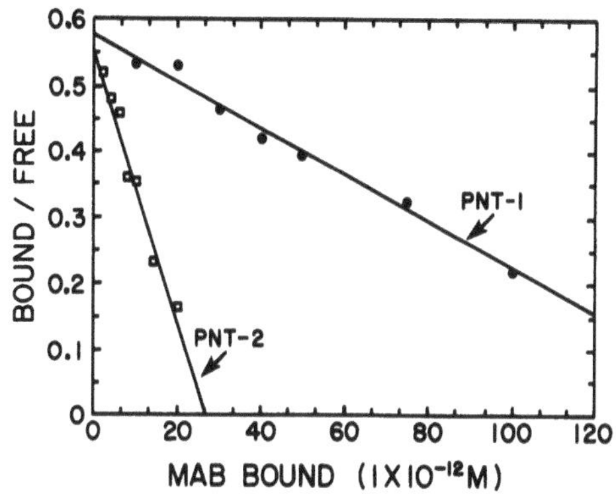

Fig. 7. Scatchard analyses of [^{35}S]PNT-1 and [^{35}PNT-2 binding to the 50,000 and 48,000-M_r sperm proteins. Proteins were extracted from sperm nuclei with 25 mM DTT, 1% cetyltrimethylammonium bromide, and were adsorbed to microtiter plates; 10-100 pmoles/well for mAb PNT-1 and 2-20 pmoles/well for mAb PNT-2. Ratios of bound/free mAb were determined as a function of bound mAb, by using a solid-phase, radioimmunoassay. MAb PNT-1: K_a = 3.53 x 10^{12} M^{-1}. MAb PNT-2: K_a = 2.08 x 10^{12} M^{-1}. [Reproduced from Bellvé et al., (1990b) by permission of the Editor,].

sperm protein (Fig. 7). The Scatchard plots for mAbs PNT-1 and PNT-2 were linear indicating a single population of binding sites were present for each antibody. By calculation, the K_a for mAb PNT-1 was 3.53×10^{12} M^{-1} and for mAb PNT-2, 2.08×10^{12} M^{-1}; both estimates representing very high binding affinities.

Comparable immunofluoresence and immunoblot patterns obtained with mAbs PNT-1 to -10, suggested the series of antibodies recognized the same protein families during spermiogenesis and epididymal maturation. Also, based on data from competitive-binding, solid-phase, radio-immunoassays, mAbs PNT-1 to -10 were found to recognize six distinct epitopes. These assays were undertaken by adsorbing 20 pmoles of 50,000- and 48,000-M_r sperm proteins to microtiter wells, saturating specific epitopes with mAbs PNT-1 to -10, and then probing for available sites with [^{35}S]methionine-labelled mAbs PNT-1 or PNT-2. Binding levels of [^{35}S]PNT-1 in the presence and absence of mAb PNT-1 were used as positive and negative controls. Similar estimates were made for mAb [^{35}S]PNT-2.

On this basis, it was possible to discern six distinct epitopes on the two sperm proteins. Thus, mAbs PNT-1 to -10 showed different levels of competitive binding, depending on whether mAb [^{35}S]PNT-1 or [^{35}S]PNT-2 was used as a probe (Fig. 8). One epitope was defined by mAbs 2, 6 and 7, which allowed ~30% binding by mAb [^{35}S]PNT-1 but only ~6% by mAb [^{35}PNT-2. A second epitope was evident from mAbs 3, 4 and 5, which competed effectively for both probes. The third, fourth and fifth were apparent from mAbs 1, 8 and 9, which all inhibited binding of mAb [^{35}S]PNT-1, but permited partial binding of mAb [^{35}S]PNT-2 to

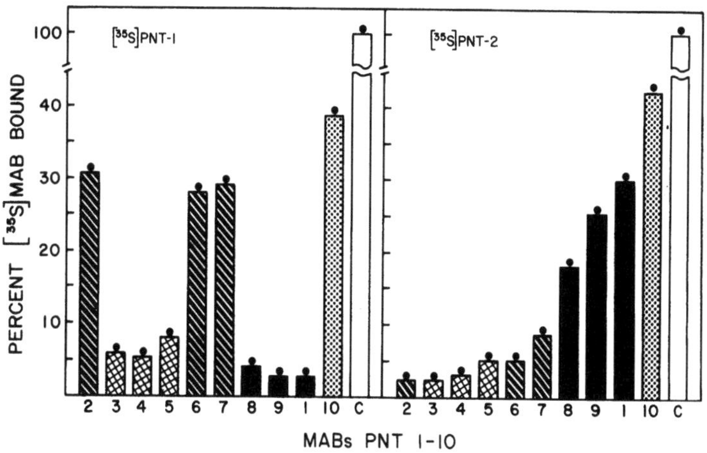

Fig. 8. Competitive binding of mAbs PNT-1 to PNT-10 to epitopes on the 50,000- and 48,000-M_r sperm proteins. The two proteins were affinity purified and 20 pmoles adsorbed to each well of a microtiter plate. After extensive washing, the proteins were incubated with one of the mAbs PNT-1 to -10, each at a saturating concentration, for 2 hr. After removing excess mAb, the samples were incubated with mAb [^{35}S]PNT-1 or [^{35}S]PNT-2. The amount bound was quantitated by scintillation counting. Positive and negative binding of mAb [^{35}S]PNT-1 was determined in the absence and presence of bound mAb PNT-1, respectively. Comparable estimates were made for [^{35}S]mAb PNT-2. [Reproduced from Bellvé et al., (1990b) by permission of the Editor].

18, 25 and 30%, respectively (Fig. 8). The last three levels of binding proved to be different, indicating mAbs 1, 8 and 9 were directed against different epitopes. The sixth epitope was delineated by mAb PNT-10, which allowed mAb [^{35}S]PNT-1 and [^{35}S]PNT-2 to bind to ~38 and 42%, respectively.

Notably, all mAbs competed with [^{35}S]PNT-1 and/or [^{35}S]PNT-2 for binding to the two sperm proteins, and each bound to SDS-denatured proteins on immunoblots. The two observations suggest the six epitopes are clustered in a single domain, rather than being distributed throughout the protein as one would expect of conformationally-dependent epitopes.

DISCUSSION

The thecins, M_rs 80,000, 77,000 and 75,000, first appear as a cap--like structure at the apical pole of the germ cell nucleus at ~step 5 of mouse spermiogenesis. Their distribution gradually extends caudally over the nucleus, coincident with the developing acrosomic system and the underlying perinuclear theca, until covering the whole anterior pole of the nucleus in testicular spermatozoa. The epitope domain then undergoes a dynamic transition. Rather than having the high-M_r proteins, sperm of the caput epididymis contain two immunoreactive proteins of M_rs 50,000 and 48,000. Moreover, the epitopes now are located in a narrow margin along the anterio-dorsal pole, and in a new perifossal region at the caudal-ventral pole. This event occurs in the absence of a cytoplasmic pool of the low-M_r proteins, and during a period in which the cells are known to be transcriptionally and translationally inactive. Thus, the transition is likely to reflect endoproteolytic processing and redistribution, as opposed to de novo synthesis and insertion of new constituents.

The biochemical and morphological transition of the thecin epitopes at the onset of sperm maturation may involve several mechanisms. The simplest explanation would invoke proteolytic processing of the high-M_r family to form the low-Mr proteins. This would be consistent with the proteins: a) appearing and disappearing simultaneously during the onset of sperm maturation, b) binding mAbs PNT-1 and PNT-2 in stoichiometric amounts, c) co-localizing to the perinuclear theca, and d) sharing a peptide domain containing six epitopes. The morphological transition would then involve a selective masking and unmasking of accessible epitopes in the respective domains, or a lateral migration of protein molecules into the new domain in the posterior region of the perinuclear theca. Precedent exists for a redistribution of sperm components in a physiological process. During the acrosome reaction in guinea pig sperm, the PH-20 antigen migrates from the plasma membrane over the posterior head to the inner acrosomal membrane over the anterior sperm head (Cowan et al., 1987). In the mouse sperm acrosome reaction, galactosyltransferase also appears to assume a new cellular organization (Lopez and Shur, 1987). However, the transition in thecin epitopes occurs at the onset of sperm maturation, rather than being an event associated with fertilization.

The transition in thecin epitopes occurs after spermiogenesis, as the sperm move from the testes into the caput epididymis. The event therefore takes place at a time when few differentiative processes are known to occur in the sperm head (Bellvé and O'Brien, 1983; Jones, 1989). One therefore can only speculate as to a possible function for the event. It might be related to modelling of the acrosome in epididymal sperm (Fawcett and Phillips, 1969). Alternatively, it may be a prerequisite to other processes that occur during the maturation of sperm in the epididymis. Since the epitope domain is unique to haploid germ cells, the proteins presumably serve a function specific to sperm. This question may best be addressed by first undertaking molecular studies to determine the relationship between the different protein species and their cellular functions. To this end, three clones from a spermatid cDNA λgt expression library have been found to translate peptidic epitopes recognized by eight mAbs directed against the thecins. One of these clones is being sequenced to determine the primary structure of the encoded protein.

ACKNOWLEDGEMENTS

Preparation of this review was funded by NICHHD grant P50 HD 05770.

REFERENCES

BALHORN, R., GLEDHILL, B. L., & WYROBECK A. J. (1977). Mouse sperm chromatin proteins: Quantitative isolation and partial characterization. Biochemistry 16, 4074-4080.

BELLVÉ, A. R. (1979). The molecular biology of mammalian spermatogenesis. Oxford Rev. Reprod. Biol. 1, 159-261.

BELLVÉ, A. R., CAVICCHIA, J. C., MILLETTE, C. F., O'BRIEN, D. A., BHATANGAR, Y. M., & DYM, M. (1977a). Spermatogenic cells of the prepuberal mouse. Isolation and morphological characterization. J. Cell Biol. 74, 68-85.

BELLVÉ A. R., CHANDRIKA, R. & BARTH, A. H. (1990a). The peri-nuclear matrix as a structural element of the mouse spermatozoan nucleus. (Submitted)

BELLVÉ, A. R., CHANDRIKA, R., & BARTH, A. H. (1990b). Temporal expression, polar distribution and transition of an epitope domain in the perinuclear theca during mouse spermatogenesis. (Submitted).

BELLVÉ, A. R., McKAY, D. J., RENAUX, B. S. & DIXON, D. H. (1988). Purification and characterization of mouse protmaines, P1 and P2. Amino acid sequence of P2. Biochemistry 27, 2890-2897.

BELLVÉ, A. R., & MOSS, S.B. (1983). Monoclonal antibodies as probes of reproductive mechanisms. Biol. Reprod. 28, 1-26.

BELLVÉ, A. R., & O'BRIEN, D. A. (1983). The mammalian spermatozoon: Structure and temporal assembly. In: Mechanisms and Control of Fertilization. (J. F. Hartman, editor). Academic Press, New York. pp. 55-137.

COWAN, A. E., MYLES, D. G. & KOPPEL, E. (1987). Lateral diffusion of the PH-20 protein on guinea pig sperm: Evidence that barriers to diffusion maintain plasma membrane domains in mamalian sperm. J. Cell Biol. 104, 917-923.

FAWCETT, D. W., ANDERSON, W. A., & PHILLIPS, D. M. (1971). Morphogenetic factors influencing the shape of the sperm head. Develop. Biol. 26, 220-251.

FAWCETT, D. W. & PHILLIPS, D. M. (1969). Observations on the release of spermatozoa and on changes in the head during passage through the epididymis. J. Reprod. Fertil. Suppl. 6, 405-418.

IERARDI, L. A., MOSS, S. B. & BELLVÉ, A. R. (1990). The nuclear matrix of mouse spermatogenesis: Temporal synthesis of polypeptides coincident with assembly of synaptonemal complexes. (Submitted).

JONES, R. (1989). Membrane remodelling during sperm maturation in the epididymis. Oxford Rev. Reprod. Biol. 11, 285-337.

KOEHLER, J. K. (1973). Studies on the structure of the postnuclear sheath of water buffalo spermatozoa. J. Ultrastruct. Res. 44, 355-368.

LALLI, M., & CLERMONT, Y. (1981). Structural changes of the head component of the rat spermatid during late spermiogenesis. Am. J. Anat. 160, 419-434.

LONGO, F. J., KROHNE, G., & FRANKE. W. W. (1987). Basic proteins of the perinuclear theca of mammalian spermatozoa and spermatids. A novel class of cytoskeletal elements. J. Cell Biol. 105, 1105-1120.

MOSS, S. IERARDI, L. A., & BELLVÉ, A. R. (1990). The nuclear matrix of mouse spermatogenesis: Differential occurrence of the lamin proteins. (Submitted).

OKO, R., & Clermont, Y. (1988). Isolation, structure and protein composition of the perforatorium of rat spermatozoa. Biol. Reprod. 39, 673-687.

OLSON, G. E., HAMILTON, D. W., & FAWCETT, D. W. (1976). Isolation and characterization of the perforatorium of rat spermatozoa. J.

Reprod. Fertil. 47, 293-297.

OLSON, G. E., NOLAN, T. D., NOLAND, V. P., & GARBERS, D. L. (1983). Substructure of the postacrosomal sheath of bovine spermatozoa. J. Ultrastruct. Res. 85, 204-218.

PARANKO, J., LONGO, F., POTTS, J., KROHNE, G., & WERNER, W. W. (1988). Widespread occurrence of calicin, a basic cytoskeletal protein of sperm cells, in diverse mammalian species. Differentiation 38, 21-27.

PRIMAKOFF, P. & MYLES, D. G. (1984). Localized surface antigens of guinea pig sperm surface migrate to new regions prior to fertilization. J. Cell Biol. 99, 1634-1641.

PRUSLIN, F. H., & RODMAN, T. C. (1985). Proteins of demembranated mouse sperm heads. Characterization of a major sperm-unique component. J. Biol. Chem. 260, 5654-5659.

REACTIVE OXYGEN SPECIES AND HUMAN SPERM FUNCTION

R.J. Aitken
MRC Reproductive Biology Unit
37 Chalmers Street
Edinburgh
EH3 9EW

Introduction

The human spermatozoon is a slightly differentiated cell which has developed a distinctly compartmentalised structure in order to accommodate the variety of diverse biological functions required to fertilize the ovum. For example, they exhibit a capacity for at least 2 different kinds of controlled movement (progressive and hyperactivated) by virtue of a flagellar beat pattern, which utilizes ATP largely generated by glycolysis in the midpiece of the cell. They also exhibit a refined capacity for cell recognition: not only will human spermatozoa not bind to any other cell type than the ovum, but they will not bind to the ovum of any other species than another hominoid ape, such as a gibbon (Bedord, 1977). These ovum-recognition sites are located on the plasma membrane overlying the acrosomal region of the sperm head. Gamete recognition is followed sudden changes in the properties of this specific area of plasma membrane, leading to its fusion with the underlying outer acrosomal membrane in an event known as the acrosome reaction. This process involves the activation and release of acrosomal enzymes including a specific protease (acrosin) which, in concert with the hyperactivated movements of the sperm flagellum, leads to the penetration of the zona pellucida. At the time of the acrosome reaction the plasma membrane of another differentiated region of the sperm head, the equatorial segment, suddenly acquires the capacity to recognise, and fuse with, the vitelline membrane of the oocyte.

The molecular biology of this complex cascade of interactive events, summarised in Fig 1, is now the subject of ongoing research focused on a more complete understanding of the fundamental mechamisms involved in the control of human sperm function. This knowledge should ultimately find application in the development of male contraceptive agents and the diagnosis and treatment of male infertility. Paradoxically perhaps, these twin objectives are complementary since elucidation of the mechanisms responsible for male infertility, the aetiology of which is largely unexplained, may provide us with clinical models on which to base radically new approaches to contraception.

In this brief review, I shall focus on an intriguing aspect of the cell biology of the human spermatozoon, which has direct relevance to male infertility and may have long-range implications for fertility control. This property concerns the capacity of human spermatozoa to generate reactive oxygen species.

NATO ASI Series, Vol. H 45
Mechanism of Fertilization
Edited by B. Dale

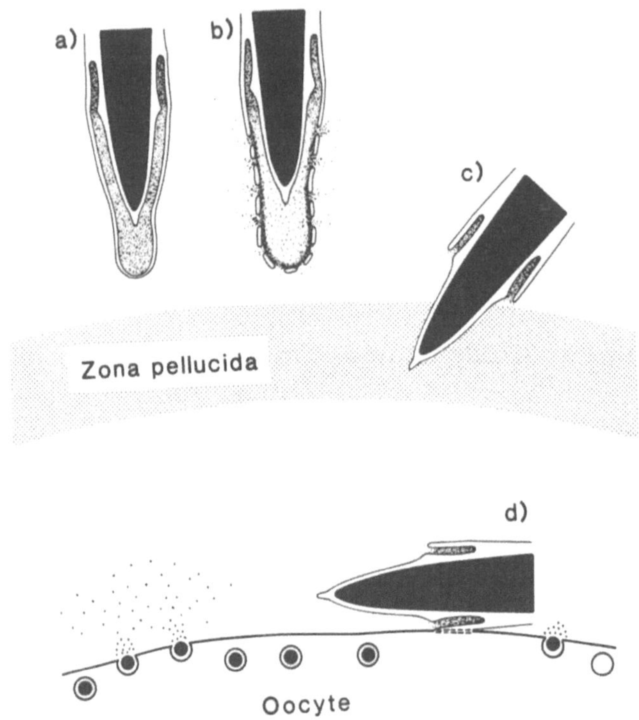

Fig 1. Progressive stages of sperm-egg interaction: a) intact capacitated spermatozoon b) acrosome reaction c) zona penetration and d) sperm-oocyte fusion.

Reactive oxygen species

The primary reactive oxygen species generated by the human spermatozoon appears to be superoxide anion, in light of the suppressive effect of superoxide dismutase on the ability of human spermatozoa to reduce ferricytochrome C (Alvarez & et al.,1987) or nitroblue tetrazolium. Similarly, superoxide dismutase will suppress the chemiluminescent signals generated by human spermatozoa in the presence of lucigenin, a membrane-impermeant probe which responds to extracellular superoxide anion.

Superoxide anion undergoes rapid dismutation within the spermatozoon, due to the presence of endogenous superoxide dismutase (Alvarez et al., 1987) so that the major reactive oxygen species being released from suspensions of human spermatozoa is hydrogen peroxide. The latter can be readily detected with sensitive spectrofluorometric techniques including the oxidation of scopoletin or homovanillic acid in the presence of horse radish peroxidase. Furthermore, peroxidase-enhanced chemiluminescent techniques can also be used to demonstrate the release of hydrogen peroxide from human spermatozoa into the extracellular space, in a manner which is readily susceptible to the scavanging effects of exogenous catalase.

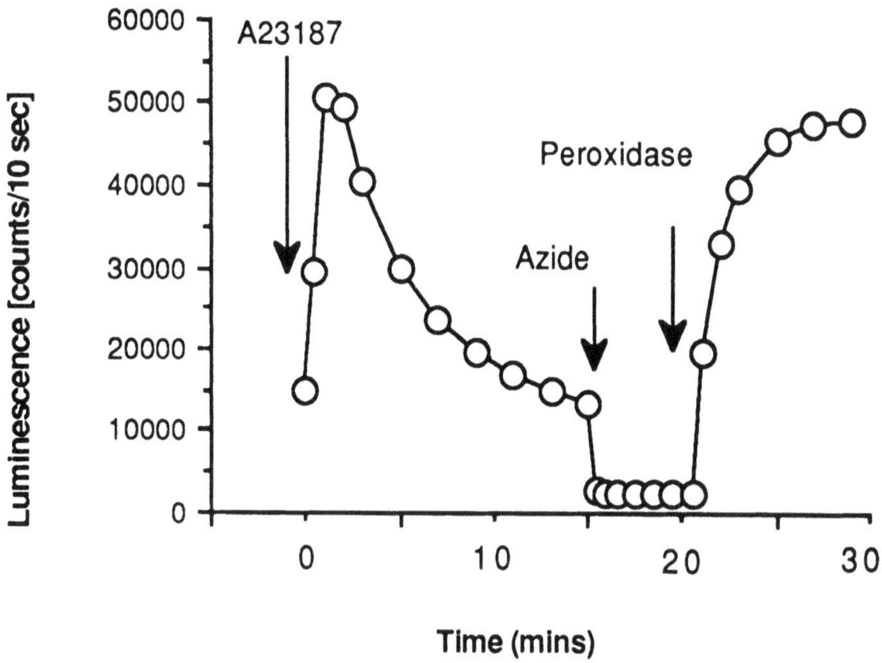

Fig 2. Addition of azide abolishes the chemiluminescent signal obtained with human spermatozoa in the presence of luminol due to the disruption of endogenous peroxidase activity. The luminol signal will return, however, if exogenous peroxidase is then added.

The luminol-induced chemiluminescent signals obtained from human spermatozoa in the absence of exogenous peroxidase appear to be due to the ability of intracellular hydrogen peroxide to oxidize luminol through the mediation of an endogenous sperm-peroxidase. If sodium azide is used to suppress this peroxidase, the luminol-dependent chemiluminescent signal is lost (Fig 2) in much the same way that suppression of myeloperoxidase in neutrophils abrogates the luminol signal obtained from these cells. If exogenous peroxidase is subsequently added to azide treated spermatozoa, the luminol signal returns (Fig 2). is unfortunate that, as yet, more definitive technique for identifying oxygen radicals such as election spin resonance, have not been applied to human spermatozoa. Nevertheless, the available spectrofluorometric and chemiluminescent data suggest that these cells exhibit the capacity for producing at least superoxide anion and hydrogen peroxide. The mechanisms responsible for the generation of these reactive oxygen species and their possible roles in the control of normal sperm function and the aetiology of male infertility are important questions which will be addressed in the following sections.

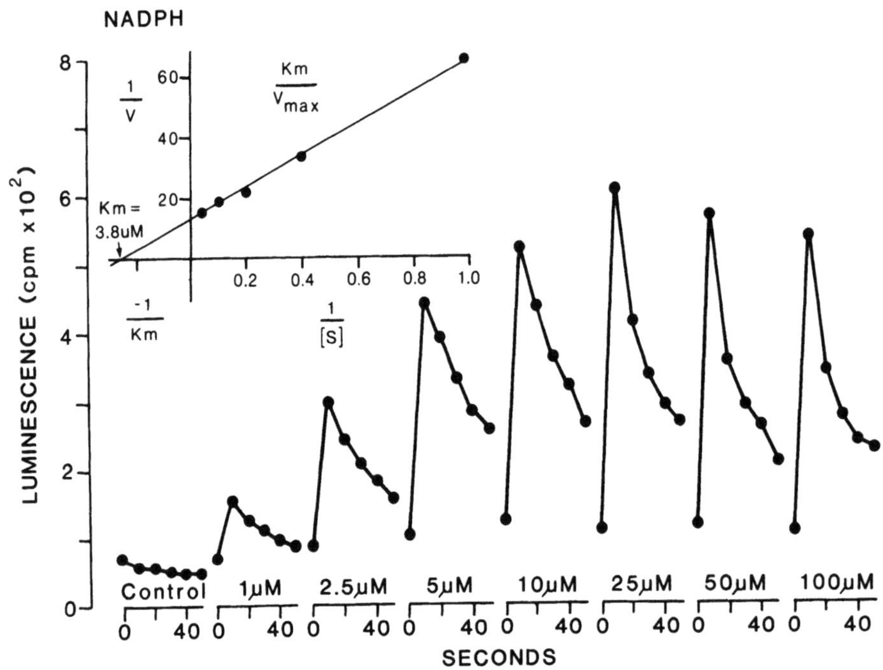

Fig 3. Induction of superoxide anion production by NADPH in triton-permeabilized human spermatozoa.

Cellular Mechanisms involved in Synthesis

The available data suggest that the electrons responsible for the reduction of molecular oxygen to superoxide anion derive from NADPH. If human spermatozoa are permeabilized by exposure to triton-X or processed as a broken-cell preparation, then a burst of reactive oxygen species can be induced by addition of either NADPH or NADH to the medium, the Km value for the former being about 4 times lower than the latter (Fig 3). The pulse of reactive oxygen species generated by exposure to NADPH is not influenced by singlet oxygen or hydroxyl radical scavengers but is profoundly influenced by scavengers of superoxide anion, such as cytochrome C.

An alternative procedure for demonstrating the role of NADPH in the generation of superoxide anion by human spermatozoa is to use an air-drying procedure to achieve permeabilization and then expose these cells to a solution containing nitroblue tetrazolium (NBT). In response to the presence of NADPH the NBT is reduced to formazan in a manner which, because of its extreme sensitivity to the presence of superoxide dismutase, is indicative of superoxide anion production.

The NADPH required for superoxide production is derived from the oxidation of glucose through the hexose monophosphate shunt. Hence, if glucose is removed from the incubation medium and replaced by 2-deoxyglucose, the production of reactive oxygen species is inhibited. Furthermore, in the presence of glucose, there is a highly significant correlation between the rate of reactive oxygen species generated by individual samples and the rate of glucose flux through the hexose monophosphate pathway, as measured by $^{14}CO_2$ production from a radiolabeled glucose precursor. Similarly, the capacity of human spermatozoa to generate reactive oxygen species, as monitored by chemiluminescence, is highly correlated with their ability to reduce NBT when air dried and exposed, not to NADPH, but to $NADP^+$ and glucose-6-phosphate, the presence of which will lead to the generation of NADPH, if the hexose monophosphate shunt is active.

Spermatozoa therefore appear to operate very much as an activated neutrophil in that the driving force for reactive oxygen species production is NADPH derived from the hexose monophosphate shunt. The spermatozoon's superoxide generating system also appears to resemble the leucocyte model in that the NADPH oxidase responsible for effecting the election transfer to molecular oxygen resides in the plasma membrane of the cell. The extent to which the biochemical nature of the NADPH oxidase in leucocytes and germ cells is similar, is currently under investigation.

Role of Reactive Oxygen Species in Sperm Pathology

One of the features of reactive oxygen species generation by washed human ejaculates, is the considerable variation in the rate of reactive oxygen species production between individuals. In Fig.4, data are presented for 11 separate individuals for the production of reactive oxygen species in response to a calcium signal generated by A23187. Within this random sample there is a 100 fold difference between the minimal and maximal response. The clinical significance of this variation in reactive oxygen species production is shown in Fig 5 in which this activity is plotted on a log scale, against the competence of the same population of spermatozoa for sperm-oocyte fusion in the zona-free hamster oocyte penetration assay. These data reveal a powerful negative correlation between reactive oxygen species production by the ejaculates and sperm function as measured in this biossay (Aitken &

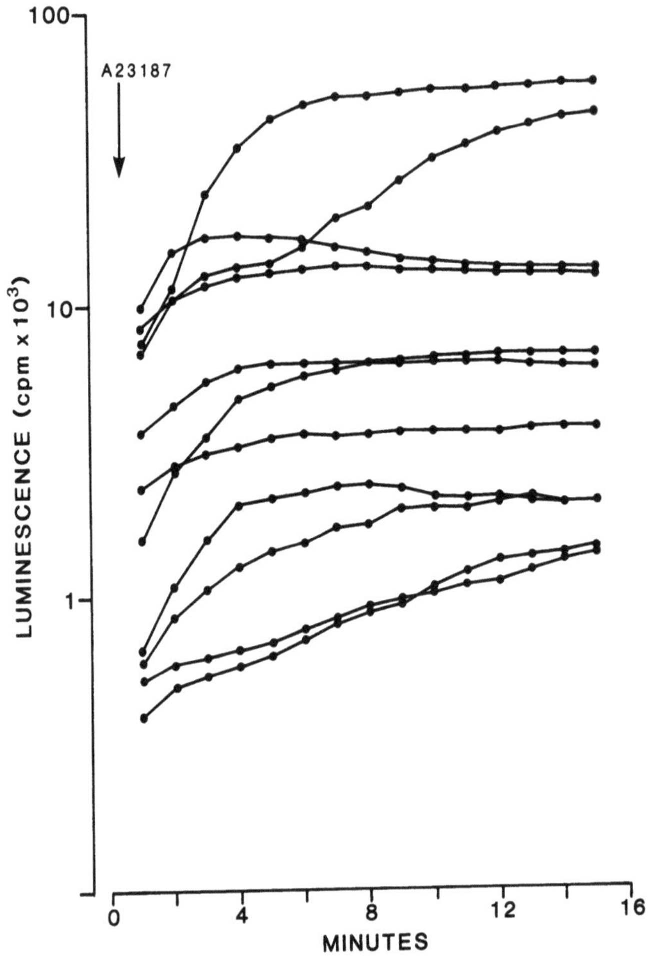

Fig 4. Reactive oxygen species production by 11 individuals in response to stimulation with A23187.

Clarkson, 1987). The mean basal levels of reactive oxygen species production are approximately 40 times greater in the samples exhibiting a defective capacity for sperm-oocyte fusion than the rest of the study population, and these levels are increased still further when such samples are treated with A23187 (Aitken & Clarkson, 1987.

Similarly, in a cohort of 74 oligozoospermic patients, 40 were found to exhibit defective sperm function, characterised by a failure to respond to A23187 with sperm-oocyte fusion, and in 17 cases this pathology was associated with the excessive generation of reactive oxygen species (Aitken et al. 1989b).

There is growing evidence that this association between the excessive generation of reactive oxygen species and defective sperm function involves peroxidative damage to the unsaturated

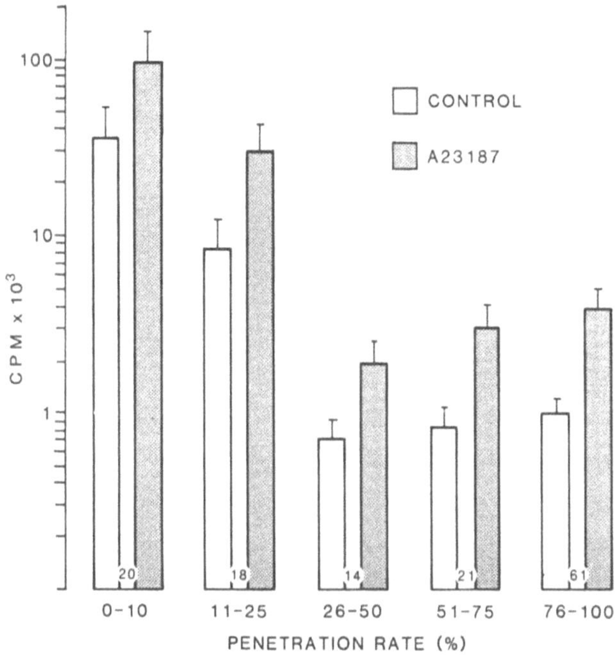

Fig 5. Relationship between reactive oxygen species generation and the capacity of human spermatozoa for sperm-oocyte fusion in response to A23187. Open bars = basal level of reactive oxygen species production; hatched bars = reactive oxygen species production in the presence of A23187 (Aitken & Clarkson, 1987).

fatty acids of the sperm plasma membrane. The human spermatozoon contains a high concentration of unsaturated fatty acids (Jones et al., 1979), 47.5% of phospholipid-bound fatty acids extracted from these cells being in the form of docosahexanoic acid, which contains 6 unsaturated double bonds per molecule. Such highly unsaturated fatty acids are susceptible to peroxidative attack because the presence of each double bond weakens the C-H bonds on the adjacent carbon atom, favouring the onset of a lipoperoxidative chain reaction which is initiated by the removal of a hydrogen atom and the creation of a lipid free radical.

The extremely small cytoplasmic space possessed by human spermatozoa also means that they are relatively deficient in the protective enzyme systems possessed by most cells. Spermatozoa apparently lack catalase activity, although they do possess both the glutathione peroxidase/glutathione reductase couple (Alvarez & Storey, 1989) and superoxide dismutase (Alvarez et.al., 1987). The activity of the latter, in particular, has been shown to correlate well with the resistance of human spermatozoa to peroxidation and their consequent ability to remain motile when cultured in albumin-free medium for prolonged periods of time in vitro (Alvarez, et. al, 1987).

The generation of both hydrogen peroxide and superoxide anion by human spermatozoa, also favours the onset of lipid peroxidation, particularly when this activity is chronically elevated, as observed within the infertile male population (Aitken et. al., 1989b). The combination of these 2 reactive oxygen species can lead to the creation of hydroxyl radicals via a Haber-Weiss reaction:

$$H_2O_2 + O_2^{-\cdot} = OH\cdot + OH^- + O_2$$

providing sufficient transition elements (iron or copper) are available to enhance the rate constant for this reaction. For seminal plasma, at least, there is recent evidence suggesting that adequate quantities of available iron may indeed be present to serve as a catalyst for hydroxyl radical formation. (Kwenang et al., 1988). Using malondialehyde production as an end-point for lipid peroxidation, we have been able to show a positive correlation between this process and the level of reactive oxygen species production by individual sperm populations, incubated in vitro in the presence of ferrous ion (Aitken et al., 1989a). The possibility that oxygen radical-induced lipid peroxidation might be a cause of defective sperm function has been demonstrated experimentally using suspensions of normal functional cells exposed to a ferrous-ion promotor system (ferrous sulphate: ascorbate). The presence of increasing amounts of promotor leads to a progressive increase in the degree of lipid peroxidation and although the motility of the spermatozoa remains unchanged, there is a corresponding decrease in the competence of these cells for sperm-oocyte fusion in the presence of A23187. In other words, the presence of promoter has achieved in the same refractory state that we observe in the infertile male population, by inducing a limited degree of lipoperoxidation in otherwise normal spermatozoa. (Aitken et al., 1989a).

That lipid peroxidation plays a causative role in the loss of sperm function observed in the above experiment is supported by the significant enhancement of sperm function achieved by adding the chain-breaking antioxidant, α-tocoperol, to the incubation medium (Aitken et al., 1989a). Such results have obvious implications for the development of therapeutic strategies, based on the use of antioxidants to treat cases of male infertility associated with peroxidative damage to the spermatozoa.

Reactive oxygen species in normal sperm function

If human spermatozoa possess a capacity for generating hydrogen peroxide and exhibit endogenous peroxidase activity, what biological purpose can such specialized systems perform? As yet, there is no definite answer to this question but a number of interesting possibilities are beginning to emerge. One intreguing observation is that when human pellucida is considerably enhanced (Aitken, et. al. 1989a). This effect is inhibited by the simultaneous presence of α-tocopherol, indicating that changes in the plasma membrane induced by peroxidation must be instrumental in effecting the observed increase in sperm-zona adhesion (Aitken et al., 1989a). In a number of other cell types whose functions are characterised by phases of increased adhesiveness, such as neutrophils or platelets, evidence has recently been obtained to indicate a mediating role for superoxide anion and lipid peroxidation (Lafuze, et, 1983, Bearpark, et.al., 1988) in the mechanism of adhesion. Similarly, peroxidation of liposomes containing unsaturated fatty acids, has been shown to induce increased vesicle adhesion and fusion (Sevanian, et. al., 1988). One consequence of lipid peroxidation, which may be involved in the mechanism of this increased adhesiveness, is the enhancement of phospholipase A_2 activity, through the creation of a multiphasic phospholipid arrangement within the plasma membrane, which enhances the binding of phospholipase A_2 to its substate (Ungemach, 1985). In addition, hydrogen peroxide is known to stimulate tyrosine kinase activity (Koshio et al., 1988) which may further enhance phospholipase A_2 since the natural inhibitor of this enzyme, lipomodulin is inactivated by tyrosine phosphorylation (Hirata et al., 1984). Activation of phospholipase A_2 should result in an enhancement in the fusogenicity of the plasma membrane through the destabilizing action of lysophospholipid. Evidence for a role for phospholipase A_2 in the induction of the acrosome reaction has been obtained in many different mammalian species (Fleming & Yanagimachi 1981; Llanos et al., 1982; Ohzu & Yanagimachi 1982), and it is possible that hydrogen peroxide plays an active role in the activation of this enzyme.

Conclusions

In the brief review, I have considered some of the evidence indicating that human spermatozoa produce reactive oxygen species and, in particular, superoxide anion and hydrogen peroxide. The biological significance of this rather specialized metabolic activity is currently unknown, although the activation of tyrosine kinase and/or phospholipase A_2 activities are possiblities currently under investigation. The clinical significance of these studies lies in the loss of sperm function observed in samples exhibiting a chronic elevation of reactive oxygen species generation. The reactive oxygen species generated in such cases appear to originate from the spermatozoa as well as from any granulocytes present in the seminal plasma. The combined release of superoxide and hydrogen peroxide from these two sources leads to a state of oxidative stress in which sperm function is suppressed as a result of the peroxidation of unsaturated fatty acids in the sperm plasma membrane. These observations are the first to define a loss of human sperm function in biochemical terms and the results have clear implications for the design of appropriate therapeutic strategies.

Reference

Aitken, R.J., Clarkson, J.S., (1987). Cellular basis of defective sperm function and its association with the genesis of reactive oxygen species by human spermatozoa. J. Reprod. Fert. 81: 459-469.

Aitken, R.J., Clarkson, J.S., Fishel S. (1989a). Generation of reactive oxygen species, lipid peroxidation and human sperm function. Biol. Reprod. (in press).

Aitken, R.J., Clarkson, J.S., Hargreave, T.B., Irvine, D.S., Wu., F.C.W. (1989b). Analysis of the relationship between defective sperm function and the generation of reactive oxygen species in cases of oligozoospermia. J. Androl. 10: 214-220.

Alvarez, J.G., Storey, B.T., (1989). Role of glutathione peroxidase in protecting mammalian spermatozoa from loss of motility caused by spontaneous lipid peroxidation. Gamete Res 23: 77-90.

Alvarez, J.G., Touchstone, J.C. Blasco, L., Storey, B.T., (1987). Spontaneous lipid peroxidation and production of hydrogen peroxide and superoxide in human spermatozoa. J.Androl. 8:338-348.

Bearpark, T., Salvemini, D., Sneddon, J.M., Vane, J.R., (1988). Endothelium -derived relaxing factor (EDRF) and superoxide anions modulate platelet adhesion to endothelial cells. J. Physiol. 339:12P.

Bedford, J.M., (1977). Sperm/egg interaction: the specificity of human spermatozoa. Anat. Rec. 188: 477-488.

Fleming, A.D., Yanagimachi, R., (1981). Effects of various lipids on the acrosome reaction and fertilizing ability of guinea pig spermatozoa with special reference to the possible involvement of lysophospholipids in the acrosome reaction. Gamete Res. 4:253-273.

Hirata, F., Matsuda, K., Notsu, S., Hattori, T., Del Carmine, R., (1984). Phosphorylation at a tyrosine residue of lipomodulin in mitogen-stimulated murine thymocytes. Proc. Nat. Acad. Sci. USA., 81: 4717-4721.

Jones, R., Mann, T., Sherins, R., (1979). Peroxidative breakdown of phospholipids in human spermatozoa, spermidical properties of fatty acid peroxides and protective action of seminal plasma. Fertil Steril, 31:531-537.

Koshio O., Akanuma, Y., Kasuga, M., (1988). Hydrogen peroxide stimulates tyrosine phosphorylation of the insulin receptor and the tyrosine kinase activity of intact cells. Biochem, J., 250: 95-101.

Kwenang, A., Kroos, M.J., Koster, J.F., Van Eijk, H.G., (1987). Iron, ferritin and copper in seminal plasma. Human Reprod. 2: 387-388.

Lafuze, J.E., Weisman, S.J., Ingraham, L.M., Butterick, C.J., Alpert, L.A., Baehner, R.L., (1983). The effect of vitamin E on rabbit neutrophil activation. In: Porter, R., Whelan, J. (eds). Biology of Vitamin E. Ciba Foundation Symposium 101, Pitman, Bath, p130.

Llanos MN., Lui W, Meizel, S., (1982). Studies of phospholipase A_2 related to the hamster sperm acrosome reaction. J.Exp. Zool 221: 107-117.

Ohzu, E., Yanagimachi, R., (1982). Acceleration of the acrosome reaction in hamster spermatozoa by lysolecithin. J.Exp. Zool. 224: 259-263.

Ungemach, F.R., (1985). Plasma membrane damage to hepatocytes following lipid peroxidation: involvement of phospholipase A_2. In Poli, G., Cheeseman, K.H., Dianzani, M.U. & Slater, T.F. (eds). Free Radicals in Liver Injury, IRL Press, Washington., p127.

Sevanian, A., Wratten, M.L., McLeod, L.L., Kim, E., (1988). Lipid peroxidation and phospholipase A_2 activity in liposomes composed of unsaturated phospholipids: a structural basis for enzyme activation. Biochem Biophys. Acta 1961: 316-327.

Fertilisation in *Fucus*: exploring the gamete cell surfaces with monoclonal antibodies

J.R. Green, J.L. Jones, C.J. Stafford and J.A. Callow

School of Biological Sciences, University of Birmingham, P.O. Box 363, Birmingham B15 2TT, U.K.

NATO ASI Series, Vol. H 45
Mechanism of Fertilization
Edited by B. Dale

1. Introduction

Fertilisation in the marine brown alga Fucus is a species-specific process involving highly differentiated sperm and eggs. These gametes provide ideal material for studying both the molecular basis of recognition and signalling during fertilisation and the molecular architecture of a natural plant protoplast surface . For the purposes of studying fertilisation the Fucus system has a number of advantages: 1) For most species, large quantities of eggs (10^6-10^7) and sperm (10^{10}-10^{11}) can be released from wild plants. 2) Fertilisation is species-specific. 3) The gametes are naked cells, so there is no interference from cell walls. 4) There is a relatively simple bioassay which can be used to measure the percentage of eggs fertilised under controlled conditions. This review will focus on the cell surfaces of the Fucus gametes since we are primarily interested in the organisation of the gamete plasma membranes and the role of cell surface molecules in recognition and fertilisation. After dealing with the morphology of the gametes, the fertilisation assay and effects of a variety of treatments on fertilisation will be discussed briefly, since these aspects have been reviewed previously (Evans et al., 1982; Callow et al., 1985). Recent studies involving the use of monoclonal antibody (MAb) techniques to study the organisation and structure of the gamete cell surfaces will then be described.

2. *Fucus* gametes

Fucus sperm are highly differentiated structures. They are drop-shaped, biflagellate and about 5 µm long (Fig. 1a). The sperm are bright orange due to carotenoid accumulation in a specialised region of the chloroplast (eyespot). The shorter, anterior flagellum bears helical rows of fine hairlike structures known as mastigonemes (Fig. 1b), (Manton and Clarke, 1951) and the tip of the anterior flagellum appears to be sticky (Friedmann, 1961). The longer posterior flagellum is smooth and lacks mastigonemes. The anterior end of the sperm extends into a proboscis (Fig. 1b). This is a flattened plasma membrane bound structure composed of an interior framework of thirteen parallel microtubules which are connected to the root of the anterior flagellum (Manton and Clarke, 1951). In motile sperm the anterior flagellum exerts rapid undulating movements whilst the posterior flagellum remains erect and the proboscis is directed forward, though its function remains unknown.

Fucus eggs (Fig. 2) are spherical (80 µm in diameter) and brown in appearance. They are non-motile but can be set into slow rotating motion by the large number of sperm that are chemo-attracted to them (Fig. 2). They are surrounded only by a plasma membrane, in comparison with for example sea urchin eggs, which have a plasma membrane surrounded by a vitelline layer that is encased in a jelly coat. The surface of the egg appears to be rough when viewed by SEM, though this could be due to the protrusion of cytoplasmic vesicles (Callow et al., 1978). The egg cytoplasm is full of these vesicles which contain polyphenols and

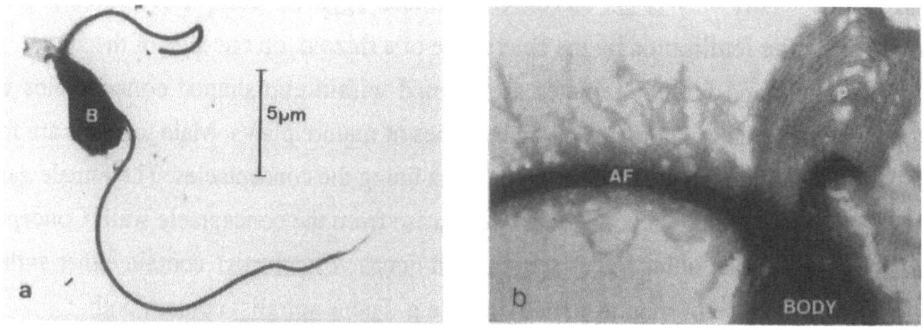

Fig 1. Negatively stained Fucus serratus sperm. a: Whole sperm. b: Close-up showing anterior flagellum (AF), proboscis (p) and body.

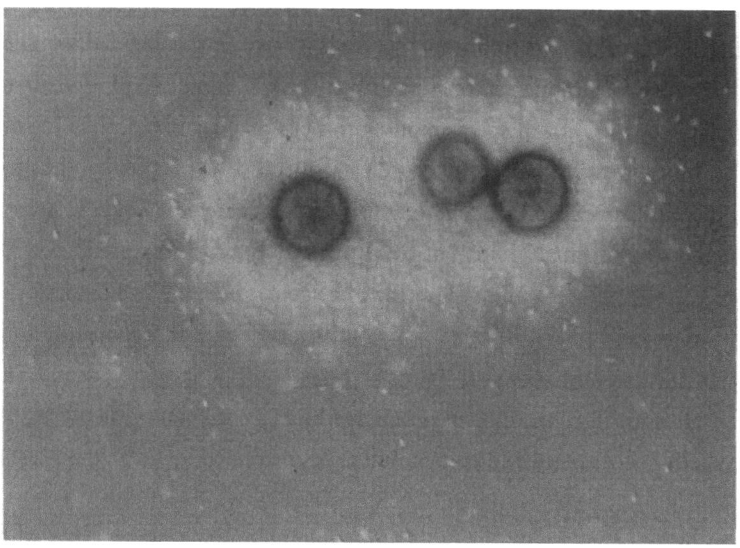

Fig 2. Fucus egg (80 μm) attracting sperm.

polysaccharides (Brawley et al., 1976). Unfertilised eggs are completely apolar; polarity occurs 12-24h after fertilisation by the emergence of a rhizoid on one side of the zyote.

The male and female gametes are formed within cup-shaped conceptacles which develop in the receptacles at the tips of the branches of mature plants. Male gametes are formed in small sac-like antheridia which develop on hairs lining the conceptacles. The female gametes are produced in large sacs called oogonia which arise from the conceptacle wall. Conceptacles of dioecious species (including Fucus serratus and Fucus vesiculosus) contain either antheridia or oogonia, whilst those of monoecious species (e.g. Fucus spiralis) contain both.

3. Fertilisation in *Fucus*

3.1. Stages in fertilisation

Sexual reproduction in marine brown algae such as Fucus is achieved by the liberation of naked gametes into the sea. The fertilisation process is species-specific and involves a series of events which can be summarised as follows: 1) Chemotaxis of sperm to eggs. 2) Species-specific recognition and sperm binding to eggs. 3) Plasmogamy i.e. fusion of sperm and eggs. 4) A block to polyspermy and the release of cell wall material by the egg.

Chemotaxis is mediated by a pheromone which is secreted by eggs to attract the sperm. This pheromone was originally isolated from eggs of Fucus serratus and is a conjugated hydrocarbon, 1,3 -trans,5 -cis-octatriene, which has been given the trivial name 'fucoserraten' (Muller and Seferaides, 1977). It is a highly active substance and concentrations as low as 10^{-6} M induce attraction of sperm of Fucus serratus and Fucus vesiculosus. These results, plus the fact that an identical compound is secreted by eggs of Fucus vesiculosus, suggests that the species specificity of sperm-egg recognition is not mediated by the pheromone (Muller and Seferaides, 1977).

Sperm make contact with the egg membrane by 'probing' the surface with the tips of their anterior flagella (Friedman,1961). Only a few sperm appear to bind to the egg surface and following specific binding, fusion of sperm and egg membranes occurs though the precise details of these events remain obscure. No evidence of phenomena comparable to the acrosome reaction in animal sperm-egg interactions has been demonstrated in Fucus.

Plasmogamy in fucoids is followed by the immediate release from the egg of the β-linked polyuronide alginic acid, which is stored pre-packaged in the large cytoplasmic vesicles. The alginic acid must act as a barrier to polyspermy and secretion of the alginate wall appears to commence at the point of entry of sperm (Evans et al., 1982). Within two minutes of mixing eggs and sperm numerous loci of wall deposition are visible. Twenty to thirty minutes after fertilisation cellulose synthesis begins and after one hour the zygote wall contains equal amounts of alginate-cellulose (Quatrano and Stevens, 1976). After 4h fucans are also present in the wall and from 6 - 24 h the proportion of the three major wall components is constant (60% alginic acid, 20% cellulose, 20% fucan). From 24 h the rhizoid becomes apparent as a protuberance on one side of the cell and the first cell division occurs about 30 h after mixing eggs and sperm (for review see Evans et al., 1982).

3.2. The fertilisation assay

An assay based on Calcofluor white staining of the newly secreted alginic acid wall was developed to quantify fertilisation in Fucus (Bolwell et al., 1977; Callow et al.,1978). The proportion of fertilised eggs can be determined by direct counting using a UV microscope. The standard assay contains 5000 eggs and fertilisation is proportional to the sperm concentration up to a saturating level of approx. 300 sperm per egg. At 150 sperm per egg, fertilisation is rapid, linear up to 5 min. and complete within 15 min. This assay permits detailed studies on the molecular basis of egg-sperm recognition. These have been based on the ability of exogenous molecules (lectins, glycosidases, polysaccharides) to perturb or inhibit the standard bioassay.

Experiments with lectins have shown that there is specificity in their binding to Fucus gametes. ConA and RCA_{120} (which bind to mannose/glucose and galactose residues respectively) bind strongly to eggs but not to sperm and they both inhibit fertilisation. Fucose-binding protein binds to both gametes and also inhibits fertilisation (Bolwell et al.,1979; Catt et al., 1982; Vithanage et al .,1983). Other lectins tested had no inhibitory activity in the fertilisation assay. The results therefore suggest a role for mannose/glucose and galactose residues on the eggs in the fertilisation process. A role for fucose residues on the egg and/or sperm also has to be considered.

A different approach involved treating gametes with a variety of carbohydrases. Pre-treatment of eggs with low concentrations of α-D-mannosidase or α-L-fucosidase, both of which remove terminal residues, inhibited fertilisation (Bolwell et al., 1979). The same carbohydrases had no effect on sperm. Other carbohydrases, including α,β- glucosidases, α,β- galactosidases and N-acetyl-glucosaminidase had no effects on eggs or sperm. These results support the notion that mannose- and fucose- containing components on the egg cell surface are involved in interactions with the sperm.

When the effects of polysaccharides were tested in the fertilisation assay, a number of heteroglycans containing mannose (e.g. yeast mannan) or an α-1-2-linked fucose (e.g. ascophyllan/fucoidan) inhibited fertilisation by binding to sperm, whereas a variety of glucans and galactans were ineffective (Bolwell et al., 1979) . Fertilisation could not be inhibited by simple mono- or disaccharides. This was strong indirect evidence for the presence of mannose- and fucose- binding proteins on the sperm cell surface with a role in sperm-egg recognition during fertilisation.

Taken together the evidence is consistent with a model for egg-sperm recognition based on mannose- and fucose- containing egg surface components interacting with complementary carbohydrate binding structures on the sperm cell surface.

Plasma membrane enriched fractions derived from Fucus serratus eggs were also found to inhibit fertilisation in a species-specific manner (Bolwell et al ,.1980). This inhibitory activity was abolished by pre-treatment of the egg fractions with α-L-fucosidase and α-D-mannosidase. ConA binding glycoproteins were isolated from solubilised egg plasma membrane fractions and they were also able to inhibit fertilisation. This was again blocked by treatment with the glycosidases. These results suggested that at least one of the important egg glycoproteins involved in the fertilisation process contains both mannose and fucose. The low levels of soluble 'receptor' isolated so far have precluded further attempts to isolate recognition molecules. The most difficult problem is the initial isolation of egg plasma membranes which are not highly contaminated with polysaccharides. In part this is due to the absence of suitable membrane markers , but this problem may be overcome with the use of monoclonal antibodies to egg cell surface components.

4. Monoclonal antibodies to the *Fucus* sperm cell surface

Although the techniques referred to in the previous section have yielded information regarding the nature of sperm-egg recognition, their usefulness in this respect is limited and clearly other approaches are needed. In order to dissect the molecular mechanisms involved in fertilisation we are now using monoclonal antibodies (MAbs) to examine the structure and functions of gamete cell surface molecules.

At the functional level, we were encouraged by the fact that a polyclonal antiserum raised against surface antigens of Fucus serratus sperm flagella, caused inhibition in the fertilisation assay in a species-specific manner (Vithanage et al., 1982). Our initial studies have used MAbs to explore the gamete cell surfaces.

4.1. Sperm cell surface organisation

One characteristic of sperm cells of many animals is that specific surface antigens are localised to particular regions (domains) of the cell (e.g. head, tail) and the molecules in these separate domains often have functions related to the fertilisation process (Primakoff and Myles, 1983; Saxena et al., 1986). MAbs were raised against cell surface antigens of Fucus serratus sperm to investigate their spatial organisation and to determine whether such antigens occur in discrete regions or are expressed over the entire cell. The antibodies have been tested for cross reactivity with Fucus serratus eggs and sperm of other fucoid species and the molecular nature of the antigens recognised has been determined (Jones et al., 1988; Jones et al., submitted).

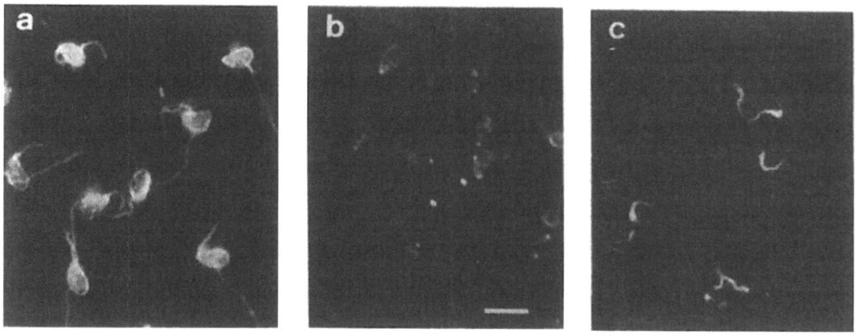

Fig 3. Immunofluorescence of Fucus serratus sperm labelled with MAbs. a: FS4, b: FS1, c: FS2. Bar: 0.5 μm.

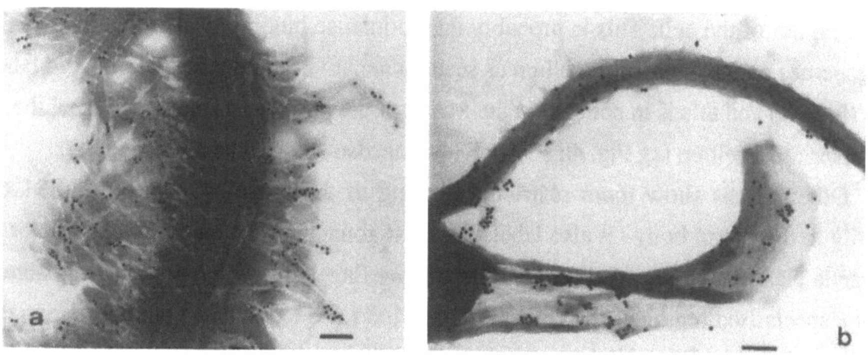

Fig 4. Immunogold labelling of negatively stained Fucus serratus sperm. a: FS7, b: FS4, Bar: 0.2 μm

Mice were immunised with live sperm and after fusion of spleen cells with myeloma cells, hybridoma supernatants were screened using indirect immunofluorescence on fixed sperm attached to slides. The sperm proved to be highly immunogenic and a large number of hybridomas secreted antibody to the sperm surface. Cell lines producing a set of twelve MAbs (designated FS 1-12) were cloned and antibody binding to sperm was examined by immunofluorescence and immunogold labelling. The results are summarised in Table 1. The MAbs fall into two broad groups comprising those antibodies which label the cell most intensely on the body region (FS1,3,4,6,8,9,10) and those which bind most intensely to the anterior region of the cell (FS 2,5,7,11,12). The results show that some surface antigens are highly concentrated in particular regions of the sperm, whereas others are more generally distributed over the cell surface. Several MAbs bind to the entire sperm, including both flagella (FS3,4,6,8,10) (Fig. 3a) and the labelling of the antibodies is concentrated on the body. Two MAbs (FS7,12) also bind to the entire sperm but their binding is highly concentrated on the anterior region of the cell. This is probably due to intense binding of these antibodies to the mastigonemes on the anterior flagellum as seen by negative staining of immunogold labelled sperm (Fig. 4a) and this is in contrast to the MAbs of the previous group which label the shaft of the anterior flagellum (eg Fig. 4b shows FS4: note also labelling of the proboscis).

Other MAbs show more restricted binding to the sperm cell surface. FS1 binds primarily to the sperm body : it also labels the mastigonemes very weakly, but does not label the flagella plasma membranes (Fig. 3b). The anterior flagellum, which bears mastigonemes, is clearly a specialised region of the cell and three MAbs (FS2, 5 and 11) bind preferentially to this region (Fig. 3c). FS11 binds to the plasma membrane of this flagellum while FS2,5 bind to the mastigonemes. FS2 and FS5 also bind weakly to the sperm body.

The results show that the surface antigens of a highly differentiated plant cell can be organised in a non-uniform manner over the cell. It appears that Fucus sperm can maintain antigens in distinct regions (anterior flagellum plasma membrane, mastigonemes, cell body) while at the same time maintaining antigens at different concentrations over the whole surface of the sperm. For mammalian sperm, antigens tend to be highly restricted to particular domains (e.g. head, tail , midpiece). In some instances antigens can be shared by adjacent domains but it is only rarely that MAbs have revealed antigens distributed over the entire cell (Primakoff and Myles, 1983; Gaunt et al.,1983). However it seems that the restriction of antigens to particular regions on the surface of Fucus sperm is less apparent than for mammalian sperm, whereas polarisation or concentration of antigens which are generally expressed over the sperm surface is more apparent. In animal cells domains are maintained by barriers to free diffusion, in the plane of the membrane, or by cytoplasmic anchors to the surface macromolecules. Such barriers exist on the surface of mammalian sperm cells (Cowan et al .,1987), and in gametes of the green alga Chlamydomonas eugametos (Musgrave et al.

Table 1. Binding patterns of MAbs FS1- FS12 to <u>Fucus serratus</u> sperm and eggs

MAb	Sperm Anterior flagellum	Sperm Body	Sperm Posterior flagellum	Egg
FS4(IgM)	++	+++	++	+
FS3(IgM)	+	++	+	-
FS6(IgG1)	+	++	+	-
FS8(IgM)	+	++	+	-
FS9(IgG2b)	+	+++	+	+
FS10(IgM)	+	+++	+	-
FS1(IgM)	+-(M)	++	-	-
FS7(IgG1)	+++(M)	++	+	-
FS12(IgG1	++(M)	+	+	-
FS5(IgM)	+++(M)	+-	-	+
FS2(IgG1)	++(M)	+-	-	+
FS11(IgM)	++	-	-	-

Mab binding to sperm was determined by indirect immunofluorescence. The intensity of fluorescence was recorded as: - absent; +- very weak; + weak; ++ intermediate; +++ strong. Antibody binding to the mastigonemes (M) on the anterior flagellum was determined by immunogold labelling of negatively stained sperm. Mabs which did not bind to the mastigonemes bound to the shaft of this flagellum. Binding of Mabs to egg vesicles was determined by ELISA.

1986), where there is a functional barrier to free diffusion of surface macromolecules, located at the flagellum-body transition zone.

4.2. Comparison of sperm and egg cell surfaces and species/genus specificity

The panel of MAbs which was raised to sperm of Fucus serratus was tested for cross reactivity with eggs and with gametes of other fucoid species using ELISA and immunofluorescence (Jones et al., 1988) The majority of the MAbs do not bind to Fucus serratus eggs (FS1,3,6,7,8,10,11,12) indicating clearly that the sperm and egg cell surfaces differ. Four of the MAbs do bind to eggs (FS2,4,5,9) and FS2 is of particular interest since immunogold labelling of egg sections shows that it binds almost exclusively to the egg plasma membrane (Fig 5). The antibody binds in patches on the egg cell surface (Stafford unpublished results) and this is similar to the binding of the lectins ConA and RCA_{120} reported previously (Catt et al., 1982). These results suggest that there may be domains on the egg plasma membrane in which particular antigens are concentrated. In addition FS2 should prove to be a useful marker for the egg plasma membrane during enrichment procedures.

Only one MAb (FS10) binds in a species-preferential manner to sperm of Fucus serratus as compared with that of Fucus vesiculosus, though nine MAbs (FS2,3,5,6,7,8,10,11,12) bind in a genus-preferential manner to sperm of these species when compared with that of Ascophyllum nodosum. These reults reflect the degree of homology between the three species, showing that a number of antigenic determinants are highly conserved whereas others show variation.

4.3. Molecular nature of sperm surface antigens

Western blotting and antigen modification procedures have been used to gain information on the antigens and epitopes recognised by this set of antibodies (Jones et al., submitted). From initial characterisation studies the MAbs can be classified into groups identical to those derived from immuno-localisation studies. The MAbs FS3,4,6,8,10, which all label the entire sperm, including both flagella, recognise glycan epitopes of a 205 kDa glycoprotein. Antibody binding is abolished by periodate, but the nature of the glycan-peptide linkage remains uncertain. We judge that this antigen is immunodominant since this set of five Mabs was initially chosen from a much larger number with the same binding specificities. Competition assays suggest that these MAbs bind to the same epitope. However, the antibodies show different specificities towards eggs and other species of brown algae and the most likely explanation is that they are binding to neighbouring or overlapping epitopes of this glycoprotien. It seems though that these different binding specificities are probably based on differences in the carbohydrate side chains of this molecule. This would be similar to the situation in Chlamydomonas eugametos in

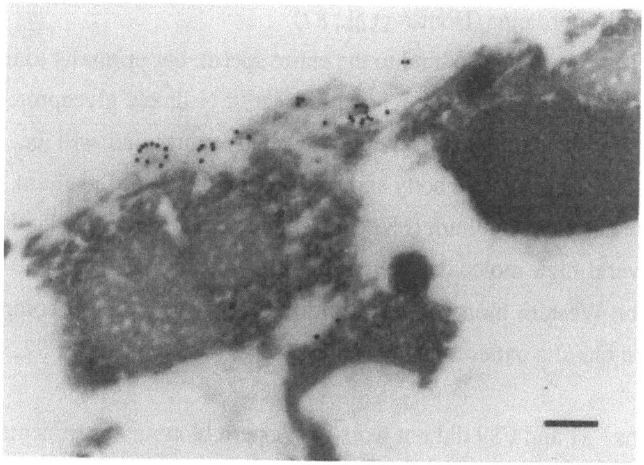

Fig 5. Immunogold labelling of egg section with FS 2. Bar: 0.5 μm.

which strain specific epitopes are correlated with the presence of certain O-methylated sugars in the isoagglutinin glycoproteins (Homan et al., 87).

The two MAbs which also bind to the entire sperm, but primarily to the mastigonemes on the anterior region (FS7,12), both recognise a set of N-linked glycoproteins ranging from 50-250 kDa. It is possible that some of these glycoproteins will be localised to the mastigonemes, some to the sperm body and some to the posterior flagellum. The three Mabs which bind to the anterior flagellum (FS2,5-mastigonemes, FS11-flagella plasma membrane) all recognise several high molecular weight proteins/glycoproteins 90-250 kDa, though the binding pattern on Western blots shows differences between FS2,5 as compared with FS11. From the Western blotting patterns and competition assays it seems that FS2,5 bind to the same epiope.

The MAbs FS1 and FS9 did not work in Western blotting experiments and also proved difficult to use in ELISA. However FS9 binding to sperm antigens could be detected in dot-blot assays, and its binding was inhibited by periodate treatment of the antigens, suggesting that it recognises a carbohydrate epitope. The antigens recognised by these two antibodies remain to be characterised.

That many of the MAbs appear to recognise carbohydrate is not surprising given previous studies in which MAbs have been produced to plant cell surface antigens. The evidence suggests that these carbohydrate epitopes can be held in distinct spatial arrays on the surface of the sperm.

5. Summary

Use of a panel of monoclonal antibodies raised to cell surface antigens of Fucus serratus sperm has shown that: A) Some antigens are distributed over the entire sperm. These include a 200 kDa glycoprotein and a set of N-linked glycoproteins. B) Some antigens have a restricted distribution- to the sperm body, to the mastigonemes or to the anterior flagellum plasma membrane. C) Some antibodies show gamete-, species- and genus- preferential binding. D) The 200 kDa glycoprotein is an immunodominant antigen. It is likely that differences in the carbohydrate side chains of this molecule are responsible for the binding characteristics of the antibodies referred to in (C).

References

Bolwell, G.P., Callow, J.A., Callow, M.E. and Evans, L.V. (1977) Cross fertilisation in fucoid seaweeds. Nature **268**, 626-627

Bolwell, G.P., Callow, J.A., Callow, M.E. and Evans, L.V.(1979) Fertilisation in brown algae. II Evidence for lectin sensitive complementary receptors involved in gamete recognition in Fucus serratus. J.Cell Sci. **36**, 19-30

Bolwell, G.P., Callow, J.A., and Evans, L.V. (1980) Fertilisation in brown algae. III. Preliminary characterisation of putative gamete receptors from eggs and sperm of Fucus serratus. J.Cell Sci. **43**, 209-224

Brawley, S.H., Wetherbee,R. and Quatrano, R.S. (1976) Fine-structural studies of the gametes and embryo of Fucus vesiculosus L. (Phaeophyta) I. Fertilisation and pronuclear fusion. J.Cell Sci. **20**, 233-254

Catt, J.W., Vithanage, H.I.M.V., Callow, J.A., Callow, M.E. and Evans, L.V. (1982) Fertilisation in brown algae. V. Further investigations of lectins as surface probes. J.Cell Sci. **147**, 127-133

Callow, J.A., Callow, M.E. and Evans, L.V. (1985) Fertilisation in Fucus. In ' Biology of Fertilisation' Vol 2, pp 389-407. Academic Press.

Cowan, A.E., Myles, D.G. and Koppel, D.E. (1987) Lateral diffusion of the PH-20 protein on guinea pig sperm: evidence that barriers to diffusion maintain plasma membrane domains in mammalian sperm. J.Cell Biol. **104**, 917-923

Evans, L.V., Callow, J.A. and Callow, M.E. (1982) The biology and biochemistry of reproduction in Fucus. In 'Progress in phycological research' F.E.Round and D.Chapman, eds., Vol.I, pp 67-110. Elsevier, Amsterdam.

Friedmann, I. (1961). Cinemicrography of spermatozoids and fertilisation in Fucales. Bull.Res. Counc. Isr. **10D**, 73-83

Gaunt,, S.J., Brown, C.R. and Jones, R. (1983) Identification of mobile and fixed antigens on the plasma membrane of rat spermatozoa using monoclonal antibodies. Exp.Cell Res. **144**, 275-284

Homan, W.L., van Kalshoven, .H., Kolk, A.H.J., Musgrave, A., Schuring, F. and van den Ende, H. (1987) Monoclonal antibodies to surface glycoconjugates in Chlamydomonas eugametos recognise strain-specific O-methyl sugars. Planta **170**, 328-335

Jones, J.L., Callow, J.A. and Green, J.R. (1988) Monoclonal antibodies to sperm surface antigens of the brown alga Fucus serratus exhibit region-, gamete-, species- and genus-preferential binding. Planta **176**, 298-306

Manton, I. and Clarke, B. (1951) An electron microscope study of the spermatozoid of Fucus serratus. Ann. Bot. **60**, 461-471

Muller, D.G. and Seferiadis, K. (1977) Specificity of sexual chemotaxis in Fucus serratus and Fucus vesiculosus (Phyaeophyceae). Z. Pflanzenphysiol. **84**,85-94

Musgrave, A., de Wildt, P., van Etten, I., Pijst, C., Scholma, C., Kooyaman, R., Homan, W. and van den Ende, H. (1986) Evidence for a functional membrane barrier in the transition zone between the flagellum and cell body of Chlamydomonas eugametos gametes. Planta **167**, 544-553

Primakoff, P.and Myles, D.G. (1983) A map of the guinea pig sperm surface constructed with monoclonal antibodies. Dev. Biol. **98**, 417-428

Quatrano, R.S. and Stevens, P.T. (1976) Cell wall assembly in Fucus zygotes. I. Characterisation of the polysaccharide components. Plant Physiol. **58**, 224-231

Saxena, N.K., Russell, L.D., Saxena, N. and Peterson, R.N. (1986) Immunofluorescence antigen localisation on boar sperm plasma membranes: monoclonal antibodies reveal apparent new domains and apparent redistribution of surface antigens during sperm maturation and at ejaculation. Anat. Rec. **214**, 238-252

Vithanage, H.I.M.V., Catt, J.W., Callow, J.A., Callow, M.E. and Evans, L.(1982) Fertilisation in brown algae IV. Appearance of sperm specific antigens on fertilised eggs. J.Cell Sci. **60**, 103-108

GAMETE INTERACTIONS AND THE INITIATION OF EGG ACTIVATION IN SEA URCHINS

Frank J. Longo[1], Susan Cook[1], David H. McCulloh[2], Pedro I. Ivonnet[2] and Edward L. Chambers[2]

[1]Department of Anatomy,
The University of Iowa,
Iowa City, IA 52242
and
[2]Department of Physiology and Biophysics,
The University of Miami School of Medicine
Miami, FL 33101

The earliest perceivable response of echinoid eggs to the fertilizing sperm is a transient depolarization, the activation or fertilization potential (Steinhardt et al., 1971; Jaffe, 1976; Chambers and de Armendi, 1979). This change in electrical activity involves the appearance of sperm associated ion channels, which depolarize the plasma membrane, as well as the opening of voltage-dependent calcium channels (Chambers and de Armendi, 1979). These changes constitute Phase 1 (Lynn et al., 1988; Chambers, 1989). In voltage clamped eggs, sperm which induce Phase 1 either enter or fail to enter the egg. If sperm penetration occurs, the inward current of Phase 1, initiating Phase 2, continues to increase; if sperm penetration fails to occur the inward current is abruptly severed. During Phase 2 a large, rapid and transient increase in intracellular free calcium (Steinhardt and Epel, 1974; Steinhardt et al., 1977; Whitaker and Steinhardt, 1982; Jaffe, 1983) propagates in the form of a wave from the point of gamete interaction to the opposite pole of the egg (Eisen et al., 1984; Swan and Whitaker, 1986; Yoshimoto et al., 1987). This wave of increased intracellular calcium is initiated following a latent period (Phase 1) of approximately 12 sec and stimulates the egg from its quiescent state to proliferation and embryogenesis (Chambers, 1989). Activation of the egg as evidenced by cortical granule exocytosis and fertilization membrane elevation begins during Phase 2. Neither the depolarization (Lynn and Chambers, 1986) nor the calcium influx (Schmidt et al., 1982) that occur during Phase 1 are necessary for the activation of the egg.

In sea urchins, elevated calcium concentrations lead to the production of inositol triphosphate (Ins 1,4,5 P_3) which, in turn, induces further calcium release from intracellular stores (Berridge and Irvine,

NATO ASI Series, Vol. H 45
Mechanism of Fertilization
Edited by B. Dale

1984; Whitaker and Irvine, 1984; Whitaker and Aitchison, 1985; Swann and Whitaker, 1986; Ciapa and Whitaker, 1986). Injection of inositol triphosphate into sea urchin eggs induces the calcium wave and activation (Whitaker and Irvine, 1984; Swann and Whitaker, 1986); however, inositol triphosphate injected eggs do not undergo mitosis or cleave, presumably since they lack centrosomes normally contributed by the sperm (Brandriff et al., 1975). Other contributions of the sperm to the egg at fertilization include its mitochondrion, plasma membrane and nucleus. Although the incorporation of the sperm nucleus into the egg and its association with the maternal genome are important in establishing diploidy and genetic variance of the embryo, it and other sperm components do not appear to be essential for initiation of cell proliferation, as parthenogenetic embryos can be produced from unfertilized eggs by inducing the calcium wave and the formation of centrosomes (Brandriff et al., 1975). What then is the crucial interaction between the sperm and egg at fertilization that leads to egg activation?

Sperm-Egg Interactions and Egg Activation

The following terms and their definitions have been employed here. **Gamete attachment**: A physical interaction of the gametes directly observable in living specimens, involving the adherence of the sperm to the egg. **Gamete binding**: Specific molecular event involving, for example, a receptor-ligand interaction between the sperm and egg. **Sperm-egg plasma membrane fusion**: A process whereby the continuity of the gamete plasma membranes is established. This would follow gamete attachment and/or binding and would result in the adjoining of the sperm and egg cytoplasms. **Gamete continuity**: Sperm and egg are electrically coupled and share diffusible components but their plasma membranes have not necessarily fused. In light of these definitions, the processes of attachment, binding, gamete continuity and sperm-egg plasma membrane fusion have been postulated as having a role in different phases of the fertilization potential or its counterpart, the activation current (Chambers and de Armendi, 1979; Shen, 1983; Schatten and Hulser, 1983; Lynn and Chambers, 1984; Lynn et al., 1988). At present it remains unclear which interactions are obligatory prerequisites for egg activation.

The lack of a precisely defined zero-time, indicating the beginning of gamete interactions, and the absence of techniques which would permit the identification of the fertilizing sperm on a single egg have hampered previous attempts to define the time of sperm-egg plasma membrane fusion, as well as other related, early events of fertilization. However, recent observations using Hoechst dye transfer of paraformaldehyde-fixed sea urchin eggs (Hinkley et al., 1986) and electron microscopic examination of electrophysiologically recorded eggs clamped at -20 mV and fixed with glutaraldehyde (Longo et al., 1986) have attempted to resolve when gamete continuity and sperm-egg plasma membrane fusion occur during early chronological events of sperm-egg interactions in the sea urchin, *Lytechinus variegatus*. Using eggs with only one attached sperm, the onset of abrupt inward current (I_{on}), recorded with a microelectrode inserted into the egg cytoplasm, served as a well-defined zero time. From 1 to 5 sec after I_{on} the tip of the sperm acrosomal process was found in contact with the vitelline layer, separated by a distance of 50 to 150Å from the egg plasma membrane (Longo et al., 1986). Continuity between the sperm and egg plasma membranes occurred approximately 5 sec following I_{on} and resulted in the adjoining of the egg and sperm cytoplasms. Similar results were also obtained by the transfer of Hoechst dye when eggs were fixed after this time (Hinkley et al., 1986). Dehiscence of the cortical granules and fertilization cone formation were initiated subsequent to gamete membrane fusion, approximately 6 to 8 and 10 sec after I_{on}, respectively (Longo et al., 1986).

In light of these observations and their possible significance to our understanding of sperm-egg interactions, the following aspects need to be considered with respect to interpretations concerning how egg activation is initiated. Sperm entry is not inhibited in eggs voltage clamped at -20 mV and eggs exhibit a conductance increase following approximately 90% of sperm attachments (Chambers, 1989; McCulloh, 1989). The development of such eggs occurs normally and synchronously with control specimens (Lynn and Chambers, 1986); only recently, however, has the timing of early events, e.g., sperm-egg fusion, in clamped vs. nonclamped eggs been tested and found to be similar in both instances (see below). That the membrane fusion process is preserved faithfully upon application of fixative is also critical to an interpretation of results obtained with voltage clamped eggs. The immediate cessation of sperm motility, and the abrupt, marked change of inward current within 1 sec after addition of fixative suggest that fixation occurs rapidly. Knowledge of the actions of the paraformaldehyde (Hinkley et al., 1986) and glutaraldehyde (Longo et al., 1986) fixatives employed in studies to determine the chronology of early fertilization events is important to an interpretation of the resultant observations and inferences as to what aspect of gamete interaction leads directly to the triggering of the activation potential/current (Phase 1) and egg activation (Phase 2).

As far as we are aware, studies have not been published examining how rapidly aldehyde fixatives stop membrane fusion when added to isolated cells. In addition, it is unclear how well intermediate stages of membrane fusion are preserved during fixation. Miniature endplate potentials recorded intracellularly at single neuromuscular junctions and resulting from exocytotic fusion of synaptic vesicle membranes with the plasma membrane continued to be recorded for up to 1 min following the addition of glutaraldehyde to the intact muscle (Hubbard and Laskowski, 1972; Clark, 1976; Heuser, 1976; Smith and Reese, 1980). This suggests that synaptic vesicle- plasma membrane fusion was not immediately halted by the addition of fixative. In these experiments, however, no information was available on how rapidly the external medium was replaced by fixative, nor was it possible to assess the role of diffusion barriers in the intact muscle tissue. In addition, conventional fixation techniques have serious short comings when applied to lipids because they may lead to membrane artifacts, such as lipid extraction and redistribution, particularly in fusing membrane systems (Peterson and Rubin, 1970; Scott, 1976; Hasty and Hay, 1978; Poste et al., 1978; Chandler and Heuser, 1979). Recently developed low temperature techniques, i.e., the combination of freeze substitution and low temperature embedding, potentially are capable of overcoming these limitations (Weibull et al., 1984; Verkleij et al., 1985; Knoll et al., 1988). However, such protocols are difficult, if not impossible, at the present time to apply to studies of early fertilization events as discussed here. As an alternative, we used solutions containing osmium to fix eggs at varying intervals after the onset of the electrically recorded activation current (see below).

In both the ultrastructural observations of electrically recorded eggs (Longo et al., 1986) and the dye transfer studies of Hinkley et al. (1986), it is presumed that the fusing sperm and egg plasma membranes are sufficiently stable during fixation and subsequent processing that their specific morphological relationship to one another is retained. To what degree this is true is unknown. Studies have shown that the membrane bilayer can remain in a labile state during fixation with glutaraldehyde (Chandler and Heuser, 1979; Eddy and Shapiro, 1976) because these structures are insufficiently stabilized against secondary changes during dehydration and polymerization (Ruthmann, 1970). This may, in part, be due to the fact that this fixative, in contrast to osmium tetroxide, fails to react with the double bonds of phospholipids which are an integral part of biological membranes. The possibility exists that during the period up to 5 sec, membrane changes associated with gamete fusion are transitional, unstable or tenuous and are not maintained or are reversed when fixative is added. An increase of capacitance, indicative of the establishment of gamete continuity, followed by a loss of capacitance (McCulloh and Chambers, 1986a and b) suggests that reversal of gamete continuity can occur in eggs clamped at negative membrane potentials even in the absence of fixatives. Presently, however, there is no evidence that sperm-egg fusion plasma membrane is reversed during the period 1 to 4 sec after I_{on} and in the absence of such data we contend that membrane fusion occurs approximately 5 sec after I_{on}. Therefore the observations of Longo et al. (1986) and Hinkley et al. (1986) suggest that the manner by which the sperm stimulates the egg to initiate I_{on} and events characteristic of Phase 1 (Chambers, 1989) involves a mechanism other than membrane fusion.

To correlate further the status of the sperm and egg plasma membranes with the electrical activity of the egg we have examined electrically recorded eggs fixed with 2% OsO_4 and 2% glutaraldehyde in sea water. Our rationale for employing osmium is that in addition to reacting with the double bonds of aliphatic and aromatic side groups in proteins, this agent readily reacts with the double bonds of unsaturated lipids and, therefore, stabilizes membrane components (Ruthmann, 1970). These analyses have been performed on nonclamped eggs and eggs clamped at different voltages that affect sperm entry and/or egg activation. Fixed specimens have been examined with high-voltage and conventional transmission electron microscopy.

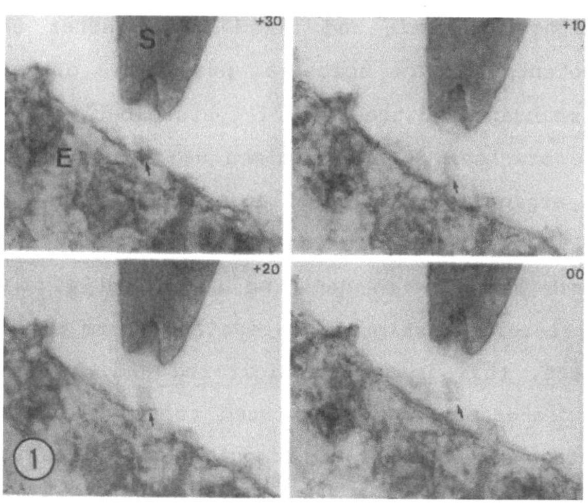

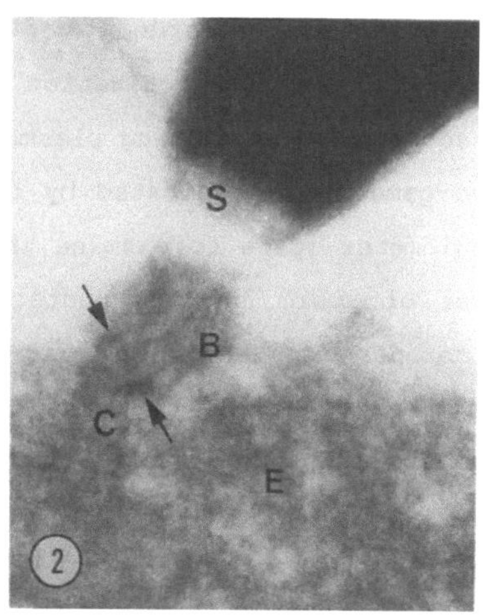

Figure 1. High voltage electron micrograph of a thick section (0.5 µm) depicting an interacting sperm (S) and egg (E) fixed with osmium teroxide, 4 sec after I_{on}, and rotated on a goniometer stage. The arrows point to the tip of the acrosomal process. The figures in the upper right of each micrograph indicate the degrees of rotation. Note the space between the acrosomal process and the egg surface when the specimen was rotated to 00.

Figure 2. Thick section showing a fused sperm and egg as viewed with high voltage electron microscopy. The narrow channel (C) connecting the sperm (S) and egg (E) is surrounded by bindin (B) and some dense aggregates (arrows) characteristic of newly fused gametes.

Results of high voltage electron microscopic examinations of eggs clamped at -20 mV (32 specimens examined) are consistent with previous observations utilizing conventional transmission electron microscopy (Longo et al., 1986). Gamete components were well preserved particularly in the area of sperm-egg interaction. In the specimen (fixed at 4 sec after I_{on}) depicted in figure 1, the acrosomal process appeared to be in contact with the egg surface. However, when the preparation was rotated on a goniometer stage, an intervening space could be seen separating the tip of the acrosomal process from the egg surface. In specimens that had fused, i.e., the plasma membranes and cytoplasms of the egg and sperm were confluent; the narrow channel connecting the two gametes was composed of the acrosomal process surrounded by bindin (Fig. 2). Midway along the exterior of the channel was an accumulation of dense material which was a consistent feature of fixed gametes observed with high voltage electron microscopy; its nature and function are unknown. In specimens prepared at later time intervals, i.e., greater than 10 sec after I_{on}, actin filaments were apparent in the developing fertilization cone.

Depicted in figure 3 is an egg inseminated while its membrane potential was recorded (nonvoltage clamped) and fixed 4 sec after the initiation of the activation potential. The acrosomal process is closely associated with the egg plasma membrane and the area of contact between the two gametes is indicated by the brackets. This specimen was rotated in a goniometer stage to examine the organization of the plasma membranes in the area of sperm and egg contact. Such a preparation, in which the plasma membranes are seen in transverse section, is depicted in figure 4. In order to determine the structural relationship of the egg and sperm plasma membranes in the area of contact, this image was digitized and analyzed with a Gould IP 8500 imaging processor. We speculated that the plasma membranes along the area of contact might share morphological relationships similar to that characterized for vesicle-plasma membrane fusion (Palade activation. They suggest that the egg is activated by the diffusion of an activator substance from the sperm.

Gamete plasma membrane fusion, as determined by electron microscopy, occurs during the shoulder (Phase 1) of the activation current (Longo et al., 1986; data presented here). The conductance increase responsible for generation of I_{on} and the shoulder of the activation current (Phase 1) is localized to the vicinity of the attached sperm (McCulloh and Chambers, 1986), and, as indicated by observations presented here, may result from (1) the opening of channels in the egg plasma membrane in response to a component released from or located on the surface of the sperm or (2) channels in the sperm plasma membrane. In regards to possibility (1) immunochemical studies have shown that bindin is present on the surface of the acrosomal process, which defines the point of gamete contact. Although sea urchin eggs immersed in a suspension of bindin show no morphological evidence of having undergone an activation response (Vacquier and Moy, 1977), a purified jet of purified bindin (Vacquier, 1983) directed onto the surface of unfertilized eggs induces increases in conductance and inward current (Lynn, Chambers and Vacquier, unpublished observations). An acrosomal protein from Urechis sperm, which is comparable to bindin of sea urchin, activates and initiates the development of Urechis eggs (Gould et al., 1986).

An alternative explanation is that the initiation of the conductance increase may come about by the inclusion of channels from the sperm plasma membrane into the egg plasma membrane. This would require an electrical connection between the cytoplasms of the sperm and egg either due to fusion of the gamete membranes or to the presence of channels in the membranes at the site of sperm-egg contact. The increase in capacitance demonstrated in patch-clamp studies on sea urchin eggs occurs simultaneously with the conductance increase at the initiation of Phase 1 (McCulloh and Chambers, 1986a and b). The capacitance increase indicates the occurrence of electrical continuity between the gametes. Whether or not these changes reflect the fusion of the sperm and egg plasma membranes has not been verified and is doubtful based on data presented here. Observations by Schackman et al., (1984) indicate that the conductance of the sperm plasma membrane is high following the acrosome reaction. When the sperm membrane is inserted in parallel with the egg plasma membrane of low conductance (Chambers and de Armendi, 1979), a high conductance membrane would be formed. If electrical continuity is established prior to sperm-egg plasma membrane fusion, such a change might involve a transitional structural digitized model generated a trimodal profile. Model 3 depicts a situation in which the membranes are in an intermediate state of fusion, i.e., the outer leaflets have coalesced and the inner leaflets are in apposition to one another. A densitometric scan through this fusion intermediate showed a bimodal profile. A fourth model which is not presented in figure 5 would be one in which the plasma membranes had fused and established cytoplasmic continuity with one another.

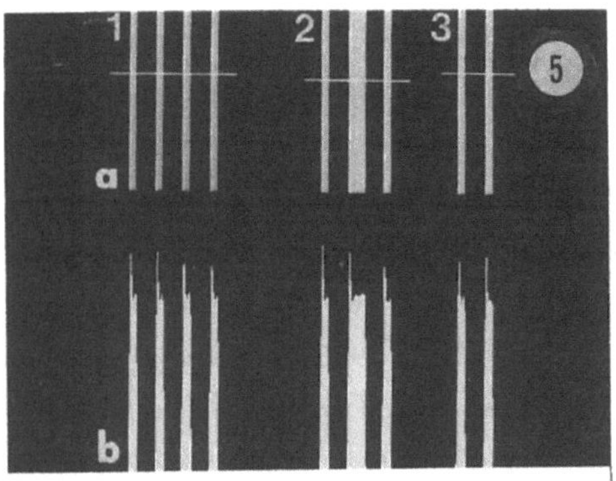

Figure 5. Models (a) and their densitometric scans (b) based on the association and fusion of unit membranes as proposed by Palade and Bruns (1968). 1, Sperm and egg plasma membranes in apposition to one another with an intervening space. The individual lines represent inner and outer leaflets of the plasma membranes. 2, Sperm and egg plasma membranes in contact with one another. The thick line represents the adjoined outer leaflets of the sperm and egg plasma membranes. 3, Fusion intermediate depicting the inner leaflets of the sperm and egg plasma membranes following the fusion of the outer leaflets. See text for details.

Figure 6. Profiles resulting from densitometric scans of the digitized image shown in figure 4. Scans through the sperm and egg plasma membranes and the region of contact between the acrosomal process and the egg surface are indicated by the lines marked s, e and c, respectively. See text for details.

Densitometric scans through the digitized, negative representation of the specimen depicted in figure 4, specifically, areas comprising the egg and sperm plasma membranes in regions where no contact between the gametes was observed, generated bimodal profiles (Fig. 6). Bimodal profiles were also generated in the area where the tip of the acrosomal process was in contact with the egg surface. The presence of a bimodal profile along the region of contact between the acrosomal process and the egg plasma membrane agrees with previous findings depicting an intermediate stage of unit membrane fusion (Palade and Bruns 1968) and suggests that for this particular specimen, the fusion process had not been completed. How these observations relate to our present concepts of what is believed to occur during the fusion of phospholipid bilayers is difficult to predict at the present time for several reasons. (1) A one-to-one relationship of what is seen in the electron microscope as the unit membrane and the conformation of phospholipid bilayers has not been established, i.e., what part of the unit membrane structure is equivalent to a specific component of the phospholipid bilayer is unclear. (2) Contemporary models of membrane fusion, many based on results of in vitro lipid systems involve some type of micelle intermediate which has not, to our knowledge, been reported for fusing membranes viewed with the electron microscope (Wilschut, 1989). Based on these considerations, the present observations emphasize the need for further experiments and a clarification of our understanding of membrane fusion as a general phenomenon.

Relationship of the Egg and Sperm Plasma Membranes at the Site and Time of Egg Activation

What then might be the nature of the association of the sperm and egg plasma membranes that leads to egg activation? Since sperm-egg fusion occurs 5 sec after the initiation of the activation potential or current this change in electrical activity of the egg may be a result of a non-fusion event, e.g., a receptor-ligand type of interaction. One of the difficulties with a hypothesis involving a plasma membrane receptor is that although many sperm can bind to the egg surface, the wave of calcium release is only initiated at the point of interaction with the successful or fertilizing sperm. This implies that simple attachment of the sperm to the egg does not necessarily lead to receptor occupancy (binding) and activation (Epel, 1989). Investigations by Kline et al. (1988) and Jaffe (1989) suggest that in amphibian and sea urchin eggs the fertilizing sperm interacts with a receptor on the egg surface to regulate the production of inositol triphosphate and diacylglycerol, which in turn are mediated by G protein probably coupled to phospholipase C (Turner and Jaffe, 1989; Turner et al., 1986, 1987; Epel, 1989). The net result is a cascade of events (receptor occupancy, G-protein and phospholipase C activation, and production of Ins 1,4,5 P_3 and diacylglycerol leading to egg activation). This suggestion has been challenged by Whitaker et al. (1989) who, analyzing the effects of injected GTP analogs into sea urchin eggs, concludes that G-proteins are not involved in the initiation of egg

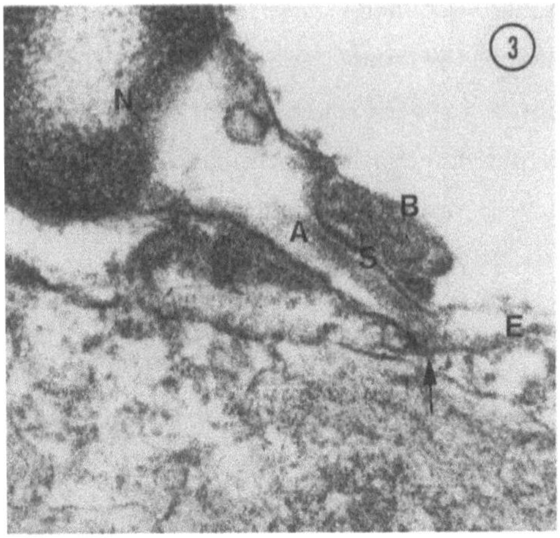

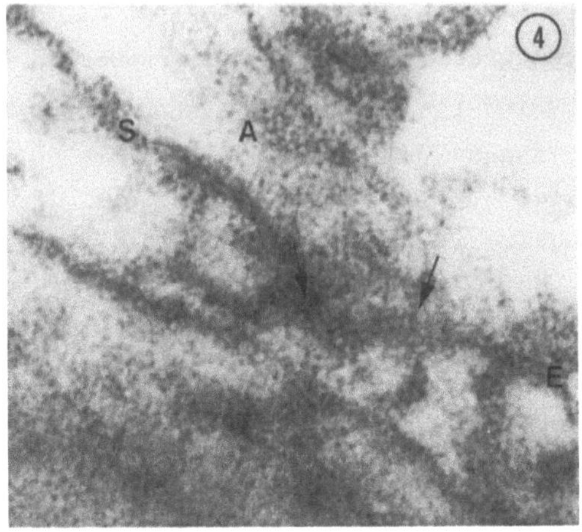

Figure 3. The area of sperm-egg interaction in a specimen from which the membrane potential was recorded (nonclamped). The egg was fixed with osmium tetroxide 4 sec after the initiation of the activation potential and viewed with conventional transmission electron microscopy. N, sperm nucleus; A, acrosomal process; B, bindin; S and E, sperm and egg plasma membranes, respectively. The arrow indicates the point of contact between the acrosomal process and the egg surface.

Figure 4. Same preparation shown in figure 3 at higher magnification and rotated in a gonometer stage to clarify images of the sperm and egg plasma membranes. A, acrosomal process; S and E, sperm and egg plasma membranes, respectively. The arrows indicate the area of contact between the acrosomal process and the egg surface.

and Bruns, 1968). Models depicting the sperm and egg plasmalemma as unit membrane structures in various stages of interaction and fusion as suggested by the investigations of Palade and Bruns (1968) were drawn, digitized as negative representations of electron density and densitometrically scanned (Fig. 5). Profiles depicting the inner (cytoplasmic) and outer (extracellular) leaflets of the different models were then ascertained and compared to scans of digitized negative representations of the image shown in figure 4 (see Figure 6). In figure 5, model 1 depicts the plasma membranes closely apposed to one another, separated by a narrow extracellular cleft. A densitometric scan of such a digitized model generated a quadramodal profile where each peak represented a plasma membrane leaflet. Model 2 represents two plasma membranes in contact with one another with no intervening space. Scans of such a alteration in the apposed sperm and egg plasma membranes at the site of contact. The putative channels in such a membrane might not be resolved by electron microscopy.

Cross-species fertilization experiments have shown that the voltage required to block fertilization is characteristic of the sperm species (Jaffe et al., 1982, 1983; Iwao and Jaffe, 1989), indicating that the voltage dependence of fertilization is contributed by the sperm membrane, i.e., the sperm may contain a positively charged protein that is inserted into the egg plasma membrane to initiate the activation and fusion. It has been suggested that the insertion of a charged protein into the lipid bilayer could be regulated by the electrical field across the membrane (Iwao and Jaffe, 1989). In Lytechinus protein insertion would require the equivalent of 5 to 10 electron charges to traverse the electric field of the egg plasma membrane (McCulloh, 1989).

Egg activation by the fertilizing sperm has been interpreted as a consequence of the injection of an activator substance from the sperm into the egg following egg gamete plasma membrane fusion (Dale et al., 1985; Whitaker et al., 1989). The present observations indicate that the onset of electrical activity of the inseminated egg (the initiation of the activation potential or activation current; i.e., events of Phase I) occurs prior to the completion of sperm-egg plasma membrane fusion. Sperm-egg plasma membrane fusion, however, precedes the onset of Phase 2 events, which occur in association with egg activation. At present there is no conclusive evidence supporting the notion that initiation of Phase 1 of the activation potential/activation current occurs simultaneously with sperm-egg plasma membrane fusion. Moreover, the activator substance presumable derived from the sperm has not been identified.

Conclusions

The data presented here demonstrate that in the sea urchin Lytechinus variegatus fusion of the egg and sperm plasma membranes follows the onset of electrical activity of the egg (i.e., the initiation of the activation potential/current) by approximately 5 sec. However, sperm-egg plasma membrane fusion is completed prior to the initiation of activation at the end of the latent period. Data are reviewed consistent with the notion that the onset of electrical activity of the egg, i.e., the initiation of the activation potential or its counterpart, the activation current, occurs prior to completion of gamete plasma membrane fusion and may be due to the presence of channels in a transitional membrane structure. This transitional state is followed by sperm-egg plasma membrane fusion, resulting in the continuity of the gamete plasmalemmas. Since fusion of the gamete plasma membranes occurs well before the end of Phase 1 (equal in duration to the latent period), prior to the initiation of the cortical granule reaction and elevation of the fertilization membrane, this suggests that events accompanying and/or following gamete membrane fusion are involved in activation (Chambers, 1989; Whitaker et al., 1989). It is anticipated that the results of further correlative ultrastructural studies, employing electrically recorded eggs, clamped at different voltages (Chambers, 1989; McCulloh, 1989) will further resolve the chronology of sperm-egg interactions leading to egg activation. Changes in the sperm and egg plasma membranes at the site of gamete interaction are believed to involve fundamental alterations in phospholipid structure and function that may be common to other cells. Investigations of such phospholipid changes will, no doubt, have a bearing on possible mechanisms attending interactions of the sperm and egg which lead to the initiation of embryonic development.

ACKNOWLEDGEMENTS

This article is dedicated to the memory of a friend and outstanding researcher of fertilization processes, Dr. Alberto Monroy. The authors' investigations reported here have been supported by grants NSF PCM-83-16864 (E.L.C.) and DCB-87-11787 (E.L.C.), NIH HD-191267 (E.L.C.) and NIH HD-22085 (F.J.L.).

REFERENCES

Berridge MJ, Irvine RF (1984) Inositol triphosphate, a novel second messenger in signal transduction, Nature 312:315-318

Brandriff B, Hinegardner RT, Steinhardt R (1975) Development and life cycle of the parthenogenetically activated sea urchin embryo. J Exp Zool 192:13-24

Chambers EL (1989) Fertilization in voltage-clamped sea urchin eggs. In: Mechanisms of Egg Activation. Nuccitelli, Cherr, GN, Clark WH, Jr., (eds). Plenum Press, New York, pp 1-18

Chambers EL, DeArmendi J (1979) Membrane potential, action potential and activation potential of eggs of the sea urchin, Lytechinus variegatus. Exp. Cell Res., 122:203-218

Chandler DE, Heuser J (1979) Membrane fusion during secretion. Cortical granule exocytosis in sea urchin eggs as studied by quick-freezing and freeze-fracture. J. Cell Biol., 69:521-538

Ciapa B, Whitaker MJ (1986) Two phases of inositol polyphosphate and diacylglycerol production of fertilization. FEBS Lett., 195:137-140

Clark AW (1976) Changes in the structure of neuromuscular junctions caused by variations in osmotic pressure. J. Cell Biol., 69:521-538

Dale B, DeFelice LJ, Ehrenstein G (1985) Injection of a soluble sperm fraction into sea urchin eggs triggers the cortical reaction. Experientia, 41:1068-1070

Eddy EM, Shapiro BM (1976) Changes in the topography of the sea urchin egg after fertilization. J. Cell Biol., 71:35-48

Eisen A, Kiehart DP, Wieland SJ, Reynolds GT (1984) Temporal sequence and spatial distribution of early events of fertilization in single sea urchin eggs. J. Cell Biol., 99:1647-1654

Epel D (1989) Arousal of activity in sea urchin eggs at fertilization. In: The Cell Biology of Fertilization. H. Schatten and G. Schatten, ed. Academic Press, New York. pp. 361-385

Gould M, Stephans JL, Holland LZ (1986) Isolation of protein from Urechis sperm acrosomal granules that binds sperm to eggs and initiates development. Dev. Biol. 117:306-318

Hasty DL, Hay ED (1978) Freeze-fracture of the developing cell surface. J. Cell Biol., 78:756-768

Heuser J (1976) Morphology of synaptic vesicle discharge and reformation at the frog neuromuscular junction. In: Motor Innervation of Muscle. S. Thesleff, ed. Academic Press, New York, pp. 51-115

Hinkley RE, Wright BD, Lynn JW (1986) Rapid visual detection of sperm-egg fusion using the DNA-specific fluorochrome Hoechst 33342. Dev. Biol., 118:148-154

Hubbard JI, Laskowski MB (1972) Spontaneous transmitter release and Ach sensitivity during glutaraldehyde fixation of rat diaphragm. Life Sci., 11:781-785

Iwao Y, Jaffe LA (1989) Evidence that the voltage-dependent component in the fertilization process is contributed by the sperm. Dev. Biol., 134:446-451

Jaffe LA (1976) Fast block to polyspermy in sea urchin eggs is electrically mediated. Nature, 261:68-71

Jaffe LA (1983) Sources of calcium in egg activation: A review and hypothesis. Dev. Biol., 99:256-276

Jaffe LA (1989) Receptors, G-proteins and egg activation. In: Mechanisms of Egg Activation. R. Nuccitelli, G.N. Cherr, W.H. Clark, Jr., eds. Plenum Press, New York. pp. 151-155

Jaffe LA, Gould-Somero M, Holland LZ (1982) Studies of the mechanism of the electrical polyspermy block using voltage clamp during cross-species fertilization. J. Cell Biol., 92:616-621

Kline DE, Simoncini L, Mandel G, Maue RA, Kado RT, Jaffe LA (1988) Fertilization events induced by neurotransmitters after injection of mRNA in Xenopus eggs. Science, 241:464-467

Knoll G, Burger KNJ, Bron R, van Meer, G, Verkley A (1988) Fusion of liposomes with the plasma membrane of epithelial cells: Fate of incorporated lipids as followed by freeze fracture and autoradiography of plastic sections. J. Cell Biol., 107:2511-2521

Longo FJ, Lynn JW, McCulloh DH, Chambers EL (1986) Correlative ultrastructural and electrophysiological studies of sperm-egg interactions of the sea urchin, Lytechinus variegatus. Dev. Biol., 118:155-166

Lynn JW (1989) Correlations between time-dependent and cytochalasin B affected sperm entry in voltage-clamped sea urchin eggs. In: Mechanisms of Egg Activation. R. Nuccitelli, G.N. Cherr and W.H. Clark, Jr., eds. Plenum Press, New York. pp. 43-60

Lynn JW, Chambers EL (1984) Voltage clamp studies of fertilization in sea urchin eggs. I. Effect of clamped membrane potential on sperm entry, activation and development. Dev. Biol., 102:98-109

Lynn JW, McCulloh DH, Chambers EL (1988) Voltage clamp studies of fertilization in sea urchin eggs. II. Current patterns in relation to sperm entry, nonentry and activation. Dev. Biol., 128:305-323

McCulloh DH (1989) Sperm entry in sea urchin eggs: Recent inferences concerning its mechanisms. In: Mechanisms of Egg Activaiton. R Nuccitelli, GN Cherr and WH Clark, Jr., eds. Plenum Press, New York, pp. 19-42

McCulloh DH, Chambers EL (1985) Localization and propagation of membrane conductance changes during fertilization in eggs of the sea urchin, Lytechinus variegatus. J. Cell Biol., 101:230a

McCulloh DH, Chambers EL (1986a) When does the sperm fuse with the egg? J. Gen. Physiol., 88:38-39a

McCulloh DH, Chambers EL (1986b) Fusion and "unfusion" of sperm and egg are voltage dependent. J. Cell Biol., 103:236a

Palade GE, RR Bruns (1968) Structural modulations of plasmalemmal vesicles. J. Cell Biol., 37:633-649

Petersen JA, Rubin H (1970) The exchange of phospholipids between cultured chick fibroblasts as observed by autoradiography. Exp. Cell Res., 60:383-392

Poste G, Porter CW, Paphadjopoulos D (1978) Identification of a potential artifact in the use of electron microscopic autoradiography to localize saturated phospholipids in cells. Biochim. Biophys. Acta, 510:256-263

Ruthmann, A (1970) Methods in Cell Research. Cornell Univ. Press, Ithaca, N.Y.

Schackman RW, Christen R, Shapiro BM (1984) Measurement of plasma membrane and mitochondrial potentials in sea urchin sperm. J. Biol. Chem., 259:13914-13922

Schatten G, Hulser D (1983) Timing the early events during sea urchin fertilization. Dev. Biol., 100:244-248

Schmidt T, Patton C, Epel D (1982) Is there a role for the Ca^{2+} influx during fertilization of the sea urchin egg? Dev. Biol. 90:284-290

Scott RE (1976) Plasma membrane vesiculation: A new technique for isolation of plasma membranes. Science, 194:743-745

Shen SS (1983) Membrane properties and intracellular ion activities of marine invertebrate eggs and their changes during activation. In: Mechanism and Control of Animal Fertilization. J.F. Hartmann, ed. Academic Press, New York. pp. 213-267

Smith JE, Reese TS (1980) Use of aldehyde fixatives to determine the rate of synaptic transmitter release. J. Exp. Biol., 89:19-29

Steinhardt RA, Epel D (1974) Activation of sea urchin eggs by calcium ionophore. Proc. Natl. Acad. Sci. USA, 71:1915-1919

Steinhardt RA, Lundin L, Mazia D (1971) Bioelectric responses of the echinoderm egg to fertilization. Proc. Natl. Acad. Sci. USA, 68:2426-2430

Steinhardt RA, Zucker RS, Schatten G (1977) Intracellular calcium release at fertilization in the sea urchin egg. Dev. Biol., 58:185-196

Swann K, Whitaker MJ (1986) The part played by inositol triphosphate and calcium in the propogation of the fertilization wave in sea urchin eggs. J. Cell Biol., 103:2333-2342

Turner PR, Jaffe LA (1989) G-proteins and the regulation of oocyte maturation and fertilization. In: The Cell Biology of Fertilization. H Schatten and G Schatten, eds. Academic Press, New York. pp. 279-318

Turner PR Jaffe LA, Fein A (1986) Regulation of cortical granule exocytosis by inositol 1,4,5-triphosphate and GTP binding protein. J. Cell Biol., 102:70-76

Vacquier VD (1983) Purification of sea urchin sperm bindin by DEAE-cellulose chromatography. Anal. Biochem., 129:497-501

Vacquire VD, GW Moy (1977) Isolation of bindin: The protein responsible for adhesion of sperm to sea urchin eggs. Proc. Natl. Acad. Sci. USA, 74:2456-2460

Verkleij AJ, Humbel B, Studer D, Müller M (1985) 'Lipidic particle' systems as visualized by thin-section electron microscopy. Biochim. Biophys. Acta, 812:591-594

Weibull C, Villiger W, Carlemalm E (1984) Extraction of lipids during freeze-substitution of Acholeplasma laidlawii-cells for electron microscopy. J. Microsc. 135:213-216

Whitaker MJ, Steinhardt RA (1982) Ionic regulation of egg activation. Q. Rev. Biophys., 15:593-666

Whitaker M, Irvine RF (1984) Inositol 1,4,5-triphosphate microinjection activates sea urchin eggs. Nature, 312:636-639

Whitaker MJ, Aitchison J (1985) Calcium-dependent phosphoinositide hydrolysis is associated with exocytosis in vitro. FEBS Lett., 182:119-124

Whitaker M, Swann K, Crossley I (1989) What happens during the latent period at fertilization. In: Mechanisms of Egg Activation. R. Nuccitelli, G.N. Cherr, and W.H. Clark, Jr., eds. Plenum Press, New York. pp. 157-171

Wilsahut J (1989) Intracellular memebrane fusion. Cur. Opin. Cell Biol., 1:639-647

Yoshimoto Y, Iwamatsu T, Hirano K, Hiramoto Y (1987) The wave pattern of free calcium released upon fertilization in medaka and sand dollar eggs. Dev. Growth Differ., 28:583-596

SELF-INCOMPATIBILITY IN *BRASSICA*: THE NATURE AND ROLE OF FEMALE GLYCOPROTEINS

H.G. Dickinson and M.J.C. Crabbe*

Department of Botany, School of Plant Sciences, University of Reading, Whiteknights, Reading, RG6 2AS

*Department of Microbiology, School of Animal and Microbial Sciences, University of Reading, London Road, Reading, RG1 5AQ

NATO ASI Series, Vol. H 45
Mechanism of Fertilization
Edited by B. Dale

Summary

The roles played by stigmatic glycoproteins in the self-incompatibility (SI) mechanism in *Brassica oleracea* have been investigated. In a physiological study, the suppression of protein synthesis was demonstrated to overcome SI, but also affect a number of other pollination-related events, such as regulated hydration of the pollen grain. Interestingly, inhibition of glycosylation appears to affect SI alone, leaving other processes to proceed normally. When stigmatic glycoproteins are used in bioassays, it is clear that they can regulate pollen tube growth in accordance with the S (incompatibility) genes present. However, the multiplicity of glycoproteins possessed by each genotype, and their presence in quantity indicates that they are unlikely to be simple intercellular signals. Database searches involving concensus sequences from the family of stigmatic glycoproteins have revealed strong homologies with domains in several classes of animal proteins, principally the von Willebrand factor and type VI collagen. These polypeptides, which are also heavily glycosylated, are held to be involved in protein-protein interactions and extracellular matrix formation - processes which may also be involved in the initial stages of pollination. Detailed modelling of these sequences indicate that, despite small differences in amino acid composition, these domains are strikingly similar in their three dimensional molecular architecture.

Introduction

A number of complex interactions take place on, or within, the pistil of flowering plants which determine the success of fertilisation. In general these interactions fall into two classes; the first are interspecific, and ensure that only pollen of the same, or very closely related species produces a tube which grows down to the ovary. The second class of interactions are the so-called self-incompatibility (SI) mechanisms, which prevent fertilisation by self pollen and thus encourage outbreeding. To consider interspecific barriers to fertilisation as "mechanisms" is probably philosophically incorrect, in that in the vast majority of cases speciation occurs by plant populations 'growing apart' through a spectrum of small physiological and morphological differences. At some point in this continuum of divergence, intercrossing becomes no longer possible. However, evidence from interspecific crosses within the Brassicas suggests that some matings between particular species are prevented by reactions which closely resemble those normally associated with SI (Lewis and Crowe, 1958).

SI itself is widespread amongst the flowering plants, and is held to have been responsible for their rapid advancement (Whitehouse, 1950). Genetical research has established that there are two ways by which the SI can be regulated, gametophytically, where the genes of the pollen grain control its compatibility with respect to the stigma (East and Mangelsdorf, 1925), and sporophytically, where pollen compatibility is determined by the genes of the mother plant (Bateman, 1952). Much of the research into gametophytically controlled SI has been carried out in *Nicotiana* (Anderson *et al.*, 1986), where self-pollen tubes are typically inhibited within the style, whilst sporophytically controlled SI has been principally studied in *Brassica* (Nasrallah *et al.*, 1988) where pollen inhibition is on the stigma surface.

In recent attempts to establish a molecular basis for this highly efficient form of intercellular communication, glycoproteins have been found to be associated with the expression of SI in female tissues, situated either in the style (gametophytic), or stigma (sporophytic). Interestingly, these glycoproteins, which are heavily glycosylated and relatively constant in structure, are present in comparatively large quantities which is unusual for signalling molecules. It has been hypothesised that these glycoproteins may perform at least one other, more basic, function in pollination and have only recently assumed their role in SI (Dickinson, 1990). Such a view is supported by the recent findings in both *Brassica* (Nasrallah *et al.*, 1988) and *Nicotiana* which indicate that glycoproteins active in SI belong to a larger family of similar proteins expressed in the pistil.

Despite the attraction of this hypothesis we have no firm indication as to what this alternative role in pollination might be. Its discovery is, however, important since its modification may form a part of the SI mechanism itself. Further, if this function includes interaction with the pollen, the component of the grain involved may well also serve as the male determinant of SI.

The nature and synthesis of stigmatic glycoproteins

The stigmatic papilla cells of *Brassica* synthesise a variety of polypeptides, many of them glycosylated. One family of glycoproteins has, however, been the focus of particular attention in that a considerable body of evidence suggests that its members are associated with SI (Nasrallah *et al.*, 1988). First, plants possessing similar S-alleles are characterised by at least one member of this glycoprotein family. Second, members of the family are missing from plants exhibiting self-compatibility. Third, these glycoproteins are first apparent within the stigmatic papillae at the time when the cells become capable of discriminating between self and cross pollen. Finally, SI can be overcome by protein synthesis inhibitors, which can be shown to suppress the synthesis of these glycoproteins. The involvement of female glycoproteins in SI is not restricted only to *Brassica*, but has also been demonstrated in *Nicotiana* (Jahnen *et al.*, 1989), *Solanum* (Kirch, 1989), *Papaver* (Franklin-Tong *et al.*, 1989) and *Antirrhinum* (A. McCubbin and H.G. Dickinson, in preparation).

Despite this wealth of circumstantial evidence, the fact that each plant can contain a number of S-associated glycoproteins means that we remain ignorant as to which particular polypeptide mediates SI. In a series of elegant experiments spanning the last decade Nasrallah and his coworkers (1989) have cloned genes encoding a number of these glycoproteins, and have classified them as S-locus linked glycoproteins (SLG) - polypeptides held to be associated with the expression of SI, and S-locus 'related' glycoproteins (SLR) - proteins believed not to be involved in SI. Trick *et al.* (1990), also studying putative S-genes in *Brassica*, have demonstrated several domains of the gene to be heavily conserved within the family. Sequences from putative SLGs and SLRs are now available from a range of sources and comparison reveals a number of clearly defined domains within these genes, which encode a protein of some 400 amino acid residues in length (Dickinson, 1989) (see Figure 1). Firstly, the amino terminus of the molecule is characterised by a signal peptide followed by a heavily glycosylated region of the molecule which varies between particular SLGs, and some SLRs. This region extends approximately to residue number 250, after which there is heavily conserved sequence encoding an unusually large number of cysteine residues.

<u>The part played by stigmatic glycoproteins in SI</u>

Despite the very persuasive circumstantial evidence, there are as yet no direct data linking the glycoproteins with the operation of SI. Perhaps the most conclusive results are from *in vitro* bioassays, and evidence from our, and other (Singh and Paolillo, 1989) laboratories suggests that this approach can produce valid data, although earlier attempts gave highly equivocal results (Ferrari and Wallace, 1975). Recent systems based on the growth kinetics of pollen tubes have produced consistent, and dramatic results. Our results indicate that carboxymethyl cellulose-separated high pI female glycoproteins can regulate pollen tube extension in accordance with the compatibility of the "pollination". We have yet to identify which particular glycoprotein is involved. Interestingly, while developing the media for the bioassay, the stigmatic glycoproteins emerged as exerting a

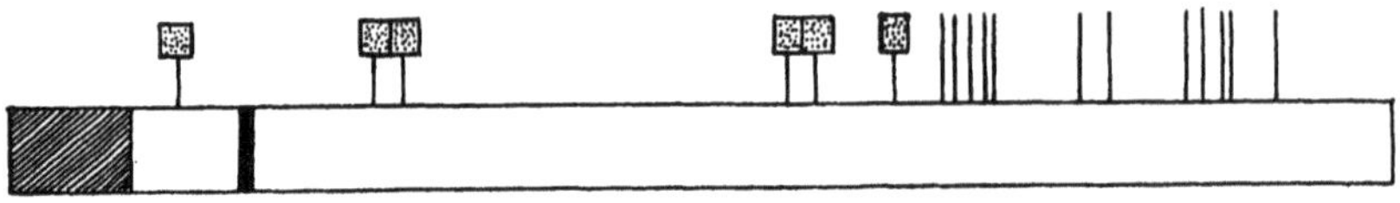

Figure 1. Schematic representation of 'generalised' SLG, derived from Dickinson 91989). The hatched region at the NH_2 terminus indicates the signal sequence, the vertical black bar a deletion common to all SLGs and the upper hatched boxes potential glycosylation points. The vertical lines at the carboxy terminus represent conserved cysteine residues.

highly promotional effect on pollen tube growth - irrespective of the S-gene being carried by the pollen (Elleman *et al.*, 1989). Singh and Paolillo have also reported similar observations (1989).

In a parallel series of studies, the physiology of the stigmatic papilla cells have been investigated. Pulse labelling experiments have shown that newly-incorporated amino acid is first transported into the cell wall, and from thence back into the cytoplasm and vacuole (Roberts *et al.*, 1984). This surprising result supported previous work on pollen adhesion (Stead *et al.*, 1980) which suggested that inhibition of stigmatic protein synthesis somehow affects the adhesion of pollen grains to the papillar surface. This work also demonstrated adhesion to be decreased by the enzymatic removal of protein from the papillar surface. Interestingly, following a period of some 120 mins after enzymatic treatment, 'normal' levels of adhesion are restored. This dynamic character of the papilla cell and its surface was also emphasised by Sarker *et al.* (1988) who showed prevention of protein synthesis to overcome both SI and the regulated hydration of the pollen grain. Although treatment with inhibitors permits self pollen tubes to grow and penetrate the stigmatic papillae, development ceases shortly thereafter for the inhibitor also prevents the production of the proteins required for continued pollen tube growth. More evidence to support a linkage between SI and pollen hydration has come from studies of immature stigmas, which are incapable of discriminating between self and cross pollen. Indeed, 'bud pollination' is a method commonly used by plant breeders to produce self seed. In addition to being incompetent with regard to SI, these immature papilla cells contain low levels of glycoprotein, and hydrate mature pollen at a rate equivalent to that of mature papillae treated with protein synthesis inhibitors. It would thus seem that these cells are missing a component (or components) capable of regulating both SI and controlled hydration. The possibility thus arises that hydration and SI could be mediated by the same determinant, perhaps a glycoprotein, and to test this hypothesis attempts were made to devise treatments capable of dissociating one effect from the other. The most successful of these involved the inhibitor tunicamycin, which prevents glycosylation of polypeptide chains. Mature stigmas treated with this inhibitor hydrated pollen grains at a normal rate, but were completely incapable of rejecting self pollen (Sarker *et al.*, 1988). While this experiment indicates that glycosylation of polypeptides is required for SI to function, it may not be concluded that S-allele specificity resides in the organisation of sugar residues of the glycosyl groups decorating these polypeptides; the results suggest only that glycosylation is required for one or more parts of the response.

Although the early studies point to newly-synthesised protein being deposited first in the wall and then in the cytoplasm, until recently there has been no evidence to suggest that the SLG/SLR follow such a pathway. However, immunological studies by the Cornell group (M.E. Nasrallah, personal communication) confirm that SLG, synthesised in the ER of the papilla cytoplasm is transported to the cell wall. Present studies are directed towards investigating its residence time in the wall, and fate on return to the cytoplasm.

Functions of stigmatic glycoproteins during pollination, other than in SI

The presence of relatively large quantities of the SLG/SLR group of glycoproteins in the stigma, combined with the presence of SLRs in self-compatible lines suggests that whatever their role in SI, these polypeptides may perform other, more general functions in pollination. Such a view is supported by their ability to promote pollen development *in vitro* (Elleman *et al.*, 1989), and the possibility that a single glycoprotein regulates both SI and controlled hydration (Sarker *et al.*, 1988).

Using amino acid sequences currently available for SLGs and SLRs, a search of databases was made to discover any similarities between the

Brassica glycoproteins and other polypeptides of which the structure and function is better known. Once significant domains had been identified within these molecules, it was proposed to study the 3D conformation of molecule in these regions using a powerful IBM-PC-computer-based modelling program (Crabbe and Appleyard, 1989) with the aim of demonstrating precise similarities between structures. Further, it was anticipated that these structures could also be used to predict some of the properties of the male molecule with which they interact during pollination. The results of this study will be published in detail elsewhere (M.J.C. Crabbe and H.G. Dickinson, in preparation) but the principal findings merit mention here. The database search revealed considerable homology between a section of *Brassica* SLG and a group of animal glycoproteins including the precursor to von Willebrand factor (Verweij *et al.*, 1986) and collagen VI (Chu *et al.*, 1989) (see Table 1). Not only is there some similarity in sequence and glycosylation sites, but there are close homologies between the cysteine-rich domain which characterise all these polypeptides. In this respect, the similarity between the *Brassica* glycoproteins and collagen VI is particularly striking. This is a significant finding, as both collagen VI and the von Willebrand factor are involved in protein-protein binding, and perhaps in forming extracellular matrices - both of which could form an important part of the pollen stigma interaction.

Although sequence homologies are interesting, relatively few small differences in amino acid residues are required to generate very different 3D structures. Using the molecular modelling and energy minimisation program, the cysteine-rich domains of the von Willebrand factor precursor, collagen VI and the *Brassica* glycoproteins were studied with respect to cysteine S-S distances (see Figure 2) and for smaller domains within the region, 3D architecture (see Figure 3). Interestingly, the differences in amino acids between the polypeptides conferred very few differences on the

Table 1. Homologies between the cystein-rich region of S-allele glycoproteins (given by the consensus sequence starting at residue 382) and two sections of Von Willebrand factor (VW1 starting at residue 700; VW2 starting at residue 354), and type VI collagen (COLL-VI, starting at residue 786). The secondary structure predictions (h, helix, t or T, turn, where T is a higher probability than t) are derived from Chou-Fasman and Garnier-Osguthorpe-Robson algorithms. The beta-turn (beta-hairpin) sequences are from Immunoglobulin Fab NEW light chain (starting at residue 89) and protease A from *S. griseus* (starting at residue 31). The sources of these sequences and their involvement in beta-turns are set out in Sibanda *et al.* (1989).

```
CON.SEQ.(382): IVDRSIGLK--ECEKRCLSDCNCTAFANADIRNGGTGC
(PRED)       :    tt       hhhhhhh     hhhhhhhh  TT
VW1 (700)    : LTCRSLSYPDEECNEACLEGCFC
(PRED)       :    tt       hhhhhhh
VW2 (354)    :             CQERCVDGCSC
(PRED)       :              hhhhhh
COLL-VI (786):            CREKKCPDYTCPITFANPADI
BETA-TURNS:
FAB L.CH.(89): SYDRSLRV
PROT. A (31) :                               ITTGGSRC
```

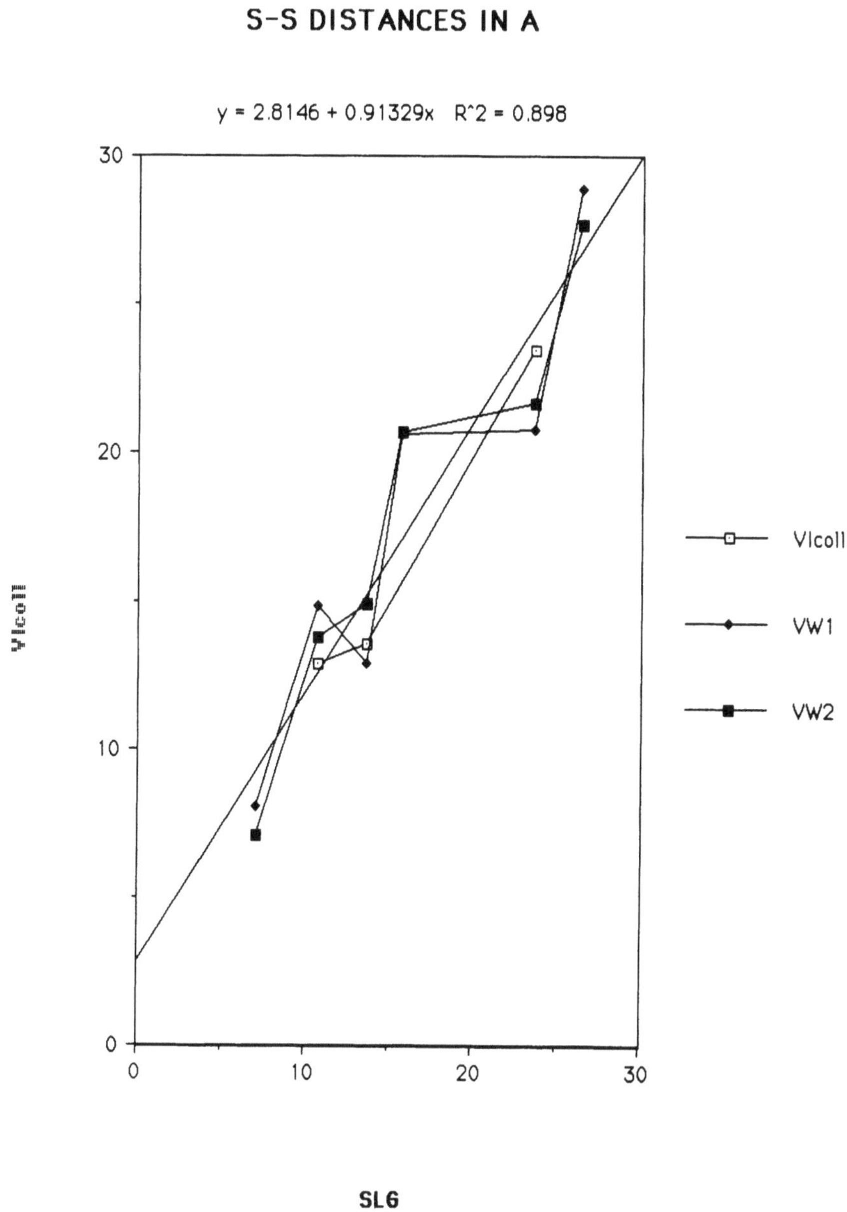

Figure 2. Predicted S-S distances in Angstroms (A) after modelling and energy minimisation for the SLG cysteine-rich consensus sequence shown in Table 1, compared with similar sequences in Von Willebrand factor (VW1 and VW2) and type VI collagen (VIcoll). The linear regression equation (illustrated graphically by the continuous straight line) is shown above the graph, with a correlation coefficient of 0.898.

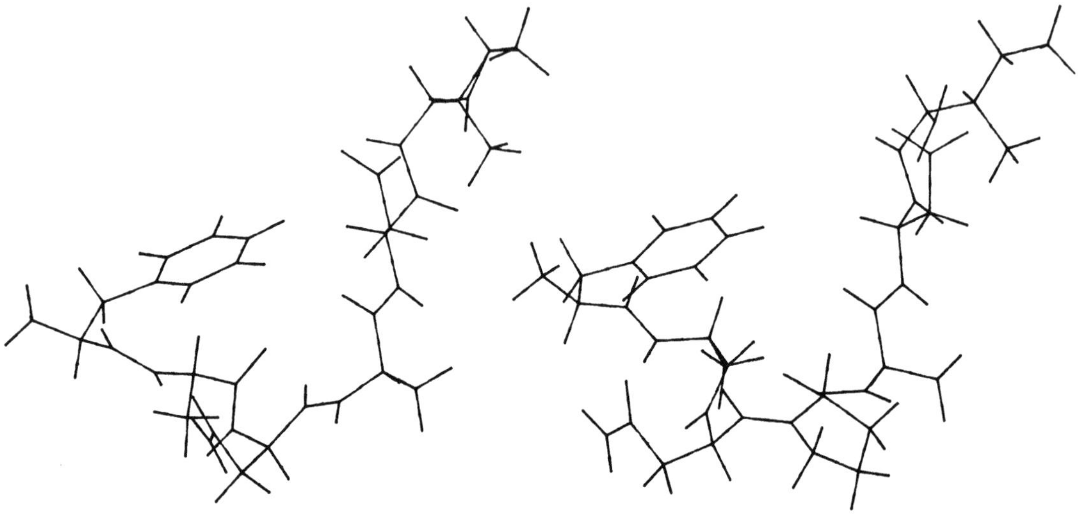

FANADI FANPADI

Figure 3. Comparison between the molecular architecture of a sequence (FANADI) in the conserved cysteine-rich region of the consensus SLG, and a homologous sequence (FANPADI) in the same domain of type VI collagen. Whilst the addition of the proline residue has 'lengthened' the assembly somewhat in the type VI collagen, the general conformation is strikingly similar. Structure predictions were made using a computer-based molecular modelling programme (Crabbe and Appleyard, 1989).

molecular architecture and a domain in the SLG and type VI collagen emerged as possessing a number of striking similarities (see Figure 3).

Despite the undoubted powers of the molecular modelling program, the quantity of data involved means that only relatively small domains can be modelled. This focus on detail renders it difficult to detect the larger, and sometimes more significant features of molecular architecture. Fortunately, recent studies by Sibanda *et al.* (1989), based primarily on crystallographic data, have identified sequences characterising β turns, or hairpins, which are recognised as important features of protein structure. The cysteine rich domain of the *Brassica* glycoprotein, and its flanking regions, have been surveyed for putative β -hairpin turn sequences, and two have been identified. Surprisingly, these apparently 'bracket' the cysteine-rich region and, although they do not contain exactly the amino acid residues reported by Sibanda *et al.* for the 'standard' types of β turn, molecular modelling reveals the key residues to be present, and that the appropriate hydrogen bonds and intermolecular dimensions can exist to ensure that β turns are formed. The *Brassica* SLRs and SLGs therefore emerge as possessing a domain, bracketed by β turns, strongly homologous with animal proteins known to bind protein and form matrices. Clearly, more work will be required to establish whether the protein binding properties of the *Brassica* glycoprotein are similar to those of collagen VI, which forms complex multimers (Chu *et al.*, 1989), or whether it binds to specific sites on other polypeptides.

Interestingly, database searches using the sequences encoding putative S-genes of *Nicotiana* have revealed striking homology with the gene for *Aspergillus* T_2 ribonuclease (McClure *et al.*, 1989). In a series of physiological studies, and using an *in vitro* bioassay, McClure *et al.* (1989) have also demonstrated the stylar glycoprotein encoded by those genes to possess ribonuclease activity. While SI in *Nicotiana* is gametophytically controlled, and while there is virtually no sequence homology between the genes encoding *Aspergillus* T_2 ribonuclease and the *Brassica* SLG, this finding of a 'general biochemical function' for a putative S-gene product - rather than a specific signal-receptor interaction - is nevertheless very significant.

The part played by glycoproteins in the pollen stigma interaction in *Brassica*

The SLG/SLR stigmatic glycoproteins are thus emerging as multifunctional agents, capable of mediating SI, hydration, and perhaps other roles including pollen grain adhesion and promotion of tube growth. However, these functions cannot be carried out in isolation and require the participation of a male determinant. At present, our ignorance as to the component of the pollen active in SI is complete. The fact that the compatibility of the pollen with respect to the stigma is sporophytically controlled suggests that this determinant is either synthesised in the grain prior to meiosis, or is somehow deposited on the grain by the sporophyte, prior to its release. It has been known for some years that cruciferous pollen bears a complex coating derived from the tapetum (Dickinson and Lewis, 1973) and, since the tapetum is sporophytic tissue, the search for the male determinant has been focussed on this layer. Significantly, the first physiological studies of extracted coatings (Dickinson and Lewis, 1975) showed it capable of stimulating some of the stigmatic responses normally associated with SI. Experiments of this type are, however, difficult to carry out and often produced equivocal results.

The search for the male determinant would be much facilitated were something known of its chemistry. Early models, such as the dimer hypothesis proposed by Lewis (1963), predicted the male partner in the interaction to resemble the female determinant, and a biologically active dimer to be formed between the two. As well as there being a number of conceptual problems with this hypothesis, no firm evidence to support it has yet emerged from any study so far reported. In gametophytic SI, where the pollen tube grows through a stylar domain rich in female SLG, it has reasonably been proposed that a traditional signalling interaction may take place, with a receptor on the pollen tube surface being the male partner (Clarke *et al.*, 1985) although the very recent evidence indicating that the SLG in *Nicotiana* has a powerful ribonuclease activity (McClure, 1989) does little to support such a view. We believe the situation to be markedly different in sporophytic SI. In a hypothesis devised to explain both the evolution of sporophytic SI from gametophytic, and the differences in structure and physiology that characterise the two systems, the male determinant is proposed to be carried in the pollen coating, and to be chemically similar (although not identical), to the female (Dickinson, 1990). There are now two powerful pieces of evidence supporting this inference; firstly, transcripts present in anthers at the microspore stage of pollen development - when the tapetum is laying down the coating - display some homology with probes raised from S-specific domains of the SLG (C. Dumas, personal communication). Second, a new method for removing the pollen grain coating uncontaminated with gametophytic products shows it to contain high concentrations of a polypeptide with a pI of about 9, similar to that of the SLG (Elleman, 1989). Further characterisation of this polypeptide is currently under way.

Unfortunately, the application of modern cell and molecular biological methods to SI in *Brassica* has served to complicate, rather than simplify matters. However, with our present knowledge of the structure and potential functions of the female glycoprotein, an increasing number of clues to the identity of the male determinant, and extensive physiological knowledge of the SI response, the stage is now becoming set for the molecular resolution of this sophisticated intercellular signalling system.

Acknowledgements

The authors thanks are especially due to Virginie Guyon for her valuable assistance with the molecular modelling work whilst visiting Reading on the EC ERASMUS Studentship scheme. Much of the work reported has been funded through Research Grants awarded by the UK Agricultural and Food Research Council. We also acknowledge Simon Hiscock for drawing Figure 1 and Valerie Norris for help in preparing the manuscript.

References

Anderson, M.A., Cornish, E.L., Mau, S.L., Williams, E.G., Hoggart, R., Atkinson, A., Bonig, I., Gregor, B., Simpson, R., Roche,P.J., Haley, J.D., Penscaow, J.D., Niall, H.D., Tredegar, G.W., Cochlan, J.P., Crawford, R.J. and Clarke, A.E. 1986. Cloning of a cDNA for a stylar glycoprotein associated with the expression of self-incompatibility in *Nicotiana alata*. Nature **321**, 38-44.

Bateman, A.J. 1952. Self-incompatibility systems in angiosperms I. Theory Heredity **6**, 285-310.

Chu, M.L., Pan, T.C., Conway, D., Kuo, H.J., Glanville, R.W., Timpl, R., Mann, K. and Deutzmann, R. 1989. Sequence analysis of x 1 (VI) and x 2 (VI) chains of human type VI collagen reveals internal triplication of globular domains similar to the A domains of von Willebrand factor and two x 2 (VI) chain variants that differ in carboxy terminus. EMBO J. **8**, 1939-46.

Clarke, A.E., Anderson, M.A., Basic, T., Harris, P.J. and Mau, S.L. 1985. Molecular basis of cell recognition during fertilisation in higher plants. J. Cell Sci. Suppl. 2, 261-285.

Crabbe, M.J.C. and Appleyard, J. 1989. Desk-top Molecular Modeller Version 1.2. Oxford University Press (Oxford, UK).

Dickinson, H.G. 1990. Self-incompatibility in flowering plants. Bioessays (in the press).

Dickinson, H.G. and Lewis, D. 1973. The formation of the tryphine coating the pollen grain of *Raphanus* and its properties related to the self-incompatibility system. Proc. R. Soc. Lond. B. **184**, 149-165.

Dickinson, H.G. and Lewis, D. 1975. Interaction between the pollen grain coating and the stigmatic surface during compatible and incompatible intraspecific pollinations in *Raphanus*. In: The biology of the male gamete, eds J.G. Duckett and P.A. Racey. Biol. J. Linn. Soc. Suppl. I, pp. 165-175.

East, E.M. and Mangelsdorf, A.J. 1925. A new interpretation of the hereditary behaviour of self-sterile plants. Proc. Natnl. Acad. Sci. USA 11, 166-183.

Elleman, C.J., Sarker, R.H., Aivalakis, G., Slade, H. and Dickinson, H.G. 1989. Molecular physiology of the pollen-stigma interaction in *Brassica*. In: Plant Reproduction: from floral induction to pollination, E. Lord and G. Bernier eds. Am. Soc. Pl. Physiol. Symp. Ser. 1, 136-145.

Ferrari, T.E. and Wallace, D.H. 1975. Germination of *Brassica* pollen and expression of incompatibility *in vitro*. Euphytica 24, 757-765.

Franklin-Tong, V.E., Ruuth, E., Marmey, P., Lawrence, M.J. and Franklin, F.C.H. 1989. Characterisation of a stigmatic component from *Papaver rhoeas* L. which exhibits the specific activity of a self-incompatibility (S-) gene product. New phytol. 112, 307-315.

Jahnen, W., Batterham, M.P., Clarke, A.E., Moritz, R.L. and Simpson, R.J. 1989. Identification, isolation and n-terminal sequencing of style glycoproteins associated with self-incompatibility in *Nicotiana alata*. The Plant Cell 1, 493-499.

Kirch, H.H., Uhrig, H., Lohspeich, F., Salamini, F. and Thompson, R.D. 1989. Characterisation of proteins associated with self-incompatibility in *Solanum tuberosum*. Theor. Appl. Genet. 78, 581-588.

Lewis, D. 1965. A protein dimer hypothesis on incompatibility. In: Genetics Today, ed. S.J. Geerts. Proc. XI Int. Congress Genet. 1963, pp. 657-663. Pergamon (Oxford, UK).

Lewis, D. and Crowe, L.K. 1958. Unilateral incompatibility in flowering plants. Heredity 12, 233-256.

McClure, B.A., Haring, V., Ebert, P.R., Anderson, M.A., Simpson, R.J., Sakiyama, F. and Clarke, A.E. 1989. Style self-incompatibility gene products in *Nicotiana alata* are ribonucleases. Nature 342, 955-957.

Nasrallah, J.B., Yu, S.M. and Nasrallah, M.E. 1988. Self-incompatibility genes of *Brassica oleracea*: expression, isolation and structure. Proc. Natnl. Acad. Sci. USA, **85**, 5551-5555.

Roberts, I.N., Harrod, G. and Dickinson, H.G. 1984. Pollen-stigma interactions in *Brassica oleracea*: ultrastructure and physiology of the stigmatic papillar cells. J. Cell Sci. **66**, 241-253.

Sarker, R.H., Elleman, C.J. and Dickinson, H.G. 1988. Control of pollen hydration in *Brassica* requires continued protein synthesis, and glycosylation is necessary for interspecific incompatibility. Proc. Natnl. Acad. Sci. USA, **85**, 4340-4344.

Sibanda, B.L., Blundell, T.L. and Thornton, J.M. 1989. Conformation of hairpins in protein structures; a systematic classification with applications to modelling by homology, electron density fitting and protein engineering. J. Mol. Biol. **206**, 759-777.

Singh, A. and Paolillo, D.J. 1989. Towards an *in vitro* bioassay for the self-incompatibility response in *Brassica oleracea*. Sex Plant Reprod. 2, 277-280.

Stead, A.D., Roberts, I.N. and Dickinson, H.G. 1980. Pollen stigma interactions in *Brassica oleracea*: the role of stigmatic proteins in pollen grain adhesion. J. Cell Sci. 42, 417-423.

Trick, M., Larsen, K. and Flavell, R.B. 1990. A basis for S-allele specificity suggested by closely-related S-allele sequences. In the press.

Verweij, C.L., Diergaarde, P.J., Hart, M. and Pannekoek, H. 1986. Full length von Willebrand factor (vWF) cDNA encodes a highly repetive protein cnsiderably larger than the mature vWF subunit. EMBO J. **5**, 1839-1847.

Whitehouse, H.L.K. 1950. Multiple allelomorph incompatibility of pollen and tyle in the evolution of the angiosperms. Ann. Bot. N.S. **14**: 198-216.

INDUCTION OF THE ACROSOME REACTION IN STARFISH

M. Hoshi, T. Amano*, Y. Okita**, T. Okinaga and T. Matsui***
Department of Life Science, Faculty of Science, Tokyo Institute of Technology, O-okayama, Meguro-ku, Tokyo 152, Japan

SUMMARY

When starfish spermatozoa reach the jelly coat of homologous eggs, they immediately undergo the acrosome reaction that is a prerequisite for fertilization. Three organic components of the egg jelly are responsible for this phenomenon; namely, an extremely large sulfated glycoprotein named acrosome reaction-inducing substance (ARIS), a group of sulfated steroid saponins named Co-ARIS, and a sperm activating oligopeptide(s) (SAP). ARIS induces the acrosome reaction species-specifically in high Ca^{2+} or high pH sea water, but it requires Co-ARIS for the induction in normal sea water. The acrosome reaction induced by ARIS and Co-ARIS is not accompanied with a transient increase in the intracellular pH (pH_i), which is generally accepted to be inevitable for triggering the acrosome reaction, and proceeds slower than the jelly-induced acrosome reaction. All the three, but nothing else, are required to reconstruct the acrosome reaction-inducing ability of the egg jelly.

ARIS and Co-ARIS cooperatively stimulate verapamil- and maitotoxin-sensitive Ca^{2+} channels in the spermatozoa and increase the intracellular Ca^{2+} ($Ca^{2+}{}_i$) to an extent sufficient for eventually inducing the acrosome reaction. SAP stimulates Na^+/H^+ exchange systems in the spermatozoa and transiently increases the pH_i, which facilitates them to undergo the acro-

Present address:
*Department of Pharmacology, Tokyo Metropolitan Institute of Gerontology, Itabashi-ku, Tokyo 173, Japan; **Suntory Pharmatech Center, Chiyoda-machi, Gunma 370-05, Japan; ***Institute for Comprehensive Medical Science, Fujitagakuen Health University School of Medicine, Aichi 480-11, Japan

NATO ASI Series, Vol. H 45
Mechanism of Fertilization
Edited by B. Dale

some reaction. When spermatozoa reach the jelly coat, they immediately undergo the acrosome reaction in consequence of concurrent increases in the $Ca^{2+}{}_i$ and pH_i.

INTRODUCTION

Fertilizing spermatozoa in most animals have to undergo the acrosome reaction in response to signals from eggs or their surroundings to ensure a spatio-temporally favorable exposure of their devices essential for penetration through the egg coats and for subsequent fusion with the egg plasma membrane. Although much is known about morphological, physiological and biochemical changes in conjunction with the acrosome reaction (for reviews see Dan, 1967; Tilney, 1985), it is still unclear for most animals where the fertilizing spermatozoa undergo the acrosome reaction. It is, however, well established in starfish that they undergo the acrosome reaction upon encountering the jelly coat (Dale et al., 1981; Ikadai and Hoshi, 1981a). We therefore attempted to isolate from the starfish jelly coat a signal molecule(s) for triggering the acrosome reaction.

This paper reviews the chemical structures and biological activities of egg-jelly components responsible for triggering the acrosome reaction in the starfish, Asterias amurensis. Most conclusions presented below are also confirmed in another starfish, Asterina pectinifera.

EGG-JELLY COMPONENTS

The jelly coat contains two macromolecular components; one is a high mannose glycoprotein of about 80 KDa. Although it has quite interesting saccharide structures (Endo et al., 1987), so far it is of no effect upon spermatozoa. The other is a highly sulfated, fucose-rich glycoprotein of an extremely high molecular weight. This glycoprotein does serve as a key molecule for the induction of the acrosome reaction as shown below and is named acrosome reaction-inducing substance(ARIS). ARIS contains fucose, galactose, xylose and N-acetylgalactosamine in the molar ratio of 5.9:3.6:1.6:1.0. The major product obtained by limited acid hydrolysis is (Fuc_5,Gal_3,Xyl_1) ► Fuc ► suggesting the presence of an unusual sugar chain (Okinaga et al., unpublished data).

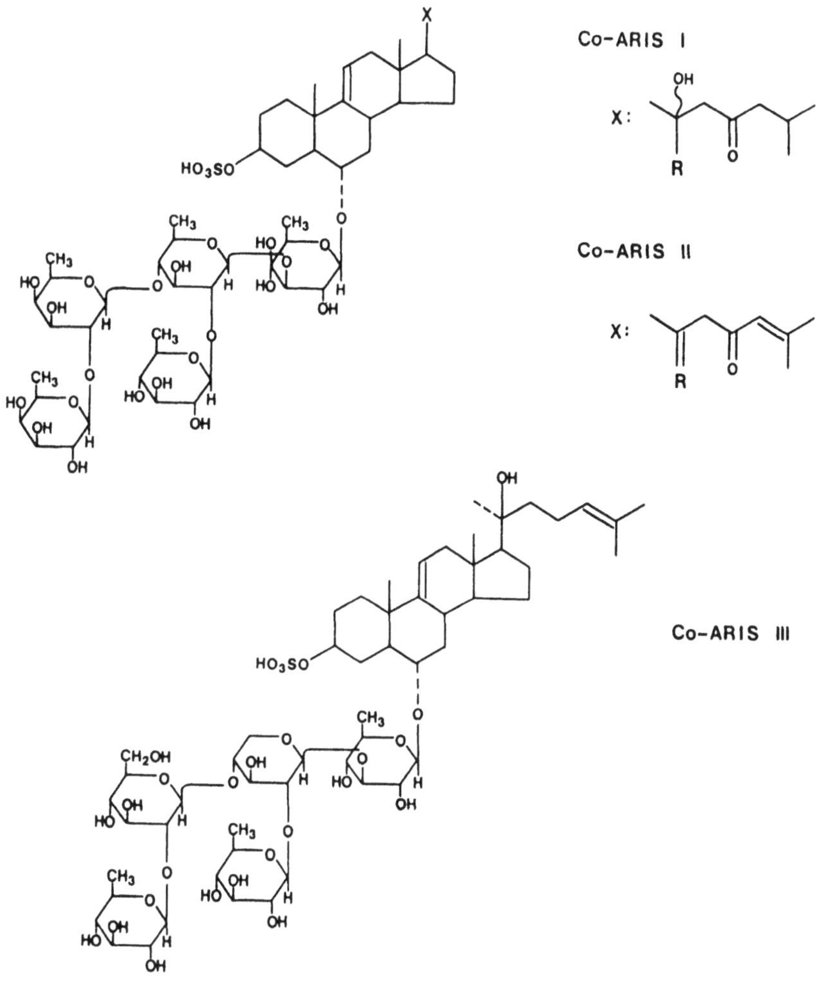

Fig. 1. Structures of major Co-ARIS substances. Co-ARIS I and II are expressed in the hydrated form at 6-deoxy-xylo-hexos-4-ulose. (From Fujimoto et al., 1987)

The activities of ARIS are species-specific, pronase-resistant and ascribable mainly to the saccharide and sulfate moieties (Matsui *et al.*, 1986a, b). This allows us to substitute a pronase digest of ARIS (P-ARIS) for intact ARIS. *Asterina* ARIS has a significantly different sugar composition.

Diffusible organic components of the egg jelly consists of steroid saponins and oligopeptides. It is known that sulfated saponins in the jelly-coat agglutinate spermatozoa irreversibly (Uno and Hoshi, 1978). Besides sperm agglutinins, the egg jelly contains sulfated saponins that serve as a co-factor for ARIS (Co-ARIS) as described below. Fig. 1 shows the structures of major Co-ARIS substances (Nishiyama *et al.*, 1987a; Fujimoto *et al.*, 1987). They appear to act synergistically on spermatozoa in natural circumstances. The jelly coat seems to contain Co-ARIS of a total concentration roughly optimum to trigger the acrosome reaction (Nishiyama *et al.*, 1987b). The activity of Co-ARIS is very much dependent upon the steroid side chain and sulfate moiety. It is, however, not so much dependent upon fine structures of the saccharide chains. The action of Co-ARIS is not very much species-specific (Nishiyama *et al.*, 1987b).

The jelly coat contains sperm-activating peptides (SAPs) (for a review, see Suzuki, 1989). Although such peptides have not yet been isolated in starfish, the data to be shown below strongly suggest that a SAP(s) participates in the physiological induction of the acrosome reaction.

ACROSOME REACTION

Egg jelly induces the acrosome reaction.

The acrosome reaction is induced to some extent by raising the intracellular Ca^{2+} ($Ca^{2+}{}_i$) with a calcium ionophore, A23187, or a calcium channel activator, maitotoxin. If sperm pH_i is concurrently increased by the addition of monensin,

NH_4Cl or a partially purified fraction of SAP, much more spermatozoa undergo the acrosome reaction (Matsui et al., 1986a; Okita et al., unpublished data). The acrosome reaction is also induced by simply increasing both external Ca^{2+} and pH. These data suggest that starfish spermatozoa undergo the acrosome reaction if both $Ca^{2+}{}_i$ and pH_i are concurrently increased.

Egg jelly triggers the acrosome reaction quickly, dose-dependently and species-specifically in starfish. As found in sea urchins, the jelly-induced acrosome reaction depends very much upon external cations, especially Ca^{2+} and H^+, susceptible to Ca^{2+}-channel antagonists (verapamil, diltiazem and nicardipine) as well as K^+-channel antagonists (quinidine and 4-aminopyridine), facilitated by an increase in pH_i. It is accompanied by an abrupt uptake of external Ca^{2+}, a Na^+-dependent transient increase in pH_i and a remarkable but transient increase in cAMP. (Matsui et al., 1986a, c) It is generally accepted therefore that the egg jelly induces the acrosome reaction by stimulating Ca^{2+}-channels and Na^+/ H^+ exchangers in the sperm plasma membrane (Schackmann and Shapiro, 1981; Tilney, 1985; Hoshi et al., 1986, 1988).

When the egg jelly is thoroughly depleted of diffusible organic components, it loses the ability to induce the acrosome reaction in normal sea water, but not that in high Ca^{2+} or high pH sea water. This observation led the hypothesis that two, at least, organic components in the jelly coat are necessary to trigger the acrosome reaction in normal sea water (Ikadai and Hoshi, 1981a; Matsui et al, 1986a). Indeed, ARIS and Co-ARIS are purified from the egg jelly as shown above.

ARIS requires Co-ARIS to trigger the acrosome reaction in normal sea water.

ARIS (or P-ARIS) is the sole jelly component that induces the acrosome reaction in high Ca^{2+} or high pH sea water. It requires Co-ARIS to induce the acrosome reaction in normal sea water. The acrosome reaction induced by ARIS and Co-ARIS is accompanied by increases in Ca^{2+} uptake and the level of cAMP, and it is sensitive to verapamil and diltiazem like the jelly-induced acrosome reaction. It is different, however, from the jelly-induced one in two points. First, these changes described above proceed about 1 min slower in the spermatozoa treated with ARIS/Co-ARIS. Secondly, a Na_+-dependent transient pH_i increase accompanying the jelly-induced acrosome reaction, has never detected in the acrosome reaction induced by ARIS/Co-ARIS (Fig. 2). Similarly, in 50 mM Ca^{2+} sea water, P-ARIS can induce the acrosome reaction without increasing the pH_i. Conversely, in normal sea water a fraction containing SAP and Co-ARIS (Fraction M_8), and in Ca^{2+}-deficient sea water the egg jelly, increases the pH_i without inducing the acrosome reaction. This activity is destroyed by pronase digestion both in the egg jelly and in Fraction M_8. Pure Co-ARIS substances do not have such activity at all (Matsui et al., 1986c; Hoshi et al., 1988).

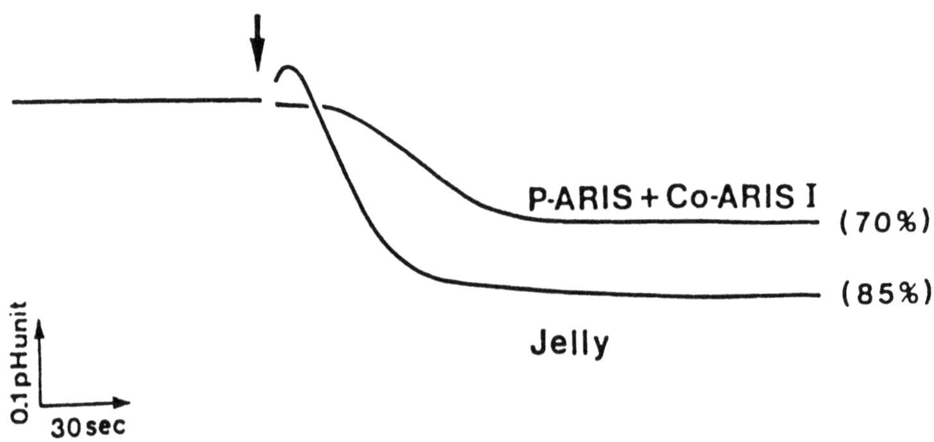

Fig. 2. Changes in sperm pH_i upon the acrosome reaction. The arrow indicates the addition of the egg jelly or its components. Values in the parentheses show the percentage of acrosome-reacted sperm. Jelly, 50 μg sugar/ml; P-ARIS, 25 μg sugar/ml and Co-ARIS I, 200 μM. (From Hoshi et al., 1986)

Biological roles for SAP

SAP is originally discovered as a substance that keeps sperm motility in slightly acidic sea water (for a review, see Suzuki, 1989). It is not obligatory for the induction of acrosome reaction as mentioned above. Nevertheless, basing upon the following account, we have proposed that it contributes much to the physiological induction of the acrosome reaction (Hoshi et al., 1986; 1988). It has very recently been confirmed in a sea urchin that SAP I is indeed involved in the acrosome reaction (Yamaguchi et al., 1989).

Fraction M_8 containing Co-ARIS substances and SAP is far more effective than a pure Co-ARIS as a co-factor for ARIS. In fact, in the presence of a sufficient amount of P-ARIS, the concentration required for 50% acrosome reaction was 5 μ g/ml for Fraction M_8, 24 μ g/ml for Co-ARIS II and III and 90 μ g/ml for Co-ARIS I (Matsui et al., unpublished data). When Fraction M_8 is digested with pronase to which Co-ARIS is resistant, the apparent activity as a co-factor for ARIS is much reduced with a concomitant loss of SAP activities (Hoshi et al., 1986, 1988; Matsui et al., 1986b, c; Nishiyama et al., 1987b). Furthermore, replacement of pure Co-ARIS with Fraction M_8 appreciably reduces the concentration of P-ARIS required for the full induction of the acrosome reaction (Matsui et al., 1986b, c; Nishiyama et al., 1987b). Combinations of P-ARIS with Fraction M_8, but not with a pronase digest of it nor with Co-ARIS, perfectly mimic the jelly in the manner and way of inducing the acrosome reaction (Hoshi et al., 1988). These data suggest that SAP play an important role in the physiological induction of the acrosome reaction (Hoshi et al., 1986, 1988).

PRETREATMENT EFFECTS

The acrosome reaction induced by the egg jelly or its components requires external Ca^{2+}. When spermatozoa are incubated in Ca^{2+}-deficient sea water for more than a half minute

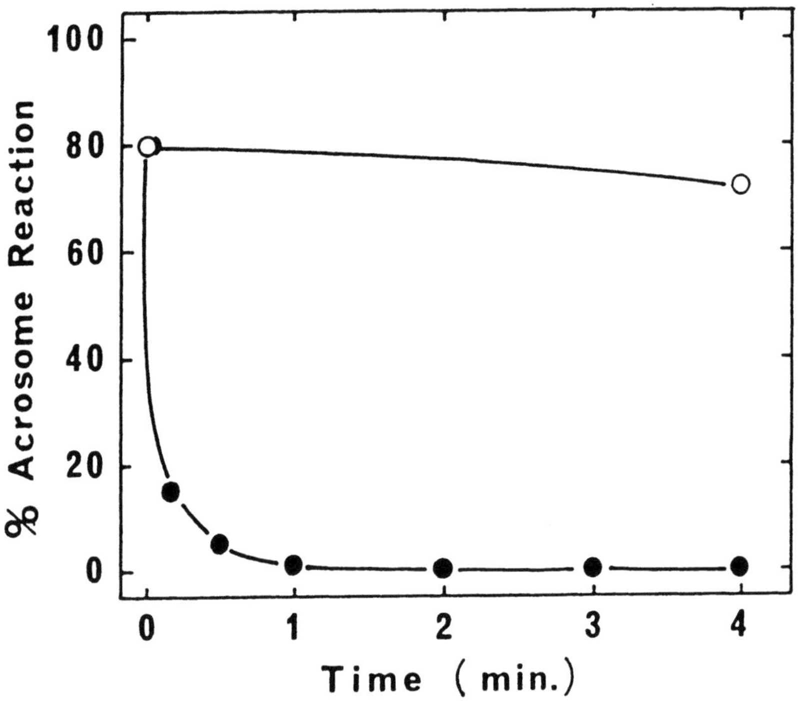

Fig. 3. Effects of the egg jelly in Ca^{2+}-deficient sea water on the acrosome reaction. Spermatozoa were treated with 50 μ g sugar/ml jelly in Ca^{2+}-deficient sea water for a given period and then Ca^{2+} was added to give a normal concentration of 10 mM (—●—), or they were treated first with Ca^{2+}-deficient sea water, and then the egg jelly and Ca^{2+} were sufficiently added to give a concentration of 50 μ g sugar/ml and 10 mM, respectively (—○—). Points are means for two experiments. (From Matsui _et al._, 1986b)

with an enough amount of the egg jelly to trigger the acrosome reaction in normal sea water, they do not undergo the acrosome reaction or the ionic changes in cytoplasmic even after Ca^{2+}, fresh jelly or the both are sufficiently fortified (Fig. 3). They seem to lose their ability to react to the jelly coat _in situ_, too. Similarly, if they are treated with P-ARIS (ARIS) and Fraction M_8 not simultaneously but sequentially in either order with an interval over minutes, they do not undergo the acrosome reaction and become unresponsive to the jelly. Fraction M_8, but not a pronase digest nor Co-ARIS, has such effects of pretreatment suggesting that an oligopeptide(s), presumably SAP, has the ability to make spermatozoa unresponsive to the jelly (Matsui et al., 1986b). This effect of Fraction M_8 is not attributable directly to its ability to raise the pH_i, because spermatozoa are still responsive to the jelly even after the pH_i has significantly been raised by incubation in alkaline sea water.

The spermatozoa that become unresponsive to the jelly by pretreatments can undergo the acrosome reaction like intact ones in response to a calcium ionophore, A23187, plus monensin (Matsui et al., 1986b) and to a Ca^{2+} channel activator, maitotoxin, plus Fraction M_8 or alkaline sea water (Nishiyama et al., 1986; Okita et al., unpublished data). The acrosome reaction induced by maitotoxin in combination with Fraction M_8 or alkaline sea water is also susceptible to verapamil. These data suggest that the pretreatment effects are due to an irreversible change(s) in the steps not later than the stimulation of Ca^{2+} channels and Na/H^+ exchangers. It is also suggested that verapamil and maitotoxin act on the same channel.

The pretreatment effect of P-ARIS is species-specific and depends mainly upon the saccharide and sulfate moieties just like its ability to induce the acrosome reaction in the presence of Co-ARIS, at high Ca^{2+} or at high pH (Matsui et al., 1986b). It is, therefore, presumable that the two apparently different effects of ARIS upon spermatozoa actually result from a single action of ARIS. This may also be true for the effects of SAP.

PROPOSED MECHANISM

ARIS, Co-ARIS and SAP seem to be taken by the spermatozoa (Matsui et al., 1986b). It is the best conceivable for the present that spermatozoa have specific receptors in the plasma membrane for ARIS and SAP but not for Co-ARIS. Co-ARIS may infiltrate or be inserted into the sperm plasma membrane to some extent, because it is a saponin potentially having cytolytic effects on spermatozoa (at unphysiologically high concentrations it actually damages spermatozoa) and because steroid side chains contribute to their activity much more than saccharide chains (Nishiyama et al., 1987b). It is favorable for this assumption that ARIS and presumably SAP, but not Co-ARIS, show pretreatment effects. It is also favored by our preliminary data suggesting that Co-ARIS enhances the P-ARIS binding to homologous spermatozoa (Kontani et al., unpublished data).

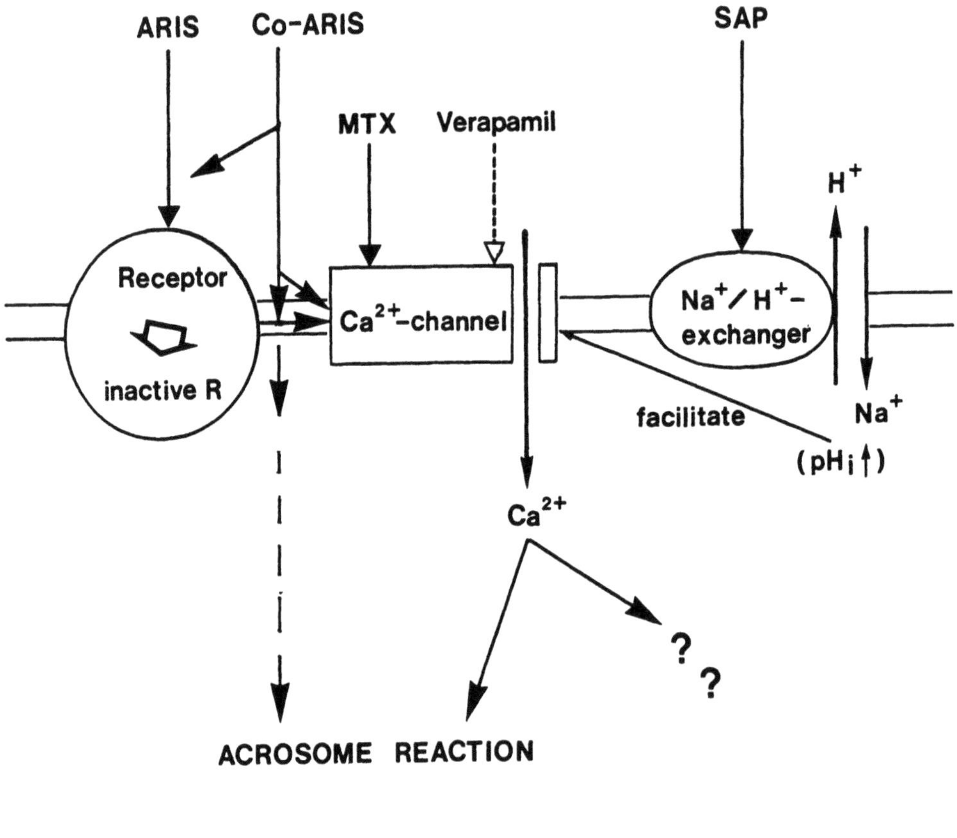

Fig. 4. Outline of the proposed mechanism. For details see the text. (Modified from Amano, 1990)

Taking account of the data presented above together, we guess that the binding of ARIS to its receptor raises Ca^{2+}_i by stimulating verapamil- and maitotoxin-sensitive Ca^{2+}-channels in the spermatozoa. This receptor could be the channel itself. Co-ARIS may contribute for the Ca^{2+}_i increase by modulating the binding of ARIS and/or the succeeding activation of the channel. Whereas SAP raises pH_i by stimulating Na^+/H^+ exchange systems. It may also contribute to some extent for raising Ca^{2+}_i as suggested in sea urchin (Schackmann and Chock 1986). It is also our guess at the moment that the pretreatment with ARIS mainly affects the coupling or signal transduction between the ARIS binding and the Ca^{2+}-channel activation. Pretreatments with SAP may affect that step and/or the ARIS binding itself, which is not a direct result of the increased pH_i. These hypotheses are now under experimental examination in our laboratory.

ACKNOWLEDGMENTS

Work in the authors' laboratory was supported in part by the grants from the Ministry of Education, Science and Culture of Japan, the Naito Foundation, Yamada Science Foundation, the Foundation for the Promotion of Research on Medical Resources, and the Institute of Developmental and Reproductive Biology in Yamagata.

REFERENCES

Amano T (1990) The histone degradation and acrosome reaction induced in starfish spermatozoa by egg jelly components. Dissertation to be submitted for the degree of Dr. Sci. in the Graduate School, Tokyo Institute of Technology.

Dale B, Dan-Sohkawa M, De Santis A and Hoshi M (1981) Fertilization of the starfish Astropecten aurantiacus. Exp Cell Res 132: 505-510.

Dan JC (1967) Acrosome reaction and lysins. In "Fertilization, Vol. 1", Metz CB and Monroy A (eds). pp. 237-293. Academic Press, New York.

Endo T, Hoshi M, Endo S, Arata Y and Kobata A (1987) Structures of the sugar chains of a major glycoprotein present in the egg jelly coat of a starfish, Asterias amurensis. Arch Biochem Biophys 252: 105-112.

Fujimoto Y, Yamada T, Ikekawa N, Nishiyama I, Matsui T and Hoshi M (1987) Structure of acrosome reaction-inducing steroidal saponins from the egg jelly of the starfish, Asterias amurensis. Chem Pharm Bull 35: 1829-1832.

Hoshi M, Matsui T, Nishiyama I, Fujimoto Y and Ikekawa N (1986) Egg-jelly components responsible for the induction of acrosome reaction. In "Advances in Invertebrate Reproduction, Vol. 4", Porchet M, Andries J-C and Dhainaut A (eds). pp. 275-282. Elsevier Science Publishers, Amsterdam.

Hoshi M, Matsui T, Nishiyama I, Amano T and Okita Y. (1988) Physiological inducers of the acrosome reaction. In "Regulatory Mechanisms in Developmental Processes", Eguchi G, Okada TS and Saxen L (eds). pp. 19-24. Elsevier Scientific Publishers Ireland, Dublin.

Ikadai H and Hoshi M (1981a) Biochemical studies on the acrosome reaction of the starfish, Asterias amurensis. I. Factors participating in the acrosome reaction. Dev Growth Differ 23: 73-80.

Ikadai H and Hoshi M (1981b) Biochemical studies on the acrosome reaction of the starfish, Asterias amurensis. II. Purification and characterization of acrosome reaction-inducing substance. Dev Growth Differ 23: 81-88.

Matsui T, Nishiyama I, Hino A and Hoshi M (1986a) Induction of the acrosome reaction in starfish. Dev Growth Differ 28: 339-348.

Matsui T, Nishiyama I, Hino A and Hoshi M (1986b) Acrosome reaction-inducing substance purified from the egg jelly inhibits the jelly-induced acrosome reaction in starfish: an apparent contradiction. Dev Growth Differ 28: 349-357.

Matsui T, Nishiyama I, Hino A and Hoshi M (1986c) Intracellular pH changes of starfish sperm upon the acrosome reaction. Dev Growth Differ 28: 359-368.

Nishiyama I, Matsui T, Yasumoto T, Oshio S and Hoshi M (1986) Maitotoxin, a presumed calcium channel activator, induces the acrosome reaction in mussel spermatozoa. Dev Growth Differ 28: 443-448.

Nishiyama I, Matsui T and Hoshi M (1987a) Purification of Co-ARIS, a cofactor for acrosome reaction-inducing substance, from the egg jelly of starfish. Dev Growth Differ 29: 161-169.

Nishiyama I, Matsui T, Fujimoto Y, Ikekawa N and Hoshi M (1987b) Correlation between the molecular structure and the biological activity of Co-ARIS, a cofactor for acrosome reaction-inducing substance. Dev Growth Differ 29: 171-176.

Schackmann RW and Chock PB (1986) Alteration of intracellular $[Ca^{2+}]$ in sea urchin sperm by the egg peptide speract. J. Biol Chem 261: 8719-8728.

Schackmann RW and Shapiro BM (1981) A partial sequence of ionic changes associated with the acrosome reaction of Strongylocentrotus purpuratus. Dev Biol 81: 145-154.

Suzuki N (1989) Sperm activating peptides from sea urchin eggs. In "Bioorganic Marine Chemistry, Vol. 3", Scheuer PJ (ed). pp. 47-70. Springer Verlag, Berlin.

Tilney LG (1985) The acrosome reaction. In "Biology of Fertilization, Vol. 2", Metz CB and Monroy A (eds). pp. 157-213. Academic Press, Orlando.

Uno Y and Hoshi M (1978) Separation of the sperm agglutinin and the acrosome reaction-inducing substance in egg jelly of starfish. Science 200: 58-59.

Yamaguchi M, Kurita M and Suzuki N (1989) Induction of the acrosome reaction of Hemicentrotus pulcherrimus spermatozoa by the egg jelly molecules, fucose-rich glycoconjugate and sperm-activating peptide I. Dev Growth Differ 31: 233-239.

NEW DATA AND CONCEPTS IN ANGIOSPERM FERTILIZATION

C.M. GUILLUY, T. GAUDE, C. DIGONNET- KERHOAS, A. CHABOUD, Ph. HEIZMANN and C. DUMAS

Reconnaissance Cellulaire et Amélioration des Plantes

LA INRA 879. Université Cl. Bernard - Lyon I

43 Bd du 11 Nov 1918

69622 Villeurbanne Cédex. - France

NATO ASI Series, Vol. H 45
Mechanism of Fertilization
Edited by B. Dale

INTRODUCTION

The process of fertilization, which has been defined as a multistep phenomenon, starting with the interaction between pollen and stigma and ending with the fusion of the gametes, constitutes a tremendous complex system in higher plants in comparison with lower plants or animals. In these last ones, most of the research in fertilization has employed the gametes of invertebrates, particularly the sea urchin. The reason for this choice was due to their easy availability as well as the minimal requirements needed for their maintenance. In addition, large quantities of gametes (male and female) can be obtained and externally fertilized (reviewed in Longo, 1989).

By contrast, in angiosperms, both types of gametes are housed in complex paternal and maternal sporophytic tissues (see Knox, 1984). The first contact between the two partners is the final step of the random transport of pollen grains, termed pollination. This contact induces the beginning of very precise but still largely unknown recognition events (see Gaude and Dumas, 1987) Such events occur either between sporophyte-sporophyte, sporophyte-gametophyte, gametophyte-gametophyte or gamete-gamete (see Clarke *et al.*, 1985). In addition, the whole multistep phenomenon is basically internal, thus very difficult either to follow step per step or to analyze at the molecular level.

The aim of this paper is to point out the recent progress achieved in angiosperms, especially in pollen quality, fertilization, male-female recognition and self-incompatibility, as wall as to mind about the model systems we have to use for further research.

1. POLLEN QUALITY

The pollen grain is the male gametophyte. Each grain is an accessible, isolated multicellular organism which assures the production and the dissemination of sperm cells until their union with the female gametes during the double fertilization. Pollen

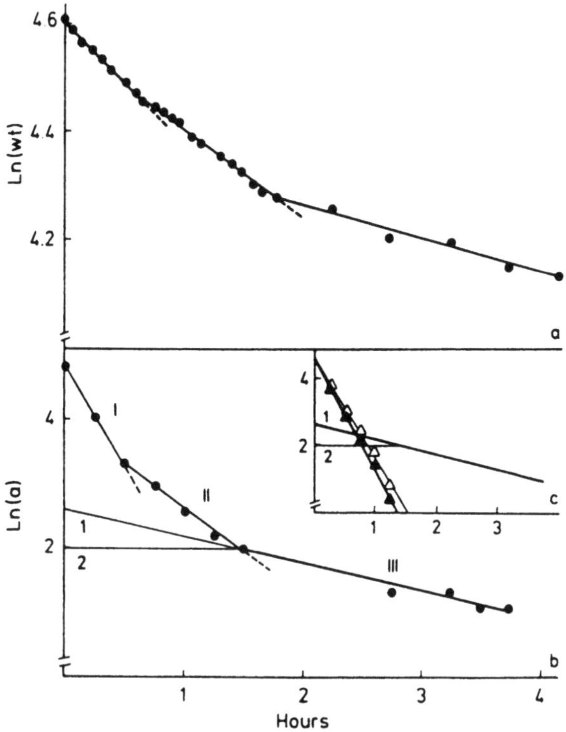

Figure 1:(a-c). Water release from the pollen of *Cucurbita pepo* var Seneca during a natural ageing (24°C , 40% RH)(from Digonnet-Kerhoas *et al.*, 1989).

a- Gravimetric study showing the weight loss.
Ln (wt): logarithm of the weight

b- ^{1}H NMR study illustrating the proton loss.
NMR spectrum is obtained at 80 MHz using a Fourier transform method. The three straight decay lines may figure three water fluxes I, II, III corresponding to three water fractions. If we consider that the fraction III leaves the pollen grains continuously since the beginning of the dehydration process, we can extrapolate the data and draw the line 1. But, if the fractions III remains stable until the others fractions have left the pollens, we can draw the line 2.
Ln (a): logarithm of integration of ^{1}H NMR peak

c- Interpretation of the ^{1}H NMR results. If we subtract the signal of the water flux represented by the line 1 from the signals of the two others fluxes I and II (b), these latter appear as a single flux that we denote ▲. Thus, two straight lines of decay are obtained after elimination of the influence of the flux III : the lines 1 and ▲. In the same way, we obtain the single flux Δ when we subtract the value of the line 2 from the fluxes I and II. We can conclude for the occurence of two majors fluxes of water which are released from *C. pepo* pollen during ageing.
In (c), legends for the axes are the same as in (b).

quality has be defined as the ability for pollen to fertilize a receptive and compatible pistil.

1.1 Pollen quality tests

Pollen quality has to be monitored at each step of the fertilization program: hydration, germination, style penetration, pollen tube growth and gamete fusion. The tests currently used for pollen viability permit testing the capacity of pollen grains to achieve some or all of the sequences of the fertilization program: *in vitro* germination, *in vivo* germination and seed set, which is the only method available to test pollen fertilization capacities (Kerhoas and Dumas, 1988).

One indirect method allows a characterization of the vegetative plasma membrane: the fluorochromatic reaction (F.C.R.), which takes a few minutes to perform (see Heslop-Harrisson *et al.*, 1984). In addition, the evaluation of the ATP content of the pollen grain has been shown to correlate with the pollen viability (Roeckel, personal communication).

1.2 Pollen longevity and storage

The last review on this subject has been established by Towill in 1985. In fact, there is a close correlation between pollen longevity and water content which is species dependent. Pollen grains with low water content at anthesis are long life pollen and can be stored easely at low temperature. Pollen grains characterized by a high water content are short life pollen . They can be stored at ultra-low temperature if they may sustain a dehydration process (Barnabas *et al.*, 1988).

1.3 Pollen quality, a multidisciplinary approach

In this part of the paper, we present the studies on pollen quality in two different species: *Cucurbita pepo* containing 45% H_2O *and Zea mays* containing 57-60% H_2O at anthesis. These pollen grains are stress sensitive. After harvest, pollen grains dehydrate and loose their viability (Gay *et al.*, 1987; Kerhoas *et al.*, 1987).

A multidisciplinary approach to the study of pollen viability permits to get an understanding of the loss of viability of hydrated pollen. This viability loss has been characterized by nuclear magnetic resonance study of the proton (Fig.1), and freeze fracture examination coupled with scanning electron microscopy (SEM) in *Z. mays* and *C. pepo* pollen (Kerhoas *et al.*, 1987; Digonnet-Kerhoas *et al.*, 1989). The quantity of protons giving a signal during pulse NMR experiments was very low when the pollen grains were judged to be dead according to the fluorochromatic test. Freeze-fracture replicas of this dead *C. pepo* pollen showed that the plasma membrane had become detached from the intine surface. The SEM study indicated that the *C. pepo* pollen-wall is not distorted and cannot follow the protoplast retraction which occurs during dehydration. On the contrary, *Z. mays* pollen is able to change its shape with water content and the plasma membrane remains contiguous to the intine in the dehydrated pollen. The gel-phase microdomains in the plasma-membrane of the dehydrated maize pollen grain could account for the viability loss. Hoekstra *et al.* (1989) suggest that the presence of sucrose is a key factor in preserving the membrane from the gel-phase in pollen grains which are not sensitive to dehydration.

Finally, pollen quality seems to be correlated to the water content but also to the ability to withstand a dehydration process. The presence of relatively high concentrations of sucrose for membrane integrity and the occurence of a pollen wall which can distort seem to characterize this ability. However, all these experiments have been performed at the pollen population level and further research is necessary to define some parameters at the individual pollen grain and gamete levels. And, because the process of fertilization requires two interacting partners similar experiments are in progress to define female quality.

2. FERTILIZATION OCCURING IN A COMPATIBLE SITUATION

Signaling and recognition events are involved at various

levels in the pistil, from the stigma surface to the female gametes (Dumas *et al.*, 1988). But almost none of them is known, even in a compatible situation (see Knox *et al.* , 1986). In order to solve some of these events and to get assessment of the pistil receptivity, several methods have been performed such as the use of aniline blue fluorochrome (ABF, see Dumas and Knox, 1983) and the *in vitro* pollination technique (Dupuis and Dumas, 1989).

2.1 Which model system should we use ?

We need a system which allows a strict control of pollen quality (see § 1) and pistil quality. In addition, further experiments need viable isolated sperm and egg cells. Till now, it seems that two systems could be used in this view: *Plumbago zeylanica* (see Russell, this issue) and *Zea mays* . According to the recent progress done in maize, it seems for us that the second one could be used for such a purpose especially at the gamete level. The preliminary step was to well defined an *in vitro* pollination system as a model for studying fertilization in maize (Dupuis and Dumas, 1989). The objective of this current research was to determine precisely the kinetics of the *in vitro* fertilization by the means of autoradiography with ^{32}P-radiolabelled pollen. At 28°C, the double fertilization takes place between 4 and 7 h after *in vitro* pollination of 4.5 cm-long silks.

2.2 Fusion of gametes

The pollen tube bearing the male germ unit (MGU: physical association between two sperm cells (SC) and the vegetative nucleus, see Dumas et al., 1984b) reaches the ovule where the double fertilization occurs. The ultimate transformation of the double fertilization is the association of the corresponding group of chromosomes derived from pronuclei, one of maternal (egg cell) and the other of paternal origin (1st SC), another one of maternal (2 fused pronuclei from the central cell) and paternal origin (2nd SC). The investigators have pointed out that the cytoplasmic DNA

inheritance (plastid, mitochondria) is still an enigma in angiosperms. It is generally supposed that cytoplasmic DNA is maternally inherited (Mogensen, 1988) with some exceptions (see Lee *et al.*, 1988; Roeckel *et al.*, in press).
Using serial sections and computer aid reconstruction and quantification, it has clearly been demonstrated by Russell (1983, 1985) that fertilization appears to be targeted in that each sperm cell is preprogramed to fuse with a particular female target cell.

2.3 Is an in vitro intergametic fusion a dream for plant biologists?

Recently, this MGU concept has been verified by its *in vitro* isolation (Matthys-Rochon *et al.*, 1987), and an enriched fraction of sperm cells obtained (Dupuis *et al.*, 1987) and partially characterized (Wagner *et al.*, 1989a). On the other side, the embryo sac and female target cells have been isolated and cytologically characterized (Wagner *et al.*, 198b). The way for the *in vitro* gamete manipulations in maize seems to be opened (Wagner *et al.*, 1990).

3. MALE-FEMALE RECOGNITION AND SELF-INCOMPATIBILITY

The fertilization process of flowering plants is controlled by recognition mechanisms which allow the pistil to discriminate between the different types of pollen that it may receive. During intraspecific matings, self-pollen, i.e., pollen belonging to the same genotype as that of the pistil, is generally rejected, while allopollen is accepted. This process, named self-incompatibility, enforces outbreeding and constitutes a quite original recognition system in flowering plants since it operates, in opposition to the immune system of vertebrates, to reject "self" while accepting "not-self" (see Gaude and Dumas, 1987). Many self-incompatibility systems of angiosperms are controlled by a single locus, the S locus (reviewed in De Nettancourt, 1977).

Genetics of self-incompatibility (SI) and cellular events of pollen-pistil interactions have extensively been studied these last ten years (see Clarke and Knox, 1980; Gaude and Dumas, 1987, for reviews). More recently, data on the molecular basis of SI has been obtained, mainly on two model systems: Solanaceae and Cruciferae (see Ebert *et al.*, 1989). The cruciferous plant *Brassica* provides an ideal experimental model for investigating cell recognition in flowering plants because of its well-determined genetic control of SI, and has been chosen for this reason in the present study.

In Cruciferae, the male-female recognition mechanisms related to self-incompatibility are governed by the monofactorial S-locus with at least fifty alleles (Ockendon, 1985). Identity of alleles in pollen and stigma leads to the incompatible reaction which results in the rejection of pollen grain on the stigma surface. Pollen acceptance or rejection may then be considered as the result of a dialogue between products of the same S alleles (Dumas *et al.*, 1984a). New data has recently been reported on the characterization of stigma S locus specific glycoproteins (SLSGs) in the cruciferous plant *Brassica* (Nasrallah *et al.*, 1985a; Takayama *et al.*, 1986a,b; Isogai *et al.*, 1987). The stigma S molecules appear exclusive to the stigmatic tissue, are able to bind the lectin concanavalin A (Con A), possess similar molecular masses, and accumulate in the papillar cells during stigma development. The amino-acid sequences of four S glycoproteins of *Brassica oleracea* L. have recently been deduced from complementary DNA clones (cDNA) encoding respectively the S_6, S_{13}, S_{14} and S_{22} alleles (Nasrallah *et al.*, 1987; Lalonde et coll., 1989). In *Brassica campestris* L., a species very closely related to *Brassica oleracea*, Takayama *et al.* (1987) have also reported partial amino-acid sequences of S_8, S_9 and S_{12} glycoproteins, determined by gas phase microsequencing analysis. By contrast to the bulk of information relative to the female partner, little is known about the nature of pollen S molecules (Gaude and Dumas, 1987). In this context, our laboratory is particularly interested in investigating the molecular basis of the expression of the S locus in the male partner. In this article, we present some attempts to identify and characterize the pollen S proteins and their corresponding DNA coding sequences.

The strategy we have been developing for this purpose is based

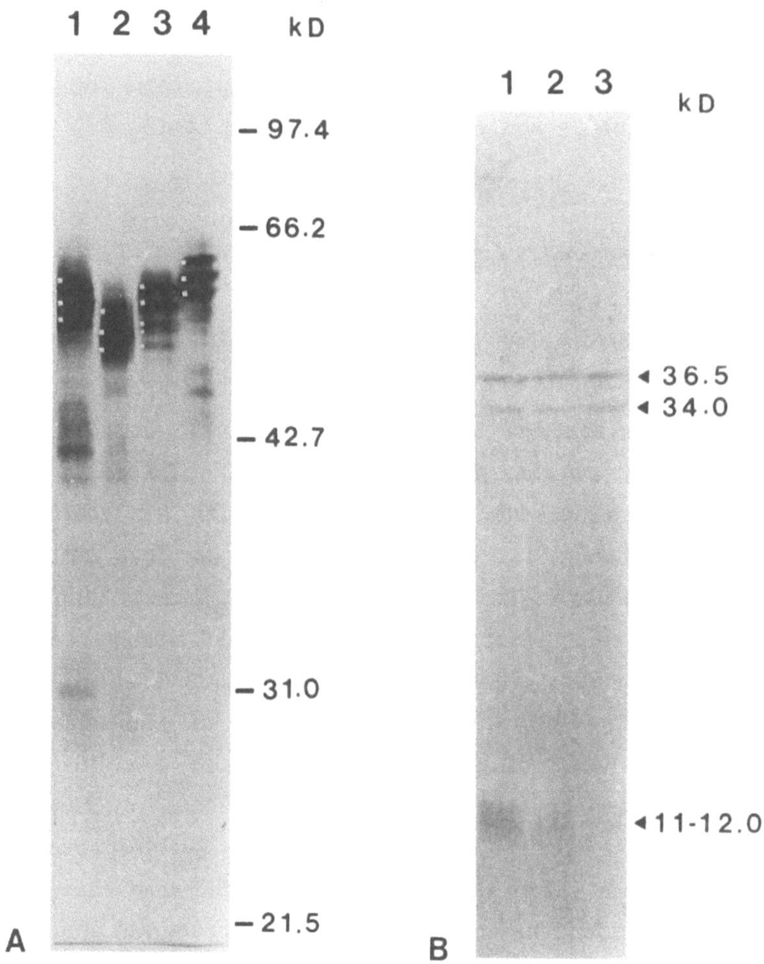

Figure 2: **A**. Immunoblot on nitrocellulose of stigma proteins separated by SDS-PAGE (10% gel) and probed with MAb 97-53 (1:1000 dilution). Lanes 1 to 4: stigma proteins (10 μg/lane) from S_6S_6, $S_{13}S_{13}$, $S_{14}S_{14}$ and $S_{22}S_{22}$ genotypes of *B. oleracea*. Stigma extracts possess several MAb-binding components whose major antigenic bands correspond to the SLSGs (white dots).

B. Immunoblot on nitrocellulose of pollen proteins separated by SDS-PAGE (15% gel) and probed with MAb 97-53 (1:100 dilution). Lanes 1 to 3: pollen proteins (40 μg) from S_6S_6, $S_{14}S_{14}$, and S_6S_{14} genotypes. Three MAb-binding proteins are detected in pollen extracts whatever the S-allele of the plant.

on the fact that the self-incompatibility response results from an identity of S alleles in pollen and stigma, and on the presumption that some structural homologies between S products of the same allele (at the protein and mRNA levels) should be found.

3.1 Translational level

We have raised a monoclonal antibody (MAb) directed against the S_6 stigma glycoprotein ($SLSG^6$) of *Brassica oleracea* and used it as a probe to detect any antigenic (structural) homology between stigma and pollen proteins (Gaude *et al.*, 1988). This antibody (MAb 97-53) revealed to present a very high affinity for the antigenic solution and possessed a broad specificity to various stigma SLSGs (Fig.2A). This broad specificity reflects a common antigenicity shared between the stigmatic S-molecules that was expected with regard to the high structural homologies previously described (Nasrallah *et al.*, 1987). Moreover, because of the wide binding ability of the MAb, the amino-acid residues involved in the antigenic determinants are unlikely to belong to the regions responsible for the specificity of each stigma S-glycoprotein.

When used on electroblots of SDS-PAGE separated pollen proteins, MAb 97-53 was able to detect three antigenic bands (Fig.2B). Though this cross-reaction must be considered carefully as this can be attributed either to partial epitope identity (relevant interactions), or to irrelevant interactions with no structural homology among antigens, the MAb-binding proteins detected in pollen constitute the only putative candidates for the S-specific molecules of the male gametophyte. An interesting point of our study is that no cross-reaction between pollen and stigma proteins of similar molecular mass was observed. This suggests that the S-specific substances in pollen may have different molecular structures from S-glycoproteins in the stigma. Lastly, even though we cannot say for certain that the cross-reacting material in pollen is associated with the S-gene expression, the MAb technology appears a promising technique in the study of the self-incompatibility at the molecular level.

3.2 Molecular genetic level

In the literature, the strategy proposed for cloning DNA sequences encoding stigma SLSGs has been based on three points: 1) S specific glycoproteins have been identified for several alleles; 2) their synthesis is stigma specific and 3) it represents in some strains, 5% of total protein synthesis, one day before anthesis (Nasrallah *et al.*, 1985a). The cDNA libraries were constructed from stigma mRNA extracted one day before anthesis and SLSG clones were isolated by differential hybridization (Nasrallah *et al.*, 1985b; Trick and Flavell, 1989).

We first hoped to use the same strategy to isolate male S-associated sequences, but no male S allele-specific glycoproteins were detectable through comparative analysis of pollen glycoproteins extracted from different S-lines (Gaude T, unpublished data). In fact, recent results suggest that putative pollen S-associated products occur in much lower concentration than stigma ones (Gaude *et al.*, 1988; Nasrallah *et al.*, 1986).

An alternative strategy has been developed based on the hypothesis of homology between male and female S-associated molecules. Our purposes are to use nucleic probes encoding stigma S-products in order to identify homologous sequences expressed in male tissues, and to characterize the synthesis regulation of such sequences throughout the male development.

By using Polymerase Chain Reaction (PCR), stigma specific S-product probes were produced without the cloning step. Oligonucleotides complementary to the 3' boundaries of the S_6 stigma glycoprotein cDNA sequence published by Nasrallah *et al.* (1987) were synthesized and annealed as primers to genomic DNA. The Taq DNA polymerase primer extension gave an amplification product of 1215 pb in size which corresponds to the size of the target sequence (Fig.3, A), knowing that the gene is intronless (Nasrallah *et al.*, 1988).

The S-specificity of the PCR product was demonstrated by RNA dot-blot analysis: a strong hybridization of this probe was obtained on mature self incompatible stigma RNA (one day before anthesis), while no signal was detected with leaf RNA (Fig. 3 B-1

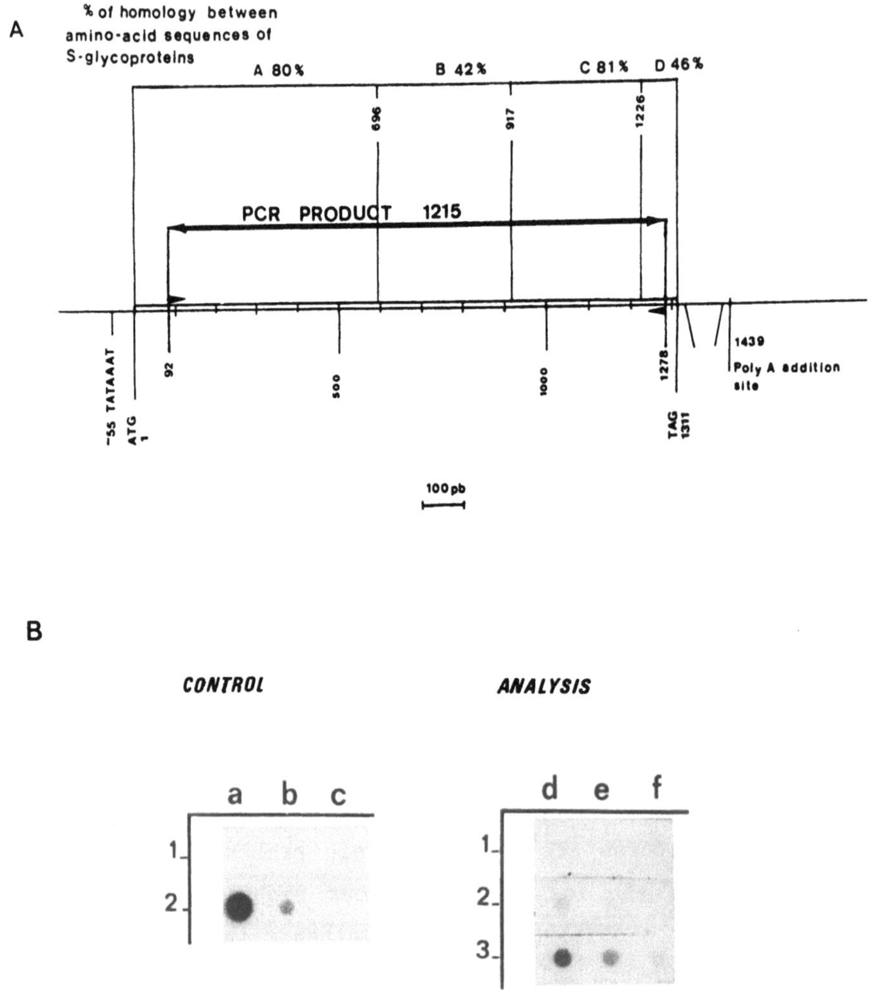

Figure 3: RNA DOT-BLOT ANALYSIS USING PCR PROBE

A- Localisation of the PCR product used as a probe
B- RNA dot-blot analysis:
CONTROL: 1-Leaf RNA, a:200ng; b:20ng; c:2ng. **2-S16-Stigma RNA (one day before anthesis),** a:200ng; b:20ng, c:2ng.
ANALYSIS: 1-Tricellular pollen stage, mRNA, d:100ng; e:10ng; f:1ng. **2-Bicellular pollen stage, mRNA,** d:500ng; e:100ng; f:10ng. **3-Microspore stage, total RNA,** d:10μg; e:1μg; f:100ng.

and -2, a,b,c). A further control was also made by using the S-associated BS 29-2 and the S-related BS 29-1 clones, supplied by Dr M. Trick, as probes (Trick and Flavell, 1989); hybridization signals were obtained only with the BS 29-2 clone. This reveals that the PCR product homology is limited to the S-gene, and so, that this fragment results exclusively from S-gene and not from S-related gene amplification (data not shown).

The RNA dot-blot analysis made with the PCR probe shows the presence of S-gene homologous sequences in self-incompatible anther tissue (Fig.3 B-1, -2, -3 d,e,f). The hybridization signal is lower than in the stigma and decreases during the male development. The PCR probe detects S-homologous sequences at immature stages, in microspore and bicellular pollen (Fig.3 B-2, -3 d, e, f). Thus, as it is generally admitted, putative male S-products seems to be produced by the anther stamen at early development stages and to get integrated in the pollen wall when stamen degenerates.

cDNA libraries have been constructed from microspore anther RNA and the isolation of sequences detected by PCR probe is in progress. Putative male S-gene(s) will be characterized in order to find out the extent of homology between male and female S-products. Once the S molecules of both partners has been identified, it will be possible to carry out the determination of the molecular interaction occuring during the male-female recognition event.

CONCLUSION

The process of fertilization in angiosperms is becoming a more and more exciting area because of recent emerging data and concepts.

The first concerns the isolation and characterization of S specific gene and S related gene controlling the self-incompatibility from the female side. But, the putative S genes and S products from the male side remain to be identified.

The second is a real possibility for investigators to build up an *in vitro* intergametic fusion in the near future. This prospect tool could be used to prove the putative preprogramed fertilization and to analyze the cytoplasmic DNA inheritance by the use of specific probes.

In addition, such an area is a typical interface between basic research: "*cell recognition is a root of biology* " and, applied research. This last aspect includes both the manipulation of breeding systems with S genes and plant transformation *via* the germ cell route.

REFERENCES

Barnabas B, Kovacs G, Abranyi A, Pfahler P (1988) Effect of pollen storage by drying and deep-freezing on the expression of different agronomic trait in maize *(Zea mays* L.). Euphytica 39: 221-225

Clarke AE, Knox RB (1980) Plants and immunity. Dev Comp Immunol 3: 571-589

Clarke AE, Anderson MA, Bacic T, Harris PJ, Mau SL (1985) Molecular basis of cell recognition during fertilization in higher plants. J Cell Sci S2: 261-285

Digonnet-Kerhoas C, Gay G, Duplan JC, Dumas C (1989) Viability of *Cucurbita pepo* pollen: biophysical and structural data. Planta 179: 165-170

Dumas C, Knox RB (1983) Callose and determination of pistil viability and incompatibility. Theor Appl Genet 67: 1-10

Dumas C, Knox RB, Gaude T (1984b) Pollen-pistil recognition: New concepts from electron microscopy and cytochemistry. Int Rev Cytol 90: 239-272

Dumas C, Knox RB, McConchie CA, Russell SM (1984a) Emerging physiological concepts in fertilization. What's New in Plant Physiol 15: 17-20

Dumas C, Bowman RB, Gaude T, Guilluy CM, Heizmann P, Roeckel P, Rougier M (1988) Stigma and stigmatic secretion reexamined. Phyton 28:193-200

Dupuis I, Roeckel P, Matthys-Rochon E , Dumas C (1987) Procedure to isolate viable sperm cells from corn (*Zea mays* L.) pollen grains. Plant Physiol 85: 876-878

Dupuis I, Dumas C (1989) *In vitro* pollination as a model for studying fertilization in maize (*Zea mays* L.) Sex Plant Reprod 2

Ebert PR, Anderson MA, Bernatzky R, Altschuler M, Clarke AE (1989) Genetic polymorphism of self-incompatibility in flowering plants. Cell 56: 255-262

Gaude T , Dumas C (1987) Molecular and cellular events of self-incompatibility. *In* KL Giles, J Prakash, eds, Pollen: Cytology and Development, Int Rev Cytol Vol 107. Academic Press, Orlando, pp 333-366

Gaude T, Nasrallah ME, Dumas C (1988) Use of monoclonal antibodies to determine putative S-products of pollen in *Brassica oleracea*. Heredity 61: 317-318

Gay G, Kerhoas C, Dumas C (1987) Quality of a stress sensitive *Cucurbita pepo* L.pollen. Planta 171: 82-87

Heslop-Harrison J, Heslop-Harrison Y, Shivanna KR (1984) The evaluation of pollen quality and a further appraisal of the fluorochromatic (FCR) test procedure. Theor Appl Genet 67: 367-375

Hoekstra FA, Crowe LM, Crowe JH (1989) Differential desiccation sensitivity of corn and *Pennisetum* pollen linked to their sucrose contents. Plant Cell Environ 12: 83-91

Isogai A, Takayama S , Tsukamoto C, Ueda Y , Shiozawa H , Hinata K, Okazaki K , Suzuki A (1987) S-locus-specific glycoproteins associated with self-incompatibility in *Brassica campestris*. Plant Cell Physiol 28: 1279-1291

Kerhoas C, Gay G, Dumas C (1987) A multidisciplanary approach to the study of the plasma membrane in *Zea mays* pollen during a controlled dehydration. Planta 171: 1-10

Kerhoas C, Dumas C (1988) Pollen quality in *Zea mays* as a pre-requisite for sperm cell isolation and pollen transformation. *In* HJ Wilms, CJ Keijzer, eds, Plant sperm cell as tools for biotechnology. Pudoc Wageningen, p 97

Knox RB (1984) Pollen pistil interactions. *In* HF Linskens, J. Heslop-Harrison, eds, pp 508-608

Knox RB, Williams EG, Dumas C (1986) Pollen, pistil and Reproductive function in Crop plants. *In* J. Janick, ed, Plant Breeding Rev. The Avi Publ. Co., Westport USA, Vol. 4, pp 8-79

Lalonde BA, Nasrallah ME, Dwyer KG, Chen CH, Barlow B, Nasrallah JB (1989) A highly conserved *Brassica* gene with homology to the S-locus-specific glycoprotein structural gene. Plant Cell 1: 249-258

Lee DJ, Blake TK, Smith SE (1988) Biparental inheritance of chloroplast DNA and the existence of heteroplasmic cell in alfalfa. Theor Appl Genet 76: 545-549

Longo FJ (1989) Fertilization. Chapman and Hall. Outline Studies in Biology. London New York. 183p.

Matthys-Rochon E, Vergne P, Detchepare S, Dumas C (1987) Male germ unit isolation from three tricellular pollen species *Brassica oleracea* , *Zea mays* and *Triticum aestivum* . Plant Physiol 83: 464-466

Mogensen HL (1988) Exclusion of male mitochondria and plastids during syngamy as a basis for maternal inheritance. Proc Natl Acad Sci 85: 2594-2597

Mogensen HL Wagner VT, Dumas C. Quantitative, three-dimensional ultrastructure of isolated corn (*Zea mays*) sperm cells. Theor Appl Genet (in press)

Nasrallah JB, Doney RC , Nasrallah ME (1985a) Biosynthesis of glycoproteins involved in the pollen-stigma interaction of incompatibility in developing flowers of *Brassica oleracea* L. Planta 165: 100-107

Nasrallah JB, Kao TH , Goldberg ML , Nasrallah ME (1985b) A cDNA clone encoding an S locus-specific glycoprotein from *Brassica oleracea*. Nature 318: 263-267

Nasrallah ME, Nasrallah JB (1986) Molecular biology of self-incompatibility in plants TIG Sept: 239-244

Nasrallah JB, Kao TH, Chen CH, Goldberg ML, Nasrallah ME (1987) Amino-acid sequence of glycoproteins encoded by three alleles of the S locus of *Brassica oleracea*. Nature 326: 617-619

Nasrallah JB, Su-May Yu, Nasrallah ME (1988) Self-incompatibility genes of *Brassica oleracea*: expression, isolation and structure. Proc Natl Acad Sci 85: 5551-5555

Nettancourt De D (1977) Incompatibility in Angiosperms. Springer Verlag, Berlin , pp 1-230

Ockendon DJ (1985) Genetics and physiology of self-incompatibility in *Brassica*. *In* I Sussex, A Ellingboe, M Crouch, R Malmberg, eds, Plant Cell/Cell Interactions. Cold Spring Harbor Laboratory, New York, pp 1-6

Roeckel P, Matthys-Rochon E, Chaboud A, Russell SD, Dumas C. Sperm cells. *In* S Blackmore, RB Knox, eds, Microsporogenesis. Academic Press (in press)

Russell SD (1983) Fertilization in *Plumbago zeylanica* : gametic fusion and fate of the male cytoplasm. Am J Bot 70: 416-434

Russell SD (1985) Preferential Fertilization in *Plumbago zeylanica*: ultrastructural evidence for gamete-level recogniton in an angiosperm. Proc Natl Acad Sci 82: 6129-6132

Takayama S, Isogai A , Tsukamoto C , Ueda Y, Hinata K , Okazaki K , Suzuki A (1986a) Isolation and some characterization of S-locus specific glycoproteins associated with self-incompatibility in *Brassica campestris*. Agric Biol Chem 50: 1365-1367

Takayama S, Isogai A, Tsukamoto C , Ueda Y , Hinata K , Okazaki K , Koseki K, Suzuki A (1986b) Structure of carbohydrate chains of S-glycoproteins in *Brassica campestris* associated with self-incompatibility. Agric Biol Chem 50: 1673-1676

Takayama S, Isogai A, Tsukamoto C , Ueda Y , Hinata K , Okazaki K , Suzuki A (1987) Sequences of S-glycoproteins, products of the *Brassica campestris* self-incompatibility locus. Nature 326: 102-105

Towill LE (1985) Low temperature and freeze-/vacuum drying preservation of pollen. *In* KK Kartha, ed, Cryopreservation of plant cells and organs, CRC Press, Inc Boca Raton Florida, p 171

Trick M, Flavell RB (1989) A homozygous S-genotype of *Brassica oleracea* expresses two S-like genes. Mol Gen Genet 218: 112-117

Wagner VT, Dumas C, Mogensen HL (1989a) Morphometric analysis of isolated *Zea mays* sperm. J Cell Sci 93: 179-184

Wagner VT, Song YC, Matthys-Rochon E, Dumas C (1989b) Observations on the isolated embryo sac of *Zea mays* L. Plant Sci 59: 127-132

Wagner VT, Dumas C, Mogensen HL (1990) Comparative three-dimensional study on the position of the female germ cells as a prerequisite for corn (*Zea mays*) transformation. Theor Appl Genet (in press)

STRUCTURE AND FUNCTION OF EGG-ASSOCIATED PEPTIDES OF SEA URCHINS

Norio Suzuki
Noto Marine Laboratory
Kanazawa University
Ogi, Uchiura
Ishikawa 927-05
Japan

The sea urchin egg is surrounded by a gelatinous extracellular matrix called the jelly coat through which spermatozoa must pass before reaching the plasma membrane of the egg during fertilization (Fig.1). Lillie (1913) observed that the jelly coat, dissolved in sea water, is able to cause a transitory activation of sperm motility and sperm agglutination. Thereafter, many investigators have shown that soluble factors associated with sea urchin eggs stimulate the motility and respiration of sea urchin spermatozoa (Carter, 1931; Cohn, 1918; Gray, 1928; Rothschild, 1956a, 1956b). Hathaway (1963) reported that the factors are diffusible in dialysis, heat-stable, alcohol-soluble and non-volatile. Ohtake (1976a, 1976b) demonstrated that the respiration of sea urchin spermatozoa could be reproducibly stimulated by partially purified egg jelly factors obtained from the sea urchin *Pseudocentrotus depressus* if extracellular pH was maintained at acidic values. Kopf *et al* (1979) and Hansbrough and Garbers (1981a) then found that the egg jelly of the sea urchin *Strongylocentrotus purpuratus* contains a factor which elevates the respiratory rate and cyclic AMP and cyclic GMP concentrations of *S. purpuratus* spermatozoa. Suzuki and co-workers (1981) have purified from the egg jelly of the sea urchin *Hemicentrotus pulcherrimus* a decapeptide able to stimulate sperm respiration whose sequence is Gly-Phe-Asp-Leu-Asn-Gly-Gly-Gly-Val-Gly. The same peptide was also purified from the egg jelly of *S. purpuratus* (Garbers *et al.*, 1982).

The jelly coat of the sea urchin has been shown to induce the acrosome reaction in sea urchin spermatozoa, which was first described in detail by Dan (1952, 1954, 1956). The jelly coat contains a polysaccharide-protein complex which can be separated into a sialoglycoprotein and a fucose-sulfate glycoconjugate (Decker *et al.*, 1976; Isaka *et al.*, 1970; Ishihara *et al.*, 1973). The fucose-sulfate glycoconjugate has been considered to be responsible for induction of the acrosome reaction (Garbers *et al.*, 1983; Kopf and Garbers, 1980; SeGall and Lennarz, 1979). Yamaguchi and coworkers (1988) have reported that sperm-activating peptide-free macromolecular fraction as

NATO ASI Series, Vol. H 45
Mechanism of Fertilization
Edited by B. Dale

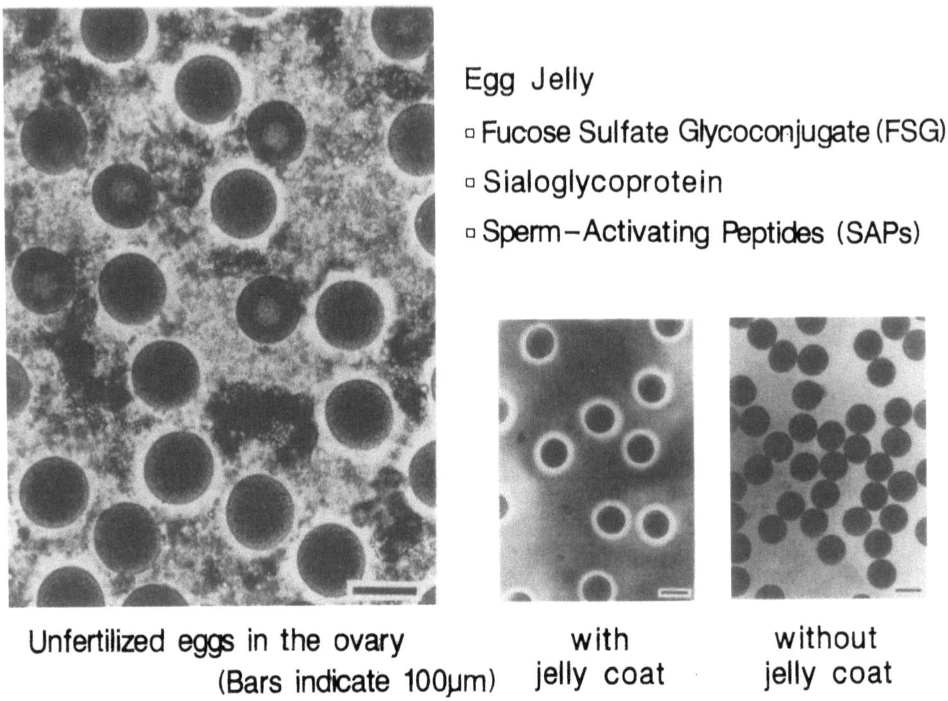

Fig. 1. Unfertilized eggs and egg jelly molecules of *Hemicentrotus pulcherrimus*. To visualized the jelly coat, a drop of 10-fold diluted indian ink was added to the specimens.

well as sperm-activating peptide-free fucose-rich glycoconjugate prepared from the egg jelly of *H. pulcherrimus* was less active than crude jelly in induction of the acrosome reaction and that addition of a synthetic sperm-activating peptide I (SAP-I) increased the rate of the acrosome reaction to those seen with crude jelly. In this paper, I describe the structure and function of sperm-activating peptides.

1. Purification and Structure of Various Peptides

Since the jelly coat of sea urchins is made of highly viscous large molecular weight glycoproteins, it is quite difficult to handle intact jelly coat with common biochemical methods. To analyze the components of the jelly coat unfertilized eggs are treated with acidified sea water (pH 5.0 to 5.5), which dissolves the jelly coat. The centrifuged supernatant 'soluble egg jelly' was then mixed with a 2-fold volume of 95% ethanol to extract peptide and then centrifuged. The resultant supernatant was concentrated under reduced pressure at 50°C and lipids were removed by chloroform extraction.

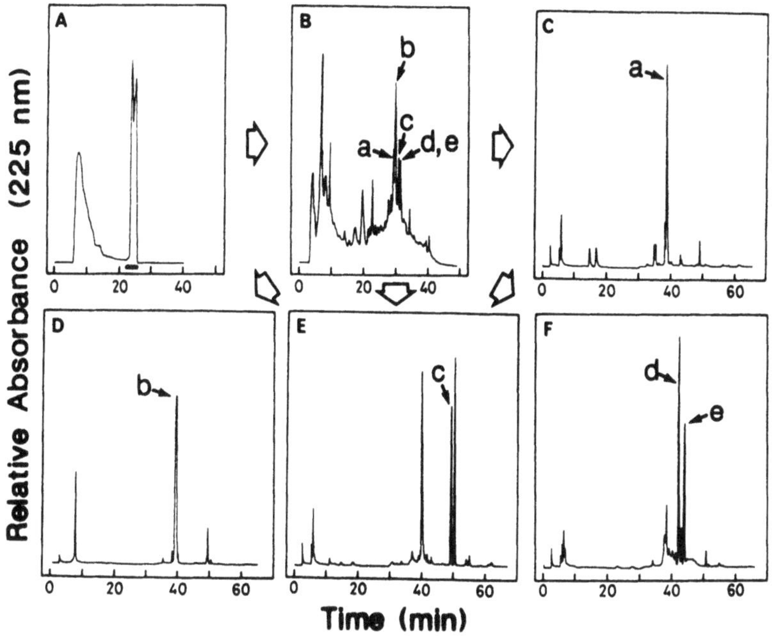

Fig. 2. Purification of sperm-activating peptides from the egg jelly of *Hemicentrotus pulcherrimus*. A. Program I; B. Program II; C-F. Program III. Peak a:Gly-Phe-Ser-Leu-Asn-Gly-Gly-Gly-Val-Gly; Peak b:Gly-Phe-Asp-Leu-Asn-Gly-Gly-Gly-Val-Gly; Peak c:Ser-Phe-Ala-Leu-Gly-Gly-Gly-Gly-Val-Gly; Peak d: Gly-Phe-Asp-Leu-Thr-Gly-Gly-Gly-Val-Gly; Peak e:Gly-Phe-Ser-Leu-Ser-Gly-Ser-Gly-Val-Asp.

The water layer was then lyophilized and the residue, dissolved in deionized and distilled water, was used for peptide purification. Sperm-activating peptides were purified by sequential HPLC on a reverse-phase column (Suzuki *et al.*, 1987b, 1988a). In general, separations were carried out using a combination of the following programs: Program I: Flow rate is 9.9 ml/min, the column (Shimadzu C-8 Prep, particle size 15μm, 20 x 250 mm) is equilibrated with 5% acetonitril (ACN) in 0.1% trifluroacetic acid (TFA) in water and eluted for 15 min with equilibration solvent, followed by elution with 60% ACN in 0.1% TFA in water for the next 15 min. Program II: Flow rate is 1.0 ml/min, the column (Unisil C-8, particle size 5μm, 4.6 x 250 mm) is equilibrated with 10% ACN in 0.1% TFA and eluted for 10 min with equilibration solvent, followed by a linear gradient of ACN from 10% to 50% in 0.1% TFA over a 50 minute time period. Program III: Flow rate is 1.0 ml/min, the same column used in Program II is equilibrated with 5% ACN in 5 mM sodium phosphate (pH 5.7) and eluted for 20 min with the equlibration solvent, followed by a linear gradient of ACN from 5% to 30% in 5 mM sodium phosphate (pH 5.7) for over a 40 minute time period.

Table 1 Structure of various sperm-activating peptides

Order Diadematoida

Diadema setosum

Gly-Cys-Pro-Trp-Gly-Gly-Ala-Val-Cys (SAP-IV)
Gly-(X)-Pro-(X)-Gly-Gly-Ala-Val-

Order Arbacioida

Glyptocidaris crenularis
Ser-Ala-Lys-Leu-Cys-Pro-Gly-Gly-Asn-Cys-Val
Lys-Leu-Cys-Pro-Gly-Gly-Asn-Cys-Val (SAP-IIB)
Leu-Cys-Pro-Gly-Gly-Asn-Cys-Val
Ser-Phe-Lys-Leu-Cys-Pro-Gly-Gly-Gln-Cys-Val
Lys-Leu-Cys-Pro-Gly-Gly-Gln-Cys-Val
Leu-Cys-Pro-Gly-Gly-Gln-Cys-Val

Arbacia punctulata
Cys-Val-Thr-Gly-Ala-Pro-Gly-Cys-Val-Gly-Gly-Gly-Arg-Leu-NH_2 (SAP-IIA)

Order Echinoida

Lytechinus pictus
Gly-Phe-Asp-Leu-Thr-Gly-Gly-Gly-Val-Gln
Phe-Asp-Leu-Thr-Gly-Gly-Gly-Val-Gln

Tripneustes gratilla
Gly-Phe-Asp-Leu-Asn-Gly-Gly-Gly-Val-Gly (SAP-I)
Gly-Phe-Asn-Leu-Asn-Gly-Gly-Gly-Val-Gly
Gly-Phe-Ser-Ile-Gly-Gly-Gly-Gly-Val-Gly
Gly-Phe-Asp-Leu-Gly-Gly-Gly-Gly-Val-Gly
Gly-Phe-Ser-Leu-Gly-Gly-Gly-Gly-Val-Gly
Gly-Phe-Gly-Leu-Gly-Gly-Gly-Gly-Val-Gly
Gly-(Br-Phe)-Asn-Leu-Asn-Gly-Gly-Gly-Val-Gly
Gly-(Br-Phe)-Asp-Leu-Asn-Gly-Gly-Gly-Val-Gly

Pseudoboletia maculata
Gly-Phe-Ala-Leu-Asp-Gly-Val-Asn
Gly-Phe-Ala-Leu-(X)-Gly-Val-Gly
Gly-Phe-Ala-Leu-Asp-Gly-Val-Gly

Pseudocentrotus depressus
Gly-Phe-Asp-Leu-Asn-Gly-Gly-Gly-Val-Gly (SAP-I)
Gly-Phe-Asp-Leu-Thr-Gly-Gly-Gly-Val-Gly
Gly-Phe-Ala-Leu-Gly-Gly-Gly-Gly-Val-Gly

Strongylocentrotus nudus
Gly-Phe-Asp-Leu-Asn-Gly-Gly-Gly-Val-Gly (SAP-I)
Gly-Phe-Ser-Leu-Ser-Gly-Gly-Gly-Val-Gly
Gly-Phe-Ala-Leu-Gly-Gly-Gly-Gly-Val-Gly
Gly-Phe-Ser-Leu-Gly-Gly-Gly-Gly-Val-Gly
Gly-Phe-Asp-Leu-Thr-Gly-Gly-Gly-Val-Gly

Strongylocentrotus purpuratus
Gly-Phe-Asp-Leu-Asn-Gly-Gly-Gly-Val-Gly (SAP-I)
Gly-Phe-Ala-Leu-Gly-Gly-Gly-Gly-Val-Gly
Gly-Phe-Ser-Leu-Thr-Gly-Gly-Gly-Val-Gly

(continued Table 1)

Hemicentrotus pulcherrimus
Gly-Phe-Asp-Leu-Asn-Gly-Gly-Gly-Val-Gly (SAP-I)
Gly-Phe-Asp-Leu-Thr-Gly-Gly-Gly-Val-Gly
Gly-Phe-Ser-Leu-Asn-Gly-Gly-Gly-Val-Ser
Gly-Phe-Ala-Leu-Gly-Gly-Gly-Gly-Val-Gly
Gly-Phe-Ser-Leu-Ser-Gly-Ser-Gly-Val-Asp

Echinometra mathaei (type A)
Gly-Tyr-Ser-Leu-Ser-Gly-Gly-Ala-Val-Asp
Gly-Phe-Asn-Leu-Ser-Gly-Gly-Gly-Val-Gly
Gly-Phe-Ser-Leu-Ser-Gly-Gly-Gly-Val-Gly
Gly-Phe-Asp-Leu-Thr-Gly-Gly-Gly-Val-Gly

Echinometra mathaei (type B)
Gly-Tyr-Ser-Leu-Ser-Gly-Gly-Gly-Val-Asp
Gly-Tyr-Asn-Leu-Asn-Gly-Asp-Arg-Ile-Asp
Gly-Phe-Ser-Leu-Ser-Gly-Gly-Gly-Val-Gly
Gly-Phe-Asp-Leu-Thr-Gly-Gly-Gly-Val-Gly

Anthocidaris crassispina
Gly-Phe-Asp-Leu-Thr-Gly-Gly-Gly-Val-Gly
Gly-Phe-Asp-Leu-Ser-Gly-Gly-Gly-Val-Gly
Gly-Phe-Ser-Ley-Ser-Gly-Ser-Gly-Val-Gly

Heterocentrotus mammillatus
Gly-Thr-Leu-Pro-Thr-Gly-Ser-Gly-Val-Ser
Gly-Phe-Glu-Met-Gly-Gly-Thr-Gly-Val-Gly
Gly-Tyr-Asn-Leu-Gly-Gly-Gly-Gly-Ile-Asp
Gly-Phe-Gly-Leu-Ser-Gly-Gly-Gly-Ile-Gly

Order Clypeasteroida

Clypeaster japonicus
Asp-Ser-Asp-Ser-Ala-Gln-Asn-Leu-Ile-Gly (SAP-III)
Asp-Ser-Asp-Ser-Ala-His-Leu-Ile-Gly
Asp-Ser-Asp-Ser-Ala-Phe-Leu-Ile-Gly

By using the procedure, in the last nine years, we purified forty-one different sperm-activating peptides from the egg jelly of fourteen sea urchin species distributed over four taxonomic orders (Fig.2 and Table 1) (Nomura *et al.*, 1983; Shimomura *et al.*, 1986b; Suzuki *et al.*, 1981, 1984b, 1988a, 1988b; Yoshino *et al.*, 1989a, 1989b). Considering the structure and biological specificity, the peptides could be classified in four groups. Before obtaining two sperm-activating peptides from *Diadema setosum* egg jelly, we have given a specific name to each group of peptides. However, it is difficult to find a good name for the peptides from *D. setosum* egg jelly using the same way, in which we have named speract (sperm-activating peptide) resact (respiration-activating peptide) and mosact (motility-activating peptide). It is possible that we obtain more different sperm-activating

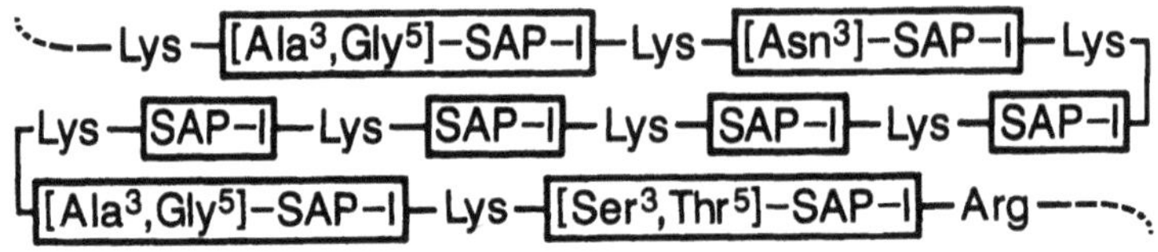

Fig. 3. A putative partial structure of sperm activating peptide precursor protein of *Strongylocentrotus purpuratus*.

peptides from the egg jelly of sea urchins in different taxonomic orders specific for spermatozoa of a given species. These peptides as well as peptides with established structures should have essentially the same biological activities toward spermatozoa of the species. Therefore, we need a generalized name for the peptides and I proposed sperm-activating peptide as the general name, which is abbreviated as SAP (Suzuki, 1989). Thus, for example, speract and [Ser^5]-speract will be called sperm-activating peptide I, which will be abbreviated as SAP-I, and [Ser^5]-sperm-activating peptide I, which will be abbreviated as [Ser^5]-SAP-I. In this nomenclature, resact becomes SAP-II and mosact SAP-III. A peptide from *D. setosum* will be SAP-IV.

As shown in Table 1, a diversity of peptide structures exists, and even within the same species, many variants of the same basic peptide structure have been isolated. Since the egg jelly has been collected from many individuals of a given species, it has not been clear wether or not the multiple peptide structures within the species is due to individual variation, a diverse number of peptides encoded within a single mRNA or multiple genes. Ramarao *et al* (1989) have isolated cDNA clones coding for SAP-I and determined the DNA sequences. The deduced amino acid sequence of the putative precursor protein contained multiple SAP-I and SAP-I-like structures (Fig.3). This shows that peptide diversity within a given species is explained by the synthesis of variants from a single mRNA.

2. Effects of Peptides on Sea Urchin Spermatozoa

Respiration and motility of sea urchin spermatozoa are highly dependent on pH of suspending medium (Fig.4). Below pH 6.0, sea urchin spermatozoa are practically immotile and do not respire. Decreased respiration rates and motility of spermatozoa due to acidification of sea water can be reversed by the addition of sperm-activating-peptides (Fig.4). The stimulated respira-

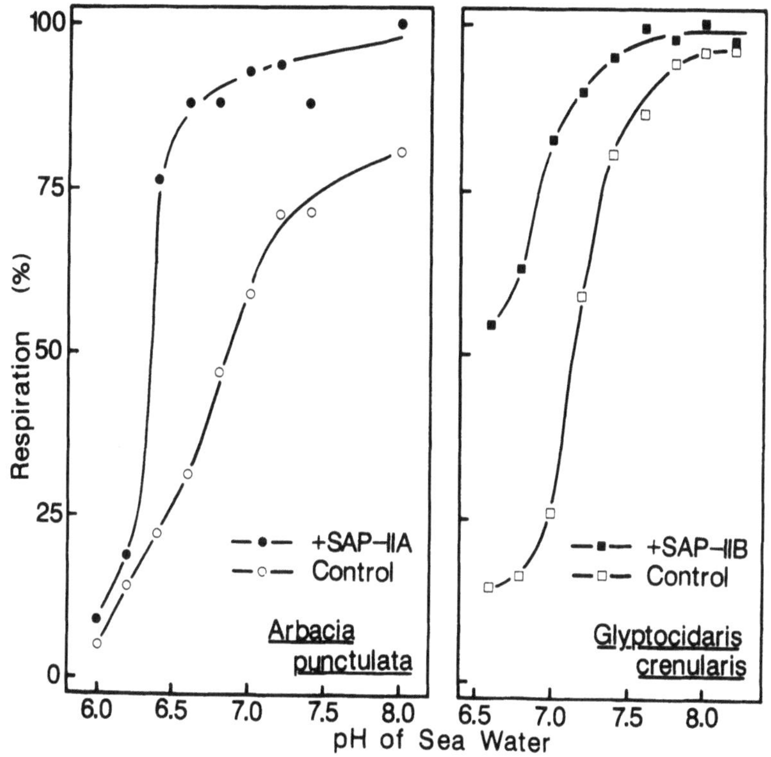

Fig. 4. pH dependence of sperm respiration with or without a sperm-activating peptide. Sperm respiration was expressed as the percentage of respiration with a sperm-activating peptide (1.36×10^{-7}M of SAP-IIA or 3.7×10^{-7}M of SAP-IIB) at pH 8.0

tion rate do not exceed that of spermatozoa in normal sea water, pH 8.2, with exception of some cases. SAP-IIA-stimulated respiration rate of spermatozoa from a given individual of *Arbacia punctulata* exceeded that of spermatozoa in any pH of sea water. This may explain why some investigators obtained respiratory stimulation effects of soluble egg jelly on spermatozoa from certain species of sea urchins even in normal sea water. However, to obtain reproducible respiratory stimulation by sperm-activating peptides, it is recommended to use slightly acidic sea water (pH 6.6 to 6.8) (Ohtake, 1976a).

Sperm-activating peptides are specific at the ordinal level (Table 2) (Suzuki *et al.*, 1982, 1988c) and stimulate sperm respiration one half-maximally at about 10-100 pM. Respiratory stimulation induced by sperm-activating peptides is dependent on the concentration of external Na^+ (Hansbrough and Garbers, 1981b; Suzuki *et al.*, 1984a). Approximately 50 mM Na^+ was required for half-maximal respiratory responses to peptides. Monensin, an ionophore

Table 2 Specificity of respiratory stimulation by sperm-activating peptides

	Spermatozoa used			
	D. setosum	*G. crenularis*	*H. pulcherrimus*	*C. japonicus*
SAP-IV	+	-	-	-
SAP-IIB	-	+	-	-
SAP-IIA	-	±	-	-
SAP-I	-	-	+	-
SAP-III	-	-	-	+

The respiration-stimulating activity is expressed as + when a peptide stimulated sperm respiration half-maximally at less than 5 nM. When the half-maximal stimulation occurred at between 5 and 500 nM, ± was used. - means no respiratory stimulation.

that catalyzes an electro-neutral Na^+/H^+ exchange across cell membrane stimulated sperm respiration and motility. Repaske and Garbers (1983) have reported that SAP-I triggered Na^+/H^+ exchange across the *S. purpuratus* sperm plasma membrane and raised an intracellular pH.

It has been known that sea urchin spermatozoa possess enzymes which involve in cyclic nucleotide metabolism (Garbers and Hardman, 1975; Garbers *et al.*, 1974, 1978; Sano, 1976; Swarup and Garbers, 1982; Wells and Garbers, 1976). An egg jelly constituent fucose-sulfate glycoconjugate causes 100-fold increases in sea urchin sperm cyclic AMP concentrations within 1 min, but fails to elevate cyclic GMP concentrations (Garbers *et al.*, 1983). Sperm-activating peptides, however, cause transient increases in cyclic GMP as well as cyclic AMP within a few seconds in both acidic and normal sea water (Garbers *et al.*, 1982; Suzuki *et al.*, 1984a, 1988b). The increases in cyclic GMP are explained by transient activation of the membrane form of guanylate cyclase which is a major protein of sperm tail plasma membrane (Bentley and Garbers, 1986a, 1986b; Bentley *et al.*, 1986a, 1986b; Garbers, 1988; Sano, 1976). It is noted that when a synthetic NH_2-terminal fragment (Cys-Val-Thr-Gly-Ala-Pro-Gly) and a synthetic CO_2-terminal fragment (Cys-Val-Gly-Gly-Gly-Arg-Leu-NH_2) of SAP-IIA were simultaneously added to *A. punctulata* spermatozoa, cyclic GMP concentrations were elevated at the same relative concentrations as observed with SAP-IIA although the CO_2-terminal fragment alone had only 0.1% of the cyclic GMP elevating activity of SAP-IIA and the NH_2-terminal fragment did not elevate cyclic GMP (Shimomura and Garbers, 1986).

Guanylate cyclase of *A. punctulata* spermatozoa could be radiolabeled when the spermatozoa were incubated in normal sea water with ^{32}P-orthophosphate (Ward and Vacquier, 1983). When the spermatozoa are treated with soluble egg jelly or SAP-IIA, the guanylate cyclase loses its ^{32}P-label and shifts its relative mobility on SDS polyacrylamide gels from 160 kDa to 150 kDa (Suzuki *et al.*, 1984b; Ward and Vacquier, 1983; Ward *et al.*, 1985a, 1985b). Similar mobility shifts are commonly observed in sea urchin spermatozoa treated with a specific sperm-activating peptide (Suzuki *et al.*, 1987b, 1988a, 1988b; Yoshino *et al.*, 1989a, 1989b). The loss of phosphate from guanylate cyclase results in a large decrease in the specific activity (Ramarao and Garbers, 1985). Loss of phosphate, the mobility shift and decrease in activity of guanylate cyclase are all Na^+-dependent, Ca^{2+}-dependent and inducible with an ionophore monensin (Vacquier, 1986; Ward *et al.*, 1985a, 1986). Therefore, it appears that a sperm-activating peptide causes an initial activation of the guanylate cyclase (Garbers, 1988; Ramarao and Garbers, 1988).

As I described above, sperm-activating peptides exhibit several biochemical and physiological activities toward sea urchin spermatozoa. These include acceleration of sperm motility and respiration, transient elevation of cyclic GMP concentrations, activation, dephosphorylation and inactivation of the membrane form of guanylate cyclase. In addition to these, two more biological roles for specific sperm-activating peptides have been identified. When SAP-IIA was injected *A. punctulata* sperm suspension, the spermatozoa showed a striking chemotactic response (Ward *et al.*, 1985b). They changed their swimming behaviour and formed a cluster in the area of microinjected SAP-IIA. The chemoattraction to SAP-IIA was species specific and dependent on concentration and the presence of Ca^{2+}. The chemotaxis of animal spermatozoa to eggs or egg secretions from female reproductive system is a widespread phenomenon (Miller, 1985a, 1985b). However, the chemical nature of animal sperm chemoattractants have until recently unknown. SAP-IIA is the first egg-derived molecule of known structure shown to be a chemoattractant of animal spermatozoa.

The jelly coat has been shown to induce the acrosome reaction which is accompanied by net Na^+ and Ca^{2+} influx, and H^+ and K^+ efflux (Cristen *et al* ., 1983; Darszon *et al.*, 1986; Guerrero and Darszon, 1989; Schackmann and Shapiro, 1981; Schackmann *et al.*, 1978, 1981). These ionic changes are also induced by sperm-activating peptides (Lee and Garbers, 1986; Schackmann and Chock, 1986). Yamaguchi *et al* (1988) have recently reported that sperm-activating peptide-free macromolecular fraction prepared from the egg jelly of the sea urchin *H. pulcherrimus* was less active than crude jelly in

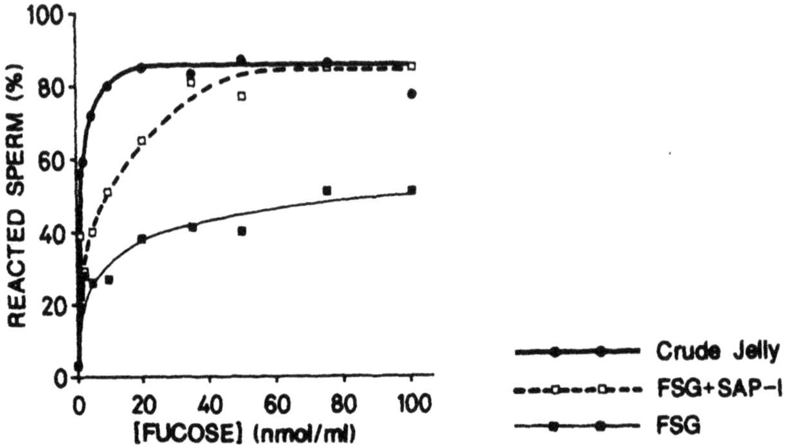

Fig. 5. Dependence of the acrosome reaction on the concentration of crude jelly, fucose-rich glycoconjugate and fucose-rich glycoconjugate with SAP-I.

induction of the acrosome reaction and that addition of synthetic SAP-I increased the rate of the acrosome reaction to those seen with crude jelly. This was also confirmed by experiments using a fucose-rich glycoconjugate purified from *H. pulcherrimus* egg jelly and synthetic SAP-I (Fig.5) (Yamaguchi *et al* ., 1989). SAP-I increased induction of the acrosome reaction with the fucose-rich glycoconjugate in a concentration-dependent manner. The half-maximal increase occurred at 4×10^{-10}M of SAP-I. This value corresponds to the concentration at which SAP-I half-maximally stimulates sperm respiration. These results suggest that SAP-I promotes induction of the acrosome reaction by acting as a specific co-factor of fucose-rich glycoconjugate.

3. Peptide Receptors on Sperm Plasma Membrane

Smith and Garbers (1983) first described that *S. purpuratus* spermatozoa had approximately 6,000-8,000 receptors/cell specific for SAP-I. Suzuki *et al* suggested that the receptors exclusively localize on the sperm tail of *H. pulcherrimus*. Using a radioiodinated synthetic SAP-I analog (Gly-Gly-Gly-Tyr-Asp-Leu-Asn-Gly-Gly-Gly-Val-Gly) or SAP-IIA analog (Gly-Gly-Gly-Tyr-Cys-Val-Thr-Gly-Ala-Pro-Gly-Cys-Val-Gly-Gly-Gly-Arg-Leu-NH_2) with respiration-stimulating abilities equal to SAP-I or SAP-IIA, Bentley and Garbers (1986a, 1986b) and Bentley *et al* (1986a, 1986b) demonstrated that SAP-I or SAP-IIA receptor exists on a plasma membrane vesicle prepared from *S. purpuratus* or *A. punctulata* spermatozoa. Dangott and Garbers (1984, 1987) successfully crosslinked ^{125}I-SAP-I analog to intact *S. purpuratus* spermatozoa with a bifunctional cross-linking agent disuccinimidyl suberate. A single radiolabeled band was detected on SDS polyacrylamide gels with an apparent molecular weight of 77,000. The cross-linking reaction was specific and the radioiodinated SAP-I analog failed to cross-link to spermatozoa of unreactive species. Recently, Dangott and co-workers (1989) purified the receptor protein from *S. purpuratus* spermatozoa using DEAE-Sepharose chromatography and preparative gel electrophoresis. They digested the purified protein with *Staphylococcus aureus* V8 protease and isolated peptide fragments. One of the fragments yielded the amino acid sequence Val-Ser-Ala-Pro-Phe-Asp-Leu-Glu-Ala-Pro-Phe-Ile-Ile-Asp-Gly-Ile. Subsequently, they isolated a clone containing DNA sequences encoding open frame of 532 amino acids that included the above peptide sequence. The deduced amino acid sequence suggests that the protein contains a 26-residue amino-terminal signal peptide, a large extracellular domain relatively rich in cysteine that includes a four-fold repeat at about 115 amino acids, a single membrane-spanning region, and only 12 amino acid residues extending into the cytoplasm.

Identification of SAP-IIA receptor was carried out by the same cross-linking method as SAP-I receptor using ^{125}I-SAP-IIA analog and *A. punctulata* spermatozoa. Shimomura *et al* (1986a) provided evidence that the cross-linked protein was the same molecule as the membrane form of guanylate cyclase. Singh and co-workers (1988) isolated cDNA clone corresponding to the guanylate cyclase. The cDNA clone would encode for 986 amino acids with a 21-amino acid signal sequence and one hydrophobic transmembrane domain. Amino acid sequences of SAP-I receptor protein and guanylate cyclase do not show significant identity. The relationship between the two proteins remains to be solved.

References

Bentley JK, Garbers DL (1986a) Retention of the speract receptor by isolated plasma membranes of sea urchin spermatozoa. Biol Reprod 34:413-421

Bentley JK, Garbers DL (1986b) Receptor-mediated responses of plasma membranes isolated from *Lytechinus pictus* spermatozoa. Biol Reprod 35: 1249-1259

Bentley JK, Shimomura H, Garbers DL (1986a) Retention of a functional resact receptor in isolated sperm plasma membrane. Cell 45:281-288

Bentley JK, Tubb DJ, Garbers DL (1986b) Receptor-mediated activation of spermatozoan guanylate cyclase. J Biol Chem 262:14859-14862

Carter GS (1931) Iodine compounds and fertilization. II. The oxygen consumption of suspensions of sperm of *Echinus esculeutus* and *Echinus miliaris*. J Exp Biol 8:176-193
Cohn EJ (1918) Studies of the physiology of spermatozoa. Biol Bull 34:167-218
Cristen R, Schackmann RW, Shapiro BM (1983) Interaction between sperm and sea urchin egg jelly. Dev Biol 98:1-14
Dan JC (1952) Studies of the acrosome. I. Reaction to egg-water.and other stimuli. Biol Bull 103: 54-66
Dan JC (1954) Studies of the acrosome. III. Effects of calcium deficiency. Biol Bull 107:335-349
Dan JC (1956) The acrosome reaction. Int Rev Cytol 5:365-393
Dangott LJ, Garbers DL (1984) Identification and partial characterization of the receptor for speract. J Biol Chem 259:13712-13716
Dangott LJ, Garbers DL (1987) Further characterization of a speract receptor on sea urchin spermatozoa. Ann NY Acad Sci 513:274-283
Dangott LJ, Jordan JE, Bellet RA, Garbers DL (1989) Cloning of the mRNA for the protein that crosslinks to the egg peptide speract. Proc Natl Acad Sci USA 86:2128-2132
Darszon A, Soto JG, Lievano A, Sanchez JA, Trejo ADI (1986) Ionic channels in the plasma membrane of sea urchin sperm. In Latorre R(ed)Ionic Channels in Cells and Model System, Plenum Press, New York London, pp 291-305
Decker GL, Joseph DB, Lennarz WJ (1976) A study of factors involved in induction of the acrosome reaction in sperm of the sea urchin, *Arbacia punctulata*. Dev Biol 53:115-125
Garbers DL (1988) Signal/transduction mechanisms of sea urchin spermatozoa. ISI Atlas of Science: Biochemistry/1988, pp 120-126
Garbers DL, Hardman JG (1975) Factors released from sea urchin eggs affect cyclic nucleotide metabolism. Nature(London) 257:677-678
Garbers DL, Hardman JG, Rudolph FG (1974) Kinetic analysis of sea urchin sperm guanylate cyclase. Biochemistry 13:4166-4171
Garbers DL, Watkins HD, Tubb DJ, Kopf GS (1978) Regulation of spermatozoan cyclic nucleotide metabolism by egg factors. Adv Cyclic Nucleotide Res 9:583-595
Garbers DL, Watkins HD, Hansbrough JR, Smith AC, Misono KS (1982) The amino acid sequence and chemical synthesis of speract and of speract analogues. J Biol Chem 257:2734-2737
Garbers DL, Kopf GS, Tubb DJ, Olson G (1983) Elevation of sperm adenosine 3':5'-monophosphate concentrations by a fucose-sulfate-rich complex associated with eggs: I. Structural characterization. Biol Reprod 29: 1211-1220
Gray J (1928) The effect of egg secretions on the activity of spermatozoa. J Exp Biol 5:362-365
Guerrero A, Darszon A (1989) Egg jelly triggers a calcium influx which inactivates and is inhibited by calmodulin antagonists in the sea urchin sperm. Biochim Biophys Acta 980:109-116
Hansbrough JR, Garbers DL (1981a) Speract. Purification and characterization of a peptide associated with eggs that activates spermatozoa. J Biol Chem 256:1447-1452
Hansbrough JR, Garbers DL (1981b) Sodium-dependent activation of sea urchin spermatozoa by speract and monensin. J Biol Chem 256:2235-2241
Hathaway RR (1963) Activation of respiration in sea urchin spermatozoa by egg water. Biol Bull 125:486-498
Iasaka S, Hotta K, Kurokawa M (1970) Jelly coat substances of sea urchin eggs. Exp Cell Res 59:37-42

Ishihara K, Oguri K, Taniguchi H (1973) Isolation and charactrization of fucose sulfate from jelly coat glycoprotein of sea urchin egg. Biochim Biophys Acta 320:628-634

Kopf GS, Garbers DL (1980) Calcium and fucose-sulfate-rich polymer regulate sperm cyclic nucleotide metabolism and the acrosome reaction. Biol Reprod 22:1118-1126

Kopf GS, Tubb DJ, Garbers DL (1979) Activation of sperm respiration by a low molecular weight egg factor and 8-bromoguanosine 3',5'-monophosphate. J Biol Chem 254:8554-8560

Lee HC, Garbers DL (1986) Modulation of the voltage-sensitive Na^+/H^+ exchange in sea urchin spermatozoa through membrane potential changes induced by the egg peptide speract. J Biol Chem 261:16026-16032

Lillie FR (1913) Studies on fertilization. V. The behaviour of the spermatozoa of *Nereis* and *Arbacia* with special reference to egg extracts. J Exp Zool 14:515-574

Miller RL (1985a) Demonstration of sperm chemotaxis in Echinodermata: Asteroidea, Holothuroidea, Ophiuroidea. J Exp Zool 234:383-414

Miller RL (1985b) Sperm chemo-orientation in the metazoa. In Metz CBJr, Monroy A(eds) Biology of Fertilization vol 2, Academic Press, New York, p 275-337

Nomura K, Suzuki N, Ohtake H, Isaka S (1983) Structure and action of sperm activating peptides from the egg jelly of the sea urchin, *Anthocidaris crassispina*. Biochem Biophys Res Commun 117:147-153

Ohtake H (1976a) Respiratory behaviour of sea-urchin spermatozoa. I. Effect of pH and egg water on the respiratory rate. J Exp Zool 198:303-312

Ohtake H (1976b) Respiratory behaviour of sea-urchin spermatozoa. II. Sperm-activating substance obtained from jelly coat of sea-urchin eggs. J Exp Zool 198:313-322

Ramarao CS, Garbers DL (1985) Receptor-mediated regulation of guanylate cyclase activity in spermatozoa. J Biol Chem 260:8390-8396

Ramarao CS, Garbers DL (1988) Purification and properties of the phosphorylated form of guanylate cyclase. J Biol Chem 263:1524-1529

Ramarao CS, Burks DJ, Garbers DL (1989) Molecular cloning of the complementary DNA for an egg peptide that stimulates spermatozoa. (submitted to J Biol Chem)

Repaske DR, Garbers DL (1983) A hydrogen ion flux mediate stimulation of respiratory activity by speract in sea urchin spermatozoa. J Biol Chem 258:6025-6029

Rothschild L (1956a) The physiology of sea-urchin spermatozoa. Action of pH, dinitrophenol, dinitrophenol + versene, and usnic acid on O_2 uptake. J Exp Biol 33:155-173

Rothschild L (1956b) The respiratory dilution effect in sea-urchin spermatozoa. *Vie Milie* 7:405-415

Sano M (1976) Subcellular localizations of guanylate cyclase and 3',5'-cyclic nucleotide phosphodiesterase in sea urchin sperm. Biochim biophys Acta 172:20-30

SeGall GK, Lennarz WJ (1979) Chemical characterization of the component of the jelly coat from sea urchin eggs responsible for induction of the acrosome reaction. Dev Biol 71:33-48

Schackmann RW, Eddy EM, Shapiro BM (1978) The acrosome reaction of *Strongylocentrotus purpuratus*. Dev Biol 65:483-495

Schackmann RW, Shapiro BM (1981) A partial sequence of ionic changes associated with the acrosome reaction of *Strongylocentrotus purpuratus*. Dev Biol 81:145-154

Schackmann RW, Cristen R, Shapiro BM (1981) Membrane potential depolarization and increased intracellular pH accompanying the acrosome reaction of

sea urchin sperm. Proc Natl Acad Sci USA 78:6066-6070
Schackmann RW, Chock PB (1986) Alteration of intracellular [Ca^{2+}] in sea urchin sperm by the egg peptide speract. J Biol Chem 261:8719-8728
Shimomura H, Dangott LJ, Garbers DL (1986a) Covalent coupling of a resact analogue to guanylate cyclase. J Biol Chem 261:15778-15782
Shimomura H, Suzuki N, Garbers DL (1986b) Derivatives of speract are associated with the eggs of *Lytechinus pictus* sea urchins. Peptide 7:491-495
Singh S, Lowe DL, Thorpe DS, Rodrigues H, Kuang WJ, Dangott LJ, Chinkers M, Goeddlel DV, Garbers DL (1988) Membrane guanylate cyclase is a cell-surface receptor with homology to protein kinases. Nature(London) 334:708-712
Smith AC, Garbers DL (1983) The binding of an ^{125}I-speract analogue to spermatozoa. In Lennon DLF, Stratman FW, Zhalten RN(eds)Biochemistry of Metabolic Processes, Elsevier/North-Holland, New York, p 15-28
Suzuki N (1989) Sperm-activating peptides from sea urchin egg jelly. In Scheuer PJ(ed)Bioorganic Marine Chemistry vol 3, Springer, Berlin, Heidelberg, New York, p 47-70
Suzuki N, Nomura K, Ohtake H, Isaka S (1981) Purification and the primary structure of sperm-activating peptides from the jelly coat of sea urchin eggs. Biochem Biophys Res Commun 99:1238-1244
Suzuki N, Hoshi M, Nomura K, Isaka S (1982) Respiratory stimulation of sea urchin spermatozoa by egg extracts, egg jelly extracts and egg jelly peptides from various species of sea urchins: Taxonomical significance. Comp Biochem Physiol 72A:489-495
Suzuki N, Ohizumi Y, Yasumasu I, Isaka S (1984a) Respiration of sea urchin spermatozoa in the presence of a synthetic jelly coat peptide and ionophores. Develop Growth Differ 26:17-24
Suzuki N, Shimomura H, Radany EW, Ramarao CS, Bentley JK, Garbers DL (1984b) A peptide associated with eggs causes a mobility shift in a major plasma protein of spermatozoa. J Biol Chem 259:14874-14879
Suzuki N, Kurita M, Yoshino K, Yamaguchi M (1987a) Speract binds exclusively to sperm tails and causes an electrophoretic mobility shift in a major sperm tail protein of sea urchins. Zool Sci 4:641-648
Suzuki N, Kurita M, Yoshino K, Kajiura H, Nomura K, Yamaguchi M (1987b) Purification and structure of mosact and its derivatives from the egg jelly of the sea urchin *Clypeaster japonicus*. Zool Sci 4:649-657
Suzuki N, Kajiura H, Nomura K, Garbers DL, Yoshino K, Kurita M, Tanaka H, Yamaguchi M (1988a) Some more speract derivatives associated with eggs of sea urchins, *Pseudocentrotus depressus, Strongylocentrotus purpuratus, Hemicentrotus pulcherrimus* and *Anthocidaris crassispina*. Comp Biochem Physiol 89B:687-693
Suzuki N, Yoshino K, Kurita M, Nomura K, Yamaguchi M (1988b) A novel group of sperm-activating peptides of the sea urchin *Glyptocidaris crenularis*. Comp Biochem Physiol 90C:305-311
Suzuki N, Yoshino K, Kurita M, Yamaguchi M, Amemiya S (1988c) Taxonomical significance of respiratory stimulation of sea urchin spermatozoa by egg associated substances. In Burke RD, Mladenov PV, Lambert P, Parsley RD (eds) Echinoderm Biology, Balkema AA, Rotterdam, p 213-218
Swarup G, Garbers DL (1982) Phosphoprotein phosphatase activity of sea urchin spermatozoa. Biol Reprod 26:953-960
Vacquier VD (1986) Activation of sea urchin spermatozoa during fertilization. TIBS 11:77-81
Yamaguchi M, Niwa T, Kurita M, Suzuki N (1988) The participation of speract in the acrosome reaction of *Hemicentrotus pulcherrimus*. Develop Growth Differ 30:159-167
Yamaguchi M, Kurita M, Suzuki N (1989) Induction of the acrosome reaction of

Hemicentrotus pulcherrimus spermatozoa by the egg jelly molecules, fucose-rich glycoconjugate and sperm-activating peptide I. Develop Growth Differ 31:233-239

Yoshino K, Kajiura H, Nomura K, Takao T, Shimonishi Y, Kurita M, Yamaguchi M, Suzuki N (1990) A halogenated amino acid-containing sperm-activating peptide and its related peptides isolated from the egg jelly of sea urchins, *Tripneustes gratilla, Pseudoboletia maculata, Strongylocentrotus nudus, Echinometra methaei* and *Heterocentrotus mammillatus*. Comp Biochem Physiol (in press)

Yoshino K, Kurita M, Yamaguchi M, Nomura K, Takao T, Shimonishi Y, Suzuki N (1990) A species specific sperm-activating peptide from the egg jelly of the sea urchin *Diadema setosum*. Comp Biochem Physiol (in press)

Ward GE, Vacquier VD (1983) Dephosphorylation of a major sperm membrane protein is induced by egg jelly during sea urchin fertilization. Proc Natl Acad Sci USA 80:5578-5582

Ward GE, Garbers DL, Vacquier VD (1985a) Effects of extracellular egg factors on sperm guanylate cyclase. Science 227:768-770

Ward GE, Brokow CJ, Garbers DL, Vacquier VD (1985b) Chemotaxis of *Arbacia punctulata* spermatozoa to resact, a peptide from the egg jelly layer. J Cell Biol 101:2324-2329

Ward GE, Moy GW, Vacquier VD (1986) Phosphorylation of membrane-bound guanylate cyclase of sea urchin spermatozoa. J Cell Biol 103:95-101

Wells JN, Garbers DL (1976) Nucleotide 3',5'-monophosphate phosphodiesterases in sea urchin sperm. Biol Reprod 15:46-53

SPERM- EGG INTERACTION IN BIVALVES

F.Rosati,R.Focarelli
Department of Evolutionary Biology
Via Mattioli 4
53100 Siena
Italy

Introduction

Fertilization is the process by which the gametes from two individuals of the same species fuse together to give rise to a new individual. Mechanisms operate at many levels to ensure the outcome of this process. However, the moment when the gametes recognize each other and interact is certainly one of the most important.

The gametes are highly differentiated cells having special structures for this interaction process. Despite the apparent variability of the gametes, according to group and species, there is a similar plan of organization of the structures involved in the interaction process in all the gamete types so far studied. The eggs are generally enclosed by an extracellular coat and another acellular or cellular envelope. The apical region of the sperm cell is characterized by an acrosomal apparatus which, on interaction with the egg, undergoes a series of changes, called acrosomal reaction. In the last decade a rigorous molecular approach to fertilization has begun to unravel the mechanisms by which eggs communicate with spermatozoa. In both vertebrates and invertebrates sperm surface receptors for the egg have been identified or strongly inferred (Shur, 1989; Wassarman,1988; Rosati, 1985; Garbers, 1989). In addition to surface receptors, the spermatozoon has been shown to contain guanine nucleotide regulatory proteins, calmodulin (a receptor for $Ca2^{+}$), cyclic AMP dependent protein kinase (the receptor for cyclic AMP) and other molecules found in signal transduction pathways in other cells (see Garbers 1989). The role of the cellular and acellular coat of the eggs in the interaction with the sperm has also been extensively studied in some species of invertebrates and vertebrates and the importance of the inner egg coat, namely the vitelline coat (or zona pellucida in mammals) in this process is widely recognized (Ruiz-Bravo and Lennarz, 1989; Garbers, 1989; Rosati, 1985; Wassarman, 1988; Yanagimachi, 1988;). In particular a glycoprotein named Zp3 and fucosyl containing glycoproteins have

NATO ASI Series, Vol. H 45
Mechanism of Fertilization
Edited by B. Dale

been identified in mouse (see Wassarman, 1988) and in tunicates (see Rosati, 1985) respectively to be the egg coat component involved in this process. These glycoproteins were given the name "sperm receptors" and more recently "ligands" or "effectors" as a consequence of the role they play (Saling, 1989; Garbers, 1989). Different conclusions have been reached regarding the status of the spermatozoon when it interacts with this coat i.e whether the acrosome reaction occurs before or as consequence of this interaction. Very recently it was suggested that the sperm-egg interaction has two separate sequential steps, i.e two different bindings at least in the mouse (Saling, 1989). The first step involves the receptors on the plasma membrane of the sperm and the ZP3 component of the zona pellucida as ligand and has been termed "primary binding"; the second involves acrosomal components and ZP2, another component of the zona pellucida, and is known as "secondary binding". In this context it has recently been suggested that in the sea urchin the interaction between the bindin, exposed after the acrosomal reaction, and the vitelline coat, is analogous to secondary binding (Garbers, 1989). In this case the inducer of the acrosome reaction appears to be a fucose sulfate polymer present in the outer acellular jelly coat. However this also raises the question of whether or not the ligands of the receptors on the sperm surface of the different species are homologous; i.e whether the oocyte itself synthesizes the molecules involved in the first binding which triggers the acrosomal reaction or wheter this important function is fulfilled by other cells. In mammals and tunicates it has been clearly shown that the oocyte itself produces the components of the vitelline coat and zona pellucida (Bleil and Wassarman, 1980; Rosati et al., 1982). Little is known, however, about the cellular origin of the jelly coat components. Although there may be differences in the interaction process in different species, only by comparing analogous stages and analogous structures of this process can the similarities and differences of the various models be correctly evaluated. A rigorous revision of the terminology used up to now for the structures and events involved in the interaction process would be useful for this purpose.

In this paper we report the results of a study of two egg models of bivalve molluscs and their interaction with sperm.

Sperm-egg interaction in a freshwater bivalve

I Polarized site of sperm entrance

Whereas external fertilization is the rule in bivalve molluscs, in the group of Unionacea the eggs are internally fertilized i.e. the spermatozoa are released into the water by the male and drawn into the inhalant siphon by the female. The sperm encounter the eggs in the suprabranchial chamber or water tube of the gills where development begins and reaches the larval stage. When the eggs of *Unio elongatulus*, the species studied by us, break free from the ovary in which they have been compressed, they immediately swell and become spheroid. Their diameter is about 150 µm. A vitelline coat 0.3 µm thick surrounds the egg but there is no jelly coat as in other species of molluscs. The vitelline coat forms a characteristic protrusion, which we termed "crater" at the vegetal pole (Focarelli et al., 1988). A circular area around this crater is completely different from the rest of the vitelline coat being contracted in concentrically disposed wrinkles (Figs.1 and 2).

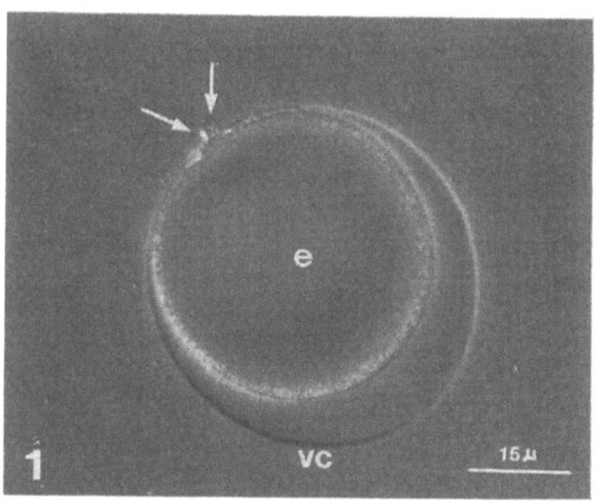

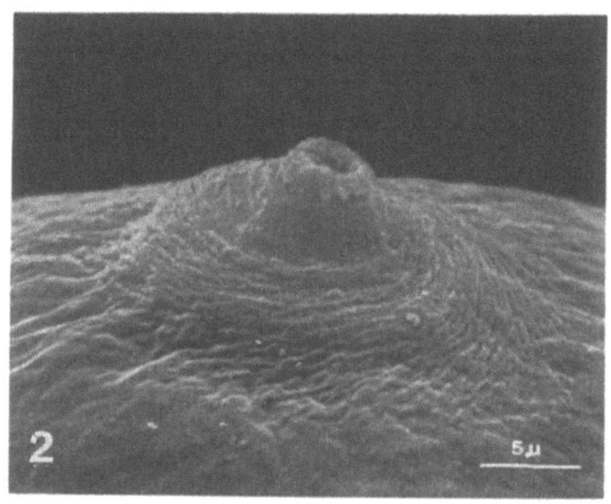

Fig.1 Interference contrast micrograph of an unfertilized egg of *Unio elongatulus*. The vitelline coat (vc) surrounding the egg (e) shows, at the vegetal pole, a conical protrusion, the crater(arrows).
Fig.2 Scanning electron micrograph of the vegetal region of the unfertilized egg of *Unio* showing the crater and the wrinkled area around it.

It has been suggested that the protruding crater acts as micropyle (Lillie 1901). However when the eggs are inseminated and the process followed by interferential phase microscopy it is evident that spermatozoa bind to and enters the vitelline coat only in the area around the crater (Figs. 3 and 5). As the spermatozoon enters, vacuolar material in the egg erupts explosively and the crater becomes wider and flatter (Fig.4) (Focarelli et al., 1988). Immediately after this, the wrinkled area seems to spirale around the crater dragging it down (Fig 6) (unpublished results).

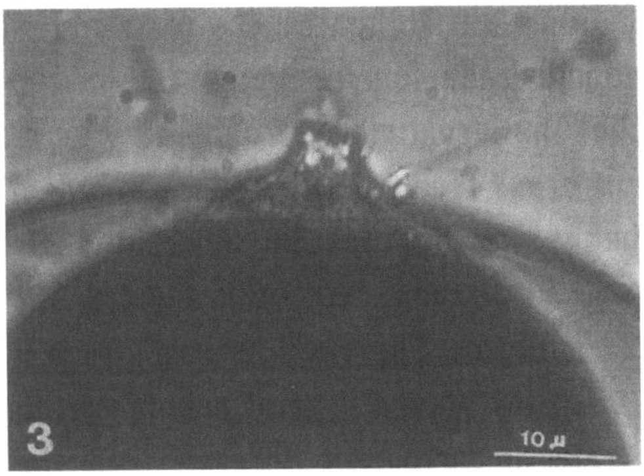

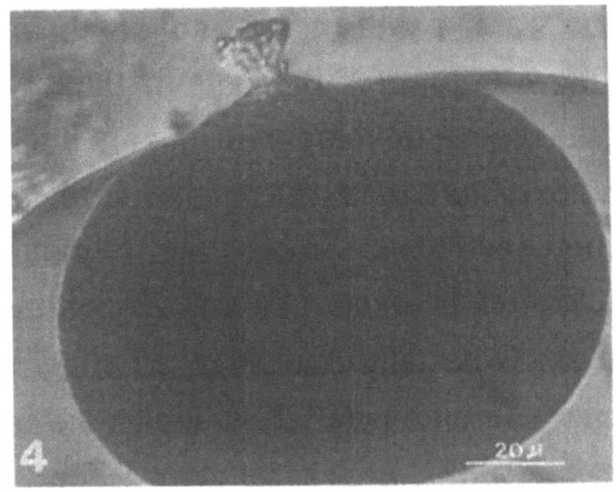

Fig.3 Interference contrast micrograph of an egg a few seconds after insemination. A spermatozoon bound to the vitelline coat in the region around the crater is clearly visible.
Fig.4 Interference contrast micrograph of an inseminated egg showing the explosive ejection from the crater of vacuolar material.

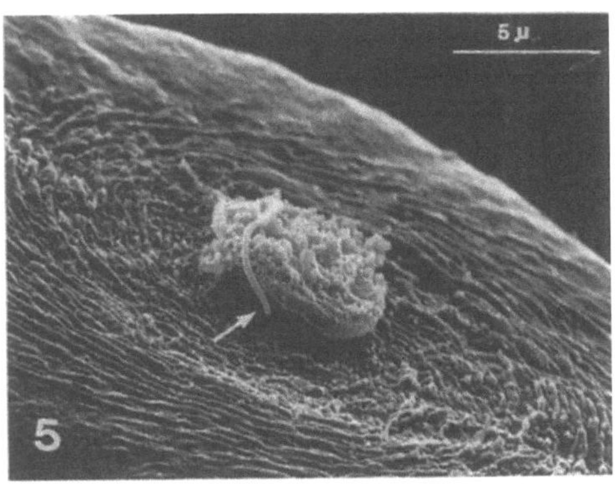

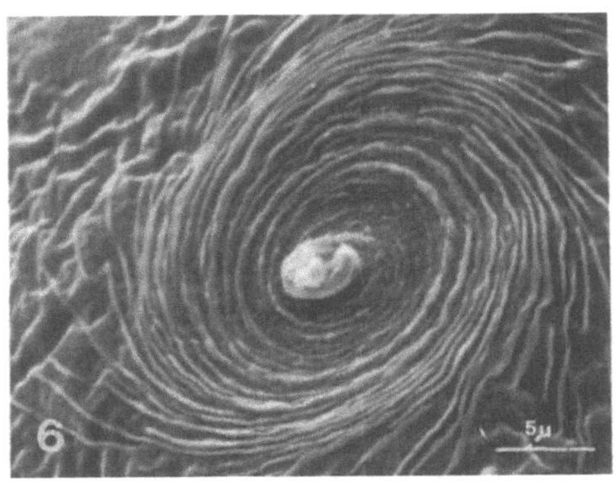

Fig.5 Scanning electron micrograph of a spermatozoon penetrating the vitelline coat (arrow) at the foot of the crater which now appears flatter and wider.
Fig.6 Scanning electron micrograph showing the spiral shape of the wrinkled area in a fertilized egg.

When tested with lectins the wrinkled area of the unfertilized egg is the only point of the vitelline coat strongly positive to LTA (Lotus tetragonolubus agglutinin) which recognizes fucosyl residues (Fig.7) (Focarelli et al., 1988). The same lectin also interferes with the binding of the sperm to the vitelline coat.

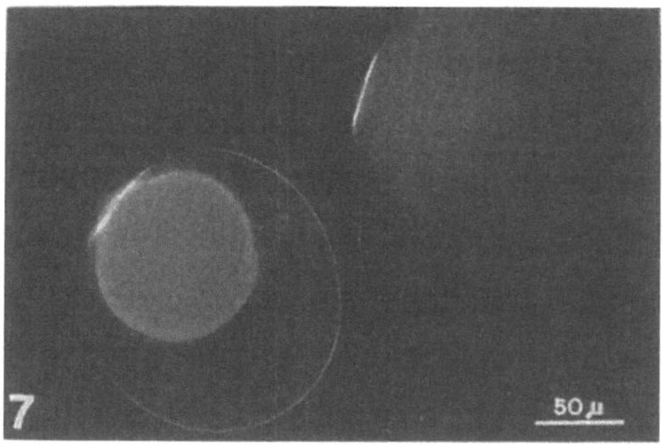

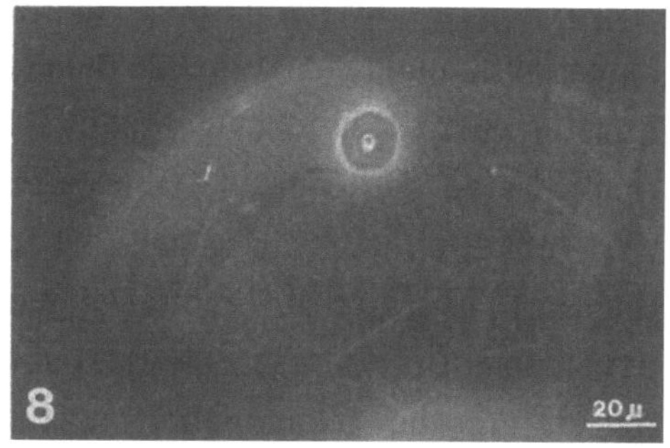

Figs.7 and 8 Fluorescence micrographs showing the localization of the LTA binding sites in the vitelline coat of two unfertilized (Fig.7) and a fertilized egg (Fig. 8)

After fertilization the LTA-positive area shrinks to about 1/3 of its initial size and shows narrow, spirally arranged zones of greater positivity (Fig.8).

The parallel between specific lectin sites and sperm binding capacity in the wrinkled area suggested to us that the sperm receptors (which we now propose to call ligands of the sperm receptors, according to the terminology of Saling,1989) are fucosyl-containing glycoproteins (Focarelli et al., 1988 and Focarelli et al., 1989). It is interesting that both in ascidians as in mammals fucose has been indicated as the key component of vitelline coat or zona pellucida glycoproteins involved in the interaction.

II Differentiation of the vitelline coat

Morphological observations have shown that the vitelline coat is synthesized by the growing oocyte and that the differentiation of the wrinkled area is strictly related to the formation of the crater in the young oocyte (Focarelli et al 1989). The position of the oocyte within the stroma of the ovary seems determinant in this process. In an early stage of development a surface bleb is formed in a site opposite to the area of attachment of the oocyte. Later the bleb loses its terminal portion, leaving behind its foot. The vitelline components in this region, which later become the crater region, are synthesized and accumulated in a completely different way from the rest of the oocyte; indeed, synthesis and accumulation continue here, and the vitelline coat remains closely apposed to the oocyte plasma membrane even when a wide perivitelline space is formed in the rest of the oocyte (Focarelli et al.,1989). LTA treatment of sections of the ovaries seems to indicate that fucosyl-containing glycoproteins are synthesized in the growing oocyte and first accumulated along the entire surface of the vitelline coat (Fig.9). Later, in conjunction with the development of the vitelline coat, they disperse into the rest of the oocyte except the crater region. In this zone they remain concentrated and continue to accumulate even in later stages of development (Focarelli et al., 1989) (Fig.10). The specific localization of many microtubules in the region of crater differentiation has suggested that they can play a role in the development of the asymmetrical shape of this oocyte and as a consequence in the differential accumulation of the vitelline coat components (Focarelli et al.,1989).

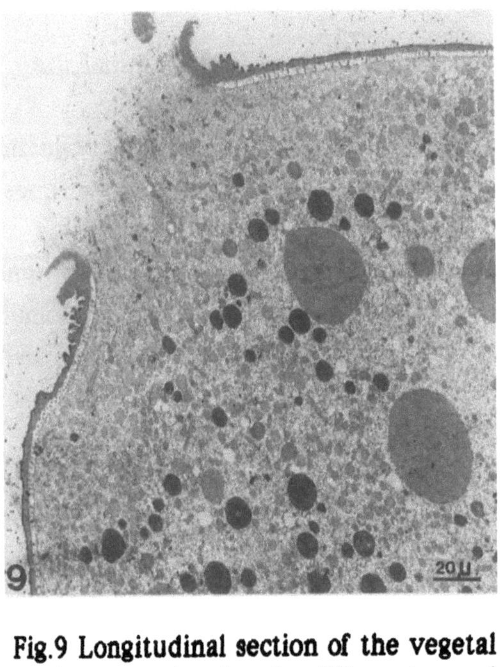

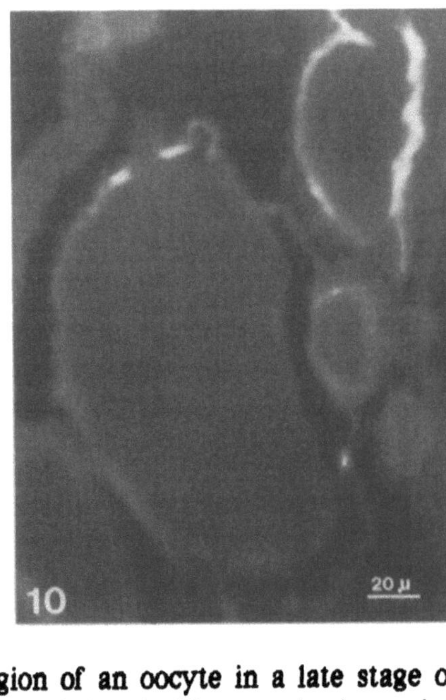

Fig.9 Longitudinal section of the vegetal region of an oocyte in a late stage of development showing the different morphology of the vitelline coat in the region around the crater.
Fig.10 High magnification of a section of ovary incubated with LTA-FITC showing strong labeling in the crater region.

III Biochemical characterization of the vitelline coat

We also performed a preliminary biochemical characterization of the vitelline coat. When proteins were extracted from the vitelline coats of the intact eggs with appropriate buffers and subjected to SDS-PAGE two main and some minor bands were detected. The two major representative peptides have an apparent molecular weight of 230 and 190 Kd respectively (Fig.11) (unpublished results). Many different bands were detected in the regions corresponding to these major components when the vitelline coats were isolated and then subjected to SDS PAGE (Focarelli et al.1989). Work is now in progress to analyze the rationale of this different pattern. In the presence of reducing agents (2-mercaptoethanol) the electrophoretic profile of the two main peptides remain unchanged (Fig.11). When the electrophoretic pattern was blotted and treated with different lectins conjugated with peroxidase, the 230 Kd component was strongly positive to all the lectins tested (WGA, DBA, ConA, LTA) whereas the other one was only slightly positive to LTA (unpublished results).

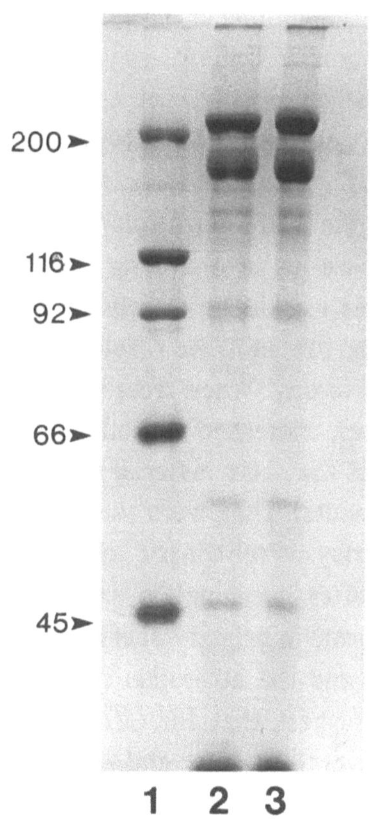

Fig.11 SDS PAGE electrophoresis of proteins extracted from vitelline coats of *Unio elongatulus* in the presence (Lane 2) or absence (Lane 3) of 2-mercaptoethanol. Lane 1: High molecular weight standards.

When the solubilized components of the vitelline coats were run on a 4% slab gel under non denaturating conditions (native) three species were resolved (unpublished results). Each of them was electroeluted and then separated by SDS-PAGE (unpublished results). Surprisingly, all of them contained both 230 and 180 Kd components. Although further studies are needed to assess this point, these preliminary results seem to suggest that the vitelline coat is made up of three different heterodimers composed by two principal glycoproteins. Work is underway to analyze the functional role of these glycoproteins.

Sperm-egg interaction in a marine bivalve

We also studied the egg and interaction process in *Mytilus galloprovincialis*. It was already known that the binding of the sperm to the vitelline coat occurred after exposure of the acrosomal content, suggesting that the inducer of the acrosomal reaction was the outer jelly coat as in the sea urchin (Longo and Anderson, 1969). As a first approach to identifying the egg molecules involved in the different steps of the recognition process we treated the oocyte with different lectins in view of the well known role of sugar residues in the process. In the course of this study we found that long filamentous protrusions (termed by us "vitelline coat spikes") could be clearly detected using the lectin DBA (Dolichos biflorus agglutinin) (unpublished results). Then we tried to detect these structures by electron microscopy. When cross sections were carefully observed the spikes were clearly seen enmeshed in jelly. In some sections it was clear that the spikes are made of the same material as the vitelline coat from which they arise (unpublished results). They were the only structures also positive to DBA in electron microscopy. Preliminary observations of the sperm-egg interaction a few seconds after insemination using living or fixed eggs, suggest that the spikes may play a role in primary sperm binding. Since there is binding between the vitelline coat and the acrosomal components of the sperm (Longo and Anderson,1969) we suggest that the *Mytilus* vitelline coat contains the components responsible for both primary and secondary binding like the mouse zona pellucida.

Conclusions

Fertilization is an event unquestionably unique; however we are now learning that it is made up of a series of steps some of which are commonly used by a wide variety of cells. In the past decade the progress in identifying the individual steps involved in the overall process has been outstanding. A detailed description of the sequential molecular events that occur during the first moment of sperm-egg interaction has been also achieved at least in mammals. Comparison between analogous moments as well as analogous and homologous molecular components involved in these first steps of sperm-egg communication in a diversity of animal models, should be useful to assess wheter a common molecular mechanism underlies this process.

REFERENCES

Bleil JD, Wassarman PM (1980) Synthesis of zona pellucida proteins by denuded and follicle-enclosed mouse oocytes during culture in vitro. Proc Natl Acad Sci USA 77:1029-1033

Focarelli R, Renieri T, Rosati F (1988) Polarized site of sperm entrance in the egg of a freshwater bivalve, Unio elongatulus. Dev Biol 127:443-451

Focarelli R, Rosa D, Rosati F (1989) (to be published) Differentiation of the vitelline coat and the polarized site of sperm entrance in the egg of Unio elongatulus (Mollusca,Bivalvia). J Exp Zool

Garbers DL (1989) Molecular basis of fertilization. Ann Rev Biochem 58:719-742

Longo FJ, Anderson E (1969) Cytological aspects of fertilization in the lamellibranch Mytilus edulis. I. Polar body formation and development of the female pronucleus. J Exp Zool 172:69-96

Rosati F, Cotelli F, De Santis R, Monroy A, Pinto MR (1982) Synthesis of fucosyl-containing glycoproteins of the vitelline coat in oocytes of Ciona intestinalis (Ascidia). Proc Natl Acad Sci U.S.A. 79:1908-1911

Rosati F (1985) Sperm-egg interaction in Ascidians. In: Metz CB and Monroy A (eds) Biology of Fertilization, Vol 2. Academic Press, London, p. 361

Ruiz-Bravo N, Lennarz WJ (1989) Receptors and membrane interactions during fertilization. In Schatten H and Schatten G (eds) The molecular biology of fertilization, Academic Press, San Diego, p 21

Saling P (1989) (to be published) Mammalian sperm interaction with extracellular matrices of the egg. In Oxford Review of Reproductive Biology

Shur BD (1989) Galactosyltransferase as a recognition molecule during fertilization and development. In: Schatten H and Schatten G (eds) The molecular biology of fertilization, Academic Press, San Diego, p 38

Wassarman PM (1988) Zona pellucida glycoproteins. Ann Rev Biochem 57:415-442

Yanagimachi R (1988) Mammalian fertilization. In: Knobil E and Neill J (eds) Physiology of reproduction, Vol 1. Raven Press, New York, p 135

GAMETE INTERACTION IN ASCIDIANS:SPERM BINDING AND PENETRATION THROUGH THE VITELLINE COAT

R. De Santis
Stazione Zoologica "A.Dohrn" di Napoli
Villa Comunale
80121 Napoli
Italy

M.R. Pinto
Institute of Protein Biochemistry and Enzymology
Via Toiano, 2
80072 Arco Felice, Napoli
Italy

NATO ASI Series, Vol. H 45
Mechanism of Fertilization
Edited by B. Dale

Despite the wealth of information available in different species, gamete interaction is still one of the most challenging problems in biology. The general consensus is that the encounter of the gametes triggers a cascade of events that culminates in sperm-egg fusion and involves many different molecules and mechanisms (De Santis and Pinto, 1988; Wassarman, 1987). Various models that allowed different approaches, have been used in attempts to clarify the steps of this highly complex process.

In this context, the study of gamete interaction in the ascidians has proved to be particularly suitable for the analysis of the early events of fertilization. In fact, in ascidians individual steps of this pathway can be dissected out when the egg envelopes are manually or chemically removed (Rosati and De Santis, 1978). Here we briefly review the results of studies carried out in ascidians to identify the molecules and the mechanisms involved in sperm-egg binding and sperm penetration through the vitelline coat.

SPERM BINDING TO THE VITELLINE COAT

Species-specific recognition and binding between the egg and the spermatozoon, which consists in a molecular match between highly specialized components of the gamete surfaces (De Santis et al., 1983; Hoshi et al., 1985; De Santis and Pinto, 1987; Casazza et al., 1988), is a preliminary and essential step of the gamete interaction (Rosati and De Santis, 1978).

Studies carried out in ascidians demonstrated that the vitelline coat (VC), the extracellular coat which surrounds the egg, is a vanguard for interaction with the spermatozoa (De Santis and Pinto, 1988). The receptors for the species specific recognition and binding are a product of the oocyte (Rosati et al., 1982) and are assembled on the vitelline coat which provides a rigid and rather stable structure in the interaction with the spermatozoa. Binding occurs between fibrils emerging from the vitelline coat and the plasma membrane at the tip of the sperm head (De Santis et al.,

1980; Honegger, 1986). In Ciona intestinalis, fucosyl residues localized on these fibrils have a key role in sperm binding (Rosati and De Santis, 1980; Pinto et al., 1981). In Phallusia mammillata and in Ascidia nigra, N-acetylglucosamine residues have the same function (Honegger, 1982; Lambert, 1986; Honegger, 1986). The sperm receptors have been isolated and partially characterized in Ciona intestinalis using fucose as a marker. Fucosyl proteins, a glycoprotein fraction of the egg VC, possess the properties of a sperm receptor: they inhibit fertilization and binding by competing with the sperm receptors on the VC and they induce sperm activation and the acrosome reaction (De Santis et al., 1983). Fucosyl proteins isolated from the ovaries were used for the biochemical and functional characterization of the receptors. This component is a high molecular weight glycoprotein complex (> 10^7) with a protein-carbohydrate ratio of 2:1. Exhaustive proteolytic digestion of fucosyl proteins yields high molecular weight glycopeptides (>4×10^5), that contain N-acetylgalactosamine, fucose, galactose and rhamnose. These glycopeptides retain sperm binding activity. However, only the intact receptor induces sperm activation and the acrosome reaction (De Santis and Pinto, 1987). Whether these are functions exerted by the polypeptide component alone, has yet to be investigated. This receptor shares many of the biochemical characteristics of the sperm receptor of the Strongylocentrotus purpuratus egg, which is a high molecular weight proteoglycan-like molecule (Rossignol et al., 1984). The biological activity is more similar to zona pellucida glycoprotein of the mouse egg (ZP3), which serves as both receptor for spermatozoa and inducer of the acrosome reaction (Wassarman, 1987).

The counterpart of these receptors on the spermatozoon and the mechanisms underlying the sperm-egg interaction have been studied in several ascidians. The data available support the hypothesis that glycoproteins on the egg envelope interact with enzymes of the sperm surface by forming an enzyme-substrate complex (Hoshi, 1984). Indeed, an α-L-fucosidase isolated from Ciona spermatozoa is involved in sperm binding: this enzyme would interact with the fucosyl glycoproteins of the vitelline coat in the very early stages of binding (Hoshi et al., 1985). In Phallusia mammillata, where N-acetylglucosamine is functional in sperm binding, it has been found that a ß-D-N-acetylglucosaminidase retains the highest activity among sperm glycosidases (Hoshi, 1984).

SPERM PENETRATION THROUGH THE VITELLINE COAT

After binding, spermatozoa undergo the acrosome reaction (De Santis et al., 1980; Honegger, 1986; Fukumoto, 1988) and pass through the vitelline coat, leaving the single mitochondrion outside (Lambert and Epel, 1979; Lambert and Lambert, 1981; Lambert and Lambert, 1984). In the genus Ascidia, ß-D-N-acetylglucosaminidase, the sperm surface VC-binding site, remains over the mitochondrion, thus providing an anchoring point throughout sperm penetration (Lambert, 1986; Lambert, 1989).

It seems likely that spermatozoa cross the egg coats by means of lytic agents, "lysins" (Hoshi, 1985). In *Ciona intestinalis* and in *Phallusia mammillata*, in which an acrosome reaction has been described, it has been inferred that these enzymes are contained in the acrosomal vesicles (Rosati et al., 1985; Honegger, 1986).

In many ascidian species, trypsin- and chymotrypsin-like protease activities, functional in sperm penetration, have been demonstrated by using a variety of natural and synthetic protease inhibitors (Hoshi, 1985). These experiments have revealed an interesting subclass-related distribution of these activities: in Pleurogona ascidians, both trypsin- and chymotrypsin-like activities are necessary, while in Enterogona only the chymotrypsin-like activity is required for the penetration through the VC (Hoshi, 1985; Yokosawa et al., 1987). These observations have been confirmed and extended by results of *in vitro* fertilization assays using specific synthetic substrates.

Most of the biochemical characterization of "lysins" has been carried out in *Halocynthia roretzi*. In this species two trypsin-like enzymes, spermosin and acrosin, and a chymotrypsin-like enzyme are involved in a "lysin system" that regulates sperm penetration (Hoshi et al., 1981; Sawada et al., 1984). The timing of action of these enzymes has also been analyzed using specific protease substrates in fertilization assays: there is evidence that spermosin and the chymotrypsin-like enzyme function at an early stage of the process of penetration through the egg investment, while acrosin functions at a later stage (Sawada et al., 1986). Furthermore, purified spermosin and acrosin are not able to digest the VC of ascidian intact eggs, while the partially purified preparation of the sperm chymotrypsin-like enzyme, still contaminated with spermosin and acrosin,exerts a lytic action on the coat (Sawada et al., 1984). Hence, the role of ascidian acrosin and spermosin and their functional relationship with the chymotrypsin-like enzyme has yet to be elucidated. A further biochemical and functional characterization of these enzyme activities in different species of ascidians could also help to ascertain whether the distribution of trypsin- and chymotrypsin-like proteases has an evolutionary meaning.

All these data point to the complexity of the mechanism of sperm penetration through the egg VC: in fact, it seems likely that spermatozoa cross the egg coat by means of several lysins acting cooperatively and sequentially as a system.

CONCLUDING REMARKS

The aim of this paper was to present the results of studies on two relevant events of fertilization in ascidians, namely sperm-egg recognition and sperm penetration through the egg VC. We expressly neglected other problems, peculiar to these species, in order to focus attention on problems of more general interest.

From data obtained in many species of ascidians it is possible to draw a general scheme of the early events of gamete interaction of these organisms. However, there are still many obscure gaps in the puzzling process leading to sperm-egg fusion. This applies also to gamete interaction in other organisms, e.g. sea urchins and mammals. These missing links are likely to remain until we have models that allow the scrutiny of individual steps of this complex process, and models that lend themselves to biochemical and molecular analysis as well as morphological and functional examination. The prospects are more promising for one of the unsolved problems that have not been touched upon in this review, the intriguing question of self-sterility in ascidians. The colonial ascidian Botryllus and the solitary ascidian Ciona intestinalis have proved to be a good tool for the study of this topic and have provided interesting insights into genetic and functional aspects respectively.

The techniques molecular biology may contribute to the understanding of the molecular mechanisms underlying self-recognition, and could provide a new tool to elucidate sperm-egg interaction.

REFERENCES

Casazza G, De Santis R, Pinto MR (1988) Plasma membrane glycoproteins of Ciona intestinalis spermatozoa that interact with the egg. Dev Growth and Differ 30:147-158

De Santis R, Jamunno G, Rosati F (1980) A study of the chorion and of the follicle cells in relation to sperm-egg interaction in the ascidian Ciona intestinalis. Dev Biol 74: 490-499

De Santis R, Pinto MR, Cotelli F, Rosati F, Monroy A, D'Alessio G (1983) A fucosyl-glycoprotein component with sperm receptor and sperm activating activities from the vitelline coat of Ciona intestinalis eggs. Exp Cell Res 148: 508-513

De Santis R, Pinto MR (1987) Isolation and partial characterization of a glycoprotein complex with sperm receptor activity from Ciona intestinalis ovary. Dev Growth and Differ 29:617-625

De Santis R, Pinto MR (1988) The pathway of sperm-egg interaction in ascidians: biology and chemistry. Zool Sci 5:919-924

Fukumoto M (1988) Fertilization in ascidians: apical processes and gamete fusion in Ciona intestinalis spermatozoa. J Cell Sci 89:189-196

Honegger TG (1982) Effect on fertilization and localized binding of lectins in the ascidian, Phallusia mammillata. Exp Cell Res 138:446-451

Honegger TG (1986) Fertilization in ascidians: studies on the egg envelope, sperm and gamete interactions in Phallusia mammillata. Dev Biol 118: 118-128

Hoshi M, Numakunai T, Sawada H (1981) Evidence for participation of sperm proteinases in fertilization of the solitary ascidian, Halocynthia roretzi: effects of protease inhibitors. Dev Biol 86:117-121

Hoshi M (1984) Roles of sperm glycosidases and proteases in the ascidian fertilization. In Advances in invertebrate reproduction. Engels W et al (eds) Elsevier Science Publishers, vol 3, p 27-40

Hoshi M (1985) Lysins. In Biology of fertilization. Metz CB et al (eds) Academic Press Inc, vol 2, p 431-462

Hoshi M, De Santis R, Pinto MR, Cotelli F, Rosati F (1985) Sperm glycosidases as mediators of sperm-egg binding in the ascidians. Zool Sci 2: 65-69

Lambert CC, Epel D (1979) Calcium-mediated mitochondrial movement in ascidian sperm during fertilization. Dev Biol 69:296-304

Lambert CC, Lambert G (1981) The ascidian sperm reaction: Ca^{2+} uptake in relation to H^{+} efflux. Dev Biol 88:312-317

Lambert CC, Lambert G (1984) The role of actin and myosin in ascidian sperm mitochondrial translocation. Dev Biol 106:307-314

Lambert CC (1986) Fertilization-induced modification of chorion N-acetylglucosamine groups blocks polyspermy in ascidian eggs. Dev Biol 116:168-173

Lambert CC (1989) Ascidian sperm penetration and the translocation of sperm surface glycosidases. J Exp Zool 249:308-315

Pinto MR, De Santis R, D'Alessio G, Rosati F (1981) Studies on fertilization in the ascidians. Fucosyl sites on vitelline coat of Ciona intestinalis. Exp Cell Res 132: 289-295

Rosati F, De Santis R (1978) Studies on fertilization in ascidians. I Self-sterility and specific recognition between gametes of Ciona intestinalis. Exp Cell Res 112:111-119

Rosati F, De Santis R (1980) The role of the surface carbohydrates in sperm-egg interaction in Ciona intestinalis. Nature 283:762-764

Rosati F, Cotelli F, De Santis R, Monroy A, Pinto MR (1982) Synthesis of fucosyl-containing glycoproteins of the vitelline coat in oocytes of Ciona intestinalis (Ascidia). Proc Natl Acad Sci USA 79:1908-1911

Rosati F, Pinto MR, Casazza G (1985) The acrosomal region of the spermatozoon of Ciona intestinalis: its relationship with the binding to the vitelline coat of the egg. Gamete Res 11:379-389

Rossignol DP, Earles BJ, Decker GL, Lennarz WJ (1984) Characterization of the sperm receptor on the surface of eggs of Strongylocentrotus purpuratus. Dev Biol 104:308-321

Sawada H, Yokosawa H, Ishii S (1984) Purification and characterization of two types of trypsin-like enzymes from sperm of the ascidian (Protochordata) Halocynthia roretzi. J Biol Chem 259:2900-2904

Sawada H, Yokosawa H, Numakunai T, Ishii S (1986) Timing of action of sperm proteases in ascidian fertilization. Experientia 42:74-75

Wassarman PM (1987) The biology and chemistry of fertilization. Science 235:553-560

Yokosawa H, Numakunai T, Murao S, Ishii S (1987) Sperm chymotrypsin-like enzymes of different inhibitor-susceptibility as lysins in ascidians. Experientia 43:925-927

CELLULAR AND MOLECULAR ELEMENTS OF MAMMALIAN FERTILIZATION

Paul M. Wassarman
Department of Cell and Developmental Biology
Roche Institute of Molecular Biology
Roche Research Center
Nutley, New Jersey 07110
USA

INTRODUCTORY REMARKS

The concept of receptor-mediated gamete adhesion is widely accepted today to account for species specificity observed during fertilization in animals. Although this concept dates back more than 70 years to F.R. Lillie (Farley, 1982), macromolecules involved in gamete adhesion have been identified only recently. Here, I review some aspects of research on the mouse egg receptor for sperm, a zona pellucida glycoprotein called ZP3, that has been studied extensively at both the cellular and molecular level. Such studies have revealed that ZP3 regulates early events of the fertilization pathway in mice, including gamete recognition and adhesion (binding), and sperm exocytosis (acrosome reaction).

FERTILIZATION IN MICE

The fertilization pathway in mice consists of several steps that must occur in a compulsory order prior to fusion of sperm and egg to form a zygote (Gwatkin, 1987; Wassarman, 1987a,b,1988a; Yanagimachi, 1988). The steps include: I. Loose attachment and then tight species-specific binding of acrosome-intact sperm to the unfertilized egg extracellular coat, or zona pellucida (ZP). Binding occurs between sperm receptors in the ZP and egg-binding proteins in plasma membrane overlying the anterior region of the sperm head. II. Completion of the acrosome reaction (AR) by sperm bound to the ZP.

NATO ASI Series, Vol. H 45
Mechanism of Fertilization
Edited by B. Dale

The acrosome is a large lysosome-like vesicle that underlies plasma membrane at the anterior region of the sperm head. The AR is an exocytotic event involving multiple point fusions between outer acrosomal membrane and plasma membrane to produce hybrid vesicles. It results in exposure of inner acrosomal membrane and its associated contents. III. Penetration of the ZP by bound, acrosome-reacted sperm. This probably involves digestion of a path through the ZP by a sperm proteinase (acrosin) associated with inner acrosomal membrane. IV. Fusion of a single acrosome-reacted sperm with egg plasma membrane (fertilization) to form a zygote. Fusion occurs between plasma membrane remaining at the posterior region of the sperm head and egg plasma membrane.

Within minutes after gamete fusion, the ZP is altered such that it becomes refractory to both sperm binding and penetration; the slow, or secondary block to polyspermy. These changes, which constitute the zona reaction, take place subsequent to the cortical reaction. The latter involves fusion of cortical granules, small, membrane-bound, lysosome-like vesicles located in the egg cortex, with egg plasma membrane. As a result, cortical granule contents (including various enzymes) are deposited into the ZP, and these alter dramatically the nature of the extracellular coat.

Even this brief description of fertilization in mice illustrates that the ZP is a principal player in the fertilization process. Recognition and adhesion of male and female gametes is regulated in part, both before and after fertilization, by ZP components. In addition, sperm undergo exocytosis (AR) while bound to the ZP, implicating a ZP component as inducer of this event.

ZP3 IS A PRIMARY SPERM RECEPTOR

ZP3 is an 83,000 M_r glycoprotein present in more than a billion copies in the mouse egg extracellular coat, or zona pellucida (ZP) (Bleil and Wassarman, 1980a; Wassarman *et al.,* 1985a, 1986a,b, 1989; Wassarman, 1988a,b, 1989a). ZP3 is one of 3 glycoproteins (ZP1-3) that constitute the ZP. It consists of a 44,000 M_r polypeptide chain, 3 or 4 complex-type, asparagine-linked (N-linked) oligosaccharides, and an undetermined number of serine/threonine-linked (O-linked) oligosaccharides. The heterogeneous appearance and relatively low pI of ZP3 on SDS-gels is due to its oligosaccharides, not to polypeptide chain.

The ZP3 polypeptide chain consists of 402 amino acids (Kinloch *et al.*, 1988; Ringuette *et al.*, 1988; Wassarman *et al.*, 1989), is rich in proline and serine plus threonine residues, is not particularly hydrophilic or hydrophobic, and contains little α-helix forming potential. It is likely that the polypeptide chain consists of stretches of extended chain (5-29 residues) interupted by short stretches of reverse turns or coils (4-10 residues). There are 6 potential N-linked glycosylation sites along the polypeptide chain, although a maximum of 4 sites are actually used. Computer searches of protein sequence data bases suggest that ZP3 has a unique polypeptide chain.

ZP3 purified from oocyte or egg ZP behaves in a manner expected of a *bona fide* sperm receptor (Bleil and Wassarman, 1980b, 1983, 1986; Wassarman *et al.*, 1985a,b, 1986a,b, 1989; Wassarman, 1987a,b, 1988a,b, 1989a). For example, it binds to the head (not midpiece or tail) of acrosome-intact (not acrosome-reacted) sperm and prevents them from binding to eggs *in vitro.* On the other hand, ZP3 purified from embryo ZP (indistinguishable from egg ZP3 on SDS-gels) neither binds to sperm nor prevents them from binding to eggs *in vitro.* In addition, oocyte and egg ZP3, but not embryo ZP3, induce sperm to undergo the AR *in vitro.* The differential behavior of egg and embyro ZP3 is to be expected, since free-swimming sperm bind to eggs, but not to embryos. There are tens-of-thousands of sites for ZP3 binding on plasma membrane surrounding the head of each acrosome-intact sperm. Whereas relatively low concentrations (nanomolar) of purified ZP3 prohibit binding of sperm to eggs *in vitro*, significantly higher concentrations are required to induce the AR.

ZP3 OLIGOSACCHARIDES MEDIATE GAMETE ADHESION

Removal of O-linked oligosaccharides from purified oocyte or egg ZP3 by alkaline-borohydride hydrolysis, results in inactivation of the glycoprotein as both sperm receptor and AR-inducer *in vitro* (Florman and Wassarman, 1985; Wassarman *et al.*, 1985a, 1986a,b, 1989; Wassarman, 1989b). On the other hand, the released ZP3 oligosaccharides themselves behave in a manner expected for a sperm receptor (see above). A specific size-class of these oligosaccharides (about 3,900 M_r), representing less than 10% of total ZP3 O-linked oligosaccharides, accounts entirely for the glycoprotein's ability to bind to sperm. However, neither the oligosaccharides nor small ZP3 glycopeptides (1,500-6,000 M_r) induce sperm to undergo the AR *in vitro.* indicating a role for ZP3 polypeptide chain in sperm exocytosis (Florman *et al.,* 1984; Florman and Wassarman, 1985).

A galactose residue, located in α-linkage at the non-reducing terminus of ZP3 O-linked oligosaccharides, serves as an essential determinant for binding of the glycoprotein to sperm (Bleil and Wassarman, 1988; Wassarman, 1989b). Either removal of the residue with α-galactosidase or conversion of its C-6 hydroxyl to an aldehyde with galactose oxidase, renders ZP3 (or purified ZP3 oligosaccharides) unable to bind to sperm. These results are not unexpected, since interactions between carbohydrate-binding proteins and sugar ligands are stabilized by hydrogen bonds and van der Waals contacts, with the former providing the major contribution to binding (Quiocho, 1986). Sugar hydroxyls participate extensively in hydrogen bonding, perhaps, accounting for the inhibitory effect of galactose oxidase.

Involvement of a terminal α-galactose in sperm receptor function may be related to the finding that oligosaccharides with repeating N-acetyllactosamine units (lactosaminoglycans) are often associated with developmentally regulated glycoconjugates (Pink, 1980; Feizi *et al.,* 1981; Schwarting and Yamamoto, 1988). In some instances, such sequences are terminated by a galactose residue in α-linkage with the galactose of the N-acetyllactosamine unit. In this connection, LA4, a monoclonal antibody that recognizes an epitope consisting of a terminal galactose residue in α-linkage with a penultimate galactose (structure based on type 2 [galactose (β1-4)-N-acetylglucosamine] lactoseries sequence) (Dodd and Jessell, 1985; Jessell and Dodd, 1985), binds exclusively to ZP3 on immunoblots containing ZP glycoproteins (Shalgi *et al.,*1989; Wassarman, 1989b).

ZP3 INDUCES THE SPERM ACROSOME REACTION

The AR is an example of receptor-mediated, signal-transduced, cellular exocytosis (Kopf and Gerton, 1989; Wassarman, 1989a). It has an absolute requirement for Ca^{++}, involves a Ca^{++}-dependent phospholipase, guanine nucleotide-binding proteins (G proteins), and altered cyclic nucleotide metabolism, and is characterized by Na^{+} and Ca^{++} influx and H^{+} efflux through plasma membrane surrounding the sperm head. The latter involves an ATP-dependent H^+ pump and leads to an increase in intracellular pH. Massive lateral displacement of transmembrane glycoproteins on the sperm surface, overlying the acrosomal vesicle, is apparently an early step in the AR (Aguas and Pinto de Silva, 1989).

As indicated above, purified oocyte or egg ZP3 (not embryo ZP3; $ZP3_f$) induces sperm to undergo the AR *in vitro.* ZP3 is as effective as ionophore A23187 in inducing the AR and is inhibited by pertussis toxin, phorbol esters, and diacylglycerol, agents that interfere with signal transduction in somatic cells (Kopf and Gerton, 1989; Wassarman, 1989a). The AR-inducing function of ZP3 is dependent on both the glycoprotein's O-linked oligosaccharides and polypeptide chain (see above). These and other findings suggest that induction of the AR by ZP3 may depend on multivalent interactions between the glycoprotein and a sperm plasma membrane component (egg-binding protein) and may result in aggregation (capping or patching) of the sperm component (Wassarman, *et al.,* 1985a; Leyton and Saling, 1989; Wassarman, 1989a).

ZP3 IS A STRUCTURAL GLYCOPROTEIN

In addition to its other functions, ZP3 serves as a structural glycoprotein during assembly of the ZP. ZP3, along with ZP1(200,000 M_r dimer) and ZP2 (120,000 M_r), is synthesized continuously by oocytes during their growth phase (2-3 weeks), the time in oogenesis when the ZP first appears (Bleil and Wassarman, 1980c; Salzmann *et al.,* 1983; Wassarman *et al.,* 1985a, 1986a,b, 1988b, 1989; Chamberlin *et al.,* 1989; Kinloch *et al.,* 1989; Kinloch and Wassarman, 1989; Wassarman, 1989a). The ZP3 gene is expressed and ZP3 is synthesized and secreted only by growing oocytes, not by other cell types, including follicle cells. Thus, ZP3 is an example of oocyte-specific and, therefore, sex-specific gene expression during mammalian development. Nascent ZP3 and ZP2 form dimers (about 180,000 M_r) that polymerize to give rise to long (several μm) ZP filaments (about 7 nm in width) (Greve and Wassarman, 1985; Wassarman *et al.,* 1985a, 1986a,b, 1989; Wassarman, 1989a; J. Greve.and P. Wassarman, unpublished results). Consequently, ZP3 and ZP2 are present periodically (every 15 nm or so) along the filaments. ZP filaments are crosslinked by ZP1 to give rise to a 3-dimensional matrix. The matrix is supported by non-covalent bonds between the 3 ZP glycoproteins. Results of recent studies strongly suggest that long ZP filaments can be assembled *in vitro* from ZP2-ZP3 dimers and higher oligomers (M. Vazquez and P. Wassarman, unpublished results).

ZP3 FROM OTHER MAMMALS

Glycoproteins that are functionally analogous to mouse ZP3 have been identified in ZP of other mammals. For example, porcine ZP3 (pZP3; 55,000 M_r) and hamster ZP3 (hZP3; 56,000 M_r) have been purified and shown to participate in gamete interactions in a manner similar to that described for mouse ZP3 (Yurewicz *et al.*, 1987; Moller *et al.*, 1989). Recently, the entire primary structure of hZP3 polypeptide chain (400 amino acids) was determined by sequencing genomic clones and found to be about 80% identical to that of mouse ZP3 (R. Kinloch, B. Ruiz-Seiler, and P. Wassarman, unpublished results). Work with bovine, rabbit, and human ZP also indicates the presence of a ZP3-like glycoprotein component in these ZP (O'Rand and Fisher, 1987; Florman and First, 1988; Cross *et al.*, 1988; Florman *et al.*, 1989). Therefore, it is likely that results of experiments carried out with mouse ZP3 will apply to other mammals as well.

SUMMARY AND CONCLUDING REMARKS

The ZP is a unique organelle that regulates gamete interactions during fertilization in mammals. Although the ZP is relatively simple in terms of composition, the multifunctional nature of its constituent glycoproteins enables the extracellular coat to perform its various tasks during development. In mice, ZP3 is responsible for gamete recognition and adhesion (sperm receptor function), as well as for induction of sperm exocytosis (AR-inducer function). The former is attributable solely to the glycoprotein's O-linked oligosaccharides, and the latter to the glycoprotein's O-linked oligosaccharides and polypeptide chain. Therefore, gamete recognition and adhesion, and sperm exocytosis, are examples of carbohydrate-mediated events during mouse development. This conclusion probably applies to gamete interactions during fertilization in many other mammals as well. Whether or not differences in sperm receptor oligosaccharide structure account for species specificity of gamete adhesion remains to be determined. The tremendous structural diversity of oligosaccharides is certainly compatible with such a role. Future studies of the mechanism of action of ZP3 during the fertilization process will undoubtedly reveal more surprising features of this unusual glycoprotein.

ACKOWLEDGEMENTS

It was an honor for me to participate in a symposium on fertilization dedicated to the memory of my late friend, Alberto Monroy. I am very grateful to the organizers of the NATO Advanced Research Workshop for enabling me to participate in this informative and stimulating meeting, held in such a beautiful setting. It is also a great pleasure for me to acknowledge past and present members of my laboratory for their valuable experimental and conceptual contributions to the research summarized here. This research was supported in part by the NICHD, NSF, Rockefeller Foundation, and Hoffmann-La Roche Inc.

REFERENCES

Aguas AP, Pinto de Silva P (1989) Bimodal redistribution of surface transmembrane glycoproteins during Ca^{++}-dependent secretion (acrosome reaction) in boar spermatozoa. J. Cell Sci 93:467-479.

Bleil JD, Wassarman PM (1980a) Structure and function of the zona pellucida: Identification and characterization of the proteins of the mouse oocyte's zona pellucida. Devel Biol 76:185-203.

Bleil JD, Wassarman PM (1980b) Mammalian sperm-egg interaction: Identification of a glycoprotein in mouse egg zonae pellucidae possessing receptor activity for sperm. Cell 20:873-882.

Bleil JD, Wassarman PM (1980c) Synthesis of zona pellucida proteins by denuded and follicle-enclosed mouse oocytes during culture in vitro. Proc Natl Acad Sci, USA 77:1029-1033.

Bleil JD, Wassarman PM (1983) Sperm-egg interactions in the mouse: Sequence of events and induction of the acrosome reaction by a zona pellucida glycoprotein. Devel Biol 95:317-324.

Bleil JD, Wassarman PM (1986) Autoradiographic visualization of the mouse egg's sperm receptor bound to sperm. J Cell Biol 102:1363-1371.

Bleil JD, Wassarman PM (1988) Galactose at the nonreducing terminus of O-linked oligosaccharides of mouse egg zona pellucida glycoprotein ZP3 is essential for the glycoprotein's sperm receptor activity. Proc Natl Acad Sci, USA 85:6778-6782.

Chamberlin ME, Ringuette MJ, Philpott CC, Chamow SM, Dean J (1989) Molecular genetics of the mouse zona pellucida. In: The Mammalian Egg Coat: Structure and Function (Dietl J, ed), Springer-Verlag, Berlin. pp.1-17.

Cross NL, Morales P, Overstreet JW, Hanson FW (1988) Induction of acrosome reactions by the human zona pellucida. Biol Reprod 38:235-244.

Dodd J, Jessell TM (1985) Lactoseries carbohydrates specify subsets of dorsal root ganglion neurons projecting to the superficial dorsal horn of rat spinal cord. J Neurosi 5:3278-3294.

Farley J (1982) Gametes and Spores. Ideas About Sexual Reproduction. Johns Hopkins Univ Press, Baltimore.

Feizi T, Kapadia A, Gooi HC, Evans MJ (1981) Human monoclonal antibodies detect changes in expression and polarization of Ii antigens during cell differentiation in early mouse embryos and teratocarcinomas. In: Teratocarcinoma and Cell Surface (Muramatsu T, Ikawa Y, eds), North-Holland Biomedical Press, Amsterdam. pp.167-181.

Florman HM, Bechtol KB, Wassarman PM (1984) Enzymatic dissection of the functions of the mouse egg's receptor for sperm. Devel Biol 106:243-255.

Florman HM, First NL (1988) The regulation of acrosomal exocytosis. I. Sperm capacitation is required for induction of acrosome reactions by the bovine zona pellucida in vitro. Devel Biol 128:453-463.

Florman HM, Tombes RM, First NL, Babcock DF (1989) An adhesion-associated agonist from the zona pellucida activates G protein-promoted elevations of internal Ca^{++} and pH that mediate mammalian sperm acrosomal exocytosis. Devel Biol 135:133-146.

Florman HM, Wassarman PM (1985) O-Linked oligosaccharides of mouse egg ZP3 account for its sperm receptor activity. Cell 41:313-324.

Greve JM, Wassarman PM (1985) Mouse egg extracellular coat is a matrix of interconnected filaments possessing a structural repeat. J Mol Biol 181:253-264.

Gwatkin RBL (1977) Fertilization Mechanisms in Man and Mammals. Plenum Press, New York.

Jessell TM, Dodd J (1985) Structure and expression of differentiation antigens on functional subclasses of sensory neurons. Phil Trans Roy Soc Lond B, Biol Sci 308:271-281.

Kinloch RA, Roller RJ, Fimiani CM, Wassarman DA, Wassarman PM (1988) Primary structure of the mouse sperm receptor's polypeptide chain determined by genomic cloning. Proc Natl Acad Sci, USA 85:6409-6413.

Kinloch RA, Roller RJ, Wassarman PM (1989) Organization and expression of the mouse sperm receptor gene. In: Developmental Biology, UCLA Symp Mol Biol, vol 125 (Davidson E, Ruderman J, Posakony J, eds), Alan R. Liss, New York, in press.

Kinloch RA, Wassarman PM (1989) Profile of a mammalian sperm receptor gene. The New Biologist, in press.

Kopf GS, Gerton GL (1989) The mammalian sperm acrosome and the acrosome reaction. In: Elements of Mammalian Fertilization (Wassarman PM, ed), CRC Press, Boca Raton, in press.

Leyton L, Saling P (1989) Evidence that aggregation of mouse sperm receptors by ZP3 triggers the acrosome reaction. J Cell Biol 108:2163-2168.

Moller CC, Bleil JD, Kinloch RA, Wassarman PM (1989) Structural and functional relationships between mouse and hamster zona pellucida glycoproteins. Devel Biol, in press.

O'Rand MG, Fisher SJ (1987) Localization of zona pellucida binding sites on rabbit spermatozoa and induction of the acrosome reaction by solubilized zonae. Devel Biol 119:551-558.

Pink JRL (1980) Changes in T-lymphocyte glycoprotein structures associated with differentiation. Contemp Top Mol Immunol 9:89-113.

Quiocho FA (1986) Carbohydrate-binding proteins: Tertiary structures and protein-sugar interactions. Annu Rev Biochem 55:287-316.

Ringutte MJ, Chamberlin ME, Baur AW, Sobieski DA, Dean J (1988) Molecular analysis of cDNA coding for ZP3, a sperm binding protein of the mouse zona pellucida. Devel Biol 127:287-295.

Salzmann GS, Greve JM, Roller RJ, Wassarman PM (1983) Biosynthesis of the sperm receptor during oogenesis in the mouse. Eur Mol Biol Org J 2:1451-1456.

Schwarting GA, Yamamoto M (1988) Expression of glycoconjugates during development of the vertebrate nervous system. BioEssays 9:19-23.

Shalgi R, Bleil JD, Wassarman PM (1989) Carbohydrate-mediated sperm-egg interactions in mammals. In: Proc World Congress on In Vitro Fertilization (Ben-Rafael Z, ed), Plenum Press, New York, in press.

Wassarman PM (1987a) The biology and chemistry of fertilization. Science 235:553-560.

Wassarman PM (1987b) Early events in mammalian fertilization. Annu Rev Cell Biol 3:109-142.

Wassarman PM (1988a) Fertilization in mammals. Scientific Amer 256(Dec):78-84.

Wassarman PM (1988b) Zona pellucida glycoproteins. Annu Rev Biochem 57:415-442.

Wassarman PM (1989a) Profile of a mammalian sperm receptor. Development 106, in press.

Wassarman PM (1989b) Role of carbohydrates in receptor-mediated fertilization in mammals. In: Carbohydrate Recognition in Cellular Function, Ciba Found Symp., no. 145 (Bock G, Harnett S, eds), John Wiley and Sons, Chichester. pp.135-155.

Wassarman PM, Bleil JD, Florman HM, Greve, JM, Roller RJ, Salzmann GS, Samuels FG (1985a) The mouse egg's receptor for sperm: What is it and how does it work? Cold Spring Harbor Symp Quant Biol 50:11-19.

Wassarman PM, Florman HM, Greve JG (1985b) Receptor-mediated sperm-egg interactions in mammals. In: Biology of Fertilization, vol 2 (Metz CB, Monroy A, eds), Academic Press, New York. pp.341-360.

Wassarman PM, Bleil JD, Florman HM, Greve JM, Roller RJ, Salzmann GS (1986a) The mouse egg's extracellular coat: Synthesis, structure, and function. In: Gametogenesis and the Early Embryo (Gall JG, ed), Alan R. Liss, New York. pp.371-388.

Wassarman PM, Bleil JD, Florman HM, Greve JM, Roller RJ, Salzmann GS (1986b) Nature of the mouse egg's receptor for sperm. In: The Molecular and Cellular Biology of Fertilization (Hedrick JL,.ed), Plenum Press, New York. pp.55-78.

Wassarman P, Bleil J, Fimiani C, Florman H, Greve J, Kinloch R, Moller C, Mortillo S, Roller R, Salzmann G, Vazquez M (1989) The mouse egg receptor for sperm: A multifunctional zona pellucida glycoprotein. In: The Mammalian Egg Coat: Structure and Function (Dietl J, ed), Springer-Verlag, Berlin. pp.18-37.

Yanagimachi R (1988) Mammalian fertilization. In: The Physiology of Reproduction,vol 1 (Knobil E, Neill JD, eds), Raven Press, New York. pp.135-185.

Yurewicz EC, Sacco AG, Subramanian MG (1987) Structural characterization of the Mr=55,000 antigen (ZP3) of porcine oocyte zona pellucida. J Biol Chem 262:564-571.

MICROMANIPULATION IN THE STUDY OF SPERM-EGG INTERACTIONS

Ralph B.L. Gwatkin
ReproGene
25460 Bryden Road
Beachwood, OH 44122, USA

Abstract

Micromanipulation can assist fertilization and answer basic questions in gamete biology. This chapter describes several methods of assisted fertilization and the new information they provide. Micromanipulation is shown to overcome the antifertility action of a monoclonal antibody to ZP3, the zona pellucida glycoprotein carrying the sperm receptor, and to allow successful in vitro fertlization with mouse sperm carrying one t-complex. When applied to mouse caput epididymal sperm, the proportion of fertilized eggs that develop to blastocyst is markedly increased.

Introduction

Micromanipulation of living cells goes back a century (see Chambers, 1940), but apparently was first applied to mammalian eggs in 1942 (Nicholas and Hall, 1942; see also Lin, 1971). With the advent of in vitro fertilization (IVF) it became apparent that some types of gamete infertility might be overcome by assisting the sperm to enter the egg, using various micromanipulative techniques (Table 1). These

NATO ASI Series, Vol. H 45
Mechanism of Fertilization
Edited by B. Dale

Table 1

ASSISTED FERTILIZATION METHODS

METHOD	DEVELOPERS
Vitelline Injection	Uehara & Yanagimachi, 1976
Zona Drilling	Gordon & Talanksy, 1986
Zona Cutting	Tsunoda *et al*, 1986
Perivitelline Injection	Laws-King *et al*, 1987 Lassalle & Testart, 1987
Partial Zona Dissection	Malter & Cohen, 1989

techniques not only have potential value in the treatment of human infertility, but they can also be expected to answer basic questions of fertilization, e.g. whether interaction of the sperm with the zona pellucida, the oolemma or with homologous cytoplasm are necessary for subsequent development and how it is affected by blocking the action of specific gene products with antibodies or antisense mRNAs.

Microinjection of the Sperm into the Vitellus

The most obvious means of assisting fertilization is to by-pass both the zona pellucida and the egg plasma membrane by injecting the sperm chromatin into the vitellus, but there are problems with this approach. The egg may be incompletely activated. The potentially selective hurdles of the zona pellucida and egg plasma membrane are absent. The vitellus may be damaged and the chromatin itself may be affected by the sonication procedure used to isolate the sperm nuclei and remove acrosomes (Martin et al. 1988). Sperm-into-vitellus injection was first developed by Uehara and Yanagimachi (1976), who injected hamster and human sperm nuclei into hamster eggs and found that both formed pronuclei. Thus, the cytoplasmic factors controlling pronucleus formation were shown by this micromanipulation to be non-species specific. Later micromanipulation studies by Thadani (1979 & 1980) confirmed this result for the mouse up to first cleavage and also showed that sperm pronucleus formation was dependent on breakdown of the egg germinal vesicle. Complete rabbit development resulting in live births has been reported by Hosoi et al (1988). Round-headed (acrosomeless) human sperm have been shown to form pronuclei when injected into the vitellus of hamster eggs (Lanzendorf, 1988), suggesting that such normally infertile sperm are capable of supporting early development. However, to my knowledge no human pregnancies have been achieved with the sperm-into-vitellus approach.

Subzonal Microinjection

Penetration of the egg plasma membrane with the micropipet may only partially activate the egg. To achieve full activation the sperm may be injected into the space between the vitelline membrane and the zona pellucida. Such perivitelline, or subzonal, insertion does not require sonication of the sperm, although it is presumed that the acrosome reaction must be induced to remove potentially damaging enzymes and to ensure that a fusible region of the sperm plasma membrane is exposed. Barg et al (1986) failed to fertilize mouse eggs with sperm induced to acrosome react with the ionophore A23187. However, these sperm were probably damaged, because Mann (1988) injected sperm that had simply been incubated for 2 hr in culture medium M2 (Quinn et al, 1982) subzonally and obtained 25% fertilized eggs, of which 54% developed into fetuses, or were born, following embryo transfer. Laws-King et al (1987) reported that subzonal insertion of human eggs with human sperm, synchronously reacted by the addition of Ca^{2+} after capacitation in Ca^{2+}-free, Sr^{2+}-containing, medium (Mortimer et al, 1986), resulted in pronucleus formation. Early cleavage was observed by Lassalle and Testart (1987). One human pregnancy has been reported (Ng et al, 1988).

Although perivitelline introduction of the sperm has the advantage over vitelline injection that I have mentioned, it is difficult to ensure monospermy. The sperm may be damaged by the micromanipulation and slower, possibly less fertile, sperm may be selected.

Zona-drilling

An alternative is to open an artificial "micropyle" in the zona, small enough to prevent an excessive loss of the cortical granule exudate and ensure a normal, or nearly normal, zona reaction to block polyspermy, but sufficient to allow the fertilizing sperm to enter (Fig. 1). This

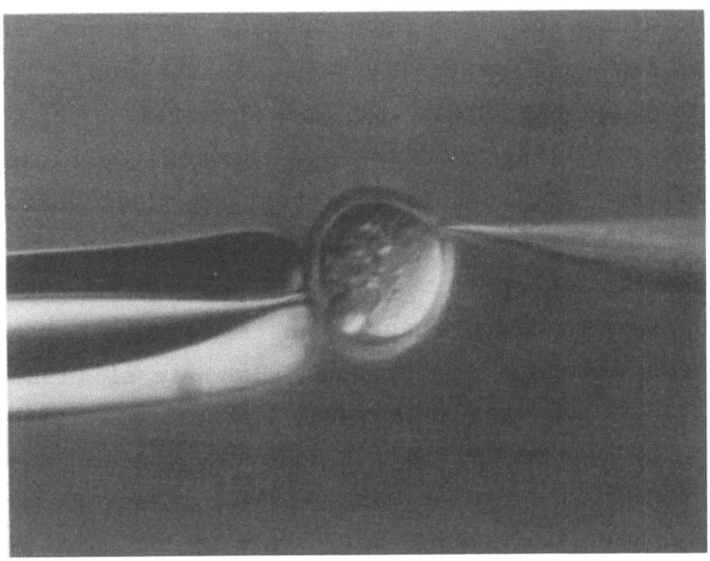

Fig 1. Mouse egg, held on holding pipet (left), is being drilled with acidified Tyrode's solution in micropipet (right).

approach was realized by Gordon and Talansky (1986) who used a stream of acidified Tyrode's solution (at pH 2.3) to dissolve an aperture in the zona. With mouse eggs, they observed only 2% polyspermy at 10^6 sperm/ml and 15% fertilization at 10^4 sperm/ml, a concentration at which there was no fertilization with undrilled eggs. Fetuses were obtained from zona-drilled eggs, following embryo transfer.

Several problems have been encountered with zona-drilling. The acidified Tyrode's may lead to temporary withdrawal of the vitelline microvilli (Cummins, personal communication) reducing sperm fusion. It is difficult to avoid over-drilling which may lead to premature hatching of the blastocyst (Depyere et al, 1988) and it is possible that cytotoxic lymphocytes may invade the cleaving embryo in the oviduct. An alternative is to drill the hole by some other means, e.g. by thrusting the micropipet through the zona (partial zona dissection, Fig. 2, Malter & Cohen, 1989) or by cutting it with a sharp microneedle (Tsunoda, 1986).

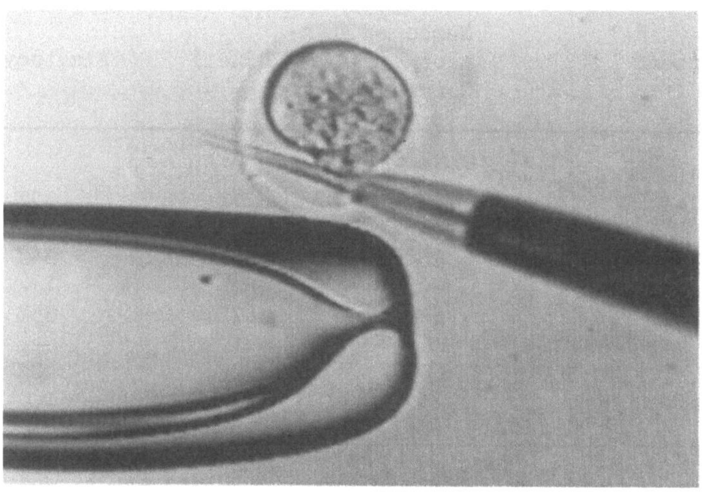

Fig 2. Partial zona dissection is achieved by thrusting the micropipet twice through the zona.

The application of partial zona dissection to human eggs in an IVF program has resulted in one reported pregnancy (Malter & Cohen, 1989). Gordon et al (1988) reported no pregnancies from zona-drilled eggs after 3 transfers. More data will need to be obtained to establish which procedure is optimal.

Micromanipulation Studies with Anti-ZP3 Antibody

We initiated our investigation of micromanipulation-assisted fertilization by confirming the mouse studies of John Gordon's laboratory and showed that zona-drilling was effective even when penetration of the zona was blocked by a monoclonal antibody to the protein core of the zona glycoprotein, ZP3 (see Table 2, Conover & Gwatkin, 1988).

Table 2

IVF OF DRILLED AND CONTROL OOCYTES
EXPOSED TO MONOCLONAL ANTIBODY TO ZP-3
PRIOR TO INSEMINATION

Treatment	Number of Oocytes	2-Cell	Blastocyst
Drilled	42	72%	32%
Drilled + ZP-3 mAb	55	65%	34%
Control	64	62%	35%
Control + ZP-3 mAb	64	2%	0%

Pooled data from 6 expts. Rat anti-mouse ZP3 mAb is from East et al. (1985). Used at 10-20 μg/ml. From Conover and Gwatkin (1988)

Since the protein core is thought to be responsible for the sperm acrosome reaction it would appear that sperm-zona interaction is not necessary for subsequent blastocyst formation.

Micromanipulation with t-mutant Sperm

We then studied sperm carrying two t complexes. These sperm are known to be unable to fuse with zona-free oocytes, although they do acrosome react (McGrath & Hillman, 1980). As expected, zona-drilling did not allow fertilization to occur. However, sperm with only one t complex, which exhibit a reduced progressive motion and undergo hyperactivation sooner than nonmutant sperm (Olds-Clarke, 1987), did not fertilize *in vitro* unless the zona was drilled (Table 3, Ahmad et al, 1989). The reason is unclear. Possibly a delay in fertilization by these mutant sperm may be circumvented by accelerating gamete contact via drilling.

Caput Epididymal Sperm

Sperm undergo final maturation for fertilization as they pass from the caput to the cauda of the epididymis (see Orgebin-Crist and Fournier-Delpech, 1982). These changes appear to involve the acquisition of vigorous progressive motility and an increased ability to bind to the zona. At least some caput sperm appear to be capable of functional fertilization, because in men bilateral obstruction of the epididymides can be treated by connecting the caput to the vas deferens in an operation known as vasoepididymostomy (Shoysman & Bedford, 1986; Silber, 1989). Uehara and Yanagimachi (1977) observed that relatively few caput sperm injected into the vitellus of hamster eggs formed pronuclei, but these eggs may not have been fully activated by the procedure, as I have emphasized. When we examined the fertilizing ability of mouse caput sperm *in vitro* we found that although 11% of the eggs fertilized by these sperm developed to the 2-cell stage, only 7.7% developed to blastocyst compared with 48% of those fertilized by cauda

Table 3

EFFECT OF ZONA DRILLING ON THE ABILITY OF T/t & t/t SPERM TO FERTILIZE

Treatment		Number of Oocytes			
		Total	2PN	Total	2Cell
Not drilled	T/t	24	0	28	0
	t/t	24	0	-	-
Drilled	T/t	22	7 (31%)	32	25 (78%)
	t/t	24	0	-	-
Not drilled/ No sperm	T/t	-	-	15	0
Drilled/ No sperm	T/t	-	-	15	0

2-5 x 10^6 sperm ($T/t^{w5/w71}$ & t^{w5}/t^{w71}) per ml. Oocytes (T/t^{w5}). From Ahmad et al (1989)

sperm (Table 4). Zona drilling improved slightly the fertilizing ability of caput sperm to 16.7%, but had a dramatic effect on the proportion (now 50%) that went to blastocyst. Partial zona dissection was ineffective. Thus caput sperm are capable of fertilizing eggs and many are capable of developing at least to the blastocyst stage once an aperture is made in the zona. We have noted that with zona drilling the vitelline membrane blebs out to the zona

Table 4

ASSISTED FERTILIZATION WITH CAUDA & CAPUT EPIDIDYMAL SPERMATOZOA

Manipulation	Number of Ova	2-Cell (%)	Blastocysts per 2-cell (%)	Blastocysts per ovum (%)
		Sperm from <u>Cauda</u> Epididymis		
None	3	98 (32.6)	47/98 (48)	47/300 (15.6)[a,b]
Zona Drilling	87	57 (65.5)	43/57 (75)	43/87 (50)[a]
Partial Zona Dissection	42	19 (45.2)	11/19 (58)	11/42 (26.2)[b]
		Sperm from <u>Caput</u> Epididymis		
None	116	13 (11.2)	1/13 (7.7)	1/116 (<1)
Zona Drilling	144	24 (16.7)	12/24 (50)	12/144 (8.3)[c]
Partial Zona Dissection	23	1 (4.3)	0 (0)	0/23 (0)

a, $p < 0.001$, b, $p = 0.09$ (not significant), c, $p = 0.006$
Sperm Concentration, 10^4-10^5/ml. Unpublished data of Wazzan, Gwatkin and Thomas

surface, making it available for fusion with even poorly motile sperm. In the partially zona dissected eggs, however, the aperture is small and the elastic zona tends to reseal. Without micromanipulation, or with a resealing aperture, penetration of the zona may be delayed and this could explain why few of the eggs fertilized with caput sperm in this way are capable of further development. Delayed first cleavage, suggestive of delayed sperm entry, has been noted in rabbit ova fertilized by sperm from the proximal corpus of the epididymis when compared to those of the distal corpus (Orgebin-Crist and Jahad, 1977). Alternatively, zona drilling might act by selecting more fertile sperm.

Future Research

There are many aspects of sperm-egg interaction now open to experimentation through micromanipulation. The ability to by-pass egg integuments by micromanipulation and to place the spermatozoon (or several spermatozoa) into the egg cytoplasm at predetermined intervals should allow us to define precisely the role of egg maturation and aging in successful fertilization. It should also be possible to inject individual sperm components (or antibodies to them) into mammalian eggs to evaluate the role of these sperm components in fertilization. The role of various growth factors and oncogene products in fertilization might also be determined in this way.

More work is also needed on the lines of research highlighted in this chapter. Studies are needed to define the basic mechanism of in vitro fertilization failure with t-mutant sperm. Studies are also needed to determine whether the improvement we observed in development following zona drilling is due to an acceleration of gamete fusion. This could be settled using the DNA probe, H33342.

On the technical side, electron microscope studies are needed to define aperture size and condition of the underlying plasma membrane following drilling. Alternative drilling techniques and egg activating agents need to be developed.

Assisted fertilization methods will need to be evaluated for their ability to produce live young following embryo transfer and for their value in the treatment of human male infertility.

References

Ahmad T, Conover JC, Quigley MM, Collins RL, Thomas AJ, Gwatkin RBL (1989) Failure of spermatozoa from T/t mice to fertilize in vitro is overcome by zona drilling. Gamete Res 22:369-373

Barg PE, Wahrman MZ, Talansky BE, Gordon JW (1986) Capacitated, acrosome reacted but immotile sperm, when microinjected under the mouse zona pellucida, will not fertilize the oocyte. J Exp Zool 237:365-374

Chambers R (1940) The micromanipulation of living cells. In "The cell and protoplasm" pp 20-30. Ed by FR Moulton, Sci Press, Wash, DC

Conover JC, Gwatkin RBL (1988) Fertilization of zona-drilled mouse oocytes treated with a monoclonal antibody to the zona glycoprotein, ZP3. J Exp Zool 247:113-118

Depyere HT, McLaughlin KJ, Seamark RF, Warnes GM, Matthews CD (1988) Comparison of zona cutting and zona drilling as techniques for assisted fertilization in the mouse. J Reprod Fert 84:205-211

East IJ, Guylas BJ, Dean J (1985) Monoclonal antibodies to the murine zona pellucida protein with sperm receptor activity: Effects on fertilization and early development. Dev Biol 109:268-273

Gordon JW, Grunfeld L, Garrisi GJ, Talansky BE, Richards C, Laufer N (1988) Fertilization of human oocytes by sperm from infertile males after zona pellucida drilling. Fertil Steril 50:68-73

Gordon JW, Talansky BE (1986) Assisted fertilization by zona drilling: A mouse model for correction of oligospermia. J Exp Zool 239:347-354

Hosoi Y, Miyake M, Utsumi K, Iritani A (1988) Development of rabbit oocytes after micromanipulation of spermatozoa. Proc 11th International Congress on Animal Reprod and AI Abstract 331

Lassalle B, Courtot AM, Testart J (1987) In vitro fertilization of hamster and human oocytes by microinjection of human sperm. Gamete Res 16:69-78

Lanzendorf S, Maloney M, Ackerman S, Acosta A, Hodgen G (1988) Fertilizing potential of acrosome-defective sperm following microsurgical injection into eggs. Gamete Res 19:329-337

Laws-King A, Trounson A, Sathananthan H, Kola I (1987) Fertilization of human oocytes by microinjection of a single spermatozoon under the zona pellucida. Fertil Steril 48:637-642

Lin TP (1971) Egg micromanipulation, Ch 10 (pp 157-171) in "Methods in mammalian embryology" Ed by JC Daniel, Jr., WH Freeman & Co, San Francisco
Malter HE, Cohen J (1989) Partial zona dissection of the human oocyte: a nontraumatic method using micromanipulation to assist zona pellucida penetration. Fertil Steril 51:139-148
Mann JR (1988) Full term development of mouse eggs fertilized by a spermatozoon microinjected under the zona pellucida. Biol Reprod 38:1077-1083
Martin RH, Ko E, Rademaker A (1988) Human sperm chromosome complements after microinjection of hamster eggs. J Reprod Fert 84:179-186
McGrath J, Hillman N (1980) Sterility in mutant (t^{Lx}/t^{Ly}) male mice. III. In vitro fertilization. J Embryol Exp Morph 59:49-58
Mortimer D, Curtis EF, Dravland JE (1986) The use of strontium-substituted media for capacitating human spermatozoa: an improved sperm preparation method for the zona-free hamster egg penetration test. Fertil Steril 46:97-103
Ng S-C, Bongso A, Ratnam SS, Sathananthan H, Chan CLK, Wong PC, Hagglund L, Anandakumar C, Wong YC, Goh VHH (1988) Pregnancy after transfer of sperm under zona. The Lancet 2:790
Nicholas JS, Hall BV (1942) Experiments on developing rats: II The development of isolated blastomeres and fused eggs. J Exp Zool 90:441-458
Olds-Clarke P (1987) Hyperactivation is accelerated, but capacitation is normal in sperm from t^{w32}/+ mice. Biol Reprod 36 (Suppl 1):54
Orgebin-Crist M-C, Jahad N (1977) Delayed cleavage of rabbit ova after fertilization by young epididymal spermatozoa. Biol Reprod 16:358-362
Orgebin-Crist M-C, Fournier-Delpech S (1982) Evidence for maturational changes during epididymal transit. J Androl 3:429-433
Quinn P, Barros C, Whittingham DG (1982) Preservation of hamster oocytes to assay the fertilizing capacity of human spermatozoa. J Reprod Fertil 66:161-168
Schoysman RJ, Bedford JM (1986) The role of the human epididymis in sperm maturation and sperm storage as reflected in the consequences of epididymovasostomy. Fertil Steril 46:293-299
Silber, SJ (1989) Role of epididymis in sperm maturation. Urology 33:47-51
Thadani VM (1979) Injection of sperm heads into immature rat oocytes. J Exp Zool 210:161-168
Thadani VM (1980) A study of hetero-specific sperm-egg interactions in the rat, mouse and deer mouse using in vitro fertilization and sperm injection. J Exp Zool 212:435-453

Tsunoda Y, Yasui T, Nakamura K, Uchida T, Sugie T (1986) Effect of cutting the zona pellucida on the pronuclear transplantation in the mouse. J Exp Zool 240:119-125
Uehara T, Yanagimachi R (1976) Microsurgical injection of spermatozoa into hamster eggs with subsequent transformation of sperm nuclei into male pronuclei. Biol Reprod 15:467-470
Uehara T, Yanagimachi R (1977) Behavior of nuclei of testicular caput and cauda epididymal spermatozoa injected into hamster eggs. Biol Reprod 16:315-321

HUMAN SPERM-OOCYTE FUSION

A. Henry Sathananthan†, S.C. Ng, A.O. Trounson, S.S. Ratnam and T.A. Bongso.
Monash Medical Centre, Melbourne and National University Hospital, Singapore.

INTRODUCTION

Sperm-oocyte fusion is the key event of fertilization in the human, as well as in animals. It is a classic example of spontaneous membrane fusion between 2 cells without the intervention of agents such as viruses or chemicals. The process of gamete fusion has been extensively studied in various animals, including mammals (see reviews, Yanagimachi, 1981, 1988a,b; Bedford, 1983; Monroy, 1985; Longo, 1989), but there is little information about gamete fusion in the human (Soupart, 1980; Sathananthan and Chen, 1986; Sathananthan et al, 1986a, 1987). Yanagimachi (1988a,b) has comprehensively reviewed sperm-egg fusion in animals. This chapter will concentrate on what is known in the human with reference to recent reviews of gamete membrane fusion in mammals. The recent impact of IVF and new technologies of assisted reproduction such as microinsemination sperm transfer (MIST) beneath the zona, using a micromanipulator (Ng et al, 1988, 1989a) have further helped us understand the events that lead up to gamete fusion and microfertilization (Trounson, 1987; Laws-King et al, 1987; Sathananthan et al, 1989).

The first detailed reports of human monospermic and polyspermic fusion with oocytes in vitro were by Sathananthan and Chen (1986) and

† Lincoln School of Health Sciences, La Trobe University, 625 Swanston Street, Carlton, 3053, Victoria, Australia.

NATO ASI Series, Vol. H 45
Mechanism of Fertilization
Edited by B. Dale

Sathananthan et al (1986a). Although Soupart and Strong (1975) published a paper on polyspermic interaction in zona-denuded oocytes, the process of gamete membrane fusion was not clearly defined as sperm with intact acrosomes were also thought to fuse with the egg. Later a vitrified frozen oocyte showed evidence of single sperm fusion after insemination in vitro (Sathananthan et al, 1987), very much like that seen during normal monospermic penetration (Sathananthan and Chen, 1986). We have also seen sperm-egg membrane fusion after MIST of poor quality sperm into the perivitelline space (PVS) from a man with severe oligoasthenoteratozoospermia. This technique has already resulted in successful pregnancies (Ng et al, 1988, 1989a). All these studies have given us some insights to the process of membrane fusion involving about 8 different sperm cells and 4 mature oocytes. The process seems to be basically similar during both monospermic or polyspermic interaction. However, since the zona was partly denuded (Fig. 1) in the study of polyspermic fertilization (Sathananthan et al, 1986a) the mode of penetration of some sperm was significantly different to that observed in the other studies, when the zona was kept intact. Figures 2-17 are electron micrographs showing the various stages of sperm penetration through the egg vestments, sperm fusion with the oocytes and sperm incorporation.

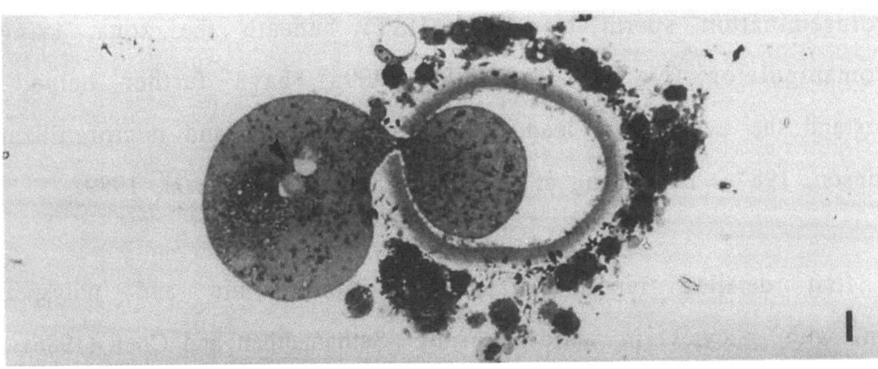

Fig. 1 Zona-punctured ovum showing naked bleb of the egg which was penetrated by many sperm. Various stages of sperm fusion and incorporation were seen by TEM in this bleb x 246 (Reproduced from Sathananthan et al, 1986a).

PREREQUISITES FOR GAMETE FUSION

Sperm capacitation and oocyte maturation

The prerequisites for successful sperm-oocyte fusion in vitro are obtaining a mature oocyte and capacitated or acrosome-reacted sperm for insemination. During in vitro fertilization (IVF), the oocytes are usually matured in culture for 3 to 8 hours after recovery, either by laparoscopy or ultrasound. Mature oocytes have a single polar body and have arrested maturation at metaphase II of meiosis, when they are ovulated. If the oocyte is immature (at the germinal vesicle stage or at metaphase I), reduced sperm penetration may occur but subsequent cleavage will not eventuate (Lopata et al, 1985). We are now investigating sperm incorporation into immature oocytes after multiple sperm injection into the PVS to determine whether normal gamete fusion and sperm nuclear decondensation occurs.

Sperm capacitation is essentially a physiological process whereby sperm acquire the capacity to penetrate through the egg vestments, particularly the zona pellucida. However, freeze-fracture replicas of mammalian sperm plasma membrane do show structural changes associated with intrinsic membrane particle proteins (see Yanagimachi, 1988a,b). Sperm capacitation in vitro usually begins spontaneously in the culture medium of choice during the sperm washing, centrifuging, layering and incubating process. This takes less than an hour after semen collection. The most motile sperm are usually harvested for insemination in the IVF procedure. The culture media used routinely are Whittingham's T_6, Ham's F_{10} and Menezo B_2 (see Trounson, 1989). Capacitated sperm show hyperactivated motility and eventually undergo the acrosome reaction (AR). In vivo, sperm capacitation most likely occurs in the oviduct, prior to fertilization. The oviductal secretions, cumulus cells and zona are capable of inducing sperm capacitation and the AR (see Meizel, 1985; Wassarman et al, 1985; Yanagimachi, 1988a,b). Both hyperactivated motility and the release of acrosomal enzymes during the AR seem to play important roles in the penetration of the cumulus and the zona of the oocyte, prior to fusion. Immotile sperm are incapable of penetrating the zona and have to be manipulated or injected into the PVS to fertilize the egg (Ng et al, 1987; Bongso et al, 1989).

In a time-course study of early penetration of human oocytes during IVF, we found that both acrosome-intact (Fig. 2) and acrosome-reacted sperm were located between cumulus cells or were at the surface of the zona, one hour after insemination (Chen and Sathananthan, 1986). Some sperm with intact acrosomes were firmly bound to the substance of the zona by their plasma membranes (Fig. 3), while others were beginning the AR by vesiculation of their surface membranes (Fig. 4). Sperm-oocyte membrane fusion had occurred two hours after insemination and sperm nuclear decondensation was evident three hours after insemination. However, only 10-30% of the sperm had reacting acrosomes at the surface of oocytes within two hours of semen collection or one hour after insemination. Sperm do not always bind to the zona during IVF, where large numbers are used for insemination (Sathananthan, 1984; Sathananthan et al, 1982). They seem to enter the zona directly and at all angles, especially if the cumulus is not present. This is the case when the cumulus is breached or removed by hyaluronidase, before insemination. However, in mice zona glycoproteins have sperm binding properties before and after the AR (Wassarman et al, 1985). Similarly, there appear to be zona receptors on the sperm surface (see Yanagimachi, 1988b).

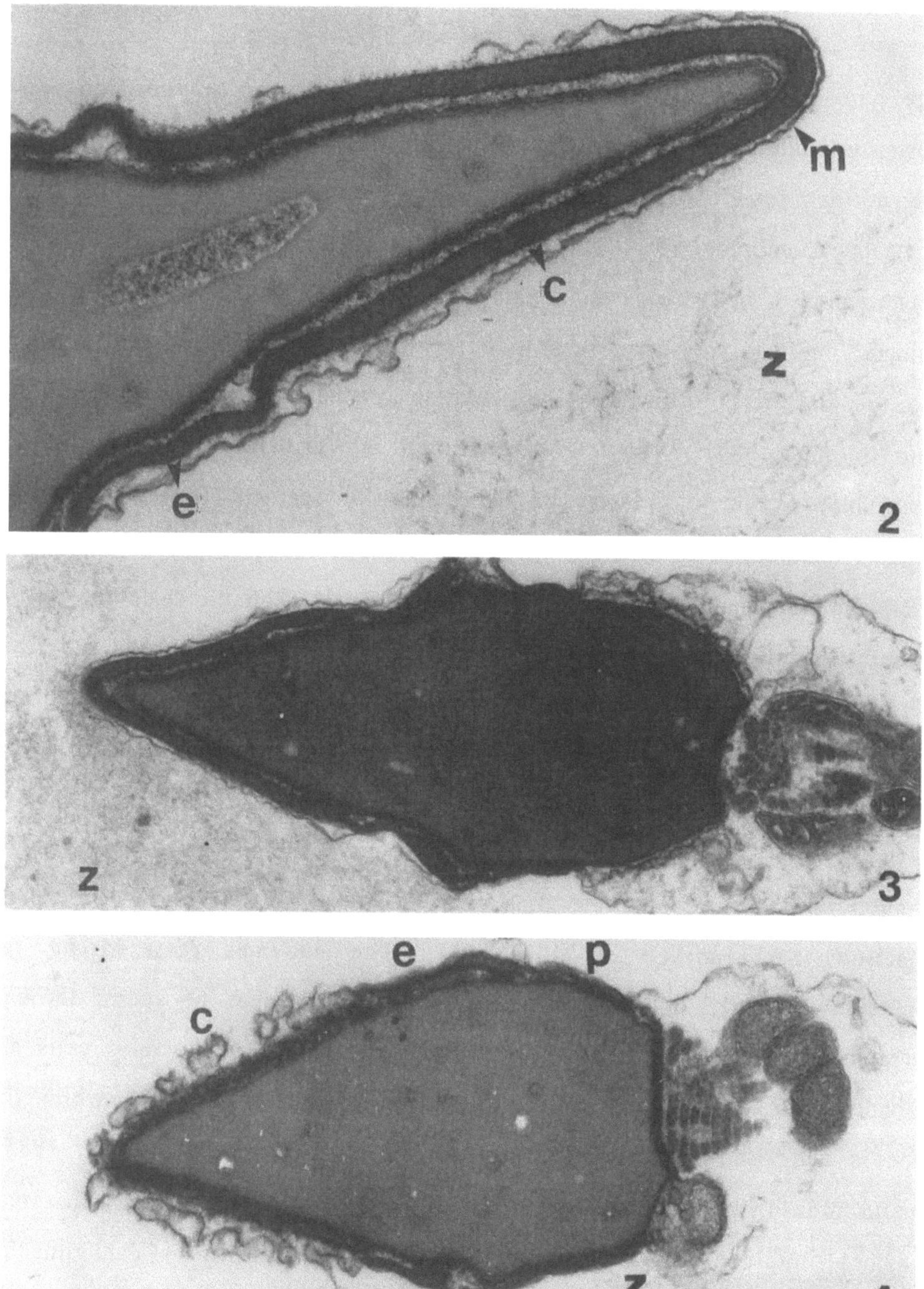

Figs. 2-4 Sperm at the surface of the zona of human oocytes. Acrosome-intact (Fig. 2), sperm firmly bound to zona (Fig. 3) and acrosome-reacting (Fig. 4) sperm are evident. c = acrosome cap, e = equatorial segment, m = plasma membrane, p = postacrosomal region, z = zona. x 57,400, x 29,274, x 29,274 (Reproduced from Chen and Sathananthan, 1986).

Acrosome-reaction prior to fusion

The human acrosome has 2 well defined segments - an anterior cap and a posterior equatorial segment (Fig. 2). Classically, vesiculation involves the intermittent fusion of the sperm plasma membrane (PM) and the outer acrosome membrane, (Fig. 4). This process is akin to exocytosis, and has been reported in various mammals (see Yanagimachi, 1988a,b). In the human, the AR was often seen to occur in 2 stages (Fig. 5), the acrosome cap reacting first, followed by vesiculation of the equatorial segment (Chen and Sathananthan, 1986; Sathananthan et al, 1982, 1986b, 1989). Invariably, a spur or vestige of the equatorial segment covered with PM remains intact (Fig. 6) after completion of the AR (Soupart and Strong, 1974; Chen and Sathananthan, 1986; Sathananthan et al, 1986a). Once the AR is completed the inner acrosome membrane (IAM) is exposed and this membrane is the first to make contact with egg plasma membrane (oolemma), when the sperm emerges from the zona into the PVS (Fig. 7). Unlike in lower vertebrates (see Yanagimachi, 1988a), the IAM is incapable of fusing with the oolemma. Partially reacted sperm were found in the cumulus, at the surface of the zona or even in the PVS (Fig. 5) after sperm injection (Sathananthan et al, 1989). The enzymes released by the acrosome evidently digest a pathway for the movement of sperm through the cumulus and the thickness of the zona (Fig. 6). The cumulus cells are embedded in a gelatinous matrix composed of hyaluronic acid while the zona is composed of three glycoproteins in mice (Wassarman et al, 1985). Hyaluronidase and acrosin are thought to be the principal acrosomal enzymes involved in sperm penetration of egg-vestments (Yanagimachi, 1988b). The stages of the AR in the human are depicted in Fig. 8. The AR does not always seem to occur in the classical manner described above. Acrosome-swelling and vesiculation within the acrosome have been reported during the AR of human sperm (Nagae et al, 1986). We have also observed this process occasionally at the surface of oocytes and more often

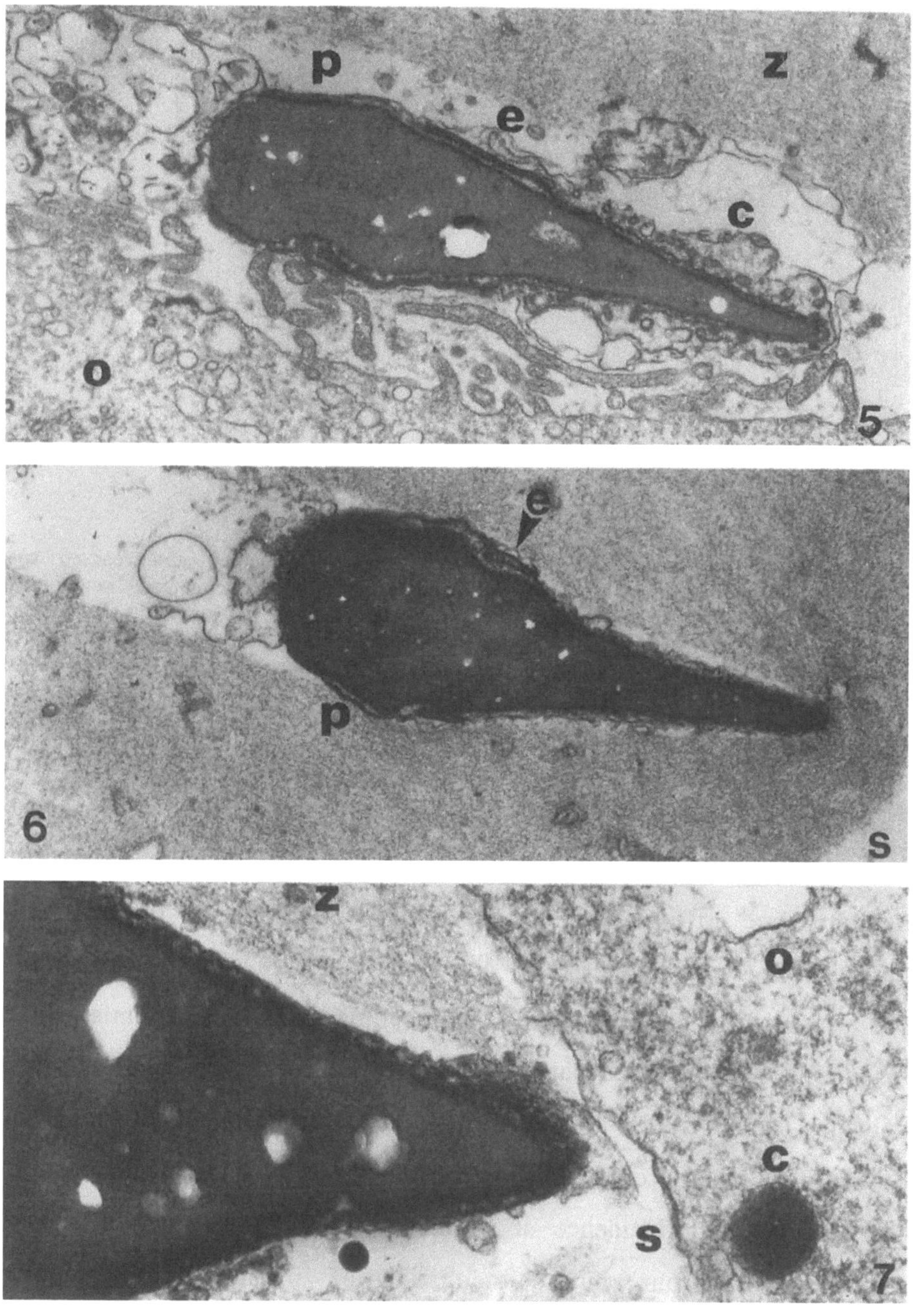

Figs. 5-7 Sperm in PVS and penetrating zona. Sperm micro-injected into PVS with a partially-reacted acrosome (Fig. 5) and fully acrosome-reacted sperm penetrating the zona (Figs. 6, 7) c = acrosomal cap, e = equatorial segment or vestige, o = ooplasm, s = PVS, z = zona. x 22,386, x 22,386, x 57,400 (Figs. 6 and 7 are reproduced from Chen and Sathananthan, 1986).

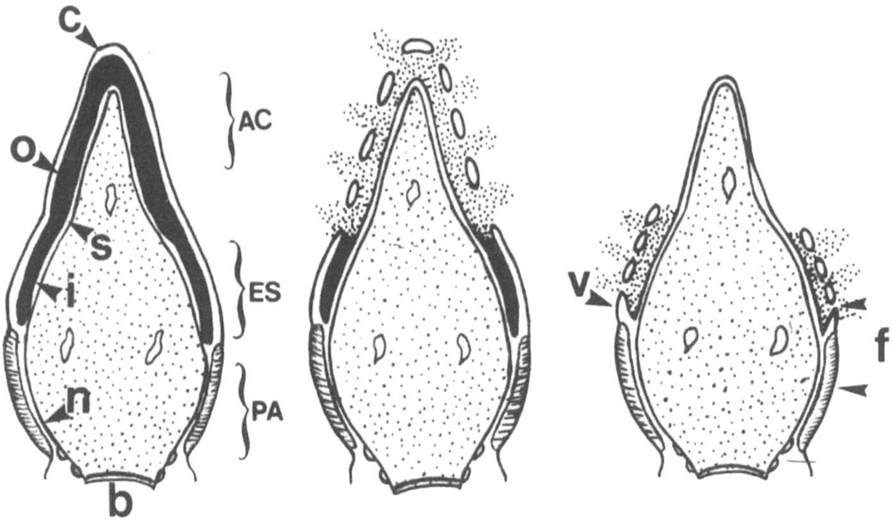

Fig. 8 Stages of the acrosome reaction: intact, partially-reacted and reacted sperm. Note vesiculation of acrosome cap (ac) followed by that of the equatorial segment (es). The fusogenic zone (f) extends, from the equatorial vestige (v) over the anterior half of the postacrosomal region (pa). c = plasma membrane, b = basal plate, i = inneracrosome membrane, n = nuclear envelope, o = outer acrosome membrane, s = subacrosomal space (modified from Sathananthan et al, 1986b).

when sperm are capacitated in the culture medium and examined as pellets. Hence there are differences in the mechanics of the AR but the end result is apparently the same - release of acrosomal enzymes and exposition of the IAM. It is well established that extracellular Ca^{2+} is essential for both acrosome vesiculation and later for the fusion of sperm and egg plasma membranes (Monroy, 1985; Meizel, 1985; Yanagimachi, 1988a,b). The AR induced in vitro is very likely to be the true physiological reaction on account of the high incidence of fertilization (~90%) recorded in most IVF centres. Apart from routine incubation of sperm, various innovative methods are being used to improve capacitation and induce the AR, such as Percoll gradient, Ficoll separation and strontium incubation in the absence and presence of Ca^{2+} (Mortimer et al, 1986). These methods, though more time-consuming, are proving invaluable for the induction or synchronization of the AR in poor-quality sperm in conjunction with sperm injection,

(Laws-King et al, 1987 Ng et al, 1988, 1989a). The significance of the AR in microfertilization has been reviewed by Sathananthan and Trounson (1989) and Sathananthan et al (1989). There is no doubt, whatsoever, that only acrosome reacted can penetrate the zona and are capable of fusing with the oocyte (Sathananthan and Chen, 1986; Sathananthan et al, 1986a,b). Even capacitated acrosome-intact sperm with hyperactivated motility are incapable of fusion (Fig. 9) and swim actively in the PVS after sperm injection, sometimes shearing the inner zona (Sathananthan et al, 1989). Although the sperm PM and oolemma are brought very close to each other, no fusion eventuates. The moment vesiculation of the acrosome is completed membrane fusion is initiated almost simultaneously (Figs. 10, 11).

SPERM-OOCYTE MEMBRANE FUSION

It must be realized that to visualize a single sperm fusing with an egg is a matter of chance and one has to painstakingly examine serial sections of many oocytes, 1 to 3 hours after insemination at 30 minute intervals (Sathananthan and Chen, 1986). We used routine methods of IVF, fixation and processing for transmission electron microscopy (TEM) and obtained 12 oocytes from 6 women who became pregnant by IVF. The oocytes were sectioned from pole to pole to obtain just one sperm fusing with an egg. The events that occur soon after fusion, namely pronuclear formation, cortical granule exocytosis and abstriction of the second polar body are easier to detect by TEM (Lopata et al, 1980), but finding a sperm tail axoneme in the ooplasm is still a difficult task.

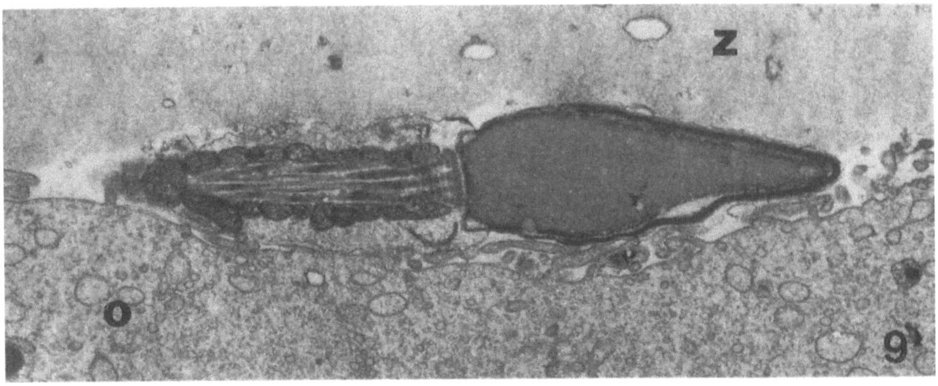

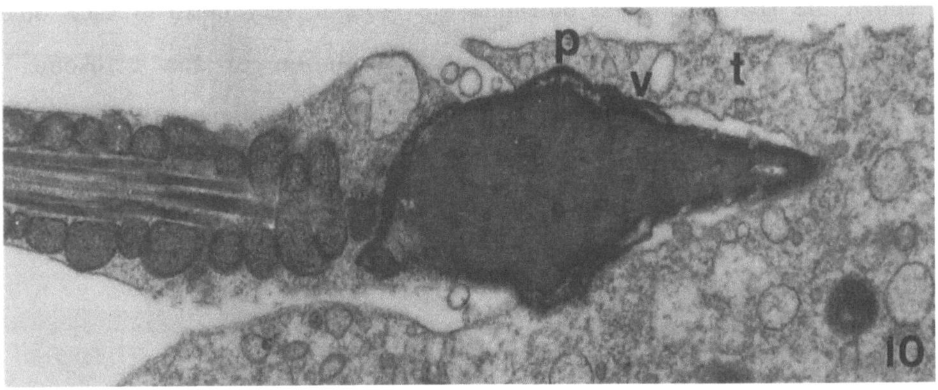

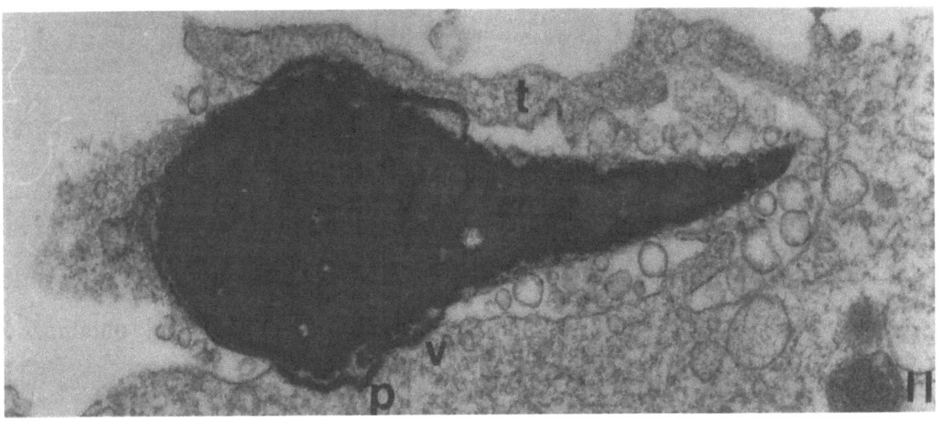

Fig. 9 Acrosome-intact sperm lying passively in the PVS after MIST. **Figs. 10-11** Serial sections of a sperm undergoing initial sperm-egg membrane fusion with a polyspermic ovum. A tongue-like process (t) engulfs the anterior region of the spermhead. Its acrosome has vesiculated (Fig. 11). The fusogenic zone extends from the equatorial vestige (v) to the anterior half of the postacrosomal region (p). o = ooplasm, z = zona x 17,220, x 22,386, x 29,274 (Figs. 10 and 11 are reproduced from Sathananthan et al, 1986a).

For clarity of visualization of the early events leading up to membrane fusion let's examine polyspermic interaction first and then consider monospermic penetration. In this experiment, the zona was partially denuded with fine needles in Menezo B_2 medium containing sperm capacitated for 4 hours, in order to promote polyspermic penetration (Sathananthan et al, 1986a). The portion of the egg within the zona was not penetrated, while the naked region showed various stages of sperm fusion, incorporation and nuclear decondensation (Fig. 1). This experiment was remarkably similar to that carried out in 1957 by Yanagimachi with a herring egg (see Yanagimachi, 1988a). During polyspermic interaction the initial contact between the sperm and egg seem to be made between microvilli of the egg and the IAM or midsegment of the spermhead. However, it was not determined with certainty whether the equatorial vestige or postacrosomal segment (PAS) were the first to make contact with the oolemma. At least, 5 sperm were seen fusing with the egg at different locations on the oolemma, while 13 others were already incorporated into the ooplasm and showed various stages of nuclear decondensation. Examination of serial sections of sperm undergoing fusion clearly show that the fusogenic zone of the spermhead extends from the equatorial vestige of the acrosome to the anterior half of the PAS (Figs. 10, 11). The sperm is completing the AR by vesiculation and is being incorporated into the egg tangentially. Gamete fusion has elicited a phagocytic response from the egg and a tongue-like extension of the cortical ooplasm, much larger than a microvillus, has engulfed the anterior region of the spermhead and has fused with its midsegment. The IAM has been exposed and the internalized stretch of oolemma is found outside this membrane. The picture of monospermic fertilization, also visualized in serial sections, show that membrane fusion is at a more advanced stage (Fig. 12) than that observed during polyspermic interaction. Membrane fusion has already taken place in the PAS posterior to the equatorial vestige and the limits of the fusogenic

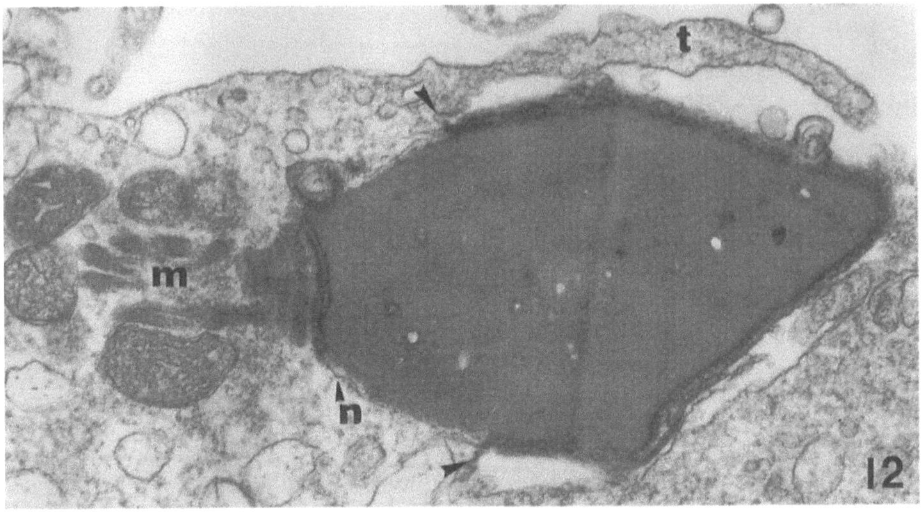

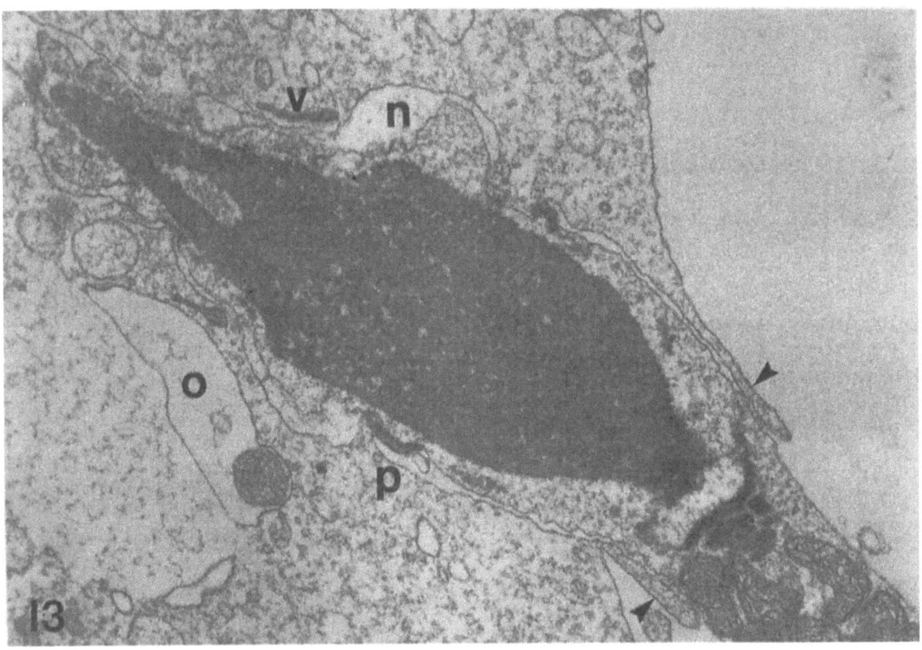

Fig. 12 Monospermic sperm-egg membrane fusion. The oolemma has fused with the postacrosomal region posterior to the equatorial vestige (arrows). The midpiece (m) and tail has been incorporated first and the head is engulfed by a process of the egg (t). **Fig. 13** Sperm penetration in a polyspermic ovum where membrane fusion has already occurred in the anterior head region, while the posterior region is engulfed by processes of the egg (arrows). Chromatin has begun to decondense and the nuclear envelope (n) is inflating. o = pocket of internalized oolemma, p = posterior part of postacrosomal region, v = equatorial vestige. x 29,274 (Reproduced from Sathananthan and Chen, 1986 and Sathananthan et al, 1986a).

midsegment is not as clearly demarcated as for polyspermic interaction. In both cases of monospermic fertilization encountered (Sathananthan and Chen, 1986; Sathananthan et al, 1987) the sperm were being incorporated tangentially but tail first. The midpiece and principal piece that were more deep-seated, had already been incorporated into the ooplasm and were devoid of PM. Sperm penetrating the egg perpendicularly in the absence of the zona, head first, also show evidence of membrane fusion along the fusogenic midsegment (Fig. 13). Phagocytic processes of the ooplasm extend toward the midpiece, which eventually fuse with the sperm PM in that region (Sathananthan et al, 1986a). Most of the tail is also incorporated progressively by membrane fusion in zipper-like fashion (Fig. 14) or occasionally by phagocytosis when the PM persists. Intermingling of considerable lengths of sperm PM with the oolemma could be envisaged both during monospermic and particularly polyspermic fusion. This has been visualized in animals by labelling the sperm surface with antibodies and then locating them on the egg-surface after fusion (see Yanagimachi, 1988a,b; Longo, 1989). Direct evidence of sperm PM incorporation into the oolemma has been provided in our study of polyspermic interaction. Abnormal sperm with nuclear defects were seen fusing with the oocyte (Sathananthan et al, 1986a) and later undergoing chromatin decondensation in the ooplasm (Sathananthan et al, 1989).

The general process of sperm-egg fusion conforms to that reported for eutherian mammals (Yanagimachi, 1981; Bedford, 1983), where the fusogenic PM is localized in the equatorial segment and/or the PAS. There seems to be some controversy with regard to the fusogenic zone of the spermhead in the hamster. Earlier studies in rodents indicated that the PM of the PAS was the first to fuse with the oolemma (see Bedford et al, 1979, 1983; Yanagimachi, 1981, 1988a,b). Bedford and co-workers subsequently showed that the PM covering the equatorial segment of the acrosome was the first to fuse with the oolemma. When fusion occurs in the human, both

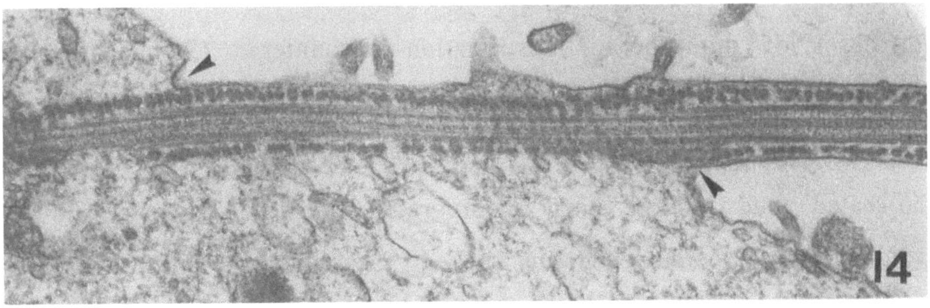

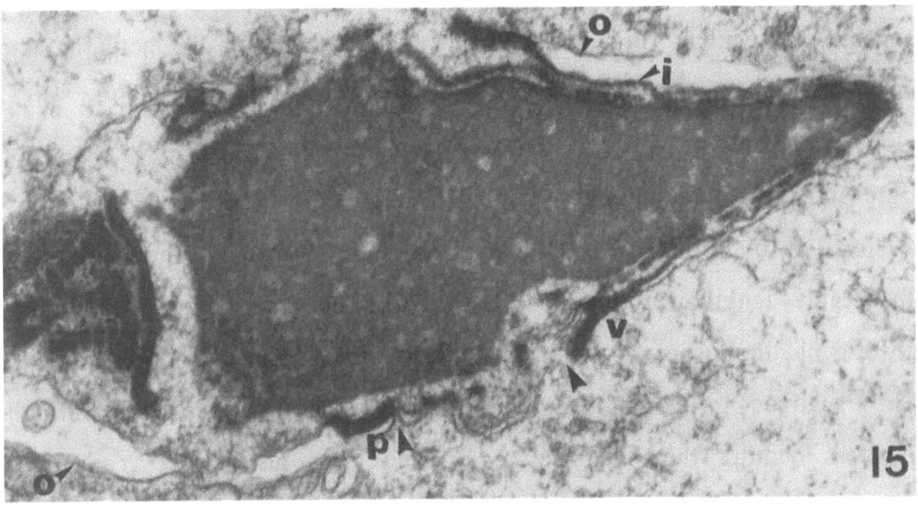

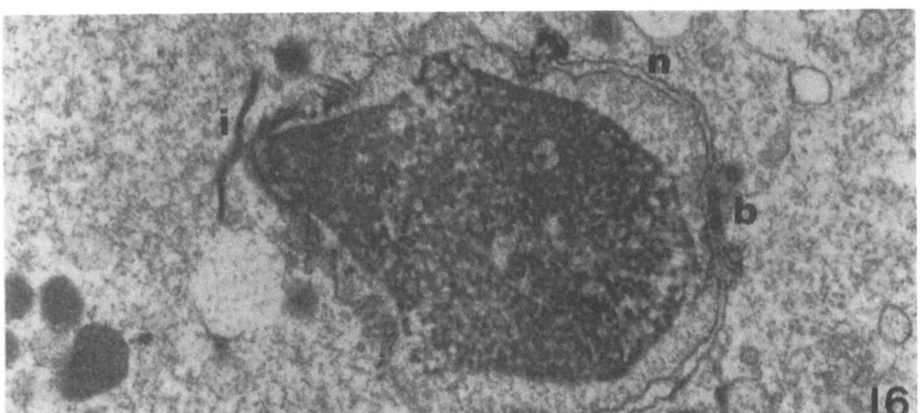

Fig. 14 Incorporation of a spermtail in zipper-like fashion. The sperm cell membrane is continuous with the oolemma (arrows). **Fig. 15** Sperm incorporated into ooplasm. The fusogenic zone is demarcated between arrows. **Fig. 16** Sperm chromatin decondensation. The nuclear envelope (n) has inflated and chromatin is coarsely granular. b = basal plate, i = inner acrosome membranes, o = internalized oolemma, p = posterior part of postacrosomal region, v = equatorial vestige. x 22,386, x 36,162, x 22,386 (Figs. 15 and 16 are reproduced from Sathananthan, 1986a).

the acrosomal cap and equatorial segment have vesiculated, completing the AR. What remains intact is a small inconspicuous equatorial vestige when the sperm enters the PVS (Soupart and Strong, 1974; Chen and Sathananthan, 1986; Sathananthan et al, 1982, 1986b). The fusogenic PM over the PAS is clearly more extensive than that over the equatorial vestige. Thus the major part of the fusogenic zone is the PM over the PAS (Fig. 11). This has been also confirmed recently with poor quality sperm seen fusing with the oocyte, after multiple sperm injection into the PVS (unpublished data). However, we have to examine earlier interactions between sperm and egg to determine which of these segments make initial contact with microvilli of the oocyte, although there are indications that both regions do so during polyspermic interaction. Microvilli were seen to make non-specific contact with all regions of the spermhead when sperm are microinjected into the PVS. No fusion was observed when the equatorial segment was intact in partially-reacted sperm (Fig. 5). Therefore it is conceivable that there are species differences in the events that lead up to gamete fusion in eutherian mammals.

POSTFUSION EVENTS

Once the spermhead is incorporated in the ooplasm the nuclear envelope inflates, sperm chromatin decondensation begins and the remnants of the equatorial vestige and posterior region of the PAS persist for some time indicating the limits of the fusogenic midsegment of the spermhead (Sathananthan et al, 1986a). The internalized oolemma and the IAM remain intact for a while but do not fuse with each other (Fig. 15). Eventually, the IAM is intermittently disrupted by evaginations of the inflating nuclear envelope (Fig. 13) and remnants of the IAM are occasionally found close to

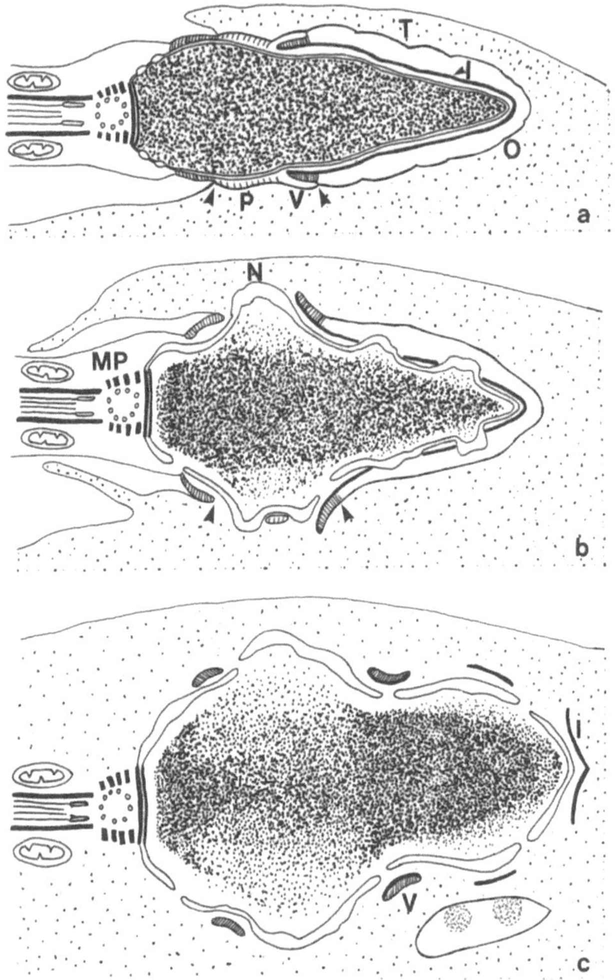

Fig. 17 Stages of sperm fusion and incorporation in the human visualized during monospermic and polyspermic fertilization (drawn from electromicrographs). a. Initial fusion between the oolemma and sperm plasma membrane extending from the equatorial vestige (V) to the anterior region of the postacrosomal segment (P) shown between arrowheads. The front of the spermhead with exposed innner acrosome membrane (I) has been engulfed by a tongue-like process of the cortical ooplasm (T). The chromatin is not expanded. b. Fusion has progressed in the midsegment of the spermhead, and the region posterior to this zone is being engulfed by two processes of cortical ooplasm extending to the midpiece (MP). The nuclear envelope (N) has inflated, and chromatin decondensation has commenced. The inner acrosome membrane is interrupted, but the internalized oolemma is still intact. The equatorial vestige and the intact region of the postacrosomal segment mark the fusogenic midsegment (arrowheads). c. Spermhead fully incorporated within the ooplasm showing further expansion of chromatin. The inflated nuclear envelope will eventually vesiculate and disappear. The oolemma has disappeared around the spermhead, but a pocket of this membrane, containing released products of cortical granules, is shown on the right. Remnants of the inner acrosome membrane, equatorial vestige (V), and posterior region of postacrosomal segment are shown. Reproduced from Sathananthan et al (1986a).

the expanded sperm nucleus (Fig. 16). The sperm nuclear envelope gradually vesiculates and looses its identity while a new envelope is assembled mainly by elements of the smooth endoplasmic reticulum of the oocyte and the male pronucleus is formed (Lopata et al, 1980; Sathananthan and Chen, 1986; Sathananthan et al, 1986a,b). The sperm basal plate, centrioles and mitochondria remain closely associated with the male pronucleus, which has now assumed a spherical form. The internalized oolemma gradually disappears but pockets of this membrane, associated with released products of cortical granules may persist (Fig. 13). Unidentified filaments of uniform thickness were associated with the incorporated sperm tail during monospermic fusion (Sathananthan and Chen, 1986). The foregoing events of gamete fusion are summarized in figure 17, based on TEM observations on both monospermic and polyspermic sperm interaction with oocytes.

The events that occur after fertilization are well documented both in vivo and in vitro (Zamboni, 1971; Gwatkin, 1977; Edwards, 1980; Soupart, 1980; Lopata et al, 1980; Sathananthan and Trounson, 1982, 1985; Dvorak et al, 1984; Sathananthan et al, 1986b, 1989) and will be dealt with only briefly. Sperm-egg membrane fusion triggers the instantaneous release of cortical granules and the abstriction of the second polar body to complete oocyte maturation. The male and female pronuclei are formed and migrate toward each other but do not conjugate or fuse. The fertilized ovum is thus activated and eventually undergoes cleavage after the paternal and maternal chromosomes are brought together and pair during syngamy. A strong primary block to polyspermy is established at the level of the inner zona after cortical granule release and zona interaction (Sathananthan and Trounson, 1982). There is also evidence of a secondary block at the oolemma when the zona is by-passed (Sathananthan et al, 1989), because multiple sperm injection into the PVS more often results in monospermy and not polyspermy (see Ng et al, chapter in this book). An electrical block to polyspermy may be operative (Jaffe and Cross, 1986), though this has not been conclusively established in mammals including the rabbit, which has no zona-block (see Yanagimachi, 1988b).

IMMOTILE SPERM INCORPORATION

Our observations on monospermic sperm interaction, where sperm were incorporated tail first, indicated that sperm motility is not essential for human gamete fusion although it is required for zona penetration (Sathananthan and Chen, 1986; Sathananthan et al, 1986b). This prompted us to experiment with immotile sperm from patients with Kartegener's syndrome (Ng et al, 1987; Bongso et al, 1989). These sperm lack dynein arms and are completely immotile, though the majority have a normal head structure. Aitken et al (1983) showed that such sperm are capable of fusing with zona-free hamster eggs, which was confirmed later by Ng et al (1987). It is evident from these studies that immotile sperm incorporation is accomplished mainly by a phagocytic response from the oocyte when sperm are brought into close contact with the oolemma. Sperm injection into the PVS is an effective way of treating such patients (Bongso et al, 1989) and has been extended to treat patients with very poor quality sperm, eventuating in successful pregnancies (Ng et al, 1988, 1989a). The role of sperm motility in sperm-egg fusion in mammals has been reviewed by Gaddum-Rosse (1985) and Yanagimachi (1989a).

SPERM INCORPORATION IN EMBRYOS

Recent TEM studies on sperm injection into the PVS of 1-16 cell human embryos reveal that sperm are sequestered into vacuoles within embryonic cells and remain more or less intact (Ng et al, 1989b). Sperm chromatin decondensation does not occur nor is there evidence of sperm-egg membrane fusion. There is some evidence of degeneration of incorporated sperm cells. Thus the phagocytic response of the oocyte seems to persist in early embryonic cells while gamete membrane fusion does not occur. In the hamster, sperm are incorporated into embryonic cells up to the 4-cell stage and undergo limited sperm chromatin condensation (Usui and Yanagimachi, 1976). However, during polyspermic interaction many supernumerary sperm fuse with the oocyte after the first sperm is incorporated and they show various stages of sperm fusion and chromatin decondensation as described above (Sathananthan et al, 1986a).

MECHANICS OF MEMBRANE FUSION

Yanagimachi (1988a) has proposed an interesting model of mammalian sperm-oocyte membrane fusion in his comprehensive review, which is both provocative and stimulating. He postulates that there exist complementary molecules on the sperm and oocyte plasma membranes that fit into one another and also fusogenic proteins on the sperm PM which insert into a 'quiescent' oolemma during fusion. Egg membrane lipid perturbance caused by sperm molecules and the phase separation of the membrane lipids through the action of Ca^{2+} leads to fusion of the 2 membranes. The electrical potential of the oolemma and an influx of Ca^{2+} appear to play regulatory roles in membrane fusion. Ca^{2+}, in particular, seems to play a key role in membrane fusion but also in cortical granule exocytosis (Monroy, 1985; Longo, 1989). It is obvious that the midsegment of the sperm PM posterior to and including the equatorial vestige undergoes some chemical modification after the AR is completed and becomes fusogenic. It has also been postulated that there could be sperm receptors on the oolemma that might play a role in sperm-egg fusion (Ng et al, chapter in this book), though these may not be entirely species specific. After fusion, there would logically be a large increase in the surface area of the oocyte due to incorporation within the oolemma of sperm PM, as well as membranes that delimit cortical granules, discharged at fertilization (see Sathananthan et al, 1986b; Longo, 1989). However, the volume of the fertilized ovum seems to remain more or less the same after fusion. Excess PM is evidently accommodated by second polar body abstriction and intensive microvillus activity at the oocyte surface, especially in the region of polar body abstriction. Internalization of the oolemma, observed in this study, and micropinocytotic activity (endocytosis) during cortical granule exocytosis (see Sathananthan, 1984; Sathananthan et al, 1986b) could also contribute to reducing surface area of the oocyte at fertilization.

In conclusion, more research needs to be done to probe the molecular aspects of gamete membrane fusion. The possible existence of sperm receptors on the oolemma has to be investigated. Another desirable avenue of research relates to the electrical regulation of sperm-egg fusion (see Jaffe and Cross, 1986), since there is evidence of a block to polyspermy at the oolemma in the human (Sathananthan et al, 1989). IVF and the new technologies of assisted reproduction that are being developed, which include sperm injection, partial zona dissection, zona-drilling or zona-cracking followed by insemination (see Ng et al, 1989a), will help us further unravel some of the mysteries of gamete membrane fusion.

REFERENCES

Aitken RJ, Ross A, Lees MM (1983) Analysis of sperm function in Kartegener's syndrome. Fertil steril 40 : 696 - 698

Bedford JM (1983) Form and function of eutherian spermatozoa in relation to the nature of egg vestments. In : Beier HM, Lindner HR (eds) Fertilization of the human egg In Vitro Springer-Verlag, Berlin, p133

Bedford JM, Moore HDM, Franklin LE (1979) Significance of the equatorial segment of the acrosome of the spermatozoon in eutherian mammals. Exp Cell Res 119:119-126

Bongso A, Sathananthan H, Wong PC, Ratnam SS, Ng SC, Anandakumar C, Ganatra S (1989) Human fertilization by microinjection of immotile sperm. Human Reprod 4: 175-179

Chen C, Sathananthan AH (1986) Early penetration of human sperm through the vestments of human eggs in vitro. Arch Androl 16: 183-197

Dvorak M, Tesarik J, Kopechny V (1984) Ultrastructure of human fertilization In: Van Blerkom J, Motta PM (eds) Ultrastructure of reproduction, Martinus Nijhoff, Boston, p. 176

Edwards RG (1980) Conception in the human female. Academic Press, London.

Gaddum - Rosse P (1985) Mammalian gamete interactions : What can be gained from observations on living eggs? Am J Anat 174: 347-356

Gwatkin RBL (1977) Fertilization mechanisms in man and mammals, Plenum Press, New York.

Jaffe LA, Cross NL (1986) Electrical regulation of sperm - egg fusion. Annu Rev Physiol 48: 191-200

Laws-King A, Trounson A, Sathananthan H, Kola I (1987) Fertilization of human oocytes by micro injection of a single spermatazoon under the zone pellucida. Fertil Steril 48: 637-642

Longo FJ (1989). Egg cortical architecture. In : Schatten H, Schatten G (eds) The cell biology of fertilization, Academic Press, San Diego, p105

Lopata A, Sathananthan AH, McBain JC, Johnston WIH, Spiers AL (1980) The ultrastructure of the preovulatory human egg fertilized in vitro. Fertil Steril 33:12-20

Lopata A, Nayudu P, Jones G, Abramczuk J (1985) The quality of human embryos obtained by IVF. In : Testart J, Frydmann R (eds) Human in victro fertilization. Actual problems and prospects, Elsevier, Armsterdam, p171

Meizel S (1985) Molecules that initiate or help stimulate the acrosome reaction by their interaction with the mammalian sperm surface. Am J Anat 174 : 285-302

Monroy A (1985) Processes controlling sperm-egg fusion. Eur J Biochem 152 : 51-56

Mortimer D, Curtis EF, Dravland JE (1986) The use of strontium-substituted media for capacititing human spermatozoa : an improved sperm preparation method for zona-free hamsters egg penetration test. Fertil Steril 46 : 97-103

Nagae T, Yanagimachi R, Srivastava PN, Yanagimachi H (1986) Acrosome reaction in human spermatozoa. Fertil Steril 45 : 701-707

Ng SC, Sathananthan AH, Edirisinghe WR, Kum Chue JH, Wong PC, Ratnam SS, Sarla G (1987),. Fertilization of a human egg with sperm from a patient with immotile cilia syndrome : case report. In : Ratnam SS, Teoh ES, Anandakumar C (eds) Advances in Fertility and Sterility, vol 4, Parthenon, Lancaster, p71

Ng SC, Bongso A, Ratnam SS, Sathananthan AH, Chan CL, Wong PC, Hagglund L, Anandakumar C, Wong YC and Goh VHH (1988) Pregnancy after transfer of multiple sperm under the zona. Lancet 2 : 790.

Ng SC, Bongso A, Sathananthan AH, Ratnam SS (1989a) Micromanipulation : Its relevance to human in vitro fertilization. Fertil Steril (in press)

Ng SC, Sathananthan AH, Bongso TA, Tok VCN, Ho JCK, Ratnam SS (1989b) Sperm transfer by micromainpulation into early human embryos (in review)

Sathananthan AH (1984) Ultrastructural morphology of ferilization and early cleavage in the human. In: Trounson A, Wood C (eds) In vitro fertilization and embryo transfer. Churchill Livingstone, Edinburgh, p131

Sathananthan AH and Trounson AO (1982) Ultrastructure of cortical granule release and zona interaction in monospermic and polyspermic human ova fertilized in vitro. Gamete Res 6: 225-234

Sathananthan AH and Trounson A (1985) The human pronuclear ovum: fine structure of monospermic and polyspermic fertilization in vitro. Gamete Res 12:385-398

Sathananthan AH and Chen C (1986) Sperm-oocyte membrane fusion in the human during monospermic fertilization. Gamete Res 15:177-186

Sathananthan AH and Trounson A (1989) The microinjection technique and the role of the acrosome reaction in microfertilization. In : Ben-Rafael Z (ed) In vitro fertilization and alternate assisted reproduction, Plenum, New York (in press)

Sathananthan AH, Trounson AO, Wood C, Leeton JF (1982) Ultrastructural observations on the penetration of human sperm into the zona pellucida of the human egg in vitro. J Androl 3:356-364

Sathananthan AH, Ng SC, Edirisinghe R, Ratnam SS, Wong PC (1986a) Sperm-egg interaction in vitro. Gamete Res 15:317-326

Sathananthan AH Trounson A and Wood C (1986b) Atlas of fine structure of human sperm penetration eggs and embryos cultured in vitro. Praeger Scientific, Philadelphia

Sathananthan AH, Trounson A, Freemann L (1987) Morphology and fertilizability of frozen human oocytes. Gamete Res 16: 343-354

Sathananthan AH, Ng SC, Trounson A, Bongso A, Laws-King A, Ratnam SS (1989) Human micro-insemination by injection of single or multiple sperm : ultrastructure. Human Reprod 4:574-383

Soupart P (1980) Fertilization. In : Hafez ESE (ed) Human reproduction: conception and contraception. Harper & Row, New York, p453

Soupart P, Strong PA (1974) Ultrastructural observations on human oocytes fertilized in vitro. Fertil Steril 25:11-44

Soupart P, Strong PA (1975) Ultrastructural observations on polyspermic penetration of zona pellucida free human oocytes inseminated in vitro. Fertil Steril 36:523-537

Trounson A (1987) Microfertilization. Proc Fifth World Congress In Vitro Fertilization, Norfolk, Virginia (abstract)

Trounson A (1989) Fertilization and embryo culture. In : Wood C, Trounson A (eds) Clinical in vitro fertilization, 2nd ed, Springer-Verlag, London, p32

Usui N, Yanagimachi R (1976) Behavior of hamster sperm nuclei incorporated into eggs at variours stages of maturation, fertilization and early development. J Ultrastructure Res 57: 276-288

Wassarman PM, Florman HM, Greve JM (1985) Receptor mediated sperm-egg interactions in mammals. In: Metz CB, Monroy A (eds) Fertilization, Vol 2, Academic Press, New York, p341

Yanagimachi R (1981) Mechanisms of fertilization in mammals. In Mastroianni L Jr, Biggers JD (eds) Fertilization and embryonic development in vitro. Plenum Press, New York, p 81

Yanagimachi R (1988a) Sperm-egg fusion. In: Duzgunes N, Bronner F (eds) Vol 32, Membrane fusion in fertilization, cellular transport and viral infection, Academic Press, New York, p 3

Yanagimachi R (1988b) Mammalian fertilization. In: Knobil E, Neill J et al (eds) The physiology of reproduction, Raven Press, New York, p 135

Zamboni L (1971) Fine morphology of mammalian fertilization. Harper and Row, New York

MICRO-INSEMINATION SPERM TRANSFER (M.I.S.T.) INTO HUMAN OOCYTES AND EMBRYOS

Soon-Chye Ng, Ariff Bongso, Sathananthan AH* and Shan S. Ratnam

Department of Obstetrics & Gynaecology,
National University of Singapore,
National University Hospital,
Lower Kent Ridge Road,
SINGAPORE 0511.

INTRODUCTION

The severely oligozoospermic patient usually has a combination of multiple spermatozoal defects (WHO, 1987), with very poor chances of spontaneous conception. Even in-vitro fertilization (IVF) with embryo replacement (ER) has not resulted in improved fertilization in such cases. In the series of 31 patients reported by Yovich and Stanger (1984) 3 groups were defined - severe oligozoospermia ($<5 \times 10^6$ motile sperm/ml), moderate oligozoospermia ($6-12 \times 10^6$ motile sperm/ml), and abnormal sperm morphology (>60% atypical forms). Fertilization rates for 164 oocytes were 41.2%, 77.4% and 48.7% respectively, compared with 85.0% for the controls. In the series published by Mahadevan et al (1985), a total of 129 oocytes from 45 patients with at least one abnormal sperm parameter were studied. When the sperm density was $<5 \times 10^6$/ml, the fertilization rate was only 17%, compared to 54% when the density was $10-19 \times 10^6$/ml. When the density was $5-9 \times 10^6$/ml, the fertilization rate was 27%. More significant was the effect of motility on fertilization; sperm with less than 30% forward progressive motility had almost no fertilization. While fewer than 200 spermatozoa reach the site of fertilization in-vivo (Ahlgren, 1975), the in-vitro situation is sub-optimal. Fertilization rate decreases from 84.7% to 48.0% when inseminating concentrations reduce from 500,000/ml to 250,000/ml motile spermatozoa (Diamond, 1985) although Craft et al (1981) have reported successful fertilization with 10,000 motile spermatozoa. Wolf and co-workers (1984) reported that fertilization rates of human pre-ovulatory oocytes decreased from 60.0% at 500,000 motile sperm/ml to

* Lincoln School of Health Sciences, La Trobe University, and Monash Medical Centre, Melbourne, AUSTRALIA.

NATO ASI Series, Vol. H 45
Mechanism of Fertilization
Edited by B. Dale

28.6% at 10,000 motile sperm/ml (with maximum fertilization of 80.8% at 25,000 motile sperm/ml).

When the number of abnormal semen parameters are taken into consideration, fertilization decreased from 67.0% (43 men, 182 oocytes) for one abnormal parameter, and 58.2% (14 men, 67 oocytes) for two abnormal parameters, to 7.1% (4 men, 14 oocytes) for three abnormal parameters (De Kretser et al, 1985). It is the latter group that is oligoasthenoteratozoospermic. In a larger study Yates et al (1989) reported that there is still a significant reduction in fertilization in triple sperm defects (35.1%, 61 men in 116 cycles), when compared to single sperm defects (61.3%, 121 men in 206 cycles) and double sperm defects (48.9%, 98 men in 177 cycles). When the sperm density is very low (<5.0 x 10^6/ml), it is usually combined with low motility and high abnormal forms (WHO, 1989), unless the low sperm density is due to an obstructive cause. Such a combination is to be expected when there is seminiferous tubular failure, usually of an idiopathic nature.

In a retrospective study of 1067 IVF attempts in Norfolk, Oehninger et al (1988) reported that 5% of them resulted in failed fertilization. Male factor was thought to be the cause in 15.3% of these 52 failed fertilization cases on initial classification. However, on re-evaluation using strict criteria for spermatozoa morphology, spermatozoa anomalies was thought to be the cause in 61.5%, while another 13.4% had a combination of spermatozoa and oocyte anomalies. Unfortunately, there was no similar study of spermatozoa patterns using the strict criteria for patients who were successful in fertilization.

Tubal embryo transfer (Wong et al, 1988), otherwise known as pronuclear stage transfer (PROST) and tubal embryo stage transfer (TEST) (Yovich, Yovich and Edirisinghe, 1988), does not offer much hope for such patients as the spouses' oocytes still need to be fertilized before the transfer into the Fallopian tubes.

Hence, there has been keen interest in developing a micro-manipulative method to introduce spermatozoa directly into the oocyte, as such a technique requires very minimal spermatozoa. At this point, I would like to introduce some new terms: assisted fertilization, micro-insemination, sperm transfer, and micro-injection. "Assisted

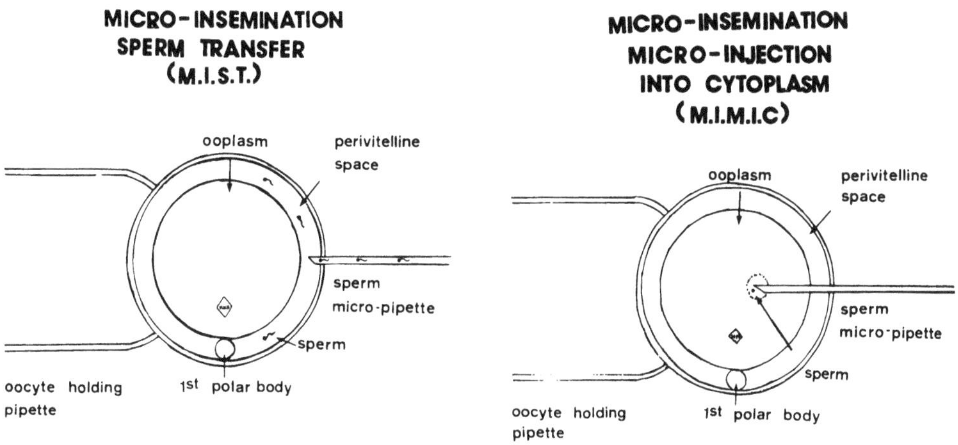

Figure 1: Micro-insemination of spermatozoa into oocytes, a schematic representation.

fertilization" is when fertilization is assisted by micro-manipulation techniques. The term "micro-insemination" has been proposed (R.Yanagimachi, 1988, personal communication), in contrast to "macro-insemination" which refers to vaginal and in-vitro insemination. The term "micro-fertilization" was first used by Alan Trounson (1987) to describe introduction of spermatozoa into the oocyte. However, "micro-insemination" is more appropriate because insemination is the process of adding sperm, while fertilization is the result of sperm union with egg (Yanagimachi, 1988, personal communication). In micro-insemination, sperm are introduced either singly or in multiples into the perivitelline space (pvs) under the zona ("sperm transfer") or directly into the cytoplasm ("sperm micro-injection") [figure 1]. We propose the following acronyms: MIST for Micro-Insemination Sperm Transfer, and MIMIC for Micro-Insemination Micro-Injection into Cytoplasm (Ng et al, 1990).

Zonal procedures, such as zona-drilling (Gordon et al, 1988), zona-cracking (Odawara and Lopata, 1989) and partial zona-dissection (Cohen at al, 1988), are also methods of assisted fertilization. However, macro-insemination is used, as the manipulated oocytes are then transferred into the insemination well/droplet.

This review will concentrate on MIST, as this is the more successful of the 2 micro-insemination techniques.

INDICATIONS FOR MICRO-INSEMINATION

The indications for micro-insemination in the human are:

1. **Severe oligozoospermia**:
When the sperm count is reduced to less than 5 million/ml, there is a significantly reduced chance of fertilization by normal IVF procedures (see above). However, recently Yovich and co-authors (1987) described pronuclear stage transfer (PROST) and tubal embryo stage transfer (TEST) for oligozoospermia in which transfer of pronuclear zygotes or cleaving embryos into the Fallopian tube was carried out. Though this improved the pregnancy rate following fertilization, the fertilization process still requires the initial semen sample to be more than 5 million/ml.
2. **Severe multiple sperm factors (oligoasthenoteratozoospermia)**:
Many severe male factors have a combination of at least 3 sperm defects because of seminiferous tubular failure (see above). But it must be realised that the potential for conception is dependent on severity. Mild forms of oligoasthenoteratozoospermia do not even require IVF. However, severe forms have very poor prognosis in terms of treatment and realisation of pregnancy. Such patients have previously relied on artificial insemination with donor sperm. Now, micro-insemination offers some hope. In borderline situations, it should be demonstrated that there is repeated failed fertilization in-vitro.
3. **Immotile sperm (total asthenozoospermia)**:
Immotile sperm in the semen sample may be caused by environmental or congenital factors, eg ciliary dyskinesis (Rossman et al, 1981), with its three major variants (see Zamboni, 1987). Environmental problems, eg drug therapy, usually result in decreased motility, and this can be improved by removal of the drug or introduction of additives into the sperm washing proce-dures (see Bongso, Ng and Ratnam, 1989). Congenital problems are more difficult to treat and there is the ethical issue of whether propagation of such genes is desirable. The immotile cilia syndrome (ciliary dyskinesis), an

autosomal recessive condition (Palmblad, Mossberg & Afzelius, 1984), is one such problem and is compatible with normal life span provided preventive therapy is instituted for the respiratory complications. Hence, for this genetic disease we feel that micro-insemination is justified.

4. **Inability to penetrate egg investments**:
While this has not been described conclusively in the literature, there are many instances where there is failed fertilization in spite of oocytes being mature and spermatozoa of good motility (Chia et al, 1984). As discussed above, the report by Oehninger and co-workers (1988) stressed that inspite of critical reappraisal of the causes of failed fertilization in 52 patients, 6 (11.5%) were unexplained. The development of a zona-penetration test (Overstreet, 1983) may shed more light onto this defect. In the same manner, the "hemi-zona" assay will help in the under-standing of sperm penetration through the zona (Burkman et al, 1988). "Spontaneous" hardening of the zona pellucida has been described in mouse oocytes cultured in-vitro (De Felici & Siracusa, 1982), as demonstrated by increasing resistance to solubilization by chymotrypsin. The authors later presented data suggesting that follicular fluid contains factors (especially glycosaminoglycans) that prevent such hardening (De Felici, Salustri & Siracusa, 1985). Alterations of the surface structure of the zona (related to maturity of the oocyte) have been linked to sperm-binding capacity to the zona (Familiari et al, 1988). For micro-insemination to be justified in this condition, it should be demonstrated clearly that there is failed fertilization on repeated IVF attempts in all the oocytes. Specific sperm abnormalities that do not allow attachment of sperm to oolemma, eg round-headed, acrosome-less sperm (Lalonde et al, 1988), might be another condition that justifies sperm injection. Lanzendorf et al (1988) have recently reported that MIMIC into hamster oocytes with such sperm can result in sperm head decondensation.

CHROMOSOMAL ABNORMALITIES IN HUMAN OOCYTES

For human IVF, patients are usually stimulated to increase the number of oocytes available, and hence to increase the number of embryos transferred and therefore the chance of pregnancy (Gronow et al, 1985).

Since stimulation results in the production of multiple oocytes (much above the normal 1 oocyte per cycle) there has been concern regarding their chromosomal and genetic normalcy. The current literature is summarized in Table 1. Wramsby and co-workers (Wramsby, Fredga and Liedholm, 1987) reported karyotyping of 23 oocytes from 17 patients after clomiphene-citrate stimulation. An aneuploidy rate of 34.8% (8/23) was obtained. Hyperploidy was seen in 2 oocytes, while the rest were hypohaploid. There were 4 oocytes with very low chromosome numbers (1-5); these spreads could have been due to artificial loss during the preparations. Clinically, the use of clomiphene citrate has not been associated with an increased congenital abnormality rate. In a study of 1034 pregnancies conceived after clomiphene citrate, congenital malformation rate was only 2.3%, comparable to the control of 1.7% for spontaneous conceptions and 1.9% for pregnancies conceived after human menopausal gonadotropin and human chorionic gonadotropin (Kurachi et al, 1983). Another study of chromosomal complements from spare oocytes was reported by Martin et al (1986). Fifty oocytes were fixed 5-54 hours after collection. Normal haploid states were seen in 34 (68%). Hypo-haploidy was seen in 14 (28%), hyper-haploidy in 1 (2%), and structural abnormalities in 2 (4%). This study showed a high hypo-haploid state which may be artifactual. Oocytes that failed to fertilize in-vitro have also been karyotyped. In our own series of 302 failed fertilized oocytes, 251 were analysable. Normal complements were found in 192 (76.6%), while 33 (13%) were found to be hypo-haploid, and 20 (8%) were hyper-haploid (Bongso et al, 1988). Pellestor and Sele (1988) also karyotyped 201 oocytes that failed to fertilize after IVF from 87 women. Of these, 188 were in metaphase II. The aneuploidy rate was 18.6%. They classified these oocytes according to the indications for IVF; the aneuploidy rate was higher in idiopathic subfertility (22.5%) than in male subfertility (10.2%). Plachot et al (1986) reported on the karyotypes of 24 oocytes that remained unfertilized after 42 hours. There were 3 hypo-haploid, 4 hyper-haploid and 1 structural abnormality. In a later series, Plachot et al (1988a) reported higher aneuploidy rates (44.3%) for metaphase II oocytes that were still not fertilized 42 hours after insemination. Hypo-haploidy was seen in 14, hyper-haploidy in 17, and 2 had structural abnormality in a series of 70 interpretable karyotypes. Wramsby (1988) also reported 30 of 53 (56.6%) analyzable spreads from metaphase II oocytes that failed to fertilize in-vitro to have aneuploidy. Hypo-haploidy was seen in the majority (27),

TABLE 1: INCIDENCE OF CHROMOSOMAL ABNORMALITIES IN HUMAN OOCYTES

Authors	No. eggs	No. inter-pretable	Normal n=23	Hypo-haploid	Hyper-haploid	Structural anomalies
A. Oocytes that were not inseminated						
Martin et al (1986)	50	50	34	14	1	2
%				(28.0)	(2.0)	(4.0)
Wramsby et al (1987)	23	22	10	8	2	-
%				(36.4)	(9.1)	
TOTAL	73	72	44	22	3	2
%				(30.6)	(4.2)	(2.8)
B. Oocytes that failed to fertilize in-vitro						
Plachot et al (1986)	39	24	17	3	4	1
%				(12.5)	(16.7)	(4.2)
Bongso et al (1988)	302	251	192	33	20	1
%				(13.1)	(8.0)	(0.4)
Plachot et al (1988a)	120	70	37	14	17	2
%				(20.0)	(24.3)	(2.3)
Plachot et al (1988b)	608	316	234	----- 76 -----		-
%				(24.1)		
Wramsby (1988)	76	52	18	27	3	-
%				(51.9)	(5.8)	
TOTAL	1145	713	498	77*	44*	4*
%				(19.4)	(11.1)	(1.0)

* excludes Plachot et al (1988b)

while 3 were hyper-haploid and 2 had structural anomalies. However, in a multicentric study, Plachot et al (1988b) reported only a 26% chromosomal anomaly rate in a multicentric study involving 3 centres and 608 oocytes that failed to fertilize after 42 hours.

INCIDENCE OF CHROMOSOMAL ABNORMALITIES IN SPERMATOZOA

Rudak et al (1978) were the first to analyse sperm chromosomes using zona-free hamster oocytes. Of the 60 they analysed, 2 were hypohaploid while one was hyperhaploid (Table 2). In a study on 1000 human male pronuclear chromosomes complements from 33 normal donors, Martin et al (1983) reported an abnormality rate of 8.5% (5.2% aneuploidy and 3.3% structural abnormalities). The incidence of sperm chromosomal aberrations from normal donors (many with proven fertility) with chromosome aberrations varies between 6.6% and 14.3%, depending on the series (see Templado et al (1988)). However, the distribution of the different types of anomaly is quite variable. Structural defects account for 1.4 to 13.0%. Hypohaploid metaphases vary from 0.5 to 7.1%, while hyperhaploidy varies from 0.5 to 5.7%. There has been no reported studies on the sperm chromosomes of oligozoospermic men (Martin RH, personal communication, 1989). In a study of chromosome abnormalities in human spermatozoa from four healthy men of proven fertility, Brandiff et al (1984) reported that of 909 karyotypes 6.5% had structural abnormalities, 0.6% were hyperhaploid, 1.0% were hypohaploid, and 3.9% had chromosomal gaps. In another study, Kamiguchi and Mikamo (1986) reported on 1091 spermatozoal complements from 4 donors. There were 0.45% hyper-haploidy, 0.45% hypohaploidy, and 13.0% structural abnormalities. In the largest study on 1582 spermatozoal chromosome complements from 30 men of proven fertility, Martin et al (1987) reported a mean of 10.4% spermatozoal chromosome abnormality in individual men. The means of numerical and structural abnormalities were 4.7% and 6.2% respectively. Hence, the incidence of chromosomal anomalies in human spermatozoa (8-12%) is lower than the 20-21% reported in oocytes (Wramsby, 1988; Plachot et al, 1988; Bongso et al, 1988). This has an important bearing on micro-insemination, because if the resulting embryo is chromosomally abnormal the problem may have been the oocyte.

While it may not have been clearly demonstrated that chromosomal anomalies may be higher in patients with oligoasthenoterato-zoospermia, it remains one of the main concerns. To our knowledge there has been no haploid sperm karyotypes of men with normal peripheral karyotypes but abnormal semen analysis. However, it has been reported by Skakkebaek, Byrant and Philip (1973) that there were no structural abnormalities detected in studies of meiotic chromosomes from 74 subfertile men, though they noted a decrease in metaphase II/metaphase I ratio from meiotic chromosomes of these men. The authors concluded that "the chiasmatic frequency, the non-pairing of sex chromosomes and homologous chromosomes and the occurence of polyploid cells in metaphase II were the same in the controls and the infertile men."

An earlier report of karyotypes on testicular biopsies from subfertile men (McIlree et al, 1966) also showed that when the peripheral karyotype was normal, there was no significant increased abnormality rates in the testicular meiotic cells.

The incidences described above are for haploid spermatozoal chromosomes. This is very different for men with abnormal peripheral karyotypes. In a survey of male patients attending a subfertility clinic, Chandley and co-workers (1975) reported that 2% of them were chromosomally abnormal; this frequency rose to 6% among men with a mean spermatozoal count of less than 20 million per ml, and was 15% among azoospermic men. Earlier studies by Kjessler (1972) also demonstrated that the majority of chromosomally abnormal subfertile men are oligozoospermic and azoospermic. More recently, Retief et al (1984) reported similiar figures with chromosome abnormalities in 14.1% of azoospermic and 5.1% of oligozoospermic men (spermatozoal densities of less than 10 million/ml).

Unfortunately there is no animal model for oligoterotoastheno-zoospermia, or for that matter oligozoospermia alone. However, there seems to be interesting work on the genetic aspects of spermatogenesis (see Bartlett et al, 1989).

An important question now is whether poor spermatozoal quality is the product of micro-environment changes in the testis (including the epididymis), or is it the result of chromosomal defects? Existing evidence seem to suggest that severe spermatozoa problems are more due to

TABLE 2: CHROMOSOMAL ABNORMALITIES IN HUMAN SPERMATOZOA

Authors	No. pt	No. sperm	Normal n=23	Hypo-haploid (%)	Hyper-haploid (%)	Structural anomalies (%)
Rudak et al (1978)	1	60	57	2 (3.3)	1 (1.7)	-
Martin et al (1983)	33	1000	915	27 (2.7)	24 (2.4)	33 (3.3)
Brandiff et al (1984)	4	909	836	5 (0.6)	9 (1.0)	59 (6.5)
Kamiguchi & Mikamo (1986)	4	1091	941	5 (0.5)	5 (0.5)	142 (13.0)
Martin et al (1987)	30	1582	1419	53 (3.4)	20 (1.3)	98 (6.2)
TOTAL	72	4642	4168	92 (2.0)	59 (1.3)	332 (7.2)

micro-environment alterations, especially in the epididymis, than to chromosomal abnormalities per se. Epididymal necrospermia has been recently described in 4 patients by Wilton et al (1988) in which spermatozoal degeneration and death occurred either during epididymal passage or storage or both. Moreover, some spermatozoal defects are shown to be transient, eg acrosomal hypoplasia (Sauer, Bustillo and Serafini, 1989). In a review of the testicular ultrastructure of 243 subfertile patients, Martinova, Kantcheva and Tzvetkov (1989) showed that the disturbances affect the cytoplasm more than the nucleus, with significant stability of synaptonemal complexes even in the severely injured testes. More work needs to be done in the areas of the epididymal and seminiferous tubular physiology and patho-physiology.

TRANSMISSION OF ABNORMAL GENETIC TRAITS

In addition to structural chromosomal abnormalities there is the theoretical risk of transmission of abnormal genetic traits. However, there is usually a relevant clinical history. Patients where the genetic trait does not affect fertility may pose possible problems; but such patients do not present themselves in subfertility clinics. Other than the genetic abnormalities that result in low or absent motility of the sperm (eg ciliary dyskinesis), such patients are not likely to be candidates for micro-insemination. Congenital hypothalamic hypogonadism (eg Kallman's syndrome) may not respond well to therapy, and may present with severe oligozoospermia; hence such patients may request for micro-insemination. Ethical issues are therefore raised. While the technique is able to help such men have their own genetic offspring, the question is whether one should try. If the offspring is able to have a normal life-span with treatment, then we feel that micro-insemination could be offered. The immotile cilia syndrome (ciliary dyskinesis), an autosomal recessive condition is one such condition. It is compatible with normal life span provided preventive therapy is instituted for the respiratory complications. Hence, for this genetic disease we feel that micro-insemination is justified (Bongso et al, 1989a).

LIMITATIONS

There are severe limitations to the widespread application of micro-insemination at present. It requires expensive equipment and training; it is a very time-consuming procedure; and, at the moment, yields low fertilization rates. Moreover, the procedure is still very much a research procedure. Optimal conditions for its successful application is still not determined. There are also conflicting reports with animal models on its possible dangers.

SPERMATOZOAL RECEPTORS ON THE OOLEMMA MEMBRANE:

There has been increasing indirect evidence for some sort of spermatozoal receptor resident on the oolemma membrane. Earlier studies with zona-free oocytes suggested that there is some form of block to excessive polyspermy (Pavlok & McLaren, 1972; Toyoda & Chang, 1968; Niwa & Chang, 1975). More recently there is a report that 97% of human sperm nuclei from donors with poor penetration of zona-free hamster eggs (<10%)

underwent decondensation after direct cytoplasmic injection into hamster oocytes (Lanzendorf et al, 1987). The first direct evidence for sperm receptors on the oolemma membrane came from a study with murine gametes by S.C.Ng and D.Solter (unpublished data, 1987). We transferred between 3-5 motile spermatozoa into oocytes before and after ethanol activation (Cuthbertson, 1983) of the oocytes. Of 101 oocytes in which there was no ethanol activation, 32 (31.7%) developed 2 pronuclei and a second polar body at 8 hours. There was also a very low incidence of polyspermy, with only 2 eggs with 3 pronuclei, and none with more than 3 pronuclei, at 8 hours (Table 3). The fertilization (monospermy and polyspermy) of oocytes was not significantly different when the oocytes were exposed to ethanol 1-2 hours after the multiple spermatozoal transfer. However, when the oocytes were exposed to ethanol 1/2 to 1 hour before spermatozoal transfer, only 8 of 60 oocytes were fertilized (Table 3). The data supports the hypothesis that spermatozoal receptors or spermatozoal blocks were present on the oolemma of zona-intact oocytes because the polyspermy rates were low inspite of transfer of 3-5 spermatozoa; moreover, ethanol activation prior to spermatozoal transfer probably resulted in removal of spermatozoal receptors because of the low fertilization rate. The presence of spermatozoal receptors on the oolemma may explain the low fertilization in zona-free oocytes, with the possible exception of golden hamster oocytes (Yanagimachi, 1984). In this species, it is possible that the spermatozoal receptors are not so specific. Interestingly, it was reported recently that heteroantibodies found on human sera reacted against the hamster oolemma, though these oolemma antigens were distinct from antigens present on the surface of human spermatozoa (Bronson & Cooper, 1988).

These spermatozoal receptors are probably modified by the release of cortical granule contents into the perivitelline space, thus making them unable to bind to other spermatozoa present on the oolemma membrane (Figure 2). Polyspermy is explained by binding of more than 1 spermatozoon onto the multiple receptors around the same time before the cortical granules are discharged (see Ng et al, 1990).

The presence of such spermatozoal receptors on the oolemma membrane allows multiple spermatozoa to be transferred under the zona pellucida. They may also exert some sort of selection on the type of spermatozoon that finally fuses with the oocyte when there are multiple spermatozoa in the perivitelline space.

TABLE 3: FERTILIZATION OF MOUSE OOCYTES AFTER TRANSFER OF 3-5 SPERMATOZOA INTO THE PERI-VITELLINE SPACE.

Ethanol activation	n	2PN and 2nd PB	3PN and 2nd PB	>3PN and 2nd PB
Nil	101	32	2	0
After sperm transfer	137	43	1	0
Before sperm transfer	60	8	0	0

EFFICIENCY OF SPERM TRANSFER

Single spermatozoal transfer in the mouse was recently reported by Mann (1988) in which 25% (55/211) of mouse oocytes "injected" with a single spermatozoon into the pvs had 2 pronuclei and a second polar body at 8 hours after the sperm transfer. About half of those there were fertilized developed to live fetuses or young after transfer into pseudopregnant recipients (54%, or 26/48). The fertilization efficiency of the sperm transfer was lower than that for in-vitro fertilization under similar conditions (78%). In another study Lacham and co-workers (1989) investigated the conditions to improve acrosome reaction in the

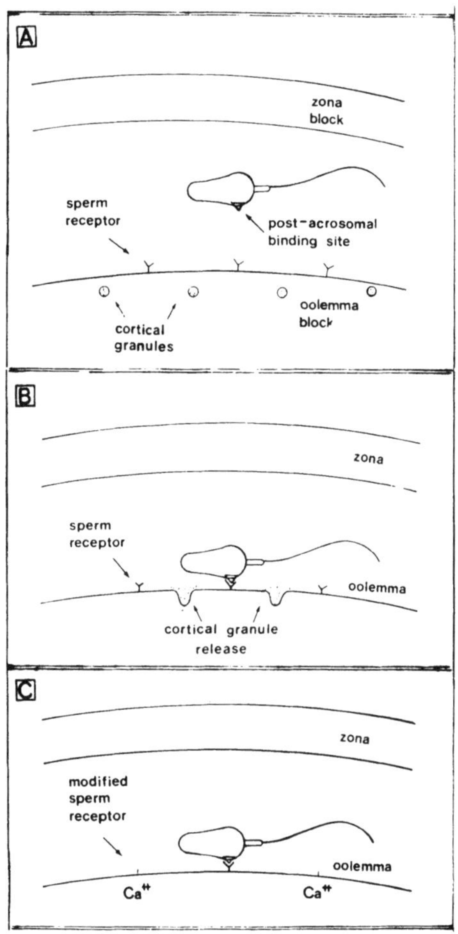

Fig 2: Schematic diagram of spermatozoal receptors on oolemma

mouse sperm before single sperm transfer into mouse oocytes. Acrosome reaction was induced by incubation in T6 medium for 30 minutes, 2 hours and 6 hours, and in 12mM dibutryl guanosine 3,5-cyclic monophosphate (dbcGMP) with 10mM imidazole for 30 minutes. Acrosomal status of spermatozoa selected for micro-insemination were as follows: T6 incubation for 30 minutes (54 ± 7.9%), for 2 hours (58 ± 5.6%), for 6 hours (73 ± 9.1%), and in cGMP (81 ± 15.1%). However, fertilization after sperm transfer was much lower than the acrosome-reacted rates of the spermatozoa - 36%, 34%, 29% and 43% respectively. Blastocyst rates were high, at 73%, 96%, 92% and 96% respectively. TEM of the unfertilized eggs revealed that the majority of the spermatozoa under the zona were not acrosomally-reacted, though some were reacted or reacting. Yamada, Stevenson & Mettler (1988) transferred single motile mouse spermatozoon into mouse oocytes after exposing the spermatozoa to dibutryl guanosine 3,5-cyclic monophosphate and imidazole. They reported 19.6% (41/209) fertilization rate with treated sperm, compared to only 5.3% (8/150) fertilization rate with untreated sperm. Again the fertilization efficiency was low, compared to 79.3% (680/858) fertilization in IVF with untreated spermatozoa, but surprisingly only 1.6% (6/368) with treated sperm. In the data from Ng and Solter (unpublished, 1987) only 31 of 101 oocytes (30.7%) that had multiple spermatozoal transfer without ethanol exposure developed to 2-cells. This was significantly lower than the 120 2-cell embryos that developed from 134 oocytes (89.6%; $p<0.001$) fertilized in-vitro within 2 hours of ovulation. Lassalle, Courtot & Testart (1987) reported multiple sperm transfer into human oocytes; 3-5 sperm were transferred into 7 oocytes and 3 fertilized, all monospermic. When the number of sperm transferred was increased to 10-12 sperm per egg, 2 of 3 oocytes fertilized but they were polyspermic. It is possible that with high sperm numbers, more than one sperm could attach onto their receptors on the oolemma before the receptors are inactivated, possibly by cortical granule release.

Hence in the mouse, fertilization after transfer of single or multiple spermatozoa into the peri-vitelline space is less efficient than fertilization in-vitro. It must be realised that these are "normal" spermatozoa. In the human situation, micro-insemination sperm transfer (MIST) is used only for very poor spermatozoal quality, in which fertilization in-vitro has very low chance of success. Hence, even low efficiencies allow some fertilization (see below).

TECHNIQUE OF MIST

Because the semen quality is very poor, the entire semen sample is used and spermatozoa are collected by the Ficoll entrapment procedure (Cummins & Breen, 1984). This procedure improves sperm samples from World Health Organization (WHO) grade 1 to WHO 2 motility with concommitant improvement in fertilization (Bongso et al, 1989b). After liquefaction, the semen was centrifuged to pellet the spermatozoa which is then mixed with 0.5 ml of 5% Ficoll in PBS. This mixture is then introduced gently under T6 medium (0.1 ml, and 0.5 ml respectively) and incubated in 5% CO_2, 5% O_2 and 90% N_2 (5:5:90) at 37oC for 1-2 hours. The motile spermatozoa are collected from the supernatant. The spermatozoa are then concentrated by centrifugation in a micro-centrifuge (MSE, Landsborough, UK) at 6,500 rpm (3,352 X g) for 10 minutes, and the supernatant is removed. Ficoll solution (5% in PBS, 0.1 ml) is added and the suspension is kept at 4oC until use. Recently, we have further modified the method by allowing a further 1-2 hours of incubation before micro-centrifugation to improve spermatazoa acrosome reaction, and suspend the final spermatazoa solution in T6 medium with 20 mM HEPES and 10% heat-inactivated human serum (HS) instead of 5% Ficoll in PBS. The final spermatozoal suspension is introduced into the spermatozoa micropipette (Drummond, Broomall, Pennsylvania, USA) from the rear end and is displaced into the tip with an Eppendorf Microinjector 5242 (Eppendorf, Hamburg, FRG). The spermatozoa micro-pipette is prepared with the Narisinghe micro-puller (#PB-7, Tokyo, Japan) and the Narisinghe micro-grinder (#EG-4, Tokyo, Japan). The external diameter of the tip is 10-15 um and the internal diameter is 8-12 um. The angle of the bevel for the spermatozoa micro-pipette is between 25 and 35°. The holding pipette for the oocyte is a Pasteur pipette drawn out on an ethanol flame. Both spermatozoa and holding pipettes are bent over the ethanol flame to allow horizontal displacement over the microscope stage.

The oocytes are collected trans-vaginally under ultrasonic guidance. They are then incubated in 5:5:90 gas mixture at 37°C in T6 medium with 10% heat-inactivated HS (Ng et al, 1985). Four to 5 hours after the oocytes are collected, the cumulus-corona complex is removed manually by micro-pipetting following exposure to 0.1% hyaluronidase in T6 medium. Only the metaphase II oocytes are manipulated, within 1 hour. Each oocyte is manipulated one at a time and is transferred into T6 medium with 20 mM

HEPES and 10% heat-inactivated HS just before micro-manipulation (Zeiss motorized micro-manipulator MR with the Zeiss IM35 inverted microscope, Carl Zeiss, Oberkochen, FRG). The Zeiss microscope has a warm stage set at 37°C. The oocyte is then directly punctured by the micropipette and between 7 and 10 motile spermatozoa are introduced into the peri-vitelline space. After the micro-insemination, the oocyte is washed 3 times in T6 medium before incubation overnight in 5:5:90 at 37°C. The oocyte is checked for pronuclei 14-18 hours later. It has been reported by us that micro-insemination has resulted in the normal fertilization ultrastructurally (Sathananthan et al, 1989). This is dealt with by Dr AH Sathananthan in another chapter in this book.

CLINICAL TRIAL

A clinical trial was conducted to evaluate the usefulness of this procedure for patients with severe oligoasthenoteratozoospermia who would otherwise have to resort to artificial insemination with donor spermatozoa (Ng et al, 1989a). It was undertaken with full consent from the patients after explanation of the risks involved, especially to the resulting embryo. The study was also approved by the National University of Singapore, the National University Hospital and the Singapore Science Council. Patients were included into the sperm transfer study when their husbands' final sperm concentration is very poor (<1.0 X 10^6/ml) or poor concentration and motility (<2.5 X 10^6/ml with sluggish or no forward motility). These patients were separated into 2 groups: (1) initial semen analysis with severe oligozoospermia (<5.0 X 10^6/ml); (2) initial semen analysis revealed oligozoospermia but not severe enough to consider sperm transfer, but resultant sperm parameters described above after swim-up on the day of the oocyte recovery necessitated their inclusion. The males had original semen analyses with motile sperm densities of between 0.07 and 2.01 X 10^6/ml (for group 1) and between 1.20 and 21.0 X 10^6/ml (for group 2); and total sperm densities of between 0.5 and 5.3 X 10^6/ml (for group 1) and between 12.9 and 76.0 X 10^6/ml (for group 2). Viable and normal forms were also low. Semen collected on the day of sperm transfer yielded very low motile sperm densities (Table 4). For MIST, the entire semen sample was used and sperm were prepared by the Ficoll entrapment procedure.

The females (29 patients, with a mean age of 31.5 and range from 24 to 36 years of age [for group 1], and a mean age of 34.0 and range from 28 to 41 years of age [for group 2]) were stimulated with one of the following 3 regimes: 1. Clomiphene + human menopausal gonadotropin (hMG), 2. Purified follicle-stimulating hormome (FSH) + hMG combination, 3. Gonadotropin-releasing hormone agonist (Buserelin) + Metrodin. They were monitored with daily serum estradiol (E2) and ultrasound scanning. Human chorionic gonadotropin (hCG) at 5,000 or 10,000 IU were administered when at least 2 follicles reached an average of 16 mm in the presence of satisfactory E2 (>300 pg/ml/follicle ≥ 14 mm). The eggs (total of 257) were collected trans-vaginally under ultrasonic-guidance 34 hours from hCG administration. They were then incubated in 5:5:90 gas mixture at 37oC in T6 medium with 10% heat-inactivated human serum (HS). Four to 5 hours after egg collection, the cumulus-corona complex were removed after exposure to 0.1% hyaluronidase in T6 medium. There were 20 oocytes at metaphase I, 10 at prophase I and 195 were at metaphase II; the rest were subjected to IVF. Only the metaphase II oocytes were manipulated, within 1 hour.

MIST was performed on 195 metaphase II oocytes and 16 (8.2%) were damaged (Table 5). Eight oocytes had 2 pronuclei (9 4%) in group 1, while 33 of 110 oocytes had 2 pronuclei in group 2 (30.0%). In-vitro fertilization was also used as controls only when the sperm concentration and motility was thought to be possible for fertilization; for group 1, none fertilized in 7 oocytes (2 patients), and for group 2 five of 26 fertilized (5 patients, 19.2%). Though the fertilization following MIST was better (30.0% versus 19.2%), this was not significant. In spite multiple sperm transfer, polyspermy was seen only in 2 oocytes in group 1 (2.3%). Parthenogenetic activation was seen in 2 (2.3%) and 6 (5.5%) in groups 1 and 2 respectively. In group 1, a total of 15 embryos and zygotes were replaced in 9 patients, with 2 pregnancies; the numbers are more than the 8 fertilized because 7 embryos were healthy looking the next day. In group 2, 29 zygotes/embryos were replaced in 12 patients, with 1 resulting pregnancy.

The first pregnancy was reported in 1988 (Ng et al, 1988). The patient is a 35 year old Chinese female with 9 years of subfertility. Her 36 year old husband had severe sperm problems with an atrophic right testis and a left varicocoele ligated 3 years ago. His semen counts varied between 1.7 and 5.5 million/ml in the past year; motility was occasional and normal forms less than 30%. She was stimulated with Metrodin after Buserilin down-regulation. Seven eggs were obtained, but on removal of the cumulus and corona after hyaluronidase exposure 5.5 hours later, only 5 were in metaphase II while the other 2 had germinal vesicles. The husband produced his semen 2 hours after the oocyte collection; 1.5 ml of semen was obtained with a count of 2 million/ml motile sperms and 12 million/ml total sperms (Makler chamber). The entire specimen was used for the Ficoll entrapment procedure and a final count of 0.3 million/ml motile sperm in 0.25 ml of supernatant was obtained. Between 7 and 10 motile spermatozoa were introduced into the perivitelline space. Of the 5 metaphase II oocytes, one was damaged and one of the remaining four had 2 pronuclei and 2 polar bodies at 18 hours after the sperm transfer. The pronuclear egg was then transferred into her left Fallopian tube under laparoscopic vision. She was confirmed pregnant. A karyotyping of amniotic cells after amniocentesis showed 46XY. She delivered a healthy baby boy weighing 3.356 kg on 20th May 1989 by LSCS. This is the first such delivery by micro-insemination. In the second pregnancy the husband also had severe oligoasthenoteratozoospermia (3 ml of semen with a count of 1.1 $X10^6$/ml very sluggishly motile sperm and 4 $X10^6$/ml total sperm). After sperm preparation, a final count of 1 $X10^6$/ml motile sperm in 0.20 ml of medium was obtained. Of 10 oocytes collected, 6 were subjected to MIST while the remaining 4 were inseminated in-vitro. None of the latter 4 fertilized, while 2 monospermic zygotes and 1 dispermic zygote were obtained after MIST. The monospermic zygotes were transferred into the left Fallopian tube at pronuclear stage. Unfortunately, she developed a left tubal ectopic pregnancy and an intra-uterine pregnancy which aborted. As the tubal pregnancy was treated conservatively with Methrotrexate, no tissue was available for karyotyping; the missed abortion revealed a 46 XX karyotype. The third pregnancy was from group 2. She was a 41 year old female, and the husband was 48 years old, with oligoasthenoteratozoospermia. His previous sperm densities were between 15.6 and 17.8 $X10^6$/ml total sperm, with motilities of 12 - 16% and viability of only 26%. On the day of the oocyte collection, his sperm density was only 1.1 $X10^6$/ml motile sperm, and 3.1 $X10^6$/ml total sperm. After the Ficoll entrapment and micro-centrifugation, only 1.0 X 10^6/ml total sperm were obtained. The wife was stimulated with FSH:hMG combination regime and had 8 oocytes. After removing the corona and cumulus, 7 were at metaphase II. Four zygotes with 2 pronuclei were obtained, and 1 other had 1 pronucleus. The 4 normospermic zygotes were replaced by TET. She became pregnant but the ultrasound at 8 weeks ammenorrhoea revealed a gestational

TABLE 4: MICRO-INSEMINATION SPERM TRANSFER: SPERM PARAMETERS.

	Severe Oligozoospermia	Final Oligozoospermia
Number of cases	15	14
Average age	31.5	34.0
Previous SA:		
Density ($X10^6$/ml)	2.64 ± 1.31	37.76 ± 55.84
Motility (%)	23.00 ± 18.12	30.44 ± 16.45
Fresh SA:		
Motile Density ($X10^6$/ml)	2.28 ± 3.79	3.84 ± 3.03
Total Density ($X10^6$/ml)	6.13 ± 8.88	15.90 ± 18.71
Post-washing:		
Motile Density ($X10^6$/ml)	0.34 ± 0.29	0.50 ± 0.65
Post-microcentrifugation:		
Motile Density ($X10^6$/ml)	1.18 ± 0.99	1.69 ± 1.98

sac without a fetal heart. Karyotyping of the chorionic villi again revealed normal karyotype (46,XX).

MIST INTO HUMAN EMBRYOS

MIST was used to investigate whether blastomere membranes or early human embryos are capable of fusing with the sperm as in the metaphase II oocyte (Ng et al, 1989b). This is important because presence of supernumary spermatozoa after multiple sperm transfer may result in abnormal incorporation of such spermatozoa into the embryo. Between 10-30 sperm were transferred into 11 donated human embryos between pronuclear and 10-cell stages. After culture for 6-24 hours in vitro, the embryos were fixed for transmission electron microscopy (TEM). Both acrosome-intact and acrosome-reacted sperm were located in the PVS and between blastomeres. Sperm-blastomere membrane fusion was not observed, even at 2 pronuclear stage. Sperm heads incorporated into blastomeres were often located in membrane-bound vesicles (both acrosome-intact and acrosome-reacted).

TABLE 5: MICRO-INSEMINATION SPERM TRANSFER: FERTILIZATION

	Severe Oligozoospermia	Final Oligozoospermia
Metaphase II eggs	85	110
Sperm transferred	6.00 ± 3.33	6.81 ± 2.81
Accidental MIMIC	16	23
PN1	2 (2.4%)	6 (5.5%)
PN2	8 (9.4)	33 (30.0)
PN3+	2 (2.4)	0
Unfertilized	65 (75.6)	63 (57.3)
Damaged	8 (9.4)	8 (7.3)
Concomitant IVF		
Number of eggs	7	26
Fertilized	0	5 (19.2)
MIST embryos transferred		
Number of embryos	15	29
Number of patients	9	12
Pregnancies (pts)	2 (22.2)	1 (8.3)

Acrosome-reacted sperm heads were lying passively in vacuoles or were undergoing degenerative changes at their surfaces. Sperm chromatin decondensation was not observed in any of the sperm heads that were detected in the blastomeres. The results clearly show that sperm are unable to undergo membrane fusion with early embryonic cells. Further, sperm nuclei are incapable of expanding their chromatin to form typical male pronuclei following MIST into early human embryos. However, the phagocytic response of the oocyte seems to be retained by the early embryonic cells. (See Figures 3-6).

CONTROVERSIAL AREAS AND FUTURE DIRECTIONS

The presence of spermatozoal receptors on the oolemma membrane needs to be confirmed. Immuno-histogical investigations would be important. If

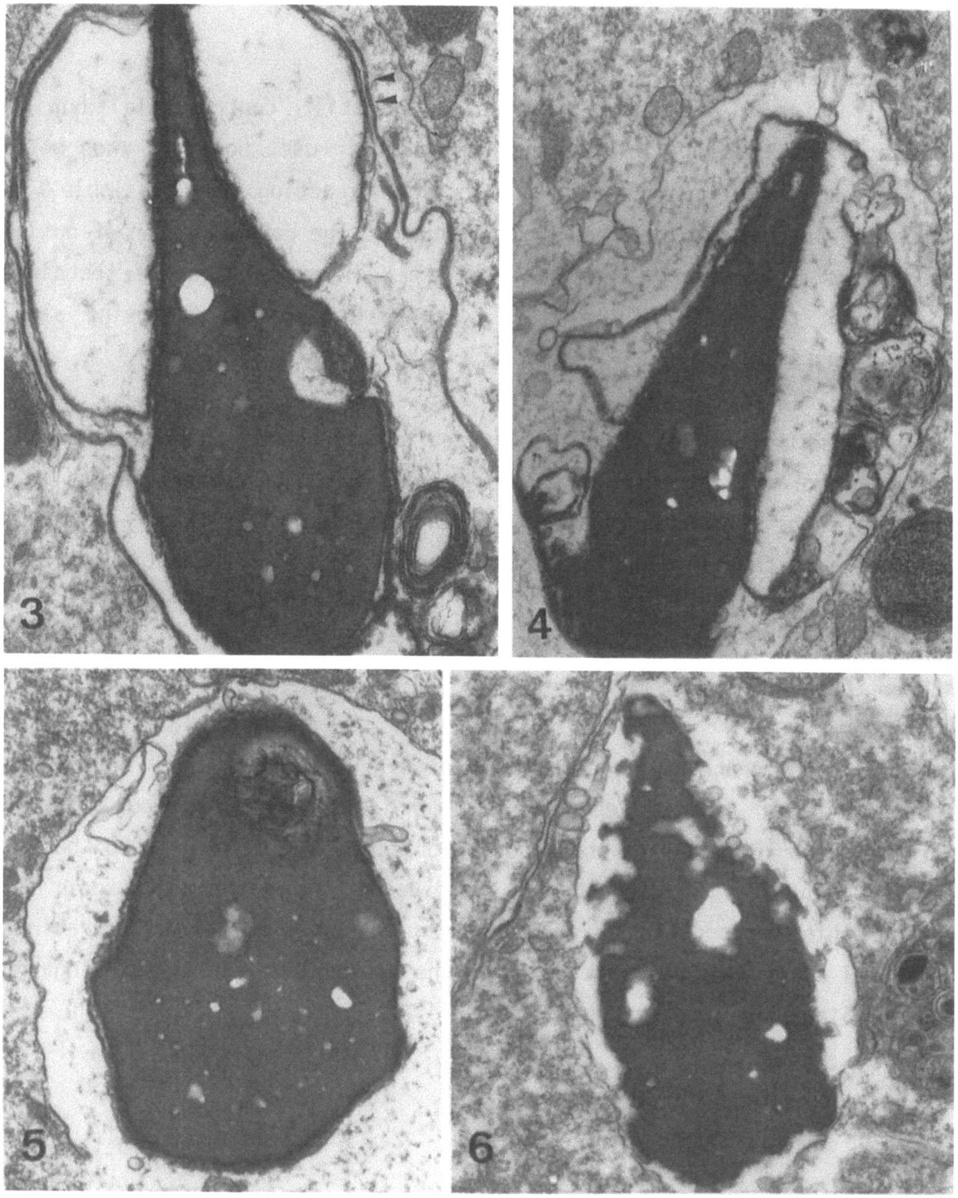

Figures 3 & 4: Acrosome-swollen sperm in blastomeres of a 8-cell embryo developed from a 6-cell embryo after MIST. The plasma and outer acrosomal membranes (arrow-heads) are dilated and complex surface membrane configurations are seen in one spermhead (Fig 4). Figures 5 & 6: Acrosome-reacted spermheads in vacuoles within 8-cell and 16-cell blastomeres developed from 6-cell and 10-cell embryos after MIST. X 29,750. (Reproduced from Ng et al, 1989b)

they are present, the receptor kinetics would need to be understood. They may have a very important role in contraception.

Transfer of multiple spermatozoa is still controversial but the extremely low rates of polyspermy justify their use. However, when better methods of inducing physiological acrosome reaction are available, the number of spermatozoa may need to be reduced. The determination of optimal conditions is necessary. Moreover, there may be combination of techniques to improve the efficiency of MIST, eg. electro-fusion following single sperm transfer.

The main cause of poor spermatozoa quality needs to be determined - is it chromosomal, genetic, or environmental? Based on current facts, it is unlikely to be chromosomal (structural or numerical). It is more likely to be related to the micro-environment of the testis, with a genetic basis.

ACKNOWLEDGEMENTS

We are grateful to the clinicians, scientists and technicians involved in our Assited Reproductive Techniques program without whom our work in micro-insemination would not be possible. We wish also to thank Ms Harjeet Kaur for her secretarial assistance, Mr Anthony Khoo for his photographic reproductions, and Ms Veronica Tok for her illustrations. Financial assistance was obtained from the National University of Singapore and the Singapore Science Council.

REFERENCES:

Ahlgren M (1975) Sperm transport to and survival in the human fallopian tube. Gynecol Invest 6:206-214

Bartlett J, Jockenhovel F, Nieschlag E (1989) New approaches to the pathophysiology of male infertility. Meeting report. Int J Androl 12:240-249

Bongso TA, Ng SC, Ratnam SS, Sathananthan AH, Wong PC (1988) Chromosome anomalies in human oocytes failing to fertilize after insemination in vitro. Human Reproduction 3:645-649

Bongso A, Ng SC, Ratnam SS (1989) Laboratory IVF technology. In: C Chan, WC Cheng, SL Tan (eds). Advances in Perinatal and Reproductive Medicine, Singapore: MacGraw Hill, p.191-205

Bongso TA, Sathananthan AH, Wong PC, Ratnam SS, Ng SC, Anandakumar C, Ganatra S (1989a) Human fertilization by microinjection of immotile sperm. Human Reprod 4:175-179

Bongso TA, Ng SC, Mok H, Lim MN, Teo HL, Wong PC, Ratnam SS (1989b) Improved sperm density, motility and fertilization rates following Ficoll treatment of sperm in a human in vitro fertilization program. Fertil Steril 51:850-854

Brandiff BL, Gordon L, Ashworth G, Watchmaker A, Carrano A, Worybek (1984) Chromosome abnormalities in human sperm: comparison among four healthy men. Hum Genet 66:193-201

Bronson RA, Cooper GW (1988) Detection in human sera of antibodies directed against the hamster egg oolemma. Fertil Steril 49:493-496

Burkman LJ, Coddington CC, Franken DR, Kruger T, Rosenwaks Z, Hodgen GD (1988) The hemizona assay (HZA): development of a diagnostic test for the binding of human spermatozoa to the human zona pellucida to predict fertilization potential. Fertil Steril 49:688-697

Chandley AC, Edmond P, Christie S, Gowans L, Fletcher J, Frackiewicz A, Newton M (1975) Cytogenetics and infertility in man. Pt 1. Karyotype and seminal analysis. Ann Hum Genet 39:231-254

Chia CM, Sathananthan H, Ng SC, Law HY, Edirisinghe WR (1984) Ultrastructural investigation of failed in vitro fertilisation in idiopathic subfertility. 18th Singapore-Malaysia Congress of Medicine, Singapore. (abstract) p.52

Cohen J, Malter H, Fehilly C, Wright G, Elsner C, Kort H, Massey J (1988) Implantation of embryos after partial opening of oocyte zona pellucida to facilitate sperm penetration. Lancet 2:162

Craft I, McLeod F, Bernard A, Green S, Twigg H (1981) Sperm numbers and in-vitro fertilization. Lancet 2:1165-1166

Cummins JM, Breen TM (1984) Separation of progressively motile spermatozoa from human semen by "sperm-rise" through a density gradient. Aust J Lab Sci 5:15-20

Cuthbertson KSR (1983) Parthenogenetic activation of mouse oocytes in vitro with ethanol and benzyl alcohol. J Exp Zool 226:311-314

De Kretser DM, Yates CA, McDonald J, Leeton JF, Southwick G, Temple-Smith PD, Trounson AO, Wood EC (1985) The use of in vitro fertilization in the management of male infertility. In: Rolland R, Heineman MJ, Hillier SG, Vemer H (eds). Gamete quality and fertility regulation. Excerpta Medica, Amsterdam, pp.213-223 De Felici M, Salustri A,

Siracusa G (1985) "Spontaneous" hardening of the zona pellucida of mouse oocytes during in vitro culture. II. The effect of follicular fluid and glycosaminoglycans. Gamete Res 12:227-235

De Felici M, Siracusa G (1982) "Spontaneous" hardening of the zona pellucida of mouse oocytes during in vitro culture. Gamete Res 6:107-113

Diamond MP, Rogers BJ, Vaughn WK, Wentz AC (1985) Effect of the number of inseminating sperm and the follicular stimulation protocol on in vitro fertilization of human oocytes in male factor and non-factor couples. Fertil Steril 44:499-503

Familiari G, Nottola SA, Micara G, Aragona C, Motta PM (1988) Is the sperm-binding capacity of the zona pellucida linked to its surface structure? A scanning electron microscopic study of human in vitro fertilization. J In Vitro Fertilization Embryo Transfer 5;134-143

Gordon JW, Grunfeld L, Garrisi GJ, Talansky BE, Richards C, Laufer N (1988) Fertilization of human oocytes by sperm from infertile males after zona pellucida drilling. Fertil Steril 50:68-73

Gronow MJ, Martin MJ, McBain JC, Wein P, Speirs AL, Lopata A (1985) Aspects of multiple embryo transfer. Annals N Y Acad Sci 442:381-386

Kamiguchi Y, Mikamo K (1986) An improved, efficient method for analyzing human sperm chromosomes using zona-free hamster ova. Am J Hum Genet 38:724-751

Kjessler B (1972) Facteurs genetiques dans la subfertile male humaine. In: Fecondite et Sterilite du male. Acquisitions recentes. Paris; Masson. p.205-225

Kurachi K, Aono T, Minagawa J, Miyake A (1983) Congenital malformations of newborn infants after clomiphene-induced ovulation. Fertil Steril 40:187-189

Lacham O, Trounson A, Holden C, Mann J, Sathananthan AH (1989) Fertilization and development of mouse eggs injected under the zona pellucida with single spermatozoa treated to induce the acrosome reaction. Gamete Res 23:233-243

Lalonde L, Langlais J, Antaki P, Chapdelaine A, Roberts KD, Bleau G (1988) Male infertility associated with round-headed acrosomeless spermatozoa. Fertil Steril 49:316-321

Lanzendorf SE, Mayer JF, Swanson J, Acosta A, Hamilton M, Hodgen GD (1987) The fertilizing potential of human spermatozoa following microsurgical injection into oocytes. 5th World Congress IVF & ET, Norfolk, Virginia, April 5-10, 1987, abstract pp-103

Lanzendorf SE, Maloney MK, Ackerman S, Acosta A, Hodgen GD (1988) Fertilizing potential of acrosome-defective sperm following microsurgical injection into eggs. Gamete Res 19:329-337

Lassalle B, Courtot AM, Testart J (1987) In vitro fertilization of hamster and human oocytes by microinjection of human sperm. Gamete Research 16:69-78

Mahadevan MM, Leeton JF, Trounson AO, Wood C (1985) Successful use of in vitro fertilization for patients with persisting low-quality semen. Annals N Y Acad Sci 442:293-300

Mann JR (1988) Full term development of mouse eggs fertilized by a spermatozoon microinjected under the zona pellucida. Biol Reprod 38:1077-1083

Martin RH, Balkan W, Burns K, Rademaker AW, Lin CC, Rudd NL (1983) The chromosome constitution of 1000 human spermatozoa. Hum Genet 63:305-309

Martin RH, Mahadevan MM, Taylor PJ, Hildebrand K, Long-Simpson L, Peterson D, Yamamoto J (1986) Chromosomal analysis of unfertilized human eggs. J Reprod Fertil 78:673-678

Martin RH, Rademaker AW, Hildebrand K, Long-Simpson L, Peterson D, Yamamoto J (1987) Variation in the frequency and type of sperm chromosomal abnormalitites among normal men. Hum Genet 77:108-114

Martinova Y, Kantcheva L, Tzvetkov D (1989) Testicular ultrastructure in infertile men. Arch Androl 22:95-98
McIlree ME, Price WH, Court Brown WM, Tulloch WS, Newsam JE, Maclean N (1966) Chromosome studies on testicular cells from 50 subfertile men. Lancet 2:69-71
Niwa K, Chang MC (1975) Requirement of capacitation for sperm penetration of zona-free rat eggs. J.Reprod.Fertil. 44:305-308
Ng SC, Ratnam SS, Law HY, Edirisinghe WR, Chia CM, Rauff M, Wong PC, Yeoh SC, Anandakumar C, Goh HHV (1985) Fertilisation of the human egg and growth of the human zygote in vitro: the Singapore experience. Asia-Oceania J. Obstet Gynecol 11:533-537
Ng SC, Bongso TA, Ratnam SS, Sathananthan AH, Chan CLK, Wong PC, Hagglund L, Anandakumar C, Wong YC, Goh VHH (1988) Pregnancy after transfer of multiple sperm under the zona. Lancet 2;790
Ng SC, Bongso A, Sathananthan AH, Ratnam SS (1989a) Micro-insemination of human oocytes. 6th World Congress on In Vitro Fertilization and Alternate Assisted Reproduction. Jerasulem, Israel. April 2-7, 1989 In "In Vitro Fertilization and Alternate Assisted Reproduction". New York: Plenum Press. In press
Ng SC, Sathananthan AH, Bongso TA, Ratnam SS, Tok VCN, Ho JKC (1989b) Micro-insemination sperm transfer (MIST) into early human embryos. Gamete Res in review
Ng SC, Bongso TA, Sathananthan AH, Ratnam SS (1990) Micro-manipulation: its relevance to human IVF (Review) Fertil Steril 1990; in press
Odawara Y, Lopata A (1989) A zona opening procedure for improving in vitro fertilization at low sperm concentrations: a mouse model. Fertil Steril 51:699-704.
Oehninger S, Acosta AA, Kruger T, Veeck L, Flood J, Jones HW Jr (1988) Failure of fertilization in in vitro fertilization: the "occult" male factor. J In Vitro Fertilization & Embryo Transfer 5:181-187
Overstreet JW (1983) The use of the human zona pellucida in diagnostic tests of sperm fertilizing capacity. In: In vitro fertilization and embryo transfer. Proceedings Serono Clinical Colloquia on Reproduction no.4. Ed: Crosignani PG, Rubin BL. London: Academic Press, p.145-166
Palmblad J, Mossberg B, Afzelius BA. (1984) Ultrastructural, cellular, and clinical features of the immotile-cilia syndrome. Ann Rev Med 35:481-492
Pavlok A, McLaren A (1972) The role of cumulus cells and the zona pellucida in fertilization of mouse eggs in vitro. J Reprod Fertil 29:91-97
Pellestor F, Sele B (1988) Assessment of aneuploidy in the human female by using cytogenetics of IVF failures. Am J Hum Genet 42:274-283
Plachot M, Junca A-M, Mandelbaum J, de Grouchy J, Salat-Baroux J, Cohen J (1986) Chromosome investigations in early life. I. Human oocytes recovered in an IVF programme. Human Reprod 1:547-551
Plachot M, de Grouchy J, Junca A-M, Mandelbaum J, Salat-Baroux J, Cohen J (1988) Chromosomal analysis of human oocytes and embryos in an in vitro fertilization program. Ann N Y Acad Sci 541:384-397
Retief AE, Van Zyl JA, Menkveld R, Fox MF, Kotze GM, Brusnicky J (1984) Chromosome studies in 496 infertile males with a sperm count below 10 million/ml. Hum Genet 66:162-164
Rossman CM, Forrest JB, Less RMKW, Newhouse AF, Newhouse MT (1981) The dyskinetic cilia syndrome: abnormal ciliary motility in association with abnormal ciliary ultrastructure. Chest 80:860-865
Rudak E, Jacobs PA, Yanaginachi R (1978) Direct analysis of the chromosome constitution of human spermatozoa. Nature 274:911-913

Sathananthan AH, Ng SC, Trounson AO, Bongso TA, Laws-King A, Ratnam SS (1989) Human micro-insemination by injection of single or multiple sperm: ultrastructure. Human Reprod 4:574-583

Sauer MV, Bustillo M, Serafini P (1989) Transient acrosomal hypoplasia of spermatozoa and male fertility. Arch Androl 22:95-98

Skakkebek NE, Bryant JI, Philip J (1973) Studies on the meiotic chromosomes in infertile men and controls with normal karytypes. J Reprod Fertil 35:23-36

Templado C, Benet J, Genesca A, Navarro J, Caballin MR, Miro R, Egozcue J (1988) Human sperm chromosomes. Human Reprod 3:133-138

Toyoda Y, Chang MC (1968) Sperm penetration of rat eggs in vitro after dissolution of zona pellucida by chymotrypsin. Nature 220:589-591

Trounson A (1987) Micro-fertilization. Plenary lecture 56, 5th World Congress In Vitro Fertilization, Norfolk, Virginia. April 5-10

Wilton LJ, Temple-Smith PD, Baker HWG, de Kretser DM (1988) Human male infertility caused by degeneration and death of sperm in the epididymis. Fertil Steril 49:1052-1058

Wolf DP, Byrd W, Dandekar P, Quigley MM (1984) Sperm concentration and the fertilization of human eggs in vitro. Biol Reprod 31:837-848

Wong PC, Bongso A, Ng SC, Chan CLK, Hagglund L, Anandakumar C, Wong YC, Ratnam SS (1988) Pregnancies after human tubal embryo transfer (TET): a new method of infertility treatment. Singapore J Obstet Gynecol 19:41-43

World Health Organization Task Force on the diagnosis and treatment of infertility (1987) Towards more objectivity in diagnosis and management of male infertility. (Prepared by Comhaire FH, de Kretser D, Farley TMM, Rowe PJ) Int J Androl Suppl 7:1-53

Wramsby H (1988) Chromosome analysis of preovulatory human oocytes and oocytes failing to cleave following insemination. Annals N Y Acad Sci 541:228-236

Wramsby H, Fredga K, Liedholm P (1987) Chromosome analysis of human oocytes recovered from preovulatory follicles in stimulated cycles. New Engl J Med 316:121-124

Yamada K, Stevenson AFG, Mettler L (1988) Fertilization through spermatozoal microinjection: significance of acrosome reaction. Human Reprod 3:657-661

Yanagimachi R (1984) Zona-free hamster eggs; their use in assessing fertilizing capacity and examining chromosomes of human spermatozoa. Gamete Research 10:187-232

Yates CA, Trounson AO, de Kretser DM (1989) Male factor infertility and in vitro fertilization. 4th International Congress in Andrology, Florence, Italy, 14-18 May 1989 Abstract, p.263

Yovich JL, Blackledge DG, Richardson PA, Matson PL, Turner SR, Draper R (1987) Pregnancies following pronuclear stage tubal transfer. Fertil Steril 48:851-857

Yovich JL, Stanger JD (1984) The limitations of in vitro fertilization from males with severe oligospermia and abnormal sperm morphology. J In-Vitro Fertilization & Embryo Transfer 1:172-179

Yovich JL, Yovich JM, Edirisinghe WR (1988) The relative chance of pregnancy following tubal or uterine transfer procedures. Fertil Steril 49:858-864

Zamboni L (1987) The ultrastructural pathology of the spermatozoon as a cause of infertility: the role of electron microscopy in the evaluation of semen quality. Fertil Steril 48:711-734

ABNORMALITIES OF THE HUMAN EMBRYONIC ZONA PELLUCIDA

Jacques Cohen
The Center for Reproductive Medicine and Infertility
Cornell University Medical Center
HT-306, 505 East 70 Street
New York, New York 10021
U.S.A[1]

The mammalian zona pellucida

The mammalian oocyte becomes surrounded by an acellular glycoprotein coat called the zona pellucida (ZP) during growth of the ovarian follicle. It is unclear whether the ZP originates from the oocyte, the follicular cells surrounding it, or both. The ZP plays an important role during both fertilization and preimplantation development. It contains species-specific sperm receptors that function during fertilization and participates in a secondary block to polyspermy following fertilization. It is not until the expanded blastocyst stage of development that embryos hatch from their ZP, allowing them to implant into the endometrial lining. Almost all studies of the mammalian ZP deal with molecular aspects of gamete recognition events (Dietl, 1988). Very few investigators study the ZP following sperm-egg fusion, despite its many important biological functions at that time.

The ZP of the mouse egg consists of three structural proteins ZP1, ZP2 and ZP3. ZP3 also functions as the sperm receptor. The amount of ZP3 diminishes following fertilization and this is accompanied by a molecular change (Wassarman et al, 1989). This may -in part- explain the increased ZP resilience following fertilization, a phenomenon referred to as "zona hardening". Zona hardening also occurs in the human ZP. Dissolution of the ZP using acidic tyrode's medium requires a longer

[1]Data obtained at Reproductive Biology Associates, suite 330-D, 993 Johnson ferry Road, Atlanta, Georgia 30342, U.S.A.

NATO ASI Series, Vol. H 45
Mechanism of Fertilization
Edited by B. Dale

exposure in the human zygote than in the unfertilised egg (Malter and Cohen, 1989a).

The hatching process and the embryonic zona pellucida

The molecular structure of the mammalian embryonic ZP at the time of hatching is largely unknown but it appears that the ZP matrix alters with time. Evidence suggests that hatching may occur via one of three mechanisms; (I) uterine enzymes digest the zona from the outside, (II) embryonic proteases digest the zona from the inside and (III) the zona is thinned through increased pressure from the expanding blastocyst. Evidence for these three routes exist in the mouse, whereas embryonic proteases appear to be absent in several other species (Kane, 1983). Zona thinning appears to be essential for hatching as it is observed in most mammalian species. In the rabbit, hatching only occurs when there is a minimal number of cells and ZP rupture appears to be associated with physical expansion. Rhesus monkey blastocysts also require a minimal diameter for hatching to occur (Boatman, 1987).

Human embryos are usually transferred at the cleaved embryo stage, due to increased embryonic demise when culture is extended. Only one in four fully expanded blastocysts derived from in vitro fertilised ovarian human oocytes will hatch (Fehilly et al, 1985; Lindenberg, 1989). This provides evidence that human blastocysts are capable of hatching in the absence of uterine lysins, but it also demonstrates that zonae are often resilient to dissolution. The incidence of blastocyst formation and hatching can be increased by culturing human embryos on monolayers of reproductive tract cells (Wiemer et al, 1989a,b; Lindenberg, 1989). This finding supports the argument that many in vitro fertilised embryos are deficient in the amount of zona lysins needed to initiate and complete hatching.

The human zona pellucida appears bilayered when it is exposed to acid Tyrode's solution (Gordon et al, 1988; Malter and Cohen, 1989a). The outside layer is thick and is digested within seconds, the inner layer is thin, but very resilient and its dissolution may require several minutes. Structural change in the embryonic zona is especially obvious when one attempts to penetrate it with a beveled, sharp enucleation or biopsy needle.

The function of the zona pellucida during embryo cleavage seems to be primarily of a physical nature: to prevent dispersal of the blastomeres and avoid direct contact between the embryo and foreign cells such as the epithelial lining of the reproductive tract, leucocytes, spermatozoa, or the cells of other embryos (Modlinski, 1970; Bronson and McLaren, 1970; Moore et al, 1968). In addition, the zona facilitates passage of the embryo through the fallopian tube. The ZP no longer appears to be essential for continued normal development in vitro once compaction has occurred (Modlinski, 1970; Bronson and McLaren, 1970; Trounson and Moore, 1974). For precompacted embryos, however, the zona must be present and relatively intact (Moore et al, 1969). Thawed human cleaved embryos with large gaps in their zonae do not develop into viable fetuses. This may be due to embryonic cell loss through the zona as a result of uterine contractions. Similar results were obtained by Nichols and Gardner (1989) who transferred mouse embryos with openings in their zonae into the oviduct. However, oviductal/tubal and intrauterine transfer of precompacted mouse and human embryos with small holes or incisions in their zonae following zona drilling with acid Tyrode's medium or mechanical opening using partial zona disssection (PZD) can result in implantation (Gordon and Talansky, 1986; Malter and Cohen, 1989a).

Impairment of the human hatching process

One can only speculate about the causes of implantation failure following in vitro fertilization techniques in the human, however four factors are thought to play key roles. First, the follicular environment and endometrial receptivity which are both under endocrinological control. Secondly, the conditions to which gametes and embryos are exposed. Embryos which are cocultured on a layer of fetal bovine uterine fibroblasts develop faster and produce fewer extra-embryonic fragments. In addition, more blastocysts are obtained when conventional culture techniques are replaced by co-culture methods, indicating that conventional in vitro conditions may be suboptimal and inhibiting growth of human embryos (Wiemer et al, 1989a and b). Thirdly, embryonic implantation can be influenced by the replacement procedure and fourthly, many early conceptuses are genetically abnormal. Approximately 20 to 40% of all human IVF embryos appear to have chromosome abnormalities.

We postulate here that a fifth important factor may be causing embryonic demise following IVF. A substantial number of apparently healthy IVF embryos are impaired in their ability to hatch from the zona pellucida even if these embryos are replaced into the fallopian tube or the uterus immediately following cleavage. Below we outline the rationale for such a hypothesis. We will also describe our use of micromanipulation to promote hatching following IVF by introducing an artificial gap in the ZP (assisted hatching).

Table 1. High zona pellucida thickness variation (< 10% refers to even shaped ZP and >25 % to uneven ZP with thin and thick areas) in human embryos

number of embryos replaced	incidence of implantation		
	zona thickness variation		
	< 10%	10-25%	>25%
1	0/2 (0%)	0/3 (0%)	1/5 (20%)
2	0/5 (0%)	3/12 (25%)	5/17 (29%)
3	0/14 (0%)	5/17 (29%)	18/38 (47%)

An important observation leading to incorporation of "Assisted Hatching" in our IVF protocol was the finding that some cleaved human embryos had uniform zonae, while others had thin patches (Cohen et al, 1988 and 1989a). Cleaved embryos with thinned areas in their zonae appear to implant more frequently than embryos with uniform zonae (Table 1). This finding is consistent and provides a significant factor in predicting embryonic implantation.

Recently, it was demonstrated that the zonae of 25% of human embryos actively thin between the zygote and four-cell stages (Wright et al, 1990). However, most embryos have not thinned their zonae at the time of replacement, two days following oocyte collection (Table 2). Consequently, it may be postulated that the ability of many human embryos to undergo hatching could be enhanced by an artificial thinning of the zona.

Table 2. Zona thinning of human zygotes and cleaved embryos

zona thickness variation higher (+) or lower (-) than 25% per embryo		proportion of embryos
zygote	cleaved embryo	
(-)	(-)	39/88 (44%)
(-)	(+)	25/88 (28%)
(+)	(-)	4/88 (5%)
(+)	(+)	20/88 (23%)

Micromanipulation of the human ZP

Although introducing an incision in the zona pellucida is a simple procedure, it may have profound consequences for embryonic development. For instance, microorganisms, viruses and cytotoxins present in the insemination suspension, could invade and infect the embryo via the artificially produced gap in the zona. Blastomeres may be lost through large holes, possibly causing embryonic death or vesiculation (Talansky and Gordon, 1988). Immune cell invasion through gaps in the zona may cause embryonic death, prior to the formation of tight junctions (Willadsen, 1979). Although it is difficult to monitor such events following embryo replacement, a trial with special attention to some of these problems was conducted in PZD patients (Cohen et al, 1990a). Every other patient received corticosteroids and antibiotics for four days in order to counteract the potential effects of immune cell invasion. The rational for corticosteroid administration was to reduce the number of intrauterine immune cells. Seven of the eight pregnancies obtained following PZD, occurred in the experimental group and immunosuppression is now routinely applied to patients whose oocytes or embryos are micromanipulated.

Disruption of the zona by micromanipulation may not only lead to loss of embryonic tissue but may also have profound consequences on the hatching process, a prerequisite to implantation. Both the timing and morphology of the hatching process in vitro of mouse and human PZD embryos are altered (Malter and Cohen, 1989b). The zona does not thin as it does normally during blastocyst expansion, and hatching occurs earlier and at a higher frequency (Table 3). Though the sample size is still small, it appears that human PZD embryos implant better than zona-intact embryos. The incidence of implantation of PZD-embryos in patients who received corticosteroids and antibiotics was 28% (Cohen et al, 1990a). The majority of the pregnancies resulted in twinning.

Assisted Hatching

A total of 99 IVF patients who had fresh embryos for replacement consented to have "Assisted Hatching" performed during three trial periods between May and October 1989 (Cohen et al, 1990b). Patients were not selected on the basis of age, number of embryos, response to follicular recruitement or etiology. Patients who had PZD performed for severe male factor treatment were obviously excluded from this study, as the ZP of their embryos were already opened. All patients received methylprednisolone and tetracycline as described elsewhere (Cohen et al, 1990b).

All embryos allocated for replacement were micromanipulated in alternate patients. This resulted in an experimental group (patients whose embryos were micromanipulated) and a control group (patients whose embryos had intact ZP). The laboratory staff checked to ensure that micromanipulation was performed without patient preference or any other possible bias. Patients were informed of the purposes and protocol of the study. They were not informed whether the actual micromanipulation had been applied to their embryos. The physician performing the embryo replacement was not aware of this either.

The procedure for "Assisted Hatching" closely follows that of the PZD technique described previously, except that embryos were not pretreated with hyaluronidase and sucrose (Malter and Cohen, 1989a). Excessive numbers of corona cells were removed with hypodermic needles. Embryos were micromanipulated in 5 to 10 µl droplets of sucrose-free culture medium under oil in glass well slides. The oil, culture medium, and slide were kept at 37 °C prior to micromanipulation. Oil and media were preequilibrated with 5% CO_2, 5% O_2 and 90% N_2. Micromanipulation was performed between 5 and 200 minutes prior to replacement. Only one embryo was pipetted into the well slide at a time. Each micromanipulation procedure required 1 to 5 minutes. All embryos allocated for replacement were micromanipulated. Embryo selection criteria have been described elsewhere (Cohen et al, 1988a). A maximum of only two or three embryos were replaced routinely, however four embryos were replaced in instances of excessive extracellular fragmentation or other morphological abnormalities. The ZP of a total of 144 fresh two to eight-cell embryos were thus micromanipulated (Table 3). All embryos appeared intact following micromanipulation and none of the blastomeres were damaged. The actual micromanipulation procedure is described elsewhere (Cohen et al, 1990b).

Positive βhCG was confirmed in 17/51 (34%) of the control patients who received zona intact embryos (Table 3), and 24/48 (50%) of the "Assisted Hatching" patients. The incidence of clinical pregnancy increased from 26% in the control group to 46% per embryo replacement in the experimental group, a significant improvement ($p<0.05$; chi-square test). Moreover, embryonic implantation increased from 13 to

Table 3. Assisted Hatching of fresh and freeze-thawed human embryos (trials 1, 2 and 3; 1989)

status of the embryo	definition of pregnancy	zona pellucida intact	zona pellucida with incision
		(incidence of implantation)	
fresh	biochemical (per patient)	17/51 (34%)	24/48 (50%)
fresh	fetal heart beat (per patient)	13/51 (26%)	22/48 (46%)
fresh	per embryo	16/129 (13%)	32/144 (22%)
thawed	biochemical (per patient)	6/23 (26%)	7/19 (37%)
thawed	fetal heart beat (per patient)	6/23 (26%)	7/19 (37%)
thawed	per embryo	6/43 (14%)	8/35 (23%)

22% ($p<0.05$). Half the pregnancies in the experimental group are either twin or triplet pregnancies.

The same procedure was performed in patients who received freeze-thawed embryos. However, this part of the work was performed at the end of our investigation, and it was felt that every consenting patient who returned for cryopreserved embryos should have "Assisted Hatching" (Table 3). The results of these micromanipulations can therefore not be compared with a statistically acceptable control group. However, we have compared the results with those of 23 patients who did not have "Assisted Hatching" performed and who were not aware of the procedure during the same time interval (Table 3). Again, implantation per embryo increased markedly, from 14% to 23%.

We conclude that approximately one quarter of all IVF embryos have the ability to implant based on the results of PZD embryos replaced in immunosuppressed patients and the "Assisted Hatching" trials presented above. This high rate of implantation is not necessarily jeopardized by embryo cryopreservation. It is postulated that a substantial number of IVF embryos are unable to breech the zona at the time of hatching and that many can be rescued by opening their ZP several days earlier. Whether "Assisted Hatching" can be applied easily in other IVF programs needs to be awaited. Only 7% of human embryos with holes in their zonae implanted when we applied micromanipulation at the IVF program in Cornell in 1989. This latter series of investigations was not performed as part of a prospective randomised trial. However, the implantation rate of zona-intact embryos in this program normally exceeds 10%, indicating that gaps in the zona may also have an adverse effect on implantation. This may have been due to many differences between culture methods and replacement procedures between the two IVF programs.

Abnormalities of implantation following ZP micromanipulation

A number of alternate hatching mechanisms have been proposed which may explain the findings presented above (Cohen et al, 1990b). Firstly, micromanipulated embryos may hatch earlier and implant in a more synchronous uterus. Hatching in vitro usually occurred one day earlier when "Assisted Hatching" was performed in two-cell mouse embryos (Malter and Cohen, 1989b). Moreover, in vitro cultured rhesus and baboon blastocysts shed their zonae 2 or 3 days later than their in vivo counterparts (Boatman, 1987). Opening the zona pellucida at the four-cell stage in the human may help the embyos to hatch earlier and correct for the "in vitro lag".

Secondly, it is likely that hatching occurs more frequently following micromanipulation. Three of four PZD-blastocysts left for observation in culture commenced with hatching, prior to expansion of the blastocyst and without zona pellucidae thinning (Malter and Cohen, 1989b). The embryos obviously followed the route of least resistence and expelled through the artificial gap. This mechanism also explains the increased frequency of dizygotic twinning in the "assisted hatching" experiments.

Application of "assisted hatching" may also increase the formation of blighted ova and monozygotic twins. The absence of zona thinning and expansion during expulsion of the blastocyst through the artificial opening may cause constriction of the trophoblast and the inner cell mass. Approximately 30% of drilled blastocysts in the mouse will become trapped in the zona forming a figure-8 configuration (Talansky and Gordon, 1988; Malter and Cohen, 1989b). The human blastocyst appears much more fragile than that of the mouse and the integrity of the trophoblast and the inner cell mass may be jeopardised during expulsion through a narrow opening. Consequently, the blastocyst may be dissected into several parts, or the inner cell mass may split and adhere to different areas of the trophoblast. One of the twin pregnancies following "assisted hatching" was identical. Two fetuses with fetal heart activity were seen in a single sac. Monozygotic twinning is rare in mammalian species. It has only been found in the human and the armadillo. The ZP in these species may be very resilient, causing a narrow hatch, possibly dissecting the trophoblast or inner cell mass during expulsion.

Is an abnormal hatching process in the human strictly an in vitro pathology, or is it a natural phenomenon - an early selection process comparable to that of sperm selection in the reproductive tract or embryonic selection caused by errors of fertilization and chromosome crossing-over? Hypothetically, an abnormal hatching process in the human in vivo - compared to other mammalian species - may be routine. The natural occurrence of blighted ova and various types of monozygotic twins (and even triplets!) may be supportive evidence. On the other hand, many human IVF embryos may be abnormal due to suboptimal culture conditions, inhibiting hatching and prolonging the preimplantation period. This latter observation explains the increase in implantation following ZP micromanipulation in some studies.

Acknowledgements: I am grateful to Henry Malter, Klaus Wiemer, Graham Wright,

Sharon Wiker and Michael Tucker for their support of these studies. Mina Alikani is acknowledged for editing the manuscript.

References

Boatman DE (1987) In vitro growth of non-human primate pre-and peri-implantation embryos. In:The mammalian preimplantation embryo. Bavister BD (ed) Plenum Press. New York, London

Bronson RA, McLaren A (1970) Transfer to mouse oviduct of eggs with and without the zona pellucida. J Reprod Fertil 22: 129-136

Cohen J, Wiemer KE, Wright G (1988) Prognostic value of morphological characteristics of cryopreserved embryos: a study using videocinematography. Fertil Steril 49: 827-834

Cohen J, Inge KL, Suzman M, Wiker S, Wright G (1989) Videocinematography of fresh , and cryopreserved embryos: a retrospective analysis of embryonic morphology and implantation. Fertil Steril 51: 820-827

Cohen J, Malter H, Elsner C, Kort H, Massey J, Mayer MP (1990a) Immunosuprression supports implantation of zona pellucida dissected human embryos. Fertil Steril, in press

Cohen J, Wright G, Malter H, Elsner C, Kort H, Massey J, Mayer MP, Wiemer KE (1990b) Impairment of the hatching process following in vitro fertilization in the human and improvement of implantation by assisting hatching using micromanipulation. Hum Reprod, in press

Dietl J (Ed) (1989) The mammalian egg coat. Springer-Verlag, Berlin, Heidelberg, New York

Fehilly CB, Cohen J, Simons RF, Fishel SB, Edwards RG (1985) Cryopreservation of cleaving embryos and expanded blastocysts in the human: a comparative study. Fert. Steril 44: 638 - 644

Gordon JW, Talansky BE (1986) Assisted fertilization by zona drilling: a mouse model for correction of oligospermia. J Exp Zool 239: 347-354

Gordon JW, Grunfeld L, Garrisi BJ, Talansky BE, Richards C, Laufer N (1988) Fertilization of human oocytes by sperm from infertile males after zona pellucida drilling. Fertil Steril 50: 68-73

Kane MT (1983) Variability in different lots of commercial bovine serum albumin affects cell multiplication and hatching of rabbit blastocysts in culture. J Reprod Fertil 69: 555-558

Lindenberg, Hyttel P, Sjogren A, Greve T (1989) A comparative study of attachment of human, bovine and mouse blastocysts to uterine epithelial monolayer. Hum.Reprod 4: 446- 456

Malter HE, Cohen J (1989a) Partial zona dissection of the human oocyte: a nontraumatic method using micromanipulation to assist zona pellucida penetration. Fertil Steril 51: 139-145

Malter HE, Cohen J (1989b) Blastocyst formation and hatching in vitro following zona drilling of mouse and human embryos. Gam Res 24: 67-80

Modlinski JA (1970) The role of the zona pellucida in the development of mouse eggs in vivo J Embryol Exp Morphol 23: 539-551

Moore NW, Adams CE, Rowson LEA (1968) Developmental potential of single blastomeres of the rabbit egg. J Reprod Fertil 17: 527-533
Moore NW, Polge C, Rowson LEA (1969) The survival of single blastomeres of pig eggs transferred to recipient gilts. Aust J Biol Sci 22: 979-988
Nichols J, Gardner RL (1989) Effect of damage to the zona pellucida on development of preimplantation embryos in the mouse. Hum Reprod 4: 180-187
Talansky BE, Gordon JW (1988) Cleavage characteristics of mouse embryos inseminated and cultured after zona pellucida drilling. Gam Res 21: 277-278
Trounson AO, Moore NW (1974) The survival and development of sheep eggs following complete or partial removal of the zona pellucida. J Reprod Fert 41: 97-108
Wasserman P, Bleil J, Fimiani C, Florman H, Greve J, Kinloch R, Moller C, Mortillo S, Roller R, Salzmann G, Vazquez M (1989) The mouse egg receptor for sperm: a multifunctional zona pellucida glycoprotein. In: The Mammalian egg coat. Dietl J (ed) Springer-Verlag Berlin Heidelberg New York
Wiemer KE, Cohen J, Amborski GF, Munuyakani L, Wiker S, Godke RA (1989a) In vitro development and implantation of human embryos following culture on fetal uterine fibroblast cells. Hum Reprod 4: 595-600
Wiemer KE, Cohen J, Wiker SR, Malter HE, Wright G, Godke RA (1989b) Coculture of human zygotes on fetal bovine uterine fibroblasts: embryonic morphology and implantation. Fertil Steril 52: 503-508
Willadsen SM (1979) A method for culture of micromanipulated sheep embryos and its use to produce monozygotic twins. Nature (Lond) 277: 298-301
Wright G, Wiker S, Elsner C, Kort H, Massey J, Mitchell D., Toledo, A. and Cohen, J. (1990) Observations on the morphology of human zygotes, pronuclei and nucleoli and implications for cryopreservation. Hum Reprod, in press

THE ROLES OF INTERMEMBRANE CALCIUM IN POLARIZING AND ACTIVATING EGGS

Lionel F. Jaffe
Marine Biological Laboratory
Woods Hole, MA 02543
USA

This paper first reviews the role of calcium in initiating tip growth in plant cells; emphasizing symmetry breaking in fucoid eggs, visualization of intermembrane calcium with chlorotetracycline, and the suppression of calcium gradients by injected calcium buffers. It then presents a new model of the path of calcium in the fertilization of deuterostomes, particularly of marine invertebrates and mammals. In this model, calcium leaks steadily across the inner acrosomal membrane of the acrosome-reacted and then fused sperm, thence beneath the oolemma whence it is slowly pumped by a cisternum of the e.r. Finally, this cisternum overloads and rapidly releases calcium to detonate a calcium wave.

POLARIZATION OF FUCOID EGGS

I can best introduce this topic by quoting Speksnijder et al. (1989b): "The localization of rhizoidal tip growth in the fucoid egg is a prototype of the symmetry-breaking problem (Prigogine and Stengers, 1984). The tip's locus can be determined by various external stimuli that may leave quite different traces in the egg; however, all are thought to be amplified by the same positive feedback loop. This inner loop is thought to include an influx of calcium ions into the nascent tip, and this influx, in turn, is thought to raise the free calcium there to a level far above the general cytosolic one

NATO ASI Series, Vol. H 45
Mechanism of Fertilization
Edited by B. Dale

(Jaffe, 1968). The evidence for this hypothesis includes preferential $^{45}Ca^{++}$ influx into the dark and future growth pole of photopolarizing eggs (Robinson and Jaffe, 1975), an early electrical current entry there (Nuccitelli, 1978) . . . preferential tip formation towards a calcium ionophore source (Robinson and Cone, 1980)" as well as permanent inhibition of outgrowth by calcium buffer injection—a study presented in detail by Speksnijder et al. (1989b) and briefly discussed and reviewed below.

Early versions of this theory envisaged a free calcium gradient extending across the whole polarizing egg from the nascent growth region. However, a different view is suggested by the dramatic CTC or chlorotetracycline images of Kropf and Quatrano (1987) in outgrowing _Pelvetia_ fucoid eggs (Fig. 1a) a well as those of Meindl (1982) in _Micrasterias_ desmids well before outgrowths appear (Fig. 1b).

All of these in _Micrasterias_ as well as some of these in _Pelvetia_ show the illustrated restriction of fluorescence to a fine line, which seems to represent a thin layer just below the plasma membrane. It is true that some of the CTC images in _Pelvetia_ also show distinct punctate or even vesicular components that must be farther below the surface. Moreover, Brownlee and Wood have measured high calcium within the tips of growing _Fucus_ rhizoids using electrodes that were surely well below the surface. Nevertheless, these striking CTC images suggest that the primary site of high calcium within the growing or even nascent growth zones of these cells is just beneath their plasmalemmas.

I have, therefore, come to imagine that these sites lie in narrow _intermembrane spaces_ between the plasmalemma and various subsurface vesicles such as secretory vesicles and cisternae of the e.r. Presumably CTC fluorescence directly indicates high calcium within subsurface vesicles in growth zones (Chandler and Williams, 1978; Caswell, 1979, pp. 159–165).

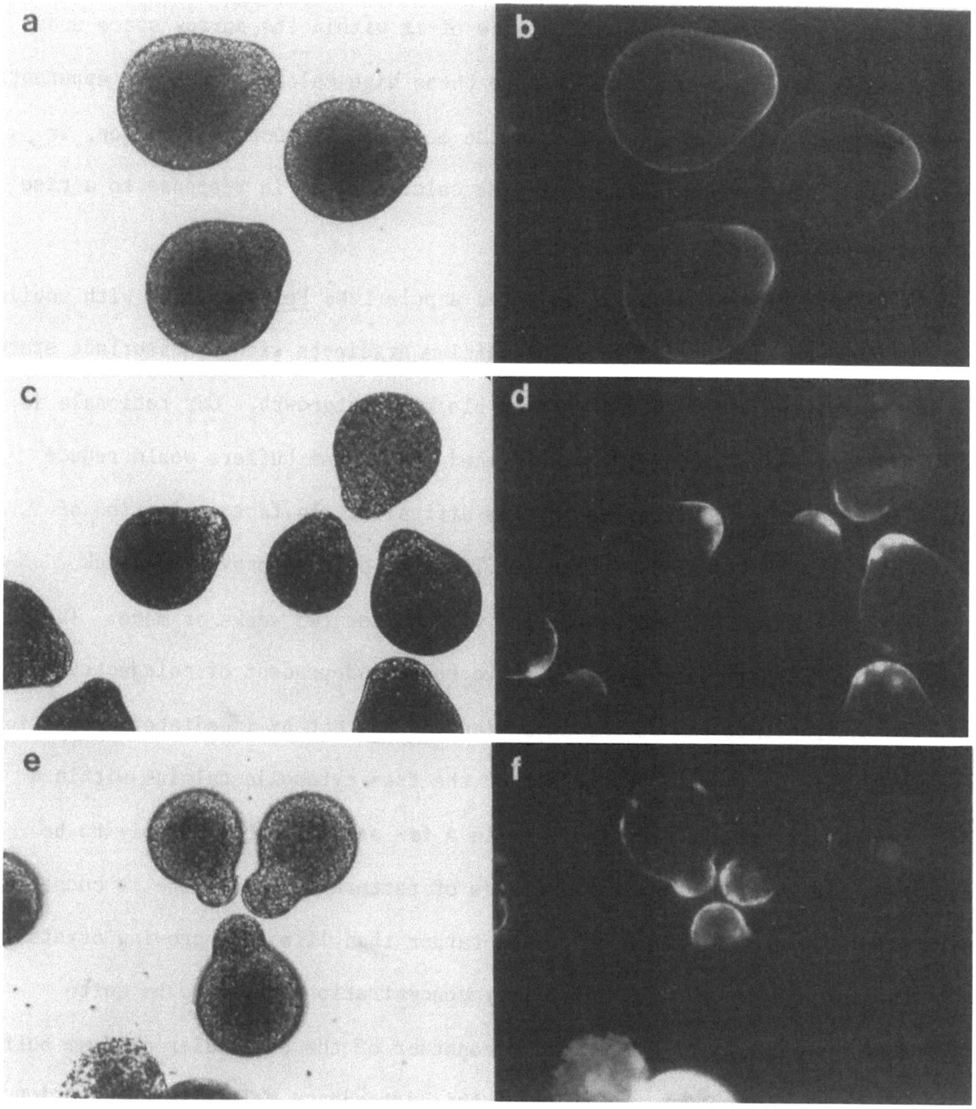

Figure 1a) Chlorotetracycline images of newly formed or nascent growth tips in plant cells. A rhizoidal outgrowth from a germinating Pelvetia egg (from Kropf and Quatrano, 1987, Fig.2a).

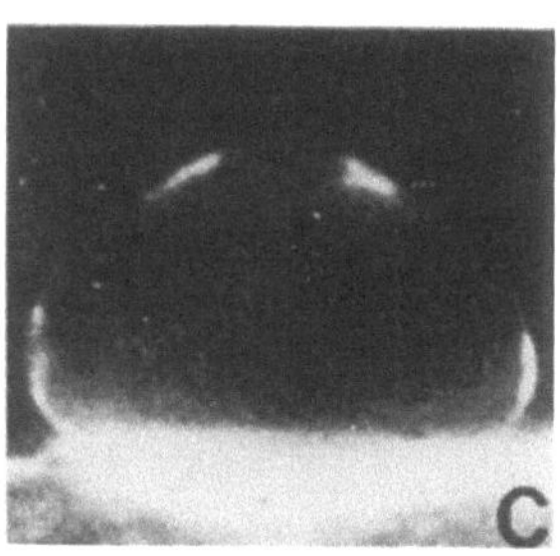

Figure 1b) Prospective lobes of developing Micrasterias cells (from Meindl, 1982).

However, I would also assume that where calcium is high within a subsurface vesicle it is likewise high just outside of it within the narrow space under the plasmalemma. I assume this because these high calcium zones are apparently produced by an influx of calcium from the medium. In such a situation, it seems safe to assume that intravesicular calcium rises in response to a rise in intermembrane calcium.

With this view in mind, we injected unpolarized *Pelvetia* eggs with mobile calcium buffers in hopes of reducing calcium gradients within subsurface spaces of the nascent growth zone and thereby blocking outgrowth. Our rationale is shown in Fig. 2a. Essentially we imagined that these buffers would reduce these gradients by facilitating calcium diffusion. In fact, injection of critical concentrations of various (BAPTA-type) buffers proves to block outgrowth formation (as well as cell division) for two weeks or more. The critical inhibitory concentrations prove to be independent of coinjected calcium showing that the buffer injections do not act by immediately resetting cytosolic calcium levels. Since all of the free cytosolic calcium within a *Pelvetia* egg is believed to turn over in a few seconds, this is only to be expected. In these dynamic early stages of pattern formation one is concerned with life as a growing flame or vortex rather than life as a growing crystal!

However, these critical inhibitory concentrations prove to be quite dependent upon the K_D or dissociation constant of the particular calcium buffer used (Fig. 2b). Moreover, the form of this dependence well fits an equation which shows to what degree various calcium buffers speed calcium diffusion within the cytosol. Note that while calcium movement through the cytoplasm of metabolizing cells is slowed to exceedingly (indeed, immeasurably) low values by sequestration into various organelles, the work of Donahue and Abercrombie on calcium movement through metabolically poisoned cytoplasm shows it to diffuse through the cytosol at rates only slightly below those found in salt solutions. Movement across the subsurface space should be by such true diffusion, and so we used Donahue and Abercrombie's data in or analysis.

In any case, I take the striking fit of our data to facilitated diffusion theory as strong evidence that these buffers do indeed act on tip growth initiation in fucoid eggs by facilitating calcium diffusion, that the nascent growth zone within fucoid eggs contains a high calcium region with a concentration of the order of 10 μM, and that this high calcium region is needed to initiate local growth. Because the effective buffers are of the BAPTA type, we informally refer to their inhibitory action as "baptism." One way to further test our analysis would be to try to rescue "baptized" eggs with local applications of calcium ionophores. If our analysis is correct, then some degree of local outgrowth might be restored by artificially providing extra local calcium carriers.

Moreover, baptism should serve as an effective general test for the action of calcium gradients as well as an indicator of their approximate size. This method has already served to indicate that local rises in free calcium to levels of the order of 10 μM are needed to accomplish cell division in fucoid eggs as well as nuclear envelope breakdown in sea urchin eggs (Speksnijder et al., 1989b; Twigg et al., 1988). Because it suppresses the cell cycle, there are certain limitations to its applicability to the analysis of developmental phenomena. However, it should be applicable to the analysis of various examples of differentiation without cleavage. To cite just one example, the famous, unequal, micromere-generating fourth division at the vegetal pole of sea urchin eggs involves a movement of pigment granules away from this pole in some species. Since this pigment withdrawal can occur without cleavage (Belanger and Rustad, 1972), one could use baptism to determine whether calcium gradients are needed for these pigment movements and thus gain an indication of whether they are needed for the profound local differentiation involved in micromere formation.

ACTIVATION OF EGGS ON THE VERTEBRATE LINE

Introduction: Ruling Out the Calcium Bomb Model

A decade ago, Gilkey et al. discovered the remarkable calcium wave which traverses the activating medaka fish egg. Its analysis helped establish the theory--introduced by Dalcq in the 1920s--that sperm activate eggs by raising their calcium levels. Moreover, it led to several new propositions (Jaffe, 1980, 1983):

1. This rise occurs by essentially different mechanisms in deuterostomes (i.e., animals on the vertebrate line) and protostomes.

2. a) It is propagated through deuterostome eggs by a traveling calcium explosion. b) The latter is mediated by calcium-induced calcium release.

3. In contrast, this calcium rise is at least begun in protostome eggs by the simultaneous, voltage-gated entry of calcium through the whole surface.

4. However, in deuterostomes it is begun or "detonated" by calcium injected into the egg by the sperm.

Propositions 1 and 3--which mainly bear on protostome eggs--have since received some further support as well as some very interesting extensions (Gould and Stephano, 1989). Proposition 2a--that all deuterostome eggs are activated by calcium waves has received considerable further support and seems to be widely accepted (Speksnijder et al., this volume). However, proposition 2b--that these waves are propagated by calcium-induced calcium release--has been seriously challenged: As an alternative, it has been suggested that these waves are propagated by the alternating release of calcium by inositol trisphosphate and the generation of inositol trisphosphate by calcium (Swann and Whitaker, 1986). For reasons touched upon by Speksnijder et al. in this volume I still feel that calcium-induced calcium release remains the more likely mechanism.

Finally, proposition 4--the so-called calcium bomb model of wave detonation--has received little consideration since 1980. Here I will reconsider and considerably modify it. It still seems reasonable to assume that the fusing tip of the sperm contains free calcium at a high enough concentration to detonate a calcium explosion via regenerative calcium release. However, Whitaker et al. have recently reconsidered the so-called latent period in the fertilization of sea urchin eggs, particularly in the light of what they rightly call the elegant experiments of Shen and Steinhardt (1984). Their analysis (of L. pictus at 16-18°C) indicates that at least seven seconds elapse between initial sperm-egg fusion and the initiation of the calcium wave. Thereafter, there appears to be a roughly constant probability of initiation with the process occurring in half the eggs after seventeen seconds. According to the calcium bomb theory this should be compared with the naturally operative mechanism of calcium-induced calcium release in cardiac muscle. According to Fabiato (1983, 1985) this mechanism only works if calcium rises within 0.1 seconds or less. Since the latent period in urchin fertilization is about a hundred times longer, this comparison effectively rules out the original calcium bomb model.

Two Modes of Regenerative Calcium Release

Fabiato and other investigators of calcium-induced or regenerative calcium release have emphasized a fast mode of such release which is triggered by a rapid rise in cytosolic calcium and is naturally operative in cardiac and perhaps other muscles. However, a second, slow mode of such release is also known and has been reviewed by Reuben et al. (1974) as well as Fabiato (1983, p. C3; 1985, p. 286).

In contrast to fast regenerative release, the slow mode is triggered by a rise in calcium within the endoplasmic reticulum, in turn induced by a relatively steady rise in calcium outside of this compartment. In experimental situations this "overload" or "supernormal accumulation" of calcium within the reticulum takes the order of 10 to 100 seconds rather than 10 to 100 milliseconds. An especially illuminating example of this phenomenon is described on page 281 of Reuben et al. (1974). In this report they studied the contractile responses of mechanically skinned crayfish muscle fibers to changes in external free calcium: "Upon adding 10 µm Ca [to a low calcium solution] a transient tension (ca. 0.8 g) occurred after a latent period of 30 seconds and similar responses were spontaneously repeated at about one minute intervals. The delay was only 6 seconds after addition of 50 µM Ca."

I will now propose an egg activation model based upon the phenomenon of slow (as well as fast) regenerative calcium release in muscle cells. In doing so, I will emphasize the gametes of some lower forms, since they are best studied, and then consider modifications needed for mammalian gametes. While the sperm's intermembrane spaces are somewhat different in the mammal than in lower forms, a fundamentally similar mechanism can be both imagined and to some extent even supported.

The Early Path of Calcium During Fertilization in Deuterostome Invertebrates

Figure 3a is a diagram of the proposed path in a deuterostome invertebrate such as an echinoderm or ascidian. The essence of this model is that after the acrosome reaction (i), calcium continues to leak into the sperm and does so via the newly formed acrosomal tubule (ii). It then steadily diffuses into the sperm head whence it is pumped out into the medium as well as the mitochondrion. The effective sperm fuses above a subsurface cisternum of the egg's e.r. Thence the calcium influx is diverted into the intermembrane space between this cisternum and the egg's plasma membrane (iii). During the latent period it is steadily pumped into this trigger region of the e.r. (iv). In the later steps (v, not illustrated), the trigger region of the e.r. becomes overloaded and rapidly releases calcium to detonate the first, activating calcium wave. In forms with postactivation waves, this sequence of slow and then fast regenerative calcium release is then repeated. A significant accessory proposition is that the influx of calcium into the acrosomal tubule goes through voltage-gated channels.

The most important piece of overall evidence for such a regenerative model remains the remarkable fact that many deuterostome eggs can be prick-activated, i.e., they can be activated simply by being pricked with a needle in a calcium-bearing medium (Jaffe, 1985, p. 128). Evidence for each step of this model will now be briefly considered.

i. The idea that the acrosomal reaction is triggered by a large influx of calcium is well based and generally accepted (Schackmann, 1989). My only new proposition is that the influx is localized in the acrosomal region. Such localization would be expected on the basis of efficiency alone. It is further supported by some recent immunocytochemical evidence that critically regulated calcium channels are localized in the acrosomal region of bull sperm (San Agustin et al., 1987).

ii. By following $^{45}Ca^{++}$ uptake, Schackmann et al. have clearly demonstrated that rapid calcium uptake continues into acrosome-reacted sea urchin sperm for up to 40 minutes and long after a fertilizing sperm would have fused with an egg (see Schackmann et al., 1978, Fig. 3,; Schackmann et al., 1981, Fig. 3). The latter data yields what seems to be the best estimate of the uptake rate after the acrosome reaction, namely, 0.5 aM/sperm/sec. Confidence in these data are strengthened by similar findings in mammalian sperm. Data on $^{45}Ca^{++}$ uptake in both guinea pig and bull sperm also yield steady rates of calcium uptake after the acrosome reaction of about half an attomole per sperm per second (see Singh et al., 1978, Fig. 1; San Agustin et al., 1987, Fig. 5); but where does this influx go?

The total free calcium in the head of an S. purpuratus sea urchin sperm (outside of the mitochondrion) can be estimated to be about 4×10^{-3} attomoles.* Dividing this by the above influx rate shows that this free calcium is turned over and should reach a steady state in a small fraction of a second. Moreover, from the X-ray microanalysis data of Cantino et al. (1983, Figs. 5d, 7d) one can estimate that after the first ten seconds or so, the mitochondrion continues to take up calcium at about 0.1 aM/sec. While this is a rather crude estimate, it accounts for only a fifth of the influx rate and should suffice to show that most of it is pumped out into the medium.

Thus the data indicate that a steady flux of about 0.5 aM/sec flows through an S. purpuratus sperm after the acrosome reaction. I will now go well beyond the data and postulate that the acrosome-reacted sperm is highly polarized so that all or most of the calcium leaks in across the inner acrosomal membrane into the acrosomal tubule, while most of it is pumped out somewhere else in the main part of the sperm head. This assumption cannot be tested by any direct evidence at this time and is only based on considerations of simplicity and efficiency. I would also second the suggestion of Shapiro et al. (1985) that the calcium channels in sea urchin sperm are voltage-gated and opened by (the experimentally demonstrated) depolarization of the plasmalemma. The chief basis for this suggestion is by analogy with the well-known

---

*This estimate is made by taking the free calcium in the non-mitochondrial compartment to be about 1.5 μM (Schackmann, 1989, p. 9) and multiplying it by half of this compartment's volume which is about 7 μm^3 (Cantino et al., 1983, Fig. 3).

voltage-gated calcium channels in eggs (Hagiwara and Jaffe, 1979). This assumption could also serve to rationalize the potassium dependence of sperm-egg fusion in guinea pigs (Yanigamachi, 1988, pp. 29-30).

However it is that calcium enters sperm, if it enters the acrosomal tubule and leaves elsewhere, then it should build up to very high levels within the long, fine intertubule space. Indeed, the steady state level of free calcium within the S. purpuratus tubule can be estimated to be of the order of 0.1 to 1 millimolar if one assumes that 0.5 aM/sec of calcium flows in evenly over the whole tubule, that the tubule is about one μm long by 0.05 μm wide (Schackmann et al., 1978, Fig. 1), and that calcium diffuses within it at rates measured in metabolically poisoned Myxicola axoplasm, i.e., at 5×10^{-6} cm^2/sec (Donahue and Abercrombie, 1987). Finally, two comments should be made on this very high level of inferred intratubule calcium: First, while this space is not an "intermembrane" space in the sense of one lying between two different kinds of membrane, it is of comparable dimensions. Second, these levels are so high that they might well induce fusigenic changes in the tubule's membrane (compare Fraser and Ahuja, 1988, p. 497).

Altogether then, these are the data and the arguments which underlie Figure 3aii.

iii. Fusion and diversion. My second main assumption is that the effective sperm fuses above a subsurface cisternum of the e.r., while other fusions fail to activate sperm and are ultimately reversed.

To my knowledge, there is little or no direct evidence available on this point. One indirect support for this assumption are the subsurface cisternae first found in ripe Xenopus eggs by Gardiner and Grey (1983). As I discussed some years ago (Jaffe, 1983, p. 269), "their distribution in both space and time strongly suggests that they are indeed a source of activation calcium in these eggs." In particular, these cisternae are two to three times more concentrated in the cortex of the animal as opposed to the vegetal hemisphere of Xenopus eggs. This fits the idea of a cisternal origin of the calcium wave because the eggs of another frog, Rana, are more easily activated by a needle, as well as by sperm, in their animal halves (Goldenberg and Elinson, 1980; Elinson, 1975).

Another observation which is compatible with the assumption of special activation sites is a curious finding of Shen and Steinhardt (1984, p. 1439): They "were . . . surprised to observe a low rate of polyspermy after the application of a [brief] window of negativity. Despite the presence of [more than two dozen] sperm, all presumably blocked at the same point, nearly all fertilizations were monospermic . . ." One could interpret this as indicating that sperm are effective in activating sea urchin eggs only when they happen to bind to a small, special fraction of the egg's surface, namely, that which is underlaid by subsurface cisternae.

Given this assumption, much or most of the calcium leaking into the acrosomal tubule would be diverted into the intermembrane space after fusion. From there it would surely be pumped into the underlying cisternum.

iv. Loading and detonation. What I have called slow regenerative calcium release is a poorly studied phenomenon. Nevertheless, there are several indications that the e.r. may have to be loaded with the order of 100 mM net Ca^{++} before it will be so overloaded as to release calcium without a sudden rise in cytosolic calcium concentration. In particular, an X-ray microanalysis study of the terminal cisternae or TC of frog muscle showed that 70 mM/Kgm dry TC was released by tetanus (Somlyo et al., 1981). Could an influx of calcium via the acrosomal tubule raise a cisternum's calcium to such high levels in the six seconds available? A simple calculation indicates that the answer is yes.

Consider a subsurface cisternum in a sea urchin egg with the profile shown in Luttmer and Longo (1985, Fig. 6). Assume that this profile represents a equidiametric disc half filled with water. This would be about three attoliters. Assume that half of the calcium entering an attached sperm flows into this space. Dividing 0.3 attomoles/sec by 3 attoliters yields an influx of 100 mM/sec.

This simple calculation shows that influx through the sperm might well overload the detonating cisternum in a single second; yet the latent period to be accounted for is at least six seconds. So the proposed process should be fast enough; but could it be slow enough? I believe so, especially when one considers that the "detonating cisternum" must be connected to the rest of the e.r. and that in the long times available calcium would be expected to diffuse out of it and into this connected e.r. Such losses might substantially slow filling of the cisternum and extend the latent period. Furthermore, most fusing sperm would not be expected to be so well centered over the responding cisternum as to fill it without lateral losses. Indeed, variability in the distance between the fusion point and its target cisternum would form one plausible source of the extremely high variability in latent periods observed by Shen and Steinhardt.

This "calcium leak" theory predicts a significant requirement for external calcium during the latent period; so let us examine the conflicting literature on this point. The paper of Sano and Kanatani (1980) on "jelly-treated sperm of the sea urchin, Hemicentrotus . . . [showed that] . . . the percentage of fertilization decreased concomitant with the reduction in . . . external calcium . . . 50% at 40 uM calcium and almost 0% at less than 10 μM." Since the sperm were transferred to low calcium media before reaching the eggs, one cannot tell to what extent calcium was required for penetration of the vitelline layer, for fusion, and/or activation. Nevertheless, this report is clearly consistent with our model.

The other four papers on the subject in sea urchins all claim that these eggs do not require external calcium for fertilization. Nevertheless, a close examination of them reveals flaws which suggest that the straightforward report of Sano and Kanatani remains the most reliable: 1. The results of Takahashi and Sugiyama (1973) are clearly attributable to calcium contamination. 2. The full paper of Schmidt et al. (1982), as well as the abstracts by Chambers (1980) and by Chambers and Angeloni (1981), leave open the serious possibility of insufficient mixing of sperm and EGTA chelator to everywhere reduce free calcium enough in the time available. The difficulty arises because the reaction between calcium and EGTA is relatively slow. According to Tsien (1980, p. 2397), "at physiological pH [and room temperature] EGTA . . . takes up to seconds to buffer a Ca^{++} transient." The authors of these three papers appear to have been unaware of this fact and reported no precautions to leave enough time for the reaction with EGTA. In the case of the report by Schmidt et al., this difficulty was compounded by beginning the mixing not at room temperature but at 10°C and continuing at 5°C <u>and</u> by the necessary use of a twenty-fold higher sperm concentration in the text experiment than was needed in a natural seawater control. With this protocol, one could imagine that enough sperm found themselves in a microenviroment containing just enough calcium, just long enough to effect fertilization.

Such problems aside, the role of calcium influxes during the latent period might best be explored with calcium channel blockers instead of calcium reduction. Tests of the application of such blockers during the latent period have not been reported. However, the ability of 0.1 to 0.5 μM uranyl ions to block sea urchin egg activation during the latent period may well prove to arise from their action as calcium channel blockers.

The remarkable ability of uranyl ions to block fertilization during the latent period was discovered by Baker and Presley in 1969. They correctly inferred that it revealed a well-defined intermediate step in the fertilization process. They also reported that "when uranyl ions are applied just before the period of fertilization membrane elevation, we have often found eggs with developing male pronuclei but apparently intact cortical granules and no obvious fertilization membrane." This indicates that uranyl ions do not act to reverse gamete fusion but rather to block or localize egg activation. More recently Shen and Steinhardt (1984) have carefully confirmed the Baker-Presley experiment and further shown that uranyl ions act throughout the latent period in each egg, however long or short this period is in that egg.

To my knowledge, none of these investigators have discussed the physiological mechanism of uranyl ion action; nor is this mechanism understood. Nevertheless, there are several substantial bits of evidence suggesting that uranyl or UO_2^{++} ions can act by somehow blocking calcium movement across membranes. The first, which is the mere fact that uranyl ions can allow sperm entry without an exocytotic wave, indicates interference with calcium entry or at least the immediate calcium action. The second is in an old paper by Hagiwara and Takahashi (1967, p. 592) who briefly report that "UO_2^{++} and La^{+++} suppressed the [barnacle muscle calcium] spike potential completely at concentrations of 2 and 1 mM respectively when the Ca^{++} concentration was 42 mM . . . the order of binding . . . was La^{+++}, UO_2^{++} > Zn^{++}, Co^{++}, Fe^{++} > Mn^{++} > Ni^{++} > Ca^{++} > Mg^{++} . . . " The third is a report that 0.2 mM UO_2^{++} causes 70% inhibition of $^{45}Ca^{++}$ uptake into inside out _Mycobacterium_ phlei vesicles. Moreover, a Lineweaver-Burk plot indicates that UO_2^{++} is a competitive inhibitor of Ca^{++} uptake by these vesicles (Agarwal and Kalra, 1983).

The calcium leak theory also offers a measure of explanation for the complex effects of clamping sea urchin eggs at unnaturally negative internal voltages (Lynn et al., 1988). Such negative clamping has two quite separate effects: One can be described as a prolongation of the latent period before detonation of the calcium wave. This is indicated by the longer average times from the start of inward current to the initiation of fertilization membrane elevation as well as of phase 2 of the activation current (see their Table 1). Indeed, as the clamped voltage becomes more negative, more and more of the eggs fail to activate at all (see their Fig. 2a). A second is an increasing inhibition of sperm entry into those activated eggs that nevertheless undergo an exocytotic wave.

In terms of the calcium leak theory, prolongation of the latent period would result from slower calcium influx, in turn brought about by closure of voltage-gated channels in the sperm. However, this effect would be limited by the considerable IR drop which should develop en route to the sperm. This, in turn, would be expected to partially insulate the sperm channels from the clamp and also electrophorese some of what calcium did still enter away from the sperm head and into the egg.

For some reason, maneuvers which should somewhat slow calcium entry prove to promote nuclear entry and thus relieve the second effect of negative clamping (McCulloh et al., 1989). This phenomenon is largely outside the scope of this calcium leak theory of egg activation. However, it could be taken as further evidence that uranyl ions slow calcium entry; this assumption would help explain Baker and Presley's remarkable observation that many uranyl-treated eggs exhibit nuclear entry without exocytosis.

v. Activation and early postactivation waves. The activation and early postactivation waves in ascidian eggs are separated by intervals of 30 to 60

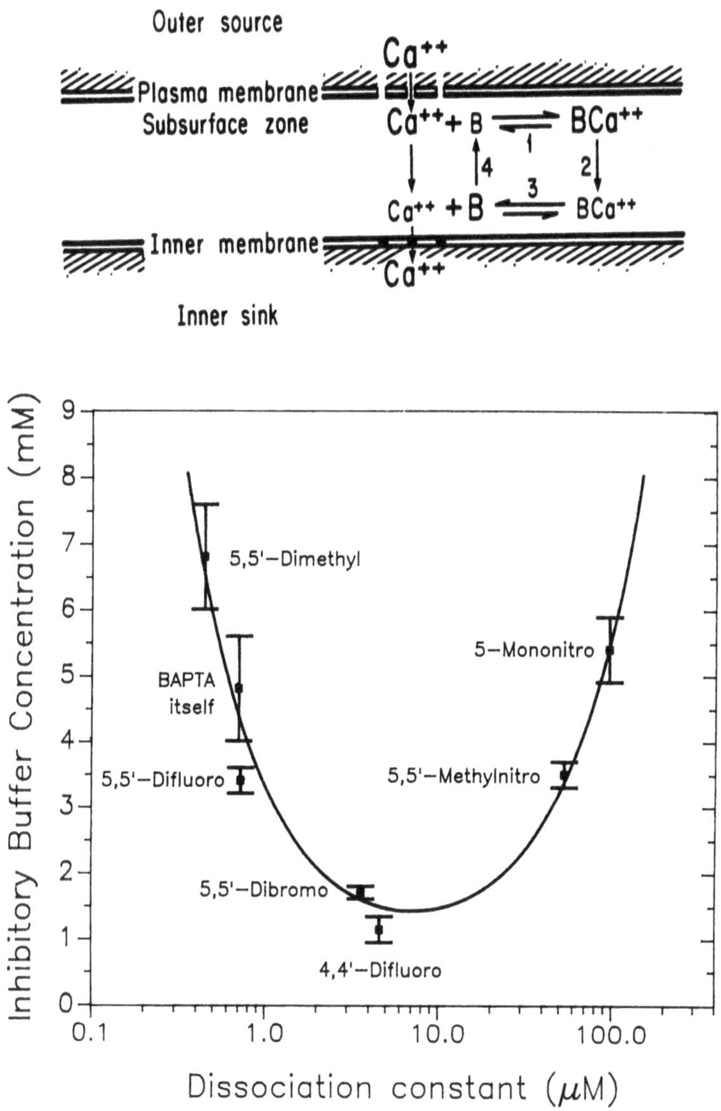

Figure 2. a) Model of how a mobile calcium buffer, B, acts to speed the steady diffusion down a gradient in an intermembrane space of the nascent growth zone of a fucoid egg. b) Mean inhibitory concentrations (± SEM) for seven BAPTA-type buffers injected into Pelvetia (fucoid) eggs. The curve is taken from facilitated diffusion theory assuming that the mean free calcium concentration in the affected region was 7 μM and that the critical buffer concentration speeded diffusion three-fold (from Speksnijder et al., 1989b).

seconds (Speksnijder et al., 1989a; Speksnijder, this volume) while those in hamster eggs are probably separated by intervals of about 60 to 90 seconds (Miyazaki et al., 1986). Since these intervals are comparable to the 10 to 30 second delay times before the activation wave in sea urchins, it seems reasonable to attribute both to the same basic mechanism, namely, the time taken to refill and overload the e.r. Moreover, in both ascidian and hamster eggs the early postactivation waves start at the site of sperm entry. Since components of the sperm's outer membrane are known to gradually intermingle with the egg's after fertilization (Yanigamachi, 1988, pp. 19-20), it seems reasonable to attribute the early initiation site to persistent calcium channels from the sperm.

Modifications of the Path of Calcium in Mammalian Fertilization

It is now well known that the structure and fusion site of mammalian sperm is substantially different from that of lower forms (Moore et al., 1983; Yanigamachi, 1988). In particular, the newly exposed inner acrosomal membrane is not everted and extended into a tubule; moreover, fusion with the egg does not occur at this tubule's tip but rather just behind the inner acrosomal membrane, at the side of the sperm, in the so-called equatorial region.

However, I would propose that a steady calcium current is set up in the acrosome-reacted mammalian sperm comparable to that in lower forms, i.e., calcium likewise leaks into the sperm across the inner acrosomal membrane and is likewise pumped out of the sperm's postacrosomal cytosol into the medium and the mitochondrion (Fig. 3b ii). Within the acrosomal process, calcium will diffuse back toward the pumps via the intermembrane (or subacrosomal) space between the inner acrosomal and nuclear membranes. Since the diffusional resistance of this space in various mammalian sperm is likely to be comparable to that of the sea urchin sperm considered above, free calcium should rise

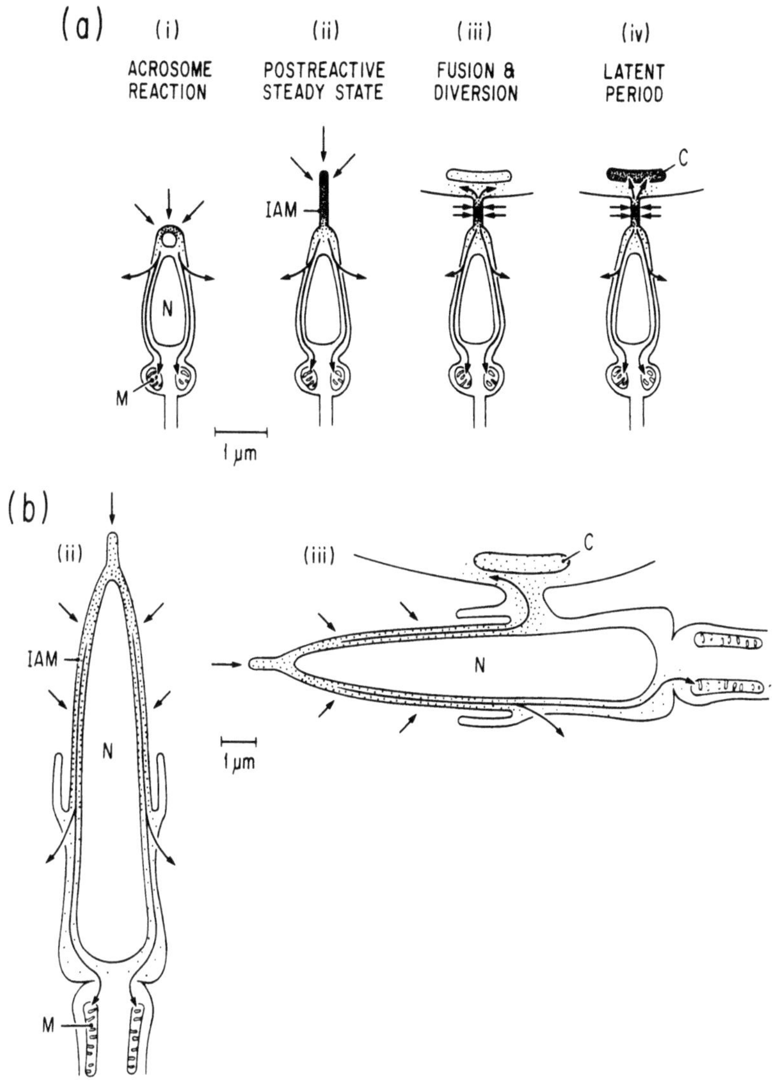

<u>Figure 3</u>. A calcium leak model of calcium's path early in deuterostome fertilization. a) Diagrams of the proposed path in sea urchin gametes. b) Modifications in mammalian gametes. Arrows indicate calcium flows; dots, calcium concentrations. Drawn roughly to scale except for widening of finest dimensions. Sea urchin sperm are drawn from Schackmann et al. (1978, Fig. 1) and Cantino et al. (1983, Fig. 1) for <u>S</u>. <u>purpuratus</u>. IAM = inner acrosomal membrane; M = mitochondrion; C = cisternum of the e.r.; N = nucleus. Mammalian sperm are drawn from Moore and Bedford (1983, Fig. 5) and San Agustin et al. (1987, Fig. 4) for the bull. The subsurface cisternae are taken from Luttmer and Longo (1985).

within this space to a comparable degree. Following fusion with the egg, calcium flowing into the acrosome should be likewise diverted into the egg (Fig. 3b iii).

Alternative Theories

One class of alternative theories involves diffusion of a trigger substance from the fertilizing sperm into the egg. Thus, Dale et al. (1985) report that injection of a soluble sperm fraction into sea urchin eggs via a patch pipette often activates them and suggest that the effective component in this experiment may be the natural trigger. More recently, Iwasa et al. (1989) have briefly reported that sea urchin sperm contain enough $InsP_3$ to activate eggs. It seems to me, however, that the main trigger mechanism is likely to be about the same throughout the deuterostomes. If this is true, then it must work for *Thyone* in which the sperm activates the egg through an acrosomal tubule 90 μm long by 0.06 μm wide (!) (Tilney, 1985). It seems obvious that neither $InsP_3$ nor any other known activator could traverse such a conduit fast enough to activate the *Thyone* egg.

However, this is not to suggest that substances within the sperm head may not play an important subsidiary role in cases--such as fish and mammals--where such a long, fine tube does not separate egg and sperm after fusion. Cyclic AMP, for example, might well play such a role. As little as 1 μM cyclic AMP enhances the rate of Ca++ accumulation into the s.r. of skinned heart muscle (Fabiato, 1983, p. C2) while various acrosome-reacted sperm raise their cyclic AMP level far about 1 μM (Garbers, 1981; Garbers et al., 1982).

A second class of alternative theories involves contact detonation, i.e., initiation of a calcium explosion without cell fusion. In particular, several investigators have been vigorously exploring the hypothesis that sperm trigger various deuterostome eggs via a receptor-G-protein-phospholipase C-$InsP_3$

sequence similar to that used by various neurotransmitters and hormones (L.A. Jaffe, 1989; Miyazaki, 1989, Nuccitelli et al., 1989).

The main difficulty with contact theories is the near lack of evidence that sperm can activate _deuterostome_ eggs without fusion. There is good evidence that sperm can activate eggs of the _protostome_ worm _Urechis_ via external (and even local) application of reasonable concentrations of a highly basic acrosomal protein (Gould and Stephano, 1989); however, the comparable protein, bindin, from sea urchin sperm does not activate their eggs (Vacquier and Moy, 1977). Furthermore, the best available evidence--in the form of electrical capacitance measurements--indicates that gamete fusion and the initiation of the fertilization current occur within milliseconds of each other in sea urchins (McCulloh and Chambers, 1986); moreover, Whitaker et al. have presented cogent arguments against the receptor-$InsP_3$ mechanism itself (Whitaker et al., 1989; Whitaker, this volume). Again, this is not to suggest that $InsP_3$ generated by contact (as well as $InsP_3$ brought in from the sperm and $InsP_3$ later regenerated by a rise in calcium) may not play a significant subsidiary role in speeding detonation of the e.r. There is good evidence that $InsP_3$ acts this way (Wakui et al., 1989).

Conclusions and Future Tests

The calcium leak model seems to provide unifying and plausible explanations for the main findings now available about the triggering of egg activation by sperm in mammals and lower deuterostomes. These include the continued calcium leak into acrosome-reacted sperm; its effectiveness through exceedingly long acrosomal tubules in some species; the apparent need for gamete fusion; the preference for sperm entry toward the animal pole; the long and variable latent period; the inhibition by uranyl ions, low calcium, and hyperpolarization of the egg; the initiation of activation and postactivation waves at the site of sperm entry; and the calcium dependence of prick activation. It also provides plausible subsidiary roles for the rises in cyclic AMP and $InsP_3$ within acrosome-reacted sperm. So this theory seems worthy of further test.

Perhaps the strongest test of the model as a whole would be to suitably extend the Baker-Presley study of uranyl inhibition. Do comparable concentrations of uranyl ions inhibit $^{45}Ca^{++}$ entry into acrosome-reacted sperm? Do known calcium channel blockers inhibit fertilization during the latent period? If so, do they inhibit $^{45}Ca^{++}$ entry into sperm and fertilization with the same concentration dependence?

Because of the particular interests in my own laboratory, I would also think of directly looking for calcium currents through acrosome-reacted sperm--particularly large-headed mammalian sperm--with our recently developed vibrating calcium electrode (Jaffe and Levy, 1987; Kuhtreiber and Jaffe, unpublished).

It is a pleasure to acknowledge helpful conversations with Michael Bedford, Laurinda Jaffe, Frank Longo, and David McCulloh as well as financial support from NSF grant DCB-8811198.

REFERENCES

Agarwal N, Kalra, VK (1983) Interaction of lanthanide cations and uranyl ion with the calcium/proton antiport systems in Mycobacterium phlei. Biochim Biophys Acta 727:285-92

Baker PF, Presley R (1969) Kinetic evidence for an intermediate stage in the fertilization of the sea urchin egg. Nature 221:488-490

Belanger AM, Rustad RC (1972) Movements of echinochrome granules during the early development of sea urchin eggs. Nature New Biology 239:81-83

Brownlee C, Wood JW (1986) A gradient of cytoplasmic free calcium in growing rhizoid cells of Fucus serratus. Nature 320:624-626

Cantino ME, Schackmann RW, Johnson DE (1983) Changes in subcellular elemental distribution accompanying the acrosome reaction in sea urchin sperm. J Exp Zool 226:255-268

Caswell AH (1979) Methods of measuring intracellular calcium. Int Rev Cytol 56:145-181

Chambers EL (1980) Fertilization and cleavage of eggs of the sea urchin Lytechinus variegatus in Ca^{2+}-free sea water. Eur J Cell Biol 22:476

Chambers EL, Angeloni SV (1981) Is external Ca^{2+} required for fertilization of sea urchin eggs by acrosome reacted sperm? J Cell Biol 91:181a

Chandler DE, Williams JA (1978) Use of chlorotetracycline as a fluorescent probe. J Cell Biol 76:371-385

Dale B, DeFelice LJ, Ehrenstein G (1985) Injection of a soluble sperm fraction into sea-urchin eggs triggers the cortical reaction. Experientia 41:1068-1070

Donahue BS, Abercrombie RF (1987) Free diffusion coefficient of ionic calcium in cytoplasm. Cell Calcium 8:437-448

Elinson RP (1975) Site of sperm entry and a cortical contraction associated with egg activation in the frog Rana pipiens. Dev Biol 47:257-268

Fabiato A (1983) Calcium-induced release of calcium from the cardiac sarcoplasmic reticulum. Am J Physiol 245:C1-C14

Fabiato A (1985) Time and calcium dependence of activation and inactivation of calcium-induced release of calcium from the sarcoplasmic reticulum of a skinned canine cardiac Purkinje cell. J Gen Physiol 85:247-289

Fraser LR, Ahuja KK (1988) Metabolic and surface events in fertilization. Gamete Res 20:491-519

Garbers DL (1981) The elevation of cyclic AMP concentrations in flagella-less sea urchin sperm heads. J Biol Chem 256:620-624

Garbers DL, Tubb DJ, Hyne RV (1982) A requirement for bicarbonate for Ca^{2+}-induced elevations of cyclic AMP in guinea pig spermatozoa. J Biol Chem 257:8980-8984

Gardiner DM, Grey RD (1983) Membrane junctions in _Xenopus_ eggs: their distributions suggests a role in calcium regulation. J Cell Biol 96: 1159-1163

Gilkey JC, Jaffe LF, Ridgway EB, Reynolds GT (1978) A free calcium wave traverses the activating egg of the medaka, _Oryzias latipes_. J Cell Biol 76:448-466

Goldenberg M, Elinson RP (1980) Animal/vegetal differences in cortical granule exocytosis during activation of the frog egg. Dev Growth Differ 22:345-356

Gould M, Stephano JL (1989) How do sperm activate eggs in _Urechis_ (as well as in polychaetes and molluscs)? In: Nuccitelli R, Cherr GN, Clarke WH (eds) Mechanisms of Egg Activation. Plenum New York pp 201-214

Hagiwara S, Jaffe LA (1979) Electrical properties of egg cell membranes. Ann Rev Biophys Bioeng 8:385-416

Hagiwara S, Takahashi K (1967) Surface density of calcium ions and calcium spikes in the barnacle muscle fiber membrane. J Gen Physiol 50:583-601

Iwasa KH, Ehrenstein G, DeFelice LJ, Russell JT (1989) Sea urchin sperm contain enough inositol 1,4,5-trisphosphate to activate eggs. J Cell Biol 109:128a

Jaffe LA (1989) Receptors, G-proteins and egg activation. In: Nuccitelli R, Cherr GN, Clarke WH (eds) Mechanisms of Egg Activation. Plenum New York pp 157-172

Jaffe LF (1968) Localization in the developing Fucus egg and the general role of localizing currents. Adv Morphogenesis 7:295-328

Jaffe LF (1980) Calcium explosions as triggers of development. Ann NY Acad Sci 339:86-101

Jaffe LF (1983) Sources of calcium in egg activation: a review and hypothesis. Dev Biol 99:265-276

Jaffe LF (1985) The role of calcium explosions, waves, and pulses in activating eggs. In: Metz CB, Monroy A (eds) Biology of Fertilization, Volume 3. Academic Press San Diego pp 127-167

Jaffe LF, Levy S (1987) Calcium gradients measured with a vibrating calcium-selective electrode. Proc IEEE/EMBS Conf 9:779-781

Kropf DL, Quatrano RS (1987) Localization of membrane-associated calcium during development of fucoid algae using chlorotetracycline. Planta 171:158-170

Lardy H, San Agustin J (1989) Caltrin and calcium regulation of sperm activity. In: Schatten H, Schatten G (eds) The Cell Biology of Fertilization. Academic Press San Diego

Luttmer S, Longo FJ (1985) Ultrastructural and morphometric observations of cortical endoplasmic reticulum in Arbacia, Spisula and mouse eggs. Develop Growth and Differ 27:349-359

Lynn JW, McCulloh DH, Chambers EL (1988) Voltage clamp studies of fertilization in sea urchin eggs II. Dev Biol 128:305-323

McCulloh DH, Chambers EL (1986) When does the sperm fuse with the egg? J Gen Physiol 88:38-39a

McCulloh DH, Ivonnet PI, Chambers EL (1989) Blockers of calcium influx promote sperm entry in sea urchin eggs at clamped negative membrane potentials. J Cell Biol 109:126a

Meindl U (1982) Local accumulation of membrane-associated calcium according to cell pattern formation in Micrasterias denticulata, visualized by chlorotetracycline fluorescence. Protoplasma 110:143-146

Miyazaki S-I (1989) Signal transduction of sperm-egg interaction causing periodic calcium transients in hamster eggs. In: Nuccitelli T, Cherr GN, Clarke WH (eds) Mechanisms of Egg Activation. Plenum New York pp 231-246

Miyazaki S-I, Hashimoto N, Yoshimoto Y, Kishimoto T, Igusa Y, Hiramoto Y (1986) Temporal and spatial dynamics of the periodic increase in intracellular free calcium at fertilization of golden hamster eggs. Dev Biol 118, 259-267

Moore HDM, Bedford JM (1983) The interaction of mammalian gametes in the female. In: Hartmann JM (ed) Mechanisms and Control of Animal Fertilization. Academic Press New York pp 453-497

Nuccitelli R (1978) Ooplasmic segregation and secretion in the Pelvetia egg is accompanied by a membrane-generated electrical current. Dev Biol 62:13-33

Nuccitelli R, Ferguson J, Jan J-R (1989) The role of the phosphatidylinositol cycle in the activation of the frog egg. In: Nuccitelli R, Cherr GN, Clarke WH (eds) Mechanisms of Egg Activation. Plenum New York pp 215-230

Prigogine I, Stengers J (1984) Order out of chaos. Bantam New York p 173

Reuben JP, Brandt PW, Grundfest H (1974) Regulation of myoplasmic calcium concentration in intact crayfish muscle fibers. J Mechanochem Cell Motility 2:269-285

Robinson KR, Cone R (1980) Polarization of fucoid eggs by a calcium ionophore gradient. Science 207:77-78

Robinson KR, Jaffe LF (1975) Polarizing fucoid eggs drive a calcium current through themselves. Science 187:70-72

San Agustin JT, Hughes P, Lardy HA (1987) Properties and functions of caltrin, the calcium-transport inhibitor of bull seminal plasma. Faseb J 1:60-66

Sano K, Kanatani H (1980) External calcium ions are requisite for fertilization of sea urchin eggs to spermatozoa with reacted acrosomes. Develop Biol 78:242-246

Schackmann RW, Eddy EM, Shapiro BM (1978) The acrosome reaction of Strongylocentrotus purpuratus sperm. Develop Biol 65:483-495

Schackmann RW, Shapiro BM (1981) A partial sequence of ionic changes associated with the acrosome reaction of Strongylocentrotus purpuratus. Develop Biol 81:145-154

Schackmann RW (1989) Ionic regulation of the sea urchin sperm acrosome reaction and stimulation by egg-derived peptides. In: Schatten H, Schatten G (eds) The Cell Biology of Fertilization. Academic Press San Diego

Schmidt T, Patton C, Epel D (1982) Is there a role for the Ca^{2+} influx during fertilization of the sea urchin egg? Develop Biol 90:284-290

Shapiro BM, Schackmann RW, Tombes RM, Kazazoglou T (1985) Coupled ionic and enzymatic regulation of sperm behavior. Curr Top Cell Regul 26:97-113

Shen SS, Steinhardt RA (1984) Time and voltage windows for reversing the electrical block to fertilization. Proc Natl Acad Sci USA 81:1436-1439

Singh JP, Babcock DF, Lardy HA (1978) Increased calcium-ion influx is a component of capacitation of spermatozoa. Biochem J 172:549-556

Somlyo AV, Gonzalez-Serratos H, Shuman H, McClellan G, Somlyo AP (1982) Calcium release and ionic changes in the sarcoplasmic reticulum of tetranized muscle: an electron-probe study. J Cell Biol 90:577-594

Speksnijder JE, Corson WD, Sardet C, Jaffe LF (1989a) Free calcium pulses following fertilization in the ascidian egg. Dev Biol 135:182-190

Speksnijder JE, Miller AL, Weisenseel MH, Chen T-H, Jaffe LF (1989b) Calcium buffer injections block fucoid egg development by facilitating calcium diffusion. Proc Natl Acad Sci USA 86:6607-6611

Swann K, Whitaker M (1986) The part played by inositol trisphosphate and calcium in the propagation of the fertilization wave in sea urchin eggs. J Cell Biol 103:2333-2342

Takahashi YM, Sugiyama M (1973) Fertilization in Ca-free sea water with egg-water-treated spermatozoa. Dev Growth Diff 15:261-267

Tilney LG (1985) The acrosomal reaction. In: Metz CB, Monroy A (eds) Biology of Fertilization Vol 2 pp 157-213

Tsien RY (1980) New calcium indicators and buffers with high selectivity against magnesium and protons. Biochem 19:2396-2404

Twigg J, Patel R, Whitaker M (1988) Translational control of $InsP_3$-induced chromatin condensation during the early cell cycles of sea urchin embryos. Nature 332:366-369

Vacquier V, Moy G (1977) Isolation of bindin: the protein responsible for adhesion of sperm to sea urchin eggs. Proc Natl Acad Sci USA 74:2456-2460

Wakui M, Potter BVL, Petersen OH (1989) Pulsatile intracellular calcium release does not depend on fluctuations in inositol trisphosphate concentration. Nature 339:317-320

Whitaker M, Swann K, Crossley I (1989) What happens during the latent period at fertilization. In: Nuccitelli R, Cherr GN, Clarke WH (eds) Mechanisms of Egg Activation. Plenum New York pp 157-200

Yanagimachi R (1988) Sperm-egg fusion. Curr Top Memb Transp 32:4-43

Yanagimachi R, Usui N (1974) Calcium dependence of the acrosome reaction and activation of guinea pig spermatozoa. Exp Cell Res 89:161-174

POLYSPERMY BLOCKS IN FUCOID ALGAE AND THE OCCURRENCE OF POLYSPERMY IN NATURE

Susan H. Brawley
Department of Biology
Vanderbilt University
Nashville TN 37235
USA

Introduction

The first block against polyspermy in eggs of fucoid algae is a sodium-dependent, electrical fast block (Robinson et al., 1981; Brawley, 1987, 1990, and in prep., see Fig. 1), similar to the electrical fast block initially discovered in the sea urchin egg by Jaffe (1976) and, subsequently, found in many animal phyla (Kline et al., 1985). At least one additional polyspermy block is present: cell wall formation, which is triggered by fertilization (Evans et al., 1982; Brawley & Bell, 1987; Brawley, 1990).

Polyspermy is lethal in the fucoid algae (Brawley, 1987, 1990), just as it is in many animals (Wilson, 1928; Jaffe & Gould, 1985). Polyspermy has been observed in other algae, including in the green algae Oedogonium (Hoffman, 1973b) and Chlamydomonas (Hans Ettel, pers. comm.) and in the red alga Bostrychia (J. West, pers. comm.); whether polyspermy is lethal in non-fucoid algae, such as these, and in higher plants is unknown. Polyspermy is often lethal in animals, because a multipolar spindle results from the inheritance of additional centrioles. The red algae and angiosperms lack centrioles, but have otherwise normal centrosomes that support a typical bipolar spindle; whether centrosomal inheritance is maternal or paternal is unknown. Additionally, the sperm of some plants are multi-flagellated; thus, the fertilized egg may contain multiple basal bodies even after normal, monospermic fertilization. Ultrastructural data (Hoffman, 1973a) from one such species, Oedogonium, suggests that the extra

NATO ASI Series, Vol. H 45
Mechanism of Fertilization
Edited by B. Dale

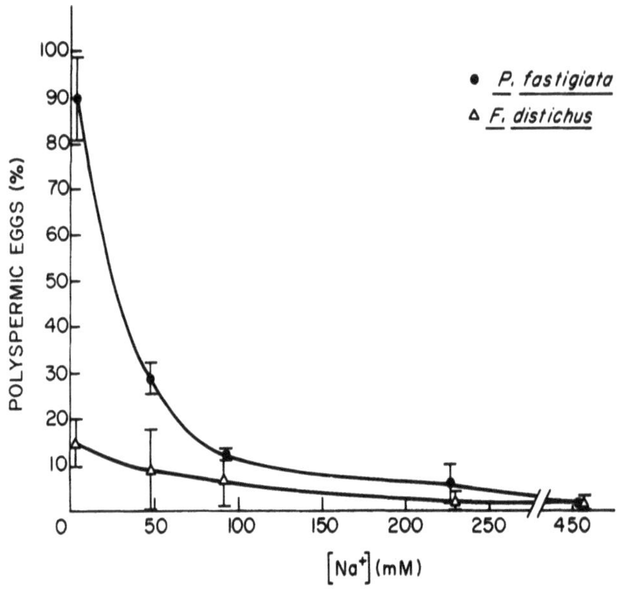

Figure 1. Percentage of polyspermic eggs as a function of external sodium concentration during fertilization. These are monoecious species, and the "natural" sperm:egg ratios were observed to be ca. 5:1 for F. gardneri (= F. distichus) and ca. 25:1 for P. fastigiata (from Brawley, 1987, with permission).

centrioles are destroyed in the zygote's cytoplasm. Polyspermy in such plants may not be lethal, because centrosome/centriole inheritance may be maternal, and/or cytoplasmic blocks to supernumerary centrioles or sperm may exist, as in "physiologically polyspermic" animals (e.g., reptiles, birds; for discussion, see Wilson, 1928; Iwao, 1989).

Natural selection walks a metaphorical circus-highwire act in species with sexual reproduction. Fertilization must be achieved via adequate production and release of gametes, but sperm/egg ratios must be low enough to prevent polyspermy (i.e., except in those species which are "physiologically polyspermic"). How common is polyspermy in nature in those species in which even one extra fertilization is lethal? Since such eggs have evolved multiple-mechanisms to block polyspermy, it would appear that polyspermy is a real threat to recruitment, with the implication that the efficiency of fertilization may be higher in nature than we have suspected. In dioecious marine organisms, such as the sea urchin, several authors (Pennington, 1985; Denny, 1988) have discussed the probability of fertilization in the turbulent sea, and consider that fertilization occurs at low efficiency under most circumstances. However, it is difficult to track gametes and zygotes of many species in nature to determine either efficiencies of fertilization or how frequently polyspermy occurs, because many organisms have planktonic zygotes and multiple larval stages that are widely dispersed.

The fucoid algal egg attaches to intertidal rocks within 4 h of fertilization, at least 12 h before the first mitosis, and there are no larval stages. These algae, therefore, offer a good system for studies of fertilization in nature. This report provides additional information on the polyspermy block in fucoid algae, and it includes preliminary information on natural levels of polyspermy in two monoecious species (Fucus gardneri, Pelvetia fastigiata) and one dioecious species (Fucus ceranoides).

Materials and Methods

Culture and electrophysiology

Pelvetia fastigiata and Fucus gardneri Silva (= F. distichus Powell) were collected and studied at Pacific Grove, CA, in the Hopkins Marine Station refuge. Fucus ceranoides was collected in the Laxey estuary, Isle of Man (U.K.). Zygotes were cultured and artificial seawaters prepared as previously described (Brawley, 1987). Fertilization was determined by staining batches of eggs with the fluorescent dye for β-linked polysaccharides, Calcofluor White (at 0.0001%); this stains cell wall material, produced within a few minutes of fertilization. Polyspermy was determined by fixing eggs in acetic-ethanol (1:3) and staining in aceto-iron-hematoxylin (Wittmann, 1965). Polyspermy was, also, determined indirectly by using a morphological indicator (Brawley, 1987; Fig. 2, this report) of polyspermy.

Unfertilized Pelvetia eggs were obtained for electrophysiology by shedding primed receptacles (Brawley, 1987) in 350 mM K^+ ASW (see below) and separating oogonia and antheridia over Nitex screen. Fresh sperm were obtained by shedding receptacles into normal seawater. Glass microelectrodes (70 - 90 MΩ) were used to impale eggs that were immobilized on polylysine-coated cover slips. Full details of the electrophysiological procedures are reported elsewhere (Brawley, 1990).

Field studies

Gametes were collected from surfaces of F. gardneri and P. fastigiata receptacles during low tide. Each receptacle was dipped into an artificial seawater in which NaCl had been substituted with KCl to increase the K^+ to 350 mM; this blocks fertilization (Robinson et al., 1981). Receptacles were selected, one per plant, along a transect line with standard ecological techniques (random numbers). In the laboratory, such eggs were washed exhaustively over Nitex screen (40 μm) 1 h after collection, then assayed for fertilization with Calcofluor White and for polyspermy by culturing in 350 mM K^+ ASW for periods less than 2 h. Zygotes of F. ceranoides were collected on artificial substrata (Seagoin' poxy putty, Permalite Plastics Corp., Newport Beach, CA) bolted to rocks in the intertidal zone.

Results

The fertilization potential

Pelvetia fastigiata has a fertilization potential (Fig. 3) similar to that of Fucus vesiculosus (Robinson et al., 1981; Brawley, 1990; Brawley, in prep.). It is rarely triggered the moment a sperm reaches the egg surface. Instead, sperm move over the surface for 10 sec or more, then attach to one spot and gyrate

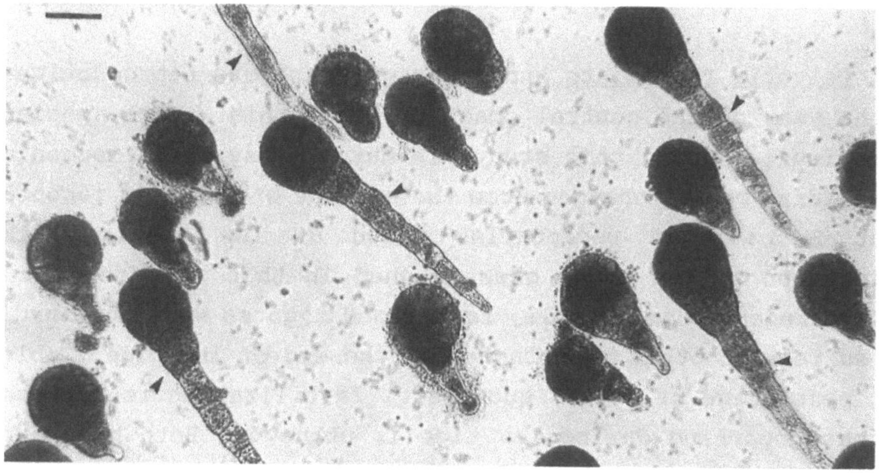

Figure 2. Polyspermic and normal (arrows) embryos of Pelvetia fastigiata at 10 d post-fertilization. Scale bar = 100 μm.

against the egg for another 10-20 sec before evoking a fertilization potential. The fertilization potential in 452 mM Na^+ (normal) artificial seawater reaches a plateau of about -25 mV from a steady resting potential of -60 mV, and the duration of the fertilization potential is about 4 min. Sperm swim away from the egg surface at 1-2 min after the beginning of the fertilization potential, and wall formation is observed with Calcofluor White by 8-10 min post-fertilization (not shown).

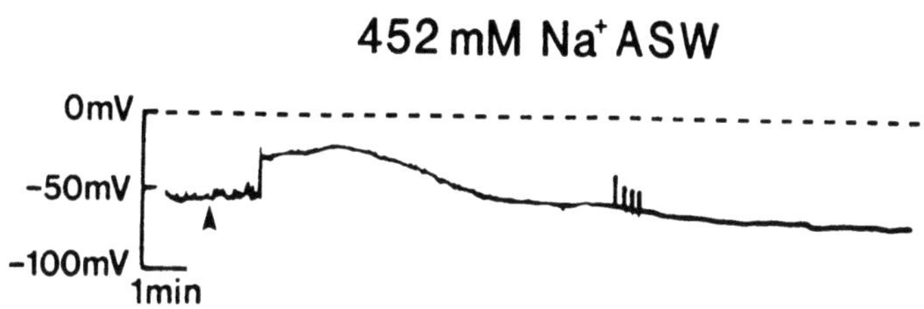

Figure 3. The fertilization potential of Pelvetia fastigiata. Sperm were added to the bath as indicated (arrow).

Fertilization at low tide in the field in monoecious species

Pelvetia fastigiata and Fucus gardneri are reproductive most of the year on the central coast of California. Both species are monoecious. During this study, in March - May, 1989, oogonia and antheridia were discharged from the surface of Pelvetia receptacles about 2 d/wk, as receptacles dried at low tide. Although desiccation killed those gametes shed in this way from the most exposed receptacles, viable, fertilized zygotes were obtained from the surface of other receptacles at the end of low tide (Table 1). What percentage of total, successful fertilization is represented by that occurring during low tide is unknown. Both hyperosmotic and hypoosmotic (rain at low tide) stress caused shedding from Pelvetia receptacles to occur, but shedding from F. gardneri was not stimulated in response to rain. Oogonia and antheridia were shed several times each week from F. gardneri receptacles at low tide. Fucus gardneri secretes much more mucilage from receptacles at low tide than does P. fastigiata, and this provides an ionic/osmotic nursery in which fertilization also occurs (see Brawley, 1990). Polyspermy, assayed on the basis of embryonic morphology, was lower in eggs of F. gardneri fertilized at low tide compared to those of P. fastigiata (Table 1).

Polyspermy in F. ceranoides in the field

Polyspermy was analyzed in Fucus ceranoides by both direct and indirect means. It was possible to collect young zygotes from this dioecious species, because fertilization must occur when plants are covered with seawater at high tide, and this is restricted to a total period of about 4 h in this estuary. This species sheds on spring tides (Brawley, in prep.), and 1,429 ± 1,099 zygotes/2.8 cm^2 of benthic surface were collected at the end of a single high tide on August 2, 1989, off 8 disks of Seagoin' poxy putty placed among two adjacent beds of F. ceranoides. Zygotes were fixed in the field from an additional 8 disks by pipetting acetic-ethanol fixative onto the disk surfaces and washing zygotes into vials. Polyspermy (Fig. 4) could be assayed directly because of the short time between fertilization and fixation of eggs, and it was 7% (n = 100 zygotes counted). Fertilization was 100% (n = 100 zygotes (eggs) counted); control tests with unfertilized eggs and zygotes on disks in the laboratory determined that eggs would not have been lost or ruptured during fixation in the field. These data for fertilization and polyspermy in F. ceranoides are typical for tides during July - August, 1989 (Brawley, in prep.).

Table 1. Fertilization and polyspermy occurring in nature at low tide in two monoecious species of fucoid algae. Values are the mean (± SD).

	Fertilization (%) (n = 100 eggs)	Polyspermy (%) (n = 100 - 250 eggs)
Pelvetia fastigiata (n = 10 receptacles) (3/3/89)	64.5 (24.2) 13 - 92	6.4 (5.8) 2 - 22
P. fastigiata (3/16/89) (n = 10 receptacles)	25.0 (22.5) 0 - 65	5.6 (4.6) 2 - 14
Fucus gardneri (3/1/89) (n = 10 receptacles)	44.8 (38.5) 25 - 100	3.5 (1.9) 2 - 6
F. gardneri (3/22/89) (n = 10 receptacles)	90.5 (14.1) 53 - 99	3.8 (2.3) 0 - 8

Discussion

This report reviews our knowledge of polyspermy and polyspermy blocks in the fucoid algae. It introduces the fertilization potential in Pelvetia fastigiata, which is very similar to that of F. vesiculosus (Brawley, 1990), and presents information on the natural occurrence of polyspermy in two monoecious and one dioecious species of fucoid algae. Polyspermy is a common (5 - 7% of all eggs) occurrence during natural fertilization despite the fact that fucoid eggs have multiple mechanisms to prevent polyspermy.

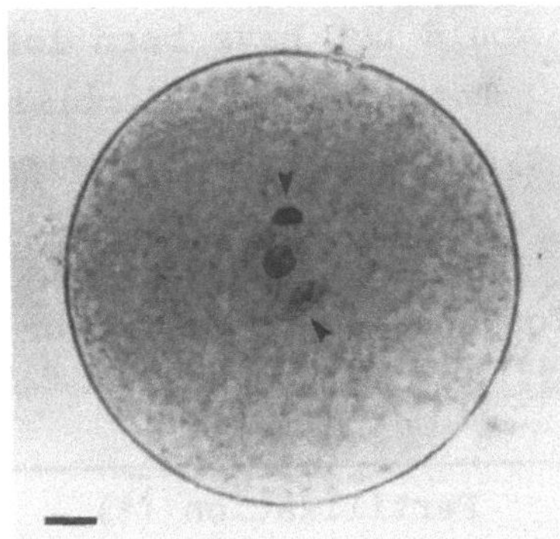

Figure 4. A polyspermic egg of F. ceranoides collected from the Laxey estuary at the Isle of Man (U.K.). Sperm pronuclei are indicated by arrows; one pronucleus has fused with the egg pronucleus. Aceto-iron-hematoxylin stain. Scale bar = 10 μm.

How do estimates of the susceptibility to polyspermy of a species compare between data collected in the laboratory versus in nature? Polyspermy in fucoid algae is sensitive both to $[Na^+]_{ext}$ and to the sperm/egg ratio, and in the dioecious species F. vesiculosus, where both factors are easily manipulated, polyspermy is low (1 - 3%) in normal Na^+ ASW until the sperm/egg ratio is increased to 1000:1 or greater (Brawley, 1990). At a sperm:egg ratio of 2000:1, 10% of eggs were polyspermic when fertilized in normal seawater in the laboratory. Only 1 - 3% of F. gardneri (= F. distichus) and P. fastigiata eggs were polyspermic when fertilized in normal ASW in the laboratory (Brawley, 1987). Thus, the percentage of polyspermic eggs found in nature in Pelvetia fastigiata and F. ceranoides is surprisingly high. In the case of those Pelvetia eggs fertilized at low tide, $[Na^+]$ may be limiting, and an unusually high sperm:egg ratio may be produced on the receptacle's surface at low tide. The same problems would confront eggs of F. gardneri, but these are fertilized in Na^+-rich mucilage (Brawley, 1990), which eliminates one of these potential problems. If laboratory work on polyspermy as a function of sperm:egg ratio in F. vesiculosus (Brawley, 1990) and F. serratus (Callow et al., 1978) can be applied to F. ceranoides, fertilization in nature in F. ceranoides may occur at quite high sperm:egg ratios, ca. 2000:1. Polyspermy is relatively common in nature in the fucoid algae, enough to be considered a factor limiting recruitment, although the conditions that permit high polyspermy also promote high levels of fertilization, a successful strategy.

Many investigators have observed shedding of oogonia and antheridia at low tide from fucoid algae, but this is the first observation that such shedding produces eggs fertilized prior to reimmersion of plants at high tide (c.f. Dawson, 1966). Monoecious species capable of such reproductive behavior should be considered "amphibious plants." Since some of the zygotes reach the substratum and attach prior to high tide, such behavior has important consequences for dispersal, recruitment and inbreeding depression in monoecious fucoid algae (also see Müller and Gassmann, 1985).

In a number of animal eggs, secretion of proteases from the egg after fertilization destroys components of the egg's surface coat that are required for sperm binding. This acts as an "intermediate" block against polyspermy, and sperm fall off the egg surface within a minute of insemination in the sea urchin (Vacquier et al., 1973) and within 1-2 min post-insemination in an ascidian egg (Lambert, 1986). A similar "intermediate" block against polyspermy might exist in the fucoid algae, since electrical repolarization of the zygote occurs prior to observation of Calcofluor White fluorescence (i.e., presence of cell wall material) in F. vesiculosus (Brawley and Bell, 1987; Brawley, 1990), Pelvetia fastigiata, and F. gardneri (Brawley, unpublished). Only in F. ceranoides (Brawley, in prep.) is cell wall material evident before repolarization of the zygote's membrane potential.

It is significant in this regard that sperm fall off and swim away from the egg surface in Pelvetia while the egg's membrane potential is still depolarized (Results and Brawley, in prep.). Of course, ultrastructural assay of the cell surface might indicate functional wall material present prior to the time at which it can be revealed with Calcofluor White staining, as suggested by several studies (Pollock, 1970; Brawley et al., 1976a,b). Combined electrophysiological and ultrastructural studies are required to assess how soon after fertilization an effective mechanical barrier against polyspermy is established.

Plasma membrane sperm receptor

Whether or not the egg plasma membrane contains a sperm receptor is a controversial point in animal systems (Kline et al., 1988). The behavior of the sperm at the fucoid egg surface suggests that a receptor is present in this system. Rarely, the sperm triggers a fertilization potential as soon as it reaches the egg, but, more commonly, it actively searches and probes the surface of the egg (see Results); this suggests that all domains of the egg plasma membrane are not suitable for sperm/egg fusion, although binding appears to occur uniformly at high sperm/egg concentration. It is important to note that sperm binding and sperm/egg fusion may involve different membrane moieties. When fertilization is prevented electrically with a current-clamp (Brawley, in prep.), sperm remain bound to the egg, and additional sperm attach to the egg's surface during such treatments.

Summary

The fucoid algae have a sodium-dependent, electrical fast block against polyspermy. Some monoecious species shed gametes at low tide, and such eggs are successfully fertilized before plants are reimmersed. Polyspermy is common in fucoid algae in nature, and it has been assayed directly in F. ceranoides. Contrary to predictions for dioecious marine animals, natural fertilization rates in this dioecious species are also high. Unanswered questions related to the polyspermy block and fertilization that should be addressed with fucoid algae and other organisms include these:

1. How common is polyspermy in other plants and is it always lethal (i.e., how are centrosomes inherited)?
2. What components of the egg's plasma membrane are involved in sperm binding and fusion?
3. Are there "intermediate" blocks against polyspermy in the fucoid algae, such as protease secretion?
4. Is ionic failure of the polyspermy block a determinant of species distributions in nature? For example, do marine species fail to penetrate freshwater environments for this reason?
5. Are high rates of fertilization common in other dioecious marine organisms, as in F. ceranoides; if so, how are they achieved?

Acknowledgements

Special thanks to David Epel and other faculty and staff of Hopkins Marine Station (Stanford University) and to Joanna Jones, Trevor Norton and other faculty and staff of the Marine Biological Station of the University of Liverpool (Port Erin, U.K.) for their hospitality during a sabbatical leave from Vanderbilt University, 1988-89. Grace Monty provided word-processing, and this work was supported by the National Science Foundation (U.S.A.).

References

Brawley SH (1987) A sodium-dependent, fast block to polyspermy occurs in eggs of fucoid algae. Dev Biol 124:390-397

Brawley SH (1990) The polyspermy block in fucoid algae. In Wiessner W (ed) Recent advances in experimental phycology, vol 5. Springer-Verlag, Berlin, in press

Brawley SH, Wetherbee R, Quatrano RS (1976a) Fine-structural studies of the gametes and embryo of Fucus vesiculosus L. (Phaeophyta). I. Fertilization and pronuclear fusion. J Cell Sci 20:233-254

Brawley SH, Wetherbee R, Quatrano RS (1976b) Fine-structural studies of the gametes and embryo of Fucus vesiculosus L. (Phaeophyta). II. The cytoplasm of the egg and young zygote. J Cell Sci 20:255-271

Brawley SH, Bell E (1987) Partial activation of Fucus eggs with calcium ionophores and low-sodium seawater. Dev Biol 122: 217-226

Callow ME, Evans LV, Bolwell GP, Callow JA (1978) Fertilization in brown algae I. SEM and other observations on Fucus serratus. J Cell Sci 32:45-54

Dawson EY (1966) Marine botany. Holt Rinehart and Winston, Inc. New York, 371 pp

Denny MW (1988) Biology and the mechanics of the wave-swept environment. Princeton University Press, Princeton, 329 pp

Evans LV, Callow JA and Callow ME (1982) The biology and biochemistry of reproduction and early development in Fucus. In Round FE, Chapman DJ (eds) Progress in phycological research, vol. 1. Elsevier Biomedical Press, Amsterdam, pp 67-110

Hoffman LR (1973a) Fertilization in Oedogonium I. Plasmogamy. J Phycol 9:62-84

Hoffman LR (1973b) Fertilization in Oedogonium II. Polyspermy. J Phycol 9:296-301

Iwao Y (1989) An electrically-mediated block to polyspermy in the primitive urodele Hynobius nebulosus and phylogenetic comparison with other amphibians. Dev Biol 134:435-445

Jaffe LA (1976) Fast block to polyspermy in sea urchin eggs is electrically mediated. Nature (London) 261:68-71

Jaffe LA, Gould M (1985) Polyspermy-preventing mechanisms. In Metz CB, Monroy A (eds) "Biology of Fertilization," vol 3. Academic Press, New York, pp 223-250

Kline D, Jaffe LA, Tucker RP (1985) Fertilization potential and polyspermy prevention in the egg of the nemertean, Cerebratulus lacteus. J Exp Zool 236:45-52

Kline D, Simoncini L, Mandel G, Maue RA, Kado RT, Jaffe LA (1988) Fertilization events induced by neurotransmitters after injection of mRNA in Xenopus eggs. Science 241:464-467

Lambert CC (1986) Fertilization-induced modification of chorion N-acetyl-glucosamine groups in the ascidian chorion, decreasing sperm binding and polyspermy. Dev Biol 116: 168-173

Müller DG, Gassmann G (1985) Sexual reproduction and the role of sperm attractants in monoecious species of the brown algae order Fucales (Fucus, Hesperophycus, Pelvetia, and Pelvetiopsis). J Plant Physiol 118:401-408

Pennington JT (1985) The ecology of fertilization of echinoid eggs: The consequences of sperm dilution, adult aggregation, and

synchronous spawning. Biol Bull 169:417-430
Pollock EG (1970) Fertilization in Fucus. Planta (Berl) 92:85-99
Robinson KR, Jaffe LA, Brawley SH (1981) Electrophysiological properties of fucoid algal eggs during fertilization. J Cell Biol 91:179a
Vacquier VD, Tegner MJ, Epel D (1973) Protease released from sea urchin eggs at fertilization alters the vitelline layer and aids in preventing polyspermy. Exp Cell Res 80:111-119
Wilson EG (1928) The cell in development and heredity, 3rd edn. Macmillan Co, New York
Wittman W (1965) Aceto-iron-haematoxylin-chloral hydrate for chromosome staining. Stain Technol 40:161-163

HOW DOES A SPERM ACTIVATE A SEA URCHIN EGG?

Michael Whitaker
Ian Crossley
Department of Physiology
University College London
Gower Street
London WC1E 6BT
UK

INTRODUCTION

Unfertilized sea urchin eggs are metabolically quiescent and arrested in interphase of the cell division cycle. Immediately after fertilization, metabolism and protein synthesis rates increase 20-fold and the cell division cycle resumes. The intracellular signal that triggers cell cycle progression at fertilization is a transient increase in intracellular free calcium concentration (Ca_i).

The Ca_i increase at fertilization is the necessary and sufficient signal that sets the egg off on the paths of early embryonic development (Whitaker & Steinhardt, 1982). The sperm's extragenetic contribution to development is to trigger the Ca_i increase at fertilization and to contribute a centrosome (Brandriff et al., 1975). The sperm's genetic contribution to development begins to be made much later, at the blastula stage (Davidson et al., 1982).

Calcium rises in the egg at fertilization in the shape of an explosive wave, beginning at the site of sperm-egg interaction (Jaffe, 1983; Eisen et al., 1984; Swann & Whitaker, 1986). The positive feedback loop that sets off and propagates the calcium explosion involves the phosphoinositide second messengers (figure 1). The calcium-releasing messenger $InsP_3$ triggers calcium release from an internal store (Whitaker & Irvine, 1984; Clapper & Lee, 1986) and the increased Ca_i stimulates further $InsP_3$ production by activating the phosphoinositidase C that hydrolyses $PtdInsP_2$ (Whitaker & Aitchison, 1985; Ciapa & Whitaker, 1986; Swann & Whitaker, 1986).

We have a clear idea, then, of the basis and the importance of the fertilization calcium wave in sea urchin eggs. We are less certain about how the sperm detonates the calcium explosion, and we shall discuss here the pros and cons of two possible mechanisms, one involving a receptor-G protein interaction, the other the introduction of an activating cytoplasmic messenger when sperm and egg fuse.

NATO ASI Series, Vol. H 45
Mechanism of Fertilization
Edited by B. Dale

THE LATENT PERIOD

There is a striking delay between the interaction of egg and sperm and the onset of the calcium wave that is easily seen in the microscope and can be precisely defined by kinetic experiments in egg populations (Allen & Griffin, 1958). It has been called the latent period. It can also be measured precisely in single eggs using electrophysiological and calcium ratio-dye methods (Whitaker et al., 1989). The first sign of sperm-egg interaction is a small, constant current (Dale et al., 1978). The sharp onset of this current precedes the onset of the calcium wave by 15 s, on average (Whitaker et al., 1989). The calcium wave itself crosses the egg in about 20 s (Whitaker & Irvine, 1984), so the latent period occupies a substantial fraction of the total egg activation-time.

The existence of a latent period suggests the accumulation of an activator that eventually triggers the calcium wave. Indeed, it is hard to think of any other explanation for it. The latency can vary substantially (5-40 s) from egg to egg (Whitaker et al., 1989). This stochastic behaviour is also characteristic of an explosive detonation. Perhaps the most interesting experiments on the latent period have been done by Lynn, McCulloh and Chambers (1984; 1988). By holding eggs

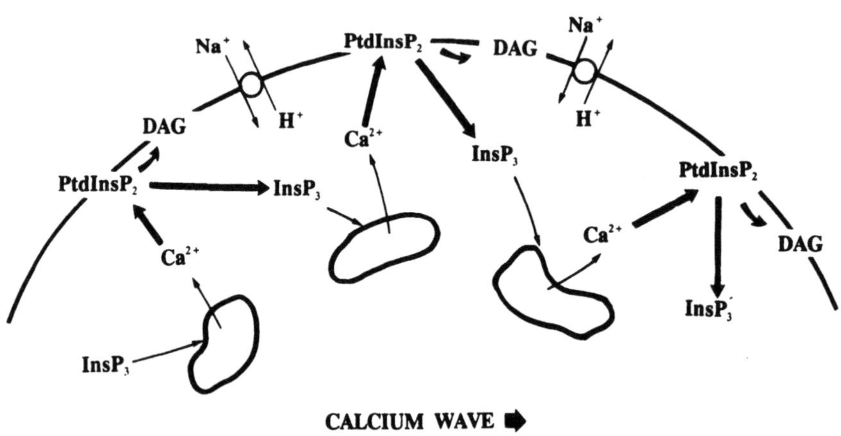

Figure 1. Phosphoinositide messengers and the fertilization calcium wave. The autocatalytic wave mechanism comprises $InsP_3$-induced calcium release and calcium-stimulated production of $InsP_3$ and DAG.

at negative membrane potentials using a voltage – clamp, they are able to prevent sperm incorporation. Sperm interact with an egg for several seconds, then detach and in some cases, swim away. The transient sperm – egg interactions are accompanied by the onset and offset of the current described above. The electrical properties of the egg membrane after these transient sperm – egg interactions are indistinguishable from those of the unfertilized egg. The most striking aspect of these experiments is that, often, an egg that has undergone one of these transient interaction with sperm will activate. In other words, the calcium wave can be triggered after sperm – egg interaction has ceased. The observations offer strong support for the idea that an activating molecule accumulates during the early stages of sperm – egg interaction. The activator persists for a time after the cessation of sperm – egg interaction. Because of the stochastic properties of the wave trigger, sometimes the activator persists long enough to set off the calcium wave; in other eggs, the activator diffuses or suffers degradation before sufficient time elapsed for the wave to be initiated. This explanation fits quantitatively, as well as qualitatively (Whitaker et al., 1989).

The existence of a latent period tells us a great deal about the basic outline of how the fertilizing sperm triggers the calcium wave, but it offers no clues about the molecular mechanism. Two hypotheses have been put forward to explain egg activation in molecular terms, one a G – protein mechanism, the other a fusion mechanism with transfer of a cytoplasmic messenger from egg to sperm (figure 2).

THE RECEPTOR – G PROTEIN HYPOTHESIS

GTP – binding proteins couple plasma membrane receptors to the effector enzymes that generate cytoplasmic second messengers (Rodbell, 1980). On this hypothesis, the sperm is an honorary hormone and triggers the calcium wave by directly stimulating phosphoinositidase C to cause $InsP_3$ production. $InsP_3$ in turn triggers the calcium wave. The latent period would be analogous to the latency that can be measured between hormone application and calcium response in mammalian cells (Marty et al., 1989).

Initial experiments offered good evidence in support of this idea. GTPγS, a thio – GTP analogue that constitutively activates GTP – binding proteins, will activate sea urchin eggs if micro – injected (Turner et al., 1986). GDPβS prevents fertilization envelope elevation, usually a sign that the calcium wave has been inhibited (Turner et al., 1986). GDPβS was also reported not to block the

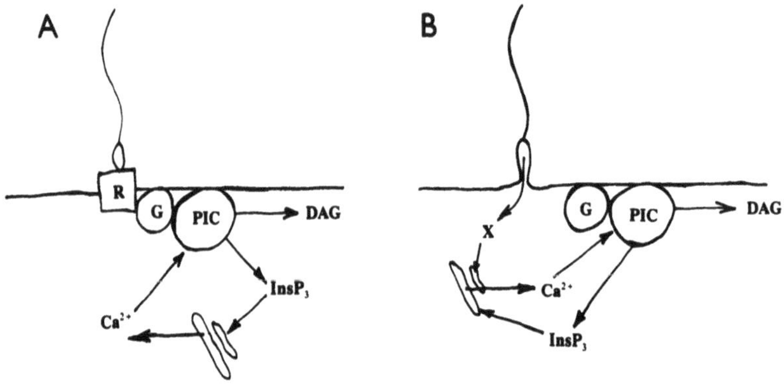

Figure 2. Two models of egg activation. In model A, the sperm triggers the calcium wave by stimulating the formation of $InsP_3$ directly, interacting with a G-protein *via* a receptor. In model B, the sperm first causes calcium release by fusing with the egg and introducing a diffusible activator, X. $InsP_3$ is produced only indirectly by the sperm, *via* the calcium-wave mechanism. The G-protein, though present, does not participate in egg activation.

fertilization envelope elevation in response to $InsP_3$ micro-injection, a result that argued that the inhibitory effects of GDPβS were exerted at a point proximal to $InsP_3$ on the signalling pathway, and therefore most probably on the G-protein. Moreover, the egg plasma membrane contains proteins that can be ribosylated by cholera and pertussis toxins, a property characteristic of GTP-binding proteins (Turner et al., 1987). These data indicate that the egg contains a GTP-binding protein that, when stimulated, will trigger the calcium wave. They also suggest that the sperm stimulates this G-protein too.

The wind has changed. Subsequent experiments, while confirming many of the above observations, have led us away from the G-protein idea. If we look, not at fertilization envelope elevation (a consequence of the calcium wave) but at the calcium wave itself, a different interpretation of the initial observations emerges. GTPγS does indeed trigger a calcium wave quantitatively identical to the fertilization wave, with a latency of 30-60 s (Swann et al., 1987), but GDPβS does not prevent the sperm from producing the calcium wave. Rather, it prevents the exocytosis that is responsible for the elevation of the fertilization envelope (Crossley & Whitaker, 1989; Swann & Whitaker, 1990). This result conflicts with the finding, mentioned above, that $InsP_3$-induced fertilization envelope elevation is unaffected by GDPβS

(Turner et al., 1986); but we cannot reproduce this result; nor can the authors of the study (L.A. Jaffe, personal communication). The upshot is that eggs contain a G-protein that can trigger the calcium wave, but there is no evidence in favour of the idea that the sperm interacts with this G-protein. In fact the GDPβS data now argue against it.

We have approached the question in a slightly different way. If both the sperm and GTPγS act in the way the hypothesis suggests, then both should produce the conjugate phosphoinositide messengers $InsP_3$ and DAG during the latent period by stimulating phosphoinositidase C via the G- protein (figure 2). Proving this is complicated by the fact that $InsP_3$ and DAG are known to be produced during the calcium wave, when the elevated Ca_i stimulates $PtdInsP_2$ hydrolysis (figure 1). We can simplify things and prolong the latent period indefinitely by micro-injecting eggs with the calcium chelator BAPTA, which prevents the calcium wave from occurring. We have measured DAG production by monitoring the activation of the egg Na/H antiporter, using a DAG kinase inhibitor to reduce any metabolism of DAG (figure 3). The antiporter is regulated by the DAG target, protein kinase C (Swann & Whitaker, 1985; Shen & Burgart, 1986; Lau et al., 1986) We find, as expected, that microinjecting GTPγS under these conditions leads to activation of the antiporter,

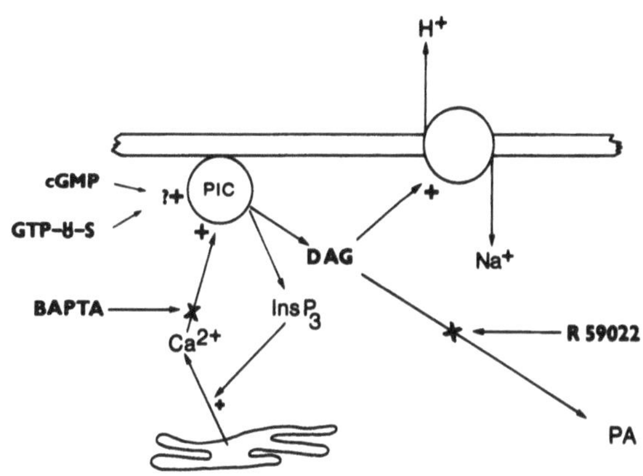

Figure 3. Measuring DAG production in a single egg. We block the calcium-wave mechanism by micro-injecting the calcium chelator BAPTA and reduce the metabolism of DAG, using the Janssen diacylglycerol kinase inhibitor, R59022. We monitor activation of the Na/H-antiporter by measuring pH_i with the fluorescent dye, BCECF.

indicating that GTPγS stimulates the production of DAG (Whalley and Whitaker, 1988). But when we fertilize eggs under these conditions, the antiporter remains inactive, despite the incorporation of ten or more sperm into the egg (Swann & Whitaker, 1990). We conclude that in the absence of the calcium wave, interaction of the sperm with the egg does not lead to the production of DAG and that DAG production at fertilization is calcium-dependent (Swann & Whitaker, 1990).

So, in attempting to confirm the receptor-G protein hypothesis, we have obtained only evidence against it. Though the egg contains a G-protein that is linked to phosphoinositidase C and that is stimulated by GTPγS, the sperm by-passes the G-protein when it triggers the calcium wave: its effects on phosphoinositide messenger production are indirect, mediated by the calcium increase itself. Abandoning the G-protein idea, we have turned instead to the possibility that the crucial activating event at fertilization is the fusion of sperm and egg.

SPERM-EGG FUSION

We had always imagined, for a variety of reasons, that sperm-egg fusion occurred at the end of the latent period, ruling it out as a participant in the activation pathway. Recent experiments by Chambers, Longo and McCulloh revived our interest in the idea that the sperm fuses with the egg, and triggers the calcium wave by introducing a diffusible activator into the egg cytoplasm (Loeb, 1915).

A first requirement for the cytoplasmic messenger idea is early cytoplasmic continuity between sperm and egg. Using a capacitance technique, McCulloh and Chambers have shown that cytoplasmic continuity is established from the very first, at the point at which the current step occurs at the beginning of the latent period (McCulloh & Chambers, 1986a;1986b). Cytoplasmic continuity may, or may not, coincide with membrane fusion defined as the continuity of egg and sperm lipid bilayers. This distinction is important from the point of view of the mechanisms of fusion and sperm incorporation (Whitaker et al., 1989). Membrane fusion, assayed ultrastructurally or by dye transfer after aldehyde or osmium fixation, occurs several seconds after the onset of the current step (Longo et al., 1986; Hinckley et al., 1986; Longo, this volume). But it is cytoplasmic continuity that determines whether or not an activating messenger can pass from sperm to egg, and cytoplasmic continuity is the earliest event that can be detected at fertilization. Moreover, the pore size at the beginning of the latent period is sufficient to allow the passage of large proteins, albeit slowly (Whitaker et al., 1989). The existence of such cytoplasmic continuity at the beginning of the latent period is the best evidence we have that the idea may be correct. It is, in fact, almost the only evidence.

CYTOPLASMIC ACTIVATORS

The other evidence in favour of cytoplasmic transfer of an activator is of two sorts. The first sort rests on the discovery of a variety of cytoplasmic messengers that will trigger the calcium wave when injected into unfertilized sea urchin eggs. These are $InsP_3$ (Whitaker & Irvine, 1984), cGMP (Swann et al., 1987) and nicotinamide nucleotide derivatives (Clapper et al., 1987). We must also include calcium ions on this list (Jaffe, 1980; Jaffe, this volume): Ca_i increases in sperm just prior to sperm–egg interaction (Trimmer et al., 1986; Guerrero & Darszon, 1989). We can exclude cAMP: although it is present in the sperm (Garbers, 1981; Jaffe, this volume), it does not activate sea urchin eggs, even when micro–injected to a final concentration of 100 μM (Karl Swann, unpublished). We consider each of the activating messenger candidates below.

$InsP_3$: $InsP_3$ is found at high concentrations in sea urchin sperm (Domino & Garbers, 1988) and is a potent activator that will trigger a stereotypic calcium wave. There are two reasons for provisionally rejecting $InsP_3$ as the diffusible activator: (i) there is barely enough $InsP_3$ in a single sperm to activate the egg if it were all

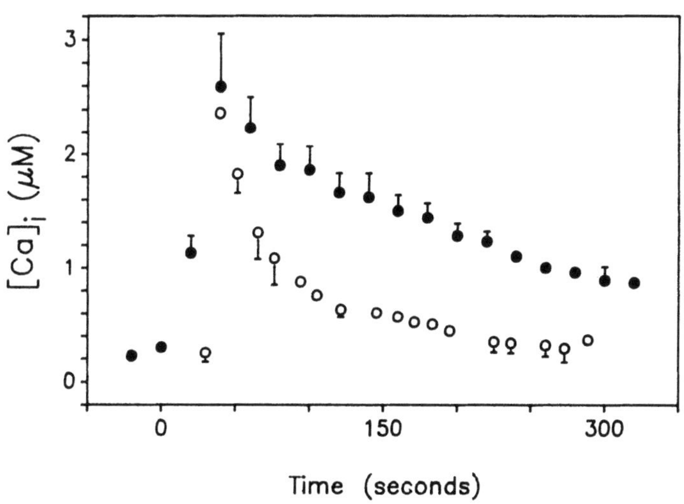

Figure 4. The calcium transient after insemination or microinjection of $InsP_3$. The $InsP_3$–induced transient is of a similar magnitude to the fertilization transient, but declines much more quickly. Mean & SEM of three experiments are shown in each case. The slower rate of rise of the fertilization calcium transient is apparent, not real, and is due to the variability in the latent period. *L. pictus.* 16°C.

introduced instantaneously (Iwasa et al., 1989), while diffusion calculations suggest that the concentration of a diffusible activator in the egg within 1 μm of the connecting pore is only 1% of the concentration in the sperm (Whitaker et al., 1989). (ii) the Ca_i transient induced by micro-injecting $InsP_3$ is shorter by 5 min than the fertilization Ca_i transient (figure 4). $InsP_3$ does not mimic the sperm Ca_i transient precisely, unlike GTPγS or cGMP (Swann et al., 1987).

cGMP: cGMP is also present in activated sea urchin sperm (Kopf & Garbers, 1979). It produces a Ca_i transient with temporal characteristics identical to the fertilization calcium transient and also resembles the sperm in producing phosphoinositide messengers indirectly, as a consequence of the Ca_i increase (Whalley & Whitaker, 1988). However, it is three orders of magnitude less potent than $InsP_3$ and, given the diffusion kinetics (Whitaker et al., 1989) would require a sperm cytoplasmic concentration of in excess of 1 M to be effective. This seems unlikely.

NAD-derivatives: These compounds are very potent calcium-releasing agents, acting at a concentration of 50 nM (Clapper et al., 1987), ten times less then $InsP_3$ (Clapper & Lee, 1986). It is not known whether NAD-derivatives are present in sperm. Micro-injection of alkaline-activated NAD causes a Ca_i transient very similar in magnitude and duration to the $InsP_3$-induced transient (Ian Crossley, unpublished results). For this reason, we do not think that an NAD-derivative is the diffusible activator.

Calcium ions: Ca^{2+} is strongly buffered by the cytoplasm (Baker, 1972) and so diffuses only slowly (Baker, 1972; Rose & Lowenstein, 1976). Ca^{2+} ions themselves are not an optimal diffusible messenger. It is unlikely that there is sufficient diffusible *internal* calcium in the sperm to trigger the calcium wave, with its characteristic latency (Jaffe, this volume). However, there are slowly-inactivating calcium channels in sperm (Guerrero & Darszon, 1989) that might provide a source of *external* calcium to supplement the sperm's internal store (Jaffe, this volume). The strongest argument against this idea is that the fertilization wave can be triggered by sperm, even under conditions in which the external calcium concentration is lower than the internal (Chambers & Angeloni, 1981; Schmidt et al, 1982). We have confirmed these results (Ian Crossley, unpublished). We observed no detectable increase in egg Ca_i due to *sperm* calcium channels during the latent period, while detectable changes in Ca_i induced by stimulating the *egg's* voltage dependent calcium channels did not trigger the calcium wave (Whitaker et al., 1989). It has also been shown that premature Ca_i increases at the site of sperm egg contact inhibit activation (Chambers & McCulloh, 1990). Calcium does not seem a likely candidate.

The other sort of evidence in favour of cytoplasmic transfer of an activating messenger comes from experiments that have shown that introducing sperm cytoplasm will activate eggs. Dale, Ehrenstein and De Felice (1985) have found that an extract made by treating sperm with distilled water will activate sea urchin eggs. It has also been demonstrated that oyster sperm (which bind to, but do not fertilize or activate sea urchin eggs under normal circumstances) will trigger the Ca_i wave when fusion is induced by treatment with the artificial fusogen polyethylene glycol (Kyozuka & Osania, 1989). In experiments with hamster eggs, Swann has shown that micro-injecting a high-molecular-weight protein factor from sperm induces the repetitive Ca_i transients that are a hallmark of fertilization in this species (Swann, 1990). This last result is particularly interesting, because the protein factor is the only parthenogenetic agent in hamster eggs that faithfully reproduces the fertilization response.

It is important to show that there are substances in sperm that can activate sea urchin eggs and to purify them. The problem remains, though, that the sea urchin egg is promiscuous: a variety of messengers can trigger the Ca_i wave. Only one of them can be the activating messenger. A minimal proof will require the demonstration that sufficient of the putative messenger is present in a single sperm to provide an activating concentration in the egg cytoplasm, given the diffusion constraints imposed by the narrow cytoplasmic channel between sperm and egg.

ACKNOWLEDGEMENTS

This work was supported by grants from the Science and Engineering Research Council, the Wellcome Trust and the Royal Society. We thank Michael Aitchison for preparing the figures.

REFERENCES

Allen RD, Griffin JL (1958) The time sequence of early events at fertilization in sea urchins. Exptl Cell Res 15:163-173

Baker PF (1972) Transport and metabolism of calcium ions in nerve. Prog Biophys Molec Biol 24:177-223

Brandriff B, Hinegardner RT, Steinhardt RA (1975) Development and life cycle of the parthenogenetically-activated sea urchin embryo. J Exp Zool 192:13-24

Chambers EL, Angeloni SV (1981) Is external calcium required for fertilization of sea urchin eggs by acrosome-reacted sperm? J Cell Biol 91:181a

Chambers EL, McCulloh DH (1990) Excitation, activation and sperm entry in voltage-clamped sea urchin eggs. J Reprod Fert (Suppl 42) (*in the press*)

Ciapa B, Whitaker MJ (1986) Two phases of inositol polyphosphate and diacylglycerol production at fertilization. FEBS Lett 195:347 – 351
Clapper DL, Lee H – C (1986) Inositol trisphosphate induces calcium release from non – mitochondrial stores in sea urchin egg homogenates. J Biol Chem 260:13947 – 13950.
Clapper DL, Walseth TF, Dargie PJ, Lee H – C (1987) Pyridine nucleotide metabolites stimulate calcium release from sea urchin egg microsomes desensitized to inositol trisphosphate. J Biol Chem 262:9561 – 9568
Crossley IB, Whitaker MJ (1989) GDPβS prevents exocytosis in sea urchin eggs at a step subsequent to the fertilization calcium transient. J Physiol 415:91P
Dale B, DeFelice, LJ, Taglietti V (1978) Membrane noise and conductance increase during single spermatozoon – egg interactions. Nature (Lond) 275:217 – 219
Dale B, De Felice LJ, Ehrenstein G (1985) Injection of a soluble sperm fraction into sea urchin eggs triggers the cortical reaction. Experientia 41:1068 – 1070
Davidson EH, Hough – Evans BR, Britten RJ (1982) Molecular biology of the sea urchin embryo. Science 217:17 – 26
Domino SE, Garbers DL (1988) The fucose – sulfate glycoconjugate that induces an acrosome reaction in spermatozoa stimulates inositol 1,4,5 – trisphosphate accumulation. J Biol Chem 263:690 – 695
Eisen A, Kiehardt DP, Wieland SJ, Reynolds GT (1984) Temporal sequence and spatial distribution of early events of fertilization in single sea urchin eggs. J Cell Biol 99:1647 – 1654
Garbers DL (1981) The elevation of cyclic AMP concentrations in flagella – less sea urchin sperm heads. J Biol Chem 256:620 – 624
Guerrero A, Darszon A (1989) Evidence for the activation of two different Ca^{2+} channels during the egg jelly – induced acrosome reaction of sea urchin sperm. J Biol Chem 264:19593 – 19599
Hinckley RE, Wright BD, Lynn JW (1986) Rapid visual detection of sperm – egg fusion using the DNA – specific fluorochrome Hoechst 33342. Devel Biol 118:148 – 154
Iwasa KH, Ehrenstein G, DeFelice LJ, Russell JT (1989) Sea urchin sperm contain enough inositol 1,4,5 – trisphosphate to activate eggs. J Cell Biol 109:128a
Jaffe LF (1980) Calcium explosions as triggers to development. Ann NY Acad Sci 339:86 – 101
Jaffe LF (1983) Sources of calcium in egg activation: a review and a hypothesis. Devel Biol 99:265 – 276
Kopf GS, Garbers DL (1979) A low molecular – weight factor from sea urchin eggs elevates sperm cyclic nucleotide concentrations and respiration rates. J Reprod Fert 57:353 – 361
Kyozuka K, Osania K (1989) Induction of cross – fertilization between sea urchin eggs and starfish sperm by polyethylene glycol treatment. Gamete Res 22:123 – 129
Lau AF, Rayson T, Humphreys TC (1986) Tumor promoters and diacylglycerol activate the Na^+/H^+ antiporter of sea urchin eggs. Exptl Cell Res 166:23 – 30
Longo FJ, Lynn JW, McCulloh DH, Chambers EL (1986) Correlative ultrastructural and electrophysiological studies of sperm – egg interaction in the sea urchin, *Lytechinus variegatus*. Devel Biol 118:155 – 166
Lynn JW, Chambers EL (1984) Voltage – clamp studies of fertilization in sea urchin eggs. I. Effect of clamped membrane potential on sperm entry, activation and development. Devel Biol 102:98 – 109
Lynn JW, McCulloh DH, Chambers EL (1988) Voltage – clamp studies of

fertilization in sea urchin eggs. II. Current patterns in relation to sperm entry, non-entry and activation. Devel Biol 128:305-323

Marty A, Horn R, Zimmerberg J, Tan YP (1989) Delay of the Ca^{2+} mobilization response to muscarinic stimulation. Soc Gen Physiol Ser 44:97-110

McCulloh DH, Chambers EL (1986a) When does the sperm fuse with the egg? J Gen Physiol 88:38a-39a

McCulloh DH, Chambers EL (1986b) Fusion and 'unfusion' of sperm and egg are voltage-dependent in the sea urchin egg *Lytechinus variegatus*. J Cell Biol 103:286a

McCulloh DH, Ivonnet PI, Chambers EL (1989) Blockers of calcium influx promote sperm entry in sea urchin eggs at clamped negative membrane potentials. J Cell Biol 109:126a

Rodbell M (1980) The role of hormone receptors and GTP regulatory proteins in membrane signal transduction. Nature (Lond) 284:17-20

Rose B, Lowenstein WR (1976) Permeability of a cell junction and the local cytoplasmic free ionized calcium concentration: a study with aequorin. J Membr Biol 28:87-119

Shen SS, Burgart LJ (1986) 1,2-diacylglycerols mimic phorbol 12-myristate acetate activation of the sea urchin egg. J Cell Physiol 127:330-340

Swann K (1990) Injection of a cytosolic factor from sperm imitates the membrane potential fertilization response in golden hamster eggs (*abstr*). J Physiol (Lond) (*in the press*)

Swann K, Whitaker MJ (1985) Stimulation of the Na/H exchanger of sea urchin eggs y phorbol ester. Nature (Lond) 314:274-277

Swann K, Whitaker MJ (1986) The part played by inositol trisphosphate and calcium in the propagation of the fertilization wave in sea urchin eggs. J Cell Biol 103:2333-2342

Swann K, Whitaker MJ (1990) Second messengers at fertilization in sea urchin eggs. J Reprod Fert (Suppl 42) (*in the press*)

Swann K, Ciapa B, Whitaker MJ (1987) Cellular messengers and sea urchin egg activation. In: O'Connor D (ed) The molecular biology of invertebrate development. Alan Liss, New York, pp 45-69

Trimmer JS, Schackmann RW, Vacquier VD (1986) Monoclonal antibodies increase intracellular Ca^{2+} in sea urchin spermatozoa. Proc Natnl Acad Sci USA 83:9055-9059

Turner PR, Jaffe LA, Fein A (1986) Regulation of cortical vesicle exocytosis by inositol 1,4,5-trisphosphate and GTP binding protein. J Cell Biol 102:70-76

Turner PR, Jaffe LA, Primakoff P (1987) A cholera toxin-sensitive G-protein stimulates exocytosis in sea urchin eggs. Devel Biol 120:577-583

Whalley T, Whitaker MJ (1988) Guanine nucleotide activation of phosphoinositidase C at fertilization in sea urchin eggs. J Physiol (Lond) 406:126P

Whitaker MJ, Aitchison MJ (1985) Calcium-dependent polyphosphoinositide hydrolysis is associated with exocytosis *in vitro*. FEBS Lett 182:119-124

Whitaker MJ, Irvine RF (1984) Inositol 1,4,5-trisphosphate microinjection activates sea urchin eggs. Nature (Lond) 312:636-639

Whitaker MJ, Steinhardt RA (1982) Ionic regulation of egg activation. Q Rev Biophys 15:593-666

Whitaker MJ, Swann K, Crossley IB (1989) What happens during the latent period at fertilization. In: Nuccitelli R, Cherr GN, Clark WH jun (eds) Mechanisms of egg activation. Plenum, New York, pp 157-171

THE BEHAVIOUR OF SPERM CELLS IN CEREAL WIDE-HYBRIDS

D.A. Laurie and M.D. Bennett[1]
The Cambridge Laboratory, Institute of Plant Science Research, Colney Lane, Norwich, NR4 7UH, U.K. [1]Jodrell Laboratory, Royal Botanic Gardens, Kew, Richmond, Surrey, TW9 3DS, U.K.

Introduction

Cereals, like most other angiosperms, differ from animals in having a life cycle which begins with a double fertilization event. When a pollen grain lands on a suitable receptive stigma it germinates, producing a pollen tube which grows through the stigma, style and ovary and delivers two haploid (1n) sperm cells to the embryo sac. One sperm cell fuses with the egg-cell (also haploid) to produce the diploid (2n) zygote while the second fuses with the two polar nuclei to produce the primary endosperm nucleus (3n). The mature endosperm serves as a food source for the embryo during germination.

In plants, it is possible to hybridize many species by simple crossing procedures (i.e. by placing pollen from one species on the stigma of another), although abnormalities of seed development usually necessitate culturing the immature embryos on nutrient medium. The plants produced from inter-specific or intergeneric crosses are commonly referred to as "wide-hybrids".

An important example is triticale, a hybrid of wheat (usually *Triticum durum*) and rye (*Secale cereale*) which is now grown on over 750 000 hectares world wide (Gregory 1987). Wide-hybrids are, however, most frequently used for transferring genes from one species into the genome of another. For instance, the short arm of chromosome 1B of wheat can be replaced with the short arm of chromosome 1R of rye, and many commercial wheat varieties now carry this 1B/1R translocation. Other examples of the transfer of agronomically useful genes to crop species by sexual hybridization are described in several recent reviews (Goodman *et al.* 1987; Gale and Miller 1987; Lange and Balkema-Boomstra 1988).

NATO ASI Series, Vol. H 45
Mechanism of Fertilization
Edited by B. Dale

Wide-hybrids also occur naturally and are an important factor in plant evolution. Their most obvious role is in the formation of allopolyploid species including the tetraploid macaroni wheats (*T. durum*), the hexaploid bread wheats (*T. aestivum*) and oats (*Avena sativa*).

In plants, most sexual wide-hybrids involve species in the same tribe, but it has recently become clear that some cereal species which are taxonomically much more diverged can be hybridized (Zenkteler and Nitzsche 1984; Laurie and Bennett 1988b; Laurie *et al.* 1989). This paper concentrates on one of these crosses, namely hexaploid wheat x maize (*Zea mays*), a hybrid which links two subfamilies of the grasses.

Cytological studies of hexaploid wheat x maize crosses have shown that zygotes have the expected F_1 combination of 21 wheat and 10 smaller maize chromosomes, and that all maize chromosomes appear to be lost in the first three cell division cycles (Laurie and Bennett 1989a). All embryos with more than four cells contained micronuclei, indicating the elimination of the maize chromosomes, suggesting that karyogamy (nuclear fusion) was essential for embryo development. It is not certain whether this is also the case for endosperm development.

The haploid wheat embryos produced can be recovered as plants by appropriate embryo rescue techniques (Laurie and Bennett 1988a; Comeau *et al.* 1988). This provides a new method of wheat haploid production and also offers the prospect of transferring maize DNA, including transposable elements, to wheat.

Fertilization typically occurs in 20% to 30% of florets when hexaploid wheat is pollinated with maize but, in contrast to "conventional" sexual wide-hybrids, there is usually a low frequency of double fertilization. For example, when the wheat variety Chinese Spring was crossed with the maize variety Seneca 60, 23% of the 343 florets examined had only an embryo, 2% had only an endosperm and only 3% had both an embryo and an endosperm (Laurie and Bennett 1988b). Other genotypes gave similar results.

It was therefore of interest to study early postpollination events in wheat x maize crosses to try and see what factors limited the frequency of fertilization to 30% and why the frequency of embryo formation was about 4-fold higher than the frequency of endosperm formation.

Pollen tube growth, fertilization and karyogamy in hexaploid wheat x maize crosses.

Ovaries from crosses between the wheat variety Chinese Spring and the maize variety Seneca 60 made in a controlled environment cabinet at 20°C were fixed in 3:1 ethanol/ acetic acid at hourly intervals between 3h and 12h after pollination and examined by light microscopy after Feulgen staining as described in Laurie and Bennett (1987). The results of an analysis of 400 ovaries, which are described in detail in Laurie and Bennett (1989b), are summarized in Fig 1 together with data from other work on the frequency of embryo and endosperm development.

In 20% of ovaries no maize pollen tube reached the embryo sac. In the remaining 80%, 68.5% showed the entry of one pollen tube into the embryo sac, 11.1% had two and 0.4% had three. In those cases where one or more pollen tubes had entered the embryo sac, four categories of aberration were identified which limited the frequency of fertilization:

1) In 1.9% of florets the pollen tube had crushed the egg-cell and polar nuclei.

2) In 26.4% of florets no sperm nuclei were visible and in a further 7.3% of florets the sperm nuclei were visible within the tip of the pollen tube. Thus, a total of 35.6% of florets showed entry of a pollen tube into the embryo sac without evidence of the release of the sperm nuclei.

3) In 16.1% of florets one or more sperm nuclei were observed in the central cell cytoplasm. Some may have been moving to the wheat egg-cell or polar nuclei at the time of fixation, but since examples of this category could still be found as late as 48h after pollination, it is likely that some sperm nuclei did not reach appropriate destinations.

4) Some embryo-sacs from material fixed 48h after pollination had one or more sperm nuclei apparently in contact with, or very close to, the polar nuclei. As there was no evidence that the frequency of endosperm formation was higher in

material fixed later than 28h after pollination (the latest time that the first division of the endosperm was seen), it is likely that failure of karyogamy also limited the frequency of endosperm formation.

Discussion

The ability of maize sperm cells to effect embryo and endosperm formation in hexaploid wheat probably reflects a conservation of intercellular and internuclear signals which are important in regulating fertilization and karyogamy. If so, the aberrations which limit the frequency of fertilization may serve to highlight three important stages in these processes.

1) Cessation of pollen tube growth and release of the sperm nuclei from the pollen tube. This may be controlled by signals within the synergid that is penetrated by the pollen tube when it enters the embryo sac.

2) Transport of the sperm nuclei to appropriate destinations within the embryo sac. Under normal circumstances one sperm cell would reach the egg-cell and the other the polar nuclei. Since the sperm cells have no structures that suggest motility, it is reasonable to suggest that they are transported by components of the cytoskeleton which are probably provided exclusively by the female parent. In wheat x maize crosses the sperm cells may remain in the central cell cytoplasm, suggesting that they cannot be recognized by the transport system, or that they have not been delivered to a correct starting point. In other cases both sperm nuclei may be associated with the egg-cell or polar nuclei, showing that the normal partitioning of one sperm to each of the female target cells could fail.

3) Karyogamy. In wheat x maize crosses, karyogamy appeared to occur with high efficiency if a sperm cell reached the egg-cell as no example of a sperm nucleus within the egg-cell was found later than 12h after pollination. However, examples of

Figure 1 (opposite). Hexaploid wheat (variety Chinese Spring) x maize (variety Seneca 60) crosses. a) Egg-cell with maize sperm nucleus (arrowed) adpressed to the nucleus. b) Polar nuclei with maize sperm nucleus (arrowed). Bars represent 10μm. c) Summary of cytological observations. *1 Data from Laurie and Bennett (1988b). *2 Data from Laurie and Bennett (1988a). All figures are percentages of the total number of florets pollinated.

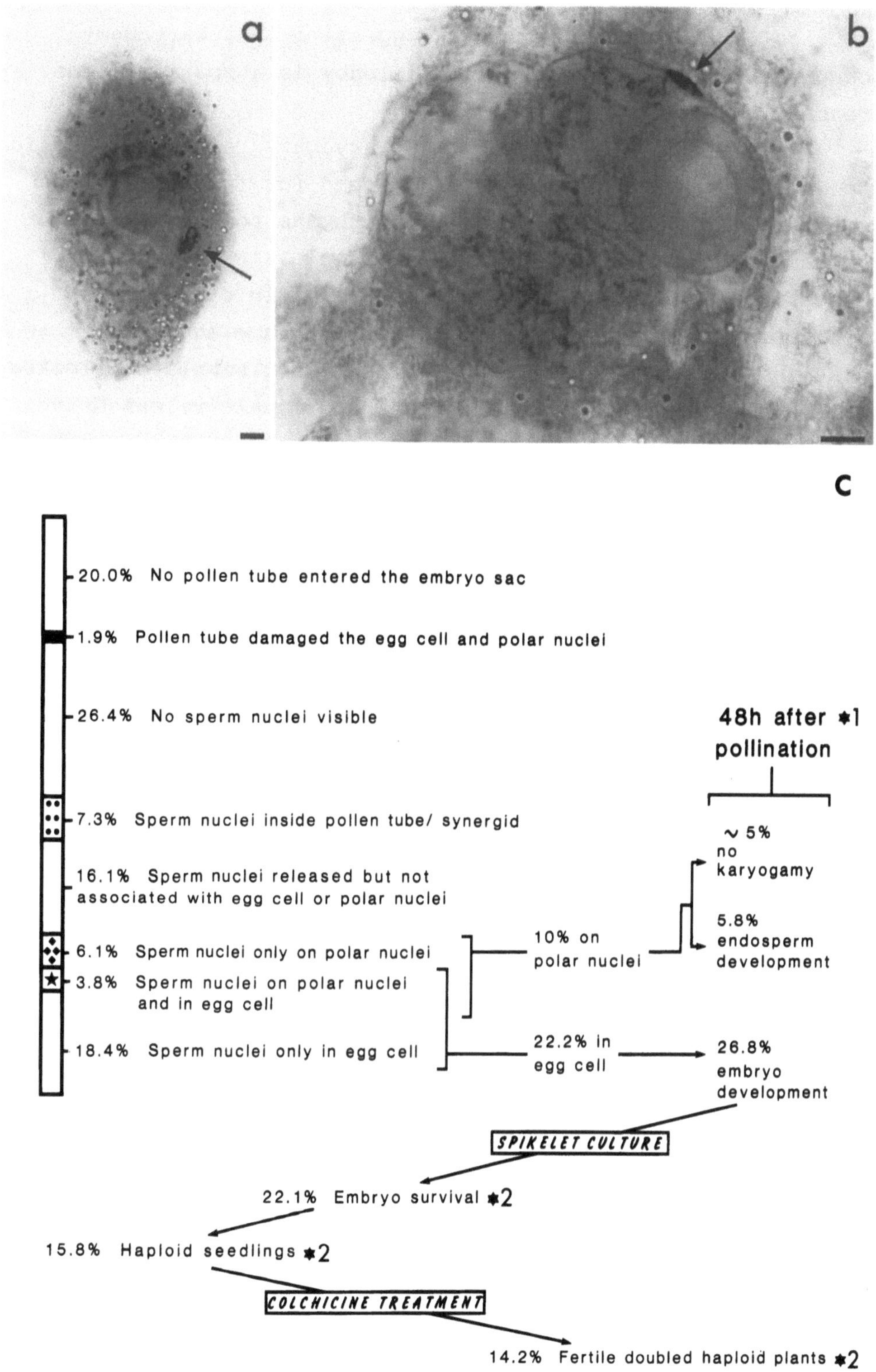
a
b
c
20.0% No pollen tube entered the embryo sac
1.9% Pollen tube damaged the egg cell and polar nuclei
26.4% No sperm nuclei visible
7.3% Sperm nuclei inside pollen tube/ synergid
16.1% Sperm nuclei released but not associated with egg cell or polar nuclei
6.1% Sperm nuclei only on polar nuclei
3.8% Sperm nuclei on polar nuclei and in egg cell
18.4% Sperm nuclei only in egg cell
48h after *1 pollination
~ 5% no karyogamy
5.8% endosperm development
10% on polar nuclei
22.2% in egg cell
26.8% embryo development
SPIKELET CULTURE
22.1% Embryo survival *2
15.8% Haploid seedlings *2
COLCHICINE TREATMENT
14.2% Fertile doubled haploid plants *2

sperm nuclei in association with the polar nuclei were seen as late as 48h after pollination, suggesting that karyogamy occurred with only about 50% efficiency if a sperm nucleus reached the polar nuclei.

Aberrations of transport and karyogamy raise the question of whether one sperm cell is programmed for fusion with the egg-cell and the other for fusion with the polar nuclei *before* release from the pollen tube. For example, in 37 of the 45 cases where only one sperm nucleus was visible within the embryo sac it was usually associated with the egg-cell. Assuming that the second sperm cell was not simply overlooked, this suggests either that the first sperm cell released from the pollen tube tended to reach the egg-cell or that a sperm cell targeted to the egg-cell was released preferentially.

Although the generality of pretargeting remains controversial, there is some evidence to suggest that it may occur in some taxa. McConchie *et al*. (1987) have claimed that the two sperm nuclei in mature unhydrated pollen grains have different morphology, which may indicate differentiation. Evidence for functional differentiation has been reported in maize, where data on the transmission of B-chromosomes suggested that the sperm nucleus carrying the B-chromosome was more likely to fertilize the egg-cell (Roman 1948). In *Plumbago* it has been shown that one sperm nucleus contains plastids and that this sperm has a high probability of fusing with the egg-cell (Russell 1985).

Overall, the results from Chinese Spring wheat x Seneca 60 maize crosses indicate that conservation of intercellular and internuclear signals may be closer for the egg-cell/ sperm cell interaction than for the polar nuclei/ sperm cell interaction.

This result appears to be consistent for this cross but a different effect is seen in crosses between the hexaploid wheat variety Highbury and Seneca 60 maize (Laurie 1989). Here, the developmental age of the floret at the time of pollination had a strong influence on the proportion of florets in which double fertilization, as opposed to the formation of only an embryo, occurred (Fig. 2). This

Figure 2. The effect of developmental age of the floret at the time of pollination on the proportion of single and double fertilization events in hexaploid wheat (variety Highbury) x maize (variety Seneca 60) crosses made at 20°C. Apart from the sample from three days before anthesis, where only four fertilization events were recorded, the percentage of fertilized florets with only an embryo (○) fell on each of the subsequent five days from 84.0% to 23.3% while the percentage showing double fertilization (●) rose from 5.5% to 70.5%. (◇) shows the percentage of florets with only an endosperm.

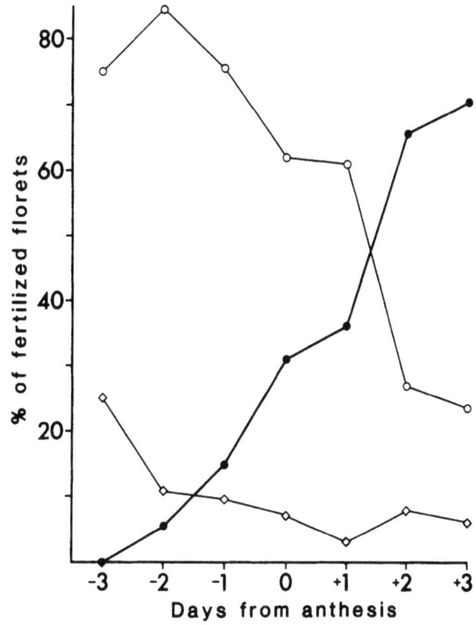

presumably reflects more efficient transport of sperm cells to the polar nuclei and, or, more efficient karyogamy. Such differences in behaviour between genotypes suggest that appropriate crosses might be used to identify genes which are important in regulating the steps of double fertilization.

Other wide-hybrids spanning comparable taxonomic distances.

It should be stressed that hexaploid wheat x maize is not the only combination of taxonomically diverse cereals that can be made. Fertilization has also been found in several others (summarized in Laurie *et al*. 1989) including tetraploid or hexaploid wheat x sorghum (*Sorghum bicolor*) or pearl millet (*Pennisetum glaucum*), barley (*Hordeum vulgare*) x maize and rye (*Secale cereale*) x maize. These crosses all resemble wheat x maize in showing elimination of the chromosomes of the male parent, but they differ in both fertilization frequency and in the ratios of single to double fertilization events (Table 1). Only crosses between hexaploid wheat and maize or sorghum had high frequencies of florets in which only an embryo was present. In hexaploid wheat x pearl millet crosses, most ovaries in which fertilization occurred had both an embryo and an endosperm. This was also the case in crosses of tetraploid wheat, barley or rye with maize. Although these results must be interpreted with

Table 1. Fertilization frequency in cereal wide-hybrids.

Cross	n	embryo only	endo-sperm only	embryo and endo-sperm	Total fert-ilized (%)
a) hexaploid wheat (Chinese Spring)					
x maize	343	80 (80.0)	8 (8.0)	12 (12.0)	29.7[1]
x sorghum	100	57 (82.6)	2 (2.9)	10 (14.5)	69.0[2]
x pearl millet	220	17 (27.0)	4 (6.3)	42 (66.7)	28.6[3]
b) tetraploid wheat (Kubanka)					
x maize	104	8 (28.6)	3 (10.7)	17 (60.7)	26.9[4]
x pearl millet	100	2 (4.2)	6 (12.5)	40 (83.3)	48.0[3]
c) barley (Sultan)					
x maize	100	4 (14.3)	8 (28.6)	16 (57.1)	28.0[1]
d) rye (Petkus Spring)					
x maize	150	4 (14.3)	10 (35.7)	14 (50.0)	18.7[5]

n is the number of florets scored. The figures in brackets are percentages of the total number of florets in which fertilization occurred. The maize, sorghum and pearl millet genotypes used were Seneca 60, S9B and Tift 23BE respectively. Superscripts refer to original sources: 1 Laurie and Bennett (1988a), 2 Laurie and Bennett (1988c), 3 Laurie (1989), 4 O'Donoughue and Bennett (1988), 5 Laurie *et al.* (1989).

caution in view of the results illustrated in Fig. 2, the results were consistent over experiments and probably represent genetically determined differences in behaviour. If so, this suggests that sperm cell/ polar nuclei interactions are particularly error prone in crosses involving hexaploid wheat.

In conclusion, it is likely that the range of sexual wide-hybrids that can be produced is greater than has previously been believed and that these, together with the crosses described above, will provide valuable opportunities for study in many areas of plant biology, including the investigation of interactions between sperm cells and the components of the embryo sac. This may provide useful insights into the processes of fertilization and karyogamy which have been described cytologically (e.g. Kiesselbach 1949; Diboll 1968; Bennett *et al.* 1973, 1975; Mogensen 1982; Van Lammeren 1986; Huber and Grabe 1989) but which otherwise remain poorly understood. These investigations would be greatly assisted if techniques for isolating sperm cells (Dupuis *et al.* 1987; Cass and Fabi 1988) could be combined with methods for isolating embryo sacs (Zhou and Yang 1985) to produce *in vitro* fertilization systems. Such systems might also be useful for transformation (Dumas *et al.* 1984) and for the production of novel wide-hybrids.

Acknowledgements: This work was funded by the United Kingdom Overseas Development Administration, project R3797.

References

Bennett MD, Smith JB, Barclay IR (1975) Early seed development in the Triticeae. Phil Trans R Soc Lond B 272: 199-227

Bennett MD, Rao MK, Smith JB, Bayliss MW (1973) Cell development in the anther, the ovule, and the young seed of *Triticum aestivum* L. var. Chinese Spring. Phil Trans R Soc Lond B 266: 38-81

Cass DD, Fabi GC (1988) Structure and properties of sperm cells isolated from the pollen of *Zea mays*. Can J Bot 66: 819-825

Comeau A, Plourde A, St. Pierre CA, Nadeau P (1988) Production of doubled haploid wheat lines by wheat x maize hybridization (Abstract). Genome 30: Supplement 1, p 482

Diboll AG (1968) Fine structural development of the megagametophyte of *Zea mays* following fertilization. Amer J Bot 55: 787-806

Dumas C, Know RB, McConchie CA, Russell SD (1984) Emerging physiological concepts in fertilization. What's New Plant Physiol 15: 17-20

Dupuis I, Roeckel P, Matthys-Rochon E, Dumas C (1987) Procedure to isolate viable sperm cells from corn (*Zea mays* L.) pollen grains. Plant Physiol 85: 876-878

Gale MD, Miller TE (1987) The introduction of alien genetic variation into wheat. In: Lupton FGH (ed) Wheat Breeding: its scientific basis. Chapman and Hall, London, New York, pp 173-210

Goodman RM, Hauptli H, Crossway A, Knauf VC (1987) Gene transfer in crop improvement. Science 236: 48-54

Gregory RS (1987) Triticale breeding. In: Lupton FGH (ed) Wheat Breeding: its scientific basis. Chapman and Hall, London, New York, pp 269-286

Huber AG and Grabe DF (1989) Double fertilization in wheat (*Triticum aestivum*): terminology and photographs. Seed Sci & Technol 17: 27-39

Kiesselbach TA (1949) The structure and reproduction of corn. Univ Nebraska Coll Agr Agric Expt Stat Res Bull No. 161

Lange W, Balkema-Boomstra AG (1988) The use of wild species in breeding barley and wheat, with special reference to the progenitors of the cultivated species. In: Jorna ML, Slootmaker LAJ (eds) Cereal breeding related to integrated cereal production. Pudoc, Wageningen, pp 157-178

Laurie DA (1989a) Factors affecting fertilization frequency in crosses of *Triticum aestivum* cv. Highbury x *Zea mays* cv. Seneca 60. Plant Breeding (in press)

Laurie DA (1989b) The frequency of fertilization in wheat x pearl millet crosses. Genome (in press)

Laurie DA, Bennett MD (1987) The effect of the crossability loci *Kr1* and *Kr2* on fertilization frequency in hexaploid wheat x maize crosses. Theor Appl Genet 73: 403-409

Laurie DA, Bennett MD (1988a) The production of haploid wheat plants from wheat x maize crosses. Theor Appl Genet 76: 393-397

Laurie DA and Bennett MD (1988b) Chromosome behaviour in wheat x maize, wheat x sorghum and barley x maize crosses. In: Brandham PE (ed) Kew Chromosome Conference III. Her Majesty's Stationary Office, London, pp 167-177

Laurie DA and Bennett MD (1988c) Cytological evidence for fertilization in hexaploid wheat x sorghum crosses. Plant Breeding 100: 73-82

Laurie DA, Bennett MD (1989a) The timing of chromosome elimination in hexaploid wheat x maize crosses. Genome (in press)

Laurie DA, Bennett MD (1989b) Early post-pollination events in hexaploid wheat x maize crosses. Sex Plant Reprod (in press)

Laurie DA, O'Donoughue LS, Bennett MD (1989) Wheat x maize and other wide sexual hybrids: their potential for genetic manipulation and crop improvement. In: Gustafson JP (ed) Gene manipulation in plant improvement II, XIX Stadler Genetics Symposium. Plenum Press, New York (in press)

McConchie CA, Hough T, Knox RB (1987) Ultrastructural analysis of the sperm cells of mature pollen of maize, *Zea mays*. Protoplasma 139: 9-19

Mogensen HL (1982) Double fertilization in barley and the cytological explanation for haploid embryo formation, embryoless caryopses, and ovule abortion. Carlsberg Res Commun 47: 313-354

O'Donoughue LS, Bennett MD (1988) Wide hybridization between relatives of bread wheat and maize. In: Miller TE, Koebner RMD (eds) Proc 7th Int Wheat Genet Symp Vol I. Inst Plant Science Res, pp 397-402

Roman H (1948) Directed fertilization in maize. Proc Natl Acad Sci USA 34: 36-42

Russell SD (1985) Preferential fertilization in *Plumbago*: Ultrastructural evidence for gamete-level recognition in an angiosperm. Proc Natl Acad Sci USA 82: 6129-6132

Van Lammeren AAM (1986) A comparative ultrastructural study of the megagametophytes in two strains of *Zea mays* L. before and after fertilization. Agric Univ Wageningen Papers 86-1

Zenkteler M, Nitzsche W (1984) Wide hybridization experiments in cereals. Theor Appl Genet 68: 311-315

Zhou C, Yang HY (1985) Observations on enzymatically isolated, living and fixed embryo sacs in several angiosperm species. Planta 165: 225-231

CALCIUM SIGNALS DURING FERTILIZATION AND OOPLASMIC SEGREGATION IN THE ASCIDIAN EGG

Johanna Speksnijder[1], Christian Sardet[2], and Lionel Jaffe
Marine Biological Laboratory
Woods Hole, MA 02543
USA

In ascidians, fertilization is followed by a dramatic reorganization of the egg cytoplasm. This process of ooplasmic segregation is of vital importance for defining the final distribution pattern of determinants as well as the bilateral symmetry of the embryo. Several lines of evidence suggest that (local) calcium fluxes may play an essential role in these segregation movements. We therefore 1) measured the changes in free cytosolic calcium at fertilization in single eggs of the ascidian <u>Phallusia mammillata</u>, 2) studied the spatial characteristics of these changes, and 3) established their possible relationship with the process of ooplasmic segregation.

Organization Of The Unfertilized Ascidian Egg

The mature unfertilized egg of ascidians displays a distinct animal-vegetal polarity. In most species, the animal pole can be recognized in the living egg by the presence of the ectoplasmic cap (Bates & Jeffery, 1987a), the polar spot (Dalcq & Vandebroek, 1937), or a clear zone surrounding the meiotic spindle (Sardet et al., 1989). This polarity is further expressed in the distribution of the subcortical mitochondria-rich

---

[1] Present address: Department of Experimental Zoology, University of Utrecht, Padualaan 8, 3584 CH Utrecht, The Netherlands.
[2] URA Biologie Cellulaire Marine CNRS Paris VI, Station Zoologique, 06230 Villefranche-sur-mer, France

NATO ASI Series, Vol. H 45
Mechanism of Fertilization
Edited by B. Dale

myoplasm, which is excluded from the animal most part of the egg (Reverberi, 1956; Mancuso, 1963; Gualtieri & Sardet, 1989). In some species, such as Styela and Boltenia, the myoplasm contains brightly coloured pigment granules, whereas in others, such pigment is absent yet the myoplasm can be made visible due to its autofluorescence (Deno, 1987) or by labelling with the mitochondria-specific dye $DiOC_2$ (Zalokar & Sardet, 1984; Sardet et al., 1989). Finally, the egg cortex contains a network of actin-filaments (Sawada & Osanai, 1981, 1985; Jeffery & Meier, 1983; Sardet et al., 1988), which is more sparse or absent in the animal pole region (Sawada, 1983, 1986, 1988).

A new aspect of polarity was recently found in the point of sperm entry (Speksnijder et al., 1989b). Although Conklin had originally described sperm entry to occur near the vegetal pole (Conklin, 1905), it subsequently became clear that both animal and vegetal fragments of various ascidian eggs can be fertilized (Ortolani, 1958; Talevi & Dale, 1986; Bates & Jeffery, 1987b). Quantitative analysis of the sperm entry point in Phallusia has shown that in fact, the animal hemisphere is the preferred area of entry (Fig. 1a). This preference for animal entry is abolished by pretreatment with cytochalasin D (Fig. 1b), which suggests that an intact actin filament network is necessary to maintain the polarity of sperm entry. This network may act to control the distribution of components involved in sperm binding, fusion or entry (Speksnijder et al., 1989b).

Despite the above described polarity, any half of the unfertilized egg can develop into a normal larva (Reverberi & Ortolani, 1962; Bates & Jeffery, 1987b). Thus relatively little developmental pattern exists in the unfertilized ascidian egg. However, after fertilization, such a pattern is formed with remarkable speed during the process of ooplasmic segregation.

The Establishment Of Developmental Pattern Following Fertilization In The Ascidian Egg

Well before first cleavage, the future embryonic axes of the ascidian embryo become apparent during a series of cytoplasmic movements known as ooplasmic segregation (Conklin, 1905; Whittaker, 1979; Jeffery, 1984). Three to five visible plasms become segregated into different regions of the egg cell, and are subsequently partioned unequally during the early cleavages. As a result, these plasms become localized in different cells of

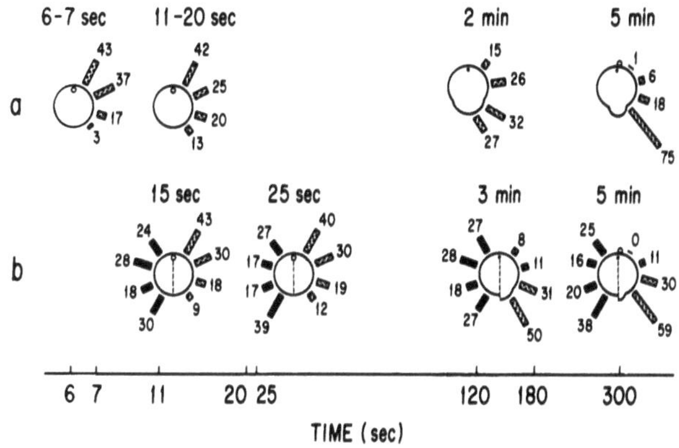

Fig. 1. Distribution of the angles between the sperm nucleus and the female chromosomes marking the animal pole (circle at top) at four times after insemination. The bars indicate frequencies of angles (expressed as percentages) between 0-45, 45-90, 90-135, and 135-180°. (a) Normal fertilization. (b) Effects of cytochalasin D (left side, black bars); controls are indicated at right side with stippled bars. (From Speksnijder et al.,1989b).

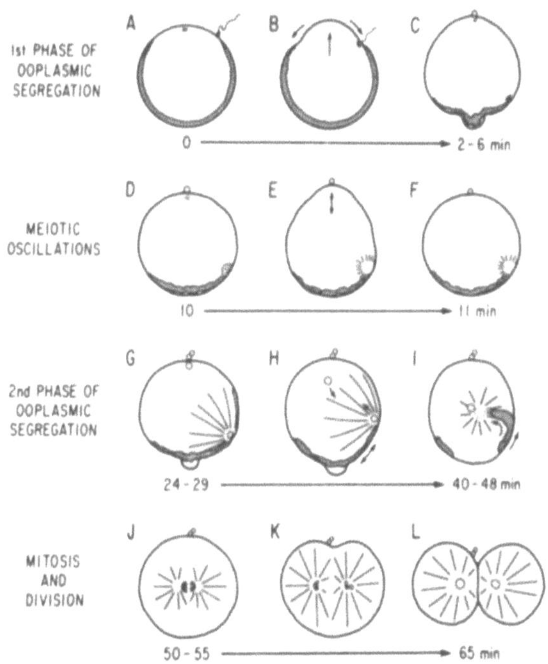

Fig. 2. See legend on next page.

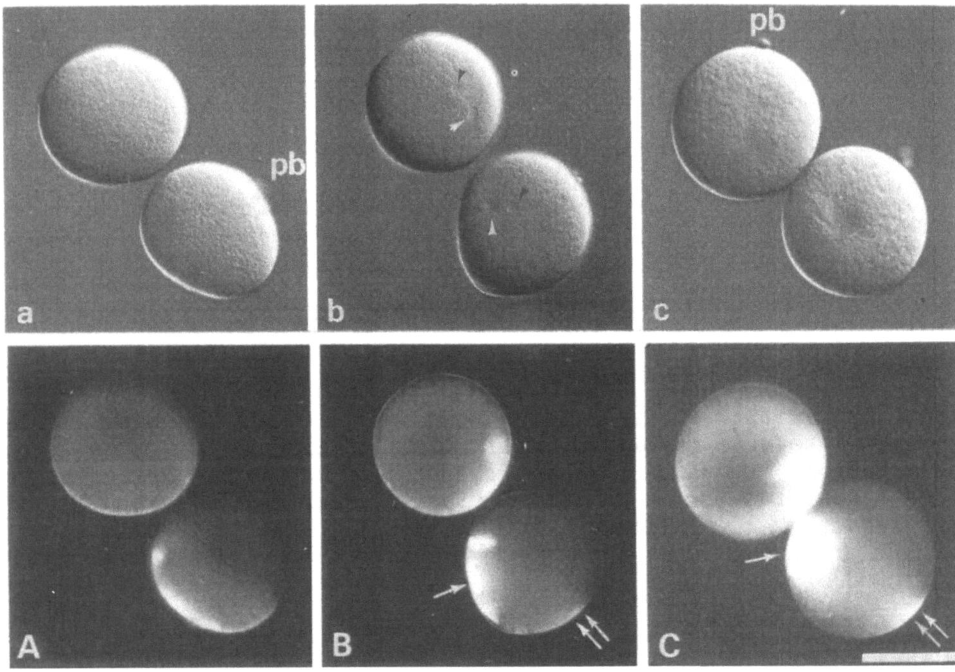

Fig. 3. The second phase of ooplasmic segregation. Two eggs are followed in differential interference contrast (a-b) and epifluorescence after staining of the mitochondria with $diOC_2$ (A-B). (A,a) The second polar body (pb) is just extruded at 20 min after fertilization and the subcortical mitochondria-rich myoplasm is still located in the vegetal hemisphere. (B,b) At 26 min, the bulk of the myoplasm (single arrow) has moved with the sperm aster (white arrowhead) toward the equator. (C,c) The male and female pronuclei have met in the centre of the egg and the myoplasm has split in two parts; the bulk is situated at the future posterior pole, whereas a small part is left behind at the anterior side (double arrows). Bar = 100 µm.

Fig. 2. Fertilization and ooplasmic segregation in the ascidian egg. During the first phase of ooplasmic segregation, the myoplasm and sperm nucleus are moved to a position more or less opposite the polar body marking the animal pole (A-C). Between formation of the first and second polar bodies, the egg surface undergoes periodic contraction cycles (D-F). After the formation of the second polar body, a second vegetal lobe is usually formed, and the bulk of the myoplasm moves with the sperm aster toward the future posterior pole. However, a small portion of the myoplasm remains in a subcortical anterior position (G-I). Finally, large mitotic asters are formed and the egg cleaves (J-L), dividing the myoplasm equally between the blastomeres (not shown). (From Sardet et al., 1989).

the 64-cell stage embryo. The most obvious of these plasms is the mitochondria-rich myoplasm. Conklin's classical account describes two distinct phases of myoplasmic segregation: during the first phase, the myoplasm concentrates in the vegetal pole region and, during the second phase, it shifts from the vegetal pole to the equatorial region of the zygote, where it marks the future posterior pole of the embryo (see Fig. 2). The myoplasm is subsequently partitioned in the blastomeres that give rise to muscle cells.

Of great embryological relevance is the observation that not all of the myoplasmic mass moves up to the equator during the second movement, but that the most anterior part gets separated from the bulk and remains located in the anterior-vegetal region of the zygote (Fig. 2,3). This region ends up in the A4.1 blastomeres of the 8-cell stage, which would explain earlier observations that descendants of blastomeres other than the B4.1 blastomeres (which receive the bulk of the myoplasm) participate in the larval muscle cell lineage (Nishida & Satoh, 1983, 1985; Nishida, 1987; Zalokar & Sardet, 1984; Meedel et al., 1987).

In recent years, the mechanisms of ooplasmic segregation have been studied in several laboratories, and it has become clear that the first phase is accompanied by a wave of cortical contraction, which sweeps from the animal to the vegetal pole in about 2 minutes at a velocity of 1.4 µm/sec (Sardet et al., 1989). This contraction wave is known to move a large complex of (sub)cortical and surface components, which includes the mitochondria-rich myoplasm, cytoskeletal structures, the nucleus of the fertilizing sperm, as well as test cells, supernumery sperm and various artificially attached surface particles (Conklin, 1905; Ortolani, 1955; Sawada & Osanai, 1981, 1985; Jeffery & Meier, 1983; Sardet et al., 1989; Speksnijder et al., 1989b). The driving force for this movement is provided by the actin filament network, as demonstrated by the inhibition of segregation and contraction with cytochalasin (Reverberi, 1975; Sawada & Osanai, 1981; Zalokar, 1984; Sawada & Schatten ,1989; Speksnijder et al., 1989b).

The second phase of ooplasmic segregation starts shortly after the formation of the second polar body, and it is during this period that the final distribution pattern of the cytoplasmic domains is established. The movement of the myoplasm to the equator and future posterior side of the embryo is closely related to the movement of the sperm nucleus and aster in the same direction, and is insensitive to cytochalasin (Sawada & Schatten,

1989). In addition, its speed is much slower than that of the contraction wave (0.1-0.4 µm/sec versus 1.4 µm/sec; Sardet et al., 1989), which suggests that motile forces other than those based on actin filaments are operative in the second phase. Microtubules are the most likely candidate for providing this force, since they are abundant and closely associated with the myoplasm, and their disruption with colchicin effectively blocks the second movement (Sawada & Schatten, 1988, 1989). This second phase of ooplasmic segregation resembles in many ways the shift of the internal cytoplasm with respect of the egg cortex that determines the dorsal-ventral axis in frog eggs (Elinson & Rowning, 1988).

From the above it is obvious that a complex series of events with important developmental consequences occur after fertilization in the ascidian egg. Many investigators have been intrigued by the factors and processes that trigger these events and, more importantly, determine their spatial characteristics. A first indication for a role of calcium fluxes came from studies using calcium ionophores, which clearly showed that treatment with these compounds, even in the absence of external calcium, causes egg activation as assessed by the occurrence of the cortical contraction wave, formation of the myoplasmic cap, or reinitiation of meiosis (Steinhardt et al., 1974; Bevan et al., 1977; Sawada & Osanai, 1981; Jeffery, 1982; Dale, 1988). This would fit the notion that a rise in cytosolic free calcium is the universal trigger for egg activation (Jaffe, 1985). Additional support for this notion was derived from the observation that microinjection of an unfertilized egg with the calcium-releasing second messenger 1,4,5-$InsP_3$ results in activation (Dale, 1988). However, the most interesting observation is that local application of calcium ionophore results in the formation of the myoplasmic cap toward the high ionophore end (Jeffery, 1982), which suggests a role of localized calcium fluxes in determining the direction of the first segregation phase. Therefore it seemed important to directly monitor the changes in free calcium that occur during fertilization and ooplasmic segregation and determine whether such changes are related, both in time and space, with the various movements in the fertilized egg. This paper describes some of the results we obtained using this approach.

The Ascidian Egg Is Activated By A Pulse Of Elevated Free Calcium

The levels of free calcium before, during, and after fertilization in the egg of *Phallusia mammillata* were monitored following injection of the calcium-specific photoprotein aequorin (Blinks, 1982). Before fertilization, low levels of luminescence are measured, from which a resting level of free calcium of about 100 nM was calculated (Speksnijder et al., 1989a). After the addition of sperm, an enormous increase in the luminescence count rates is observed (Fig. 4), which coincides with fertilization of the egg. Within 40 to 50 seconds, peak levels of luminescence of 2,000-6,000 counts/sec are reached, after which the signal gradually drops to background levels. The total duration time of the fertilization transient is 2 to 3 minutes at 21° C (mean = 2.8; n = 10), and levels of about 7 µM free calcium are reached during peak luminescence. A similar calcium pulse was found during fertilization in *Ciona intestinalis*, although even higher peak levels of calcium may be reached in this species (about 10 µM; Speksnijder et al., 1989a; Brownlee & Dale, 1989).

Surprisingly, intracellular levels of free calcium continue to oscillate after the fertilization transient has ended: a distinct series of 12 to 25 smaller postactivation pulses follows the large pulse. The peak levels of calcium reached during these pulses is about 2 to 4 µM. Although the total number of pulses is somewhat variable, the time period during which they occur is very predictable: they start within one minute after the end of the fertilization transient, and they stop shortly before the formation of the second polar body at about 22-25 minutes (Table I). Thus they coincide with the completion of meiosis.

The lectin wheat germ agglutinin (WGA), which is known to activate ascidian eggs as demonstrated by the occurrence of the cortical contraction wave and the formation of polar bodies (Zalokar, 1980), was found to induce a similar series of calcium transients. This indicates that the lectin probably activates the egg by inducing an increase in intracellular calcium.

The source of the calcium necessary for the increase in cytosolic calcium was investigated by fertilizing eggs in the absence of external calcium. It was found that this treatment did not affect development (provided that calcium was returned to the medium just before the onset of cleavage), nor the occurrence, timing and magnitude of the calcium transients. This suggests that most (if not all) of the calcium is derived from internal stores. Such stores might be provided by the extensive endoplasmatic reticulum network that exists in the Phallusia egg (Sardet et al., 1988), and which is known to contain most of the sequestered calcium (Gualtieri & Sardet, 1989).

Thus, from the above it can be concluded that transient rises in cytosolic calcium do indeed occur during natural activation via sperm as well as during artificial activation with lectins. However, in order to understand the potential role of these transients in determining the direction of ooplasmic segregation, it is clear that their spatial characteristics need to be studied.

The Pulse Of Calcium That Activates The Ascidian Egg Is A Wave

Using an imaging photon detector (Whitaker, 1985), we found that the activation pulse is in fact a wave, which starts at a point at the egg surface and subsequently fills the entire egg at a peak velocity of 8-9 µm/sec (Fig. 5). This wave starts in the animal hemisphere in about 80% of the eggs, which suggests that it is probably initiated at the point of sperm entry (Fig. 1). Since the point of sperm entry is difficult to identify directly in the living egg, we confirmed this point by fertilizing eggs preincubated with cytochalasin so as to block the cortical contraction wave which carries the sperm nucleus in vegetal direction following entry (Fig. 1b). The growing sperm aster, which is first visible about 10 minutes after fertilization, is thus expected to be located at or close by the point of entry after such treatment. Imaging of the calcium-dependent luminescence following fertilization of such cytochalasin-treated eggs revealed that the calcium wave always starts at the site where the sperm aster is subsequently formed (Speksnijder et al., in preparation), which indicates that the calcium wave is in fact initiated by the sperm at its entry point, which is usually in the animal hemisphere. Similarly, the activation wave of calcium in Ciona also starts in the animal hemisphere (Brownlee & Dale, 1989; Speksnijder et al., unpublished observations). The next step was to determine how this calcium wave relates to the contraction wave that accompanies the first phase of ooplasmic segregation.

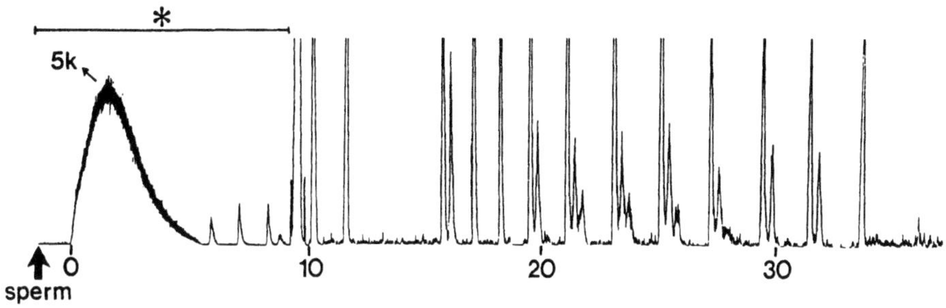

Fig. 4. Photon emission rates from an aequorin-injected Phallusia mammillata egg following insemination at 18°C. Trace with * : The 5,000 cps fertilization pulse as well as three smaller postactivation pulses. Trace after * : The remainder of the postactivation pulses shown at about 15-fold increased sensitivity. The second polar body was formed shortly after the last postactivation pulse.

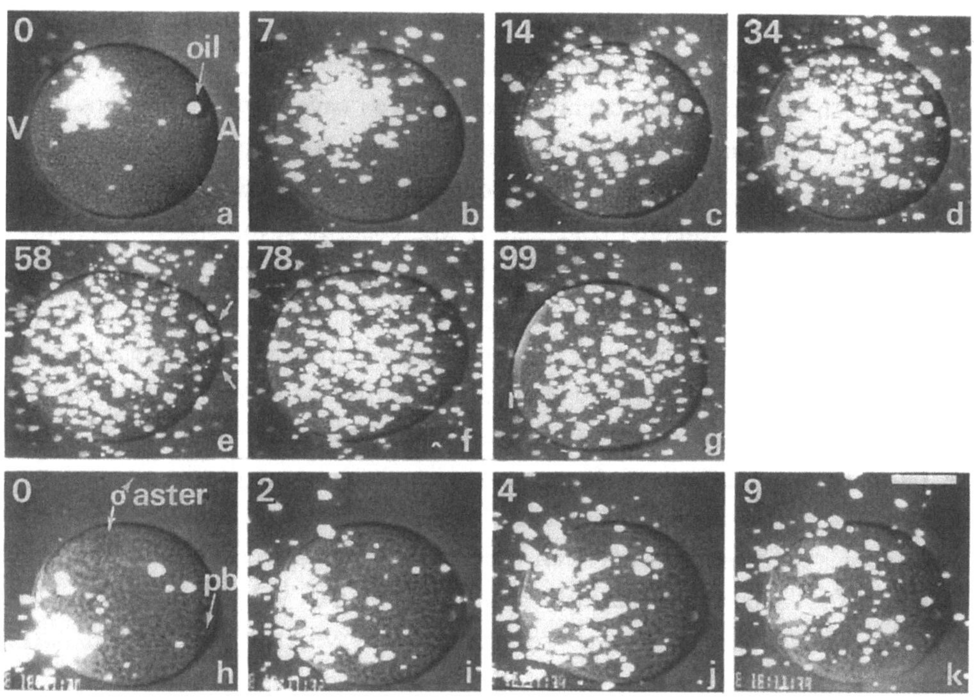

Fig. 5. Overlay of the calcium-dependent luminescent image and the differential interference contrast image of an aequorin-injected Phallusia egg. (a)-(g). The activation wave of calcium, which in this example starts just below the equator (A, animal pole; B, vegetal pole). The wave fills the egg in about 30-40 seconds, and triggers the wave of cortical contraction, which is first visible as a bulging (arrows) near the animal pole (d,e). While the contraction wave travels toward the vegetal pole, the calcium level remains elevated in the entire egg. (h)-(k) A postactivation wave that starts opposite the polar body marking the animal pole, and travels at about twice the speed of the activation wave. Times (in seconds) are indicated at the top of each picture. Bar = 50 µm.

Relationship Between The Activation Wave Of Calcium And The Contraction Wave

To elucidate this point we used an imaging set-up that allowed simultaneous observation of the calcium-dependent luminescence as well as the morphology of the aequorin-injected Phallusia egg (Speksnijder et al., in preparation). We found that there is a substantial delay between the onset of the calcium wave and the contraction wave: the latter does not start until the calcium wave has filled the entire egg, i.e. at about 30-40 seconds after its start (Fig. 5). A similar delay has been observed in the egg of Ciona (Brownlee & Dale, 1989). This probably indicates that substantial rearrangement or reassembly of the cortical contractile machinery takes place before contraction actually occurs.

The calcium wave starts at the point of sperm entry, and thus usually starts in the animal hemisphere, but it can also start below the equator (Fig. 5). In contrast, the cortical contraction wave always starts near the animal pole, and moves toward the vegetal one where the myoplasm becomes concentrated. Thus it appears that the location of the myoplasmic cap is independent of the starting position of the calcium wave, and is determined mainly by the animal-vegetal axis of the egg. This would fit the model originally proposed by Sawada (1983,1988), in which the animal-vegetal polarity in the organization of the cortical actin network determines the direction of myoplasmic segregation. A determination according to this axis is also suggested by recent experiments of Bates & Jeffery (1988), which indicate that a gradient of factors with its high point at the vegetal pole directs the first segregation phase.

However, we recently made some observations that suggest that the sperm entry point and the resulting calcium wave may well affect the final position of the myoplasmic cap relative to the vegetal pole in Phallusia. Careful examination of eggs that had completed the first segregation phase revealed that the position of the contraction pole (the end point of the contraction wave) and thus of the myoplasmic cap is rarely directly opposite the polar body that marks the animal pole. In fact, in 12/13 eggs that were oriented properly for an analysis of this phenomenon, we found that the contraction pole is 15-50° away from the antipode to the polar body (Fig. 6). In retrospect, such an asymmetry had been described previously in the egg of Ascidiella (Dalcq, 1938). This asymmetry could already be present in the unfertilized egg, as was claimed for Ascidiella

(Dalcq, 1938), and become subsequently reinforced following fertilization. However, we have not found evidence for such asymmetry in the unfertilized egg of Phallusia. Therefore we would suggest that the sperm entry point and resulting calcium wave are responsible for tilting the direction of segregation away from the vegetal pole.

In order to explain how sperm entry acts to tilt the direction of segregation, we see the following two possibilities. The first is based on observations on activation in the medaka fish egg. When this egg is prick-activated at its equator, the resulting exocytotic wave proceeds more quickly near the animal pole than near the vegetal one, with the result that it ends at a point halfway between the antipode and the vegetal pole (Yamamoto, 1961). More recently Yoshimoto et al. (1986) have confirmed these results by activating a medaka egg at its equator by microinjection of calcium and observing the resulting calcium wave directly via aequorin. If a similar reduction of calcium wave speed occurs in the egg of Phallusia, then the end point of the calcium wave would be located in the vegetal hemisphere even if the calcium wave is initiated near the equator. Therefore we suggest that the contraction pole forms at the terminus of the activating wave and the reason it would form there is because unattenuated calcium waves are thought to leave a residue of high calcium. Even though this slowing down of the calcium wave in the vegetal half and the proposed residue are not obvious in the Phallusia egg - which could be due to its much smaller size (140 µm) as compared to the medaka egg (1.2 mm) - this model is strongly supported by the observation of Jeffery (1982), that if calcium ionophore gradients are applied to ascidian (Boltenia) eggs, the myoplasm strongly tends to aggregate toward the high ionophore end. So it seems that the contraction pole can be induced to form in all or most regions of the egg if cytosolic calcium is raised there, which in turn suggests that indeed a rise in cytosolic calcium naturally occurs in the forming contraction pole and is a major factor in determining its location.

The alternative possibility is that the asymmetric position of the contraction pole results from an asymmetric contraction of the cortical actin network, which is determined by the starting position of the calcium wave. The actin network has been described as a basket with its opening toward the animal pole (Sawada, 1983, 1986,1988), a situation also found to be true for Phallusia (Sardet, unpublished observations). Contraction of this basket results in its movement toward the vegetal pole, and as a result, in the bulging of the centrally located cytoplasm toward the animal pole. Assuming that the contraction is triggered by an elevation of free calcium, then it would start first at the side of the egg where the sperm enters and where the calcium wave is initiated, and last at the side opposite the sperm entry point. In those eggs in which sperm entry does not occur exactly on either the animal or vegetal pole, the contraction of the actin network would thus be asymmetric, and the myoplasm would be segregated to a site near but not right at the vegetal pole. The location of the sperm on the opposite side of the contraction axis relative to the animal pole fits this model (Fig. 6). In addition, one would expect the contraction pole to be furthest away from the vegetal pole when sperm entry occurs near the equator. We generally found this to be true in the eggs studied. Thus from our observations we infer that the sperm entry point and resulting calcium wave somehow tilt the direction of segregation and thus help determine the final position of the myoplasmic cap. The relevance of this phenomenon for early embryonic development should be sought in the establishment of a second axis of developmental asymmetry by the sperm and resulting calcium wave. This axis might well correspond to the dorsal/ventral axis, since recent work has demonstrated that axial determinants are co-segregated with the myoplasm during the first phase of ooplasmic segregation (Bates & Jeffery, 1987b). These factors determine the future site of gastrulation and thus the dorsal side of the embryo. This asymmetry is subsequently reinforced during the second phase of ooplasmic segregation, when the larger part of the myoplasm becomes located at the future posterior side of the embryo.

The positional clues for this second part of the ooplasmic movements are much less understood. We investigated whether calcium fluxes may play a role also in this second part, and as a first step we established the spatial characteristics of the postactivation pulses of calcium that occur between the first and second phase of ooplasmic segregation.

The Periodic Postactivation Pulses Also Spread Like Waves

The 12-25 postactivation pulses of calcium, which start within 8-24 seconds after the fertilization pulse has ended (Table I, Fig. 4), also travel like waves. They spread across the egg with velocities of 13 to 24 µm/sec (mean = 18; n = 2), which is twice that of the activation wave. In most eggs, the first few postactivation waves start rather diffusely

Table 1. Timing of the early events following fertilization in the egg of Phallusia mammillata at 21° C.
Indicated are times in minutes as mean ± sem (n).

Start calcium wave	0
Calcium wave has crossed egg	0.5 ± 0.1 (3)
Start contraction & first phase segregation	0.6 ± 0.3 (6)
Contraction wave has crossed egg	2.6 ± 0.1 (4)
End fertilization pulse of calcium	2.8 ± 0.1 (10)
Start first post fertilization pulse	3.1 ± 0.2 (9)
Formation first polar body	6.0 ± 0.1 (10)
Start final post fertilization pulse	23.0 ± 0.9 (9)
Formation second polar body	23.5 ± 0.5 (11)
Start second phase segregation	23.5 ± 1.2 (4)
End second phase segregation	42.0 ± 1.1 (4)
First cleavage	54.5 ± 1.2 (10)

somewhere in the animal hemisphere or the equatorial zone. The later ones almost all start in a very small spot somewhere in the region that includes the vegetal pole and the nearby contraction pole, where at that time the myoplasm is concentrated (Fig. 5). These later waves are often followed by a contraction of the egg surface (Sardet et al., 1989). In addition,in about half of the eggs observed, 1 to 4 "out of phase" pulses are present during the last part of the pulse train (Fig. 4). These pulses start within 30 seconds (instead of the usual 60-100 seconds) after the previous one (Speksnijder et al., 1989a). Such an out-of-phase pulse behaves like an echo, i.e. its starting point is usually in the animal hemisphere and thus more or less opposite the previous pulse (not shown). The only previous report of postactivation waves of calcium concerned aequorin-loaded zona-

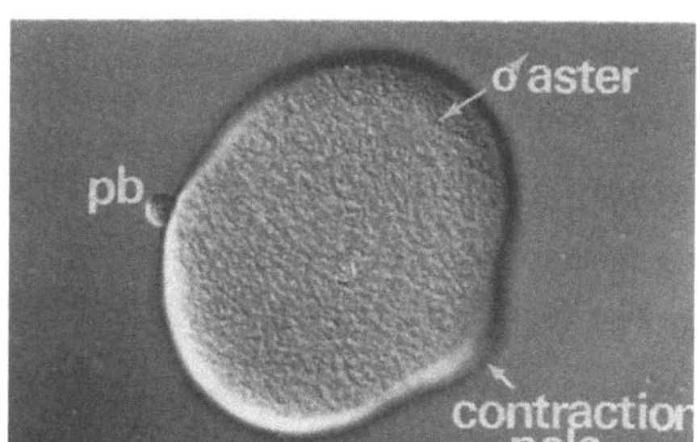

Fig. 6. Egg of Phallusia about 10 minutes following fertilization showing its contraction pole around which the myoplasm is centered well away from the vegetal pole. The sperm aster is located at the opposite side of the contraction axis relative to the polar body marking the animal pole.

free hamster eggs (Miyazaki et al.,1986). One or two such waves were seen to follow fertilization, and as in Phallusia, they began near the sperm and travelled at twice the speed of the activation wave. However, subsequent pulses were not seen as waves. In assessing the developmental significance of these postactivation waves in Phallusia, it seems important to know if they move steadily across the egg so as to add up over time to a uniform rise in free calcium or whether they are non-uniform so as to add up to a gradient. Accordingly, we integrated the later vegetally started waves and found that they usually add up to a substantial net gradient with its peak toward the vegetal end where they begin (not shown). This finding can be compared with earlier indications of high calcium levels at the vegetal end of medaka fish eggs (Jaffe, 1986).

At present, the developmental significance of these postactivation waves is unclear, and further research is required to assess their meaning. Among the early developmental events that they might affect are polar body formation, the formation or activation of vegetal factors needed for gastrulation (Bates & Jeffery, 1987b), and movement of most of the myoplasm to the future posterior pole during the second phase of ooplasmic segregation (Sardet et al., 1989).

Conclusions

As in many other organisms, fertilization in the ascidians Phallusia mammillata and Ciona intestinalis is accompanied by a large transient increase in the concentration of free cytosolic calcium (Speksnijder et al., 1986ab, 1989a; Brownlee & Dale, 1989). In the egg of Phallusia, this activation pulse starts where the sperm enters, which is usually in the animal hemisphere (Speksnijder et al, 1989b), and travels across the egg at a peak velocity of 8-9 µm/sec.

An important question is how the sperm triggers this increase in calcium, and what the mechanism of propagation is. It has been proposed that sperm may act as a source of high calcium (Jaffe, 1983) or other activator (Dale, 1988), which initiates the transient by locally introducing the calcium/activator at a trigger level. Evidence has also been presented to suggest that binding of sperm to a receptor leads to activation of a G-protein (Turner et al., 1986,1987; Miyazaki, 1988; Kline et al, 1988), which in turn stimulates the inositol lipid signal transduction pathway (Berridge, 1987). In any case, a role for $InsP_3$ in egg activation is inferred from the observation that eggs of many different organisms are activated after injection of this calcium-releasing second messenger (Whitaker & Irvine, 1984; Busa et al., 1985; Turner et al., 1986; Miyazaki, 1988).

In the unfertilized ascidian egg $InsP_3$-injection induces the cortical contraction wave (Dale, 1988), and sometimes polar body formation (Speksnijder, unpublished observations). In addition, a large pulse of calcium is induced (Speksnijder, unpublished observations). However, the spatial characteristics of this pulse are unknown. It is also not known whether production of $InsP_3$ occurs during fertilization in Phallusia, and whether it plays a role in the calcium-induced calcium release during wave propagation, as suggested for the sea urchin egg (Swann & Whitaker, 1986). The recent finding that periodic (though not necessarily traveling) pulses in pancreatic cells can be induced in the presence of a non-metabolizable $InsP_3$ analogue (Wakui et al., 1989) would favor a mechanism of direct calcium-induced calcium release (Jaffe, 1983).

However, the one question that makes the ascidian egg particularly interesting is not so much how the sperm induces the calcium wave, but if and how this wave affects embryonic pattern formation. Previous studies suggested that sperm entry may play a role in axis determination in Boltenia and Styela, but the underlying mechanisms remained unclear (Jeffery, 1984). From the observation that both in Ascidiella (Dalcq, 1938) and Phallusia, the endpoint of contraction is usually not opposite the polar body marking the animal pole, we suggest that the sperm entry point and resulting calcium wave tilt the direction of segregation such that its focal point is 15-50° away from the vegetal pole. In Styela, the focal point of ooplasmic segregation determines the site of gastrulation, even if segregation is altered such that it is foccussed on the animal pole or equatorial region (Jeffery & Bates, 1987). Thus it may well be that the site of sperm entry helps determine the site of gastrulation, which in turn becomes the dorsal side of the embryo.

The series of calcium waves that follow the activation wave in Phallusia constitute the first example of localized repetitive calcium pulses in eggs that continue embryonic development. These aequorin-injected Phallusia eggs frequently developed into normal free-swimming larvae; thus we are certain that these pulses are a natural phenomenon. The post activation waves occur during the completion of meiosis, and might therefore be associated with this process. However, the majority start in the vegetal/contraction pole area, where at that time the myoplasmic cap is located, which suggests that they may be involved in developmentally important processes such as the formation of activation of vegetal factors needed for gastrulation, or the movement of the myoplasm during the second phase of ooplasmic segregation. One way to test these possibilities would be to inhibit these calcium waves via microinjection of calcium buffers (Steinhardt & Alderton, 1988; Twigg et al., 1988; Speksnijder et al., 1989c).

Acknowledgments

This work was supported by NIH grant HD-18818 and NATO grant 86/0432.

References

Bates W.R., and Jeffery W.R. (1987a). Alkaline phosphatase expression in ascidian egg fragments and andromerogons. Dev.Biol. 119, 382-389.

Bates W.R., and Jeffery W.R. (1987b). Localization of axial determinants in the vegetal pole region of ascidian eggs. Dev. Biol. 124, 65-76.

Bates W.R., and Jeffery W.R. (1988). Polarization of ooplasmic segregation and dorsal-ventral axis determination in ascidian embryos. Dev. Biol. 130, 98-130.

Berridge M.J.(1987). Inositol trisphosphate and diacylglycerol: two interacting second messengers. Annu. Rev. Biochem. 56, 159-193.

Bevan S.J., O'Dell D.S., and Ortolani G. (1977). Experimental activation of ascidian eggs. Cell Differentiation 6, 313-318.

Blinks J.R. (1982). The use of photoproteins as calcium indicators in cellular physiology. Techniques in Cellular Physiology P126, 1-38.

Brownlee C., and Dale B. (1989). Temporal and spatial correlation of fertilization current, calcium waves and cytoplasmic contraction in eggs of Ciona intestinalis. Proc. Roy. Soc. , in press.

Busa W.B., Ferguson J.E., Joseph S.K., Williamson J.R., and Nuccitelli R. (1985). Activation of frog (Xenopus laevis) eggs by inositol trisphosphate. I. Characterization of Ca^{2+} release from intracellular stores. J. Cell Biol. 101, 677-682.

Conklin E.G (1905). The organization and cell-lineage of the ascidian egg. J. Acad. Natl. Sci. Phil. 13, 1-126.

Dalcq A.M. (1938). "Form and Causality in Early Development". Cambridge University Press, p. 108.

Dalcq A.M., and Vandebroek G. (1937). On the significance of the polar spot in ripe unfertilized and in fertilized ascidian eggs. Biol. Bull. 72, 311-318.

Dale B. (1988). Primary and secondary messengers in the activation of ascidian eggs. Exp. Cell Res. 177, 205-211.

Deno T. (1987). Autonomous fluorescence of eggs of the ascidian Ciona intestinalis. J. Exp. Zool. 241, 71-79.

Elinson R.P., and Rowning B. (1988). A transient array of parallel microtubules in frog eggs: Potential tracks for a cytoplasmic rotation

that specifies the dorsal-ventral axis. Dev. Biol. 128, 179-185.
Gualtieri R., and Sardet C. (1989). The endoplasmic reticulum network in the ascidian egg: Localization and calcium content. Biology of the Cell 65, 301-304.
Jaffe L.F. (1983). Sources of calcium in egg activation: a review and hypothesis. Dev. Biol. 99, 265-276.
Jaffe L.F. (1985). The role of calcium explosions, waves and pulses in activating eggs. In: Metz C.B., and Monroy A. (eds.) Biology of Fertilization, Vol. 3. Academic Press, San Diego, pp. 127-165.
Jaffe L.F. (1986). Calcium and morphogenetic fields. In: Calcium and the cell, CIBA Foundation Symposium 122. Wiley & Sons, Chichester, pp. 271-282.
Jeffery W.R. (1982). Calcium ionophore polarizes ooplasmic segregation in ascidian eggs. Science 216, 5 45-547.
Jeffery W.R. (1984). Pattern formation by ooplasmic segregation in the ascidian egg. Biol. Bull. 166, 277-298.
Jeffery W.R., and Meier S. (1983). A yellow crescent cytoskeletal domain in ascidian eggs and its role in early development. Dev. Biol. 96, 125-143.
Jeffery W.R., and Bates W.R. (1988). Axial determinants in ascidian eggs. In O'Connor J.D. (ed.), Molecular biology of invertebrate development. Liss, New York, p. 159-176.
Kline D., Simoncini l., Mandel G., Maue R.A., Kao R.T., and Jaffe L.A. (1988). Fertilization events induced by neurotransmitters after injection of mRNA in Xenopus eggs. Science 241, 464-467.
Mancuso V. (1963). Distribution of the components of normal fertilized eggs of Ciona intestinalis examined at the electron microscope. Acta Embryol. Morph. Exp. 6, 260-274.
Meedel T.H., Crowther R.J., and Whittaker J.R. (1987). Determinative properties of muscle cell lineages in ascidian embryos. Development 100, 245-260.
Miyazaki S.-I.(1988). Inositol 1,4,5-trisphosphate-induced calcium release and guanine-binding protein-mediated periodic calcium rises in golden hamster eggs. J. Cell Biol. 106, 345-353.
Miyazaki S.-I., Hashimoto N., Yoshimoto Y., Kishimoto T., Igusa Y., and Hiramoto, Y. (1986). Temporal and spatial dynamics of the periodic increase in intracellular free calcium at fertilization of golden hamster eggs. Dev. Biol. 118, 259-267.
Nishida H. (1987). Cell lineage analysis in ascidian embryos by intracellular injection of a tracer enzyme. III. Up to the tissue-restricted stage. Dev. Biol. 121, 526-541.
Nishida H. and Satoh N. (1983). Cell lineage analysis in ascidian embryos by intracellular injection of a tracer enzyme. I. Up to the eight cell stage. Dev. Biol. 99, 382-394.
Nishida H. and Satoh N. (1985). Cell lineage analysis in ascidian embryos by intracellular injection of a tracer enzyme. II. The 16- and 32-cell stages. Dev. Biol. 110, 440-454.
Ortolani G. (1955). I movementi corticali dell uovo di ascidie alla fecondazione. Riv. Biol. 47, 169-177.
Ortolani G. (1958). Cleavage and development of egg fragments in ascidians. Acta. Embryol. Morphol. Exp. 1, 247-272.
Reverberi G. (1956). The mitochondrial pattern in the development of the ascidian egg. Experientia 12, 55-56.
Reverberi G. (1975). On some effects of cytochalasin B on the eggs and tadpoles of the ascidians. Acta Embryol. Exp. 2, 137-158.
Reverberi G., and Ortolani G. (1962). Twin larvae from halves of the same egg in ascidians. Dev. Biol. 5, 84-100.
Sardet C., Terasaki T., Speksnijder J.E., and Jaffe L.F. (1988). The egg

cortical endoplasmic reticulum. Biol. Bull. 175, 310 (abst).
Sardet C., Speksnijder J.E., Inoué I., and Jaffe L.F. (1989). Fertilization and ooplasmic movements in the ascidian egg. Development 105, 237-249.
Sawada T. (1983). How ooplasm segregates bipolarly in ascidian eggs. Biol. Bull. Mar. Biol. Station, Ashamushi, Tohoku Univ. 17, 123-132.
Sawada T. (1986). Cortical actin filaments in unfertilized eggs of Ciona savignyi. Zool. Sci. 3, 1038 (abst.).
Sawada T. (1988). The mechanism of ooplasmic segregation in the ascidian egg. Zool. Sci. 5, 667-675.
Sawada T., and Osanai K. (1981). The cortical contraction related to the ooplasmic segregation in Ciona intestinalis eggs. Roux's Arch. Dev. Biol. 190, 208-214.
Sawada T., and Osanai K. (1985). Distribution of actin filaments in fertilized egg of the ascidian Ciona intestinalis. Dev. Biol. 111, 260-265.
Sawada T., and Schatten G. (1988). Microtubules in ascidian eggs during meiosis and fertilization. Cell Mot. Cytoskel. 9, 219-231.
Sawada T., and Schatten G. (1989). Effects of cytoskeletal inhibitors on ooplasmic segregation and microtubule organization during fertilization and early development in the ascidian Molgula occidentalis. Dev. Biol. 132, 331-342.
Speksnijder J.E., Corson D.W., Jaffe L.F., and Sardet C. (1986a). Calcium pulses and waves through ascidian eggs. Biol. Bull. 171, 488 (abst.)
Speksnijder J.E., Corson D.W., Qiu T.H., and Jaffe L.F. (1986b). Free calcium pulses during early development of Ciona eggs. Biol. Bull. 170, 542 ((abst.)
Speksnijder J.E., Corson D.W., Sardet C., and Jaffe L.F. (1989a). Free calcium pulses following fertilization in the ascidian egg. Dev. Biol. 135, 182-190.
Speksnijder J.E., Jaffe L.F., and Sardet C. (1989b). Polarity of sperm entry in the ascidian egg. Dev. Biol. 133, 180-184.
Speksnijder J.E., Miller A.L., Weisenseel M.H., Chen T.-H., and Jaffe L.F. (1989c). Calcium buffer injections block fucoid egg development by facilitating calcium diffusion. Proc. Natl. Acad. Sci. USA 86, 6607-6611.
Steinhardt R.A., Epel D., Carrol E.J., and Yanimachi R. (1974). Is calcium ionophore a universal activator for unfertilised eggs? Nature 252, 41.
Steinhardt R.A. and Alderton J. (1988). Intracellular free calcium rise triggers nuclear envelope breakdown in the sea urchin embryo. Nature 332, 364-366.
Swann K., and Whitaker M.J. (1986). The part played by inositol trisphosphate and calcium in the propagation of the fertilization wave in sea urchin eggs. J. Cell Biol. 103, 2333-2342.
Talevi R., and Dale B. (1986). Electrical characteristics of ascidian egg fragments. Exp. Cell Res. 162. 539-543.
Turner P.R., Jaffe L.A., and Fein A. (1986). Regulation of cortical vesicle exocytosisin sea urchin eggs by inositol 1,4,5-trisphosphate and GTP-binding protein. J. Cell Biol. 102, 70-76.
Turner P.R., Jaffe L.A., and Primakoff P. (1987). A cholera toxin sensitive G-protein stimulates exocytosis in sea urchin eggs. Dev. Biol. 120, 577-583.
Twigg J., Patel R. and Whitaker M. (1988). Translational control of $InsP_3$-induced chromatin condensation during the early cell cycles of sea urchin embryos. Nature 332, 366-369.
Wakui M., Potter B.V.L., and Petersen O.H. (1989). Pulsatile intracellular calcium release does not depend on fluctuations in inositol trisphosphate concentration. Nature 339, 317-320.
Whitaker M.J. (1985). An imaging photon detector for the measurement of

low-intensity luminescence. J. Physiol. (London) 365, 5P.
Whitaker M.J. & Irvine R.F. (1984). Inositol (1,4,5)- trisphosphate microinjection activates sea urchin eggs. Nature 312, 636-639.
Whittaker J.R. (1979). Cytoplasmic determinants of tissue differentiation in the ascidian egg. In: Subtelny S., Konigsberg I.R. (eds.), Determinants of spatial organization. Academic Press, New York, pp. 29-51.
Yamamoto T. (1961). Physiology of fertilization in fish eggs. Int. Rev. Cytol. 12, 361-405.
Yoshimoto Y., Iwamatsu T., Hirano K., and Hiramoto Y. (1986). The wave of free calcium upon fertilization in medaka and sand dollar eggs. Develop. Growth Differ. 28, 583-596.
Zalokar M. (1974). Effect of colchicin and cytochalasin B on ooplasmic segregation of ascidian eggs. Roux's Arch. Dev. Biol. 175, 243-248.
Zalokar M. (1980). Activation of ascidian eggs with lectins. Dev. Biol. 79, 232-237.
Zalokar M., and Sardet C. (1984). Tracing of cell lineage in embryonic development of Phallusia mammillata (Ascidia) by vital staining of mitochondria. Dev. Biol. 102, 195-205.

SOLUBLE SPERM FACTORS, ELECTRICAL EVENTS AND EGG ACTIVATION

Brian Dale
Stazione Zoologica
Villa Comunale
80121, Napoli, Italy

and

Louis J. DeFelice
Dept. of Anatomy and Cell Biology
Emory University
Atlanta
Georgia, 30322, USA

INTRODUCTION

Alberto Monroy, together with Kao, Grundfest and Tyler, were amongst the first to recognize the importance of ion fluxes through the plasma membrane in the process of egg activation (Tyler et. al. 1956). In the 1970's the electrical events at activation in sea urchin eggs were recorded (Steinhardt et al 1971; Ito and Yoshioka, 1973; Jaffe,1976 and Chambers and de Armendi,1979), and shortly afterwards measurements were extended to fish, amphibia and mammals (see Igusa et al., 1983; McCulloh et al.,1983; Nuccitelli, 1980; Talevi et al., 1985 for references). The sea urchin response was noted to be a biphasic depolarization (Ito and Yoshioka, 1973). However, neither the biophysical origin of this depolarization, nor the mechanism by which it was triggered, were known. Here we follow the progress made in this field and describe how electrical measurements have been utilized to elucidate the mechanism of egg activation.

THE FIRST ELECTRICAL EVENT OF SPERM-EGG INTERACTION LEADING TO SUCCESSFUL PENETRATION IS A SMALL STEP-LIKE DEPOLARIZATION

Close observation of the initial phase of the activation potential in sea urchin eggs showed it to be composed of discrete step-like events. Eggs were impaled for electrophysiological recording, inseminated and then examined for sperm nuclei. Each spermatozoon that entered the egg induced a small, 1-2mV, step-like depolarization (De Felice and Dale, 1979). A composite shoulder phase indicted polyspermy. In a monospermic situation, of the hundreds of attached spermatozoa, only the fertilizing spermatozoon is capable of reacting with the egg, thereby inducing a single step-like depolarization (Fig.1, Dale et al., 1978). This report was the first to show how the fertilizing spermatozoon differed from the supernumerary spermatozoa in its capacity to generate a discrete electrical event. This step, which is the earliest detectable event in

NATO ASI Series, Vol. H 45
Mechanism of Fertilization
Edited by B. Dale

the egg at fertilization is not seen in parthenogenetically activated eggs (Dale and deSantis, 1981a).

Voltage-clamp studies confirmed this sperm induced electrical event in the sea urchin (Lynn and Chambers, 1984; see also Nuccitelli et al., 1989 for general references), and a comparable event has been identified both in the ascidian (Dale et al, 1983; DeFelice and Kell, 1986) and in the anuran Discoglossus pictus (Talevi et al, 1985).

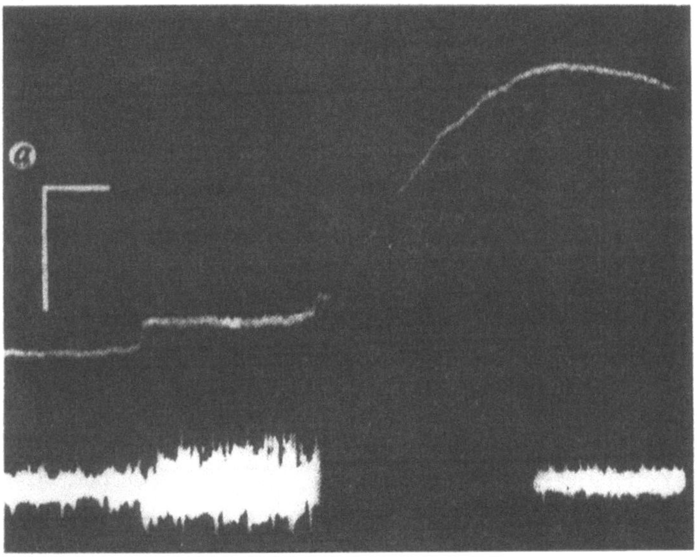

Fig.1 The fertilization potential in the sea urchin egg. Note the small step-like depolarization that precedes the bell-shaped potential by about 10sec. The lower trace, a.c coupled from 1Hz, shows the increase in voltage noise during the step. The vertical bar represents 12mV for the upper trace and 1mV for the lower trace; the horizontal bar represents 5sec. (From Dale et. al., 1978).

This small step-like electrical event lasts about 5-10 seconds at room temperature, at which time the second larger depolarization, the fertilization potential, is triggered. The delay between the two corresponds more or less to what has been described in sea urchins as the latent period (see Ginsberg, 1988 for a review), during which there is no obvious morphological change in the egg surface. Shortly after the fertilization potential in the sea urchin (Eisen, et al., 1984) and in the ascidian (Speksnijder et al., 1989; Brownlee and Dale, 1990) there is a massive release of Ca from intracellular stores that causes an increase in cytosolic Ca from 0.1μM to 10 μM. Within minutes the Ca returns to resting values. In the sea urchin the cortical reaction occurs at about the same time as the fertilization potential (Dale and deSantis, 1981a), whereas in the ascidian the Ca wave starts to propagate 5-10 seconds after the initiation of the fertilization potential and egg contraction starts some 50 sec later (Brownlee and Dale, 1990).

LARGE NON-SPECIFIC CHANNELS, GATED BY THE SPERMATOZOON, UNDERLY THE FERTILIZATION POTENTIAL

Although eggs from most animals are low conductance cells, they often possess a wide variety of both voltage-gated and chemically-gated channels. At fertilization a new population of chemical gated channels called " fertilization channels " appear in the membrane. To estimate the elementary conductance change underlying the fertilization potential in sea urchins the ratio of the change in voltage noise variance to the change in potential was calculated. Knowing the cell membrane resistance during this change we estimated that the single channel conductance was about 30-90pS (Dale et al., 1978).

Cross fertilization and Ca substitution experiments implied that the step event was post acrosome reaction and led to the hypothesis that the spermatozoon triggered egg activation by releasing a chemical factor (Dale et al., 1978). There have been several attempts to identify the activating factor, but as yet there is no conclusive evidence (Jaffe, 1980; Whitaker and Irvine, 1984; Dale et al., 1985).

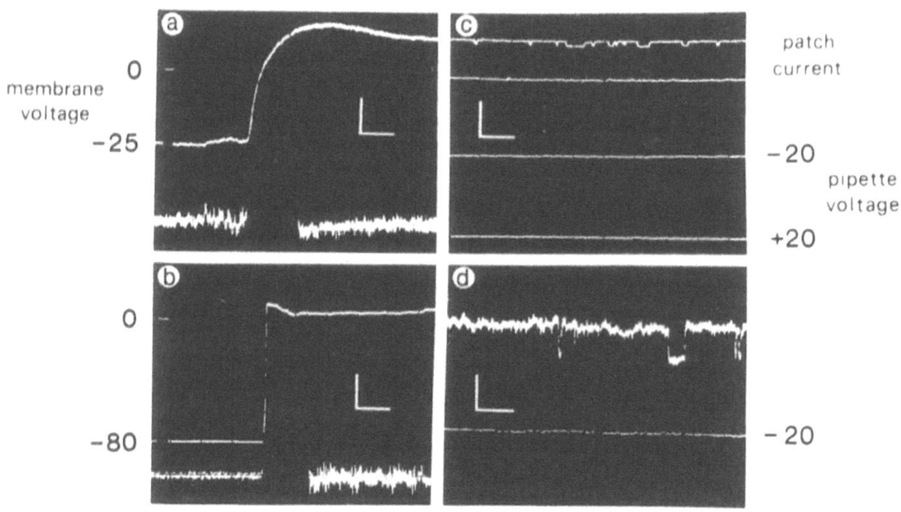

Fig.2. Voltage responses (a,b) and single channel currents (c,d) activated by sperm in ascidian eggs. Fertilization channels have a single channel conductance of about 400pS and reverse around 0mV suggesting they are non specific for ions (From Dale and Defelice, 1984).

In the ascidian *Ciona intestinalis* a small step depolarization marks the initiation of egg activation and this is followed some 5-7 seconds later by a much larger depolarization. The potential remains at a positive value for several minutes and then gradually returns to its original value (Dale et al., 1983). With the advent of the patch-clamp technique currents from single fertilization channels were first measured in 1984. Fertilization channels in Ciona were found to have a single-channel conductance of 400 pS (Fig.2., Dale and De Felice, 1984). Since the reversal potential was around 0 mV, it was suggested that these channels were not ion specific. To date these channels are amongst the largest observed in biological membranes. Whole-cell currents in ascidian eggs also studied during fertilization were shown to peak near -30 mV and approach zero near 0 mV (De Felice and Kell, 1987), supporting the single-channel data. Knowing the total conductance change at fertilization, the single channel conductance and the probability of a channel being open, we estimated that the fertilizing spermatozoon opens between 200 to 2000 fertilization channels in the egg.

FERTILIZATION CHANNELS ARE ACTIVATED LOCALLY TO THE SPERM ENTRY SITE IN THE MOSAIC ASCIDIAN EGG

One great advantage of the ascidian egg is the possibilty to manually remove the extracellular coats leaving a nude egg. Nude eggs may then be cut into small fragments and each fragment has the capability of developing into an embryo. By using the whole cell clamp technique on unfertilized and fertilized fragments and inseminating each fragment, it was found that fertilization channel precursors and voltage-gated ion channels are uniformly distributed around the ascidian egg surface (Talevi and Dale, 1986; De Felice et al., 1986; Dale and Talevi, 1989). Since fertilization currents were similar in whole eggs or fragments, irrespective of their size and global origin, it was concluded that the fertilizing spermatozoon opens a fixed number of fertilization channels limited to an area around its point of entry (Fig.3., De Felice et al., 1986). The localized ion current through these channels may regulate movements of the cytoskeleton involved in cytoplasmic segregation.

Unfortunately, for technical reasons, it is not possible to repeat similar experiments in the sea urchin. However since chemical or physical partial disruption of the egg cortex results in a reduction in amplitude of the fertilization potential (Dale et al., 1982; 1989),it appears that the channels responsible for this depolarization in the sea urchin are activated around the egg surface.

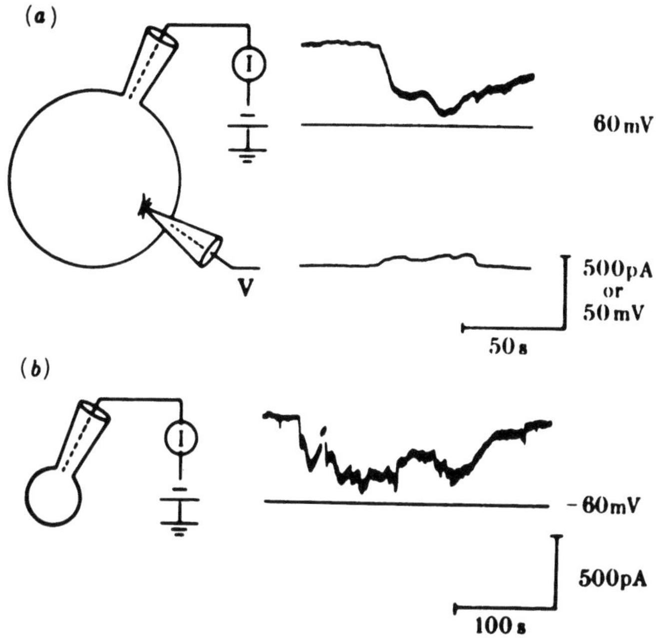

Fig.3. Since the fertilization current in a whole ascidian oocyte (a) is comparable in amplitude to that in an oocyte fragment (b), it is probable that the fertilizing spermatozoon activates a fixed number of channels localized to its site of fusion. (From DeFelice et.al. 1986).

HOW ARE FERTILIZATION CHANNELS GATED ?

Since intracellular Ca increases in eggs at fertilization, and several types of ion channel are known to be regulated by Ca, experiments were designed to determine whether fertilization channels are regulated by Ca. Raising the level of intracellular Ca in ascidian eggs, by perfusion or by loading the egg cortex (>50uM) with Ca through voltage gated channels, did not activate fertilization channels. Alternatively, eggs exposed to low-Ca sea water, perfused with the Ca chelator K-EGTA or Ca blocking agents to prevent the release of Ca from intracellular organelles, and subsequently inseminated, generated fertilization currents (Dale , 1987). Eggs exposed to the Ca ionophore A23187 were found to contract without generating a fertilization current, while microinjection of IP3 or soluble fractions of homogenized spermatozoa induced both a contraction and a fertilization current (Fig. 4., Dale, 1988). Although elevated Ca does not gate fertilization channels, it appears to be involved in the mechanism of cortical contraction (Sawada and Osanai, 1981; Dale, 1988; Brownlee and Dale, 1990). Measurements with ion selective electrodes show that the intracellular pH of ascidian eggs ranges from 7.2- 7.4 and does not vary during activation, making pH an unlikely trigger of early activation events (Russo et al., 1989).

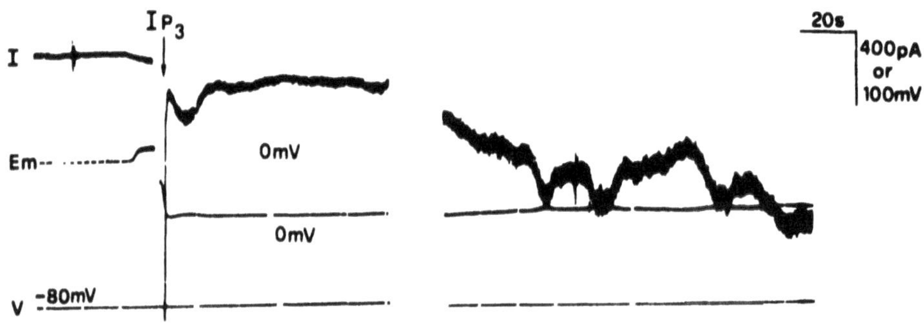

Fig.4. Whole cell currents (I) in an ascidian egg clamped at -80mV and penetrated by a second electrode containing 10μM IP_3 in KCl. The second pipette also served to verify clamp conditions by recording cell membrane potential (Em). Note the fertilization type current.(From Dale 1988).

In the anuran Discoglossus pictus the sperm entry site is a pre-determined specialized area called the animal dimple(Hibbard,1928). Discoglossus eggs as other frog eggs may be activated by pricking with a steel needle. (It should be pointed out that eggs from other animal groups are not activated upon pricking). At fertilization, the spermatozoon initiates a large regenerative depolarization that is Cl- dependent (Talevi et al, 1985); pricking elicits a comparable response. When eggs were pricked outside the dimple area, we saw a wave of contraction spread from the puncture site to the antipode. Since the activation potential was not generated until the wave reached the dimple, it appears that the channels underlying the depolarization are found in the dimple, and that they are gated by a second messenger liberated in the egg cytoplasm that spreads around the egg with the contraction wave.

WHAT STAGE OF SPERM EGG INTERACTION CORRESPONDS TO THE PRIMARY ELECTRICAL EVENT?

Owing to the rapid succession of change in the egg during activation it is difficult to distinguish individual events. To partially overcome this problem we selected the sea urchin germinal vesicle stage oocyte. These cells may be fertilized. However, since they are immature they do not give rise to the cortical reaction or other autocatalytic events seen in the egg. When an oocyte is fertilized not all of the spermatozoa are capable of penetrating the cell or of producing a fertilization cone. Over a thousand spermatozoa may attach to the oocyte, but usually not more than 10 enter, as demonstrated by the formation of fertilization cones and histological sections. If we record electrically from oocytes, successful spermatozoa give rise to an electrical depolarization and conductance increase and induce a fertilization cone some 50 sec later (Dale and Santella, 1985). Other sperm were not capable of inducing either an electrical event or a cone, whilst a third category induced a step depolarization that after several seconds spontaneously reversed. These sperm did not enter the oocyte, nor did they induce the formation of a fertilization cone.

These experiments raised the possibility that the step depolarization was the direct result of sperm-egg fusion, the conductance increase being due to the appearence of sperm channels in the newly formed syncytium. If fusion is inhibited by the ATPase inhibitor Quercetin (Eckberg and Perotti, 1983), or by removing Mg (Sano et al, 1980), spermatozoa are unable to generate electrical changes in the oocyte. Furthermore, the step event may be experimentally reversed by adding a spermicide to inseminated oocytes (Dale and Santella, 1985). Spontaneously or experimentally induced reversible sperm steps have since been seen in a variety of circumstances, (see Nuccitelli et al., 1989, for references).

Although these rather indirect experiments suggest that the step event is the result of sperm-egg fusion the question is still open to much debate. In a more direct attempt to answer this problem, Longo and colleagues voltage clamped sea urchin eggs, fertilized them and then serially sectioned them to locate the fertilizing spermatozoon. Since the authors did not detect gamete fusion until 5 sec after the step depolarization they suggested that the step is a pre-fusion event (Longo et al., 1986 ;1990). Although preliminary, McCulloh and Chambers (1986) have data that show the onset of the step event is coincident with an increase in capacitance. Since, in biological membranes capacitance is proportional to surface area, an increase in capacitance at this moment indicates gamete fusion.

In the sea urchin oocyte a step event may typically reduce the input resistance of the cell from 25Mohms to 15M ohms. By assuming the sperm membrane area is in parallel with the egg membrane area, we calculate that the sperm would have a conductance of 15nS, or in other words, it would have to contain 40 channels of 400pS.

WHAT HAPPENS DURING THE LATENT PERIOD?

In most animals there is a time delay from the moment the spermatozoon attaches to the egg surface until cortical exocytosis. This period varies from animal to animal (see review by Ginsberg, 1988), and electrophysiological recording has shown that a step depolarization marks the beginning of this period. Rothschild and Swann in 1952 suggested that a fast propagated change traverses the egg surface during the latent period. The sea urchin egg is a useful model to study the latent period for two reasons; first because the latent period is relatively long and is temperature dependent (Ginsberg, 1988; Dale et al., 1978; Dale & DeSantis, 1981a), and second because the cortical reaction may be reversibly interrupted by a mild heat shock (Allen and Hagström, 1955; Hagström and Runnström, 1959), giving rise to " partially fertilized eggs ". In these eggs, 50% of the surface may be activated, while the rest is undistinguishable from a virgin egg. Upon re-insemination spermatozoa are able to interact with the " virgin surface " of such eggs and therefore it appears unlikely that any major change has traversed this area during the latent period (Dale et al., 1989).

When fertilization occurs in the presence of the microfilament inhibiting agents cytochalasin B or D, the latent period is increased by up to 100% (Dale and deSantis, 1981b). In contrast, there is no change if the gametes are pre-exposed to these agents and subsequently fertilized in natural sea water. Together these experiments suggest that a microfilament dependent stage of sperm-egg interaction occurs during the latent period. If the preceeding arguements are correct this event is post fusion.

CONCLUDING REMARKS

The idea of a soluble sperm-borne factor that triggers the egg into metabolic activity is not new; Robertson (1912) and Loeb (1913) were among the first to design experiments in support of this hypothesis. In later models for activation emphasis was placed on the interaction of the spermatozoon with receptors located on the external surface of the egg (Lillie and Baskerville, 1922 ; Mazia et al., 1975).

The two hypotheses prevail. One school of thought points to a trigger factor in the spermatozoon that is released into the egg cytoplasm following gamete fusion (Fig 5., Dale et al, 1978; 1985, Dale, 1988; Iwasa et. al., 1988; 1989; Whitaker et al., 1989). The contrasting idea favours an externally located receptor and G-protein trans membrane transduction mechanism (Kline et al., 1988, and this volume). Most evidence supporting either of these ideas is indirect and confirmation will require the identification and characterization of the trigger elements involved.

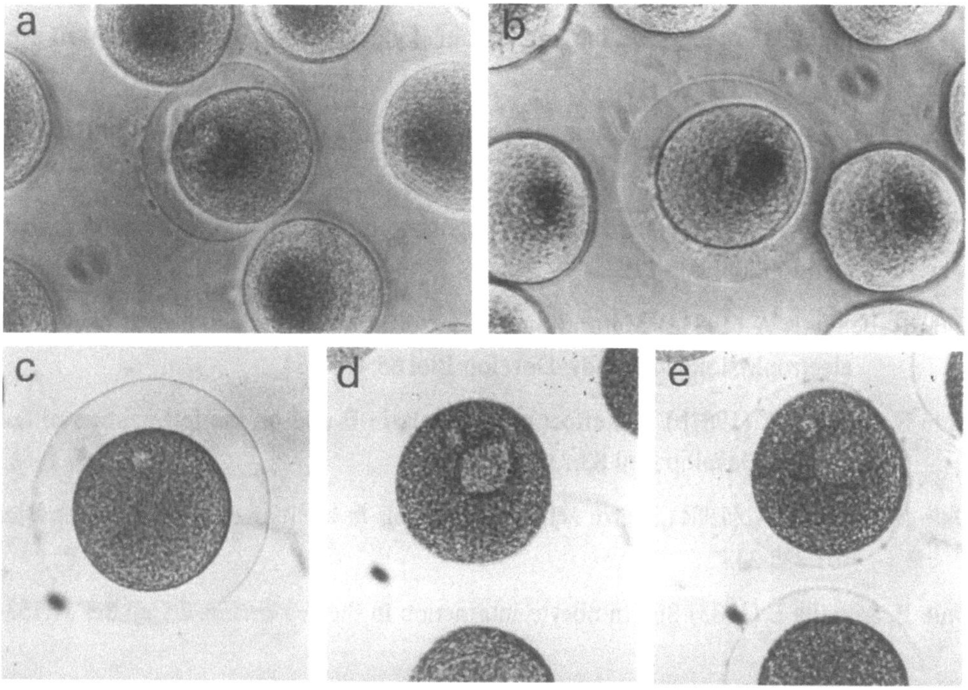

Fig.5. Pressure injection of soluble sperm extracts (in 0.5 M KCl containing EGTA) into sea urchin eggs induces the cortical reaction (a, b). A similar reaction may be induced by microinjecting a small amount of distilled water (c), but not iso-osmotic KCl (d). The egg in (d) was subsequently inseminated showing that the cortical granules remained functional following micro-injection. (From Dale et al., 1985).

As far as the former hypotheses is concerned, an active soluble factor has been successfully extracted from spermatozoa (Dale et al 1985; Dale, 1988), and although far from complete attempts to characterize this factor have begun (Iwasa et al., 1988;1989). Since the former hypothesis is based on a post fusion event, information on the timing of sperm-egg plasma membrane fusion may help to further elucidate the mechanism of fertilization.

REFERENCES

Allen R, Hagström B (1955) Interruption of the cortical reaction by heat. Exp Cell Res 9:157-167

Brownlee C, Dale B (1990) Temporal and spatial correlation of fertilization current, calcium waves and cytoplasmic contraction in eggs of Ciona intestinalis. Proc R Soc B Lond (in press)

Chambers E, de Armendi J (1979) Membrane potential, action potential and activation potential of eggs of the sea urchin, Lytechinus variegatus. Exp Cell Res 122:203-218

Dale B (1987) Fertilization channels in ascidian eggs are not activated by Ca. Exp Cell Res 172:474-480

Dale B (1988) Primary and secondary messengers in the activation of ascidian eggs. Exp Cell Res 177:205-211

Dale B, deSantis A (1981a) Maturation and fertilization of the sea urchin oocyte; an electrophysiological study. Develop Biol 85:474-484

Dale B, deSantis A (1981b) The effect of cytochalasin B and on the fertilization of sea urchins. Develop Biol 83:232-237

Dale B, De Felice L (1984) Sperm activated channels in ascidian oocytes. Develop Biol 101:235-239

Dale B, Santella L (1985) Sperm-oocyte interaction in the sea-urchin. J Cell Sci 74:153-167

Dale B, Talevi R (1989) Distribution of ion channels in ascidian eggs and zygotes. Exp Cell Res 181: 238-244

Dale B, DeFelice L, Taglietti V (1978) Membrane noise and conductance increase during single spermatozoon-egg interactions. Nature Lond 275:217-219

Dale B, DeSantis A, Hagström B (1982) The effect of nicotine on sperm-egg interaction in the sea urchin: Polyspermy and Electrical Events. Gamete Res 5: 125-135

Dale B, De Santis A, Ortolani G (1983) Electrical response to fertilization in ascidian oocytes. Develop Biol 99: 188-193

Dale B, DeFelice L, Ehrenstein G (1985) Injection of a soluble sperm fraction into sea urchin eggs triggers the cortical reaction. Experentia 41:1068-1070

Dale B, Hagström B, Santella L (1989) Partially fertilized sea urchin eggs: An electrophysiological and morphological study. Develop Growth and Diff 31:165-170

De Felice L, Dale B, (1979) Voltage response to fertilization and polyspermy in sea urchin eggs and oocytes. Develop Biol 72:327-341

De Felice L, Kell M, (1986) Sperm activated currents in ascidian oocytes. Develop Biol 119:123-128

De Felice L, Dale B, Talevi R (1986) Distribution of fertilization channels in ascidian oocyte membranes. Proc R Soc Lond 229:209-214

Eckberg W, Perotti M (1983) Inhibition of gamete membrane fusion in the sea urchin by Quercitin. Biol Bull mar biol Lab Woods Hole 164:62-70

Eisen A, Kiehart D, Reynolds G (1984) Temporal sequence and spatial distribution of early events of fertilization in single sea urchin eggs. J Cell Biol 99:1647-1654

Ginsburg A, (1988) Egg cortical reaction during fertilization and its role in block to polyspermy. Sov Sci Rev F Physiol Gen Biol 1:307-375

Hagström B, Runnström J, (1959) Re-fertilization of partially fertilized sea urchin eggs. Exp Cell Res 16: 309-314

Hibbard H, (1928) Contribution l'tude de l'ovogenèse de la fècondation et de l'histogenèse chez Discoglossus pictus. Arch Biol 32: 251-326

Igusa Y, Miyazaki S, Yamashita N, (1983) Periodic hyperpolarizing responses in hamster and mouse eggs fertilized with mouse sperm. J Physiol 340:633- 647.

Ito S, Yoshioka K, (1983) Effects of various ionic compositions upon the membrane potentials during activation of sea urchin eggs. Exp Cell Res 78:191-200

Jaffe L, (1976) Fast block to polyspermy in sea urchin eggs is electrically mediated. Nature Lond 261:68-71

Jaffe L, (1980) Calcium explosions as triggers of development. Ann NY Acad Sci 339: 86-101

Kline D, Simoncini L, Mandel G, Maue R, Kado R, Jaffe L, (1988) Fertilization events induced by neurotransmitters after injection of mRNA in Xenopus eggs Sciencs 241:464-467

Lillie R, Baskerville M,(1922) The action of ultraviolet rays on Arbacia eggs, especially as affecting the response to hypertonic sea water. Am J Physiol 61:272-288

Loeb J, (1913) "Artificial Parthenogenesis and Fertilization". University Press, Chicago.

Longo F, Lynn J, McCulloh D, Chambers E,(1986) Correlative ultrastructural and electrophysiological studies of sperm-egg interactions of the sea urchin Lytechinus variegatus. Develop Biol 118:155-166

Lynn J, Chambers E, (1984) Voltage clamp studies of fertilization in sea urchin eggs.l. Effect of clamped membrane potential on sperm entry, activation and development. Develop Biol 102:98-109

Mazia D, Schatten G, Steinhardt R, (1975) Turning on of activities in unfertilized sea urchin eggs: correlation with changes of the surface. Proc Natn Acad Sci USA 72: 4469-4473

McCulloh D, Chambers E,(1987) When does the sperm fuse with the egg. J Gen Physiol 88: 38a

McCulloh D, Rexroad C Levitan H, (1983) Insemination of rabbit eggs is associated with slow depoloarization and repetitive diphasic membrane potentials. Develop Biol 95:372-377

Nuccitelli R, (1980) The electrical changes accompanying fertilization and cortical vesicle secretion in the medaka egg. Develop Biol 76:483-498

Nuccitelli R, Cherr G, Clark W,(1989) "Mechanisms of egg activation". Plenum, New York

Robertson T, (1912) Studies on the fertilization of the eggs of a sea urchin Strongylocentrotus purpuratus by blood-sera, sperm, sperm extract and other fertilizing agents. Arch Entwick Lungs Mech 35:64-130

Rothschild Lord, Swann M, (1952) The fertilization reaction in the sea urchin. The block to polyspermy. J Exp Biol 29:469-483

Russo P, Pecorella M, De Santis A, Dale B, (1989) pH during fertilization and activation of ascidian eggs. J Exp Biol 250:329-332

Sano K, Usui N, Ueki K, Mohri H, (1980) Magnesium ion requiring step in fertilization of sea urchins. Dev Growth Differ 22:531-541

Sawada T, Osanai K, (1981) The cortical contraction related to the ooplasmic segregration in Ciona intestinalis eggs, Wilhelm Roux's Archives 190:208-214

Speksnijder J, Corson D, Sardet C, Jaffe L, (1989) Free calcium pulses following fertilization in the ascidian egg. Develop Biol 135:182-190

Steinhardt R, Lundin L, Mazia D, (1971) Biolectric responses of the echinoderm egg to fertilization. Proc natn Acad Sci USA 68:2426-2430

Talevi R, Dale B, (1986) Electrical characteristics of ascidian egg fragments. Exp Cell Res 162:539-543

Talevi R, Dale B, Campanella C, (1985) Fertilization and activation potentials in Discoglossus pictus (Anura) eggs: A delayed response to activation by pricking. Develop Biol 111:316-323

Tyler A, Monroy A, Kao C, Grundfest H, (1956) Membrane potential and resistance of the starfish egg before and after fertilization. Biol Bull 111: 153-177

Whitaker M, Irvine R, (1984) Inositol 145 triphosphate microinjection activates sea urchin eggs Nature Lond 312: 636-638

Whitaker M, Swann K, Crossley, I (1989) What Happens during the Latent period at Fertilization. In "Mechanisms of Egg Activation" pp 157-171, Eds. R. Nuccitelli, G. Cherr, W. Clark, Plenum Press, New York

CAN THE HANDLING OF OOCYTES INFLUENCE THE SUCCESS OF THERAPEUTIC IN VITRO FERTILIZATION?

M.H. Johnson, P.R. Braude, C. Vincent and S.J. Pickering
Embryo and Gamete Research Group
Departments of Anatomy and Obstetrics and Gynaecology
University of Cambridge
Cambridge CB2 3DY
U.K.

The new reproductive technologies of *in vitro* fertilization (IVF) and gamete intrafallopian transfer (GIFT) have transformed the prospects for treating the infertile successfully, as well as opening up possibilities for the preimplantation diagnosis of genetic disease. However, despite these advances, treatment using these techniques fails more often than it succeeds, births being recorded in only 10 to 20% of treatment cycles, even in the best centres (Voluntary Licensing Authority Report, 1989). It is difficult to explain these failures as being due entirely to the inherent inefficiences of the human reproductive process (Johnson *et al.*, 1990). Rather, it seems that therapeutic procedures may contribute to the substantial losses of oocytes and conceptuses that occurs. The question is which procedures and can they be changed to avoid or minimise damage? These sorts of question become increasingly acute as more invasive

NATO ASI Series, Vol. H 45
Mechanism of Fertilization
Edited by B. Dale

procedures are contemplated such as biopsy, cryopreservation, zona drilling or sperm injection. In Cambridge, we have embarked on a programme to investigate these problems, using the mouse as a model system with which to refine questions for addressing on the limited amount of human material available for research. In the process, we have found incidentally much that has improved our capacity to handle mouse oocytes more effectively.

Reductions in temperature can affect oocytes adversely and irreversibly

Mouse oocytes, cooled to temperatures below below 37oC, undergo a number of changes the severity of which depends upon the temperature and method of cooling. In practice, two situations occur therapeutically in which cooling might occur: during recovery and handling of oocytes, a transitory drop to room temperature may occur unless special precautions are taken, and during the cryopreservation of oocytes,cooling below 37oC is inevitable during the pre-freeze and post-thaw periods. We will concentrate on these two situations in our discussion.

1. Cooling

In the mouse, it is established clearly that cooling leads to major disturbances in the organisation of the microtubules of the oocyte. In control oocytes, microtubules are restricted to the second meiotic spindle, which is barrel shaped and anastral. This restriction arises because the critical concentration for tubulin polymerisation is exceeded only in the vicinity of the chromosomes, which modulate conditions locally in such a way as to permit the formation of microtubules (Maro *et al.*, 1985). The barrel shape of the spindle is determined, not by centrioles which are lacking at these stages, but by multiple foci of peri-centriolar material (PCM) at the spindle poles. In the mouse, but not the human, additional foci of PCM are evident around the oocyte cortex (Maro *et al.*, 1985; Pickering *et al.*, 1988). When mouse oocytes are cooled, the microtubules begin to disassemble, the extent of disassembly depending on the degree of cooling (Magistrini & Szollosi, 1980; Pickering & Johnson, 1987). As a result, the spindle dismantles and the chromosomes and polar pericentriolar material disperse round the cytocortex. Clearly, if such oocytes were to be fertilized, aneuploid zygotes could result from the failure of the maternal chromosomes to segregate properly. Unfortunately, even if cooled mouse oocytes are restored to 37°C, up to 45% of them do not regroup their chromosomes on the metaphase plate, but have abnormal shaped spindles with detached clusters of chromosomes.

The human oocyte seems to react even more rapidly and more permanently to cooling. Thus, if freshly recovered human oocytes are cooled to room temperature for periods of 10 to 30 minutes, the majority (c.80%) show abnormalities of spindle organisation and/or chromosomal organisation. If, even after only 10 minutes exposure to room temperature, the oocytes are restored to 37°C for 1 to 4 hours before being examined, some 70% remain abnormal (Pickering *et al.*,

1990). Transient cooling of human oocytes during retrieval, manipulation and replacement is likely to be a consistent feature of therapy, unless special precautions are taken. Such cooling-induced defects might account for part of therapeutic failure, and could account for a significant fraction of the high levels of aneuploid embryos observed in cytogenetic studies of preimplantation stages after IVF (Angell *et al.*, 1986; Veiga *et al.*, 1987; Plachot *et al.*, 1988).

Cooling has a second adverse effect on the mouse oocyte, but only if temperatures as low as 4°C are reached (Johnson *et al.*, 1988). Brief exposure to this temperature causes hardening of the zona pellucida and a reduction in sperm penetration, so reducing fertilization rates. Cooling to such temperatures is unlikely to occur unless cryopreservation is anticipated. Since successful survival after freezing requires that a cryoprotectant be present, it is important to determine whether cryprotectant itself also has direct adverse effects on the oocyte.

2. <u>Cryoprotectant</u>

Although the use of cryoprotectants such as dimethylsulphoxide (DMSO) permits survival of human oocytes during freezing and thawing, attempts to fertilise and develop such oocytes have been remarkably unsuccessful (Chen, 1986; Van Uem *et al.*, 1987). This lack of success may be due primarily to the adverse effects of the pre-freeze/post-thaw manipulations of cryoprotectant (Johnson,

1989), and studies on mouse oocytes have identified two targets that are affected.

When oocytes are added to DMSO at 37°C, a massive proliferation of microtubules occurs throughout the cytocortex of the cell, the spindle becomes increasingly disorganised, the metaphase plate is disrupted and the chromosomes disperse (Johnson & Pickering, 1987). Between 60 to 90% of oocytes are restored to apparent normality on removal of the DMSO, but in the remainder both microtubules and chromosomes are organised abnormally. This effect of DMSO is almost exactly the reverse of that described above for cooling. The DMSO evidently sequesters water, raises the concentration of free tubulin and so drives polymerisation. If cooling and cryoprotectant are combined by placing oocytes in DMSO at 4°C, the DMSO acts to stabilise the spindle and the cooling prevents excessive polymerisation. Even so, the organisation of microtubules in most oocytes is still abnormal, albeit less so than when either condition is experienced alone, and recovery rates on the removal of cryoprotectant and restoration to 37°C are improved marginally. Clearly, development of optimal protocols for the addition and withdrawal of cryoprotectant must take these observations into account if genetically abnormal zygotes are to be reduced or avoided.

The second adverse effect of placing oocytes in 1.5M DMSO at 37°C is the generation of zona hardening, as assessed by an increased resistance of the zona pellucida to penetration by spermatozoa and to digestion by chymotrypsin (Johnson, 1989). This result is reminiscent of that induced by cooling to 4oC (see above). Paradoxically, when the two adverse conditions were combined, and oocytes were placed in DMSO at 4oC, no effect on the zona was observed. Zona hardening changes of this sort occur naturally at fertilization or activation, as part of the block to polyspermy, and are thought to be due to the release of cortical granule contents. It seemed possible that premature release of cortical granules might provide an explanation for the zona hardening seen here, a suspicion borne out by two observations. First, exposure of isolated zonae to DMSO at 37oC did not cause them to become resistant to chymotrypsin digestion, suggesting that the DMSO must act upon the oocyte to stimulate the change in the zona, consistent with the requirement for cortical granule release (Vincent *et al.*, 1989a). Second, exposure of oocytes to DMSO at 37oC is associated with a reduction in the number of cortical granules in the oocyte cytocortex, coupled with evidence of their release into the sub-zonal space. How could exposure to DMSO at 37oC result in cortical granule release?

It is possible that a disrupting effect on the microfilament system of the oocyte is involved (Vincent et al., 1989b). This system is organised in control oocytes into a thin sub-cortical mesh from which short microvillous cores project into the oolemma. In the mouse, but not the human, the cytocortex adjacent to the spindle lacks microvilli and the subcortical actin mesh is much thicker (Maro et al., 1984, 1986). When oocytes are placed in DMSO at 37oC, the sub-cortical microfilament mesh becomes interrupted, the response being particularly marked in the region overlying the spindle, in which complete microfilament disassembly occurs in some oocytes. This effect of DMSO on microfilaments is reversed in most oocytes on removal of the cryoprotectant. It is interesting to note that exposure to DMSO at 4°C rather than 37°C does not cause disorganisation of microfilaments and neither does it lead to zona hardening. Thus, disruption of the sub-cortical actin mesh is associated with access of cortical granules to the oolemma; perhaps in the normal oocyte, the mesh acts to maintain their separation and prevent fusion.

Clearly, exposure to DMSO can lead to a block to fertilization of the mouse oocyte and a disorganisation of its spindle. In many oocytes, this damage is not reversed. Manipulating the conditions of exposure to DMSO can reduce some of the adverse effects and could lead to more success in oocyte cryopreservation. It will be important to analyse response to DMSO of human oocytes to see whether they are also susceptible in the same ways. However, it is already clear from these observations that the failure of oocyte cryopreservation should not suprise us.

Parthenogenetic activation

When a spermatozoon fuses with the oocyte, activation occurs, resulting in the completion of meiosis and extrusion of the second polar body. Although this process differs slightly in mice and humans (Pickering *et al.*, 1988), in both the cytoskeleton of the oocyte is involved. If the microfilament system is disrupted, polar body extrusion fails, leading to retention of an additional set of chromosomes, and thus gynogenetic triploids. Disorganisation of the spindle, such as can occur after exposure to nocodazole or colchicine, can also lead to retention of two chromosome sets. Failure of the block to polyspermy, due to inadequate release of the cortical granule contents, can lead to the incorporation of superfluous sets of paternal chromosomes and thus to androgenetic polyploidy. Counting the numbers of pronuclei that form in the zygote provides one way of assessing whether polyploidy has occurred. However, the presence of just two pronuclei does not mean that normal fertilization has occurred, since activation of mouse oocytes in the absence of a spermatozoon can occur, especially in oocytes ovulated several hours previously. This process is called parthenogenetic activation and is elicited by several diverse stimuli (Kaufman, 1983). Activation may be followed by polar body extrusion, in which case a haploid parthenogenote is generated with just one pronucleus, but many eggs do not form a second polar body and retain both sets of maternal chromosomes to form two female pronuclei and thus a diploid parthenogenote. These eggs are easily confused with fertilized oocytes, and can, moreover, develop through pre-implantation stages and implant, but die shortly thereafter (Surani, 1986). If the protocols used in therapeutic procedures, such as IVF or GIFT, cause activation of human oocytes, it will be difficult to distinguish these parthenogenotes from fertilized oocytes. The replacement of eggs or cleavage stages with no potential for development would occur.

We have recently found that, in certain circumstances, a significant level of parthenogenetic activation can occur spontaneously in human oocytes. Parthenogenesis can be detected most reliably by the immunocytochemical analysis of oocytes for the organisation of their microtubules and chromosomes. Activation can be assumed if a post-metaphase (anaphase, telophase, early interphase) spindle with segregating maternal chromosomes and no evidence of paternal chromosomes is found. In a series of over 100 fresh and aged human oocytes, we found activation rates of 20 to 30% (Pickering *et al.*, 1988). In this study, the eggs had been exposed to hyaluronidase to loosen the cumulus mass, and to acid Tyrode's solution to remove the zona pellucida.Either of these procedures could have led to activation but it was also possible that handling *per se* was responsible. Hyluronidase is used clinically in the assessment of oocyte quality, in the preparation of oocytes for insemination with poor quality spermatozoa or in preparation for sperm microinjection (Mahadevan & Trounson, 1985; Lazendorf *et al.*, 1988). Acid Tyrode's solution has been used in zona drilling or to facilitate removal of a polar body for purposes of preimplantation diagnosis of genetic disease (Gordon & Talanksky, 1986). We undertook a systematic analysis of activation rates under different conditions to show that acid Tyrode's solution was the sole responsible agent. Hyaluronidase and pronase (an alternative approach to zona removal) did not activate human oocytes (although mouse oocytes can be activated by either; Kaufman, 1983; Johnson *et al.*, 1989). Simply handling oocytes did not seem to cause significant levels of activation either (Johnson *et al.*, 1989). These results are reassuring, but do illustrate that activation is a possibility when new protocols are being developed. They also show that activation agents effective for the mouse oocyte may differ from those effective in the human. Tests must be done on human oocytes directly.

Conclusions

Our studies indicate that the human and mouse oocytes are readily disturbed in ways that impair both fertilization and developmental potential. It seems likely to us therefore that a significant part of the failure experienced in therapeutic programmes may be induced. We need more careful analyses to determine which manipulations cause irreversible damage and so to avoid this problem. It is clear that such analyses cannot come from work on mouse oocytes alone, since human oocytes show significant differences. Research on human oocytes and pre-embryos themselves will be required.

Acknowledgements

The work reported here was funded by the Medical Research Council and approved by the Interim Licensing Authority.

References

Angell RR, Templeton AA, Aitken RJ (1986) Chromosome studies in human in vitro fertilization. Hum Genet 72:333-339

Chen C (1986) Pregnancy after human oocyte crtopreservation. Lancet 1(no.8486):884-886

Gordon JW, Talansky BE (1986) Assisted fertilization by zona drilling: a mouse model for correction of oligospermia J exp Zool 239:247-354

Johnson MH (1989) The effect of fertilization of exposure of mouse oocytes to dimethylsulphoxide: an optimal protocol. J IVF Embryo Transfer 6:168-175

Johnson MH, Pickering SJ (1987) The effect of dimethylsulphoxide on the microtubular system of the mouse oocyte. Development 100:313-324

Johnson MH, Pickering SJ, George M (1988) The influence of cooling on the properties of the zona pellucida of the mouse oocyte. Hum Reprod 3:383-387

Johnson MH, Pickering SJ, Braude PR, Vincent C, Cant A, Currie J (1989) Acid Tyrode`s solution can stimulate parthenogenetic activation of human and mouse oocytes. Fert Steril in press

Johnson MH, Vincent C, Braude PR, Pickering SJ (1990) The cytoskeleton of the oocyte; its role in the generation of normal and abberrant pre-embryos. In The Establishment of Successful Human Pregnancy. Raven Press, in press

Kaufman MH (1983) Early Mammalian Development: Parthenogenetic Studies. Cambridge University Press, Cambridge

Lazendorf SE, Slusser J, Maloney MK, Hodgen GD, Veeck LL, Rosenwaks Z (1988) A preclinical evaluation of pronuclear formation by microinjection of human spermatozoa into human oocytes. Fert Steril 49:835-842

Magistrini M, Szollosi D (1980) Effects of cold and isopropyl-N-phenylcarbamate on the second meiotic spindle of mouse oocytes. J Cell Biol 22:699-707

Mahadevan MM, Trounson AO (1985) Removal of the cumulus oophorus from the human oocyte for in vitro fertilization Fert Steril 43:263-267

Maro B, Howlett SK, Webb M (1985) Non-spindle microtubule organising centres in metaphase II-arrested mouse oocytes. J Cell Biol 101:1665-1672

Maro B, Johnson M, Pickering S, Flach G (1984) Changes in actin distribution during fertilization of the mouse egg. J Embryol Exp Morphol 81:211-237

Maro B, Johnson MH, Webb M, Flach G (1986) Mechanism of polar body formation in the mouse oocyte: an interaction between the chromosomes, the cytoskeleton and the plasma membrane. J Embryol Exp Morphol 92:11-32

Pickering SJ, Braude PR, Johnson MH, Cant A, Currie J (1990) Transient cooling to room temperature can cause ireversible disruption of the meiotic spindle in the human oocyte. Fertil Steril submitted.

Pickering SJ, Johnson MH (1987) The influence of cooling on the organization of the meiotic spindle of the mouse oocyte. Hum Reprod 2:207-216

Pickering SJ, Johnson MH, Braude PR (1988) Cytoskeletal organisation in fresh, aged and spontaneously activated human oocytes. Hum Reprod 3:978-979

Plachot M, de Grouchy J, Junca AM, Mandelbaum J, Salat-Baroux J, Cohen J (1988) Chromosome analysis of human oocytes and embryos: does delayed fertilization increase chromosome imbalance? Human Reprod 3:125-127

Surani MAH (1986) Evidences and consequences of differences between maternal and paternal genomes during embryogenesis in the mouse. In: Experimental Approaches to Mammalian Embryology, J Rossant, RA Pedersen (eds) pp.401-436. Cambridge University Press, New York

Van Uem JF, Siebzehnrubl ER, Schuh B, Koch R, Trotnow S, Lang N (1987) Birth after cryopreservation of unfertilized oocytes. Lancet 1(no.8535):752-753

Veiga A, Calderon G, Santalo J, Barri PN, Egozcue J (1987) Chromosome studies in oocytes and zygotes from an IVF programme. Hum Reprod 2:425-430

Vincent C, Pickering SJ, Johnson MH (1989a) The zona hardening effect of dimethylsulphoxide requires the presence of an oocyte and is associated with a reduction in the the number of cortical granules present. J Reprod Fert submitted

Vincent C, Pickering SJ, Johnson MH (1989b) unpublished observations

Voluntary Licensing Authority for Human In Vitro Fertilization and Embryology, Fourth Report (1989) Medical Research Council, London

IS THE EGG ACTIVATION-INDUCED INTRACELLULAR pH INCREASE NECESSARY FOR THE EMBRYONIC DEVELOPMENT OF *XENOPUS LAEVIS* (ANURAN AMPHIBIAN)?

N Grandin and M Charbonneau

Laboratoire de Biologie et Génétique du Développement
URA CNRS 256; Univ. Rennes I
Campus de Beaulieu
35042 Rennes Cedex
France

NATO ASI Series, Vol. H 45
Mechanism of Fertilization
Edited by B. Dale

SUMMARY

The initiation of embryonic development in the anuran amphibian, *Xenopus laevis*, is accompanied by an increase in intracellular pH (0.3 pH unit) which starts around 5 min after fertilization. This intracellular pH (pHi) increase has no known physiological role. In the present study, we have analyzed the consequences of maintaining the pHi of fertilized eggs at the same level as that in unfertilized eggs during the early embryonic cell divisions. Among several weak acids and drugs capable of producing changes in pHi, CO_2, a weak acid, was preferred because it allowed to impose very rapid and controllable variations in pHi. Our results indicate that those embryos in which the fertilization-induced pHi increase was prevented showed a delay in development as early as the second or third division. When pHi "clamping" was maintained after the lengthening of the cell cycle was observed, this frequently resulted in a total arrest of embryonic development. On the other hand, artificial perturbation of the periodic pHi oscillations (0.05 pH unit around the plateau value, 7.7-7.8) which begin at the time of the first cell division, had no noticeable effect on embryonic development.

INTRODUCTION

During fertilization, fusion between the female and male gametes is followed within a few minutes by a burst of reactions which prepare the newly formed embryo to cell division. In various cell types, several of these reactions depend on variations in the activities of particular ions. Thus, activation of the egg of the anuran amphibian, *Xenopus laevis*, is accompanied, at the time of fertilization by a depolarization of the plasma membrane, followed by an increase in intracellular free calcium ($Ca^{2+}i$) and an increase in intracellular pH (pHi). Activation of the egg also corresponds to a reinitiation of the cell cycle, since the nucleus evolves from the metaphase II-blocked stage of meiosis to form the female pronucleus which fuses with the sperm nucleus. This process is under the control of an M-phase promoting factor (MPF), whose activity is controlled by cytoplasmic Ca^{2+} ions (Lohka and Maller, 1985), probably originating from the transient $Ca^{2+}i$ rise taking place about 1 min after egg activation (Busa and Nuccitelli, 1985). The $Ca^{2+}i$ transient has probably similar triggering effects on several other events of egg activation, such as the opening of ion channels, cortical granule exocytosis, cortical contraction and sperm nuclear decondensation.

In *Xenopus* eggs, a permanent rise in pHi has been shown to occur about 10 min after activation (Webb and Nuccitelli, 1981). No role has been attributed to this pHi increase in fertilized *Xenopus* eggs, contrary to the situation in other cells in which pHi is known to play a role in protein synthesis, DNA synthesis, or more generally in cell activation or proliferation (reviewed by Busa and Nuccitelli, 1984; Epel and Dubé, 1987). In the present paper, we have addressed the question of knowing whether the pHi increase accompanying *Xenopus* egg activation is necessary or not to the increase in the rate of protein synthesis and to the subsequent cell divisions of the embryo. A major problem in the regulation of pHi in *Xenopus* eggs is that its mechanism is still unknown. Consequently, we have used the weak acid CO_2 to prevent the normally occurring cytoplasmic alkalinization, as assessed by measuring pHi with intracellular microelectrodes, and evaluate the consequences on the rate of protein synthesis and early embryogenesis. Our results indicate that "clamping" the pHi of activated eggs to a value close to that in unactivated eggs did not prevent the activation-induced stimulation of protein synthesis. On the other hand, preventing this same pHi increase during the first 2 or 3 embryonic cell divisions produced a lengthening of the cell cycle.

MATERIALS AND METHODS

Animals, gametes and solutions

Mature gametes of *Xenopus laevis*, reared at the laboratory, were handled as described elsewhere (Charbonneau *et al*, 1986). Artificial activation, fertilization and culture of the eggs and embryos were performed in F1 medium, modified from Hollinger and Corton (1980), which contained (mM): NaCl, 31.2; KCl, 1.8; $CaCl_2$, 1.0; $MgCl_2$, 0.1; NaOH, 1.9; buffered with Hepes, 10.0, at pH 7.4 or with Capso, 10.0, at pH 9.0.

Microinjections and pHi measurements

Oviposited (unactivated jellied) eggs were dejellied by 2% cysteine in F1, pH 7.8, for about 5 min and thoroughly washed in F1. Dejellied eggs were placed in holes punched in the agarose bottom of a glass dish covered with F1. Glass capillaries (GC-150, Clark Electromedical Instruments) were pulled on a Campden micropipet puller. Each micropipet was slightly broken at the tip, to attain a diameter of 10 to 20 µm, and calibrated to deliver a volume of 40 nl, after measuring the length and diameter under a microscope. The same pipet was used for one series of experiments, including control and treated eggs. Eggs from the same experiment were injected at 5 min interval at most. Artificial activation was triggered upon microinjection. Intracellular pH was measured with microelectrodes filled with an H^+-selective resin (Fluka) fabricated as described in Charbonneau *et al* (1985). Microelectrodes were calibrated before and after each experiment.

Protein synthesis measurements

Each egg was injected with 40 nl of ^3H-leucine (1 Ci/mmol, Amersham) concentrated to deliver around 80 pmoles, in order to flood the endogenous leucine pool (Shih *et al*, 1978; Wasserman *et al*, 1982). The rate of incorporation into proteins was measured by counting the radioactivity in the TCA-precipitable material as described in Grandin and Charbonneau (1989a). The absolute rate of protein synthesis was calculated assuming that newly synthesized proteins are 10% leucine by weight (Shih *et al*, 1978; Wasserman *et al*, 1982).

CO_2 treatment

Since the physiological mechanism responsible for pHi regulation in *Xenopus* eggs and embryos is not known yet, it was necessary to compensate the activation-induced pHi increase by continuously perfusing the egg with F1 solution containing various amounts of CO_2. CO_2 is a weak acid which easily passes through plasma membranes, binds intracellular water, thus releasing into the cytoplasm HCO_3^- and H^+ ions (reviewed by Roos and Boron, 1981). Such a mechanism is actually operating in *Xenopus* eggs (Fig. 1). The advantages of using CO_2 to lower pHi, at least in *Xenopus* eggs, are four-fold (1) CO_2 acts very rapidly on pHi, on the order of a few seconds; (2) CO_2 is very easy to dose in the standard solution by counting the time of bubbling; (3) CO_2 has apparently no other effects besides those on pHi (Roos and Boron, 1981), provided pHi is not driven too acidic to levels uncompatible with most biochemical reactions; (4) the effects of CO_2 are reversible; for instance, when it is used as an anesthetic preventing egg activation, pHi returns to its initial value within 15 min after rinsing and the egg recovers its ability to activate (Grandin and Charbonneau, 1989a).

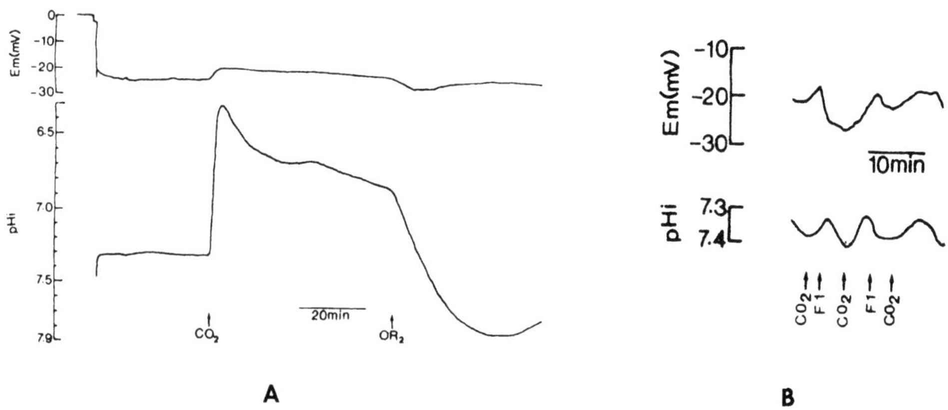

Fig. 1: Effects of CO_2 on the pHi of oocytes and eggs of *Xenopus laevis*. The physiological solution containing 10mM $NaHCO_3$ (A) or without $NaHCO_3$ (B) was gassed with CO_2 and perfused around the oocytes (A) or eggs (B) in the recording chamber (arrow CO_2). In both A and B, the upper traces represent the membrane potential (Em) and the lower traces, the intracellular pH (pHi). The Em value recorded with a second microeletrode impaled in the same cell is subtracted at the pen recorder input from the value recorded by the pH microelectrode, Em + pHi.

In A, the solution (OR_2, Wallace *et al*, 1973) was gassed with CO_2 for about 1 min, resulting in a very large acidification of the cytoplasm of the oocyte (manually isolated from *Xenopus* ovaries, stage VI, Dumont, 1972). Intracellular pH promptly returns towards its initial value upon rinsing (arrow OR_2). Note the so-called "alkaline rebond" occurring after washing off CO_2. After rinsing, CO_2 tends to leave the cell in order to restore the disrupted equilibrium due to the removal of external CO_2. Such a reaction necessitates that internal HCO_3^- ions take up H^+ ions, to form H_2O and CO_2, which produces an increase in pHi, the alkaline rebond (see Thomas, 1976).

In the present study, CO_2 was used at low doses (B), in order to control pHi and maintain it at a chosen value. Perfusion of the eggs can be done alternatively with a solution containing CO_2 and a CO_2-free solution, thus allowing the experimentator to control pHi when it is changing with cell metabolism. Note that the egg plasma membrane depolarizes at each addition of CO_2, which is due to the activation of a Cl^--dependent conductance (Charbonneau, unpublished results).

RESULTS

The activation-induced increase in intracellular pH

In *Xenopus* eggs, pHi, measured with pH-sensitive glass microelectrodes or ^{31}P-NMR, has been found to increase by about 0.3 pH unit, 10 min after fertilization, and to remain at this elevated value until at least the blastula stage (Webb and Nuccitelli, 1981; Nuccitelli *et al*, 1981). Our measurements with microelectrodes containing an H^+-selective neutral carrier-based resin confirmed these observations (Fig. 2). The pHi increase was seen to start 4 to 8 min after egg activation and to stabilize 25 to 40 min after egg activation (Fig. 2), earlier than in the study by Webb and Nuccitelli (1981) in which these events were detected respectively 10 and 60 min after egg activation.

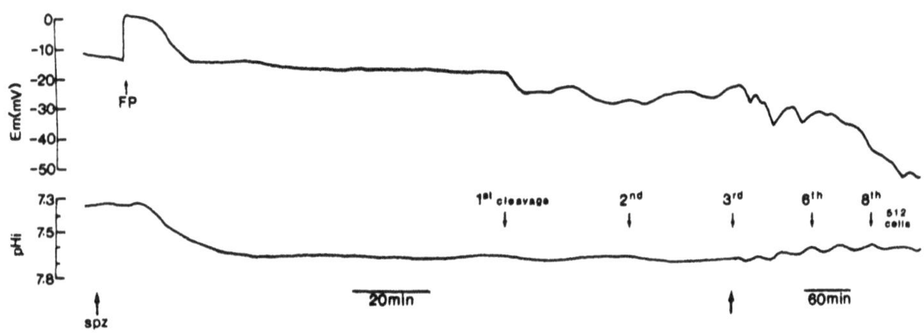

Fig. 2: Variations of pHi during fertilization and early development of *Xenopus laevis*. Around 5 min after the fertilization potential (FP), which represents the sperm-induced membrane depolarization of the egg, pHi begins to increase and reaches a stable permanent value 20 to 35 min later. After first cell division (1.5 hr at 22°C), pHi cyclically oscillates around its basal elevated value, each cycle corresponding to one cell cycle (duration 25 to 30 min). Note that the chart speed was changed at the second arrow. In the present study, pHi was 7.39 ± 0.09 S.D. (n=90) in unactivated eggs and 7.67 ± 0.07 (n=34) in activated eggs, the increase in pHi upon activation being 0.29 ± 0.04 (n=33) pH unit.

In order to evaluate the role of pHi during egg activation and embryonic development, we tried to prevent the associated changes in pHi. Changing the pH of the external solution between 6.5 and 9.5 had very little effect on the basal pHi of *Xenopus* eggs (at most 0.08 pH unit in some experiments). In addition, neither applying the potent inhibitor of the Na^+/H^+ and $Na^+/HCO_3^-/Cl^-$ exchangers, methyl propyl amiloride (gift of Dr. Pouysségur, Nice, France) or DIDS, an inhibitor of the HCO_3^-/Cl^- exchanger, had noticeable effects on basal pHi or on the increase accompanying egg activation (Fig. 3A). These three exchangers are at the origin of classical pHi regulating mechanisms operating in most cells so far studied (reviewed by Roos and Boron, 1981; Thomas, 1984). Similarly, the use of DCCD (N,N'-dicyclohexyle carbodiimide), an inhibitor of proton pumps (Solioz, 1984), did not affect pHi or physiological pHi changes when used at 1 to 100 µM (data not shown). Another hypothesis, which was tested in the present study was that the cortical granule membrane fusing with the plasma membrane upon exocytosis, which occurs between 1 and 3 min after activation, might contain proton pumps, or other pHi regulating molecules, capable of modifying pHi. For this purpose, the cortical reaction (exocytosis) was inhibited by $LaCl_3$ (which in other systems act by blocking Ca^{2+} fluxes). In this case, activation-induced pHi changes were similar to those taking place in the presence of the cortical reaction (Fig. 3C,D). Finally, we also verified that external calcium was not needed in the reaction responsible for the pHi increase (Fig. 3B).

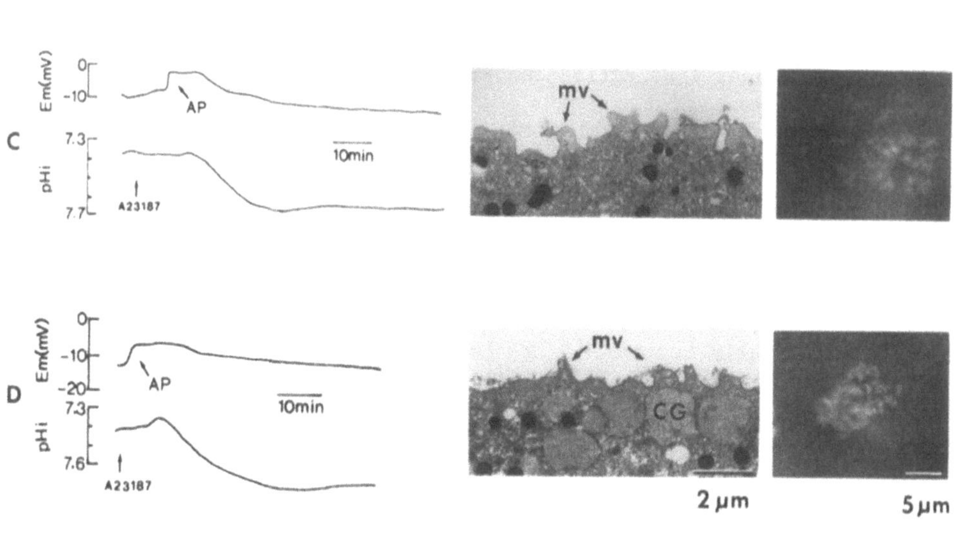

Relations between the pHi increase and the increase in the rate of protein synthesis

In other systems, pHi changes have been shown to be associated with an increased metabolic activity (reviewed by Busa and Nuccitelli, 1984). In some cases, an increase in pHi was found to stimulate the rate of protein synthesis accompanying the activation of embryonic development, as it is the case in sea urchins (Grainger *et al*, 1979). In other cases, cell cycle-dependent pHi variations were found to control protein synthesis, as in *Dictyostelium*, for instance (Aerts *et al*, 1985). In a recent study, we have demonstrated that in unactivated *Xenopus* eggs whose pHi was artificially increased with weak bases, the rate of protein synthesis was similar to that in unactivated eggs, whereas activated eggs exhibited a 2.5-fold increase in the rate of protein synthesis with respect to unactivated eggs (Grandin and Charbonneau, 1989a). This apparent independency between pHi and protein synthesis in the intact egg was confirmed *in vitro*, in a cell-free system prepared with *Xenopus* egg extracts (Grandin and Charbonneau, 1989a).

Fig. 3: Effects of various treatments on the pHi increase accompanying *Xenopus* egg activation.

A- The amiloride analog, 5-(N-methyl,N-propyl) amiloride (MPA), an inhibitor specific for the Na^+/H^+ antiporter (L'Allemain *et al*, 1984), used here at 100 µM, and DIDS (4,4'-diisothiocyanostilbene-2,2'-disulfonic acid), which specifically inhibits anion fluxes (Cabantchik *et al*, 1978) and hence the HCO_3^-/Cl^- and Na^+-coupled HCO_3^-/Cl^- exchanges (Russel and Boron, 1976), used here at 1 mM, could not prevent the egg activation-induced pHi increase. AP: activation potential.

B- Incubation of the eggs, prior to activation, in a nominally Ca^{2+}-free solution (0 Ca + 1 mM EGTA), did not prevent the occurrence of a normal pHi increase. A,B: activation with 4 µM A23187.

C,D- Treatment of the eggs with $LaCl_3$, an inhibitor of Ca^{2+} fluxes (Evans, 1983) completely inhibited the cortical reaction of exocytosis. Compare the electron micrographs of the peripheral cytoplasm of control activated eggs (in C, 30 min after prick-activation in F1) and treated eggs (in D, 30 min after prick-activation in F1 + 10 mM $LaCl_3$). Cortical granules (CG) remained present in the cortex of $LaCl_3$-treated eggs and microvilli (mv) remained short (D), a situation that normally exists before egg activation, whereas in control eggs cortical granules were extruded and microvilli elongated and thickened (C). On the other hand, nuclear cycle resumption normally occurred in $LaCl_3$-treated eggs, as attested by observing the state of the chromosomes or nucleus stained with bisbenzimide. In both $LaCl_3$-treated and control activated eggs, the nucleus evolved from metaphase II to anaphase II, about 10 min after artificial activation by the calcium ionophore A23187 (4 µM). Experimental procedures for electron microscopy were as described in Charbonneau *et al* (1986) and for bisbenzimide staining as in Grandin and Charbonneau (1989a).

In the present study, we have extended our investigation on the possible relationships between pHi and the rate of protein synthesis by measuring this latter parameter in activated eggs in which pHi was "clamped" at 7.5 with CO_2 (Fig. 4). Continuous perfusion of the eggs with CO_2 (see Materials and Methods) allowed to control the pHi and to maintain it at the desired value. When pHi was maintained at the value measured in the unactivated egg (7.40 ± 0.11 S.D., n=47) from between activation until 1 hr or 1.5 hr later, the rate of protein synthesis was similar to that in normal activated eggs (pHi: 7.69 ± 0.09, n=11). Thus, control eggs, injected with ^{3}H-leucine, synthesized the equivalent of 26.0 ± 17.0 S.D. ng protein/hr/egg (n=38; 20 females), while treated eggs synthesized the equivalent of 23.0 ± 11.0 ng protein/hr/egg (n=41; 20 females). Experiments with ^{35}S-methionine gave similar results (data not shown). In some experiments, maintaining the pHi of the embryos between 7.0 and 7.1 during an hour had no effect on protein synthesis with respect to control embryos.

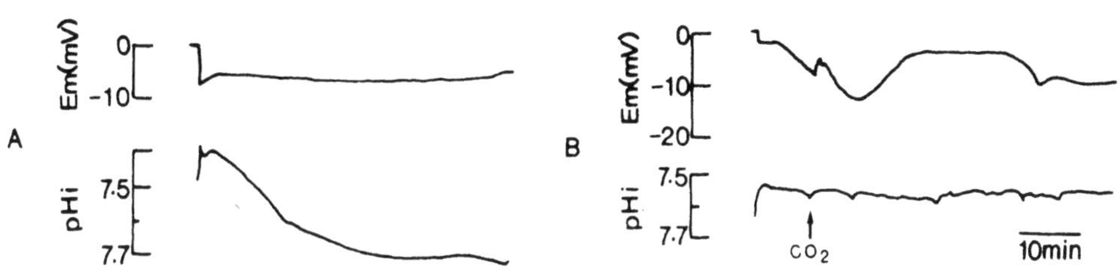

Fig. 4: Principle of the experiments designed to maintain pHi of activated or fertilized eggs at the same level as before activation. Dejellied eggs were injected with ^{3}H-leucine (see Materials and Methods). Since microinjection triggered activation, these eggs had to be rapidly transferred in the electrophysiological recording chamber and impaled with microelectrodes before pHi started to increase. As soon as pHi increased, the F1 solution was exchanged for an F1 solution containing CO_2 (arrow) and pHi maintained at its original value, that of the unactivated egg, for 1 or 1.5 hr. In such "clamped" eggs, protein synthesis, whose rate increases 2.5-fold after activation (Grandin and Charbonneau, 1989a), was similar to that in control eggs displaying an elevated pHi. On these traces, the absence of the activation potential indicates that the eggs were impaled after activation. (A: control, B: CO_2-treated).

Relations between pHi and the cell cycle

On first analysis, the values of 7.4 and 7.7 both represent optimum levels of pHi for protein synthesis. However, further increasing pHi in fertilized eggs by 0.3-0.4 pH unit lead to a partial depression of protein synthesis which resulted in an arrest of cell division (Grandin and Charbonneau, 1989b). This value of about 8.2 also corresponds to the value at which protein synthesis is depressed *in vitro*, both in egg cytoplasmic extracts and in rabbit reticulocyte lysates (Grandin and Charbonneau, 1989a,b). In the present study, the sensitivity of embryonic development to the pHi level was more precisely investigated, particularly the question of knowing whether an elevation of pHi at fertilization was required for a correct development. There again, CO_2 was used in order to keep the pHi of fertilized eggs at the same level as that in unactivated eggs (Fig. 5). In this series of experiments, pHi was 7.39 ± 0.08 (n=32) in unactivated eggs, and 7.65 ± 0.07 (n=12) in fertilized eggs. Our results indicate that the normal cell cycle was delayed when pHi was maintained at the value of the unactivated egg for at least 140 min (Table 1). Shorter periods of "pHi clamping" had no effect on embryogenesis. In *Xenopus*, the interval between fertilization and first cell division is 1.5 hr (at 22°C), while the following cell cycles last between 20 and 30 min. In our experiments, the effect on embryogenesis of maintaining a constant pHi was a function of time rather than a function of the number of imposed cell cycles.

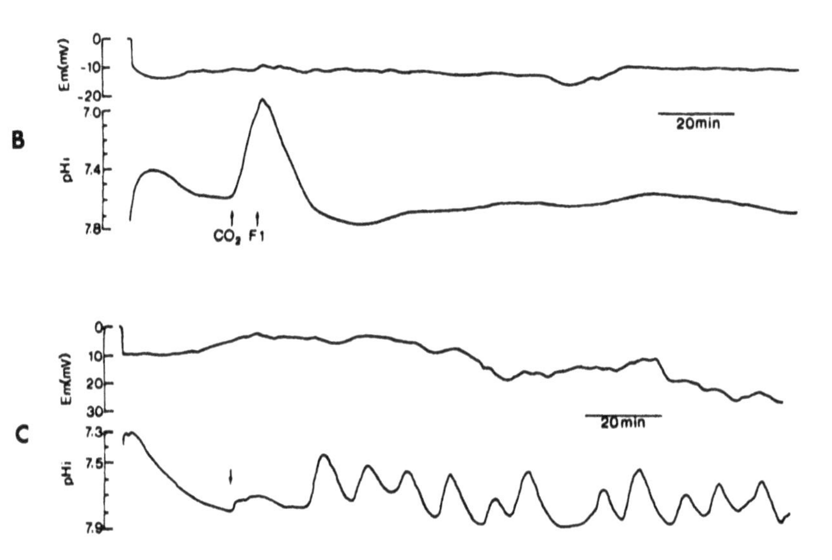

In order to rule out the possibility that cell division arrest was not only due to the prevention of the pHi increase, two sorts of controls were conducted. The first control was to verify that minor peaks accidentally produced during treatment with CO2 were not the real cause of cell division arrest. Such peaks were rare and of limited amplitude and duration. In two experiments in which pHi of fertilized eggs was purposedly driven from 7.7 to 7.0-7.1 (that is at least 0.3 pH unit more acidic than the value of the unactivated eggs) for about 15 min, there was no inhibitory effect on embryogenesis (Fig. 5B). The second control was made in order to rule out the possibility that CO2 was in some way poisoining the cell, either by itself or by allowing the accumulation of some unusual products, possibly leading to reductions in the activities of some specific enzymes. The only reliable control we could find to test this hypothesis was to perfuse fertilized eggs with CO2 in such a way that pHi oscillated around 7.7 instead of 7.4 (compare A and C in Fig. 5). In two such control experiments, cell division was not perturbed with respect to untreated fertilized eggs, eventhough the treated eggs received as much CO2 as in the experiments in which cell division was delayed.

Fig. 5: Relationships between pHi and the cell cycle were analyzed during a series of 11 experiments, using eggs from 17 females. Jellied eggs were inseminated in the recording chamber and impaled with microelectrodes 5 to 10 min after insemination, at the time the cortical contraction, indicating a successful fertilization, became visible. When pHi began to increase a few min later, perfusion with CO_2 was started (A, arrow). Control eggs run simultaneously on another set-up, in the same room, were exactly in the same conditions as CO_2-treated eggs (same females, same temperature, same number of solution exchanges). Starting at the first cell division, embryos were continuously observed and their development scored individually. Since washing off external CO_2 generated an alkaline rebond (see Fig. 1A) susceptible of interfering with cell division in a manner undistinguishable from that due to CO_2, only those events observed in the recording chamber before rinsing, and while pHi was recorded, were taken into account.

Two sorts of controls were performed. In B, an "acidic peak" was purposedly generated in order to mimic accidental displacement of pHi from the chosen value of clamp. This had no effect on cell division. In C, CO_2 was continuously perfused around the embryos, but this time the pHi level was kept oscillating around the normal elevated value. This had no effect on cell division, which rules out the possibility of some poisoining effect of CO_2.

Table 1: Relations between the level of pHi and cell division

unact. pHi [a]	fertil. pHi [b]	pHi inc.	duration (min)	develop. stage [c]	number clamped	number cont. [d]
7.47	7.90	0.35	140	no effect	7	10
7.51	7.59	0.28	175	no effect	17	20
7.44	7.70	0.24	180	no effect [e]	13	9
7.37	7.63	0.31	155	2-cell	6	6
7.39	-	-	155	4-cell	7	8
7.31	7.61	0.35	165	4-cell	12	14
7.39	7.59	0.28	180	2-cell	9	11
7.51	-	-	180	4-cell	10	9
7.52	7.71	0.29	190	4-cell	7	7
7.42	-	-	190	4-cell	6	6
7.37	7.71	0.35	190	2-cell	8	5
7.32	-	-	190	8-cell	5	6

a-pHi of the unactivated egg = pHi maintained during division
b-pHi of fertilized controls
c-Developmental stage at which lengthening of cell division was observed
d-Number of "pHi-clamped" eggs with similar effects (same female, same recording chamber). Number of control eggs (same female, same conditions in a neighbouring recording chamber)
e-In 3 other experiments, using eggs from 5 females, in which pHi was maintained at the unactivated level (7.32, 7.36 and 7.39) during 95, 100 and 105 min, there was no effect on cell division.

DISCUSSION

One of the main findings of the present study is that cell division in *Xenopus* embryos is delayed when the increase in intracellular pH accompanying egg fertilization is prevented. This represents so far the only possible role that can be attributed to pHi in *Xenopus* oocytes, eggs and embryos. The present results also demonstrate that pHi does not appear to play a role in the regulation of protein synthesis, thus reinforcing previous data obtained in *Xenopus* oocytes (Cicirelli *et al*, 1983) and eggs (Grandin and Charbonneau, 1989a).

The success of our technique employed to prevent the increase in pHi in activated or fertilized eggs is based on the use of CO_2 (see Fig. 1). We found that CO_2 was a very convenient tool to manipulate pHi, unlike sodium acetate and propionate which produced slow variations, difficult to control once initiated (data not shown). This is contrary to the results by Lee and Steinhardt, 1981) who used sodium propionate to lower pHi in *Xenopus* embryos and reported that CO_2 "caused precipitous decreases of pHi and was not useful for imposing small changes". The present data do not support such conclusions. Lee and Steinhardt (1981) have reported that lowering pHi in *Xenopus* embryos from 7.66 to about 7.3 had no effect on cell division, and that further lowering pHi resulted in an arrest of cell division (7.2) and furrow regression (7.0). Their results are contrary to ours, since we find an effect on cell division when pHi is lowered at 7.3-7.4, the pHi of the unactivated eggs. However, their results partially agree with ours, since they also found an effect at lower pHis. Data reported by Lee and Steinhardt (1981) suffer from two flaws: 1- the level of pHi before fertilization was not measured, making it impossible to establish correlations between the preexisting and clamped pHi values (our rule was to maintain pHi at the level existing before fertilization in order to eliminate egg to egg variability); 2- their technique for keeping pHi at a constant level is not, from their own statement, satisfactory (see Fig. 3 of their paper), thus rendering interpretation of their experiments difficult.

Our results suggest that the fertilization-associated pHi increase is involved in the regulation of cell division. A simple explanation would be that a certain level of pHi is needed for some biochemical reactions necessary to cell division. The most famous example is the case of phosphofructokinase which was inhibited by more than 90% when pH was lowered from 7.2 to 7.1 (Trivedi and Danforth, 1966; see other references in Roos and Boron, 1981). In the case of *Xenopus* embryos, the hypothesis of a reduction in enzymatic activities could be tested only if the specific defect resulting from pHi increase prevention was known more precisely. Another staightforward explanation to the role of pHi in cell division is that pHi might affect cell-cell coupling. Indeed, Turin and Warner (1980) have demonstrated that lowering pHi in the range 6.4-6.9 abolished current flow between blastomeres in *Xenopus*. A similar uncoupling has been observed in embryos of another amphibian, ***Ambystoma***, which was appreciable at pHis lower than 7.1 (Spray *et al*, 1981). However, it should be noted that dissociated, hence uncoupled , ***Ambystoma*** blastomeres continue to cleave (Hara, 1977). Moreover, the pHi level at which blastomeres become uncoupled is lower than that at which we observe an inhibition of cell division.

Acknowledgements: We thank Mrs. Annie Cavalier for cutting thin sections. This work was supported by grants from the INSERM, the Association pour la Recherche contre le Cancer and the Fondation Langlois.

REFERENCES

Aerts RJ, Durston AJ, Moolenar WH (1985) Cytoplasmic pH and the regulation of the *Dictyostelium* cell cycle. Cell 43: 653-657

Busa WB, Nuccitelli, R (1984) Metabolic regulation *via* intracellular pH. Am J Physiol 246: R409-R438

Busa WB, Nuccitelli, R (1985) An elevated free cytosolic Ca^{2+} wave follows fertilization in eggs of the frog, *Xenopus laevis*. J Cell Biol 100: 1325-1330

Cabantchik ZI, Knauf PA, Rothstein A (1978) The anion transport system of the red blood cell. The role of membrane protein evaluated by the use of "probes". Biochim Biophys Acta 515: 239-302

Charbonneau M, Busa WB, Grey RD, Nuccitelli, R (1985) External Na^{+} inhibits Ca^{2+}-ionophore activation of *Xenopus* eggs. Dev Biol 108: 369-376

Charbonneau M, Dufresne-Dubé L, Guerrier, P (1986) Inhibition of the activation reaction of *Xenopus laevis* eggs by the lectins, WGA and SBA. Dev Biol 114: 347-360

Cicirelli M, Robinson K, Smith LD (1983) Internal pH of *Xenopus* oocytes: A study of the mechanism and role of pH change during meiotic maturation. Dev Biol 100: 133-146

Dumont JN (1972) Oogenesis in *Xenopus laevis* (Daudin). I. Stages of oocyte development in laboratory maintained animals. J Morphol 136: 153-180

Epel D, Dubé F (1987) Intracellular pH and cell proliferation. In: Control of animal cell proliferation, vol III. Boynton AL, Leffert HM (eds) Academic Press, New-York, pp 363-393

Evans CH (1983) Interesting and useful biochemical properties of lanthanides. Trends Biochem Sci 8: 445-449

Grainger JL, Winkler MM, Shen SS, Steinhardt RA (1979) Intracellular pH controls protein synthesis rate in the sea urchin egg and early embryo. Dev Biol 68: 396-406

Grandin N, Charbonneau M (1989a) Intracellular pH and the increase in protein synthesis accompanying activation of *Xenopus* eggs. Biol Cell (in press)

Grandin N, Charbonneau M (1989b) An increase in the intracellular pH of fertilized eggs of *Xenopus laevis* is associated with inhibition of protein and DNA syntheses, and followed by an arrest of embryonic development. Exp Cell Res (in press)

Hara K (1977) The cleavage pattern of the axolotl egg studied by cinematography and cell counting. Wilhem Roux's Arch Dev Biol 181: 73-87

Hollinger TJ, Corton GL (1980) Artificial fertilization of gametes from the South African clawed frog, *Xenopus laevis*. Gam Res 3: 45-57

L'Allemain G, Franchi A, Cragoe E Jr, Pouysségur J (1984) Blockade of the

Na^+/H^+ antiport abolishes growth factor-induced DNA synthesis in fibroblasts. J Biol Chem 259: 4313-4319

Lee SC, Steinhardt RA (1981) Observations on intracellular pH during cleavage of eggs of *Xenopus laevis*. J Cell Biol 91: 414-419

Lohka MJ, Maller JL (1985) Induction of nuclear envelope breakdown, chromosome condensation, and spindle formation in cell-free extracts. J Cell Biol 101: 518-523

Nuccitelli R, Webb DJ, Lagier ST, Matson GB (1981) ^{31}P NMR reveals increased intracellular pH after fertilization in *Xenopus* eggs. Proc Natl Acad Sci USA 78: 4421-4425

Roos A, Boron WF (1981) Intracellular pH. Physiol Rev 61: 296-434

Russell JM, Boron WF (1976) Role of chloride transport in regulation of intracellular pH. Nature 264: 73-74

Shih RJ, O'Connor CM, Keem K, Smith LD (1978) Kinetic analysis of amino acid pools and protein synthesis in amphibian oocytes and embryos. Dev Biol 66: 172-182

Solioz M (1984) Dicyclohexylcarbodiimide as a probe for proton translocating enzymes. Trends Biochem Sci 9: 309-312

Spray DC, Harris AL, Bennett MVL (1981) Gap junctional conductance is a simple and sensitive function of intracellular pH. Science 211: 712-715

Thomas RC (1976) The effect of carbon dioxide on the intracellular pH and buffering power of snail neurones. J Physiol (Lond) 255: 715-735

Thomas RC (1984) Experimental displacements of intracellular pH and the mechanism of its subsequent recovery. J Physiol (Lond) 354: 3P-22P

Trivedi B, Danforth WH (1966) Effect of pH on the kinetics of frog muscle phosphofructokinase. J Biol Chem 241: 4110-4112

Turin L, Warner AE (1980) Intracellular pH in early *Xenopus* embryos: its effect on current flow between blastomeres. J Physiol (Lond) 300: 489-504

Wallace RA, Jared DW, Sega, MW (1973) Protein incorporation by isolated amphibian oocytes. III. Optimum incubation conditions. J exp Zool 184: 321-334

Wasserman WJ, Richter JD, Smith LD (1982) Protein synthesis during maturation promoting factor- and progesterone-induced maturation in *Xenopus* oocytes. Dev Biol 89: 152-158

Webb DJ, Nuccitelli R (1981) Direct measurement of intracellular pH changes in *Xenopus* eggs at fertilization and cleavage. J Cell Biol 91: 562-567

THE DISTRIBUTION AND EXOCYTOSIS OF CORTICAL GRANULES IN THE MAMMAL

D.G. Cran, AFRC, Institute of Animal Physiology and Genetics Research, Babraham Hall, Cambridge CB2 4AT, United Kingdom.

Introduction

As long as 35 years ago the observation was made that as a result of fertilization the zona pellucida was altered such that it became refractory to the ingress of supernumerary spermatozoa (Braden, Austin and David, 1954) and cortical granules were reported to disappear from the surface of the oocyte. It was a short step to suggest that substances liberated from these exocytosed organelles acted on the zona pellucida in some unknown manner to prevent the passage of sperm (Austin and Braden, 1956). Since this time it has become clear that a trypsin-like protease is involved (Gwatkin et al. 1973; Wolf and Hamada, 1977) although, while considerable progress has been made towards a full analysis of the contents of cortical granules from invertebrates such as the sea urchin, we are still far from an understanding in regard of the mammal. Further, it is clear that the granules must have a precise relationship with the oocyte plasma membrane in order that they may fuse with it, and, in addition, they must and also be capable of responding to signals resulting from sperm/oocyte fusion. It is with the distribution of the granules within the oocyte and the possible nature of intracellular signals involved in the exocytotic process that this communication will be concerned.

Distribution of Cortical Granules

In the immature oocyte the granules may be found at varying distances from the plasma membrane. However, as maturation proceeds, in those species in which this feature has been studied, there is a progressive centrifugal migration such that they come to lie within a few nanometres of the oolemma (Soupart and Strong, 1974, 1975; Cran et al, 1980; Suzuki et al, 1981; Sathananthan and Trounson, 1982; Kruip et al, 1983; Cran and Cheng, 1985). In the pig this change is temporally associated with the breakdown of the germinal vesicle. To date, however, there is little evidence to suggest that there is a direct correlation between the two events. In some species it has been reported that there is an increase in number during the final maturational phase following the LH surge (Sathananthan and Trounson, 1982; Zamboni, 1979; Cran and Cheng, 1985). These data have been based on sectioned material, and in terms of the number per unit length of plasma membrane there does appear to be an increase. However, this may not be a completely accurate representation of

NATO ASI Series, Vol. H 45
Mechanism of Fertilization
Edited by B. Dale

the total oocyte population since there may well be fluxes and losses which cannot be taken account of by such an approach. Procedures which allow the visualization of the granules in the intact oocyte would clearly overcome this problem and the papers by Lee et al, (1988), Cherr et al, (1988) and Ducibella et al (1988) where lectin binding properties have been exploited offer hopeful possibilities for the future.

Not only are there movements of the granules in a radial direction, there is also evidence that there may also be circumferential migration and loss of granules from the surface of the oocyte before fertilization. Nicosia et al. (1977) studying the distribution of the granules in the mouse, observed that in the mature cell approximately 20% of the cell surface homolateral to the spindle was totally devoid of granules. In addition, the plasma membrane in this zone was markedly smoother than in the remainder of the cell. Finally, the authors showed that sperm/oolemma fusion consistently occurred only in the villous region possessing cortical granules. Subsequently Okada et al. (1986) examining hamster oocytes showed that at the germinal vesicle stage the granules are evenly distributed but at metaphase II there is a cortical granule free zone which occupies some 8% of the oocyte surface (Fig. 1). This was attributed to a discharge of granule material into the perivitelline space and to migration of the granules. This latter observation has received support from observations based on the distribution of the lectin Lens culinaris agglutinin (LCA) (Ducibella et al. 1988). These have confirmed the previous finding that there is uniform distribution in immature oocytes.

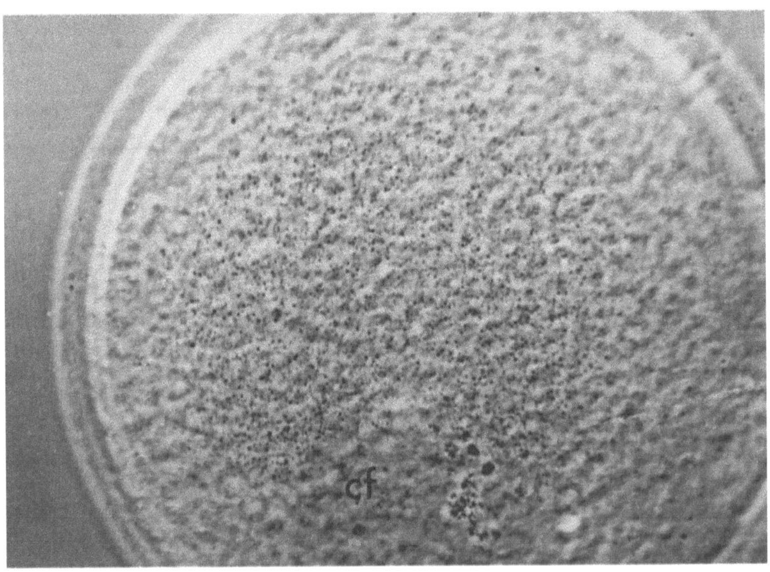

Fig. 1 Living mature hamster oocyte, slightly compressed to show the cortical granules. Note the cortical granule free area (CF) x 1500.

At 10 h following administration of hCG when the cell is at telophase/anaphase there is a marked clustering around the nucleus (Cran and Moor, in press) a region which is particularly rich in organelles, containing numerous mitochondria in addition to the cortical granules.

An understanding of the processes whereby a cortical granule free zone is formed may have important implications regarding the mechanisms of exocytosis as well as the site of fertilization. If as described above, the granule free zone results from premature loss, this raises some interesting considerations. Assuming that a particular subpopulation which is not involved in the raising of a fertilization block (and there is some evidence for at least two subpopulations (cf. Nicosia et al. 1977, Cran, 1985) is not released, why is a general block to sperm penetration not raised? Is the amount of material released insufficient, or does it not readily diffuse through the perivitelline space and thus interacts only with a small region of the zona pellucida? These observations would also indicate that there is little rotation of the oocyte since the relationship betwen the cortical granule free zone and the region of lack of penetration remains constant.

It seems likely that in order that the granules might undergo their outward movements within the cell and subsequent fusion, that there must be some mechanism, structural or otherwise, that would interact with them. Perhaps the most obvious intracellular inter-organelle relationship which might be involved is with the cytoskeleton, which has been shown on several occasions to be largely peripheral (e.g. Le Guen et al. 1989). In this regard it is perhaps relevant that subjection of the oocyte to a strong centrifugal force will displace all of the membraneous organelles with the exception of the cortical granules (Cran, 1987), (Fig. 2). However, addition of inhibitors of polymerization of tubulin and actin followed by mild centrifugation results in almost total exocytosis. It is tempting to suggest that there is a direct involvement of the peripheral skeleton and these movements. This evidence is lacking in mammalian oocytes. However, in those of the sea urchin there is a change in actin form from the F to the G type which is related to the timing of exocytosis.

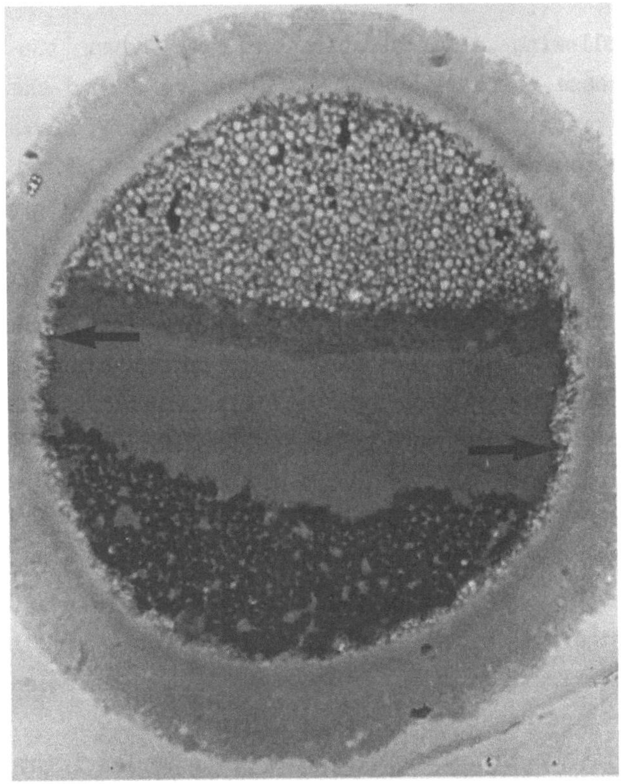

Fig. 2. Centrifuged sheep oocyte. The organelles are separated into clearly demarcated zones. At the edge of the organelle-free area cortical granules (arrows) are evident. X 850. From Cran, 1987.

Induction of exocytosis

As with many other secretory cells there is an increasing amount of evidence indicating that the induction of the loss of cortical granules is a result of some form of receptor mediated coupling. It seems quite clear that receptor mediated binding takes place between the zona pellucida and the fertilizing spermatozoa (Wassarman, 1983 inter alia), however, there is a need for the determination of the molecular nature of the interaction between sperm components and the oolemma. In this regard a recent paper by Kline et al. (1988) is of particular interest. These authors have shown that in Xenopus eggs in which receptors to serotonin and acetycholine had been artificially raised, the addition of agonists resulted in the elevation of the fertilization membrane. In addition, various sperm protein fractions will activate the eggs of the worm Direclus. These observations would suggest that some form of receptor mediated sperm/oolemma interaction is important in the initiation of the cortical reaction. However, as far as the author is aware, there is no comparable information for the mammal.

Whatever the mechanism, it is clear that some form of signal transduction system is operative to elicit the discharge of the cortical granules into the perivitelline space and thus induce the establishment of the block to polyspermy. Over the past few years there has been a burgeoning of interest in the involvement of inositol lipids in cell signalling. These studies have led to the generally accepted hypothesis that hydrolysis of membrane bound phosphatidylinositol 4,5-bisphosphate occurs after receptor activation, giving rise to two intracellular messengers, diacyglycerol (DAG) (Nishizuka, 1984) and inositol 1,4,5 trisphosphate Ins(1,4,5)P_3 (Berridge and Irvine, 1984) the latter of which is free to diffuse within the cytoplasm. Information of a biochemical nature regarding mechanisms operative are much more advanced in Xenopus and the sea urchin than in the mammal as is demonstrated by the contribution on G proteins by Whitaker in this volume. In such organisms evidence for cortical granule discharge in response to an external agent is seen by the elevation of the fertilization envelope. No such morphological change is apparent in the mammal, there being no overt structural change in the zona pellucida. Evidence for the occurrence of the cortical reaction must, therefore, rely on the presence or absence of the cortical granules or, more indirectly, on the binding and/or penetration of spermatozoa. Clearly the use of electron microscopy apart from relatively straightforward applications is too tedious to provide much useful information in this regard. The advent of the capacity to obtain numerical information using lectins alluded to above is timely.

Intracellular calcium also plays a key role in the exocytotic event. For example Turner et al. (1986) have shown that chelation of intracellular Ca^{2+} will prevent exocytosis in the sea urchin. In addition Miyazaki et al. (1986) and Igusa and Miyazaki (1986), using aequorin, have demonstrated that a wave of Ca^{2+} release spreads around the entire circumference of the cell emanating from the point of attachment of the fertilizing sperm on the oocyte plasma membrane. Miyazaki (1988) has also demonstrated that the injection of Ins(1,4,5)P_3 to give an intracellular concentration of [2nM]i and also the G-protein analogue guanosine -5'-O-(3-thiotrisphospate) (GTP-γ-S) from about [12 μM]i will provoke a release of calcium throughout the oocyte. Under these conditions a regular series of calcium transients similar to that observed after fertilization only occurs following GTP-γ-S at ~[70 μM]i. Following microinjection of Ins(1,4,5)P_3 at low dose levels there is only a single Ca^{2+} transient, a reasonably regular series only being observed at high dose levels ([6 μM]i).

Since intracellular release of Ca^{2+} is directly related to fertilization and membrane hyperpolarizations also take place as a result of microinjection of G-proteins and Ins(1,4,5)P_3 it would be tempting to suggest that there is a direct relationship between these events and the loss of the cortical granules together with the raising of the block to polyspermy. In a study examining cortical granule numbers in the hamster and the sheep Cran et al. (1988) have demonstrated that there is a pH and dose dependent effect on exocytosis resulting from the microinjection of Ins(1,4,5)P_3 with the effect being most pronounced at an alkaline pH. The lowest effective concentration was equivalent to an intracellular concentration of some 0.4 uM. In addition, microinjection of GTP-γ-S at a concentration of [32]i uM resulted in almost complete exocytosis while the non hydrolysable analogue GDP-γ-S was ineffective. In the case of GTP-γ-S the concentration required to obtain complete granule loss and to achieve a regular series of hyperpolarizations (Miyazaki, 1988) were similar. On the other hand there was approximately a 100 fold difference between the level giving a single initial hyperpolarization in response to Ins(1,4,5)P_3 and that allowing the release of granules. This disparity could be due to a number of reasons. For example, it is possible that a regular number of hyperpolarizations are necessary before full exocytosis occurs. This suggestion is supported to some extent by the observation that some 5 min elapses between microinjection of Ins(1,4,5)P_3 and full exocytosis (Cran, 1989). Alternatively there may be various unknown intermediate interactions which must take place before the final exocytotic reaction or it may be necessary to have a high local concentration. Kurasawa et al. (1989) have recently expanded information in this area by examining the effect of microinjecting Ins(1,4,5)P_3 on the capacity of mouse eggs to undergo fertilization. They have found that at low intracellular levels (5-10 nM) there is an inhibition of both binding and of the induction of the acrosome reaction and, in addition, there is a change in the electrophoretic properties of the zona protein ZP2 to ZP2f. There is, however, no effect on activation as is found with calcium releasing agents such as the ionophore A23187. In this regard the effect of Ins(1,4,5)P_3 closely resembles that occurring at fertilization and it is tempting to suggest that, in spite of the problem relating to the concentration necessary to induce a full cortical reaction that the two events are directly related.

In addition to the action of Ins(1,4,5)P_3 and its metabolites interest has also been taken of the other second messenger which results from the hydrolysis of phosphatidyl bisphosphate, DAG, which amongst other actions activates the action of protein kinase C. It would seem possible that either alone or in concert with inositol phospholipids, protein kinase C might act to induce exocytosis. Endo et al. (1987) have examined this possibility by looking at the effect of phorbol esters and DAG on fertilization and modifications of the zona pellucida of mouse eggs. They found that fertilization was indeed impaired and that this was a result of an alteration of the zona rather than of the egg itself. They further found that while binding of sperm to the zona was not impaired, penetration was. The conclusion may be drawn that the activation of protein kinase C does not result in a direct modification of the properties of the zona binding protein ZP3 but does on the other hand result in a change in that moeity which is involved in the activation of the acrosome reaction. These results are similar to those found following treatment with Ins(1,4,5)P_3 except that in the latter case there was also a modification of ZP3 to prevent binding. It remains to be seen what the interaction, if any, is between the two branches of this bifurcating second messenger system. These results however, only indicate that protein kinase C activation results in an impairment of penetration through zona modification, they do not tell us whether this is due to shedding of the cortical granules. Recent work in this laboratory has indicated that phorbol esters (activators of protein kinase C) act in a dose dependent manner to induce cortical shedding and experiments are in progress to determine whether the activation of this second messenger is a normal consequence of fertilization.

Conclusion

Much of what we know about the distributional changes of cortical granules during oocyte maturation is observational in nature. There is an indication that the peripheral cytoskeleton may be involved in the maintenance of spatial position. What is the molecular nature of this interaction, how does it change with maturational status and is it involved in the final exdocytotic events? In the mouse and the hamster a cortical granule free zone is produced during the final maturational phase of the oocyte which is brought about partly by the loss of the first polar body, by premature exocytosis, and to a lesser extent by circumferential movement. It remains to be elucidated whether there is a functional significance to this loss and how it does not result in an impediment to subsequent sperm penetration. With regard to the production of cell signals which result in the cortical reaction, it seems clear that the inositol lipids play a major role. What is the function of DAG and protein kinase C, and is there any interaction between the two pathways?

Finally there is a marked paucity of information regarding the biochemical nature of the cortical granules, changes at exocytosis and the details of the interaction with the investments of the oocyte at fertilization.

References

Austin CR and Braden AWH (1956) Early reactions of the rodent egg to spermatozoa penetration. J Exp Biol 33:358-365

Berridge MJ and Irvine RF (1984) Inositol trisphospate, a novel second messenger in signal transduction. Nature Lond 312:315-321

Braden AWH, Austin CR and David HA (1954) The reaction of the zona pellucida to sperm penetration. Aust J Biol Sci 7:394-409

Cherr GN, Drobnis EZ and Katz DF (1988) Localization of cortical granule constituents before and after exocytosis in the hamster egg. J Exp Zool 246:81-93

Cran DG (1987) The distribution of organelles in mammalian oocytes following centrifugation prior to injection of foreign DNA. Gamete Res 18:67-76

Cran DG (1988) Qualitative and quantitative structural changes during porcine oocyte maturation. J Reprod Fert 74:237-245

Cran DG (1989) Cortical granules during oocyte maturation and fertilization. J Reprod Fert Suppl 38:49-62

Cran DG and Cheng W T-K (1985) Changes in cortical granules during porcine oocyte maturation. Gamete Res 11:311-319

Cran DG and Moor RM (to be published) Programming the oocyte for fertilization. In Bavister, B Cummins, J Roldan, ERS (eds) Fertilization in Mammals. Plenum Press New York

Cran DG, Moor RM and Hay MF (1980) Fine structure of the sheep oocyte during antral follicle development. J Reprod Fert 59:125-132

Cran DG, Moor RM and Irvine RF (1988) Initiation of the cortical reaction in hamster and sheep oocytes in response to inositol trisphosphate. J Cell Sci 91:139-144

Ducibella T, Anderson E, Albertin DF, Aalberg J and Rangarajan S (1988) Quantitative studies of changes in cortical granule number and distribution in the mouse oocyte during meiotic maturation. Devel Biol 130:184-197

Endo Y, Schultz RM and Kopf GS (1987) Effects of phorbol esters and a diacyclglycerol on mouse eggs: inhibition of fertilization and modification of the zona pellucida. Dev Biol 119:199-209

Gwatkin RBL, Williams DT, Hartmann JF and Kniazuk M (1973) The zona reaction of hamster and mouse eggs: Production in vitro by a trypsin like protease from cortical granules. J Reprod Fert 32:259-265

Igusa Y and Miyazaki S (1986) Periodic increase of cytoplasmic free calcium in fertilized hamster eggs measured with calcium-sensitive electrodes. J Physiol (Lond) 377:193-205

Kline D, Simoncina L, Mandel G, Mane RA, Kado RT and Jaffe LA (1988) Fertilization events induced by neurotransmitters after injection of mRNA in Xenopus eggs. Science 241:464-467

Kruip TAM, Cran DG, Von Beneden TH and Dieleman SJ (1983) Structural changes in bovine oocytes during final maturation in vivo. Gamete Res 8:29-47

Kurasawa S, Schultz RM and Kopf GS (1989) Egg-induced modifications of the zona pellucida of mouse eggs: effects of microinjected inositol 1,4,5-trisphosphate. Dev Biol 133:295-304

Le Guen P, Crozet N, Huneau D and Gall L (1989) Distribution and role of microfilaments during early events of sheep fertilization. Gamete Res 22:411-425

Lee SH, Ahuja KK, Gilburt DJ and Whittingham DG (1988) The appearance of glycoconjugates associated with cortical granule release during mouse fertilization. Development 102:595-604

Miyazaki S (1988) Inositol 1,4,5-trisphosphate induced calcium release and GTP-binding protein mediated periodic calcium rises in golden hamster eggs. J Cell Biol 106:345-353

Miyazaki S, Hashimoto N, Yoshimoto Y, Kishimoto T, Igusa Y and Hiramoto Y (1986) Temporal and spatial dynamics of the periodic increase in intracellular free calcium at fertilization of golden hamster eggs. Dev Biol 118:259-267

Nicosia SV, Wolf DP and Inoue M (1977) Cortical granule distribution and cell surface characteristics in mouse eggs. Dev Biol 57:56-74

Nishizuka Y (1984) The role of protein kinase C in cell surface signal transduction and tumor promotion. Nature London, 308:693-698

Okada A, Yanagimachi R and Yanagimachi H (1986) Development of a cortical granule-free area of cortex and the perivitelline space in the hamster oocyte during maturation and following ovulation. J Sub Microsc Cytol 18:233-247

Sathananthan AH and Trouson AO (1982) Ultrastructural observations on cortical granules in human follicular oocytes cultured in vitro. Gamete Res 5:191-198

Souport N and Strong PA (1974) Ultrastructural observations on human oocytes fertilized in vitro. Fertil Steril 25:11-94

Souport D and Strong PA (1975) Ultrastructural observations on polyspermic penetration of zona free human oocytes inseminated in vitro. Fertil Steril 26:523-537

Suzuki S, Kitai H, Tojo R, Seki K, Oba M, Fujiwara T and Iizuka R (1981) Ultrastructure and some biologic properties of human oocytes and granulosa cells cultured in vitro. Fertil Steril 35:142-148

Turner PR, Jaffe WA and Fein A (1986) Regulation of cortical granule exocytosis in sea urchin eggs by inositol 1,4,5-trisphosphate and GTP-binding protein. J Cell Biol 102:70-76

Wassarman (1983) Oogenesis: synthetic events in the developing mammalian eggs. J F Hartmann (ed) In: Mechanisms and Control of Animal Fertilization. J F Hartmann (ed) Academic Press New York

Wolf DP and Hamada M (1977) Induction of zonal and egg plasma membrane blocks to sperm penetration in mouse eggs with cortical granule exudate. Biol Reprod 17:350-354

Zamboni L (1970) Ultrastructure of mammalian oocytes and ova. Biol Reprod (Suppl) 2:44-63

RECEPTORS, G-PROTEINS, AND ACTIVATION OF THE AMPHIBIAN EGG

Douglas Kline, Raymond T. Kado[1], Gregory S. Kopf[2], and Laurinda A. Jaffe[3]

Department of Physiology
Ponce School of Medicine
P.O. Box 7004
Ponce, Puerto Rico 00732 U.S.A.

INTRODUCTION

We have investigated the role of guanine nucleotide-binding proteins (G-proteins) in activation of the egg of Xenopus laevis by injection of activators or inhibitors of G-proteins and by assaying for pertussis toxin or cholera toxin substrates. To examine the possibility that a receptor might be involved in the activation of a G-protein in the egg membrane, we introduced exogenous G-protein-related neurotransmitter receptors into the egg membrane, added the corresponding agonists, and looked for activation events. This paper reviews some of our published and unpublished experiments to test the hypothesis that activation of the amphibian egg is mediated by receptor activation of a G-protein in the egg membrane.

CALCIUM-DEPENDENT ACTIVATION RESPONSES

Activation of the Xenopus egg by the sperm causes a number of responses. These responses include the opening of chloride channels in the egg membrane which produce the fertilization potential (Webb and Nuccitelli, 1985), exocytosis of cortical granules resulting in the elevation of the fertilization envelope (Grey et al., 1974), a transient contraction of the

---

[1]Laboratoire de Neurobiologie Cellulaire, C.N.R.S., Gif-sur-Yvette 91198, France; [2]Division of Reproductive Biology and Endocrinology, Department of Obstetrics and Gynecology, University of Pennsylvania School of Medicine, Philadelphia, PA 19104, U.S.A.; [3]Department of Physiology, University of Connecticut Health Center, Farmington CT 06032, U.S.A.

NATO ASI Series, Vol. H 45
Mechanism of Fertilization
Edited by B. Dale

pigmented hemisphere toward the animal pole (Stewart-Savage and Grey, 1982), and the resumption of meiosis (Masui, 1985). These responses are all calcium dependent since injection of a calcium chelator to prevent or suppress the rise in calcium at fertilization prevents the responses (Kline, 1988). These events which follow fertilization also occur after artificial activation and we have used them as indicators of activation. In these studies, we were particularly interested in the membrane potential change and the exocytosis since these are the first in a series of responses to a rise in calcium at fertilization.

A SIGNAL TRANSDUCTION PATHWAY INVOLVING G-PROTEINS

The responses of a variety of cells to various hormones, neurotransmitters, light and other stimuli or "signals" depends on G-proteins which act as transducers between the receptor for a given stimulus and the effector molecule that effects the response to the stimulus (Stryer and Bourne, 1986; Gilman 1987). For fertilization, the hypothesis is that sperm, through interaction with a receptor in the egg membrane, stimulate a G-protein which in turn leads to stimulation of the effector enzyme phospholipase C (phosphatidylinositol-4,5 bisphosphate phosphodiesterase; Cockcroft and Gomperts, 1985) or some other enzyme of the phosphatidylinositol pathway. Phospholipase C produces inositol trisphosphate (IP_3) and diacylglycerol (DAG) from the membrane lipid phosphatidylinositol-4,5 bisphosphate (PIP_2). IP_3 causes the release of calcium from intracellular stores (Berridge, 1987) and DAG may activate protein kinase C (Nishizuka, 1986). The substrates for protein kinase C in eggs are not completely known (recent evidence suggests that protein kinase C may not be involved in activation of the Na^+-H^+ antiporter by sperm as previously thought; Shen, 1989).

Turnover of membrane phospholipids in the sea urchin egg is indicated by measurements of increases in PIP_2, IP_3, and DAG within 15 seconds after fertilization (Turner <u>et al</u>., 1984; Ciapa and Whitaker, 1986). No such measurements have been made following fertilization of the frog egg, but microinjection of IP_3 activates <u>Xenopus</u> eggs (Busa <u>et al</u>., 1985).

A G-protein normally becomes activated when it binds GTP to replace bound GDP; the G-protein is inactivated when the GTP is hydrolysed back to GDP. Normally, the exchange of GTP for GDP is slow, but it is accelerated by the influence of a receptor molecule which has bound a specific ligand. The hydrolysis-resistant analog of GTP, GTP-γ-S (Guanosine-5'-O-(3-thiotriphosphate) can activate G-proteins in the absence of receptor stimulation because there is always some slow exchange of GTP for GDP. GTP-γ-S produces a persistent activation of G-proteins since it is hydrolysis resistant. Introduction of GDP-β-S, a metabolically stable analog of GDP, can inhibit G-proteins by competing with GTP for binding to the G-protein.

Cholera toxin (CTX) and pertussis toxin (PTX) ADP-ribosylate certain G-proteins and cause labeling if radioactive NAD is present. The toxins can also modify the function of G-proteins; cholera toxin activates certain G-proteins while PTX inhibits some G-proteins. However, not all G-proteins are sensitive to toxin modification.

INJECTION OF GTP-γ-S ACTIVATES _XENOPUS_ EGGS

Injection of GTP-γ-S at final concentrations of 50-460 μM, in 23 of 24 _Xenopus_ eggs, caused a change in membrane potential which mimicked the potential change normally produced by sperm (Fig. 1A). The change in potential (activation potential) began 1-4 min after injection, although it occasionally began in less than 10 seconds. This delay might represent the time it takes for sufficient GTP-γ-S to diffuse to the cell membrane and replace GDP bound to G-proteins. The change in membrane potential is one of the first indicators of egg activation. Injected eggs also underwent cortical granule exocytosis, as indicated by the elevation of the fertilization envelope, as well as a cortical pigment contraction like that occurring at fertilization.

In addition to using commercial GTP-γ-S (Boehringer Mannheim Biochemicals, Indianapolis, IN), we repurified the commercial preparations of this nucleotide by anion exchange chromatography on DE-52 using the volatile buffer trimethylammonium bicarbonate (Connolly _et al_., 1982). The commercial

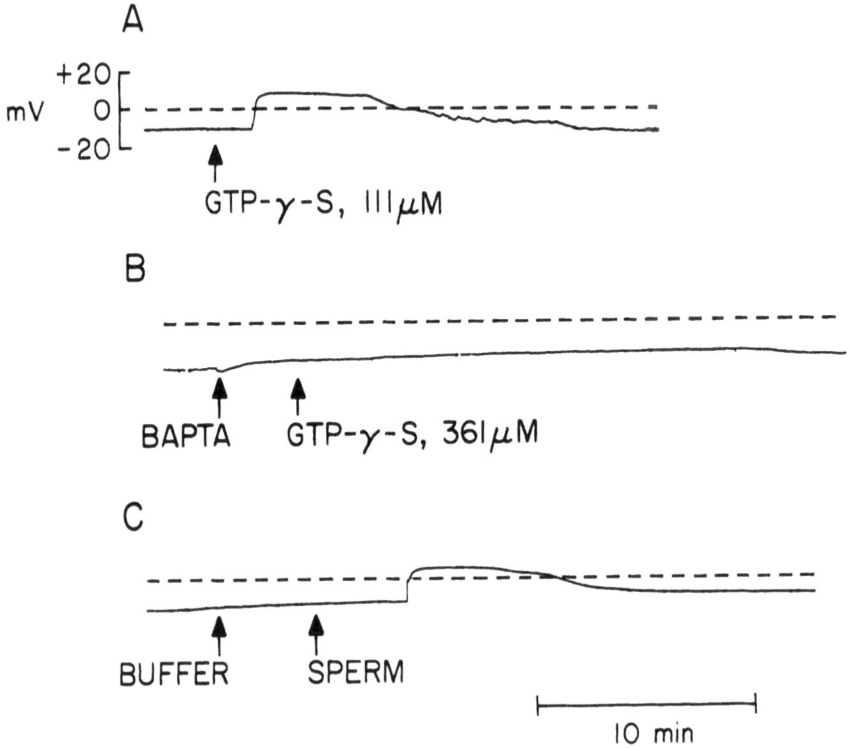

Figure 1. Activation potential initiated by injection of GTP-γ-S. Membrane potential is shown as a function of time and the dashed lines indicate 0 mV. (A) Injection of 1.0 nl of GTP-γ-S giving a final concentration of 111 μM. (b) Injection of GTP-γ-S (361 μM) after injection of the calcium chelator BAPTA (final concentration 0.21 mM). (C) Control injection of 2.7 nl buffer solution and insemination. (Kline, Kopf, and Jaffe, unpublished)

and repurified preparations were equally effective in activating eggs. To determine if activation of the egg by GTP-γ-S requires an increase in intracellular calcium, we first injected the calcium chelator BAPTA (1,2-bis)o-aminophenoxy)ethane-N,N,N′,N′-tetraacetic acid) to prevent or suppress the rise in calcium and then injected GTP-γ-S from the same injection pipet. Seven eggs were injected with 0.21 or 0.36 mM BAPTA. Injection of GTP-γ-S at a final concentration between 51 and 444 μM, after injection of BAPTA, did not activate any of the eggs (Fig. 1B).

As shown in Fig. 1C, control injections of equivalent volumes of buffer (1-12 nl, 10 eggs) or buffer with 10 mM LiCl (commercial GTP-γ-S is a tetra-lithium salt, 5 eggs) did not cause activation. Injection of GppNHP (5-guanylyl imidodiphosphate), another hydrolysis-resistant analog of GTP, did not produce an activation potential (11 eggs, final concentration 0.3 to 10.2 mM GppNHP). Although GppNHP activates G-proteins, it is less effective than GTP-γ-S in some preparations (Northup et al., 1982).

We also injected repurified GDP-β-S (guanosine 5'-O-(2-thiodiphosphate), the stable analog of GDP, which can inhibit G-protein function. None of 15 eggs injected with GDP-β-S (1.2-5.8 mM final concentration) activated in response to injection (data not shown). When the eggs were inseminated 3-14 min later, only 5/15 showed a normal fertilization potential. In 10/15 eggs, the fertilization potential was inhibited or delayed.

While these results suggest that GDP-β-S may inhibit activation, such inhibition might not be due to the specific inhibition of G-proteins for the following reasons. First, the degree of inhibition did not correlate with the amount of GDP-β-S injected nor with the time between injection and insemination. Second, GDP-β-S caused some dispersion of the pigment granules at the site of injection (not seen with control injections) which may indicate damage to the egg or some effect on the cytoskeleton. Finally, if GDP-β-S inhibits sperm-induced activation of the egg by preventing activation of a G-protein, we would expect that the inhibition could be overcome by bypassing the G-protein step and artificially increasing intracellular calcium. This was tested in 4 of the GDP-β-S-injected eggs that did not activate after sperm addition. Only 2 of 4 eggs activated in response to the calcium ionophore A23187 (6.4 µM).

Further experiments would be required to demonstrate that GDP-β-S blocks activation of the frog egg by specifically inhibiting G-proteins. It would be necessary to test whether the GDP-β-S inhibition occurs at a step prior to the production of IP_3 and if the inhibition can be overcome by subsequent injection of IP_3. Recent experiments with the sea urchin egg indicate that GDP-β-S may inhibit exocytosis at some site other than, or in addition to, the G-protein, since GDP-β-S injection appears to inhibit exocytosis but not the normal rise in calcium following fertilization (Whitaker et al., 1989). In the hamster egg, GDP-β-S may specifically inhibit G-proteins involved in activation at fertilization. In this egg, GDP-β-S blocks the calcium-dependent opening of ion channels at a step preceding IP_3 induced calcium release since injection of IP_3 overcomes the block (Miyazaki, 1988).

THE MATURE XENOPUS EGG CONTAINS A PERTUSSIS TOXIN SUBSTRATE

The demonstration that GTP-γ-S microinjection activates the frog egg suggested that the activation normally brought about by sperm-egg interaction may be regulated by a G-protein. The existence of G-proteins in the Xenopus egg was determined by examining egg membrane fractions, as well as egg cortices, for CTX and PTX substrates. ADP-ribosylation of the egg cortices and membrane fractions was carried out in the presence of [^{32}P]-NAD$^+$ as described by Kopf et al., 1986. No CTX substrates were demonstrated. However, both preparations contained a PTX substrate with an apparent molecular weight of 39,000-40,000.

PERTUSSIS TOXIN DOES NOT INHIBIT ACTIVATION INDUCED BY SPERM

To examine if activation of the frog egg could be prevented by inhibition of the G-protein by PTX, we incubated mature, ovulated eggs in PTX (4 μg/ml) for 1-3 hours. All but 1 of 41 eggs examined activated in response to sperm. To increase the time available for ADP-ribosylation by PTX, we incubated Xenopus oocytes in 4 μg/ml PTX for 3-4 days and then induced maturation by addition of progesterone. After maturation was complete, we removed the vitelline layer and applied sperm in a solution containing egg jelly water (removal of the vitelline layer and addition of jelly water are required for fertilization of eggs matured in vitro).

Nine of 9 such PTX-treated oocytes, when matured, activated and produced a fertilization potential in response to sperm. The cortical contractions and fertilization potentials produced by these PTX-treated eggs were indistinguishable from the responses recorded from 9 control-treated eggs. The 18 eggs used for this series of experiments were obtained from 4 separate animals. Additional PTX- and control-treated eggs from 2 of these animals were used to prepare plasma membranes to determine the extent of ADP-ribosylation. These membrane preparations revealed that at least 80% of the PTX substrate was modified by the PTX incubation (indicated by the fact that subsequent ribosylation in the presence of radioactive NAD was <20% of control levels). Therefore, although over 80% of the PTX substrate was ribosylated (and presumably inactivated), activation was normal.

We also tested whether or not injection of PTX at higher concentrations might inhibit activation. We injected PTX (activated by incubation with 25-50 mM dithiothreitol at 35°C for 30 min) into Xenopus oocytes about 3-5 hours after they were induced to mature by addition of progesterone. Three hours after the PTX injections (final concentration 40 µg/ml) sperm were added to the mature eggs. These PTX injections did not inhibit activation by sperm and first cleavage occurred in 14 of 14 eggs tested.

These experiments indicate that, if a G-protein is involved in activation of the frog egg, it is apparently not one which is ADP-ribosylated by PTX. Such PTX-insensitive G-proteins, thought to be involved in signal transduction, have been described (Fong et al., 1988; Matsuoka et al., 1988). The sea urchin egg contains both CTX and PTX substrates and microinjection of CTX (30 µg/ml) caused cortical granule exocytosis; PTX did not cause or inhibit exocytosis (Turner et al., 1987). The mature frog egg contains a PTX substrate but no CTX substrate. Microinjection of CTX (1 to 240 µg/ml) did not activate the mature, ovulated Xenopus egg. The function of the PTX-substrate we identified in the mature egg could be involved in an aspect of development other than fertilization. A PTX substrate is also present in immature Xenopus oocytes but conflicting results have been reported concerning the action of PTX on oocyte maturation (Sadler et al., 1984; Goodhardt et al., 1984; Mulner et al., 1985; Pellaz and Schorderet-Slatkine, 1989).

EVIDENCE FOR RECEPTOR ACTIVATION OF A G-PROTEIN

There is evidence, particularly in sea urchin and mouse, for a sperm receptor in the egg's extracellular coat which is involved in sperm binding (Ruiz-Bravo and Lennarz, 1986; Wassarman, 1987). However, it is not known, for any species, if there is a sperm receptor in the egg plasma membrane that mediates activation of the egg. To further examine the function of G-proteins in egg activation and to begin to investigate the possibility of receptor-mediated activation of such G-proteins, we introduced specific neurotransmitter receptors, which act by way of G-proteins, into the egg. The receptors that activate G-proteins belong to a structurally homologous family (Dohlman et al., 1987). We chose two neurotransmitter receptors that are known to function by way of a G-protein to activate phospholipase C in other cells; these were the type 1C serotonin (5HT) receptor and the type M1 muscarinic acetylcholine (ACh) receptor (Kline et al., 1988).

In our first experiments, we injected rat brain total poly A^+ mRNA, which contained mRNA for the serotonin receptor, into Xenopus oocytes. After two to three days of incubation, during which exogenous neurotransmitter receptors were produced, the oocytes were matured in vitro to the fertilizable egg stage by application of the maturation inducing hormone, progesterone. Application of the neurotransmitter, serotonin, to the mature mRNA-injected eggs produced a change in membrane potential similar to the fertilization potential and caused exocytosis of cortical granules. Control non-injected eggs did not respond to serotonin. Recently, we have injected a specific mRNA for the type 1C 5HT receptor (synthesized from a DNA clone provided to us by David Julius, Columbia University). Application of serotonin to mature eggs injected with this specific message produced the same results as those obtained in eggs injected with rat brain mRNA.

We also introduced a specific mRNA derived from a DNA clone of the M1 muscarinic ACh receptor. Application of ACh to mature eggs following expression of the specific M1 receptor mRNA in the oocyte produced an activation potential and cortical granule exocytosis. Control eggs not injected with mRNA did not respond to ACh.

We propose that the exogenously added receptor proteins interact with the egg's G-protein to initiate activation. These experiments suggest that there is an endogenous sperm receptor that could be related to the neurotransmitter receptors. The proposed sperm receptor is bypassed by stimulating the introduced receptors.

CONCLUSIONS AND FUTURE RESEARCH

The amphibian egg is activated by a rise in intracellular calcium. We suggest that the pathway leading to the rise in calcium involves the receptor-mediated activation of a G-protein. The evidence in support of this hypothesis includes the observation that GTP-γ-S, which activates G-proteins, activates the *Xenopus* egg. ADP-ribosylating with CTX or PTX to alter G-protein function does not activate the frog egg or inhibit activation by sperm. Thus, if a G-protein is involved in activation, it is insensitive to CTX and PTX. This is not too surprising since, in other cells, G-proteins involved in phosphoinositide metabolism may be insensitive to both CTX and PTX (Lo and Hughes, 1987). Additional evidence for a G-protein mediating activation of the egg comes from the experiments with messenger RNA injection. Specific mRNA for neurotransmitter receptors known to initiate production of IP_3 through activation of a G-protein, but which are not normally present in the egg, are able to interact with a G-protein in the egg membrane to cause several of the early activation responses normally initiated by sperm.

The results of these experiments with the *Xenopus* egg, together with those with sea urchin (Turner and Jaffe, 1989) and hamster (Miyazaki, 1988) eggs support the hypothesis that egg activation occurs through activation of a G-protein by interaction of molecules on the sperm surface with a receptor molecule in the egg membrane.

There are several other hypotheses for the mechanism of egg activation. For example, the sperm, during the process of fusion with the egg, might insert an activating molecule into the egg membrane that would trigger a signaling mechanism in the egg, or the sperm might introduce an activating molecule into the egg cytoplasm following fusion of sperm and egg membranes. Such models would not require interaction of the sperm with a receptor on the egg membrane; activation would be initiated after diffusion of the activating molecule in the plane of the membrane or through the cytoplasmic bridge or perhaps through transient gap junctions between the two gametes (Longo et al., 1986; Whitaker et al., 1989). One proposed activating substance is cGMP which has been reported to activate sea urchin eggs if injected into the egg (Swann et al., 1987). Preliminary experiments indicate that cGMP does not activate the Xenopus egg; injection of 0.3 to 0.8 mM cGMP (final concentration in the cytoplasm if uniformly distributed) did not produce a change in membrane potential in 3/3 eggs tested. Other activating substances that might be introduced through a fusion bridge to activate the egg include IP_3 and calcium. Dale et al. (1985) reported that injection of sperm extracts activate the sea urchin egg but the identity of the activating substance is not known.

Xenopus eggs can be activated by artificial stimulation of the G-protein by introduction of GTP-γ-S or introduction of exogenous receptors that activate the G-protein. Additional evidence is required to determine if sperm activate the egg through receptor-mediated activation of G-proteins. Since GDP-β-S may not be specific and PTX does not block activation, it will be necessary to use other specific inhibitors of G-proteins to test whether sperm initiate egg activation by stimulating a G-protein. These inhibitors might include antibodies against G-proteins (Yatani et al., 1988), peptides based on the C-terminal sequences of G-proteins which can inhibit the interaction between receptors and G-proteins (Hamm et al., 1988), and perhaps other specific inhibitors as they become available. It may also be possible to look for an increase in GTPase activity in the egg at fertilization, or an increase in GTP exchange; this would support the role of a G-protein in activation. Further characterization of the G-proteins will be required since there appear to be several G-proteins in eggs. These experiments may help to delineate the mechanism of activation of the egg at fertilization.

References cited

Berridge MJ (1987). Inositol trisphosphate and diacylglycerol: Two interacting second messengers. Annu Rev Biochem 56:159-193.
Busa WB, Ferguson JE, Joseph SK, Williamson JR, Nuccitelli R (1985) Activation of frog (Xenopus laevis) eggs by inositol trisphosphate. I. Characterization of Ca^{2+} release from intracellular stores. J Cell Biol 101:677-682.
Ciapa B, Whitaker M (1986) Two phases of inositol polyphosphate and diacylglycerol production at fertilisation. FEBS Lett 195:347-351.
Cockcroft S, Gomperts BD (1985) Role of guanine nucleotide binding protein in the activation of polyphosphoinositide phosphodiesterase. Nature (London) 314:534-536.
Connolly BA, Romaniuk PJ, Eckstein F (1982) Synthesis and characterization of diastereomers of guanosine 5'-O-(1-thiotriphosphate) and guanosine 5'-O-(2-thiotriphosphate). Biochemistry 21:1983-1989.
Dale B, DeFelice LJ, Ehrenstein G (1985) Injection of a soluble sperm fraction into sea-urchin eggs triggers the cortical reaction. Experientia 41:1068-1070.
Dohlman HG, Caron MG, Lefkowitz RJ (1987) A family of receptors coupled to guanine nucleotide regulatory proteins. Biochemistry 26:2657-2664.
Fong HKW, Yoshimoto KK, Eversole-Cire P, Simon MI (1988) Identification of a GTP-binding protein α subunit that lacks an apparent ADP-ribosylation site for pertussis toxin. Proc Natl Acad Sci USA 85:3066-3070.
Gilman AG (1987) G proteins: Transducers of receptor-generated signals. Annu Rev Biochem 56:615-649.
Grey RD, Wolf DP, Hedrick JL (1974) Formation and structure of the fertilization envelope in Xenopus laevis. Dev Biol 36:44-61.
Goodhardt M, Ferry N, Buscaglia M, Baulieu EE, Hanoune J (1984) Does the guanine nucleotide regulatory protein N_i mediate progesterone inhibition of Xenopus oocyte adenylate cyclase? EMBO J 3:2653-2657.
Hamm HE, Deretic D, Arendt A, Hargrave PA, Koenig B, Hofmann KP (1988) Site of G protein binding to rhodopsin mapped with synthetic peptides from the alpha subunit. Science 241:832-835.
Kline D (1988) Calcium-dependent events at fertilization of the frog egg: Injection of a calcium buffer blocks ion channel opening, exocytosis, and formation of pronuclei. Dev Biol 126:346-361.
Kline D, Simoncini L, Mandel G, Maue RA, Kado RT, Jaffe LA (1988) Fertilization events induced by neurotransmitters after injection of mRNA in Xenopus eggs. Science 241:464-467.
Kopf GS, Woolkalis MJ, Gerton GL (1986) Evidence for a guanine nucleotide-binding regulatory protein in invertebrate and mammalian sperm. J Biol Chem 261:7327-7331.

Lo WW, Hughes J (1987) Receptor-phosphoinositidase C coupling: Multiple G-proteins? FEBS Lett 224:1-3.
Longo FJ, Lynn JW, McCulloh DH, Chambers EL (1986) Correlative ultrastructural and electrophysiological studies of sperm-egg interactions of the sea urchin, Lytechinus variegatus. Dev Biol 118:155-166.
Masui Y (1985) Meiotic arrest in animal oocytes. In: Metz CB, Monroy A (eds) Biology of fertilization, Vol. 3. Academic Press, Orlando, p 189.
Matsuoka M, Itoh H, Kozasa T, Kaziro Y (1988) Sequence analysis of cDNA and genomic DNA for a putative pertussis toxin-insensitive guanine nucleotide-binding regulatory protein α subunit. Proc Natl Acad Sci USA 85:5384-5388.
Miyazaki S (1988) Inositol 1,4,5-trisphosphate-induced calcium release and guanine nucleotide-binding protein-mediated periodic calcium rises in golden hamster eggs. J Cell Biol 106:345-353.
Mulner O, Megret F, Alouf JE, Ozon R (1985) Pertussis toxin facilitates the progesterone-induced maturation of Xenopus oocyte. FEBS Lett 181:397-402.
Nishizuka Y (1986) Studies and perspectives of protein kinase C. Science 233:305-312.
Northup JK, Smigel MD, Gilman AG (1982) The guanine nucleotide activating site of the regulatory component of adenylate cyclase. J Biol Chem 257:11416-11423.
Pellaz V, Schorderet-Slatkine S (1989) Evidence for a pertussis toxin-sensitive G protein involved in the control of meiotic reinitiation of Xenopus laevis oocytes. Exp Cell Res 183: 245-250.
Ruiz-Bravo N, Lennarz WJ (1986) Isolation and characterization of proteolytic fragments of the sea urchin sperm receptor that retain species specificity. Dev Biol 118:202-208.
Sadler SE, Maller JL, Cooper DMF (1984) Progesterone inhibition of Xenopus oocyte adenylate cyclase is not mediated via the Bordatella pertussis toxin substrate. Mol Pharmacol 26: 526-531.
Shen SS (1989) Na^+-H^+ antiport during fertilization of the sea urchin egg is blocked by W-7 but is insensitive to K252a and H-7. Biochem Biophys Res Comm 161:1100-1108.
Stewart-Savage J, Grey RD (1982) The temporal and spatial relationships between cortical contraction sperm trail formation and pronuclear migration in fertilized Xenopus eggs. Wilhelm Roux's Arch Dev Biol 191:241-245.
Stryer L, Bourne HR (1986) G proteins: A family of signal transducers. Annu Rev Cell Biol 2:391-419.
Swann K, Ciapa B, Whitaker M (1987) Cellular messengers and sea urchin egg activation. In: O'Connor JD (ed) Molecular biology of invertebrate development. Alan R. Liss, New York, p 45.
Turner PR, Jaffe LA (1989) G-proteins and the regulation of oocyte maturation and fertilization. In: Schatten H, Schatten G (eds) The cell biology of fertilization. Academic Press Inc., Orlando, p 297.
Turner PR, Jaffe LA, Primakoff P (1987) A cholera toxin-sensitive G-protein stimulates exocytosis in sea urchin eggs. Dev Biol 120:577-583.

Turner PR, Sheetz MP, Jaffe LA (1984) Fertilization increases the polyphosphoinositide content of sea urchin eggs. Nature (London) 310:414-415.

Wassarman, PM (1987) Early events in mammalian fertilization. Annu Rev Cell Biol 3:109-142.

Webb DJ, Nuccitelli R (1985) Fertilization potential and electrical properties of the Xenopus laevis egg. Dev Biol 107:395-406.

Whitaker M, Swann K, Crossley I (1989) What happens during the latent period at fertilization. In: Nuccitelli R, Cherr GN, Clark WH Jr. (eds) Mechanisms of egg activation. Plenum Press, New York, p 157.

Yatani A, Hamm H, Codina J, Massoni MR, Birnbaumer L, Brown AM (1988) A monoclonal antibody to the alpha subunit of G_K blocks muscarinic activation of atrial K^+ channels. Science 214:828-831.

ELECTRICAL MATURATION OF THE MOUSE OOCYTE

Joan Murnane and Louis J DeFelice
Department of Anatomy and Cell Biology
Emory University School of Medicine
Atlanta, Ga 30322

Abstract

This article compares three kinds of oocytes from F1 hybrid mice: germinal vesicle oocytes from 10-20 day mice, GV oocytes from 12 week, superovulated mice, and first polar body oocytes from 12 week, superovulated or normal estrus mice. All oocytes have two voltage-dependent currents: an inward, inactivating current (Ca), and an outward, non-inactivating current (K). In 1.5 mM external Ca, the Ca currents increase with growth and development; the averages are -2.9, -12.4, and -13.8 uA/cm^2, for the three respect stages. Thus the increase in Ca current precedes nuclear maturation.

NATO ASI Series, Vol. H 45
Mechanism of Fertilization
Edited by B. Dale

Introduction

Neonatal germinal vesicle (NGV) oocytes grow from 20 um to 60-70 um in diameter. Ovulation begins in response to a surge of LH and a gonadotropin-induced decrease in the ability of the follicle cell to maintain meiotic arrest (Eppig & Downs 1988). At ovulation the oocyte extrudes the first polar body (FPB), where it is again arrested until fertilization. Thus the first clear sign of maturation is germinal vesicle breakdown (GVBD).

Meiotic competence, the ability of GV oocytes to resume meiosis, is acquired before puberty at about 15 days in mice, when follicular cell shave proliferated and oocytes have attained 80% of their size. In this paper these oocytes are called mature GV oocytes (MGV). Meiotic competence may depend on growth, gene expression, or other processes in the ooplasm or oocyte membrane (Erickson & Sorenson 1974; Sorenson & Wasserman 1976). Fully grown GV oocytes undergo spontaneous, hormone-independent breakdown when mechanically removed from follicles (Edwards, 1965). In contrast, smaller NGV oocytes from preantral follicles (explanted prior to the LH surge and cultured in hormone-free or hormone-supplemented medium) retain their GVs (Umezu, et al. 1978). Although oocytes do not require gonadotropins for growth, they perhaps require them in a developmental program that results in meiotic competence (Eppig 1977, 1979, 1982; Bar-Ami & Tsafriri 1981; Eppig & Downs 1984).

Mechanisms underlying attainment of meiotic competence and release from meiotic arrest are in question. In vitro GVBD requires external Ca concentration and functioning Ca channels (Paleos & Powers 1981; DeFelici & Siracusa 1982; Maruska et al. 1984; Bae & Channing 1985; Kline 1988). Recently, Ca was postulated as a mediator of LH-induced GVBD in mammalian oocytes (Preston et al. 1987). Ca-selective channels blocked by Co, La, or verapamil are found in FPB oocytes (Hagiwara & Byerly 1981; Peres 1986ab, 1987). Ca ionophore, A23187, induces GVBD in some species (Chambers et al. 1974; Steinhardt et al. 1974; Wasserman

& Masui 1975). La ions, which have a high affinity for cationic binding sites, block Ca influx, induce hyperpolarization, and prevent GVBD (Williams 1976). Oocyte viability, in vitro maturation, and amplitude of the action potential are reduced if Mg ions replace Ca (Hagiwara & Berly 1981), and verapamil, procaine, and epinephrine all perturb normal maturation in some way (Paleos & Powers 1981; DeFelici & Siracusa 1982; Maruska et al. 1984; Bae & Channing 1985). It therefore appears likely that the oocyte plasma membrane, particularly Ca ion transport, underlies some phase of maturation which prepares the oocyte for fertilization (Igusa et al. 1983).

Materials and Methods

Preparation We used two populations of cells separated by development and endogenous hormone exposure, and defined by morphological nuclear status: (1) NGV oocytes (25-65 um in diameter) from mechanically dispersed antral follicles from neonatal (10-20 day) mouse ovaries. (2) FPB oocytes (50-75 ϕm in diameter) and MGV (40-70 ìm in diameter) from minced ovaries or ampulla of the oviducts of 12 week old mice sacrificed at estrus or after superovulation. Superovulation was induced in C57BL/SJl F1 hybrid mice by injection of 5 IU of pregnant mare serum (PMSG; Sigma no: G-4877), followed 48 hours later by injection of 5 IU of human chorionic gonadotropin (hCG: Sigma CG-2). Approximately 20-30 cumulus-enclosed oocytes were collected from the ampulla of the oviduct 14-18 hours after administration of PMSG. Prior to removing the ovaries from mature mice, the stage of estrus was determined by appearance of the vaginal epithelia. Post-pubescent GV oocytes treated with enzyme often demonstrated GVBD, and FPB oocytes broke when we formed cell-attached patches. The cumulus mass, follicular cells, and zona pellucida were therefore removed mechanically. Nuclear and cytoplasmic characteristics identified cells as immature, prophase I, primary oocytes (NGV/prepuberty, and MGV/postpuberty) or mature, metaphase II, secondary oocytes (FPB). We did the dissections at 37 °C and the experiments at room temperature. Currents were normalized to surface area (uA/cm^2).

Solutions and Electrical Recording The normal bath solution (in mM) was 130 Na, 1.3 K, 1.5 Ca, 0.5 Mg, 0.5 SO4, PO4 1.3, 131.7 Cl, 5 dextrose, 10 HEPES (adjusted to pH 7.35, 270 mOsm). The whole-cell electrode contained 120 K, 0.1 Ca, 2 Mg, 124.2 Cl, 1.1 EGTA, 10 HEPES (adjusted to pH 7.4, 265 mOsm). The bath solution for the 20 mM Ca experiments contained 90 Na, 1.3 K, 20 Ca, 0.5 Mg, 0.5 SO4, 133.3 Cl, 5 dextrose, 10 HEPES (adjusted to pH 7.35, 270 mOsm). We eliminated Na current using 1 uM TTX in the bath. Mouse oocytes rarely form gigohm seals. Using electrodes with 3-5 megohm resistance, it was often necessary to wait 5-10 minutes for the patch to stabilize. Clamping the patch to -80 mV improved the seal (Peres 1987). The seals were 0.3 to 1 gigohm prior to breaking the patch with suction. We repeatedly checked the seal and series resistance compensation between protocols. We used a List PC-7 in the voltage-clamp (current-clamp) mode to record currents (potentials).

Data Analysis Using I = G(V - E) where G the leak conductance (-3.4 to -1.02 nS) and E the leak reversal potential, the theoretical leak was calculated and subtracted from each current, and the peak inward current and steady-state outward current was plotted against the test voltage (Fig. 1). The voltage protocol (A) and leak subtraction method (B-D) used to eliminate a linear background current from the nonlinear, time-variant and voltage-dependent currents. Cells held at -80 mV were stepped to -120 to 50 mV in 10 mV steps lasting 500 msec. Hyperpolarizing pulses gave a linear response, and the reversal potential of the linear leak was approximately zero in most cases.

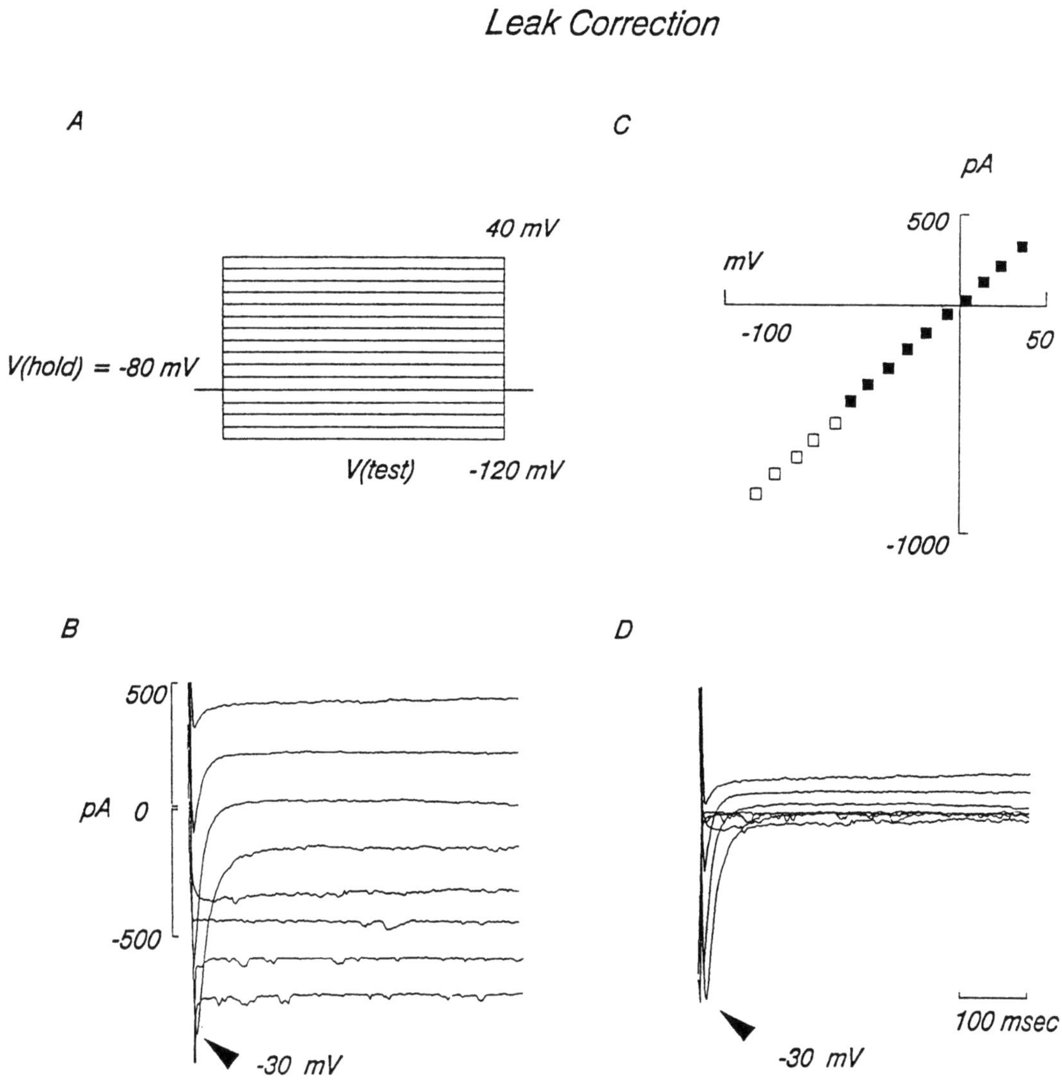

Fig. 1 Leak subtraction. (A) Cells were held at -80 mV and stepped for 500 msec to -120, -110, ... 50 mV separated by 2 seconds. (B) Typical data showing every other response. (C) Leak current determined from steady-state response at 400-450 msec. The open squares are from steps to -120 from -80 mV; closed squares are calculated -70 to 40 mV. (D) Leak-corrected currents.

Results

Figure 2 gives examples from the three kinds of oocytes we recorded from. The inward current is relatively small in NGV oocytes compared to MGV and FPB oocytes. The outward current is the same or slightly lower as the oocyte matures. As the relative proportion of outward to inward current increases, the reversal potential of the inward component shifts to the right on the voltage axis. Increasing the concentration of external Ca increases the inward current and shifts the reversal potential further to the right (Fig. 3).

Figure 4 summarizes data from 16 experiments in NGV and MGV oocytes in 1.5 and 20 mm external Ca. The FPB experiments are virtually identical to the MGV experiments. Peak inward currents are -2.95 $\pm$ 1.19 uA/cm^2 in NGV oocytes is, 12.39 $\pm$ 2.34 uA/cm^2 in MGV oocytes, and -13.76 uA/cm^2 $\pm$ 0.86 in FPB oocytes. The change represents a four-fold increase in inward current in MGV and FPB oocytes over NGV oocytes. In 20 mM external Ca, NGV oocytes have peak inward current of -10. 63 $\pm$ 0.58 uA/cm^2 and a reversal potential of near 30 mV; FPB oocytes have a peak current of -37.51 $\pm$ 11.86 uA/cm2 and reversal potentials from 30 to 50 mV. The outward current is smaller in MGV and FPB oocytes compared to NGV oocytes. In 16 experiments, the average steady-state current at 40 mV was 8.89 $\pm$ 2.07 uA/cm^2 in NGV oocytes, 5.03 $\pm$ 1.99 uA/cm^2 for MGV oocytes, and 5.53 $\pm$ 1.65- uA/cm^2 for FPB oocytes. The current has delayed activation and a reversal potential of -50 to -60 mV. 20 mM Ca significantly decreased the outward steady-state component from 6.65 $\pm$ 2.97 in NGV oocytes to 2.52 $\pm$ 0.79 in FPB oocytes.

Discussion

The kinetics, the I(V) relationships, the reversal potential, and the Ca sensitivity of the inward current suggest it is carried by

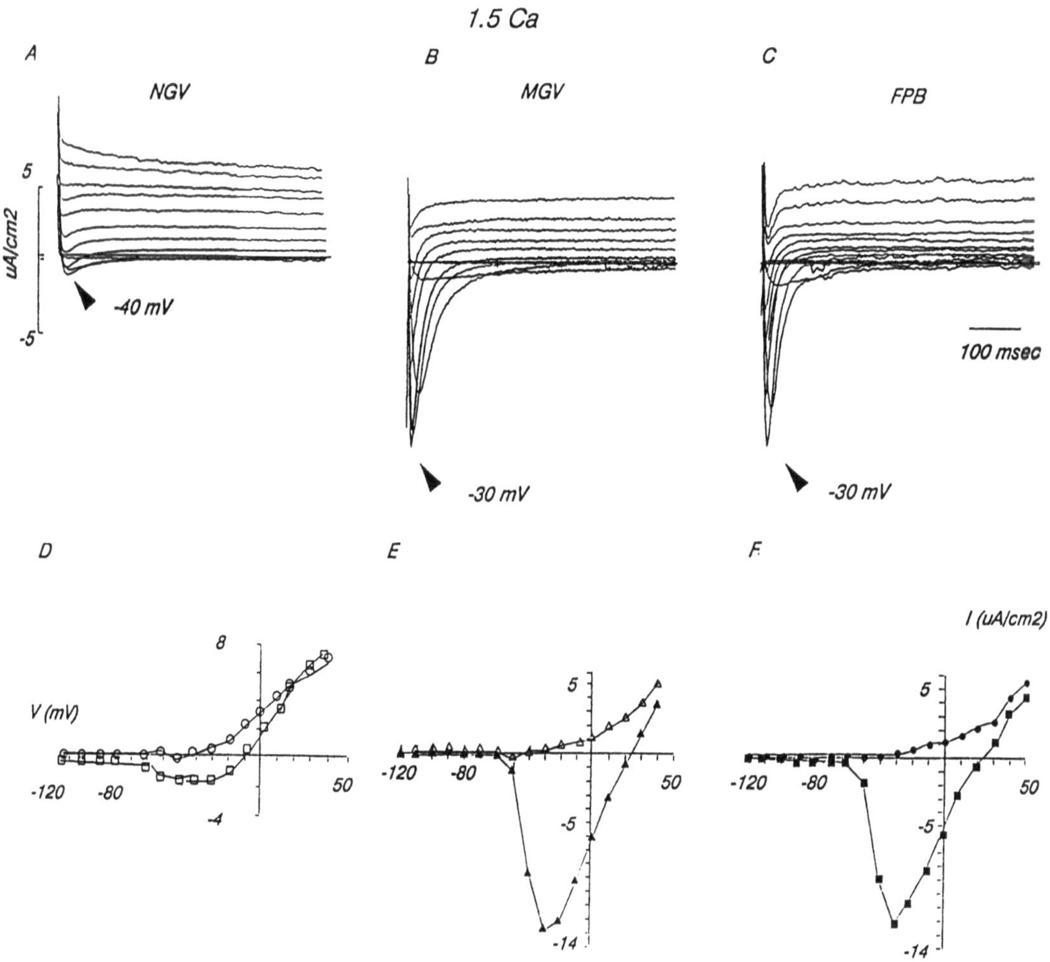

Fig. 2 Comparison of the leak-corrected, whole-cell currents in NGV, MGV, and FPB oocytes. The bottom row shows the I(V) curves of peak inward currents and plateau outward currents. As oocytes mature, the inward current increases about 4-fold, and the outward current decreases slightly.

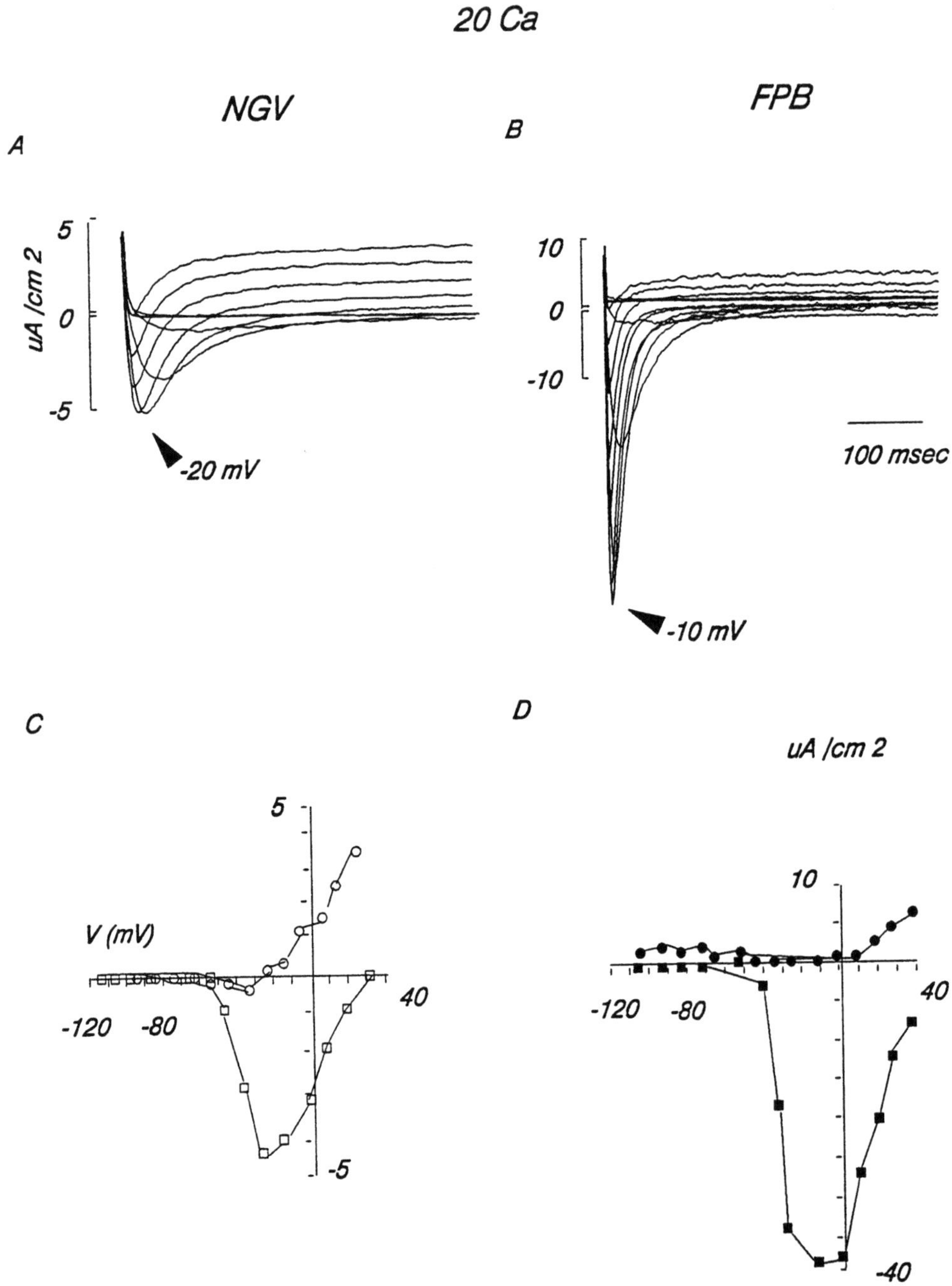

Fig. 3. Leak-subtracted currents and the I(V) relationships of NGV and FPB oocytes in 20 mM external Ca. The increase in peak inward current in NGV and FPB oocytes, in 20 Ca compared to 1.5 Ca, is accompanied by a shift in reversal potential.

I(V) Relationships

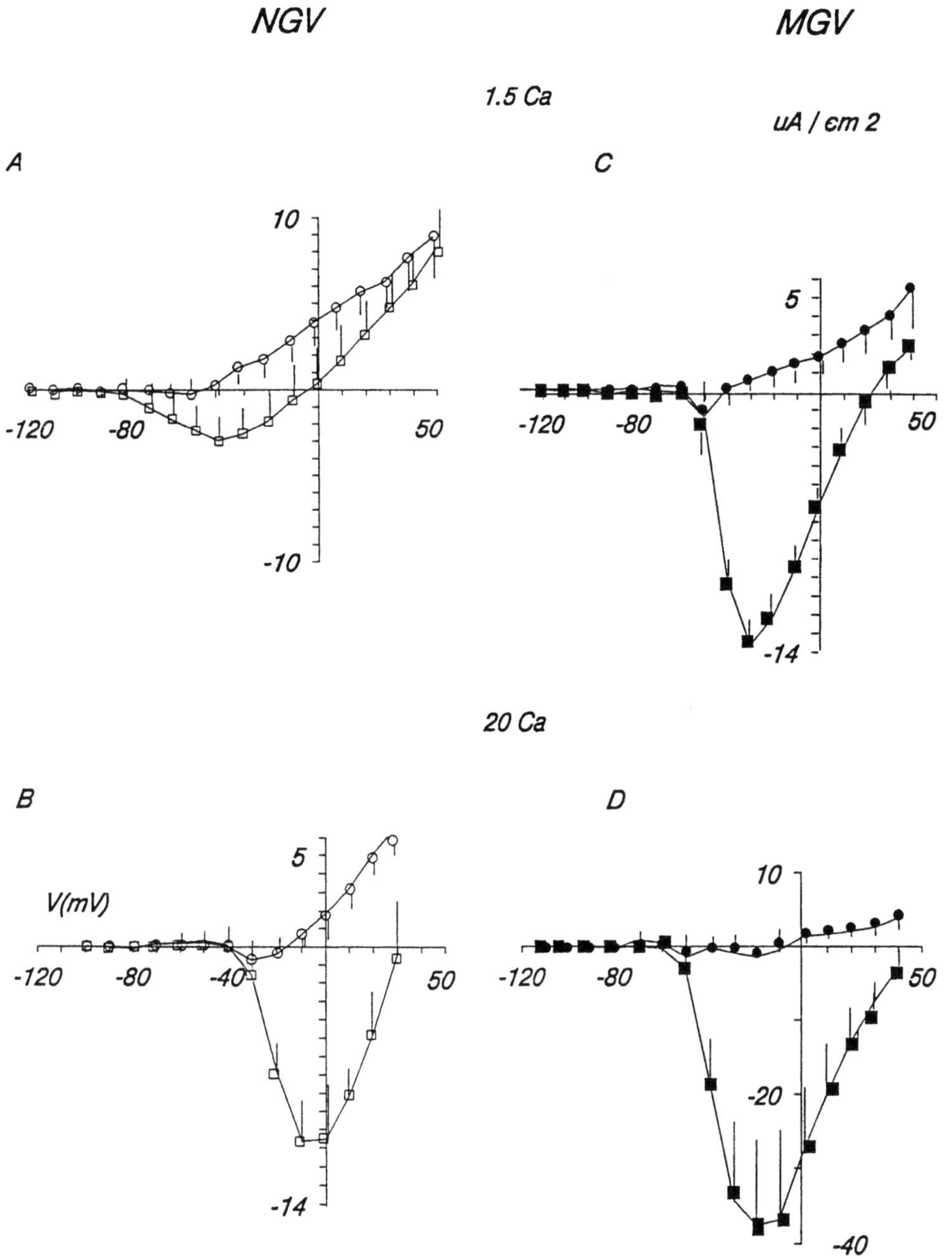

Fig 4. Summary of the I(V) relationships of NGV and MGV oocytes in 1.5 and 20 mM Ca.

Ca-selective channels similar to those found in other tissues. Our hypothesis is that the increase in Ca current in MGV/FPB oocytes over NGV oocytes reflects a greater number of these Ca channels. Similarly, kinetics, rectification, and reversal potential of the outward current suggest it is carried by K-selective channels, and its decline in MGV/FPB oocytes over NGV oocytes is a decrease in the number of K channels. The Ca and K currents, which are not separated in our experiments, compete with each other. It is an increase in a steady-state component of the Ca current in 20 mM Ca that causes a reduction in outward current. Ca and K channels have already been described in mouse oocytes (Yamashita 1982; Peres 1986ab, 1987). Our study finds that prepubescent ovarian oocytes express a much smaller Ca current and a slightly larger K current than postpubescent first polar body oocytes. Furthermore, cells defined morphologically as immature by the presence of a germinal vesicle, but isolated from antral follicles of mature mice, are electrophysiologically mature.

Not all MGV oocytes exhibit mature Ca currents. MGV oocytes from stage 7 or 8 antral follicles always exhibited large Ca currents, but ovarian oocytes from stage 5 or 6 preantral follicles exhibited large Ca currents only about half the time. The increase in Ca channels at day 20 coincides with an increase in serum LH (Dullart et al. 1975; Bar-Ami & Tsafriri 1981) and an increase in the percent of meiotically competent oocytes in antral follicles (Sorenson & Wasserman 1976). Thus gonadotropins induce nuclear maturation after membrane maturation, principally, an increase in the number of functional Ca channels). Whether the Ca current we observe indicates an indirect effect of LH on the follicular cell or a direct effect on the oocyte membrane remains to be tested.

The action of LH, oocyte maturation, GVBD, and 1st and 2nd polar body formation depends on external Ca concentration. During hormone stimulation Ca ion flux increases (Rasmussen 1982;

1986ab), and cytosolic Ca may signal cell division (Berridge 1975; Whitefield et al. 1979). Injecting Ca activates oocytes (Moreau et al. 1976; Fulton & Whittingham 1978; Jaffe 1980), and agents that perturb intracellular Ca or calmodulin metabolism inhibit maturation (Jagiello et al. 1982; Bornslaeger et al. 1984, 1986ab). Removal of internal Ca ions by microinjection of EGTA into oocytes inhibits maturation. A correlation exists between preovulatory LH surge, increase in cytosolic free Ca concentration, and GVBD in rat oocytes (Preston et al. 1987). Furthermore, sensitivity to Ca-induced maturation is acquired in parallel with the acquisition of meiotic competence, which progressively increases throughout the growth phase of the oocyte (Paleos & Powers 1981). Therefore, LH-initiated oocyte maturation may be mediated by alterations in Ca flux across the plasma membrane (Dawson & Conrad, 1972; Paleos & Powers, 1981; Maruska et al., 1984; Bae & Channing, 1985; Preston et al. 1987). Thus Ca, a ubiquitous 2nd messenger in development (Jaffe 1980; Rasmussen 1982; 1986ab), couples ligand-receptor interactions at the plasma membrane with cytosolic and nuclear responses.

Oocytes depolarize as they progress from the GV to the FPB stage (Cross et al. 1973). LH depolarizes immature oocytes in vitro (Dawson & Conrad 1972; Kline 1988), suggesting that a decrease in resting potential may be connected to meiotic maturation. Mattioli (1989) suggests that indirect action through follicular cells causes the depolarization. We also observed a depolarization from the NGV to the FPB stage. Our whole-cell clamp experiments suggest that a decrease in K channels and an increase in Ca channels are partly responsible for this depolarization. We find that Ca channels appear in the oocyte membrane before the response to gonadotropin stimulation. Our hypothesis is that this increase in the number of Ca channels, coupled with a decrease in K channels, depolarizes the oocyte and causes an influx of Ca. Only then, we suggest, is gonadotropin stimulation able to trigger maturation.

References

Bae I-H, Channing C (1985) Effects of calcium ion on the maturation of cumulus-enclosed pig follicular oocytes isolated from medium-sized Graafian follicles. Biol Reprod 33:79-87
Bar-Ami S, Tsafriri A (1981) Acquisition of meiotic competence in the rat: role of gonadotrophin and estrogen. Gamete Res 4:463-472
Berridge, MJ (1975) The interaction of cyclic nucleotides and calcium in the control of cellular activity. In:Greengard P, GA Robinson (eds) Advances in Cyclic Nucleotide Research, vol 6. Raven Press, New York pp 1-98
Bornslaeger EA, Wilde MW, Schultz RM (1984) Regulation of mouse oocyte maturation: Involvement of cyclic AMP phosphodiesterase and calmodulin. Dev Biol 105: 488-499
Bornslaeger EA, Mattei P, Schultz RM (1986a) Involvement of cAMP-dependent protein kinase and protein phosphorylation in regulation of mouse oocyte maturation. Dev Biol 114:453-462
Bornslaeger EA, Poueymirou WT, Mattei P, Schultz RM (1986b) Effects of protein kinase C activator on germinal vesicle breakdown and polar body emission of mouse oocytes. Exp Cell Res 165:507-517
Chambers EL, Pressman BC, Rose B (1974) The activation of sea urchin eggs by the divalent ionophore A23187 and X-537a. Biochem Biophys Res Commun 60:126-132
Cross MH, Cross PC, Brinster RL (1973) Changes in membrane potential during mouse egg development. Dev Biol 33: 412-416
Dawson, JE, Conrad JT (1972) The effect of human chorionic gonadotrophin and luetinizing hormone upon the membrane potential of unovulated frog oocytes. Biol Reprod 6:58-66
DeFelici M, Siracusa G (1982) Survival of isolated, fully grown mouse ovarian oocytes is strictly dependent on external Ca^{2+}. Dev Biol 92:539-543
Dullart J, Kent, Ryle M (1975) Serum gonadotrophin concentration in infantile mice. J Reprod Fertil 43: 189-192
Edwards RG (1965) Maturation in vitro of mouse, sheep, cow, pig, rehsus monkey and human ovarian oocytes. Nature London 208:349-351
Eppig JJ (1977) Mouse development in vitro with various culture systems. Dev Biol 60:371-388
Eppig JJ (1979) A comparison between oocyte grown in co-culture with granulosa cells and oocytes with granulosa cell-oocyte junctional contact maintained in vitro. J Exp Zool 209:345-353
Eppig JJ (1982) The relationship between cumulus cell-oocyte coupling, oocyte meiotic maturation, and cumulus expansion. Dev Biol 119:313-321
Eppig JJ, Downs SM (1984) Chemical signals that regulate mammalian oocyte maturation. Biol Reprod 30:1-11

Eppig JJ, Downs SM (1988) Gonadotrophin-induced murine maturation in vivo is not associated with decreased cyclic adenosine monophosphate in the oocyte-cumulus cell complex. Gamete Res 20:125-131

Erickson GF, Sorenson, RA (1974) In vitro maturation of mouse oocytes isolated from late, middle and preantral Graafian follicles. J Exp Zool 190:123-127

Fulton BF, Whittingham BG (1978) Activation of mammalian oocytes by intracellular injection of calcium. Nature London 273:149-151

Hagiwara S, Byerly L (1981) Calcium channel. Annu Rev of Neurosci 4:69-125

Igusa Y, Miyazaki S, Yamashita N (1983) Periodic hyperpolarizing responses in hamster and mouse eggs fertilized with mouse sperm. J Physiol 340:633-647

Jaffe LF (1980) Calcium explosions as triggers of development Ann NY Acad Sci 339:86-101

Jagiello G, Ducayen MB, Downey R, Jonassen A (1982) Alterations of mammalian oocyte meiosis I with divalent cations and calmodulin. Cell Calcium 3:153-162

Kline D (1988) Calcium-dependent events at fertilization of the frog egg; Injection of a calcium buffer blocks ion channel opening, exocytosis, and formation of pronuclei. Dev Biol 126:346-361

Maruska DV, Leibfreid ML, First NL (1984) Role of calcium and calcium-calmodulin complex in resumption of meiosis, cumulus expansion, viability, and hyaluronidase sensitivity of bovine cumulus-oocyte complexes. Biol Reprod 31:1-6

Mattiolli M, Galeati G, Bacci, M L (1989) (to be published) Electrophysiology of pig oocytes during maturation

Moreau MP, Dorée M, Guerrier P (1976) Electrophoretic introduction of calcium ions into the cortex of Xenopus laevis oocytes triggers meiosis reinitiation. J Exp Zool 197:443-449

Paleos GA, Powers, RD (1981) The effect of calcium on the first meiotic division of the mammalian oocyte. J Exp Zool 217:409-416

Peres A (1986a) Calcium current in mouse eggs recorded with tight-seal, whole-cell voltage-clamp technique. Cell Biol Int Rep 10:117-119

Peres A (1986b) Resting membrane potential and inward current properties of mouse ovarian oocytes and eggs. Pflugers Archiv 407:534-540

Peres A (1987) The calcium current of mouse egg measured in physiological calcium and temperature conditions. J Physiol 391:573-588

Preston SL, Parmer TG, Behrman HR (1987) Adenosine reverses calcium-dependent inhibition of follicle-stimulating hormone action and induction of maturation in cumulus-enclosed rat oocytes. Endocrin 120:1364-1353

Rasmussen H (1982) Calcium as intracellular messenger in hormone action. Adv Exp Med Biol 151:473-491

Rasmussen H (1986a) The calcium messenger system I. N Engl J Med 314:1094-1101

Rasmussen H (1986b) The calcium messenger system II. N Engl J Med 314:1164-1170
Sorenson R, Wasserman PM (1976) Relationship between growth and meiotic maturation of the mouse oocyte. Dev Biol 50:531-536
Steinhardt RA, Epel D, Carroll EF, Yanagimachi R. (1974) Is calcium ionophore a universal activator for unfertilized eggs? Nature London 252:41-43
Umezu M, Hashizume K, Masaki J (1978) Follicular and oocyte maturation during the period of ovulation in immature rats pretreated with PMS. Japan J An Reprod 24:69-72
Wasserman WJ, Masui Y (1975) Initiation of meiotic maturation in Xenopus laevis oocytes by the combination of divalent cations and ionophore A23187. J Exp Zool 193:369-375
Whitefield JF, Boyton AL, Macmanus JP, Sikorska M, Tsang BK (1979) The regulation of cell proliferation by calcium and cyclic AMP. Mol Cell Biochem 27:155-179
Williams RJP (1976) Calcium chemistry and its relation to biological function. IN:CJ Duncan (ed) Calcium in Biological Systems, Cambridge University Press, Cambridge, MA pp 1-17
Yamashita N (1982) Enhancement of ionic currents through voltage-gated channels in the mouse after fertilization. J Physiol 329:263-280

Acknowledgments

We wish to thank Antonio Peres, Mauro Mattioli, and Jacques Cohen for their criticisms of this paper and contributions to the work, and Donna Reynolds, William N Goolsby, and Michael Aiken for technical assistance. The research was supported by the NIH:HD 19770-03.

THE CONTRIBUTION OF *Discoglossus pictus* FERTILIZATION IN THE STUDY OF AMPHIBIAN SPERM-EGG INTERACTION

Campanella C.*, Talevi R., Gualtieri R., Andreuccetti P.

Dipartimento di Biologia Evolutiva e Comparata,
via Mezzocannone 8, 80134 Napoli, Italy
* Dipartimento di Scienze e Tecnologie Biomediche e di Biometria, Collemaggio 67100, L'Aquila.

KEYWORDS: Amphibians, gametes, fertilization, Discoglossus pictus.

ABSTRACT We have here raised the following questions which deserve attention in the study of amphibian fertilization: 1. How do sperm acquire motility and what is the action of the jelly coat on sperm; 2. What are the egg surface properties that unable sperm fusion; 3. How is a localized stimulus transmitted to the rest of the egg; 4. What is the role of cortical organelles at activation; 5. How is excess of sperm entrance avoided. Fertilization in Discoglossus pictus may contribute to the understanding of these mechanisms.

INTRODUCTION

Recently, amphibian gametes have provided remarkable contribution to our understanding of gametogenesis and fertilization. The control of G2-metaphase transition at maturation through an histone protein kinase, which appears to be similar from yeast to somatic cell lines (among many references : Arion et al, 1988; Dunphy et al, 1988; Labbé et al, 1988; Lohka et al, 1988; Gautier et al, 1988) has his basis in the discovery of the maturation promoting factor (MPF) in Rana pipiens oocytes by Masui and Market (1971).

The expression of heterologous m-RNA in fully grown oocytes offered the basis of elegant experimentation which suggests the presence of sperm receptors

NATO ASI Series, Vol. H 45
Mechanism of Fertilization
Edited by B. Dale

at the egg plasma membrane, of signal transduction through G-proteins and of endogenous hydrolisis of IP3 (McIntosh and Catt, 1987; Kline et al, 1988).

The finding that the vitelline envelope (VE) changes its composition following ovulation (first evidences given by Elinson, 1973; Katagiri, 1974; Grey et al, 1977; Miceli et al, 1978) thus acquiring its full ability to interact with sperm, has now been discovered in other classes, such as mammals (Oikawa et al 1988).

Interesting evolutionary interpretations of the fertilization process have been inspired by the sharp differences between urodele and anuran fertilization and the variation with respect to "orthodox patterns" in some species belonging to the most ancient families of the two orders, (Elinson, 1986).

However it is also true that many fundamental points of the fertilization process in amphibians have not been explored or require more experimentation. Among these, how do sperm acquire motility and what is the action of jelly coats on sperm; what are the egg surface properties that enable sperm fusion with the egg; how localized stimuli can be transmitted to the rest of the egg; what are the molecules involved in signal-transduction and the chain of metabolic events initiated at activation; what is the role of cortical organelles in egg activation; how excess of sperm entrance or multiple fusion with the female pronucleus is avoided.

The study of fertilization in the anuran Discoglossus pictus, belonging to the ancient superfamily Discoglossidea, has provided new data that can be useful for understanding the mechanism of fertilization. We shall first review the main features of the "Discoglossus model" and then examine some of the overmentioned points in the light of what we have learned from this species.

GENERAL FEATURES OF GAMETES AND FERTILIZATION IN *DISCOGLOSSUS PICTUS*

Many functional and morphological characteristics of gametes and gametic interaction are unique to this species, while other features are shared with anuran or urodele species. Among the features typical of D.pictus there are: 1. egg morphology (egg diameter: 1.6 mm). The animal hemisphere is indented by a jelly component, the "animal plug" (Ghiara, 1960). The center of this concavity is further invaginated and forms the cup-shaped dimple (Hibbard, 1928; Campanella, 1975) (fig 1). 2. Spermatozoa lenght and arrangement in bundles. Sperm are 2.33 mm long and are ejaculated in bundles of about 20. They are embedded in a shell made of orderly arranged filaments (Campanella et al, 1979) (figs 2 and 3).

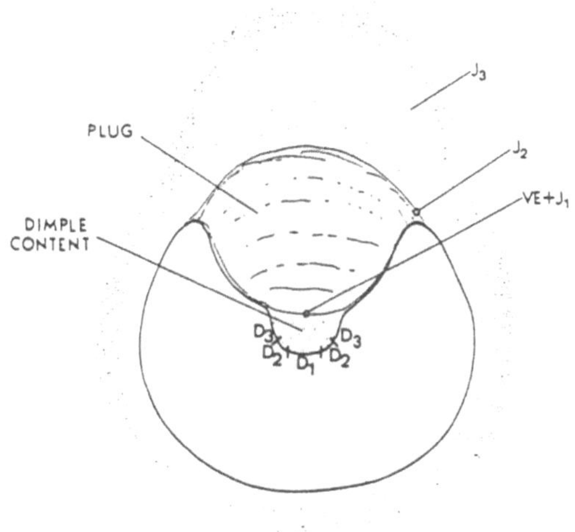

Fig 1. Schematic drawing of a longitudinal section of D. pictus egg. J1; J2; J3 indicate jelly coats; VE= vitelline envelope. The "plug" sits in the animal concavity. In the dimple, D1 is the 200 μm disk where fertilization occurs. D2 and D3 are the dimple walls.

3. Sperm behaviour at insemination. When sperm come into contact with the surface of J3 (the outermost jelly layer) the filament shell sticks to the jelly. Sperm acquire motility, come out of their shells and become embedded in the thick jelly coat facing the dimple. Indeed, they converge, through the plug, at the center of the dimple, D1, which is a disk 200 μm in diameter (Campanella and Gabbiani, 1979; Talevi and Campanella, 1988) (figs 1 and 5). D1 is the only site of the egg where fertilization occurs (Campanella, 1975; Talevi and Campanella, 1988) and has regularly arranged microvilli (fig 4). At the dimple walls (fig 1), D2 has less orderly arranged microvilli and D3 has short microvilli (Campanella et al, 1988). Moreover D1 is endowed with special properties with respect to the rest of the egg, as will be reported in the next pages.

Table I summarizes the main events occurring at fertilization in D. pictus. Cortical granules (CG) are present only in the dimple and the fertilization envelope does not lift at activation, being already far apart the dimple plasma membrane (see fig 1 and table I). What are then the signs of activation in this egg? They consist mainly in two groups of phenomena: 1) a cortical contraction wave which starts in the dimple a few minutes following fertilization and then spreads to the rest of the egg. As a consequence the dimple regresses (Talevi et al, 1985) (see table I). 2) About 15 min from insemination the concavity disappears as a result of the plug liquefaction. Similarly to urodeles, a capsular chamber is formed in which - the egg with its VE and Jl is free to rotate (see table I).

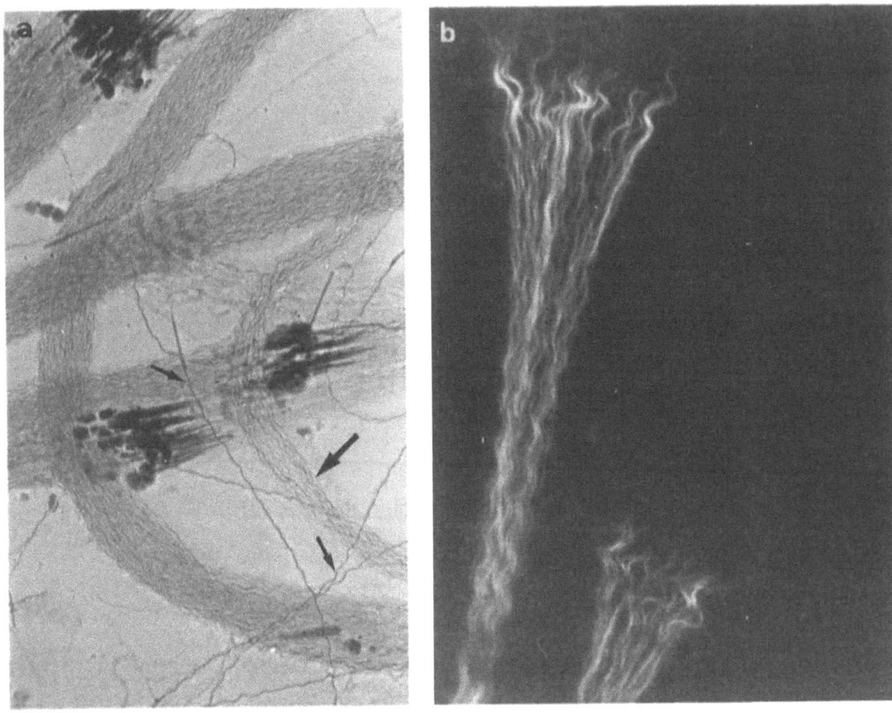

Fig 2. Sperm bundles. a: The very long threads are the heads, which are about 900 μm long (arrow), or the tails (small arrows). The strongly stained segments are the posteriormost portions of the head. The arrowheads indicate the fibrillar matrix where sperm bundles are embedded. x 1,000. b: Sperm stained with anti-actin antibody to show the sperm heads (for details see Campanella and Gabbiani, 1979). x 2,000.

Such a phenomenon is a consequence of the release of peripheral vacuole contents from the egg and involves the reduction of disulphide bonds present in the plug macromolecules (Pitari et al, 1987).

HOW DO SPERM ACQUIRE MOTILITY AND WHAT IS THE ACTION OF JELLY COAT ON SPERM

The acquisition of sperm motility or the variations in the motility pattern is linked to a change in salinity of the medium or to interaction with substances deriving from the female gamete or the genital tract. In amphibians, sperm become motile following release in the external medium (see also Morisawa's report in this volume). In D.pictus, in addition to this, contact of the bundle shell with J3 or with a polylysine coated dish triggers initiation of motility and liberation of sperm from the shell (Campanella and Gabbiani, 1979).

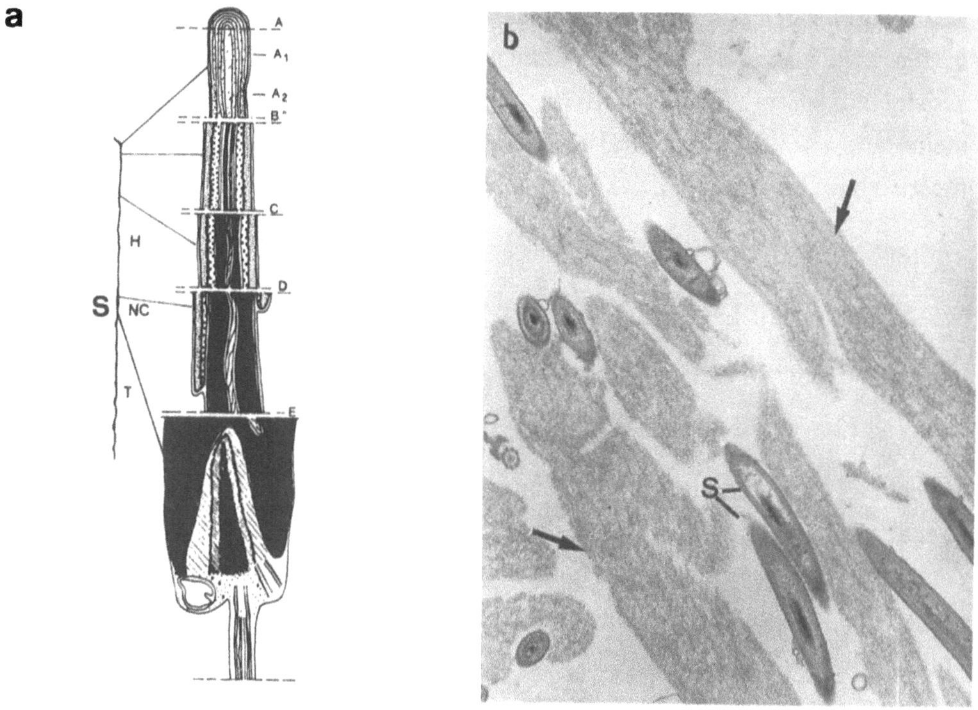

Fig 3. a: Schematic drawing of D. pictus sperm (S) where a head (H) with an apical rod, a "neck" (N) and a tail (T) are distinguishable. From A to B= apical rod; from B to E= nucleus covered by the acrosomal cap. Note the presence of an endonuclear canal, well distinguishable at the posteriormost portion of the sperm, approximately at the E level (see Gabbiani and Campanella, 1979, for details). b: TEM micrograph showing an oblique section of a sperm bundle. The arrows indicate the fibrillar matrix (the shell) where sperm (S) are embedded. x 12,000.

The shell is made of filaments which come to surround sperm in the "seminal vesicles", a wolffian duct specialization of this species (Mann et al, 1963) (fig 2). This phenomenon raises the following questions: Can external matrix or surface molecules in sperm plasma membrane function as target for environmental signalling to be transduced to the sperm cell? Can such molecules be considered ancestral precursors of mammalian epididymal products?

The mechanism of sperm penetration in amphibian jelly coat is unknown and should not involve acrosome content, as the acrosome reaction generally occurs at the VE. In particular, D.pictus sperm - whose motility expires in about 15 s- are able to penetrate the thick jelly layer facing the dimple in as little as 2-3 min. What are the biophysical and/or biochemical properties that make this penetration possible?

When sperm come into contact with the jelly, they acquire the ability to perform the acrosome reaction. Both cations (Ca, Mg) present in the jelly at relatively elevated concentration (Ishihara et al, 1984), and a 500 dalton molecule in the jelly dialyzable fraction (Katagiri, 1973) have a fertilization-supporting activity on sperm. Neither the role of such jelly components on the sperm membrane nor the biophysical-biochemical changes occurring at the acrosome reaction have been studied.

WHAT ARE THE EGG PROPERTIES WHICH ENABLE SPERM FUSION

A) Vitelline coat and egg surface

In invertebrate and mammalian fertilization, sperm-egg recognition and attachment are based on the interaction between sperm proteins and complementary glycoconjugates in the egg extracellular coat. The binding of these complementary molecules generally triggers the acrosome reaction. In a wide range of species, fucose is involved in this interaction (for reviews see Hoshi, 1984 and Macek and Shur, 1988).

In amphibians, a protease secreted by the oviducal "pars recta" modifies the coelomic oocyte VE. As a consequence the number of exposed ligands for the 32 Kdalton VE sperm lysin (Yamasaki et al, 1988) is increased following passage of the oocyte through the oviduct (Miceli et al, 1980; Takamune et al, 1986; Gerton and Hedrick, 1986). The nature of such ligands and/or sites for sperm attachment has not been investigated. However lectins, thanks to their properties of binding specific saccharides, were used to characterize the carbohydrate moieties of glycoproteins involved in sperm-egg interaction. In Bufo and Xenopus, incubation of dejelled eggs with ConA, WGA, SBA or PHA-P, inhibits fertilization (Charbonneau et al, 1986; del Pino and Cabada, 1987) and in Bufo renders the VE resistant to digestion by both proteolytic enzymes and sperm lysin (del Pino and Cabada, 1987). In Xenopus, the inhibition of fertilization could be due to an effect of lectins on sperm (Charbonneau et al, 1986).

In Discoglossus denuded egg, as a result of maturation, D1 surface acquires the ability of binding FBP. Following fertilization the affinity for such a lectin is lost. By contrast, fertilization does not change the ability of the rest of the egg surface to bind WGA and SBA (Denis-Donini and Campanella, 1977). Indeed, this work suggests that fucose residues have a role in fertilization because they are expressed on D1 during maturation and disappear following fertilization. In D.pictus, they are localized at the dimple surface. More sperimentation in D.pictus, as well as in other species should define the glycoconjugates containing fucose residues and their involvement in sperm binding at the VE and/or the egg surface.

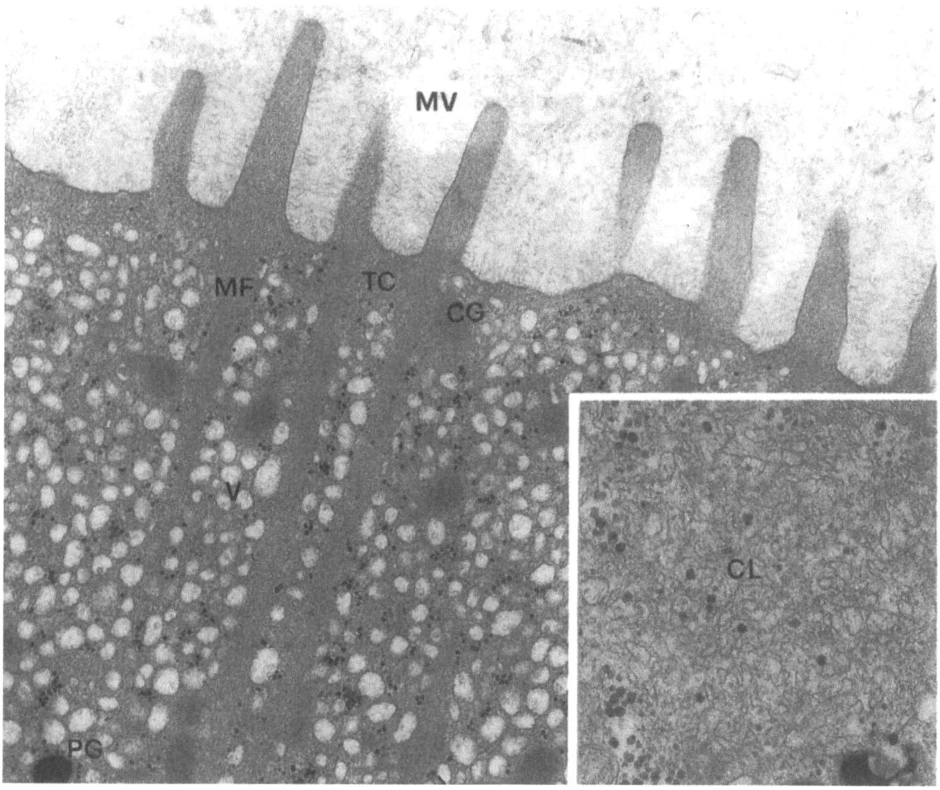

Fig 4. Unfertilized egg of D. pictus. Sperm entrance site (D1). Microfilament bundles (MF) inside microvilli (MV) are clearly discernable. The smooth endoplasmic reticulum is arranged in cortical tubular cisternae (TC), peripheral vacuoles (V) and subperipheral clusters of tubular cisternae (inset, CL). CG= cortical granule; PG= pigment granule. x 25,000; inset, x 45,000. (From Gualtieri et al., 1989 with permission of Dev. Gr. Differ.).

B) Plasma membrane-cortex

In anurans, the region where successfull sperm-egg interaction occurs is restricted to the animal hemisphere. The plasma membrane and/or cortical features that render this region capable of fusing with sperm are still unknown. In the animal half plasma membrane, lipids have a smaller coefficient of mobility (Dictus et al, 1984) and the Cl^- channels responsible for membrane depolarization at fertilization are more concentrated than in the vegetal half (Cross, 1981; Jaffe et al, 1985). Furthermore, microvilli are distributed at a higher density in the animal hemisphere than in the vegetal hemisphere (Grey et al, 1974). In D.pictus, further restriction of such a privileged area to the dimple, may give some clue for studying the prerequisites for the accomplishment of fertilization.

Table I. Summary of fertilization events in D. pictus. On the upper left side sperm penetration through the jelly envelope is schematically outlined. At the upper right side a typical fertilization potential, as recorded at 0 time fertilization, is drawn. SER= smooth endoplasmic reticulum; CL= clusters of SER; V= vacuoles; FE= fertilization envelope; VE= vitelline envelope, CG= cortical granules.

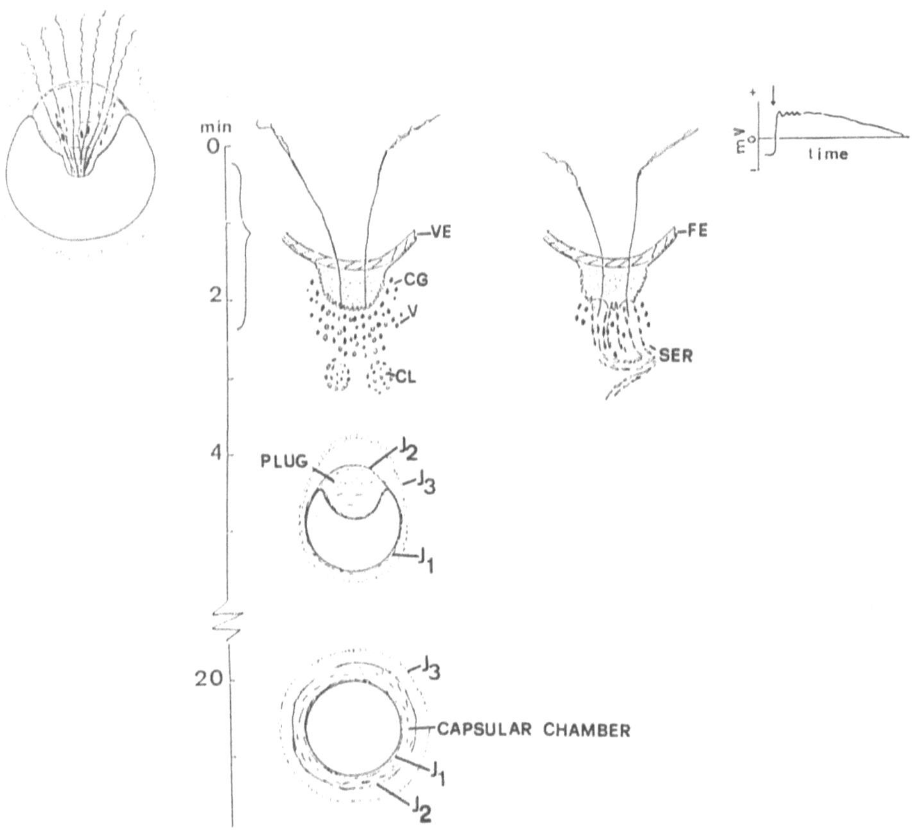

In the dimple, D1, D2 and D3 can be distinguished into three domains on the basis of their surface and cytoplasmic features (Campanella et al., 1988). The dimple regionalization includes functional characteristics: D1 is the only site in the egg, where electrical responses can be elicited by the sperm or by activating agents.

In physiological conditions, the jelly and particularly the plug make spermatozoa converge into D1 (fig. 5a). Gamete interactions occurs only in this area leading to the generation of the fertilization potential (FP), to fertilization cone formation and to the onset of development (Hibbard, 1928; Campanella, 1975; Talevi et al, 1985; Talevi and Campanella, 1988). When the plug is not in axis with the center of the dimple, sperm converge into the dimple walls, D2 and D3 (fig 5b). In these regions, sperm interaction with the plasma membrane leads to small cytoplasmic protrusions and brief depolarizations of few mV (Talevi and Campanella, 1988). A fertilization potential, similar in intensity, delay and shape to the activation potential elicited by pricking (see below) is recorded only when multiple sperm interactions occur in the D2 and D3. In these regions sperm partially penetrate into the egg where, however, degenerate; development is not initiated (Talevi and Campanella, 1988).

Studies on parthenogenetic stimuli have furnished further data on the egg response at activation (Talevi et al, 1985). Activation of the dimple by pricking or by injection with inositol 1,4,5-triphoshate (IP3) in the dimple causes an immediate depolarization, similar to the FP. A wave of contraction that reflects a propagation of $[Ca^{2+}]_i$ (Nuccitelli et al, 1988), spreads from the dimple to the antipode. When eggs are stimulated outside the dimple, a contraction wave spreads from the site of stimulus application to the antipode. In this case, the activation potential (AP) is not generated until the wave reaches the dimple (Talevi et al, 1985; Nuccitelli et al., 1988). As the Cl^- channels are Ca^{2+}-activated (Cross, 1981; Young et al, 1984; Kline, 1988), these data strongly suggest that Cl^- channels are present only in the dimple. On the other hand, by the vibrating probe a large Cl^--transported inward current has been detected only in the D1 domain, further suggesting that Cl^- channels are located only in this region (Nuccitelli et al, 1988). The above mentioned parthenogenetic experiments allow us to hypothesize that sperm may act as activating stimuli in D2 and D3 ,and that the stimulus has to be trasmitted to D1 to elicit a full FP. In the whole dimple the features which enable sperm penetration and the onset of development are probably distributed according to a gradient which has is highest intensity and threshold values in D1. In this respect in the dimple, a similar gradient of distribution concerns also cytoplasmic organelles such as CG, as well as cisternae, vacuoles and clusters of smooth endoplasmic reticulum (SER), potential sources of $[Ca^{2+}]_i$ to be released at fertilization (fig 4).

Freeze-fracture quantitative analysis of D.pictus egg membrane show that D1 has intramembranous particles with higher concentration and opposite ripartition in the two leaflets with respect to the rest of the egg membrane

(Gualtieri et al, 1989). Such a structural polarity of the egg plasma membrane settles during the oocyte maturation at the time of the first polar body extrusion (Gualtieri et al, 1988; Gualtieri et al, in press) (fig 6). Are these particles sperm receptors and/or Cl^- channels? The biochemical and functional analysis of such plasma membrane domains might offer important information on the study of sperm-egg interaction.

a

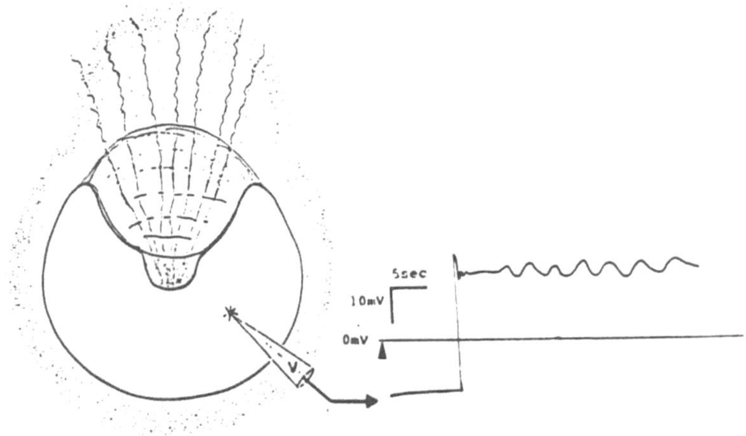

b

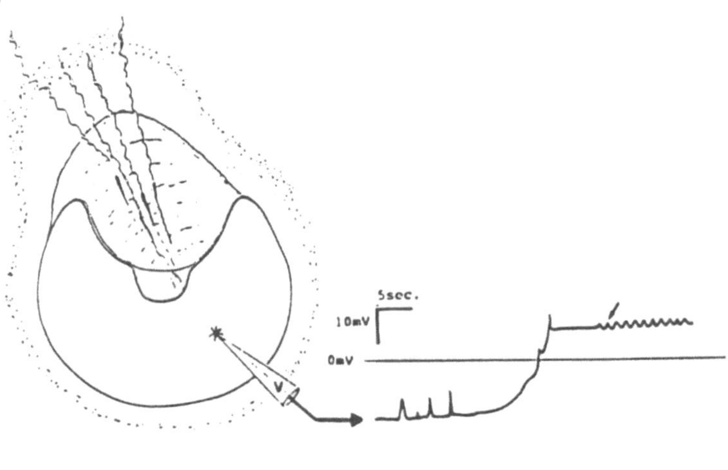

Fig 5. a: On the left, schematic drawing of an egg inseminated in physiological conditions and, on the right, the corresponding fertilization potential. The arrowhead indicates the time of sperm addition. b: On the left, schemating drawing of an egg with the plug not in axis with the center of the dimple. Upon insemination sperm converge into D2 or D3. On the right, corresponding variations of the membrane potential. Explanations in the text.

In conclusion, in D.pictus the multi-step process of fusion between sperm and egg involves specific features to be found only in the dimple. One can also suggest that specific molecules are present only in this region able to interact with sperm, to produce fertilization cones and to start development.

3) HOW IS A LOCALIZED STIMULUS TRANSMITTED TO THE REST OF THE EGG

The fine molecular events related to sperm-egg interaction and the triggering of activation are still one of the most important pitfall of the study of fertilization. Most probably sperm release a substance which is able to start PIP2 hydrolysis. In sea urchin, a sperm soluble factor is able to initiate activation when injected into the cytoplasm (Dale et al, 1985), and in Urechis a protein has been isolated from the sperm which triggers membrane depolarization upon addition to the egg (Gould and Stephano, 1987) (see articles by Dale, by Kline and by Whitaker, in this volume). In amphibians sperm contribution to gametic fusion has been investigated. It has been postulated that sperm inserts into the egg membrane either receptors which interact with G proteins or factors which directly activate such proteins (Kline et al, 1988; Iwao and Jaffe, 1989). However, it remains to be demonstrated why such putative factor cannot be effective in promoting sperm fusion at the vegetal hemisphere, or in the case of D.pictus, in the regions outside the dimple.

The indication that G proteins are involved in the activation process fits well with the knowledge that, in anurans, stimulation of the egg by sperm or parthenogenetic agents (pricking, IP3 injection etc.) is followed by an increase of cytosolic calcium which is propagated from the site of activation to the rest of the egg through a Ca^{2+} release (Busa and Nuccitelli, 1985; Busa et al, 1985; Kubota et al, 1987). Extracellular calcium is not needed during such a release, however, in its absence, activation by sperm or by pricking does not occur. Presently, there is not full explanation of such a limited and precise need for extracellular calcium.

Many events, such as ion channels opening, exocytosis and formation of pronuclei are calcium-dependent (Kline, 1988). More work is needed to understand whether the important changes which accompany activation, namely changes in pHi (Webb and Nuccitelli, 1981; Grandin and Charbonneau, 1989 and see Charbonneau, this volume) and in protein phosphorylation (Bement and Capco, 1989) are directly related to the increase in cytosolic calcium.

Among the numerous questions concerning the first stages of activation, one is whether the increase of intracellular Ca^{2+} can be considered as the first signal which travels through the cortex or the whole egg.

Fig 6. a: Replica of the germinative area in D. pictus coelomic oocytes. Insets: P and E face of two microvilli. b: Replica of the sperm entrance site (D1) in D. pictus egg. Inset: P and E face of two microvilli. The intramembraneous particle density in D1 is markedly increased with respect to the maturing oocyte. a and b, x 25,000; insets, x 60,000. CG = cortical granule. (From Gualtieri et al., 1989 with permission of Dev. Gr. Differ.).

In D.pictus egg, the contraction wave elicited by parthenogenetic stimulation, starts egg activation at its arrival in the dimple. This indicates that a subcortical mechanism of stimulus transmission can work in both direction: from the dimple to the rest of the egg, at fertilization, and from the egg regions outside the dimple to the dimple as a consequence of experimental activation. The particular organization of D.pictus egg permits the following experiment: pricking the denuded egg at the equator in two opposite sites gives rise to two distinct contraction waves in case that the stimuli are applied almost contemporaneously.

They reach the dimple after a time that varies from 5 to 15 min according to the egg clutch. However, when pricking is first conducted at one side and about 20 s later at the opposite side, the second wave never develops; the first wave regularly reaches the dimple where the activation is elicited (Campanella, unpublished). What is the nature of such an early signal which inhibits the second wave formation and travels along the egg cortex or plasma membrane? Is it present in other anuran species where it might not be depicted because of the different egg organization?

4) WHAT IS THE ROLE OF CORTICAL ORGANELLES

A) Cortical endoplasmic reticulum (SER).

Cortical constituents, such as vesicles and cisternae, have been extensively studied in Xenopus laevis eggs. They are commonly considered SER because they derive from annulated lamellae (Campanella et al, 1984; Larabell and Chandler, 1988) which in germinal as well as in somatic cells transform into rough or smooth endoplasmic reticulum (Kessel, 1985). The cisternae form a continuous network in the cortex and particularly surround CG as a shell (Grey et al, 1974), thus suggesting they might be related to CG exocytosis and be site of accumulation and release of Ca^{2+} at activation (Campanella and Andreuccetti, 1977; Andreuccetti et al, 1984). Furthermore, vacuoles are in close contact with the plasma membrane in loci, where the presence of junctions reminiscent of the triads junctions in striated muscle has been suggested (Gardiner and Grey, 1983). Biochemical analysis is needed to study the characteristics of the peripheral vacuoles and cisternae and their possible analogy with calciosomes (Volpe et al, 1988). The cortical SER anastomozes with whorls of cisternae located in the subcortical layer. Shortly following CG exocytosis, the SER rearranges, and chains of cisternae are transiently seen crossing the egg periphery where later only vacuoles are found (Campanella and Andreuccetti, 1977). The anatomical organization of such probable calcium reservoirs, the dynamic of their transformations as well as electrophysiological measurements of free Ca^{2+} following IP3 injection, suggest that Ca^{2+} release is a complex, multi-step event involving an IP3-sensitive pool and an IP3-insensitive pool (Busa et al, 1985; Parker and Miledi, 1986 ;Berridge, 1988).

In this context several specific points need explanation. It is known that Cl^- channels are activated by an increase of free calcium and that, as a consequence, the plasma membrane depolarizes. The FP is recorded always before the changes in free calcium concentrations, and this might be due to limits of the utilized tecniques (i.e. diffusion time from sites of liberation to the Ca^{2+}-selective electrode). However, it cannot be excluded that a minor release of Ca^{2+} responsible for Cl^- channels activation precedes the larger increase of cytosolic Ca^{2+}. In this respect we hypothesize that the vacuoles next to the plasma membrane are specialized in the first release of Ca^{2+} while the rest of the cortical SER is involved in the transmission of the measurable increase of Ca^{2+} as suggested by electrophysiological work (Busa et al, 1985). A second point is: How can the IP3 injection determine Ca^{2+} release in the cytoplasm where very few vacuoles (SER or calciosomes?) are found? A possible answer is suggested by somatic cells where the IP3-induced Ca^{2+} release does not appear to be the property of a single organelle but of specialized regions of rough and smooth ER and other smooth surfaced structures (Ross et al, 1989).

The function of the subcortical SER clusters has been studied in D.pictus eggs (fig 4, table I). In this species, an eightfold increase of internal Ca^{2+} has been measured at fertilization. It starts in the dimple and then propagates to the rest of the egg (Nuccitelli et al, 1988). In the dimple there are tubular cisternae close to the plasma membrane, vacuoles intermingled with CG and more deeply located cisternae clusters (Campanella et al, 1988). They have a peak of concentration in D1, and are scantly present in the rest of the egg (Campanella et al, 1986; Campanella et al, 1988). The tubular cisternae are equivalent in position and, in our opinion, in function, to X.laevis sub-plasmalemma vacuoles. Following fertilization D.pictus vacuoles become an anastomosed cortical network of SER (similar to that present in unfertilized X.laevis eggs), which migrates towards the plasma membrane. At the same time the cisternae clusters transform into whorls of cisternae which become part of the cortical network (table I). Such transformation occurs only in the dimple, but it can be induced experimentally also in those cisternae clusters located next to the point of IP3 injection. Furthermore cisternae deriving from the clusters participate in plasma membrane wound healing. It has been hypothesized that the clusters open as a consequence of a physiological or artificially induced increase of Ca^{2+} concentration and that their cisternae might constitute reservoirs of membranes (Campanella et al, 1988a and 1988b; Gualtieri et al, in press; Bracci Laudiero et al, in press).

B) Cortical cytoskeleton and territorial compartimentalization

Eggs, as most cells, are able to maintain compartimentalization at the plasma membrane as well as in the cytoplasm. How is this achieved?

In amphibian eggs, cytoskeletal proteins, such as actin, villin, prekeratin, myosin, tubulin, actin-binding protein and vimentin were depicted (Palecek et al, 1982; Gall et al, 1983; Corwin and Hartwig, 1983; Christensen et al, 1984; Tang et al, 1988). No differences have been reported in the distribution of such proteins between the animal half, the region where sperm penetrate into the egg and the vegetal half, except for a general more abundant presence in the animal hemisphere.

As previously mentioned, D.pictus eggs show an evident compartimentalization of both surface and plasma membrane constituents (sugar residues, intramembranous particles and ionic channels) and cytoplasmic territories within the dimple. In this region, 10 μm long actin microfilament bundles, appear to compartimentalize organelles (i.e. vacuoles, CG) along and between the bundles. Furthermore, in D3, which is located at the boundary of the dimple with the rest of the animal half, mitochondria are found below the bundles palisade, in significantly larger amount than in D1 and D2. Actin, myosin, alfa-actinin, tropomyosin have been immunocytologically defined in the whole egg cortex (Campanella and Gabbiani, 1980; Campanella et al, 1982; Campanella et al, in press) while spectrin has been found only in the dimple peripheral cytoplasm (Campanella et al, submitted) and a positivity to anti-gelsolin antibodies only in D1 (Campanella, unpublished results). This suggests that the cytoskeleton may play a role in organelle distribution. Furthermore, localized cytoskeletal functionality may actively contribute to CG, vacuoles and cisternae locomotion at activation, and, in general, in egg contraction. In particular, spectrin might stabilize plasma membrane domains thanks to its sites of interaction with actin and, as recently suggested (Campanella et al, submitted) with receptors or ionic channels in the plasma membrane. It would be interesting to study whether in other anurans eggs spectrin is similarly found in the region where fertilization occurs and fertilization channels are found.

6) HOW IS EXCESS OF SPERM ENTRANCE AVOIDED

In anuran eggs, fertilization is characterized by the relatively limited area of sperm interaction, the long-lasting membrane depolarization, and the exocytosis of CG at fertilization. These features seem to be the factors responsible for maintaining monospermy in eggs. In urodeles, where eggs are physiologically polyspermic, sperm entrance can occur all over the egg, the membrane does not depolarize, or even hyperpolarize at fertilization, and CG are absent (for reviews see Schmell et al, 1984; Elinson, 1986).

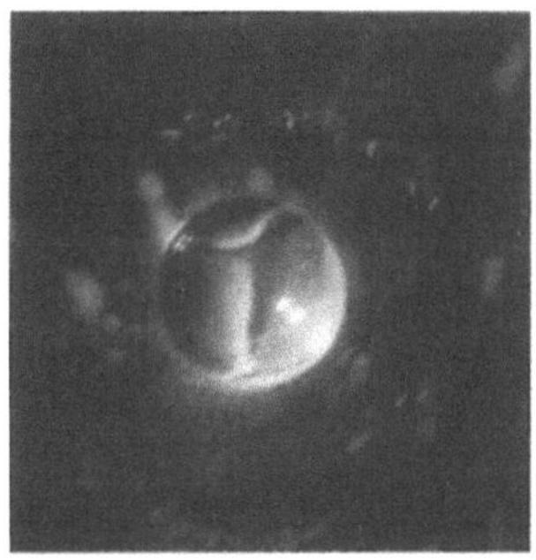

Fig 7. D. pictus cleaving eggs. Three irregular furrows are present. x 10.

Is this sharp distinction between the two orders a general rule? Are there alternative mechanisms present in anuran eggs that avoid excess of sperm entrance?

In D. pictus, a fast long-lasting depolarization as well as CG exocytosis occurs at fertilization. Yet, recent work has shown that multiple sperm penetration occurs in eggs with normal membrane depolarization and which develop into tadpoles, evidencing an interesting feature of this egg: the voltage independence of sperm penetration. Indeed, this was shown in eggs where the voltage was kept either positive through current injection or negative by bathing the eggs in high NaI Ringer (Talevi, 1989). However, the number of sperm penetrating polyspermic eggs is generally low and normal development is compatible in eggs penetrated by two up to five sperm. Moreover, in spite of the high frequency of polyspermy found (36%), not all egg clutches display polyspermy. The occurrence of this condition does not depend upon sperm "efficiency", but is correlated with characteristics of some egg clutches (Talevi, 1989). Extracellular mechanisms could play an important role in regulating sperm entrance in D. pictus egg. In this regard the jelly coat is a good candidate. The plug is made of fibrillar material, and, as previously described, conveys sperm to the D1 area; the fibrillar matrix of the plug is probably organized in a series of set routes which support sperm penetration. In polyspermic egg clutches this structure may be altered, increasing the number of sperm entering the plug and reaching the VE.

Two further blocks to sperm entrance are present in this species. In experiments of reinsemination, the newly added sperm encounter two barriers i.e. the sperm bundle shells which cover the J3 surface, and the VE which has changed into fertilization envelope as a consequence of fertilization (Campanella, unpublished results).

An important point to be still investigated in fertilization of D. pictus polyspermic clutches is the behaviour of supernumerary sperm in the dimple cytoplasm and how this can be compatible with normal development. Occasionally, upon raising tadpole in laboratory, cleaving embryos with irregular furrows have been observed (fig 7), suggesting that in polyspermic zygotes supernumerary sperm decondense, develop two asters and participate in mitosis (see also Elinson, 1986).

D. pictus is not the only species, among anurans, to display polyspermic fertilization. Recently, a high frequency of polyspermy with normal development has been found in the terrestrial breeding frog Eleutherodactylus coqui (Elinson, 1987).

With respect to urodeles, monospermy has been described in the primitive family Hinobiliae, where a fast membrane depolarization at fertilization has been reported. Furthermore in Hynobius nebulosus the outermost jelly layer constitutes the normal barrier to polyspermy (Iwao, 1989).

These new findings represent important exceptions to the schematic distinction between the two orders, and provide new data for the general understanding of amphibian fertilization. They suggest that several mechanisms can be called into action to guarantee the achievement of the final goal of fertilization: amphimissis of the female aploid genome with one paternal aploid genome.

REFERENCES

Arion D, Mejier L, Brizuela L, Beach D (1988) CDC2 is a component of the M phase-specific histone Hl kinase: evidence for identity with MPF. Cell 55:371-378

Bement WM, Capco DG (1989) Activators of protein kinase C triggers cortical granule exocytosis, cortical contraction, and furrow formation in Xenopus oocytes and eggs. J Cell Biol 108:885-892

Berridge MJ (1988) Inositol triphosphate-induced membrane potential oscillations in Xenopus oocytes. J Physiol 403:589-599

Bracci Laudiero L, Gualtieri R, Andreuccetti P Possible role of the smooth endoplasmic reticulum during the early stages of development in the painted frog Discoglossus pictus (Anura). In press on Acta Embr Morph Exp ns

Busa WB, Nuccitelli R (1985) An elevated free cytosolic Ca^{2+} wave follows fertilization in eggs of the frog Xenopus laevis. J Cell Biol 100:1325-1329

Busa WB, Ferguson JE, Joseph SK, Williamson JR, Nuccitelli R (1985) Activation of frog (Xenopus laevis) eggs by inositol triphosphate. I. Characterization of Ca^{2+} release from intracellular stores. J Cell Biol 101:677-682

Campanella C, Andreuccetti P, (1977) Ultrastructural observation on the cortical and subcortical endoplasmic reticulum and on residual cortical granules in the egg of Xenopus laevis. Dev Biol 56:1-10

Campanella C (1975) The site of spermatozoon entrance in the unfertilized egg of Discoglossus pictus (Anura): an electron microscope study. Biol Reprod 12:439-447

Campanella C, Gabbiani G (1979) Motile properties and localization of contractile proteins in the spermatozoon of Discoglossus pictus. Gamete Res 2:163-175

Campanella C, Gabbiani G (1980) Cytoskeletal and contractile proteins in coelomic oocytes, unfertilized and fertilized eggs of Discoglossus pictus (Anura). Gamete Res 3:99-114

Campanella C, Rungger-Brandle E, Gabbiani G (1982) Immunolocalization of alfa-actinin in an amphibian egg (Discoglossus pictus). In "Embryonic Development" Part B: Cellular Aspects (M Burger and R Weber eds), 45-53. Alan Liss New York

Campanella C, Andreuccetti P, Taddei C, Talevi R (1984) The modifications of the cortical endoplasmic reticulum during "in vitro" maturation of Xenopus laevis oocytes and its involvment in cortical granules exocytosis. J Exp Zool 229:283-294

Campanella C, Talevi R, Atripaldi U, Quaglia L (1986) The cortical endoplasmic reticulum and its possible role in activation of Discoglossus pictus (Anura) egg. In "Molecular and Cellular Biology of Fertilization" (JL Hedrick eds) Plenum Press New York

Campanella C, Kline D, Nuccitelli R (1988) The cortical reaction in the egg of Discoglossus pictus: Changes in the cisternae clusters at activation. In "Cell Interactions and Differentiation" (G Ghiara eds) University of Naples

Campanella C, Talevi R, Kline D, Nuccitelli R (1988) The cortical reaction in the egg of Discoglossus pictus: A study of the changes in the endoplasmic reticulum at activation. Dev Biol 130:108-119

Campanella C, Chaponnier C, Quaglia L, Gualtieri R, Gabbiani G Different cytoskeletal organization in two maturation stages of Discoglossus pictus (Anura) oocytes: Thickness and stability of actin microfilaments and tropomyosin immunolocalization. In press on Gamete Res

Campanella C, Carotenuto R, Gabbiani G Anti-spectrin antibodies stain the oocyte nucleus and the site of fertilization channels in the egg of Discoglossus pictus (Anura). Submitted to Gamete Res

Charbonneau M, Dufresne-Dubé, Guerrier P (1986) Inhibition of the activation reaction of Xenopus laevis eggs by the lectins WGA and SBA. Dev Biol 114:347-360

Charbonneau M, Grey RD, Baskin RJ, Thomas D (1986) A freeze-fracture study of the cortex of Xenopus laevis eggs. Dev Growth Differ 28(1):75-84

Christensen K, Sauterer R, Merriam RW (1984) Role of soluble myosin in cortical contractions of Xenopus eggs. Nature 310:150-151

Corwin HL, Hartwig JH (1983) Isolation of actin-binding protein and villin from toad oocytes. Dev Biol 99:61-74

Dale B, De Felice LJ, Ehrenstein G (1985) Injection of a soluble sperm fraction into sea-urchin eggs triggers the cortical reaction. Experientia 41:1068-1070

Dale B, Talevi R (1989) Distribution of ion channels in ascidian eggs and zygotes. Exp Cell Res 181:238-244

Del Pino EJ, Cabada MO (1987) Lectin binding sites in the vitelline envelope of Bufo arenarum oocytes: Role in fertilization. Gamete Res 17:333-342

Denis-Donini S, Campanella C (1977) Ultrastructural and lectin binding changes during the formation of the animal dimple in oocytes of Discoglossus pictus (Anura). Dev Biol 61:140-152

Dictus WJAG, Van Zoelen EJJ, Tetteroo PAT, Tertoolen LG, De Laat SW, Bluemink JG (1985) Lateral mobility of plasma membrane lipids in Xenopus laevis eggs: Regional differences related to animal/vegetal polarity become extreme upon fertilization. Dev Biol 101:201-211

Dunphy WG, Brizuela L, Beach D, Newport J (1988) The Xenopus cdc2 protein is a component of MPF, a cytoplasmic regulator of mitosis. Cell 54:423-431

Elinson RP (1973) Fertilization of frog body cavity eggs enhanced by treatment affecting the vitelline coat. J Exp Zool 183:291-302

Elinson RP (1986) Fertilization in amphibians: The ancestry of the block to polispermy. Intern Rev Cytol 101:59-100

Elinson RP (1987) Fertilization and aqueous development of the puerto rican terrestrial breeding frog, Eleutherodactylus coqui. J Morph 193:217-224

Gabers D, Bentley JK, Dangott LJ, Ramarao CS, Shimomura H, Suzuki N, Thorpe D (1986) Peptides associated with eggs: mechanism of interaction with spermatozoa. In "The Molecular and Cellular Biology of Fertilization" pp 315-357. JL Hedrick eds Plenum Publ Corp

Gall L, Picheral B, Gounon P (1983) Cytochemical evidence for the presence of intermediate filaments and microfilaments in the egg of Xenopus laevis. Biol Cell 47:331-342

Gardiner MD, Grey RD (1983) Membrane junctions in Xenopus eggs: Their distribution suggests a role in calcium regulation. J Cell Biol 96:1159-1163

Gautier J, Norbury C, Lohka M, Nurse P, Maller J (1988) Purified Maturation-Promoting Factor contains the product of a Xenopus homolog of the fission yeast cell cycle control gene cdc2. Cell 54:433-439

Ghiara G (1960) Ricerche intorno alla struttura microscopica, submicroscopica ed istochimica ed alle funzioni degli involucri ovulari di Discoglossus pictus Otth e di altre specie di Anfibi. Arch Zool Ital 45:9-92

Grandin N, Charbonneau M (1989) Intracellular pH and the increase in protein synthesis accompanying activation or stimulation by weak bases in Xenopus eggs. (submitted)

Greve LC, Hedrick JL (1978) An immunocytochemical localization of the cortical granule lectin in the fertilized and unfertilized eggs of Xenopus laevis. Gamete Res 1:44-61

Grey RD, Wolf DP, Hedrick JL (1974) Formation and structure of the fertilization envelope in Xenopus laevis. Dev Biol 36:44-61

Grey RD, Working PK, Hedrick JL (1977) Alteration of structure and penetrability of the vitelline envelope after passage of eggs from coelom to oviduct in Xenopus laevis. J Exp Zool 219:87-95

Gualtieri R, Andreuccetti P, Cafiero G (1988) Cell surface changes during oocyte maturation in Discoglossus pictus (Anura). Eur J Cell Biol 47(23)

Gualtieri R, Cafiero G, Andreuccetti P (1989) plasma membrane domains and the site of sperm entrance in Discoglossus pictus (Anura) eggs. Dev. Growth Diff 31(5):511-517

Gualtieri R, Cafiero G, Andreuccetti P Surface and cortex regionalization during "in vivo" maturation in Discoglossus pictus (Anura) oocytes. In press.

Hibbard H (1928) Contribution a' l'etude de l'ovogenese, de la fecondation et de l'histogenese chez Discoglossus pictus Otth. Arch Biol 32:251-326

Hoshi M (1984) Roles of sperm fucosidases and proteases in the ascidian fertilization. In "Advances in Invertebrate Reproduction" 3. Elsevier Sci Publ BV

Ishihara K, Konosono J, Kanatani H, Katagiri C (1984) Toad egg-jelly as a source of divalent cations essential for fertilization. Dev Biol 105:435-442

Iwao Y (1989) An electrically mediated block to polyspermy in the primitive urodele Hynobius nebulosus and phylogenetic comparison with other amphibians. Dev Biol 134:438-445

Iwao Y, Jaffe LA (1989) Evidence that the voltage-dependent component in the fertilization process is contributed by the sperm. Dev Biol 134:446-451

Katagiri C (1973) Chemical analysis of toad egg-jelly in relation to its sperm-capacitating activity. Dev Growth Differ 15(2):81-92

Katagiri (1974) A high frequency of fertilization in premature and mature coelomic toad eggs after enzymic removal of vitelline membrane. J Embryol Exp Morphol 31:573-581

Kessel RG (1985a) Annulatae lamellae (porous cytomembranes): With particular emphasis on their possible role in differentiation of the female gamete. In "Developmental Biology: A Comprehensive Synthesis", 1:179-232. Browder LW eds, Plenum Press New York and London

Kline D (1988) Calcium-dependent events at fertilization of the frog egg: injection of a calcium buffer blocks ion channels opening, exocytosis and formation of pronuclei. Dev Biol 126:346-361

Kline D, Robinson KR, Nuccitelli R (1983) Ion currents and membrane domains in the cleaving Xenopus egg. J Cell Biol 97:1753-1761

Kline D, Simoncini L, Mandel G, Maue RA, Kado RT, Jaffe LA (1988) Fertilization events induced by neurotransmitters after injection of mRNA in Xenopus eggs. Science 241:464-467

Kubota HY, Yoshimoto Y, Yoneda M, Hiramoto Y (1987) Free calcium wave upon activation in Xenopus eggs. Dev Biol 119:129-136

Labbe' JC, Picard A, Karsenti E, Doree M (1988) An M-Phase-specific protein kinase of Xenopus oocytes: Partial purification and possible mechanism of its periodic activation. Dev Biol 127:157-169

Larabell CA, Chandler DE (1988) Freeze-fracture analysis of structural reorganization during meiotic maturation in oocytes of Xenopus laevis. Cell Tissue Res 251:129-136

Lohka MJ, Hayes MK, Maller JL (1988) Purification of maturation-promoting factor, an intracellular regulator of early mitotic events. Proc Natl Acad Sci USA 85:3009-3013

Macek MB, Shur BD (1988) Protein-carbohydrate complementarity in mammalian gamete recognition. Gamete Res 20:93-109

Mann T, Lutwak-Mann C, Hay MF (1963) A note on the so-called seminal vesicles of the frog Discoglossus pictus. Acta Embryol Morphol Exp 6:21-24

Masui Y, Markert C (1971) Cytoplasmic control of nuclear behavior during meiotic maturation of frog oocytes. J Exp Zool 177:129-146

McIntosh RP, Catt KJ (1987) Coupling of inositol phospholipid hydrolisis to peptide hormone receptors expressed from adrenal and pituitary mRNA in Xenopus laevis oocytes. Proc Natl Acad Sci USA 84:9045-9048

Miceli DC, Fernandez SN, Raisman JS, Barbieri DF (1978) A trypsin-like oviducal proteinase involved in Bufo arenarum fertilization. J Embryol Exp Morphol 48:79-91

Miceli DC, Fernandez SN, Morero RD (1980) Effect of oviducal proteinase upon Bufo arenarum vitellin envelope. A fluorescence approach. Dev Growth Differ 22:639-643

Nuccitelli R, Kline D, Busa W, Talevi R, Campanella C (1988) A highly localized activation current yet widespread intracellular calcium increase in the egg of the frog, Discoglossus pictus. Dev Biol 130:120-132

Oikawa T, Sendai Y, Kurata S, Yanagimachi (1988) A glycoprotein of oviducal origin alters biochemical properties of the zona pellucida of hamster egg. Gamete Res 19:113-122

Palecek J, Ubbels GA, Macha J (1982) An immunocytochemical method for the visualization of tubulin-containing structures in the egg of Xenopus laevis. Histochemistry 76:527-538

Parker I, Miledi R (1986) Changes in intracellular calcium and in membrane currents evoked by injection of inositol triphosphate into Xenopus oocytes. Proc R Soc Lond 228:307-315

Pitari G, Dupré S, Amore F, Talevi R, Campanella C (1987) Ruolo dei ponti disolfuro nella dissoluzione delle gelatine nelle uova di anfibio. Atti XXXV Congresso SIB, Brescia-Gardone.

Takamune K, Yoshizaki N, Katagiri Ch (1986) Oviducal pars recta-induced degradation of vitelline coat proteins in relation to acquisition of fertilizability of toad eggs. Gamete Res 14:215-224

Talevi R (1989) Polyspermic eggs in the anuran Discoglossus pictus develop normally. Development 105:343-349

Talevi R, Dale B, Campanella C (1985) Fertilization and activation potential in Discoglossus pictus (Anura) eggs: A delayed response to activation by pricking. Dev Biol 111:316-323

Talevi R, Campanella C (1988) Fertilization in Discoglossus pictus (Anura) 1.Sperm-egg interactions in distinct regions of the dimple and occurrence of a late stage of sperm penetration. Dev Biol 130:524-535

Talevi R, Dale B (1986) Electrical characteristcs of ascidian egg fragments. Exp Cell Res 162: 539-543

Tang P, Sharpe CR, Mohun TJ, Wylie (1988) Vimentin expression in oocytes ,eggs and early embryos of Xenopus laevis. Development 103:278-287

Volpe P, Krause KH, Hashimoto S, Zorzato F, Pozzan T, Meldolesi ,J, Lew DP (1988) "Calciosome" a cytoplasmic organelle: The inositol 1,4,5 triphosphate-sensitive Ca^{2+} store of non-muscle cells? Proc Natl Acad Sci USA 85: 1091-1095

Yamasaki H, Takamune K, Katagiri C (1988) Classification, inhibition, and specificity studies of the vitelline coat lysin from toad sperm. Gamete Res 20:287-300

Ward GE, Brokaw CJ, Gabers DL, Vacquier VD (1985) Chemotaxis of Arbacia punctulata spermatozoa to resact, a peptide from the egg jelly layer. J Cell Biol 101:2324-2329

INTRACELLULAR SIGNALLING DURING FERILISATION AND POLARISATION IN *FUCOID ALGAE*

C. Brownlee,
Marine Biological Association of the UK,
The Laboratory, Citadel Hill,
Plymouth PL1 2PB UK.

Several species of marine algae show characteristics of fertilisation which are at least superficially more typical of animals than plants. The most distinctive feature is the liberation of free eggs and motile sperm into the sea. Certain members of the order Fucales produce large numbers of eggs and sperm on separate plants and the production of gametes can be controlled so that many eggs can be fertilised and induced to develop synchronously. Development of the zygote is characteristic of plant development in that a cell wall is secreted at an early stage and development occurs through the cumulative effects of polarisation, division and growth of individual cells.

SPERM ATTRACTION. Sperm are attracted to the egg by the release from the egg of a chemoattractant "fucoserratin". This has been shown to be a conjugated hydrocarbon (Muller & Gassmann, 1978, 1985) and its chemoattractant properties can be mimicked by a variety of simple hydrocarbons. The nature of the intracellular signalling systems involved in this response is unknown. Whether extracellular receptors in the sperm plasmalemma are involved is not clear. Fucoserratin is a lipid-soluble compound which will diffuse readily into the cell, possibly obviating the need for a classical second messenger system.

NATO ASI Series, Vol. H 45
Mechanism of Fertilization
Edited by B. Dale

EGG ACTIVATION. The first recorded fertilisation event in the *Fucus* egg is the fertilisation potential (Robinson, Jaffe & Brawley, 1981). This depolarisation lasts for several minutes. Its ionic basis is still unknown, but voltage-gated channels may be involved since action potentials can be elicited in the unfertilised egg by imposing inward current pulses. In animal eggs such as the sea urchin and ascidian, the fertilisation current is probably carried through channels selective for calcium and sodium (Hice & Moody, 1988; Hagiwara & Jaffe, 1979). In all animal eggs so far studied, an early fertilisation event is an elevation of cytoplasmic calcium which in deuterosomes spreads as a wave of intracellular calcium release through the cortical cytoplasm. (eg. Jaffe, 1983, 1989, also Jaffe, this volume; Brownlee & Dale, 1990). The wave is likely to propagate as a calcium-induced calcium release (see Jaffe, this volume) and can be triggered by the injection of inositol trisphosphate (Swann & Whitaker, 1986), implicating the involvement of the inositol phosphate/calcium messenger system in egg activation (see also Whitaker, this volume).

Direct evidence for a calcium wave, or indeed an elevation of cytoplasmic calcium in activating *Fucus* eggs is so far missing. There is, however, reason to suppose that a calcium wave is involved in *Fucus*. Some kind of activation wave visibly passes across fertilised eggs of the brown alga *Cytosira barbatta* (Knapp, 1931). In these eggs, roughening and contraction of the egg plasmalemma surface were observed within 1 minute of fertilisation. This appeared to spread as a wave across the egg surface. The first major structural change so far observed in the *Fucus* egg following fertilisation is the secretion of a cell wall (see Evans *et al* 1982). There is evidence to suggest that this occurs as a slow exocytotic wave across the egg (Brawley & Bell, 1987). This may be a calcium-regulated process since calcium ionophores have been reported to stimulate cell wall secretion (Brawley & Bell 1987). Replacement of sodium in the external medium with N-methylglucamine also caused egg activation. This has led to the suggestion that there is a sodium/calcium exchange mechanism in the plasmalemma. There is not yet, however any direct evidence for a sodium/calcium exchange in plant cells. The presence of calcium-binding proteins, including calmodulin in *Fucus* eggs (Brawley & Roberts, 1989) suggests that calcium-mediated signalling may occur during egg activation. However, eggs were found to lack any major calmodulin-binding activity, compared with sperm and zygotes. This calls into question the role in *Fucus* of calcium-calmodulin control of egg activation such as that postulated by Epel *et al* (1981), involving calmodulin activation of NAD kinase via the elevation of free calcium in sea urchins. Until it has been established that cytoplasmic calcium does elevate during egg activation in brown algae and that this elevation is of sufficient magnitude to release the egg from its arrested unfertilised state, speculation on the involvement of other second messengers, such as the inositol phosphate signalling pathway, the posible role of receptor-G-protein complexes in the plasmalemma or the nature of any other activating factor carried by the sperm is premature.

POLARISATION. For a few hours after fertilisation, the single celled *Fucus* zygote generates a polar axis which becomes fixed at around 12 hours. This polar axis is perpendicular to the plane of the first cell division. The direction of the polar axis can be determined by a variety of external signals, the most important of which is unilateral light. Whatever the initial polarisisng factor, the establishment of a polar axis must involve an initial asymmetry which is amplified by positive feedback to result in the formation of a stable and ultimately a fixed axis of polarity.

Figure. 1 describes the possible events involved in the perception of polarising stimuli and the development and reinforcement of intracellular gradients leading ultimately to axis fixation. In unilateral light, zygotes will germinate with the rhizoid growing away from the source of light. The presence of a specific light receptor (see below) makes it possible to speculate on the molecular mechanism by which the initial asymmetry is established and the way in which this can be amplified. However, the fact that zygotes can be polarised by a wide range of external gradients suggests that the initial step involving a specific receptor is not essential for polarisation.

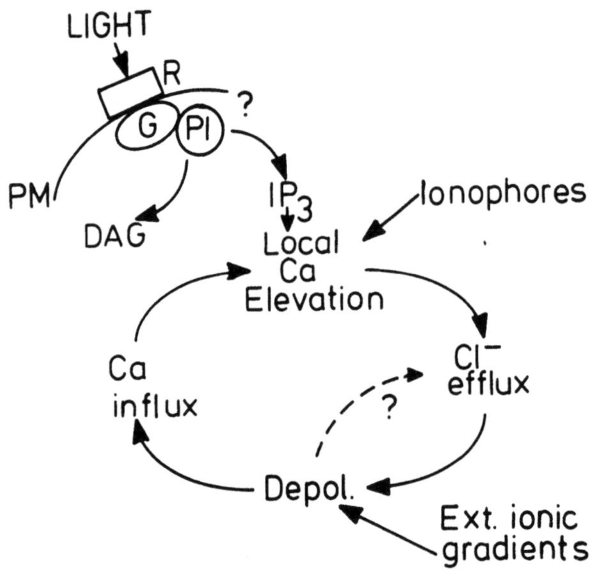

Fig. 1. Model to describe the likely events in the perception of an external gradient leading to the establishment and amplification of an intracellular gradient and axis fixation. G: G-protein, PI: phosphoinositidase, R: receptor, DAG: diacylglycerol PM: plasmamembrane.

The mechanism by which a unidirectional light stimulus can be perceived by a zygote has been more fully discussed recently (Brownlee, 1990). A membrane-associated, protein-bound flavinoid is likely to be the photoreceptor in this response (Brownlee, 1990 and references therein). This may be in the form of a G-protein-receptor complex, though evidence for the involvement of G-proteins and the inositol phosphate pathway in blue light perception is scant. Morse *et al* (1989) showed that white light stimulates inositol phospholipid turnover in *Samanea saman* pulvini, leading to increased levels of diacylglycerol and inositol polyphosphates. This response has been shown to be blue light mediated (Satter *et al*, 1981). In the *Fucus* zygote, the receptor is likely to be inactivated at high irradiance, resulting in optimal activation of the complex at the shaded side of the zygote under normal circumstances. G-protein-receptor interaction (Cockcroft & Gomperts, 1985) will stimulate the production of inositol phosphates and diacylglycerol which stimulates the phosphorylation of proteins (Berridge, 1987) and the release of calcium from intracellular stores. Calcium may in turn stimulate reversible protein phosporylation (Blowers & Trewavas, 1988; Ranjeva & Boudet, 1987). To test these possibilities, further characterisation of the blue light receptor in the polarisation response is required, together with the demonstration of the presence of G-proteins and the elucidation of their roles. Fucus zygotes are one of the few plant materials with which it should be possible to monitor changes in inositol polyphosphates and diacylglycerol and at least establish the presesnce of this signalling pathway.

Polarisation of the *Fucus* zygote in unilateral light can be separated into two processes. Initially, a labile axis of polarity is established. This can be modified by reorienting the zygotes in the light field up to 10 hours after fertilisation (Kropf & Quatrano, 1987). Formation of the labile axis most likely involves the establishment of a reversible gradient. Central to the model presented in Fig. 1 is the development of a gradient of cytoplasmic free calcium, or the localisation of elevated calcium at the future growing rhizoid tip during polarisation (Robinson & Jaffe, 1975). There is much evidence to suggest that a gradient of calcium ions is established in the polarising zygote. A gradient of cytoplasmic calcium in the growing rhizoid has been directly demonstrated using both calcium-selective microelectrodes and fluorescent indicators (Brownlee & Wood, 1986; Brownlee & Pulsford, 1989)with elevated levels at the growing tip. In the polarising zygote, there is a significant calcium component of the positiive ion influx at the site of the future rhizoid outgrowth (Robinson & Jaffe, 1975). More compelling is the demonstration that various analogues of the calcium buffer BAPTA could effectively arrest polarisation and future development of recently fertilised *Pelvetia zygotes* when applied at critical concentrations (Speksnijder *et al*, 1989). From concentrations of the buffers required to arrest development and their different Kd values, it was concluded that they were acting to facilitate the diffusion of a calcium gradient with maximal concentration around 7 micromolar. It is important to know whether abolishing calcium gradients in this manner prevents polarisation or simply prevents its expression by inhibiting rhizoid germination. It will therefore be necessary to achieve the release of zygotes from the effect of BAPTA inhibition and test for axis formation and fixation during BAPTA treatment.

Evidence against the presence of a gradient of free calcium prior to fixation of the polar axis was provided by Kropf & Quatrano (1987). Chlorotetracycline fluorescence only became localised at the rhizoid tip after fixation of the polar axis.

However, chlorotetracycline does not necessarily reflect the concentration of free calcium in the cytosol. Evidence against the role of a calcium gradient in polar axis fixation was also provided by Kropf & Quatrano (1987), who showed that almost total removal of external calcium (down to 10^{-10} M) did not prevent fixation. To resolve these problems, direct demonstration of free cytoplasmic calcium gradients during polarisation and before axis fixation is required. There is preliminary evidence to show that cytoplasmic calcium may be elevated at the shaded edge of the zygote at least during axis fixation (Brownlee, 1989). However, much more data is required, using different techniques for monitoring spatial variations in cytoplasmic calcium, such as the use of aequorin which does not compartment into organelles and vesicles. It seems that the initial steps in polarisation of the *Fucus* zygote are closly linked to the action of the photoreceptor and may involve a localised plasmamembrane change not directly involving spatial variations in calcium. This may involve components of the inositol phosphate pathway and possible localised protein phosphorylation via diacylglycerol (Fig. 1). Localised oscillations in inositol phosphate levels (Berridge & Irvine, 1989) may be involved in a sustained loacalised response.

Figure 1 suggests a possible mechanism by which an initial asymmetry can be amplified. Local elevation of calcium near the plasmalemma can cause the opening of calcium-regulated chloride channels. Such channels were first postulated to operate in the growing *Pelvetia* rhizoid by Nucitelli & Jaffe (1976). Calcium-regulated chloride channels have been described in higher plants (eg. stomatal guard cells (Schroeder, & Hagiwara, 1989)). Since the driving force for chloride is likely to be outward in the developing zygote (Weisenseel & Jaffe, 1971), the resulting chloride efflux will cause a localised depolarisation. This may be large enough to open voltage-regulated calcium channels and so allow calcium influx. Voltage-regulated anion channels have also been demonstrated in higher plants (eg. Keller *et al*, 1989). Voltage-regulated calcium channels are likely to occur in giant algae (eg. Bielby, 1984; Lunevsky, *et al* 1983). Evidence for the existence of voltage-regulated calcium channels in Fucoid algae is less clear. Depolarising the membrane potential by up to 50 mV by raising external K^+ (Weisinseel & Jaffe, 1971) does cause cytoplasmic calcium to elevate in growing rhizoids of *Fucus serratus* (Fig. 2).

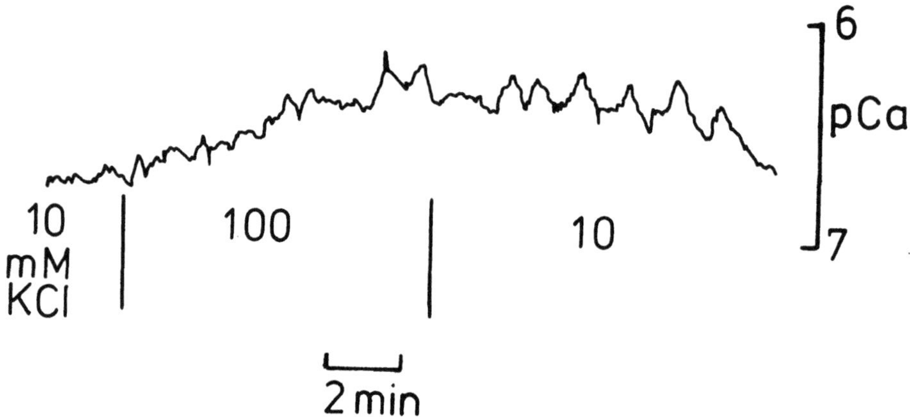

Fig. 2. Effect of increasing K^+ in the external artificial sea watter medium on average cytoplasmic free calcium levels in the *Fucus* rhizoid cell. Calcium was monitored using fura-2 microinjected into the rhizoid cell with dual wavelength excitation fluorescence microscopy (Brownlee & Pulsford, 1987; McAinsh *et al*, 1990).

However the response is somewhat slow and at this stage it is not clear whether it is due to direct regulation of plasmalemma calcium channels or to some secondary effect of elevated K^+ and depolarisation. If localised depolarisation is occuring, it should be measurable. So far, there is no convincing demonstration of a voltage gradient in the growing rhizoid, though vibrating probe studies suggest that the rhizoid apex must depolarise by up to at least 6 mV during inward current pulses (Nuccitelli & Jaffe, 1976).

Critical testing of the above hypothesis will require direct electrophysiological demonstration of calcium and voltage-regulated channels in the plasmalemma of polarised and polarising zytgotes.

Patch clamping is currently being employed with protoplasts to characterise the channel types in the *Fucus* plasmalemma and to map their distribution during polarisation in order to distinguish between polarised incorporation or local activation of symmetrically incorporated channels during polarisation.

SUMMARY

The following questions summarise the areas where more knowledge will greatly help our understanding of the process of fertilisation and early development in the *Fucus* system. This information will also provide insights into the mechanisms involved in fertilisation and early development in plants in general.

1). What is the nature of the intracellular signalling system in the attraction of sperm to the egg? 2). Is calcium involved in the activation of the *Fucus* egg. Does a wave of free cytoplasmic calcium traverse the egg and if so what is its function? 3). What is the role of the electrical changes on fertilisation? What sort of channels do they represent? 4). What are the relative roles of role of G-proteins, calcium or other activation factors, eg. GTP IP_3? 5). How is the polarising gradient set up? Is there a gradient of calcium in the polarising egg? 6). Are plasmalemma channels uniformly distributed and locally activated during polarisation and rhizoid growth, or is localised channel incorporation involved?

REFERENCES

Beilby MJ (1984) Chloride channels in *Chara*. Plant Cell Envir 7: 415-421

Berridge MJ (1987) Inositol trisphosphate and diacyl glycerol: two interacting second messengers. Ann Rev Biochem 56: 159-193

Berridge MJ, Irvine RF (1989) inositol phosphates and cell signalling. Nature 341: 197-205

Blowers DP, Trewavas AJ (1988) Phosphatidyl inositol kinase activity of a plasma-membrane associated calcium-activated protein kinase from pea. FEBS lett 88: 97-89

Brawley, SH, Bell E (1987) partial activation of *Fucus* eggs with calcium ionophore and low-sodium seawater. Devel Biol 122: 217-226

Brawley SH, Roberts DM (1989) Calmodulin-binding proteins are developmentally regulated in gametes and embryos of Fucoid algae. Devel Biol 131; 313-320

Brownlee C (1989) Visualising cytoplasmic calcium in polarised and polarising *Fucus* zygotes. Biol Bull 176(S): 14-17

Brownlee C (1990) Light and development: cellular and molecular aspects of photomorphogenesis in brown algae. In: Herring PJ, Campbell AR, Whitfield MJ, Maddock L (eds) Light and Life in the Sea. Cambridge University Press, Cambridge (In Press)

Brownlee C, Dale B (1990) Temporal and spatial correlation of fertilization current, calcium waves and cytoplasmic contraction in eggs of *Ciona intesinalis*. Proc Roy Soc (Lond) B. (In Press).

Brownlee C, Pulsford AL (1989) Visualisation of the cytoplasmic free calcium gradient in *Fucus serratus* rhizoids: correlation with cell ultrastructure and polarity. J Cell Sci 91: 249-256

Brownlee C, Wood JW 1986 a gradient of cytoplasmic free calcium in growing rhizoid cells of *Fucus serratus*. Nature 320: 624-626.

Cockcroft S, Gomperts BD (1985) Role of guanine nucleotide binding protein in the activation of phosphoinositide phosphodiesterase. Nature 314: 534-536.
Epel D, Patton C, Wallace RW, Cheung WY (1981) Calmodulin activates NAD kinase of sea urchin eggs: an early event of fertilisation. Cell 23: 534-549
Evans LV, Callow JA, Callow ME (1982) The biology and biochemistry of reproduction and early development in *Fucus*. Prog Phycol Res 1: 68-110
Hagiwara S, Jaffe LA (1979) Electrical properties of egg cell membranes. Ann Rev Biophys Bioeng 8: 385-416
Hice RE, Moody WJ (1988) Fertilisation alters the spatial distribution and the density of voltage-dependent sodium current in the egg of the ascidian *Boltenia villosa*. J Physiol 347: 301-325
Jaffe LF (1983) Sources of calcium in egg activation: a review and hypothesis. Devel Biol 99: 265-276
Jaffe LF (1989) Calcium pulses, waves and gradients in early development. In: Fiskum G (ed) Cell Calcium Metabolism. Plenum, New York London, pp 313-321.
Keller BU, Hedrich R, Raschke K (1989) Voltage-dependent anion channels in the plasma membrane of guard cells. Nature 341: 450-453
Knapp E (1931) Entwicklungsphysiologische Untersuchungen an Fucaceen-Eieren. Planta 14: 731-751
Kropf DL, Quatrano RS (1987) Localisation of membrane-associated calcium during development of Fucoid algae using chlorotetracycline. Planta 171: 158-170
Lunevsky VZ, Zherehova OM, Vastrikov IY, Beretovsky GN (1983) Excitation of Characeae cell membranes as a result of activation of calcium and chloride channels. J Membr Biol 72 43-58
McAinsh M, Brownlee C, Hetherington AM (1990) Abscisic acid-induced elevation of cytoplasmic free calcium precedes stomatal closure in guard cells of *Commelina communis*. Nature (In press)
Morse MJ, Crain RC, Cote GG, Satter RL (1989) Light-stimulated phospholipid turnover in *Samanea saman* pulvini. Plant Physiol 89: 724-727
Muller DG, Gassman G (1978) Identification of the sex attractant in the marine brown alga *Fucus vesiculosus*. Nuturwiss 65: 389-393
Muller DG, Gassmann G (1985) Sexual reproduction and the role of sperm attractants in monoecious species of the brown algae order Fucales (*Fucus, Hesperophycus, Pelvetia, Pelvetiopsis*). J Plant Physiol 118: 401-408
Nuccitelli R, Jaffe LF (1976) The ionic components of the current pulses generated by developing Fucoid eggs. Devel Biol 49: 518-531

Ranjeva R, Boudet AM (1987) Phosphorylation of proteins in plants: regulatory effects and potential involvement in stimulus/response coupling. Ann Rev Plant Physiol 38: 73-93
Robinson KR, Jaffe LF (1975) Polarising Fucoid eggs drive a calcium current through themselves. Science 187: 70-73
Robinson KR, Jaffe LA, Brawley SH (1981) Electrophysiological properties of fucoid algal eggs during fertilisation. J Cell Biol 91: 9051a (abstr).
Satter RL, Guggino SE, Lonergan TA, Galston AW (1981) The effects of blue and far red light on rhythmic leaflet movements in *Samanea* and *Albizia*. Plant Physiol 67: 965-968
Schroeder JI, Hagiwara S (1989) Nature 338: 427-430
Speksnijder JA, Miller AL, Weisenseel MH, Chen T-H, Laffe LF (1989) Calcium buffer injections block fucoid egg development by facilitating calcium diffusion. PNAS 86: 6607-6611
Swann K, Whitaker MJ (1986) The part played by inositol trisphosphate and calcium in the propagation of the fertilisation wave in sea urchin eggs. J Cell Biol 103: 2333-2342
Weisenseel MH, Jaffe LF (1972) Membrane potential and impedence of developing *Fucus* eggs. Devel Biol 27: 555-574

MECHANISM OF DORSOVENTRAL AXIS DETERMINATION IN THE ASCIDIAN EMBRYO

W. R. Jeffery, B. J. Swalla, and J. M. Venuti
Center for Developmental Biology
Department of Zoology
University of Texas
Austin, Texas 78712
USA

NATO ASI Series, Vol. H 45
Mechanism of Fertilization
Edited by B. Dale

INTRODUCTION

The dorsoventral (DV) axis of the ascidian embryo is determined during ooplasmic segregation, a series of cytoplasmic movements that occurs in two phases after fertilization. During the first phase of ooplasmic segregation (OS1), a regional cytoskeletal domain (the myoplasm) moves from the egg periphery to the vegetal pole of the zygote (Jeffery and Meier, 1983). Blastomeres inheriting the vegetal pole region have been shown to be the first to invaginate during gastrulation and may be instrumental in establishing the DV axis (Bates and Jeffery, 1987). During the second phase (OS2), the myoplasm shifts from the vegetal pole to the subequatorial region of the zygote where it forms a crescent which enters presumptive muscle cells during cleavage (Conklin, 1905). Both animal and vegetal fragments of unfertilized eggs develop normally, but after OS1 only vegetal fragments are able to gastrulate and form a DV axis (Ortolani, 1958; Reverberi and Ortolani, 1962; Bates and Jeffery, 1987). Bates and Jeffery (1987) showed that deletion of a small region including the vegetal pole of the zygote between OS1 and OS2 prevents gastrulation and DV axis formation. It has been proposed that axial determinants localized at the vegetal pole cause the vegetal cells to invaginate during gastrulation and establish the DV axis (Bates and Jeffery, 1987).

This paper summarizes our current studies on the mechanism of DV axis determination in the ascidian embryo. First, we show that there is a rotation of the internal cytoplasm with respect to the cortex during the period of axis determination. Second, we demonstrate that axis determination is sensitive to ultraviolet (UV) light. Third, we present evidence that a 30 kilodalton protein (P30) encoded by a UV-sensitive maternal mRNA is involved in establishing the DV axis. Fourth, we show that P30 is associated with the cytoskeleton. Finally, we present an hypothesis for the mechanism of axis determination in the ascidian embryo.

CYTOPLASMIC ROTATION DURING AXIS DETERMINATION

It was previously thought that both phases of ooplasmic segregation involved the concerted movement of cell surface components, cortical organelles, and underlying cytoplasm to the vegetal pole, and then to the myoplasmic crescent (see Jeffery, 1984). However, by marking specific areas of the egg surface with chalk particles before and during ooplasmic segregation, Bates and Jeffery (1987) showed that the internal cytoplasm may shift with respect to the stationary cell surface during OS2. When a chalk particle was placed on the animal hemisphere, it was translocated with the myoplasm into the vegetal region of the zygote during OS1. In contrast, a chalk granule placed on the vegetal pole

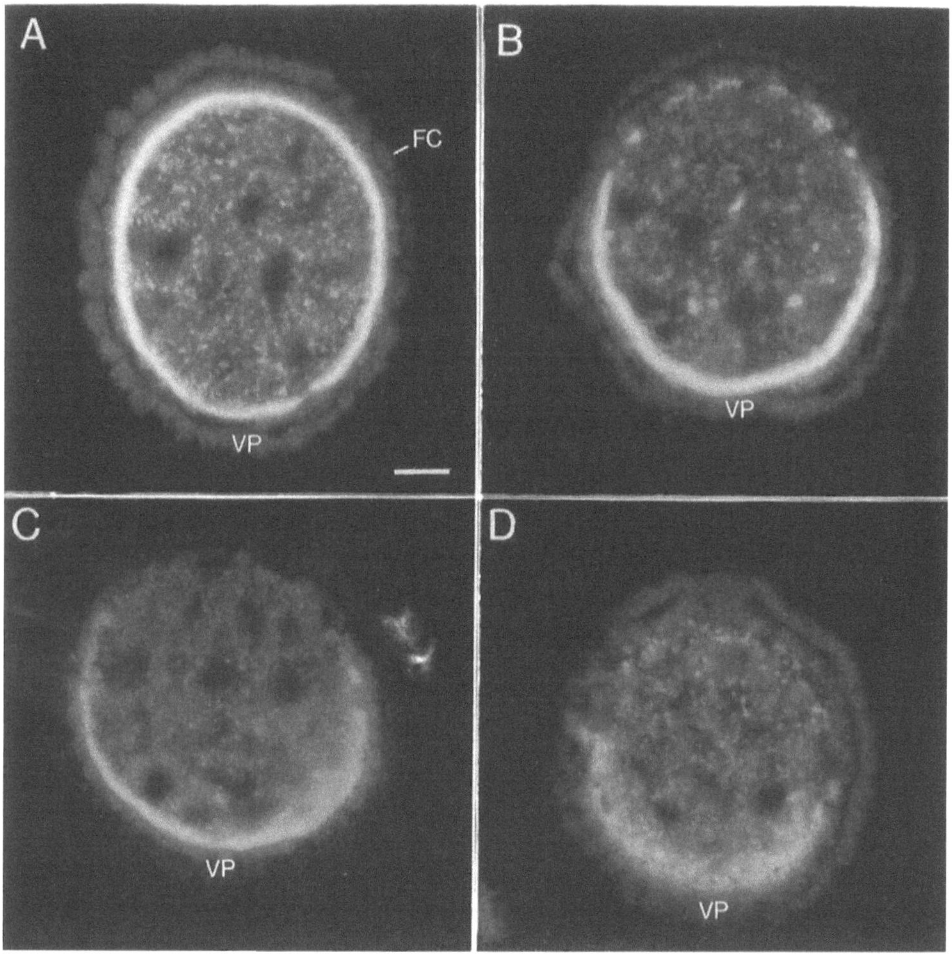

Figure 1. Sections of *Ascidia cerratodes* eggs stained with NN18 during OS1. A. An unfertilized egg. B-D. Fertilized eggs undergoing OS1. Eggs were fixed in methanol, then ethanol, and embedded in polyester wax. Sections were stained with RITC goat anti-mouse IgG after incubation with NN18. VP: approximate location of vegetal pole region. FC: follicle cells. Scale bar: 20μm.

surface after completion of OS1 remained stationary as the myoplasm shifted to the subequatorial region of the zygote during OS2.

We have continued to study cytoplasmic movements during the period of axis determination by examining the segregation of molecular components defined by monoclonal antibodies. The monoclonal antibody NN18, which recognizes a neurofilament polypeptide epitope in vertebrates (Debus *et al.*, 1983; Shaw *et al.*, 1984), reacts with the myoplasm in eggs and early embryos of many different ascidians (Swalla and Jeffery, in preparation). As shown by immunofluoresence, NN18 initially stains the peripheral region of the unfertilized egg (Fig. 1A). As myoplasm segregates during OS1,

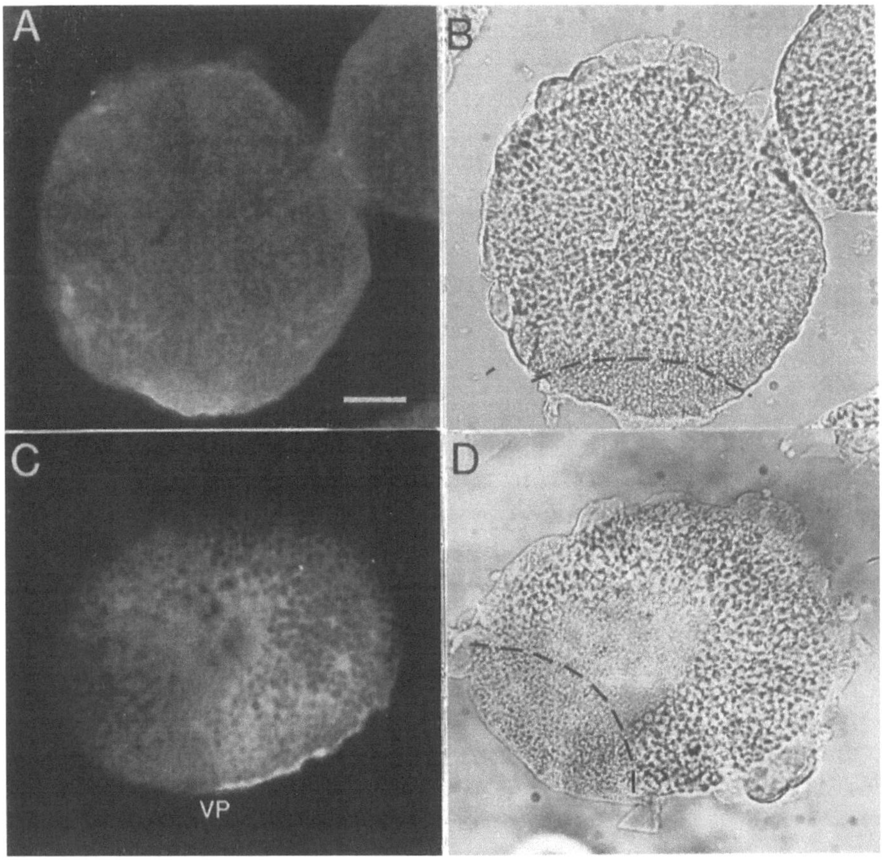

Figure 2. Sections of *Styela clava* eggs at the completion of OS1 (A, B) and OS2 (C, D) stained with IFA. Eggs were fixed in methanol, dehydrated in ethanol, and embedded in paraplast. A, C. Sections stained with primary antiserum and RITC goat anti-mouse IgG. B, C. The same sections viewed with bright field microscopy. VP: approximate location of vegetal pole region. Dashed lines: extent of myoplasm. Scale bar: 20µm.

NN18 staining was gradually restricted to the vegetal hemisphere (Fig. 1B-D). During OS2, NN18 staining shifted to the subequatorial region of the zygote with the myoplasm, and entered the presumptive muscle cells. Thus, NN18 reacts with a cytoplasmic component that shifts from the vegetal pole region into the myoplasmic crescent during the period of axis determination. The monoclonal antibody IFA recognizes an epitope shared by all classes of intermediate filaments (Pruss *et al.*, 1981). In Western blots, IFA reacts with several polypeptides similiar in molecular weight and isoelectric point to vertebrate intermediate filament proteins (Venuti and Jeffery, in preparation). As shown by immunofluorescence, IFA stained punctate components that appear near the vegetal pole of the zygote after OS1 (Fig. 2A-B). In contrast to the staining seen with NN18, the IFA-staining material remained near the vegetal pole as the myoplasm shifted to the subequatorial region during OS2 (Fig. 2C-D). These experiments and chalk marking studies (Bates and Jeffery, 1987) suggest there is a shift of myoplasmic components with respect to the cortex and egg surface during the period of DV axis determination.

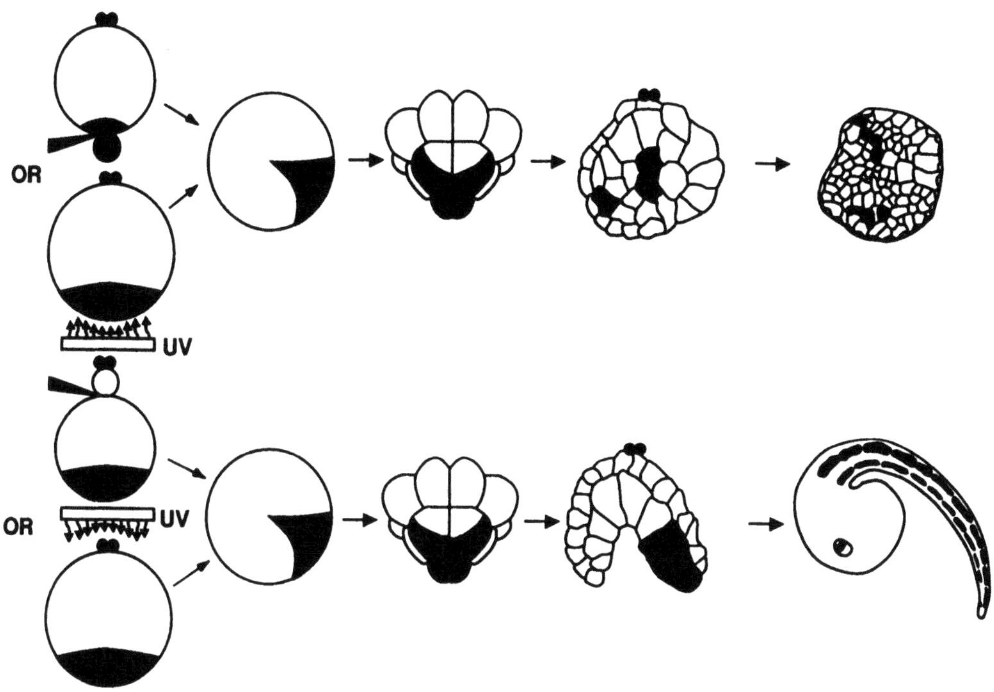

Figure 3. A summary of the effects of microsurgical deletion and UV irradiation of the animal or vegetal pole area of *Styela* eggs on gastrulation and DV axis formation. Upper diagram: Deletion or irradiation of the vegetal pole does not affect cleavage or cell differentiation but prevents gastrulation and DV axis formation. Lower diagram: Deletion or irradiation of the animal pole has no effect on gastrulation or DV axis formation. Solid areas: myoplasm-containing regions or cells.

EFFECT OF ULTRAVIOLET RADIATION ON AXIS DETERMINATION

In previous studies (Bates and Jeffery, 1987), fragments of eggs and embryos containing cell surface and underlying cytoplasmic regions were deleted microsurgically from the polar areas of *Styela* eggs before and after fertilization. When this operation was performed on unfertilized eggs, or when fragments were removed from the animal pole of zygotes between OS1 and OS2, development was unaffected. However, when fragments were deleted from the vegetal pole between OS1 and OS2, gastrulation was abolished and embryos developed without a visible DV axis. In amphibians, ultraviolet (UV) irradiation during a critical period between fertilization and first cleavage also inhibits development of dorsal features in the embryo (Malacinski *et al.*, 1975; Scharf and Gerhart, 1980). To determine if UV radiation has a similar effect in ascidians, the animal or vegetal hemisphere of *Styela* eggs was irradiated between OS1 and OS2. A UV dose of about 2.5×10^{-3} ergs mm^{-2} inhibited gastrulation and DV axis formation without affecting cleavage or differentiation of endodermal and muscle cells (see Fig. 5D). The effect of UV radiation and deletion of the polar areas of zygotes on DV axis formation is summarized in Figure 3.

To determine the timing of the UV sensitive period, eggs were irradiated with the minimal UV dose affecting DV axis determination at intervals between fertilization and first cleavage. Gastrulation and DV axis formation were inhibited only when UV was applied prior to 40 minutes after fertilization, about the time OS2 is completed. UV treatment after

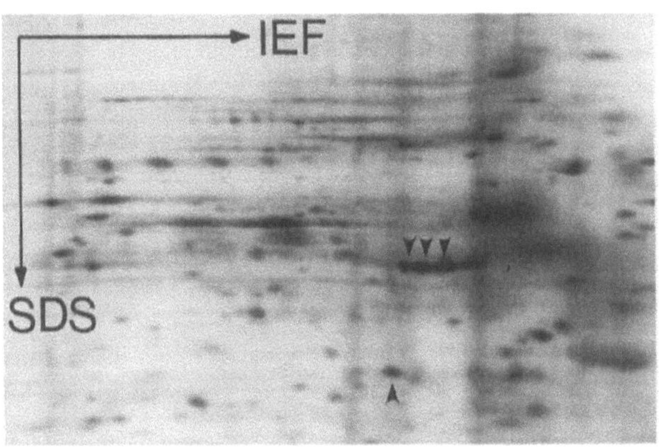

Figure 4. 2D- gel electrophoresis of proteins synthesized in the *Styela* embryo between the 2-cell and early tailbud stages. Upward-facing arrowhead: P30. Downward-facing arrowheads: the three major actin isoforms (see Tomlinson *et al.*, 1987). IEF: isoelectric focussing dimension. SDS: SDS dimension.

this time led to normal development. The results suggest that DV axis determination is sensitive to UV radiation only during a critical interval between fertilization and the completion of ooplasmic segregation.

EFFECT OF ULTRAVIOLET RADIATION ON PROTEIN SYNTHESIS

Experiments were designed to determine the molecular components involved in DV axis determination. These experiments required development of a method for UV irradiating eggs in mass. Normally development is affected only when UV radiation is applied to the vegetal hemisphere. However, when quantities of eggs were irradiated while spinning in a small chamber, UV irradiation had the same developmental consequences and effective doses as for stationary eggs exposed in the vegetal hemisphere. The effect of UV radiation on axis determination might be accompanied by changes in embryonic protein synthesis. Identification of proteins involved in DV axis formation might therefore lead to the axial determinants themselves. Thus, zygotes irradiated between OS1 and OS2 were allowed to cleave, and then incubated with [^{35}S]- methionine. Incubation with radioactive amino acids was continued until controls reached the early tailbud stage. Subsequently, proteins were extracted from the UV-irradiated embryos and controls, separated by two dimensional (2D)-gel electrophoresis, and the gels autoradiographed. Autoradiographs showed several hundred polypeptides, but the pattern of protein synthesis in normal and UV-irradiated embryos was almost identical. The major protein affected by UV was a 30kd polypeptide (P30) that sometimes consisted of two isoforms ranging between 5.5 and 6.0 in pI (Fig. 4; Fig. 5A).

The next series of experiments investigated the relationship between P30 synthesis and DV axis development. The effect of various doses of UV on P30 synthesis was examined by [^{35}S]-methionine labelling, 2-D gel electrophoresis, and autoradiography. The results showed that P30 synthesis and determination of the DV axis are affected by the same UV dose (Fig. 5). Experiments were then conducted to determine the timing of P30 synthesis. In these experiments, eggs and embryos were incubated with [^{35}S]-methionine for 2-hour intervals, and labelled proteins examined as described above. The results show that P30 synthesis occurred during a restricted period of development. Incorporation of radioactive amino acids into P30 was undetectible in unfertilized eggs and cleaving embryos before the 16-32 cell stage, corresponding in time to less than an hour before gastrulation. P30 synthesis peaked during gastrulation, continued during neurulation, and subsided by the tailbud stage. P30 synthesis was not detected during later stages of development or in adults. Finally, pulse-chase experiments were conducted to determine the stability of the newly-synthesized P30. In these experiments, [^{35}S]-methionine was added to UV irradiated embryos after first cleavage, and labelling was continued for 2 hours. Subsequently, cultures were washed free of radioactivity, aliquots of embryos

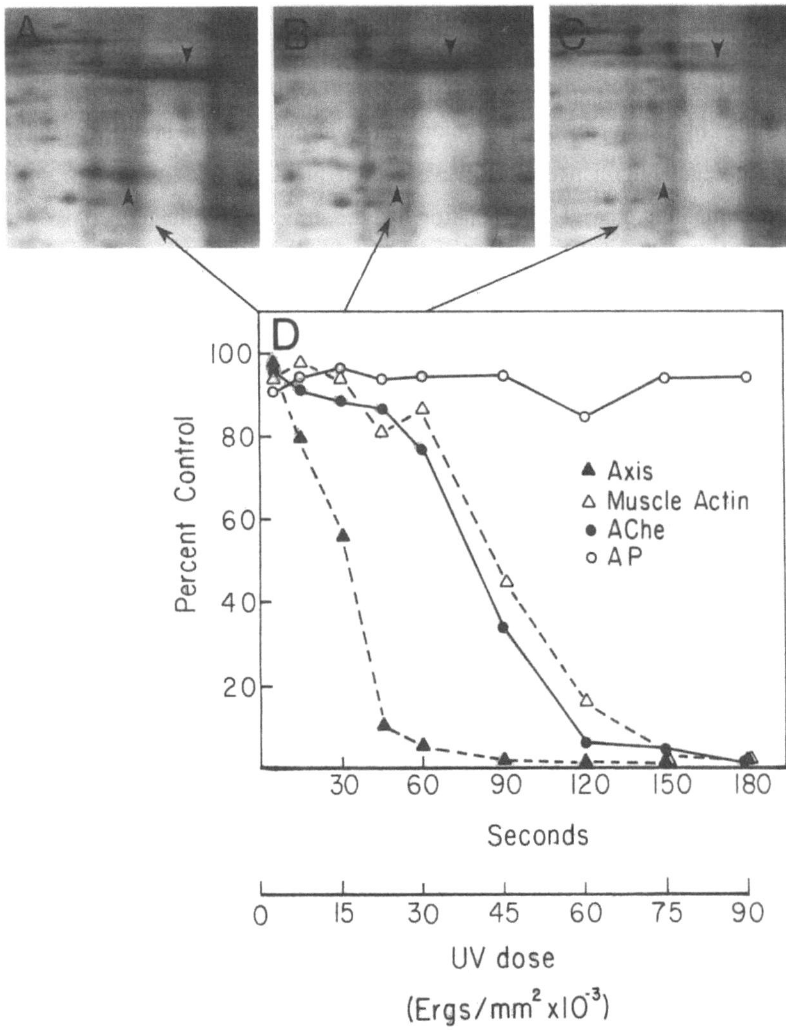

Figure 5. The effect of UV dose on P30 synthesis. A-C shows gels containing proteins extracted from zygotes irradiated with different doses of UV (diagonal lines). Only the region of the gels containing P30 is shown; other details of electrophoresis were similar to Figure 4. Upward-facing arrowhead: P30 isoforms. Downward-facing arrowhead: the most acidic actin isoform. D. The UV dose-response curve. AChe: acetylcholinesterase activity in the muscle cells. AP: alkaline phosphatase activity in the endodermal cells.

removed at various times in development, and labelled proteins examined as described above. The results showed that labelled P30 was stable until the late neurula stage when it was degraded.

In summary, the results indicate that P30 synthesis is correlated with DV axis formation: synthesis of this protein occurs during the interval between gastrulation and the tailbud stage - *precisely when the DV axis is being established* - and is abolished by the same UV dose that disrupts axis formation. P30 might therefore be a protein involved in some aspect of gastrulation or DV axis formation.

MATERNAL mRNA ENCODING P30 IS A ULTRAVIOLET-SENSITIVE TARGET

Although axis formation appears to be correlated with P30 synthesis, the latter cannot be the UV sensitive target because it is not synthesized until the 16-32 cell stage. In insect eggs, RNA can be inactivated by UV irradiation (Kalthoff, 1979). Thus, it is possible that maternal mRNA encoding P30, rather than P30 itself, is the UV-sensitive target. To determine whether P30 is encoded by a maternal mRNA, we translated RNA extracted from unfertilized *Styela* eggs in a reticulocyte lysate supplemented with [^{35}S]-methionine, the translation products were separated by 2-D gel electrophoresis, and the gels were autoradiographed. The results showed that eggs contain mRNA encoding a protein with the same molecular weight and isoelectric point as the major P30 isoform (Fig. 6A), indicating that P30 is encoded by a maternal mRNA. If maternal P30 mRNA is the UV

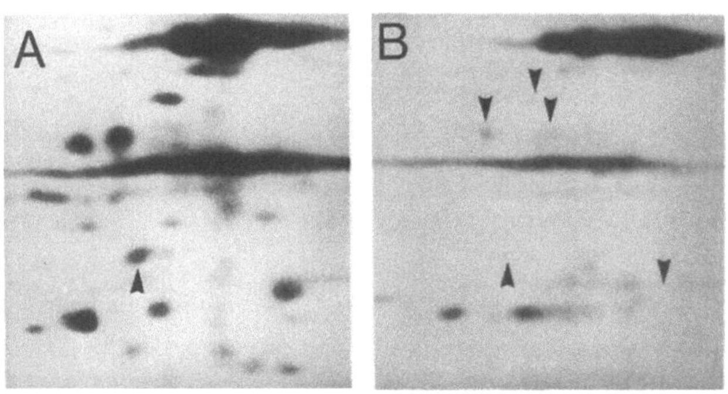

Figure 6. The effect of UV irradiation of zygotes on mRNA translation. Autoradiographs of 2D gels containing proteins synthesized *in vitro* directed by RNA from (A) control and (B) UV-irradiated zygotes. Upward pointing arrowheads: P30. Downward pointing arrowheads: other polypeptides whose translation was affected by UV irradiation. Other details are similar to Figure 4.

sensitive target, it should be possible to inactivate translational activity by UV irradiation of the zygote. To test this, *Styela* zygotes were irradiated between OS1 and OS2, their RNA was then extracted, and P30 synthesis was examined in the reticulocyte lysate. P30 (and several other polypeptides located near it in the gel) were absent from the translation products directed by RNA extracted from UV-irradiated zygotes (Fig. 6B). Thus, maternal mRNA encoding P30 is a UV-sensitive target in uncleaved zygotes. Because UV radiation probably does not penetrate deeply into the egg cytoplasm, complete inhibition of mRNA translation in UV-irradiated eggs suggests that P30 mRNA may be localized near the surface of the zygote.

P30 IS ASSOCIATED WITH THE CYTOSKELETON

Since UV affects DV axis formation in *Styela* embryos by abolishing gastrulation, and the latter is dependent on the activity of microfilaments that initiate cell shape changes in the vegetal blastomeres (W. R. Bates, unpublished results), it was reasoned that P30 might be a cytoskeletal protein. To test this possibility, we extracted [^{35}S]-methionine-labelled gastrulae with the non-ionic detergent Triton X-100, using procedures developed previously for preparing cytoskeletons from *Styela* eggs and embryos (Jeffery and Meier, 1983). After extraction, proteins were prepared and separated by 2D gel electrophoresis, and the gels autoradiographed. As shown in Figure 7, P30 was associated primarily with the cytoskeletal fraction. Experiments designed to determine the relationship between P30 and microfilaments relied on DNase I to depolymerize F actin in isolated cytoskeletons (see Jeffery and Meier, 1983). Labelled gastrulae detergent extracted in the presence of DNase I

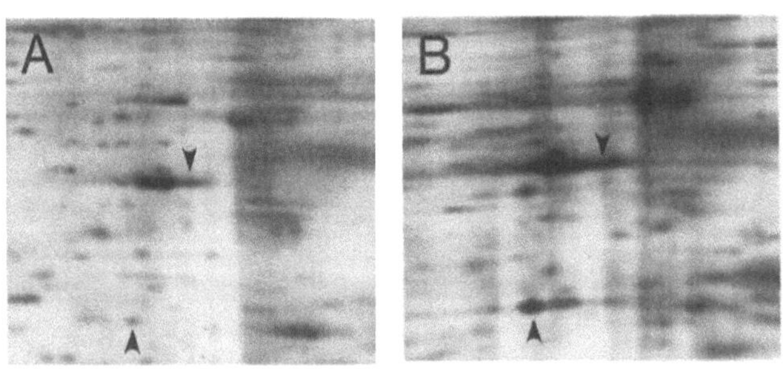

Figure 7. Association of P30 with the cytoskeletal fraction of gastrulae. A. A 2D gel containing proteins extracted from the detergent soluble fraction. B. A 2D gel containing proteins extracted from the detergent insoluble (cytoskeletal) fraction. Only the P30 region of the gels are shown; other details are similar to Figure 5A-C.

(or bovine serum albumin, as a control) were analyzed by gel electrophoresis as described above. The results showed that P30 was released from the cytoskeletal fraction by treatment with DNase I suggesting that the association of this protein with the cytoskeleton is dependent on the integrity of F-actin. Thus, P30 may be an actin-binding protein.

DISCUSSION

The results of experiments presented here lead to the following conclusions concerning DV axis determination in the ascidian embryo. First, ooplasmic segregation during the period of axis determination involves a shift of the inner cytoplasm with respect to the egg cortex. Second, DV axis determination is sensitive to UV radiation. Third, one of the UV sensitive components is maternal mRNA encoding P30, a zygotic protein that may be involved in gastrulation and DV axis establishment. Fourth, P30 is associated with the cytoskeleton.

Ooplasmic segregation during the period of DV axis determination appears to be more complex than previously believed. The results of antibody-staining and chalk marking studies (Bates and Jeffery, 1987) suggest that myoplasm is composed of two regions: a cortical and a cytoplasmic domain. The cortical domain includes the cell surface, plasma membrane, underlying actin network (Jeffery and Meier, 1983), egg cortex, and components recognized by IFA. The cytoplasmic domain consists of pigment granules, mitochondria, the deep filamentous network (Jeffery and Meier, 1983), and components recognized by NN18. Although both domains segregate to the vegetal pole region during OS1 (Jeffery, 1984), different behaviors are exhibited during OS2. The cytoplasmic domain shifts into the myoplasmic crescent region and later enters presumptive muscle cells, while the cortical domain remains near the vegetal pole and is localized in vegetal blastomeres that initiate gastrulation. Cytoplasmic rotation resembles the cortical rotation process that has been observed during axis determination in amphibian eggs (Ancel and Vitemberger, 1948; Vincent *et al.*, 1986), but in ascidians cytoplasm appears to rotate under a stationary cortex, rather than *vice versa*.

Microsurgical studies have shown that axial determinants are localized in the vegetal pole region of the zygote between OS1 and OS2 (Bates and Jeffery, 1987), and the present results show that these determinants are sensitive to UV radiation. This UV sensitivity has allowed experiments to be designed to learn more about the nature and function of axial determinants. Thus, we discovered P30, a protein whose synthesis is inhibited by UV radiation. Evidence that P30 is involved in axis determination stems from two observations. First, P30 synthesis is curtailed by the same dose of UV radiation that inhibits gastrulation and the DV axis without affecting cleavage or cell differentiation. Second, P30 is synthesized only during the interval including gastrulation and DV axis formation, and then it is degraded. P30 cannot be the UV-sensitive component itself, however, because the UV sensitivity period terminates before the embryo begins to synthesize this protein. Therefore, another component must be the UV target. Since P30 synthesis is directed by a maternal mRNA, the latter is likely to be the UV target. The UV sensitivity of P30 mRNA was demonstrated by comparing the *in vitro* translation products of RNA extracted from normal and UV-treated embryos. RNA from control embryos directed P30 translation, however, no P30 was synthesized by RNA prepared from UV treated embryos. Thus, UV radiation inactivates P30 mRNA translation, probably by affecting its secondary or tertiary structure (Kalthoff, 1979). UV-sensitive axial determinants (Elinson and Pasceri, 1989) and localized mRNA (Capco and Jeffery, 1982; Melton, 1987) are also present in the vegetal hemisphere of amphibian oocytes.

The DV axis of ascidian embryos is established during gastrulation by the invagination four large vegetal-pole blastomeres. The descendants of these and other cells continue to invaginate and migrate inside the embryo, eventually reaching a position where they may induce dorsal structures, such the brain pigment cells (Nishida and Satoh, 1989). W. R. Bates (unpublished results) has shown that invagination can be inhibited by cytochalasin, indicating that it is mediated by microfilaments. Thus, we tested whether P30 is a cytoskeletal protein by following its distribution in the soluble and cytoskeletal fractions of gastrulae extracted with Triton X-100. The results showed that P30 was located primarily in the cytoskeletal fraction, and could be released by treatment with DNase I, which depolymerizes F-actin. This suggests that P30 is an actin-binding protein.

Our results are consistent with the following hypothesis. UV-sensitive maternal mRNAs encoding proteins involved in DV axis formation segregate to the vegetal pole region with the myoplasm during OS1. These mRNAs remain in the vegetal pole region during OS2 and subsequently enter the vegetal blastomeres that initiate gastrulation. Within the vegetal cells, these mRNAs translate proteins which initiate the cell shape changes involved in gastrulation and DV axis formation, possibly by interacting with the actin cytoskeleton. We conclude that myoplasm serves dual functions during early development of the ascidian embryo: specification of muscle cells (Jeffery, 1985) and determination of the site of gastrulation and DV axis formation.

ACKNOWLEDGMENTS

This paper is dedicated to the memory of Alberto Monroy. We acknowledge support of NIH (HD-13970) and NSF (DCB-8812110) research grants (WRJ), and an NIH post-doctoral fellowship (HDO 6840; JMV).

REFERENCES

Ancel P, Vitemberger P (1948) Recherches sur le déterminisme de la symétrie bilatérale dans l'oeuf des amphibiens. Bull Biol Fr Belg Suppl 31: 1-182.

Bates WR, Jeffery WR (1987) Localization of axial determinants in the vegetal pole region of ascidian eggs. Dev Biol 124: 65-76.

Capco DG, Jeffery WR (1982) Transient localizations of messenger RNA in *Xenopus laevis* oocytes. Dev Biol 89: 1-12.

Conklin EG (1905) The organization and cell lineage of the ascidian egg. J Acad Nat Sci Phila 13: 1-110.

Debus E, Weber K, Osborn M (1983) Monoclonal antibodies specific for glial fibrillary acidic (GFA) protein and for each of the neurofilament triplet polypeptides. Differentiation 25: 195-203.

Elinson RP, Pasceri P (1989) Two UV-sensitive targets in dorsoanterior specification of frog embryos. Development 103: 511-518.

Jeffery WR (1984) Pattern formation by ooplasmic segregation in ascidian eggs. Biol Bull 166: 77-298.

Jeffery WR (1985) Specification of cell fate by cytoplasmic determinants in ascidian embryos. Cell 41: 11-12.

Jeffery WR, Meier S (1983) A yellow crescent cytoskeletal domain in ascidian eggs and its role in early development. Dev Biol 99: 408-417.

Kalthoff K (1979) Analysis of a morphogenetic determinant in an insect embryo (*Smittia Spec., Chironomide, Diptera).* Subtelny S, Konigsberg IR (eds) Determinants of Spatial Organization Academic Press New York p. 97-126.

Malacinski GM, Benford H, Chung H-M (1975) Association of an ultraviolet irradiation sensitive cytoplasmic localization with the future dorsal side of the amphibian egg. J Exp Zool 191: 97-110.

Melton DA (1987) Translocation of a localized mRNA to the vegetal pole of *Xenopus* oocytes. Nature (London) 328:80-82.

Nishida H, Satoh N (1989) Determination and regulation in the pigment cell lineage of the ascidian embryo. Dev Biol 132:355-367.

Ortolani G (1958) Cleavage and development of egg fragments in ascidians. Acta Embryol Morphol Exp 1: 247-272.

Pruss RM, Mirsky R, Raff MC, Thorpe R, Dowding AJ, Anderton BH (1981) All classes of intermediate filaments share a common antigenic determinant defined by a monoclonal antibody. Cell 27: 419-428.

Reverberi G, Ortolani G (1963) Twin larvae from halves of the same egg in ascidians. Dev Biol 5: 84-100.

Scharf SR, Gerhart JG (1980) Determination of the dorso-ventral axis in eggs of *Xenopus laevis*: Complete rescue of UV-impaired eggs by oblique orientation before first cleavage. Dev Biol 79: 181-198.

Shaw G, Debus E, Weber K (1984) The immunological relatedness of neurofilament proteins of higher vertebrates. Eur J Cell Biol 34: 130-136.

Tomlinson CR, Bates WR, Jeffery WR (1987) Development of a muscle actin specified by maternal and zygotic mRNA in ascidian embryos. Dev Biol 123: 470-482.

Vincent J-P, Oster GF, Gerhart JC (1986) Kinematics of gray crescent formation in *Xenopus* eggs: The displacement of subcortical cytoplasm relative to the egg surface. Dev Biol 113: 484-500.

THE INSECT OOCYTE: FERTILIZATION, ACTIVATION AND CYTOPLASMIC DYNAMICS

Klaus Sander
Institut für Biologie I (Zoologie)
Albert-Ludwigs-Universität
Albertstrasse 21a
D 7800 Freiburg
West Germany

Key words : Insects / Experimental Fertilization / Egg Activation / Sperm Functions / Cytoplasmic Movements / Ooplasmic Segregation / Oogenesis / Embryogenesis.

1. Abstract

The transition from oogenesis to embryogenesis has received little attention in insects so far, owing largely to internal fertilization that prevents direct observation. Fertilization is characterized by extreme sperm economy, sometimes linked to gigantic sperm length. Egg activation as a rule is not initiated by sperm entrance but by changes connected with egg deposition (deformation, water uptake etc.). Both fertilization and activation can be achieved in vitro. The ooplasmic movements prominent in many species during both oogenesis and early embryogenesis seem largely independent of cleavage nuclei but may require extranuclear sperm contributions.

2. Introduction

This contribution deals with 3 interlinked aspects of the transition from the oocyte stage to early embryogenesis in insects. While the genetic foundations of development are better known to-day in insects than in most - or perhaps all - other animal groups (see e.g. Ingham 1988, Malacinski 1989), knowledge on fertilization, egg activation and

NATO ASI Series, Vol. H 45
Mechanism of Fertilization
Edited by B. Dale

ooplasmic rearrangements lags far behind that available for some other systems, notably echinoderms and chordates (see the respective contributions in this volume) 1).

The lack of solid knowledge on insect fertilization, especially at the molecular level, is probably due to certain pecularities of insect reproduction, internal fertilization being the dominant obstacle to research. In addition, insects have very large eggs when compared to maternal body size, and consequently the female can produce only a few eggs at a time, at least by comparison with sea urchins or amphibians. The maximum number of eggs maturing simultaneously may amount to a few thousands e.g. in mayflies where an average of 4000 eggs per female was given by Hunt (1951) but most insects fall far short of such numbers. Insect spermatozoa frequently are larger, even on an absolute scale, than those of most other animals, and there are cases of sperm gigantism (megaspermy) where the spermatozoon is longer than the entire body of the adult (see e.g. Heming-van Battum & Heming 1986, Virkki et al. 1987). The record so far stands at about 15 mm sperm length (Afzelius et al. 1976). Such giant spermatozoa require considerable investments and this means of course that the numbers of sperm produced and transmitted must be low (in Drosophila about 350 - 1200 per copulation, see Sander 1985a), which fact in turn requires extremely efficient mechanisms for guiding sperm to the egg cell in order to minimize wastage (see e.g. Afzelius 1972). Indeed, in several species about one out of every two sperm stored in the female genital ducts may reach an egg cell, and mono- or oligospermy is the rule. These are some of the pecularities that make studying physiological details of fertilization in insects a technically very frustrating enterprise, at least by comparison with the more commonly used systems. On the other hand, the specific adaptations involved offer quite some fascination to the general biologist.

Egg activation in insects as a rule occurs during oviposition.It starts in many cases inside the mother's body but is completed outside, a mode which - together with the low number of eggs deposited - did not favour research on this topic either.

The dynamics of ooplasm, while sometimes being quite conspicuous, seem to vary so strongly between insect species that there is little hope for finding general rules except perhaps for the triggering role of free calcium. Thus, present knowledge is based on

1) The juvenile state of research on fertilization and egg activation in insects is documented by the lack of journals recognized as "marketplaces" by a research community. Consequently, relevant publications are scattered over a wide range of journals of different vocations (and sometimes standards). This makes spotting relevant information and references somewhat haphazard, and the author would welcome information on publications he may have overlooked.

some interesting but as yet anecdotical studies. So far these studies have hardly touched on the basic question, namely what role (if any) cytoplasmic movements play in the crucial function of early embryogenesis, namely embryonic pattern formation (Sander 1976a, Kalthoff & Rebagliati 1990).

The aim of the present report is to pull together and to discuss in some detail such data on insect fertilization and egg activation as have been published - or come to the author's attention - after completion of a review that was written half a decade ago at the request of the late Alberto Monroy (Sander 1985a). Work on ooplasmic dynamics was summarized in the reviews of Sander (1976b), Gutzeit (1986) and Van der Meer (1988, 1989) while egg activation and fertilization were also reviewed, albeit in different contexts, by Went (1982), Margaritis (1985a) and Retnakaran & Percy (1985). References listed in these reviews will as a rule not be discussed here again; so these earlier sources of information ought to be consulted as well. Together they should provide a comprehensive overview of present knowledge on the transition between oogenesis and embryogenesis in insects, and might perhaps define some rewarding problems for future research in this underdeveloped field.

3. Gamete structure and fertilization

By comparison with other animals studied, insects span an incredible range of variation in gamete and especially sperm structure. This fact may be less surprising when one considers that roughly 2/3 of all animal species are insects. Yet this structural multitude implies that the different steps of fertilization are as varied, a point that will foil almost any attempt to draw general conclusions.

While the functional aspects of insect sperm were largely neglected so far, sperm structure is known from an impressive number of insect species, and mostly at the ultrastructural level. We list here some information mainly concerning species or groups that will subsequently be considered in other contexts; unless marked otherwise, the data are taken from the comprehensive review of Jamieson (1987). Most insect sperm show some degree of motility in salt solutions and this may be helpful in negotiating the narrow ducts they have to pass in many species, but one may well doubt that free swimming movements play a general role in fertilization (for arguments see Sander 1985a), except perhaps in mayflies (Ephemeroptera) where the mode of oviposition (see e.g. Hunt, 1951) seems to require motile sperm. In the mayfly Dolania, the spermatozoa are about 25-40 µm long (Fink & Yasui, 1988). Mayfly spermatozoa differ from most other insect sperm in having a recognizable centriole, albeit made up of doublets rather than the usual triplets of microtubules. In crickets (Orthoptera), sperm length is about 1 mm (Pohlhammer 1978, McFarlane & McFarlane 1988; the value of 100 µm also given in the latter paper must be erroneous). The elongated sperm head carries an acrosomal cap from which an acrosomal filament is said to emerge (McFarlane & McFarlane 1988). Such filaments may occur in a few lower arthropods (including a juliform millipede) but so far they were not described in insects. The spermatozoon of the sawfly Athalia (Hymenoptera) is about 55µm long, with the elongate head accounting for 15µm; the tail is of helical shape and shows screw motion rather than beating or undulating (Sawa et al. 1989). Among lower dipterans, the psychodid Phlebotomus (sandfly) may have the shortest insect sperm on record (13 µm) while while the sperm of some Psychodina are distinguished by absolute immotility. Although needle-shaped and about 200 µm long, they represent merely the equivalent of a sperm head: there are no microtubular structures whatsoever and centrioles are lacking already at the spermatogonial stage. The extremely slender nucleus extends centrally through nearly the whole length of the needle, surrounded by an acrosome throughout, and its low diameter (below .1 µm) makes it impossible to detect sperm in the Psychoda egg cell by fluorescence microscopy (see below). Sciaria sperm are about 350 µm long and their single centriole sports 70 or more singlet microtubules; however, this anomaly appears trivial when compared to the really bewildering variety of structural modifications that occur in the closely related cecidomyids. Spermatozoa in the genus Drosophila are from 90 µm (Kurokawa & Hihara 1976) to more than 1 centimetre long; in some species several size classes seem to occur side by side, and may be stabilized by sperm competition (Joly & Lachaise 1989) which may favour megaspermy also in other groups (Heming-van Battum & Heming 1986). The Drosophila spermatozoon is probably the best-known of insect sperm with reference to spermiogenesis and its genetical basis (for a review see Hennig 1989) but data on function are scanty even in this genus. A recent finding is that both the acrosome and the endpiece of the tail show intense binding of concanavallin A (Perotti & Riva 1988). This is interesting because housefly sperm may enter the micropyle head first or tail first (Degrugiller & Leopold 1976), and complementary binding molecules may exist in the micropylar region of the Drosophila egg (Perotti & Riva 1988).

Sperm penetration trough the egg shell into the oocyte clearly must require special mechanisms at least in species with giant or immotile sperm, if not with insect sperm in general. To the paradigm of Drosophila fertilization as discussed and illustrated in

Sander (1985a), the detailed description of Pohlhammer (1978) on fertilization in the cricket Teleogryllus will be added here. This work was unfortunately missed not only by previous reviewers but also by subsequent authors expressly writing on fertilization in crickets (see below). The Teleogryllus female interrupts her stereotyped sequence of oviposition movements by a brief rest when sperm are being transported to an egg that is about to be deposited. This enabled Pohlhammer (1978) to study gamete union in considerable detail after dumping "resting" females into liquid nitrogen (instantly blocking any further changes), and then dissecting the respective organs. The crucial facts established are that the egg for a brief period is held in the orifice of the common oviduct (Fig. 1) while by peristaltic movements a thin strand of interwoven spermatozoa

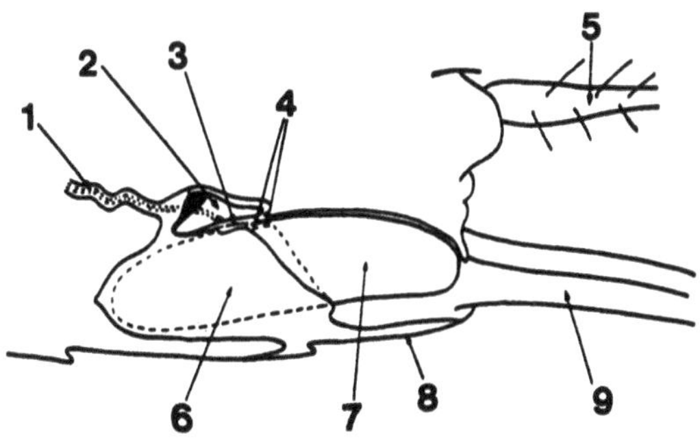

Fig. 1: Fertilization in the cricket Teleogryllus commodus (after Pohlhammer 1978). The egg is held with its anterior half in the oviduct (6) while the posterior half protrudes into the vagina (7). Sperm reach the anterior micropyle (3) and less frequently the posterior micropyles (4) via the spermathecal duct (1). The micropyles are on the ventral egg side (called dorsal by Pohlhammer because of its orientation in the maternal body). 2 = papilla carrying the orifice of the duct, 5 = circus, 8 = sternite, 9 = ovipositor. The egg is about 2.5 mm long.

is pressed out of the ductus receptaculi and towards the micropyles of the egg. Straying of spermtozoa beyond the micropylar region is largely prevented by a transverse fold thrown up in the border region between the oviduct and the vagina into which the far end of the egg projects. The two or three micropyles on the ventral egg side have each a circular outer opening that narrows down to a channel penetrating the egg shell obliquely in posterior direction. The strand of sperm is being pressed posteriorly so that some spermatozoa can get caught by the funnel-shaped micropyles (Fig. 1, 2). Active contributions of sperm motility to these events are unlikely, but once trapped the sperm head may get pushed deeper into the narrow micropylar channel and thence into the oocyte by some kind of straightening of the tail. Eggs removed 10 seconds after the onset of the female's resting phase were capable of development, implying that the sperm head had reached the oocyte. However, this time span apparently is not sufficient for the entire sperm to pass the micropyle, especially since its tail may still be entangled with other spermatozoa; the protruding end is probably shorn off when the egg leaves the vagina (Pohlhammer 1978).

Similar anatomical adaptations guiding sperm to the micropyles are found in several other insect groups (Heming-van Battum & Heming 1986; further references in Pohlhammer 1978 and Sander 1985a). In view of this body of knowledge the suspicion of

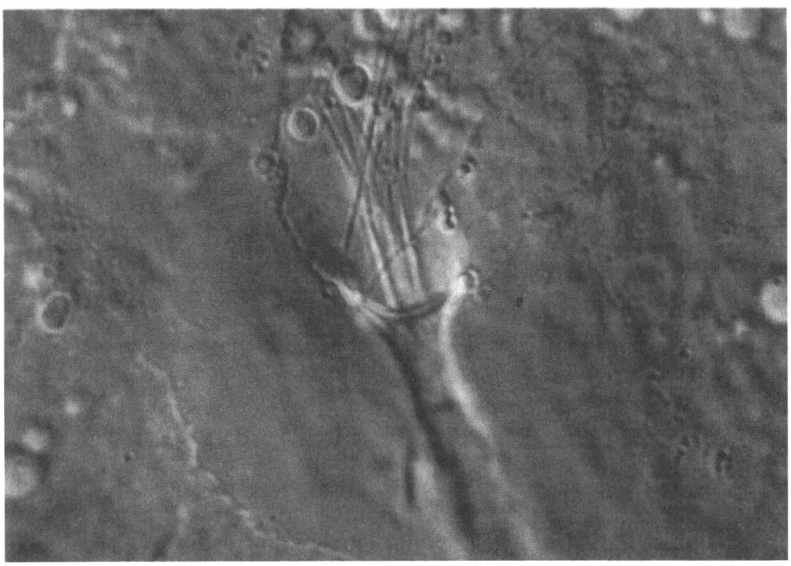

Fig. 2: Spermatozoa trapped in the funnel-shaped micropyle of Teleogryllus (from Pohlhammer 1978). The outer diameter of the slender micropylar tube is about 3 µm, its lumen about 1 µm wide.

McFarlane & McFarlane (1988) that in crickets and insects in general the sperm may not at all enter the egg through micropyles seems rather unfounded. Contrary to their claims, sperm have indeed been seen within the micropyles not only in Teleogryllus but far more than 100 years ago in Calliphora (at that time named Musca, see Sander 1985a), and since then in several other species as listed by Pohlhammer (1978) and Retnakaran & Percy (1985). That cricket sperm heads may subsequently be observed in egg regions distant from the micropyles, a finding listed by McFarlane & McFarlane (1988) in support of their views, may be due to secondary causes, among them streaming or sliding mechanisms within the ooplasm. The extremely high speed of egg deposition in the cricket, successive eggs being laid with intervals of only 2 seconds (McFarlane & McFarlane 1988), would also seem to favour sperm entrance via the micropyles rather than through the intact chorion.

Mayfly egg deposition (Hunt 1951) differs strongly from the modes discussed so far, and the circumstances suggest that the females in this group do not practice egg-by-egg fertilization in the oviduct but rather some kind of bulk exposure of eggs to sperm. In some other groups, the eggs may by fertilized in the ovary and probably before egg shell formation. This is likely to occur in Psychoda (Sander & Feddersen 1985), but the extreme slenderness of the sperm nucleus has prevented direct proof so far.

Two further aspects of insect fertilization should be mentioned briefly. In several species, the female can block sperm access to the eggs. The resulting haploid development - or rather the hemizygous/homozygous condition at the sex locus - will then determine one of the sexes, usually the male. In the honeybee, this faculty is exerted when the queen oviposits while sitting on a stretch of drone cells in the honeycomb. In another instance described recently (Putters & Van den Assem 1985), some counting mechanism seems to be involved. When the parasitic wasp Nesolynx albiclavus starts egg laying after pauses of 12 or more hours, the first 5 eggs to be laid are fertilized while the 6th egg invariantly is not. If egg laying continues without interruption, further unfertilized eggs may appear on a chance basis.

The second aspect relates to non-genetic sperm contributions to embryogenesis. While many insects are known to transmit molecular or physical signals and possibly also trophic factors with the ejaculate (see Chen 1984, Davey 1985, Boucher & Huignard 1987, Chen et al. 1988), the giant sperm might themselves be contributing not only the paternal nucleus and the centrioles (if any, see above) but also other factors, e.g. histones possibly required for chromatin functions in early development (Hennig 1989). This could provide another raison d'etre for megaspermy. Non-genetic sperm functions evidently are crucial in the several cases of pseudogamy known in insects (see Retnakaran & Percy 1985) where a sperm must enter the egg for development to begin, but then fails to unite with the female pronucleus. Commonly the sperm in these instances is credited with activating the egg (e.g. Fuyama 1984) but this view is probably too simple, the sperm at best representing just one of several factors required for the initiation of normal embryogenesis (see below).

A final topic akin to fertilization, and one better studied in insects than anywhere else (see Laugé 1985), is the origin of certain mosaic or chimeric individuals (chimeras being defined as originating from more than one zygote nucleus). Several reports document in Drosophila that chimeras can arise by double karyogamy, two haploid descendants of the oocyte nucleus fusing each with a sperm (Brand-Rosquist & Lüning 1984, Palmgren & Lake 1986). This explanation seems also to hold for part of the patchy silk moth embryos described by Ebinuma et al. (1988). On the other hand, participation of sperm nuclei can be excluded in various sawfly mosaics observed after shocking non-inseminated eggs of the sawfly Athalia (Hymenoptera) by low temperatures before and/or by high temperatures during experimental activation. The phenotypes obtained after this treatment suggest that in this species all four gonial nuclei deriving from oocyte meiosis are capable of participating in embryonic development, with or without subsequent automictic fusion (Hatakeyama et al., unpubl. results). Fusion of two meiotic products of the oocyte nucleus to constitute a diploid egg nucleus is also known to occur in a gynogenetic strain of Drosophila that produces viable offspring after pseudogamic fertilization with the defective sperm of a male sterile strain (Fuyama 1986).

4. Experimental fertilization

Experimental fertilization of insect eggs has been reported from time to time, but until recently the precaution of employing genetic makers was not taken. So some claims for in vitro fertilization are not fully convincing, especially since the diploid state of nuclei believed to prove karyogamy in these cases may have resulted from pronuclear fusion or diploidization of a haploid egg nucleus. Recently, however, in the sawfly Athalia Sawa & Oishi (1989b,c) injected genetically marked sperm bundles from the male genital tract into oocytes activated experimentally (see below), using a pigmentation mutant as the donor of either sperm or oocytes. Among the offspring, the males mostly were of the maternal genotype (they arise from unfertilized eggs - see previous section) while most of the rare females resulted from fertilization since they were heterozygous for the marker and its wildtype allele; this was revealed elegantly by studying their male progeny of which about 50% were wildtype and the other 50% mutant. Initially the yield of fertilized eggs in these experiments was around 10% at best, but when sperm were injected 20-25 min after experimental egg activation, it rose to above 20% (Sawa & Oishi 1989c). Pretreatment of the sperm with distilled water rather than salt solutions seems to increase the chances for karyogamy. Perhaps it serves to alter the sperm surface so that interaction with egg cell components can occur.

Injecting sperm into individual oocytes is of course a tedious approach. Submersion of eggs in a sperm solution would therefore be a welcome alternative. However, success seems convincing only in mayflies (Ephemeroptera), with a technique adapted from fish breeding: the gametes were stripped from both sexes and then mixed, water being added only after some minutes (Hunt 1951, Humpesch 1980a). The percentage of larval hatching was rather variable, from below 3% to above 88%, but was nonetheless taken to signal fertilization since without addition of the male products no larvae hatched at all (Hunt 1951) or hatching yields were very low (Humpesch 1980b). Yet the possibility remains that the procedures employed did not lead to karyogamy but rather to parthenogenetic egg activation by male genital tract secretions or by non-genetic sperm functions, followed perhaps by autogamy or somatic diploidization. The same problems arise with the claim of McFarlane & McFarlane (1988) for in vitro-fertilization in the cricket using sperm taken from the female's storage organs. Success at initiating nuclear division was again highly variable and in all eggs but one development came to a standstill soon after germ band formation. The single individual that hatched and reached the adult stage was a female that could have arisen by parthenogenesis. A further unresolved problem concerns the pathways by which cricket sperm could pass the covers of eggs submerged in a liquid. Penetration through plain stretches of chorion, as implied by the statements of McFarlane & McFarlane (1988) quoted above, would be singular among insects by present-day knowledge. The chorion (in contrast to the vitelline envelope, see below) is structurally complete some time before oviposition (Margaritis 1985). In view of its specific protein structure , it should be fairly resistant to any acrosomal enzymes, especially under in vitro conditions. Sperm entrance through the micropyles, on the other hand, seems unlikely in these experiments because in Teleogryllus, a species whose eggs strongly resemble those of the domestic cricket, spermatozoa swarming in Ringer's fail to be attracted by the micropyles (Pohlhammer 1978).

5. Egg cell activation

At the end of oogenesis, the insect oocyte as a rule is arrested in the first meiotic metaphase and in a state of low metabolic activity that in Drosophila can last for at least 15 days without detrimental effects on subsequent embryogenesis (Wyman 1979).

The presence of ecdysteroids is, in migratory locust, cricket and cockroach, a prerequisite for the breakdown of the germinal vesicle and the concomitant transition from prophase I to metaphase I (Lanot et al. 1989); these processes, termed meiotic reinitiation I, take place before or during the time when the oocyte moves from the follicle to the oviduct (ovulation). In the locust, meiotic reinitiation II, i.e. the initiation of subsequent meiotic stages which occurs during egg deposition, is also ecdysteroid dependent (Lanot et al. 1987). This event, by analogy with other insects, should be triggered by the stimuli that activate the oocyte, i.e. terminate its metabolically inert state. In other insect species, hormonal requirements for the initiation or completion of meiosis have not yet been recognized, while a peptide promoting ovulation is known in Drosophila (Chen et al. 1988).

Descriptions of egg activation in insects and the structural changes attendant on it are as yet rare. A notable negative finding is that so far no cortical granules nor exocytotic events related to activation were observed; on the contrary, multivesicular bodies appear peripherally in large numbers soon after egg activation in Drosophila, indicating a massive internalization of oolemma components (Mahowald et al. 1981). A change described in several species is the compaction of the vitelline envelope (see e.g. Zarani & Margaritis 1986, Fehrenbach et al. 1987). This event is accompanied by a transition from permeability to impermeability at least for water, and may establish an effective barrier for sperm that have not yet entered the egg cell by this time. The ultrastructural investigations of Mahowald et al. (1983) demonstrated that, contrary to earlier beliefs, unfertilized Drosophila eggs are being activated during deposition. The micropylar region of the vitelline envelope also changes from a liquid crystal configuration to the homogeneous appearance seen in fertilized eggs (Zarani & Margaritis 1986). On the molecular level, activation without fertilization was more recently confirmed by the observation that eggs deposited by virgin females, like fertilized eggs, start translating a maternal mRNA into the bicoid morphogen that then can be transplanted with high efficiency (Frohnhöfer & Nüsslein-Volhard 1986, Driever & Nüsslein-Volhard 1988). Translation in unfertilized eggs was also demonstrated for a maternal histone H2A mRNA in the dipteran Sciara (Ruder et al. 1987). Egg activation can lead not only to de novo protein synthesis, but also to enzymatic modification of proteins synthesized already during oogenesis. In Sciara, a cortical egg protein of 38 kD synthesized towards the end of oogenesis is modified soon after egg deposition into a 35 kD protein (Müller & Büsen 1988). This process does not require fertilization and can be mimicked in vitro using trypsin; apparently a trypsin-like protease (serine protease ?) is activated independent of fertilization merely by the act of egg deposition .

Mahowald et al.(1983) were the first to stress the fact that activation is accompanied by an extensive rehydration of the oocyte, conferring turgescence on the previously quite flaccid egg. In the cricket, the newly laid egg weighs from 20% to 80% more than the mature ovarian egg (McFarlane & McFarlane 1988), apparently due to water uptake from the substrate. Swelling of eggs immediately after deposition in humid surroundings was also documented for lower dipterans (Sander 1985b). The dramatic uptake of water may not be an epiphenomenon of activation but rather represent the triggering step itself, at least in some species habitually ovipositing in moist surroundings. This follows from the fact that oocytes taken from the mature follicles or lateral oviducts of such species can be activated by submersion in tap water or distilled water, as was observed in species belonging to the Orthoptera (McFarlane & McFarlane 1988), Hymenoptera (Sawa & Oishi 1989b), and Diptera (Sander 1985b). However, while in some species (Athalia, Psychoda) nearly all eggs so treated show signs of activation and develop to the hatching stage, the percentages of activation and/or completion of embryogenesis are lower in others.

The previously published catalogue of stimuli capable of activating insect eggs (Sander 1985) can now be extended to include brief desiccation (Sawa & Oishi 1989a), pricking the egg (Sawa & Oishi 1989a), submersion of eggs in Pringle's saline with increased concentrations of NaCl or KCl (Vinson & Jang 1987), cold shock in Pringle's solution but not in water (Vinson & Jang 1987), controlled lowering of ambient pH (Sawa & Oishi 1989a; see also Li et al. 1988 quoted therein), and tabbing the egg with a host of chlorides and organic acids (Saini et al. 1987). This array of partly or completely artificial stimuli reinforces earlier suspicions that, once physiologically ready for activation, the oocyte can be released from the blocked state either by some seemingly trivial factors linked to normal oviposition, or by a wide range of mechanical and physiological insults. Van der Meer (1988, 1989) has tried to subsume the known data under a unifying concept that embodies egg cell metabolism, cytoarchitecture and surges of intracellular free calcium. In this context it may be interesting to note that calcium action potentials have now been demonstrated in an insect oocyte, albeit (for technical reasons ?) so far only before the formation of the egg covers (O'Donnell 1985).

As mentioned above, initiation of embryogenesis is not necessarily followed by complete development up to larval hatching. Early failure of embryogenesis, occuring as a rule before or during nuclear multiplication (Mahowald et al. 1983, Wolf 1985, Sander & Feddersen 1985, Müller & Büsen 1988), indicates that activation may consist of several steps requiring each a set of specific conditions. This is most clearly revealed by experiments on the hymenopteran Venturia (Sander & Feddersen 1985) which showed that nuclear multiplication can be triggered by low ambient osmolarity (i.e. rehydration) whereas an organized blastoderm and further development ensue only if in addition the egg is either deformed temporarily or exposed to proteinase K (100 ug/ml PBS) (Sander & Feddersen 1985, their note added in proof). It should be noted that Venturia is propagating solely by parthenogenesis so that the question of paternal contributions to normal egg activation and especially its later steps does not arise.

6. Cytodynamics of the oocyte

The activated insect oocyte is a highly dynamical structure and remains so until thousands of nuclei have formed and are being separated by cell borders (blastoderm formation). However, cytodynamical events begin much earlier, in Drosophila at the latest around mid-oogenesis. At this time (stage 10) the nurse cells, sister cells of the oocyte maintaining cytoplasmic continuity among themselves and with the oocyte, start shrinking, probably by contraction of F-actin bundles. Thereby the nurse cell contents (except for the polyploid nuclei) are being pressed into the oocyte (Gutzeit 1986). This import of cytoplasm is accompanied by intense cytoplasmic movements within the oocyte that embody circular streaming and oscillations with varying directions. These movements may play a role in transporting morphogenetic factors synthesized by the nurse cells to the sites where they get of deposited or anchored in the mature oocyte (Gutzeit & Koppa 1982). The nurse cells were shown to synthesize the bcd messenger RNA that enters the oocyte and gradually gets localized at the anterior pole where it codes for an anterior morphogen (Berleth et al. 1988). The nurse cells also contain transplantable posterior morphogenetic activity (Sander & Lehmann 1988) that in normal oogenesis must be transported through the entire oocyte to the posterior pole where its presence after egg deposition was demonstrated by transplantation (Frohnhöfer et al. 1985). A similar transport mechanism must be involved in localizing the germ cell determinants in the posterior oocyte pole (Illmensee et al. 1976). The mechanisms assembling the oosome (the structure probably containing the germ cell determinants) can be blocked in the dipteran Bradysia by injection of colchicin but not by lumicolchicin, showing that cytoskeletal elements are involved in the translocation of oosomal components (Gutzeit 1985).

Once deposited, the oocyte or egg cell in several species is no less a dynamical structure than during oogenesis. Some recent data may serve to document this. The orientation of the meiotic spindles depends on a cytoplasmic process that in Athalia (Hymenoptera) and several other insects can be influenced experimentally. In normal development, the spindle during anaphase in both meiotic divisions is oriented at right angles to the oocyte surface, but instead of giving rise to polar bodies (as in other phyla) the spindles push one daughter nucleus deeper into the ooplasm; the most centrally located nucleus then turns into the egg pronucleus while the other 3 nuclei decay in the ooplasm. After exposure of the unfertilized Athalia egg to a succession of extreme temperatures (see above), the meiotic spindles fail to orient properly and this apparently permits some or all of the prospective polar body nuclei to turn into pronuclei capable of syngamy and/or cleavage mitoses (Hatakeyama et al., pers. communication). In several beetles, hymenopterans and dipterans, waves of cytoplasmic oscillations pass through the ooplasm during cleavage cycles and in Wachtliella (Diptera) these are prerequisite for releasing the chromosomal anaphase movement (which consequently also occurs in waves spreading over the embryo). Waves of calcium release are believed to trigger both these phenomena (Wolf 1985).

On the organelle level, the large cytasters (migration cytasters) involved in transporting the early cleavage nuclei of Wachtliella were observed to arise, migrate and divide also in the absence of nuclei (Wolf 1980). In Drosophila, when nuclear multiplication is blocked by aphidicolin during late cleavage, the centrosomes continue to divide and migrate (Raff & Glover 1989). They ultimately can establish a pattern resembling the distribution of the incipient blastoderm cells at the egg cell surface. No cell borders will form there in aphidicolin-treated eggs, but cytoplasmic budding can occur at the posterior pole so that pole "cells" with centrosomes but without nuclei detach. This process is suppressed by colchicin, indicating that cytoplasmic dynamics are dependent on microtubular arrays directed at this stage by the centrosomes (Raff & Glover 1989). Similar events (reviewed in Sander 1976b) have been described earlier in other insects under the names of pseudocleavage and pseudoblastoderm formation, respectively; this shows that autonomous patterning of cytoplasmic components and organelles occurs widely in early insect embryogenesis.

A last outcome of cytoplasmic dynamics to be discussed here is ooplasmic segregation. Although rarely reaching the extent seen in some other animal groups, e.g. the Tunicata, segregation is evident near the egg cell poles in lower dipterans and was shown to shift anterior morphogenetic determinants (Ripley & Kalthoff 1983) as well as symbiotic bacteria toward the egg surface (Gutzeit et al. 1985). Ooplasmic segregation can take a reproducibly aberrant course in unfertilized Psychoda eggs (see below).

How far cytoplasmic movements in the activated egg cell are required for embryonic pattern formation is an open question as yet. In some species they are so conspicious as to suggest that they must play a vital role. However, there is no proof so far that shifting of anterior determinants as described by Ripley & Kalthoff (1983) is indispensible for pattern formation. Moreover, in the few cases where mitotic waves or cytoplasmic streaming were altered experimentally, this apparently did not alter the segment patterns subsequently formed (Jung et al. 1977, Van der Meer et al. 1982, Wolf 1985).

The dynamical events observed after egg deposition and discussed in this section are cytoplasmic in the sense that they occur (probably) without a requirement for (normal) zygotic gene activity. This, however, need not mean that they are independent of syngamy. Although sperm entrance may not be required for egg activation (see above), the sperm may potentially supply the oocyte with some extrakaryotic stimuli or components indispensible for orderly development. In the absence of these, the oocyte pronucleus may replicate its DNA several times and even produce considerable amounts of daughter nuclei, yet these nuclei fail to distribute properly within the egg cell and ultimately degenerate (Fig. 3). This was observed e.g. in the house cricket (McFarlane &

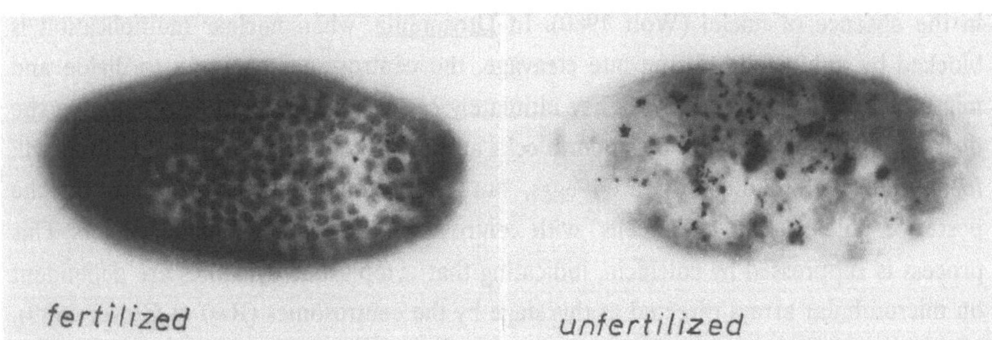

Fig. 3: Fertilized and unfertilized eggs of the dipteran Sciara coprophila, age 10h, Feulgen stained, length of the living egg is about 0.25 mm. The fertilized egg has formed a blastoderm, the unfertilized egg contains chromatin masses of varying sizes. Photographs courtesy of W. Büsen, Tübingen.

McFarlane 1988) and the dipterans Sciara (Müller & Büsen 1988) and Psychoda (see below). The symptoms suggest that the missing factor might be a sperm centriole or pericentriolar material required for cytoskeletal organization in the oocyte, but many if not most insect sperm lack typical centrioles (see Phillip 1970, Jamieson 1987), and parthenogenetic species apparently do well without. In Psychoda the immotile spermatozoa lack any traces of axionema or centrioles (Baccetti et al. 1973) so some other sperm component must be indispensible - unless the organizing stimulus is provided by the act of mating rather than by sperm components in the egg cell (Sander & Feddersen 1985).

7. Future problems and prospects

Although insects may not be the optimal systems in which to study fertilization and egg activation, some problems will be outlined here which should be studied in order to fill in gaps or resolve discrepancies, and some routes of approach where insects might offer advantages over other animal groups. In either instance, generalizing the findings may require caution in view of the terriffic degree of variation even now evident among insect species. Yet some variation must occur in other animal groups as well and the seemingly uniform sets of events discussed in textbooks may just signal that few people care to work on other than standard systems and thereby take the risk of being confronted with biological diversity.

A discrepancy that should be investigated soon concerns the descriptions of sperm entry in the crickets Acheta domesticus and Teleogryllus comodus, especially since both species are used in much other research.

Major gaps that could and should be filled in concern the union of egg and sperm and the symptoms of egg activation at all levels from the macroscopic to the molecular; species or strains where both fertilized and unfertilized eggs are capable of development might be especially rewarding in this respect. The question which role hormons play among higher insects in meiotic re-initiation and perhaps in egg activation is open to approach now. The roles of cytoskeletal elements in cytoplasmic streaming and determinant localization, and the non-genetic sperm effects possibly involved in these events, would be rewarding topics.

In almost all these instances, the use of antibodies either for localizing structures or for blocking suspected reactions or interactions should be very promising, especially since many molecules involved should be highly conserved among insects and even beyond the arthropods, so that antibodies against proteins from distant species might be tried.

Last but not least, functional studies employing Drosophila mutants along the lines that have proven so successful for analyzing spermiogenesis and embryonic pattern formation (see e.g. Hennig 1988, Nüsslein-Volhard & Roth 1989) and new methods that permit identifying localized products of previously unrecognized genes (Bellen et al. 1989, Fasano & Kerridge 1988) should be most rewarding in the long run. It is here that insects could lead the field rather than being runners-up.

Acknowledgements

The author wishes to express his thanks to Margrit Scherer for patient secretarial help, to Dieter Zissler for locating references, to Kurt Pohlhammer and Werner Büsen for kindly supplying the illustrations, and to Kugao Oishi for permission to quote unpublished data. Alberto Monroy in his friendly yet persevering ways provided the incentive that ultimately lead to this review.

References

Afzelius BA (1972) Sperm morphology and fertilization biology. In Beatty RA, Gluecksohn-Waelsch S (eds) The genetics of the spermatozoon. Edinburgh and New York. p 131

Afzelius BA, Baccetti B & Dallai R (1976) The giant spermatozoon of Notonecta. J Submicrosc Cytol 8:149-161

Baccetti B, Dallai R & Burrini AG (1973) The spermatozoon of Arthropoda. XVIII. The non-motile bifurcated sperm of Psychodidae flies. J Cell Sci 12:287-311

Bellen HJ, O'Kane, CJ, Wilson C, Grossniklaus U, Pearson RK and Gehring WJ (1989) P-element-mediated enhancer detection: a versatile method to study development in Drosophila. Genes & Development 3:1288-1300

Berleth T, Burri M, Thoma G, Bopp D, Richstein S, Frigerio G, Noll M & Nüsslein-Volhard C (1988) The role of localization of bicoid RNA in organizing the anterior pattern of the Drosophila embryo. EMBO J 7:1749-1756

Boucher L & Huignard J (1987) Transfer of male secretions from the spermatophore to the female insect in Caryedon serratus (Ol.): Analysis of the possible trophic role of these secretions. J Insect Physiol 33:949-957

Brandt-Rosquist K & Lüning KG (1984) Double fertilization in Drosophila melanogaster. Hereditas 101:69-73

Chen PS (1984) The functional morphology and biochemistry of insect male accessory glands and their secretions. Ann Rev Entomol 29:233-255

Chen PS, Stumm-Zollinger E, Aigaki T, Balmer J, Bienz M & Böhlen P. (1988) A male accessory gland peptide that regulates reproductive behavior of female D. melanogaster. Cell 54:291-298

Davey KG (1985) The female reproductive tract. In: Kerkut GA, Gilbert LI (eds) Comprehensive insect physiology, biochemistry and pharmacology. Pergamon Press Vol. I, Oxford p 15

Degrugillier ME & Leopold RA (1976) Ultrastructure of sperm penetration of house fly eggs. J Ultrastruct Res 56:312-325

Driever W & Nüsslein-Volhard C (1988) A gradient of bicoid protein in Drosophila embryos. Cell 54:83-93

Ebinuma H, Kobayashi M, Kobayashi J, Shimada T & Yoshitake N (1988) The detection of mosaics and polyploids in a hereditary mosaic strain of the silk moth, Bombyx mori, using egg colour mutants. Genet Res Camb 51:223-229

Fasano L & Kerridge S (1988) Monitoring positional information during oogenesis in adult Drosophila. Development 104:245-253

Fehrenbach H, Dittrich V & Zissler D (1987) Eggshell fine structure of three lepidopteran pests: Cydia pomonella (L.) (Tortricidae), Heliothis virescens (Fabr.), and Spodoptera littoralis (Boisd.) (Noctuidae) Int J Insect Morphol Embryol 16:201-219

Fink TJ & Yasui LS (1988) Ultrastructure of the sperm of Dolania americana Edmunds and Traver (Ephemeroptera : Behningiidae). Int J Insect Morphol Embryol 17:447-454

Frohnhöfer HG & Nüsslein-Volhard C (1986) Organization of anterior pattern in the Drosophila embryo by the maternal gene bicoid. Nature 324:120-125

Frohnhöfer HG & Nüsslein-Volhard C (1987) Maternal genes required for the anterior localization of bicoid activity in the embryo of Drosophila. Genes Dev 1:880-890

Frohnhöfer HG, Lehmann R, Nüsslein-Volhard C (1986) Manipulating the anteroposterior pattern of the Drosophila embryo. J Embryol Exp Morphol 97:169-179

Fuyama Y (1984) Gynogenesis in Drosophila melanogaster. Jpn J Genet 59:91-96

Fuyama Y (1986) Genetics of parthenogenesis in Drosophila melanogaster. II. Characerization of a gynogenetically reproducing strain. Genetics 114:495-509

Gutzeit HO & Koppa R (1982) Time-lapse film analysis of cytoplasmic streaming during late oogenesis of Drosophila. J exp Embryol Morphol 67:101-111

Gutzeit HO (1985) Oosome formation during in vitro oogenesis in Bradysia tritici (syn. Sciara ocellaris). Roux's Arch Dev Biol 194:404-410

Gutzeit HO (1986) Transport of molecules and organelles in meroistic ovarioles of insects. Differentiation 31:155-165

Gutzeit HO, Zissler D & Perondini ALP (1985) Intracellular translocation of symbiotic bacteroids during late oogenesis and early embryogenesis of Bradysia tritici (syn. Sciara ocellaris) (Diptera:Sciaridae) Differentiation 29:223-229

Heming-van Battum KE & Heming BS (1986) Structure, function and evolution of the reproductive system in females of Hebrus pusillus and Hebrus ruficeps (Hemiptera Gerromorpha Hebridae). J Morphol 190:121-168

Hennig W (1989) Spermatogenesis in Drosophila. In: Malacinski GM (ed) Developmental genetics of higher organisms. Collier Macmillan Publ. London, p 239

Humpesch UH (1980a) Effect of temperature on the hatching time of eggs of five Ecdyonurus spp. (Ephemeroptera) from austrian streams and english streams, rivers and lakes. J Anim Ecol 49:317-333

Humpesch UH (1980b) Effect of temperature on the hatching time of parthenogenetic eggs of five Ecdyonurus spp. and two Rhithrogena spp. (Ephemeroptera) from austrian streams and english rivers and lakes. J Anim Ecol 49:927-937

Hunt BP (1951) Reproduction of the burrowing mayfly, Hexagenia limbata (Serville), in Michigan. Florida Entomol 34:59-70

Illmensee K, Mahowald AP & Loomis MR (1976) The ontogeny of germ plasm during oogenesis in Drosophila. Devel Biol 49:40-65

Ingham PW (1988) The molecular genetics of embryonic pattern formation in Drosophila. Nature 335:25-34

Jamieson BGM (1987) The ultrastructure and phylogeny of insect spermatozoa. Cambridge University Press

Joly D & Lachaise D (1989) Sperm competition in Drosophila. Abstract, 11th European Drosophila Res. Conf., p. 57

Jung E, Nuss E & Wolf R (1977) Geschnürte Pimpla-Eier zeigen nur im hinteren Teilembryo Segmentausfall: Sind abgeänderte Ooplasmaströmungen die Ursache? Verh Dtsch Zool Ges 1977:307

Kalthoff K & Rebagliati M (1990) Cytoplasmic localization in insect eggs. In: Malacinski GM (ed) Cytoplasmic organization systems. McGraw-Hill Publ Comp New York. p 295

Kurokawa H and Hihara F (1976) Number of first spermatocytes in relation to phylogeny of Drosophila (Diptera: Drosophilidae). Int J Insect Morphol Embryol 5:51-63

Lanot R, Roussel JP & Thiebold JJ (1989) Ecdysteroids and meiotic reinitiation in oocytes of Periplaneta americana (Dictyoptera) and Gryllus bimaculatus (Orthoptera). J Invert Reprod Develop 15:69-74

Lanot R, Thiebold J, Lagueux M, Goltzene F & Hoffmann JA (1987) Involvement of ecdysone in the control of meiotic reinitiation in oocytes of Locusta migratoria (Insecta, Orthoptera). Devel Biol 121:174-181

Laugé G (1985) Sex determination: genetic and epigenetic factors. In: Kerkut GA, Gilbert LI (eds).Comprehensive insect physiology, biochemistry and pharmacology Pergamon Press, Oxford, Vol. I, p 295

Lehmann R & Nüsslein-Volhard C (1987) Involvement of the pumilio gene in the transport of an abdominal signal in the Drosophila embryo. Nature 329:167-170

Mahowald AP, Allis CD & Caulton JH (1981) Rapid appearance of multivesicular bodies in the cortex of Drosophila eggs at ovulation. Devel Biol 86:505-509

Mahowald AP, Goralski TJ & Caulton JH (1983) In vitro activation of Drosophila eggs. Devel Biol 98:437-445

Malacinski GM (ed) (1988) Developmental genetics of higher organisms. A primer in developmental biology. Macmillan Publ Comp. New York

Margaritis LH (1985) Structure and physiology of the eggshell. In: Kerkut GA, Gilbert LI (eds). Comprehensive insect physiology, biochemistry and pharmacolog. Pergamon Press, Oxford, Vol. 1, p 154

McFarlane C & McFarlane JE (1988) Sperm penetration and in vitro fertilization of the egg of the house cricket Acheta domesticus. Int J Invert Reprod Develop 13:171-182

Müller A & Büsen W (1988) Development of fertilized and unfertilized eggs of Sciara coprophila (Diptera): A cytological comparison. Abstract IV. Internat Congr Cell Biol Montreal p 408

Nüsslein-Volhard C & Roth S (1989) Axis determination in insect embryos. In: Cellular basis of morphogenesis. Wiley, Chichester (Ciba Foundation Symposium 144) p 37

Perotti ME & Riva A (1988) Concanavalin A binding sites on the surface of Drosophila melanogaster sperm: A fluorescence and ultrastructural study. J Ultrastruct Mol Struct Res 100:173-182
Phillips DM (1970) Insect sperm: Their structure and morphogenesis. J Cell Biol 44:243-277
Pohlhammer K (1978) Insemination of eggs in the australian cricket Teleogryllus commodus Walker (Insecta, Orthoptera). Zool Jb Anat 99:157-173
Putters FA & Van den Assem J (1985) Precise sex ratio in a parasite wasp: the result of counting eggs. Behav Ecol Sociobiol 17:265-270
Raff JW & Glover DM (1989) Centrosomes, and not nuclei, initiate pole cell formation in Drosophila embryos. Cell 57:611-619
Retnakaran A & Percy J (1985) Fertilization and special modes of reproduction. In: Kerkut GA, Gilbert LI (eds). Comprehensive insect physiology, biochemistry and pharmacology. Pergamon Press, Oxford, Vol. 1, p 231
Ripley S & Kalthoff K (1983) Changes in the apparent localization of anterior determinants during early embryogenesis (Smittia spec., Chironomidae, Diptera). Roux's Arch Dev Biol 192:353-361
Ruder FJ, Frasch M, Mettenleiter TC & Büsen W (1987) Appearance of two maternally directed histone H2A variants precedes zygotic ubiquitination of H2A in early embryogenesis of Sciara coprophila (Diptera). Devel Biol 122:568-576
Saini MS, Singh D & Kaur M (1987) Induction of artificial parthenogenesis in Athalia lugens proxima (Hymenoptera: Tenthredinidae). Entomol Genet 12:171-176
Sander K (1976a) Specification of the basic body pattern in insect embryogenesis. Adv Insect Physiol 12:125-238
Sander K (1976b) Morphogenetic movements in insect embryogenesis. In: Lawrence PA (ed) Insect development. Blackwell/Oxford, p 35
Sander K (1985a) Fertilization and egg cell activation in insects. In: Metz CH, Monroy A (eds) Biology of fertilization. Academic Press, Vol. 2, p 409
Sander K (1985b) Experimental egg activation in lower dipterans (Psychoda, Smittia) by low osmolarity. Int J Invert Reprod Develop 8:175-183
Sander K & Feddersen I (1985) Developmental failure after experimental activation of insect eggs. Int J Invert Reprod Develop 8:219-226
Sander K & Lehmann R (1988) Drosophila nurse cells produce a posterior signal required for embryonic segmentation and polarity. Nature 335:68-70
Sawa M & Oishi K (1989a) Studies on the sawfly, Athalia rosae (Insecta, Hymenoptera, Tenthredinidae) II. Experimental activation of mature unfertilized eggs. Zool Sci (Tokyo) 6:549-556
Sawa M & Oishi K (1989b) Studies on the sawfly, Athalia rosae (Insecta, Hymenoptera, Tenthredinidae) III. Fertilization by sperm injection. Zool Sci (Tokyo) 6:557-563
Sawa M & Oishi K (1989c) Delayed sperm injection and fertilization in parthenogenetically activated insect eggs (Athalia rosae, Hymenoptera). Roux's Arch Dev Biol 198 (in press)
Sawa M, Fukunaga A, Naito T, & Oishi K (1989) Studies on the sawfly, Athalia rosae (Insecta, Hymenoptera, Tenthredinidae). I. General biology. Zool Sci (Tokyo) 6:541-547
Van der Meer JM (1988) The role of metabolism and calcium in insect eggs: a working hypothesis. Biol Reviews (Cambridge) 63:107-157
Van der Meer JM (1990) Control of mitosis and ooplasmic movements in insect eggs. In: Malacinski GM (ed) Cytoplasmic organization systems. McGraw-Hill Publ Comp, New York. p 263
Van der Meer JM, Kemner W, Miyamoto DM (1982) Mitotic waves and embryonic pattern formation: No correlation in Callosobruchus (Coleoptera). Roux's Arch Dev Biol 191:355-365
Vinson SB & Jang HS (1987) Activation of Campoletis sonorensis (Hymenoptera: Ichneumonidae) eggs by artificial means. Ann Entomol Soc Am 80:486-489
Virkki N, Bruck T & Denton A (1987) Brief notes of the cytology of neotropical coleoptera. 5. Storage and activation of large sperm cells in male Alticinae. J Agric Univ Puerto Rico. 71:415-417
Went DF (1982) Egg activation and parthenogenetic reproduction in insects. Biol Rev (Cambridge) 57:319-344

Wolf R (1980) Migration and division of cleavage nuclei in the gall midge, Wachtliella persicariae. II. Origin and ultrastructure of the migration cytaster. Roux's Arch Dev Biol 188:65-73

Wolf R (1985) Migration and division of cleavage nuclei in the gall midge, Wachtliella persicariae. III. Pattern of anaphase-triggering waves altered by temperature gradients and local gas exchange. Roux's Arch Dev Biol 194:257-270

Wyman R (1979) The temporal stability of the Drosophila oocyte. J Embryol exp Morph 50:137-144

Zarani FE & Margaritis LH (1986) The eggshell of Drosophila melanogaster. V. Structure and morphogenesis of the micropylar apparatus. Can J Zool 64:2509-2519

FERTILIZATION IN CTENOPHORES

Danièle CARRE, Christian SARDET, Christian ROUVIERE
URA 671 C. N. R. S.
BIOLOGIE CELLULAIRE MARINE
UNIVERSITE PIERRE ET MARIE CURIE
STATION ZOOLOGIQUE, VILLEFRANCHE SUR MER F-06230

NATO ASI Series, Vol. H 45
Mechanism of Fertilization
Edited by B. Dale

Ctenophores constitute a small phylum of approximatly 100 species. All species are marine and most of them are large planktonic forms moving with 8 rows of beating ciliated comb plates (ctenes). The animals have a simple body plan with biradial symmetry defined by the position of the tentacles (tentacular plane) and the stomodeum (sagittal or stomodeal plane) (Harbison 1985).

At the turn of the century, ctenophore eggs and embryos were the subject of intense interest because of their mosaïc development. This is best illustrated by the fact that for example, ablation of a blastomere at the 4 cell-stage led to the disappearance of 2 of the 8 comb rows (see review by Reverberi 1971).

The relationship of the oral-aboral axis of the embryo to the first cleavage initiation site was established by Reverberi and Ortolani in the 1960's (Reverberi and Ortolani 1963, 1965). Because the cleavage initiation site was observed to often be close to the site of meiosis, it was thought that the oral-aboral axis of the embryo corresponded to the animal-vegetal axis of the egg, suggesting that, as in many species, this axis was predetermined during oogenesis. Freeman clearly showed that this was not so and observed that in many eggs in which the cleavage initiation site was situated away from polar bodies, development was normal (Freeman 1976, 1977, 1979). He could even shift at will the cleavage initiation site by displacing the zygotic mitotic apparatus. We carried these studies one step further by showing that in Beroe ovata, the location of the mitotic apparatus was, in fact, the site of sperm entry (Carré and Sardet 1984, Sardet et al. 1990).

A problem with studying ctenophore eggs has been that fertilization could not be controlled adequately. We have now achieved this control in Beroe ovata and observed all phases of fertilization. We have also studied mitosis and the successive unipolar cleavages that lead to the formation of micromeres (enriched in cortical cytoplasm) and macromeres, with restricted developmental potential (Carré and Sardet 1984, Sardet et al. 1990).

CTENOPHORE GAMETES

With the exception of 2 deep sea species, all ctenophores are hermaphrodite with visible gonads beneath the comb rows (Harbison and Miller 1986, Strathman 1987). The eggs grow rapidly (1-2 days). Spawning is controlled by light after exposure to the dark (or the reverse in some species.) (Dunlap 1974). In Beroe ovata, the species we work with, mature specimens can be obtained at Villefranche-sur-Mer, from February through June (Fig. 1). Kept in the laboratory, they will give 50-100 eggs every 2 days on average.

Sperm is emitted first in spurts, then eggs, and, if several specimens are kept in a tank together, fertilization and development will take place immediatly. There are several reasons why one would like to study Beroe ovata eggs. They are very large (1mm) but so transparent that one can observe all phases of fertilization with unusual clarity. (Carré and Sardet 1984, Carré and Sardet 1990). In addition, in these eggs, all events remain cortical. Furthermore, blastomere ablation and marking experiments have provided reliable fate maps (Reverberi and Ortolani 1963, 1965, Farfaglio 1963).

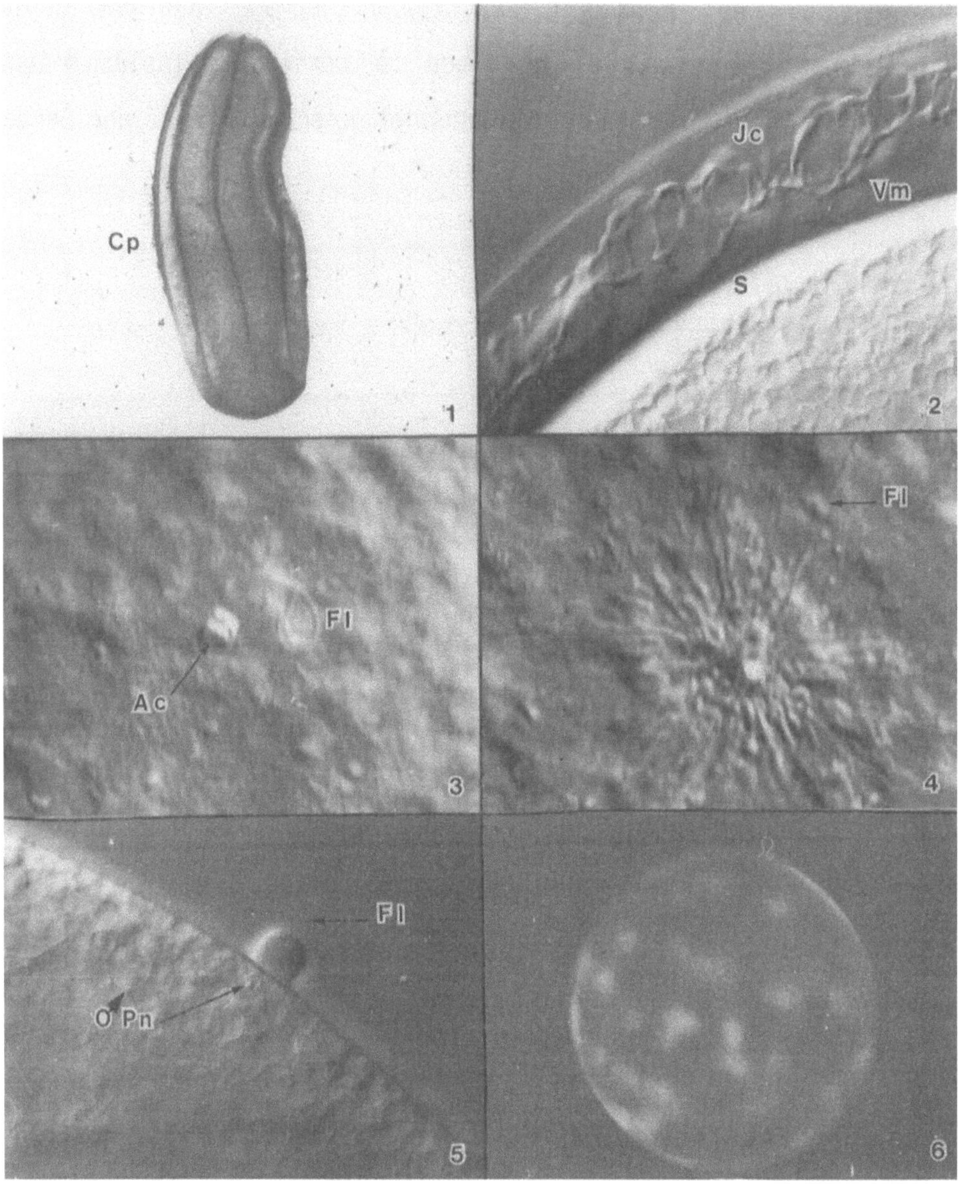

Fig 1:The ctenophore Beroe ovata showing the rows of comb plates (Cp) (x0,5).
Fig 2:The egg surface (S) vitelline (Vm) and jelly coat (Jc) (x200).
Fig 3 :A sperm with beating flagellum (F1)and acrosomal process (Ac) (x1000).
Fig 4:Entering sperm with rigid flagellum (F1) and the egg surface contraction
Fig 5:A fertilization cone (♂ Pn : sperm pronucleus) (x250).
Fig 6:Differentiations around sperm entry sites in epifluorescence (x54)

THE CONTROL OF FERTILIZATION

When freshly collected Beroe ovata are isolated and their eggs handled carefully, we observed that a large proportion of the eggs remained unfertilized. This block to self-fertilization is eventually lost with time or after the animals are kept in captivity for a while.
The block to self-fertilization is at the level of the vitelline membrane and the jelly coat of the eggs (fig 2). When these envelopes are removed the eggs can be fertilized by their own sperm. This simple operation therefore provides for the first time an opportunity to control and describe fertilization in ctenophore.

FERTILIZATION IN BEROE OVATA

Sperm entry

The egg is shed as it is completing the first meiotic division. We have observed that sperm penetration is preceeded by a modification of the spherical sperm head, a flange appearing that probably corresponds to an acrosomal process (Fig 3) (see Franc 1973 for sperm morphology in the testes). The next indication that fertilization occured is the sudden rigidification of the sperm flagellum (Fig 4). The egg reacts by a localized contraction-relaxation of the surface around the sperm entry site, that last about 2 mn (Fig 4), and the formation of a fertilization cone that disappears 20 mn later (Fig 5).
This process can be observed simultaneously or successively as up to 20 spermatozoïds can be incorporated at any position in the egg cortex,

without alteration of normal development. Each sperm pronucleus remains close to the sperm entry site.

Unlike most other eggs where a wave of cortical reorganization modify evenly the entire surface, the cortex of Beroe ovata eggs is only differenciated around each sperm pronucleus (Fig 6). These differenciations include a local tuft of microvilli, local exocytosis, and an accumulation of mitochondria and autofluorescent vesicles recruted from the surrounding cortical region (Fig 7, 8, 9). They remain around each sperm nucleus until female pronuclear fusion with one of the sperm pronucleus, and represent a convenient way to assess the degree of polyspermy and the distribution of fertilization sites (Fig 6).

The female pronuclear movements

Meiosis proceeds from the time of spawning whether or not fertilization occurs.

As soon as the female pronucleus is formed under the polar bodies, it starts moving beneath the plasma membrane,towards one of the sperm pronucleus, not necessarily towards the closest in case of polyspermy (Fig 10). The subsurface trajectory is rectilinear and as the nucleus moves at a speed of approximatly 0,5 µm/sec, it may be observed with exceptional clarity for long periods of time.

At the level of the edge of the differentiated cortical zone arround the sperm pronucleus, the female pronucleus is confronted with 2 possible alternatives :

- in cases of monospermy the female pronucleus penetrates the zone and fusion of the pronuclei takes place (Fig 13,14,15).

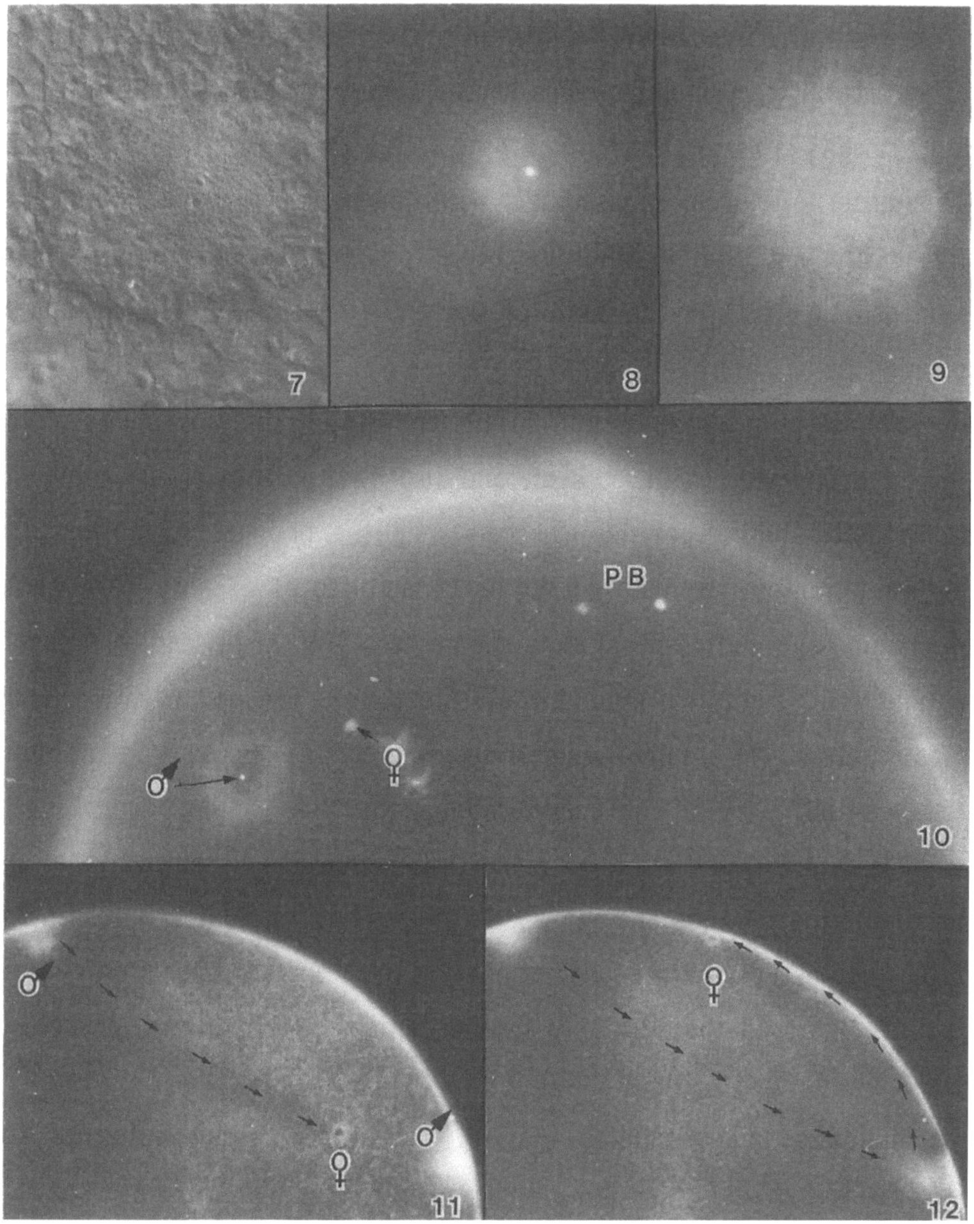

Fig 7,8,9: The zone of differentiation of the egg cortex around sperm entry site. 7 : in differential interference contrast. 8 : epifluorescent illumination in the UV spectrum after hoechst staining of DNA. 9 : epifluorescent illumination in the FITC spectrum (x450).

Fig10: Beroe Ovata egg after hoechst staining seen in epifluorescent illumination in the UV spectrum. After leaving the polar body region (Pb), the female pronucleus migrates toward a male pronucleus (x150).

Fig11,12: Two photomicrographs 20 minutes apart of the female pronucleus trajectory after exploring 2 male pronuclei (x125).

- in cases of polyspermy the female pronucleus either enters the zone or reorients to migrate towards another sperm nucleus, and eventually other ones (Fig 11, 12).

This latter situation is associated with "choosing" among sperm nuclei from different specimens, althought, at present, we have only indirect evidence that this is the basis of the puzzling behavior of the female pronucleus.

As demonstrated in sea urchin (Schatten 1982), the movements of the female pronucleus in Beroe ovata are clearly dependant on microtubules since drugs that depolymerize them stop the pronuclear motion. The cortex of unfertilized Beroe egg is constituted of a dense network of microtubules running parallel to the surface (Fig 16). Near each sperm pronucleus giant two dimensional asters several hundred microns in size develop from originally small asters (Fig 17). We have been able to observe that close to the sperm pronucleus, the female pronucleus is embedded in a dense array of astral microtubules. We suspect that the reorientation towards another sperm nucleus is associated with sensing microtubules (possibly + end) emanating from another sperm entry site. We feel however that there is a major enigma to solve with respect to the movement of the nucleus, an organelle that is known to be connected to the endoplasmic reticulum (ER). In fact, in Beroe ovata, the cortex is constituted essentially by sheets of ER and microtubules running parallel to the egg surface. Serial sectioning of migrating female pronucleus shows that there are numerous connections between the nuclear envelope and the surrounding sheets of ER (Fig 18). In light of the recent discoveries of

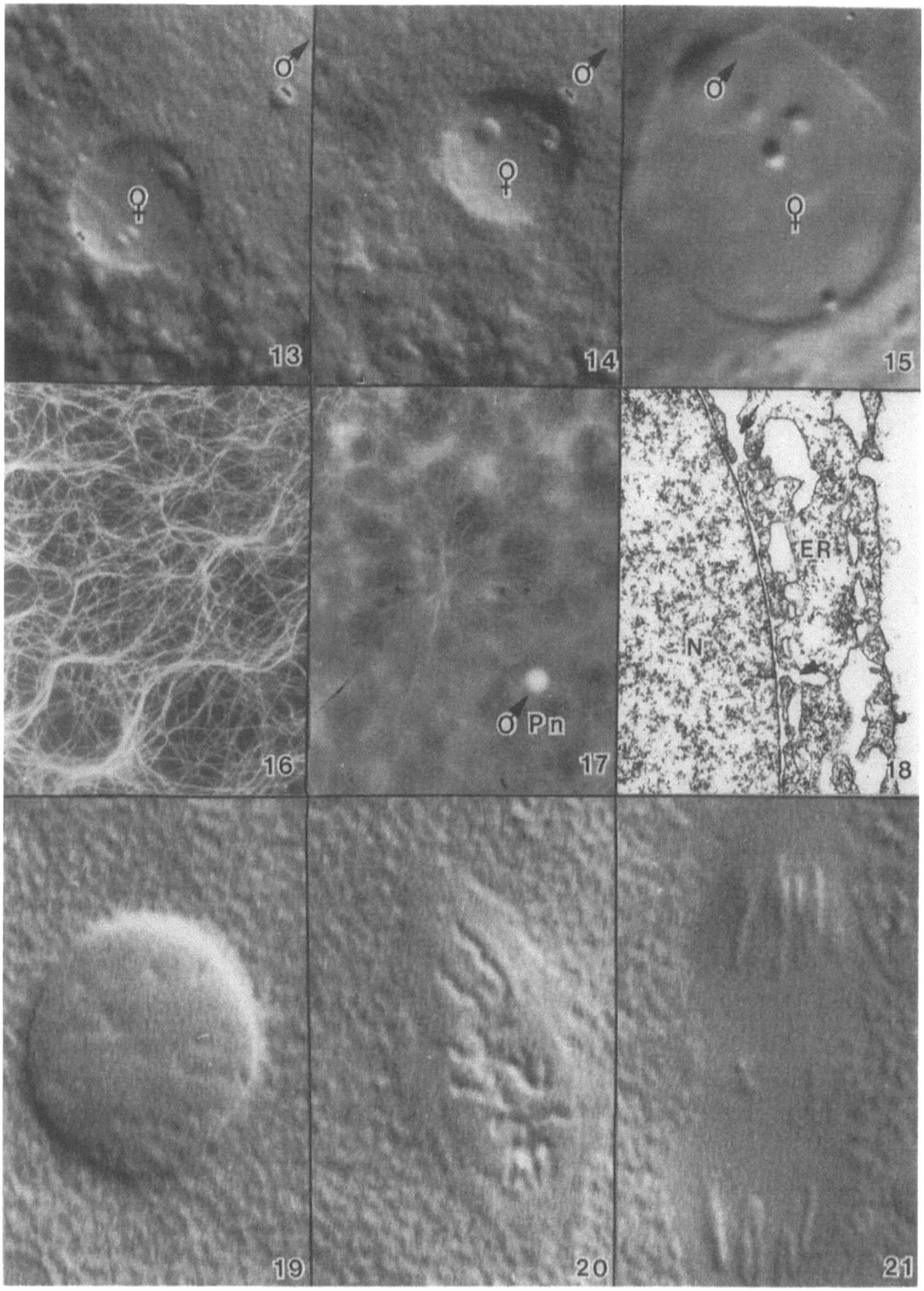

Fig13,14:Movement of the female toward the male pronucleus (x1400).
Fig15:pronuclear fusion (x3000).
Fig16:Microtubules network in the unfertilized egg.
Fig17:Astral microtubules near the male pronucleus (male Pn) (x1200).
Fig18:A migrating nucleus and connecting ER sheets arrows (x12000).
Fig 19,20,21: First mitosis in Beroe Ovata (x2000).

the physical interactions between microtubules and ER vesicles we are left with 3 possibilities for the pronuclear movements :

- the nuclear core and inner nuclear membrane move within the ER in a yet unknown microtubule dependant mechanism.
- the nucleus breaks and reforms constantly its connections with the ER.
- The ER sheets and the nucleus move together against fixed microtubules.

It is at present impossible to decide which mechanism is at work but we will attempt to distinguish among them.

Pronuclear fusion, mitosis and cleavage.

The fusion of pronuclei and mitosis take place beneath the surface, within the differenciated cortical zone, with the greatest clarity one can obtain among animal eggs (Fig 15,19, 20, 21). All other zones differentiated around supernumary sperm disappear. The first cleavage is unipolar, due to the location of the mitotic apparatus beneath the egg surface.

FERTILIZATION AND ORAL-ABORAL AXIS FORMATION

We have established that in Beroe ovata :

- sperm can enter anywhere in the cortex.
- the sperm pronucleus does not move appreciably from the site of sperm entry.

-the female pronucleus migrates to any of the male pronuclei and selects one.

-the mitotic apparatus and the cleavage furrow occur at the location of the successfull sperm entry site.

If one takes into accounts our observations and those of Reverberi, Ortolani and Freeman that showed that the oral pole of the embryo and the adult correspond to the site of initiation of the first unipolar cleavage and the location of the mitotic apparatus, we can propose that the ctenophore egg surface is totipotent and that the site of successfull sperm entry determines the oral-aboral axis of the embryo.

REFERENCES

Carré, D and Sardet, C (1984). Fertilization and early development in Beroe ovata. Dev. Biol. 105 : 188-195.

Dunlap, H (1974). Ctenophora in "Reproduction in Marine Invertebrates". Giese and Pearse ed., Acad. Press. New York : 201-265.

Farfaglio, G (1963). Experiments on the formation of the ciliated plates in ctenophores. Acta Embryol. Morphol. Exp. 6 : 191-203.

Franc, J. M (1973). Etude ultrastructurale de la spermatogenèse du cténaire Beroe ovata, J. ultra struct. Res. 42 : 255-267.

Freeman, G (1976). The effects of altering the position of cleavage planes on the process of localization of developmental potential, Dev. Biol. 51(2) : 332-337.

Freeman, G (1977). The establishment of the oral-aboral axis in the ctenophore embryo. J. Embryol. Exp. Morph. 42 : 237-260.

Freeman, G. (1979). The multiples roles which cell division can play in the localization of developmental potential, in "Determinants of spatial organization". S. Subtelny and I. R. Konigsberg ed. Acad. Press. New York : 53-76.

Harbison, G. R. (1985). On the classification and evolution of ctenophora, in. The origin and relationships of lower invertebrates. S. Conway Morris, J. D George, R. Gilson and H. M. Platt, ed., Clarendon Press, London.

Harbison, G. R. and Miller, R. (1986). Not all ctenophores are hermaphrodites. Studies on the systematics, distribution, sexuality and development of two species of Ocyropsis. Mar. Biol. 90 : 413-424.

Reverberi, G. and Ortolani, G. (1963). On the origin of ciliated plates and of the mesoderm in the ctenophores. Acta. Embryol. Morphol. Exp. 6 : 175-190.

Reverberi, G. and Ortolani, G. (1965). The development of the ctenophore's egg. Riv. Biol., (Lisbon) 58 : 113.

Reverberi, G. (1971). Ctenophores in : "Experimental embryology of marine and fresh- water invertebrates". North-Holland Publ. Amsterdam : 85-103.

Schatten, G. (1982). Motility during fertilization. Int. Rev. Cytol. 79 : 60-156.

Sardet C., Rouvière C. and Carré D. (1990). Reproduction and Development in Ctenophores in: Experimental embryology of aquatic organisms J. Marty ED. Plenum Press. In press.

THE SEA URCHIN EGG: AN OLD AND NEW MODEL

Prof. Giovanni GIUDICE
Department of Cellular and Developmental Biology "Alberto Monroy"
and Istitute for Developmental Biology, CNR
via Archirafi 30
90123 Palermo
Italy

Since this meeting is dedicated to the memory of Alberto Monroy I will touch on a few problems which represent some of the main fields of the research which are being carried out in the Department of Cellular and Developmental Biology of Palermo, which bears his name, problems which are along the lines he established many years ago, i.e. some of the mechanisms underlying the regulation of sea urchin development. One of the well known and thoroughly studied characteristics of the sea urchin, shared by the eggs of many aquatic organisms, is that of being equipped with many macromolecules and cell structures, synthesized during oogenesis, which allow this large cell to rapidly divide following fertilization, into many smaller cells, with no need for undergoing the time-and energy-consuming synthesis of the above molecules and structures. The list of these substances is a long one, so that it might be easier to describe those which are not previously stored. The storage includes histones, ribosomes and mitochondria, whose synthesis is coupled with that of DNA in somatic cells, but not in the egg and in the early embryo, where one is faced with a rapid synthesis of nuclear DNA without any appreciable synthesis of for example ribosomes and mitochondria.

The question then arises of what uncouples the synthesis of nuclei from that of mitochondria and ribosomes in the early embryos. In order to answer this question Rinaldi et al. (1,2) have centrifuged the *Paracentrotus lividus* egg thus splitting into a nucleated half and a non nucleated one. If both halves are now parthenogenetically activated and the synthesis of nuclear and mitochondrial DNA are measured, it is found that in the nucleated half a lively synthesis of nuclear DNA takes place, but no synthesis of mitochondrial DNA occurs, whereas in the non nucleated half the synthesis of mithocondrial DNA is activated, as shown by labelled precursor incorporation and by the finding in spread DNA preparations observed under the electron microscope of clear figures of replicating mitochondrial DNA. Also mitochondrial RNA and proteins start to be synthesized in the non nucleated halves and mitochondria as a whole start duplicating as shown in figure 1. The hypothesis was then made that is was the nucleus

NATO ASI Series, Vol. H 45
Mechanism of Fertilization
Edited by B. Dale

to prevent the egg mitochondria from replicating. As a proof that, if a pronucleus is introduced in the non nucleated half by fertilizing it, no mitochondrial syntheses are started. A negative control of the nucleus on the mitochondrial syntheses in the early embryo was therefore discovered. What is the mechanism through which the nucleus exerts this negative control?

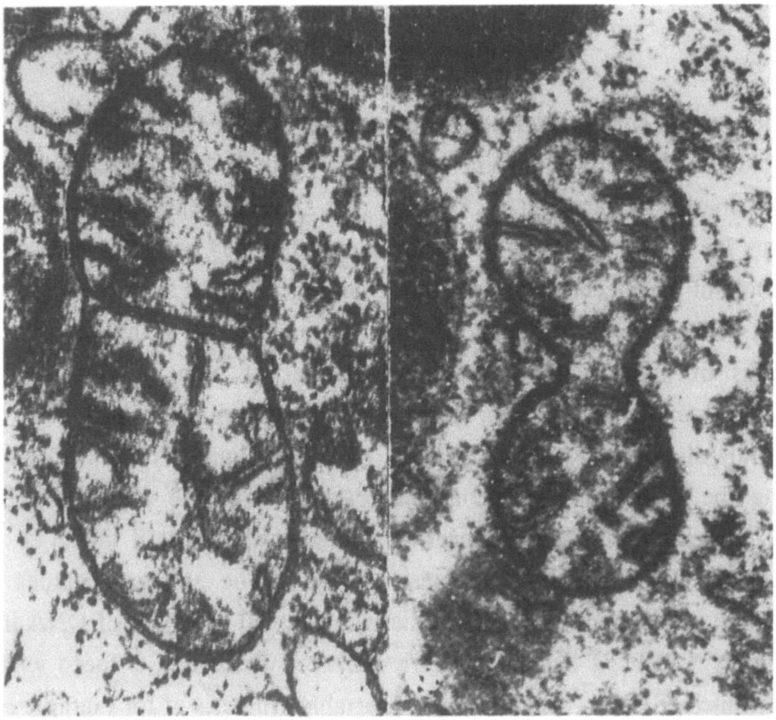

Fig. 1. Two figures of dividing mithochondria in non nucleated parthenogenetically activated eggs of the sea urchin *Paracentrotus lividus* (from Rinaldi et al. 1977. Cell. Biol. Intern. Rep., 3, 179-182)

Several hypotheses have been explored in these last years and discarded. The last experiments however point to a role of Ca^{2+} movements controlled by the nucleus in the regulation of mitochondrial syntheses. Actually if the non nucleated halves are parthenogenetically activated with ammonia, mitochondrial syntheses is activated, whereas if the parthenogenetic activation is made with the Ca^{2+} ionophore A23178, which mimics the sperm in mobilizing Ca^{2+} from compartments, no mitochondrial activation occurs. Moreover if Ca^{2+} is sequestered with Chlortetracycline in the ammonia activated halves, no mitochondrial synthesis is observed. The hypothesis was therefore put forward that internal Ca^{2+} ions are necessary for mitochondrial activation, and that nuclei cause a Ca^{2+} efflux which inactivates them.

Another advantage of the sea urchin embryo is that as shown by Giudice 1962 (4), it can be dissociated into single cells and these can be reaggregated to restitute entire embryos (figure 2). Two kind of questions have therefore been asked one concerns the mechanisms of specific cells adhesion and the other the importance of cell interactions for the metabolic changes which are observed during the early embryonic development.

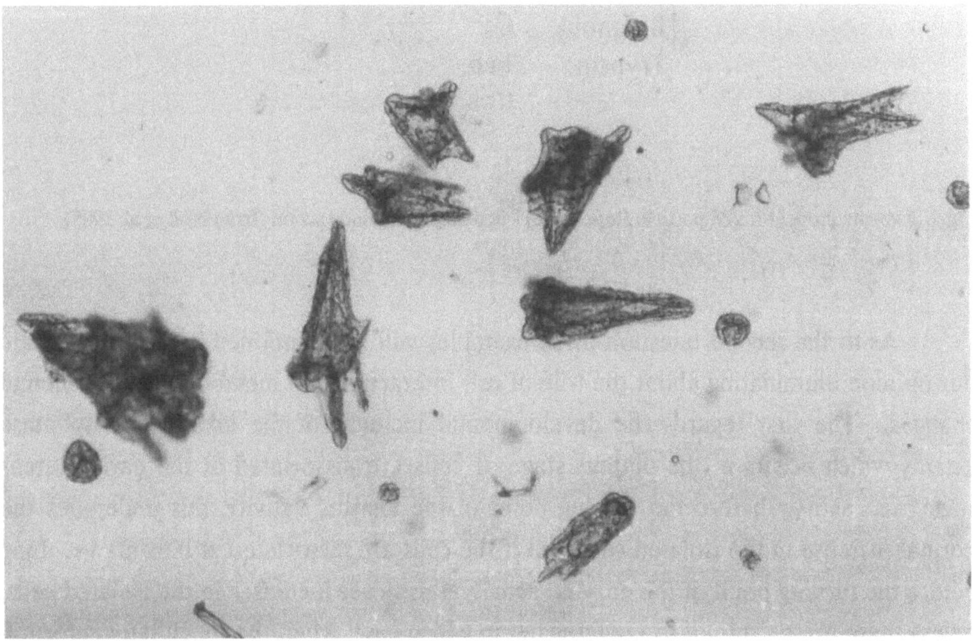

Fig 2. Pluteus-like embryos obtained after reaggregation of cells dissociated at the mesenchyme-blastula stage (from G. Giudice, 1973. Developmental Biological the Sea Urchin Embryo. Academic Press)

As to the first one, a 22S particle was discovered, which is a component of the cell surface and is able to specifically promote cell to cell adhesion and reaggregation. Experiments carried out in the last year have provided evidence that this particle is synthesized during gametogenesis presumably by the intestine and represents a major constituent of what has been considered up to recently to be yolk particles (5). A hypothesis has also been put forward that through the combination of the various protein subunits of the 22S particle, a positional code arises which specifies to the cell the position it has to occupy in the embryo context (6). (figure 3).

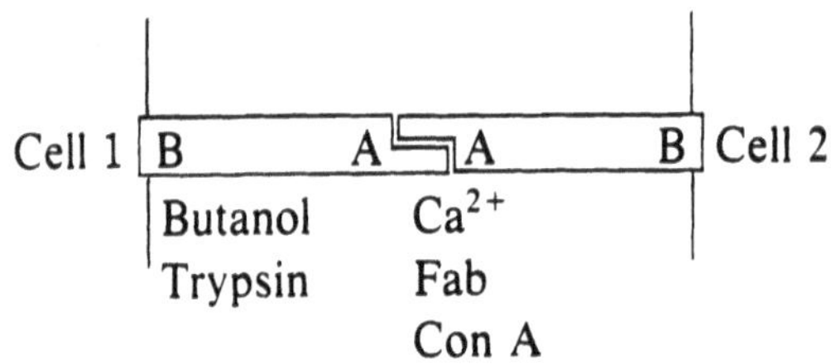

Fig 3. Two-site model for 22S particle (toposome) - dependent cell interaction (from Noll et al. 1985).

As to the second question three examples will be mentioned here, which are in our opinion illuminating about the role of cell interactions on metabolic developmental changes. The first regards the developmental increase of the alkaline phosphatase activity which occurs at the pluteus stage. If cells are dissociated at the early pluteus stage, i.e. shortly before the turning point of the enzyme activity, this undergoes the normal increase in the isolated cells, but if the cells are dissociated at blastula i.e. long before the turning point of the enzyme activity, this never increases in the isolated cells, unless these are permitted to reaggregate, in which case, when the much later reform a pluteus-like structure, the increase in alkaline phosphatase activity occurs (Figure 4). This indicates that cell interactions are needed to commit the cells to undergo the increase in alakline phosphatase activity and that once the commitment has taken place cell interactions are no longer needed for the increase in enzymatic activity to occur (7).

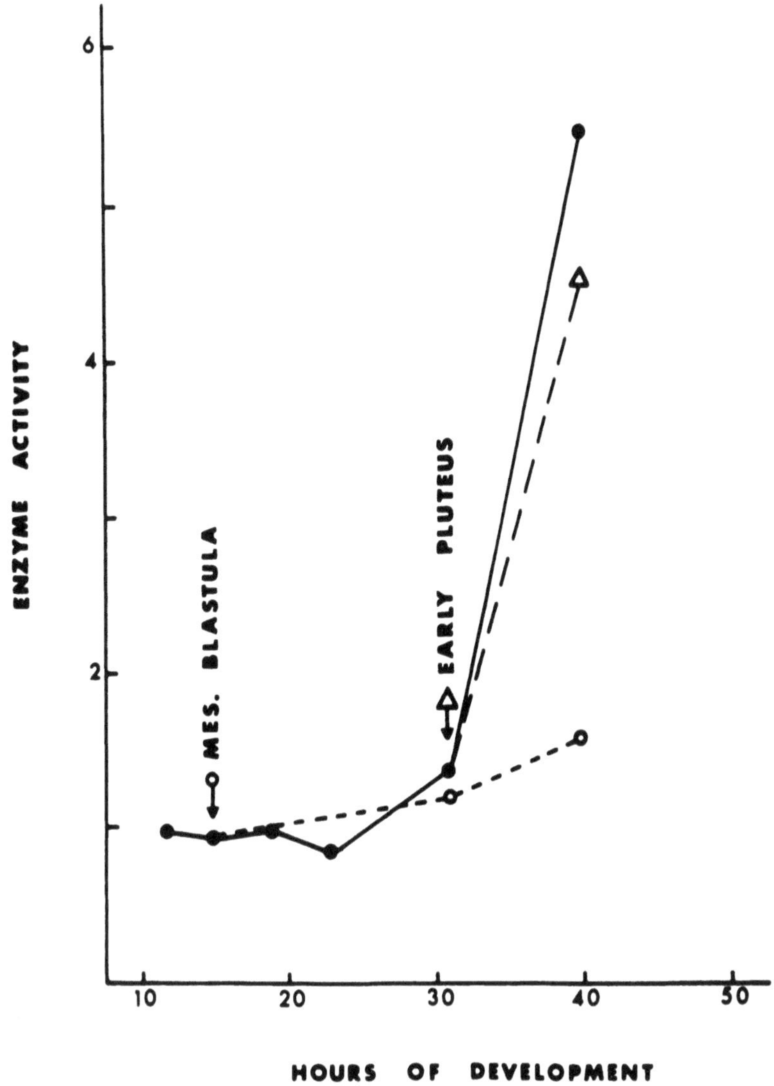

Fig 4. Alkaline phosphatase activity of entire sea urchin embryos --- cells dissociated at mesenchyme blastula o---o cells dissociated at pluteus.

A need for cell interactions is also shown for the wave of DNA synthesis which occurs at the onset of gastrulation. If the embryos are dissociated into cells shortly before that stage there DNA synthesis stops unless these are permitted to reaggregate (8). This indicates that a signal starting from the membrane informs the cells of the lack of contact and this halts DNA synthesis. As a counter proof of this if the cell surface of the dissociated cells is damaged with a mild trypsin digestion the cells resume DNA synthesis even in the absence of reaggregation (9). Treatment of the isolated cells with univalent antibodies against the plasmamembrane also result in resumal of the DNA synthesis (10).

These results do not have to be taken as an indication for the necessity of cell interactions for all changes in metabolic activities during sea urchin development, because there are cases in which the embryo seems to be endowed with an internal clock which signals the metabolic change also in the absence of cell interactions. One example of this is the ability of responding to heat shock with the synthesis of heat shock proteins. This, at least for the h.s.p. 70, arises after the hatching blastula stage, in *Paracentrotus lividus*. If the embryo is dissociated into cells at the stage of 16-32 blastomeres, the isolated cells develop the ability to synthesize h.s.p. when the entire embryo reaches the stage of hatching blastula (11).

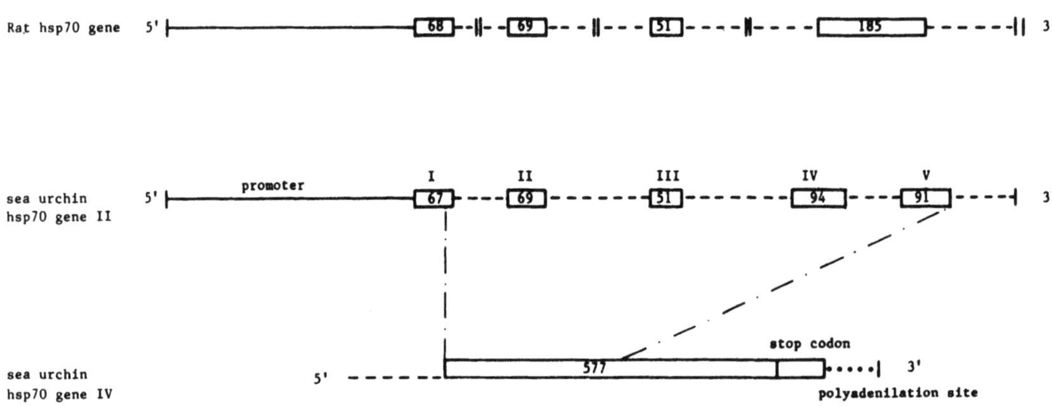

Fig 5. Schematic structure of two genomic clones for the h.s.p. 70 of *Paracentrotus lividus* (Sconzo et al. unpublished).

The sea urchin embryo has long been used as a model for the regulation of protein synthesis throughout early development. The synthesis of H2A histones, heat shock proteins, tubulins, collagen have been and are currently being studied, in Palermo. It is worth mentioning here that a cis acting element has been described which enhances the transcription of the early histone H2A gene, as shown by microinjection into Xenopus oocytes of a construct containing this sequence followed by the thymidine kinase promoter and by the CAT reporter gene. Furthermore, gel retardation and foot printing analyses showed that this sequence contains at least two upstream sequence elements which specifically bind proteins extracted from the nuclei of embryos which actively transcribe the H2A gene (12). Two genomic clones or the h.s.p.7O gene have been analysed. The sequence of the first one shows the interesting feature of being interrupted by 5 introns and of showing 4 heat shock elements in the 5' region, whereas the second one shows no introns (Figure 5). This suggests a double type of expression control, this making the sea urchin embryo once more a suitable model for the study of modern problems.

The work reported in this paper was partially supported by funds of the M.P.I. to Prof. G.Giudice.

REFERENCES

1) A.M.Rinaldi, G.De Leo, A.Arzone, I.Salcher, A.Storace & V.Mutolo. Biochemical and electron microscopic evidence that cell nucleus negatively controls mitochondrial genomic activity in early sea urchin development. Proc.Natl.Acad.Sci.76, (1979) 1916.

2) A.M.Rinaldi & G.Giudice. Nuclear-cytoplasmic interactions in early development. in "Biology of Fertilization", C.B.Metz and A.Monroy eds. ,vol.3, (1985) 367-377.

3) G.Giudice. Restitution of whole larvae from disaggregated cells of sea urchin embryos. Develop.Biol. 5 (1962) 402-411.

4) G.Giudice. (1986). The sea urchin enbryo. A Developmental Biological System. Springer Verlag.

5) M.Cervello & V. Matranga. Evidence of a precursor-product relationship between vitellogenin and toposome, a glycoprotein complex mediating cell adhesion. Cell Diff.and Develop. 26 (1989) 67-76.

6) H.Noll, V. Matranga, M.Cervello, T.Humphreys, B.Kuwasaki & D.Adelson. Characterization of toposomes from sea urchin blastula cells: A cell organelle mediating cell adhesion and epressing positional information. Proc. Natl.Acad.Sci.USA 82 (1985): 8062-8066.

7) G.Giudice & R.J.Phohl. The role of cell interactions in the control of enzyme activity during embryogenesis. Biochim. Biophys. Acta 142 (1967) 263-266.

8) G.Giudice. The sea urchin embryo. A developmental biological system. Springer Verlag Berlin-eidelberg N.Y.-Tokyo 1986.

9) M.L.Vittorelli, G.Cannizzaro & G.Giudice. Trypsin treatment or cells dissociated from sea urchin embryos elicits DNA synthesis. Cell Different. 2 (1973) 279-284.

10) M.L.Vittorelli, V.Matranga, S.Feo, G.Giudice & H.Noll. Inverse affects on thymidine incorporation in dissociated blastula cells of the sea urchin *Paracentrotus lividus* induced by butanol treatment and Fab addition. Cell Different. 9 (1980) 63-70.

11) G.Sconzo, M.C.Roccheri, M.Di Carlo, M.G.Di Bernardo & G.Giudice. Synthesis of heat shock proteins in dissociated sea urchin embryonic cells. Cell Different. 2 (1973) 279-284.

12) F.Palla, C.Casano, I.Albanese, L.Anello, F.Gianguzza, M.G.Di Bernardo, C.Bonura & G.Spinelli. Cis-acting elements of the sea urchin histone H2A modulator bind transcriptional factors. Proc.Natl.Acad.Sci.USA 86 (1989) 6033-6037.

COLLAGEN-BINDING PROTEINS IN SEA URCHIN EGGS AND EMBRYOS

E. Nakano, M. Iwata* and V. Matranga**
Department of Biology
Nagoya University
Nagoya 464-01
Japan

Gastrulation is the first important event of morphogenesis in the sea urchin embryo. Near the end of the blastula stage, certain cells in the vegetal plate lose adhesivity to each other and to the hyaline layer and migrate from the blastula wall into the blastocoelic cavity giving rise to the primary mesenchyme cells. Following migration of the primary mesenchyme cells, the cells remaining in the vegetal plate begin to invaginate into the blastocoel to form the archenteron. The tip of the archenteron advances towards the animal pole and touches the ectodermal wall to open the stomodeum. The process of gastrulation seems to be controlled by the ability of cells to adhere to other cells and to substrates.

The mechanism of cell adhesion has been extensively studied in recent decades and many cell adhesion molecules have been obtained from a wide variety of animals. Among them, fibronectin is one of the well-characterized molecules with the ability of cell adhesion.

We isolated fibronectin from the gonads of the Japanese sea urchin, using affinity chromatography on gelatin-Sepharose 4B (Iwata and Nakano 1981; 1983). Biochemical and immunological studies indicated that sea urchin fibronectin has properties similar to those of mammalian fibronectin. Recently, we introduced a new method for the preparation of fibro-

---

* Present address: Department of Physiology and Biophysics, University of Washington, Seattle, Washington 98195, USA
** Istituto di Biologia dello Sviluppo, Palermo 90123, Italy

NATO ASI Series, Vol. H 45
Mechanism of Fertilization
Edited by B. Dale

nectin from sea urchin gonads and embryos. The gonads or pelleted embryos were suspended in 6 M urea, 50 mM Tris buffer, pH 7.5, containing a cocktail of protease inhibitors (2 µg/ml aprotinin, antipain, pepstatin A and leupeptin and 1 mM benzaminidin). The suspension was homogenized and sonicated in an ice bath. The homogenate was centrifuged at 12,000 g at 4°C for 45 min. The supernatant was diluted 3 times with Tris buffer and loaded on a 4 ml column of gelatin-Sepharose 4B. The column was washed with Tris buffer and again with the same buffer containing 1 M NaCl and then eluted with 6 M urea. The solutions used for affinity chromatography were supplemented with the cocktail of protease inhibitors. The peak fractions were collected, dialyzed against Tris buffer and elctrophoresed on 6 % polyacrylamide gel containing 0.1 % SDS (SDS-PAGE).

SDS-PAGE revealed that fibronectin was present in the ovaries of all the sea urchin species examined: *Pseudocentrotus depressus*, *Anthocidaris crassispina*, *Temnopleurus toreumaticus*, *Hemicentrotus pulcherrimus*, *Paracentrotus lividus* and *Arbacia lixula*. Fibronectin was also found in the eggs and embryos of *P. depressus* and *P. lividus*. Besides fibronectin, different types of collagen-binding proteins were reported in sea urchin eggs and embryos (Nakano and Iwata 1982; Alliegro et al. 1988; Di Ferro et al. 1989). These proteins were prepared by the same method used for the preparation of fibronectin. As listed in Table 1, the molecular weights of these proteins vary considerably in different sea urchin species. Like fibronectin, the function of these collagen-binding proteins appears to be to promote cell-substrate and cell-cell adhesion of embryonic cells. In fact, both 116 kD (*Lytechinus*) and 120 kD (*Paracentrotus*) proteins promote the adhesion of dissociated blastula cells to the substrate in the centrifugal adhesion assay (Alliegro et al. 1988; Di Ferro et al. 1989).

Con A can also stimulate the adhesion of dissociated embryonic cells to the substrate (Iwata and Nakano 1984). The increase in adhesion by Con A is reduced by the addition of α-methyl-D-mannoside, suggesting that the Con A-induced adhesion is attributable to Con A-carbohydrate interaction. Scanning electron microscopic observations revealed that in the presence of Con A, dissociated cells are surrounded by the extracellular fibrous materials. The increase in cell-substrate adhesion induced by Con A is explained by the formation of such fibrous materials in the extracellular space. Sequential extraction of Con A-treated dissociated cells with Triton X-100 and urea solubilized most of the cellular components leaving the fibrous materials on the substrate. Biochemical components of the isolated fibrous materials include fibronectin and another type of collagen-binding protein, 88 kD.

Table 1. Collagen-binding proteins in sea urchins

Mol. Wt.(kD)	Name	Species	Reference
220	Fibronectin	*Pseudocentrotus depressus*	1
220	Fibronectin	*Anthocidaris crassispina*	1
220	Fibronectin	*Hemicentrotus pulcherrimus*	1
88		*Pseudocentrotus depressus*	2
88		*Anthocidaris crassispina*	2
120		*Hemicentrotus pulcherrimus*	2
120		*Dendraster excentricus*	2
116	Echinonectin	*Lytechinus variegatus*	3
120		*Paracentrotus lividus*	4
210	Fibronectin	*Paracentrotus lividus*	5
210	Fibronectin	*Arbacia lixula*	5
210	Fibronectin	*Temnopleurus toreumaticus*	5

1 Iwata and Nakano (1981); 2 Nakano and Iwata (1982);
3 Alliegro et al. (1988); 4 Di Ferro et al. (1989);
5 Present study

Numerous events during the early development of sea urchin embryos are accompanied, at least in part, by the change in cell adhesion caused by cell adhesion molecules. Since collagen-binding proteins, including fibronectin, may act as distinct adhesion molecules playing important roles in morphogenesis, their intense synthesis during the early development was expected.

We examined the incorporation of (^{35}S) methionine into the fraction of collagen-binding proteins of sea urchin eggs and embryos. During the early development of Japanese sea urchins, *Anthocidaris crassispina* and *Pseudocentrotus depressus*, there was no synthesis of fibronectin, but the synthesis of another type of collagen-binding proteins was detected. A group of these proteins was tentatively designated as 88 K protein, because the major component has a molecular weight of 88 kD (Nakano and Iwata 1982). As shown in Fig. 1, (^{35}S) methionine-labelled proteins were eluted with 8 M urea from the gelatin-Sepharose 4B column.

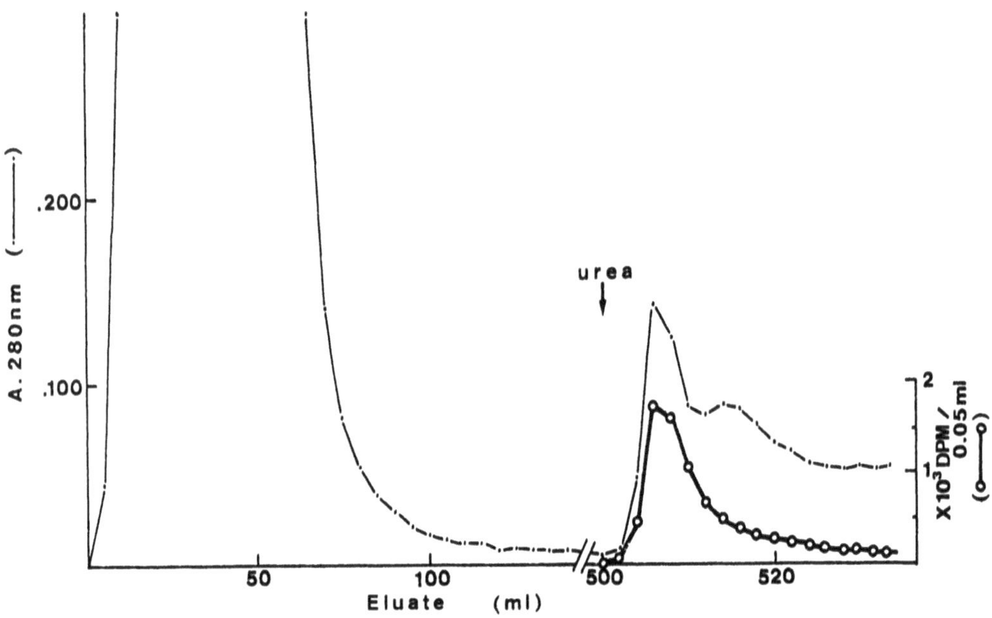

Fig. 1. Affinity chromatography of 88 K protein of *A. crassispina* on gelatin-Sepharose 4B.

SDS-PAGE of these fractions showed that the 88 K protein isolated from the embryos of *A. crassispina* and *P. depressus* were made up of three different subunits (α, β, γ) with molecular weights of 86, 88 and 90 kD, respectively. Without reduction, they existed as high molecular weight disulfide-bonded aggregates with a molecular weight of more than 300 kD. Besides the 88 K protein, no band was observed on the fluorogram (Fig. 2).

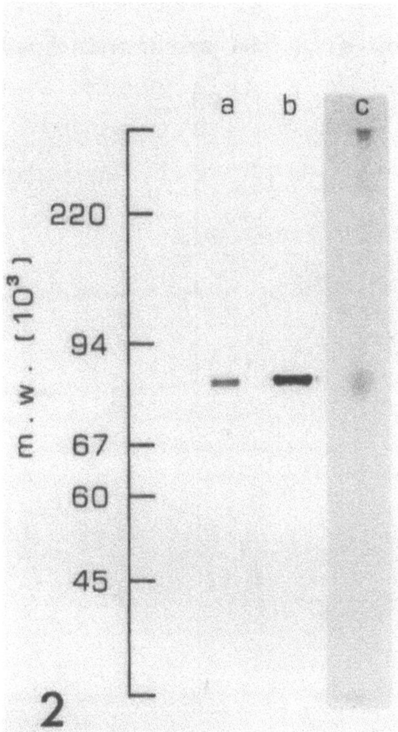

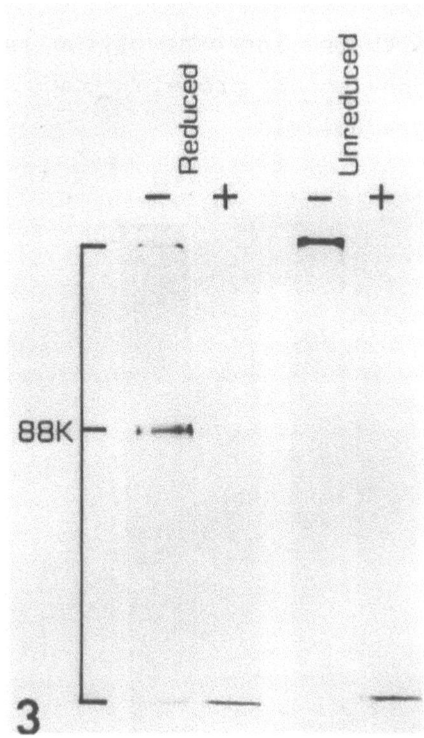

Fig. 2. SDS-PAGE of the 88 K protein of *A. crassispina*. a and b, reduced with 1 % 2-mercaptoethanol; c, unreduced. Protein concentration: a, 25 μg; b, 50 μg and c, 25 μg.

Fig. 3. Digestion of the 88 K protein of *A. crassispina* by trypsin. -, before digestion; +, after digestion.

The 88 K protein did not bind to Sepharose 4B nor to gelatin-Sepharose 4B digested with purified collagenase. This result indicated that the 88 K protein bound to gelatin, but neither to agarose nor to other minor collagenase-insensitive contaminants. The 88 K protein did not react with the antiserum against sea urchin fibronectin using the methods of Ouchterlony and ELISA. Fig. 3 shows that the 88 K protein was susceptible to digestion with trypsin (6 μg/ml, 1h). After digestion, the band of 88 K protein disappeared and was cleaved into low molecular peptides.

To analyze the accumulation of 88 K protein during the early development, fertilized eggs were cultured in filtered sea water containing (^{35}S) methionine for 24 h. Appearance and accumulation of (^{35}S) methionine-labelled 88 K protein were monitored directly with gelatin-Sepharose 4B columns under the same condition, followed by SDS-PAGE (Fig. 4). In parallel with the increase of the incorporation of (^{35}S) methionine into the total protein fraction, the 88 K protein accumulated gradually.

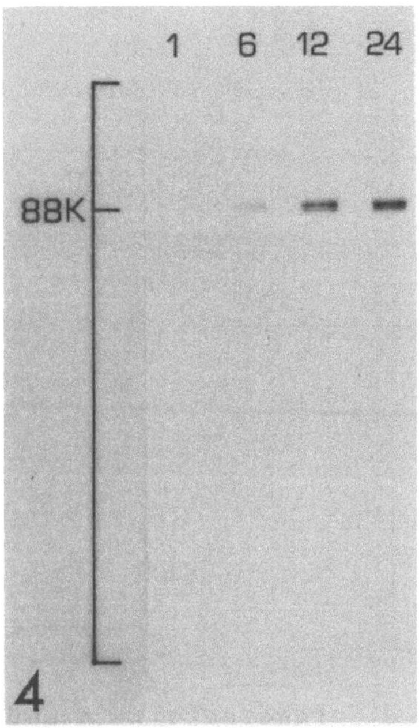

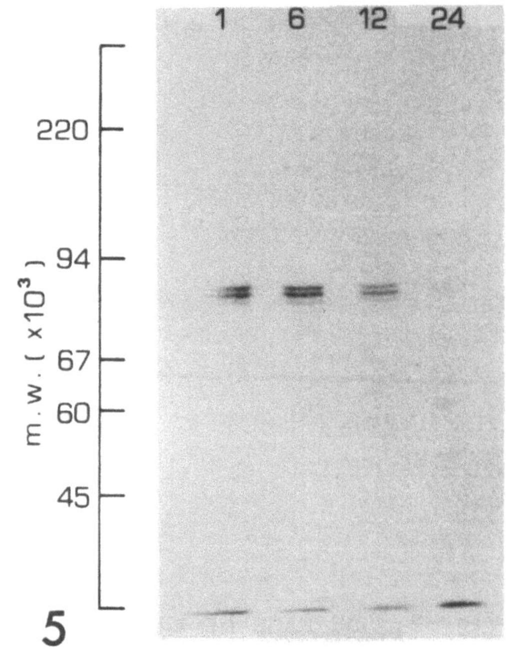

Fig. 4. Accumulation of the 88 K protein in developing embryos. Eggs of *A. crassispina* were labelled with 1 μCi/ml of (^{35} S) methionine from fertilization to the early gastrula stage. The 88 K protein was isolated from the embryos at different developmental stages: 1, 1 h (2-cell); 6, 6 h (morula); 12, 12 h (blastula) and 24, 24 h (early gastrula).

Fig. 5. Pulse labelling of the 88 K protein of *A. crassispina*. Embryos at different developmental stages were labelled with 5 μCi/ml of (^{35}S) methionine for 3 h. 1, 2-cell; 6, morula; 12, blastula and 24, early gastrula.

The rate of synthesis of 88 K protein was studied by pulse labelling, using (^{35}S) methionine at different developmental stages. Fig. 5 shows that the three subunits of 88 K protein were synthesized from the cleavage stage (1 - 4 h) to the blastula stage (12 - 15 h). Among the three subunits, the β-subunit was synthesized more intensely than the other two subunits throughout the stages examined. It should be noted that the bands of 88 K protein were not detected at the gastrula stage (24 - 27 h), though the total protein synthesis increased at this stage. When fertilized eggs were incubated with (^{35}S) methionine for 1 h immediately after fertilization, incorporation into the 88 K protein was already detected on the fluorogram, suggesting that the synthesis of 88 K protein may start within 1 h after fertilization.

These results may indicate that there is a storage of messenger RNA for the 88 K protein in the unfertilized eggs, which is translated after fertilization. As development proceeds, the synhtesis of 88 K protein ceases, while fibronectin begins to be synthesized.

Localization of fibronectin in sea urchin embryos examined by using indirect immunofluorescence with an antibody raised against sea urchin fibronectin is quite different from that of other collagen-binding proteins (Iwata and Nakano 1985). At the gastrula stage, fibronectin is localized in the primary mesenchyme cells and their adjacent regions and in the cells surrounding the blastopore at the vegetal pole. The restricted localization of fibronectin on the primary mesenchyme cell surface during migration was also reported in the sea urchin embryo (Katow et al. 1982). On the other hand, echinonectin (116 kD collagen-binding protein from *Lytechinus variegatus*) was shown by indirect immunofluorescence throughout the cytoplasm of unfertilized eggs (Alliegro et al. 1988). After fertilization, it is transported to the cell surface of ectodermal cells during gastrulation. Some fluorescence was found in the inner surface layer of archenteron. Similar localization of 120 kD collagen-binding protein was also reported in the eggs and embryos of *Paracentrotus lividus* (Di Ferro et al. 1989).

Another aspect of the role of fibronectin in sea urchin development is the induction of spicule formation. When the isolated micromeres of sea urchin embryos are cultured in sea water containing sea urchin fibronectin (0.5 mg/ml), the spicule formation is enhanced. Some spicules grow to triradiate forms, while others remain as granular or rod-like forms. The spicule fomration increases as a function of the concentration of fibronectin in a culture medium. Spicules are not formed at a concentration less than 0.1 mg/ml of fibronectin (Miyachi et al. 1984).

In conclusion, collagen-binding proteins, including fibronectin, exist in sea urchin embryos and some of them are synthesized during the early development. Comparison with genes of collagen-binding proteins in the embryos among different sea urchin species should be carried out. Further studies on collagen-binding proteins may provide an approach to analyzing the molecular basis of morphogenesis in sea urchin embryos.

References

Alliegro MC, Ettensohn CA, Burdsal CA, Erickson HP, Mc Clay DR (1988) Echinonectin: A new embryonic substrate adhesion protein. J Cell Biol 107: 2319-2327

Di Ferro D, Zito F, Cervello M, Matranga V (1989) A new collagen binding protein mediating cell-to-substratum adhesion. Cell Differ Develop 27 (Suppl): 82

Iwata M, Nakano E (1981) Fibronectin from the ovary of the sea urchin, *Pseudocentrotus depressus*. Wilhelm Roux's Arch 190: 83-86

Iwata M, Nakano E (1983) Characterization of sea-urchin fibronectin. Biochem J 215: 205-208

Iwata M, Nakano E (1984) Cell-to-substratum adhesion of dissociated embryonic cells of the sea urchin, *Pseudocentrotus depressus*. Wilhem Roux's Arch 193: 71-77

Iwata M, Nakano E (1985) Fibronectin-binding acid polysaccharide in the sea urchin embryo. Wilhelm Roux's Arch 194: 377-384

Katow H, Yamada K M, Solursh M (1982) Occurrence of fibronectin on the primary mesenchyme cell surface during migration in the sea urchin embryo. Differ 22: 120-124

Miyachi Y, Iwata M, Sato H, Nakano E (1984) Effect of fibronectin on cultured cells derived from isolated micromeres of the sea urchin, *Hemicentrotus pulcherrimus*.Zool Sci: 265 -271

Nakano E, Iwata M (1982) Collagen-binding proteins in sea urchin eggs and embryos. Cell Differ 11: 339-340

EVIDENCE WHICH SUPPORTS THE PRESENCE OF STORED MESSENGER RIBONUCLEOPROTEIN (mRNP) IN THE UNFERTILISED EGGS OF *FUCUS SERRATUS*

Alistair M. Hetherington; John Sommerville*; Andrew K. Masters and Andrew G. Mitchell

Division of Biological Sciences
Institute of Environmental and Biological Sciences
Lancaster University
Lancaster, LA1 4YQ. UK

*Department of Biology
University of St Andrews
St Andrews
Fife
Scotland UK

NATO ASI Series, Vol. H 45
Mechanism of Fertilization
Edited by B. Dale

Introduction

A considerable body of information is available concerning the physiology of fertilisation and early zygotic development in the Fucaceae (Quatrano *et al.* 1985). In contrast the events leading up to fertilisation, especially the biochemistry and molecular biology of oogonial development has received much less attention. A number of factors may have contributed to this, including the difficulty in isolating oogonia at different stages of development and also the presence of large amounts of polysaccharides and polyphenols which are known to interfere with many biochemical analyses. The present report describes our recent attempts to establish whether Fucus employs a similar strategy to the amphibian Xenopus in regulating translation during its early development.

During development Xenopus oocytes synthesise large quantities of mRNA. Only a small fraction of this mRNA is translated during oogenesis, the remainder becomes associated with protein to form complexes known as messenger ribonucleoprotein (mRNP) particles. In this form the mRNA is unavailable for translation. Derepression of translation occurs after fertilisation and the (maternal) mRNA gene transcripts are used to support the first 12 rapid synchronous cell divisions which occur from cleavage to mid-blastula. It is only after the mid-blastula stage that the embryo begins to produce noticeable quantities of zygotic mRNA and to become less dependent upon maternal transcripts (reviewed in Davidson 1986).

It was a specific object of this study to investigate whether a similar strategy is employed during early development of Fucus zygotes. This preliminary study is restricted to a biochemical investigation into the occurrence of mRNP in the unfertilised eggs.

Material and methods

Material

Fucus serratus was collected near Plymouth by Dr. C. Brownlee (Marine Biological Association of the UK). Oogonia were released into seawater, pooled and stored until required at -80°C.

Extraction and purification of mRNP

A 2 ml (packed cell volume) aliquot of unfertilised eggs was homogenised in

5 ml homogenisation buffer [50 mM HEPES pH 7.4, 30 mM EDTA, 200 units human placental ribonuclease inhibitor (Amersham International)] then, after sonication (8 x 5 sec bursts) was centrifuged (10 min, 11000 xg) and the supernatant removed. This was centrifuged again (6.5 hours, 90000 xg) and the resulting pellet resuspended in loading buffer (LB), [0.2 M NaCl, 10 mM Tris/HCl pH 7.4, 2 mM $MgCl_2$, 5 mM β-mercaptoethanol, 2% glycerol]. This was then loaded onto a pre-equilibriated oligo (dT)-cellulose (0.5 g, Type 7, Pharmacia) column and washed extensively with LB. These washes were pooled and saved. Oligo (dT)-cellulose binding material was eluted with 30% formamide and dialysed against LB for 2 hours. $MnCl_2$ was then added to an aliquot of the oligo (dT)-cellulose binding fraction (final concentration 1 mM) and 50 μCi [γ-^{32}P]-ATP added (Amersham International, specific activity of stock 2000 Ci/m mol). After 30 minutes incubation at room temperature the reaction was terminated by the addition of an equal volume of 2X SDS PAGE sample buffer.

SDS PAGE and Western Blotting

SDS PAGE of fractions from the oligo (dT)-cellulose column and phosphopolypeptides was carried out according to Laemmli (1970). Polypeptides were detected by silver staining, while [^{32}P]-phosphopeptides were detected in dried gels after autoradiography. Polypeptides were transferred to nitrocellulose membranes and probed with an antibody raised to a Xenopus mRNP protein (pp 56) as described in Cummings, Barrett and Sommerville (1989). Immunopositive material was detected with [^{125}I] protein A (Amersham International) and subsequently autoradiographed.

Results and discussion

It was originally intended to make a comparative study of protein synthesis in the unfertilised eggs and zygotes of *Fucus serratus*. With the inclusion of the transcriptional inhibitor actinomysin D in the zygotic studies it would have been possible to determine directly the contribution which maternal transcripts make to the control of early development. However, numerous attempts to investigate *in vivo* protein synthesis using [^{35}S] methionine as a radiolabelled precursor resulted in very low levels of [^{35}S] incorporation into soluble proteins (results not shown). Recent studies by Kropf, Berge and Quatrano (1989) and Kropf, Hopkins and Quatrano (1988) used [^{14}C] sodium carbonate to investigate protein synthesis in Fucus eggs and zygotes.

Interestingly, 80-90% of the proteins synthesised by unfertilised eggs and 1 day old embryos were identical and as the authors point out similar results have been reported for early embryogenesis in amphibian eggs (Brock and Reeves 1978).

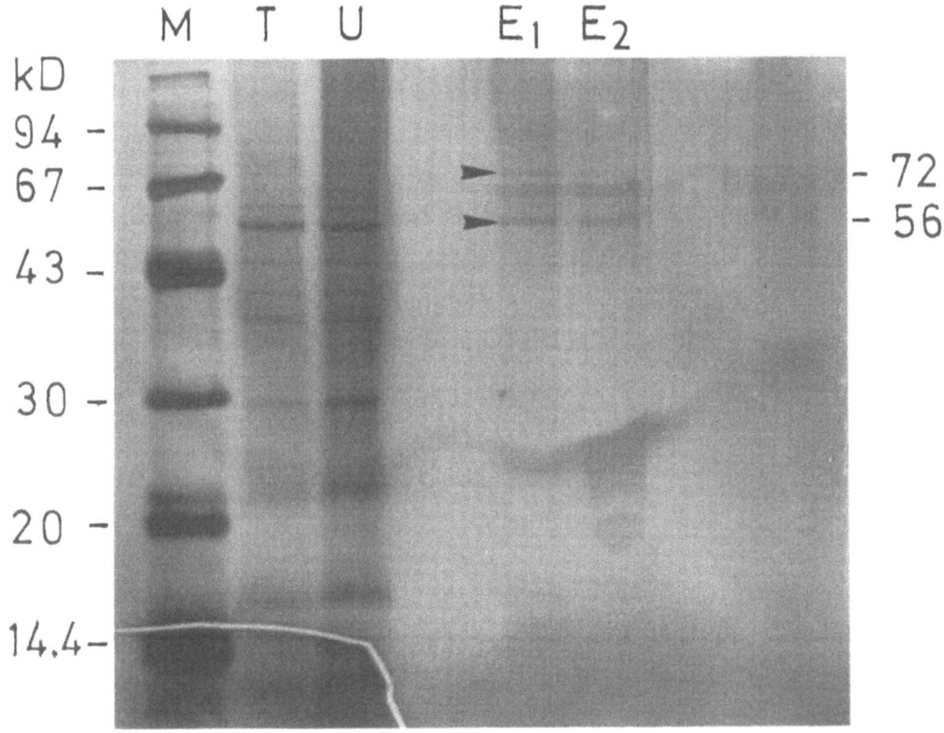

Fig. 1: Silver stained SDS PAGE (12% acrylamide) separation of Fucus proteins pelleting between 10000 and 90000 xg. Tracks:- M = molecular weight markers (Pharmacia); T = total Fucus protein pelleting between 11000 and 90000 xg; U = protein eluting in the void volume from the oligo (dT)-cellulose column; E1 and E2 polypeptides eluting from the oligo (dT)-cellulose column with 20% (E1) and 60% (E2) formamide.

However, with the failure of our *in vivo* protein synthesis studies it was necessary to use an alternative strategy in our investigation of whether maternal transcripts play a major role in the early development of Fucus. Evidence was sought for the presence of mRNP in the unfertilised eggs. A number of approaches were employed. In the first approach oligo (dT)-cellulose chromatography was used to purify mRNA binding proteins. These were then analysed by SDS PAGE. Track T of Fig. 1 shows a PAGE separation of total particle protein sedimenting between 11000 and 90000 xg, while track U is a separation of the material which elutes from the oligo (dT)-cellulose column in the void volume. The fractions separated in tracks E1 & E2 were obtained after washing the column with 20% (track E1) and 60% (track E2) formamide (in LB). These fractions represent the material which bound to oligo (dT)-cellulose. A number of polypeptides are visible in each track after silver staining. Of these it is possible that the 70 kD band (arrowed) represents the mRNA poly (A)- binding protein. From comparative studies in other organisms it is known that this protein is highly conserved in size and in RNA binding sequences (Dreyfuss *et al.* 1989). However further experimental work is required in Fucus before this identification can be any more than tentative. Interestingly, the other arrowed band in tracks E1 & E2 has a similar molecular weight (approximately 56 kD) to a major Xenopus mRNP phosphoprotein which is believed to be involved in the maintenance of the block to mRNA translation (Kick *et al.* 1987; Cummings and Sommerville 1988). This observation prompted an immunological investigation into the possible occurrence of this protein in the unfertilised eggs of Fucus.

In a separate experiment oligo (dT)-cellulose binding material was eluted with 60% formamide (in LB) and separated on SDS PAGE (in parallel with an authentic Xenopus mRNP preparation) and was then transferred to nitrocellulose membranes. The transferred material was then probed using an antibody raised against Xenopus pp56 (Cummings, Barrett and Sommerville, 1989). Immunopositive material was detected with [^{125}I] protein A. Fig. 2 illustrates the results of this experiment, in track 1 as expected the antibody recognised the partially purified Xenopus pp56. A similar although weaker reaction is seen with the Fucus material in track 2. These data provide further evidence to support the presence of mRNP particles in unfertilised Fucus eggs.

It is known that in Xenopus a protein kinase of the casein kinase II type is associated with mRNP (Cummings and Sommerville, 1988). It is also known that it is possible to correlate the activity of this enzyme with stages in development when the repression of mRNA translation is at its maximum, whereas, after fertilisation there is an increase in both phosphatase activity

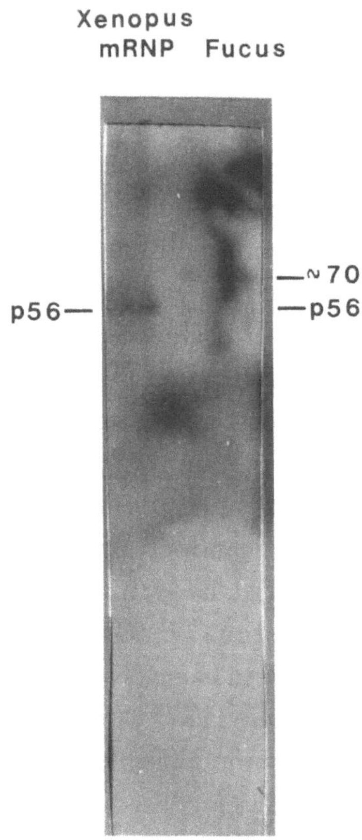

Fig. 2: Western blot of partially purified Xenopus and Fucus mRNP proteins. Proteins were transferred to nitrocellulose membranes and probed with an antibody raised to pp56 (a Xenopus mRNP phosphoprotein). Detection was by [^{125}I] protein A.

and the presence of kinase inhibitors which are associated with the derepression of mRNA translation (Cummings, LaRovere and Sommerville 1990). For this reason it is believed that there is a causal relationship between the block to mRNA translation during oogenesis and the activity of the protein kinase. It is known that higher plants contain a number of protein kinases including casein kinases (reviewed in Hetherington, Battey & Millner 1990). Given the possible central role of this enzyme in the regulation of early development in Xenopus it was decided to determine whether similar activity was present in the Fucus preparation.

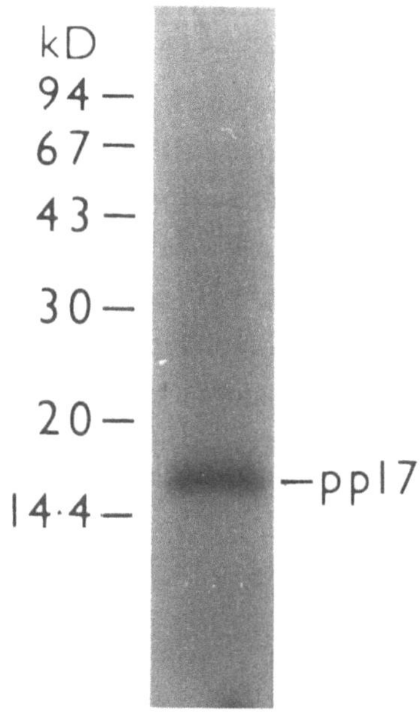

Fig. 3: Autoradiogram of SDS PAGE separation of Fucus mRNP phosphoprotein pp17. Material eluting from the oligo (dT)-cellulose column was incubated in the presence of [γ-^{32}P]-ATP and separated using SDS PAGE. The phosphoprotein was detected using autoradiography.

A fraction which eluted from the oligo (dT)-cellulose column with 60% formamide was incubated in the presence of [γ–^{32}P] ATP for 30 min. The resulting phosphopolypeptides were separated by SDS PAGE. Fig. 3 shows the resulting autoradiogram. It is obvious that a single phosphorylated polypeptide is present with a molecular weight of approximately 17 kD. Although a phosphopolypeptide of this molecular weight has not been reported in Xenopus mRNP it does bear similarities to the subunit of a higher plant calcium dependent membrane bound protein kinase which is known to autophosphorylate (Hetherington and Trewavas 1982; Blowers and Trewavas 1987). Alternatively, the phosphopolypeptide may represent a mRNP component which is a substrate but is not a subunit of the kinase or it is also possible that it results from the proteolytic digestion of a larger protein. However from Fig. 1 it is obvious that this phosphopeptide is not a major component of the Fucus mRNP fraction.

To summarise the results reported in this paper, we have identified polypeptides which bind to an oligo (dT)-cellulose column and are eluted in the same manner as Xenopus mRNP with formamide. Some of these polypeptides share other properties with Xenopus mRNP proteins including recognition by an antibody raised against a Xenopus pp56 and the presence of endogenous protein kinase activity. Future work will concentrate on determining whether the material eluting from the oligo (dT)-cellulose column shares other properties with mRNP such as a similar buoyant density and the presence of associated mRNA. Parallel studies of *in vivo* protein synthesis will attempt to establish whether maternal transcripts have a major role to play in the early development of the Fucus zygote.

Acknowledgements

AMH is grateful to The Marine Biological Association of the UK and the University of Lancaster Research Committee for the award of research grants.

References

Blowers DP and Trewavas AJ (1987) Autophosphorylation of plasma membrane bound calcium calmodulin dependent protein kinase from pea seedlings and modification of catalytic activity by autophosphorylation. Biochim. Biophys. Res. Commun. 143:691-696

Cummings A, Sommerville J (1988) Protein kinase activity associated with stored messenger ribonucleoprotein particles of Xenopus oocytes. J. Cell Biol. 107:45-56

Cummings A, Barrett P and Sommerville J (1989) Multiple modification in the phosphoproteins bound to stored messenger RNA in Xenopus oocytes. Biochim. Biophys. Acta (in press)

Cummings A, LaRovere J and Sommerville J (1990) Expression of messenger RNA-associated protein kinase during the early development of Xenopus. Development (submitted)

Davidson EH (1986) Gene activity in early development. Academic Press, New York

Dreyfuss G, Swanson MS and Pinol-Roma S (1988) Heterogeneous nuclear ribonucleoprotein partciles and the pathway of mRNA formation. Trends Biochem. Sci. 13: 86-91

Hetherington AM and Trewavas A (1982) Calcium dependent protein kinase in pea shoot membranes. FEBS Lett. 145:67-71

Hetherington AM, Battey NH and Millner PA (1990) Protein kinases, in Methods in Plant Biochemistry vol. 7A ed PJ Lea, Academic Press, New York (in press)

Kick D, Barrett P, Cummings A and Sommerville J (1987) Phosphorylation of a 60 kDa polypeptide from Xenopus oocytes blocks messenger RNA translation. Nucleic Acids Res. 15:4099-4109

Kropf DL, Hopkins R and Quatrano RS (1988) Developmental regulation of protein synthesis in Fucus distichus. in Algal Biotechnology, T Stadler, ed. Elsevier New York

Kropf DL, Berge SK and Quatrano RS (1989) Actin localisation during Fucus embryogenesis. The Plant Cell 1:191-200

Laemmli UK (1970) Cleavage of structural proteins during the assembly of the head of bacteriophage T4. Nature 227:680-685

Quatrano RS, Griffing LR, Huber-Walchli V, Doubet RS (1985) Cytological and biochemical requirements for the establishment of a polar cell. J. Cell Sci. Suppl. 1:129-141

DORSAL SPECIFICATION IN THE FERTILIZED FROG EGG

Richard P. Elinson
Department of Zoology
University of Toronto
25 Harbord Street
Toronto M5S 1A1
CANADA

NATO ASI Series, Vol. H 45
Mechanism of Fertilization
Edited by B. Dale

Eggs of most animals have asymmetric distributions of molecules and organelles which are used to specify developmental pathways in the embryo. These asymmetries are enhanced and new ones are created by cytoplasmic movements after fertilization. Development in frogs illustrates these points.

The unfertilized frog egg is radially symmetric around an animal-vegetal axis. The animal half appears to have more active cytoplasm than the yolk-rich vegetal half; nonetheless, developmental information is localized vegetally and moves animally during development. After fertilization, the cortex of the frog egg rotates relative to the cytoplasm forming the grey crescent. This microtubule-dependent rotation changes the symmetry from radial to bilateral by specifying dorsal axial development. Although dorsal specification depends on the rotation, there are factors localized in the frog oocyte required for later formation of dorsal structures. The relation between these factors and the cortical/cytoplasmic rotation is unknown.

Microtubules and the cortical/cytoplasmic rotation

The cortical/cytoplasmic rotation occurs during a phase of the first cell cycle when microtubule polymerization is favored (Elinson, 1985), and the rotation can be blocked by colchicine, colcemid, and nocodazole (Manes *et al*., 1978; Vincent *et al*., 1987). In addition to a role in firming the cytoplasm (Elinson, 1983, 1989), microtubules may also serve as tracks for the rotation of the cortex. Coincident with the rotation, a transient array of parallel microtubules forms in the shear zone between the cortex and cytoplasm (Elinson and Rowning, 1988). The parallel microtubules cover the vegetal hemisphere of the egg and are oriented parallel to the direction of rotation.

While considerable evidence suggests that the parallel microtubules are part of the rotational machinery (Elinson and Rowning, 1988), proof of this function will depend on further characterization of the mechanisms. It is not currently known whether mechanochemical ATPases, such as kinesin, are specifically associated with the parallel microtubules or whether all of the microtubules have the same polarity. The hope is that the mechanism of the cortical/cytoplasmic rotation is not as elusive as the mechanism of chromosome movement in mitosis!

The parallel microtubules can be prevented from forming by irradiating the vegetal halves of fertilized eggs with ultraviolet (UV) light (Elinson and Rowning, 1988). The cortical/cytoplasmic rotation does not occur (Manes and Elinson, 1980; Vincent and Gerhart, 1987), and the eggs develop into radial ventral embryos, lacking all dorsal and anterior structures. Dorsal development of UV-irradiated eggs can be rescued by tilting the eggs off their animal-vegetal axis (Scharf and Gerhart, 1980; Chung and Malacinski, 1980). Under these conditions, the cytoplasm moves relative to the cortex (Vincent and Gerhart, 1987), and a grey crescent forms on the up side (Elinson and Pasceri, 1989).

The rescue by gravity suggests that UV does not destroy an essential dorsal determinant in the fertilized egg, but rather prevents the cytoplasmic movement. This movement, whether natural or induced by gravity, is required for dorsal development.

Dorsal factors localized in the oocyte

The cytoplasmic movement likely activates some component differentially on the dorsal side. The nature of this event or its location in the egg (Elinson and Kao, 1989) is not known, but we have evidence of a localized component in the oocyte, required for later dorsal development (Elinson and Pasceri, 1989).

As with fertilized eggs, UV-irradiation of the vegetal half of full grown oocytes leads to radial ventral embryos (Holwill *et al*. 1987; Elinson and Pasceri, 1989). While the target in the fertilized egg appears to be the parallel microtubules, several lines of evidence suggest that the oocyte target is a factor required for the development of dorsal mesoderm (Elinson and Pasceri, 1989). Unlike UV-irradiated fertilized eggs, UV-irradiated oocytes later form parallel microtubules and undergo the cortical/cytoplasmic rotation after fertilization (Table 1).

Table 1. EFFECTS OF UV-IRRADIATION

	Xenopus prophase I or *Rana* metaphase II oocytes	*Xenopus* or *Rana* fertilized eggs
Dorsoanterior Development	Eliminated	Eliminated
Gravity rescue	No	Yes
Lithium rescue	Yes	Yes
Cortical/Cytoplasmic Rotation	Yes (*Rana*)	No
Parallel microtubules	Yes (*Rana*)	No
UV dose (J/mm^2)	3×10^{-4}	$10\text{-}16 \times 10^{-4}$

UV-irradiation of oocytes dissociates grey crescent formation, which occurs, from dorsal development, which does not occur.

Dorsal development of UV-irradiated oocytes cannot be rescued by tilting them with respect to gravity after fertilization (Table 1). Since tilting moves the cytoplasm, this result supports the idea that failure of dorsal development is due to the inactivation of a dorsal factor rather than to the inhibition of cytoplasmic movement. Dorsal structures do form when 32-cell embryos derived from UV-irradiated oocytes are treated with lithium. Lithium acts on mesoderm induction, causing animal cells to interpret a ventral mesodermal inducing signal as a dorsal one (Slack *et al.*, 1988; Kao and Elinson, 1989; Cooke *et al.*, 1989). Therefore, the UV-irradiated oocytes are able to generate a ventral mesodermal inducer but fail to produce a dorsal mesodermal inducer.

The UV dose required to inactivate the oocyte target is less than one-third the dose required to inactivate the fertilized egg target (Table 1). The oocyte target must be moved or changed during oocyte maturation, or else it would be inactivated when fertilized eggs are UV-irradiated. This is not the case since gravity rescues UV-irradiated fertilized eggs but not UV-irradiated oocytes.

There are several attractive candidates for the oocyte target. These share the properties that they are localized close to the vegetal surface of the full-grown prophase I oocyte and either move from the surface or change during maturation. For instance, both annulate lamellae (Imoh et al., 1983) and a network of intermediate filaments (Klymkowsky and Maynell, 1989) lie near the surface and break down during maturation. The functions of these organelles are not clear, although the possibilty that they have RNA associated with them has been raised (Kessel, 1983; Pondel and King, 1988).

Another candidate is Vg1 mRNA. This maternal mRNA is localized near the vegetal cortex of the full grown oocyte and disperses through the vegetal cytoplasm during oocyte maturation (Melton, 1987; Weeks and Melton, 1987). The dispersal would move most of the Vg1 RNA out of the UV target area. Vg1 mRNA codes for a transforming growth factor (TGF) like protein (Weeks and Melton, 1987), and various TGF-ß's can induce animal cells to form dorsal mesoderm (Kimelman and Kirschner, 1987; Rosa et al., 1988). The regulation of Vg1 mRNA translation and Vg1 protein processing is complex (Dale et al., 1989; Tannahill and Melton, 1989), but dorsoventral differences have yet to be detected. The effect of UV on oocytes could be explained if UV prevents the correct utilization of Vg1 mRNA and if Vg1 protein is indeed a dorsal mesodermal inducer.

A Model for Dorsal Axial Specification

The available data permits several models for dorsal development in frogs. Keeping the behavior of Vg1 mRNA in mind, I will describe one possible model (Figure 1). A UV-sensitive dorsal factor would be localized at the vegetal surface of the prophase I oocyte and would change as it disperses through the vegetal cytoplasm during oocyte maturation. Following fertilization, the cortex rotates relative to the cytoplasm due to the parallel microtubules. As a result, animal and vegetal cytoplasms would be mixed on the prospective dorsal side, and

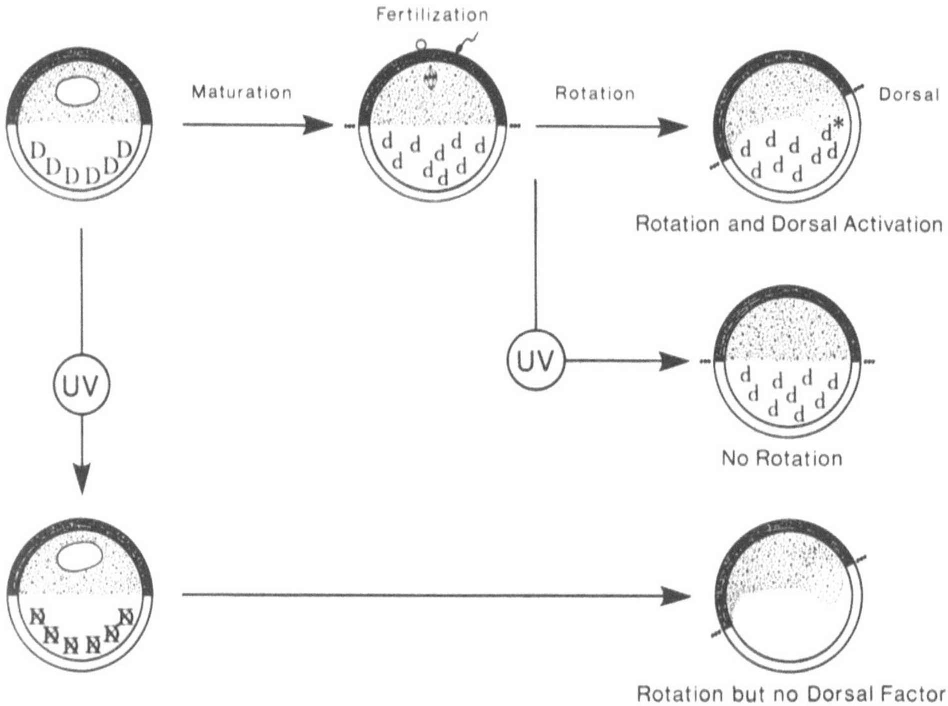

Figure 1. Model of dorsal specification. The prophase I oocyte has a dorsal factor (D) which moves away from the vegetal surface and changes (d) during maturation. The cortical/cytoplasmic rotation mixes animal and vegetal cytoplasm on the dorsal side activating the dorsal factor (d*). UV irradiation of the prophase I oocyte destroys the dorsal factor while UV irradiation of the fertilized egg prevents rotation.

this mixing would cause processing or activation of the dorsal factor. The activated dorsal factor would act like a TGF-ß molecule which would be secreted and which would induce animal cells to become dorsal mesoderm. Once formed, dorsal mesoderm functions as Spemann's organizer in generating the dorsal to ventral spectrum of mesoderm and in inducing the ectoderm to form the central nervous system.

This model presumes that a precursor, capable of becoming a dorsal mesodermal inducer, is found generally in the vegetal half and that the conditions required for its processing arise locally on the dorsal side. The local conditions need not be highly specific, but could be rather general synthetic or secretory pathways.

The Frog and the Chicken

The specification of dorsal axial development in frogs vs. chickens presents a problem. Dorsal axial specification is so fundamental that it defines the vertebrates; yet, this process is different between frogs and chickens. Axis specification in frogs involves an energy-requiring cytoplasmic rearrangement in the one-cell fertilized egg (Vincent et al., 1986), while axis specification in chickens involves a gravity-driven sloughing of cells in the multicellular embryo (Kochav and Eyal-Giladi, 1971; Eyal-Giladi and Fabian, 1980). How can such a fundamental event occur through different pathways?

The model in the previous section provides a common mechanism which can unite the frog and chicken processes. A generally distributed precursor for a dorsal mesodermal inducer would be locally processed via ordinary cellular pathways. These cellular pathways would be more active in one area of the frog egg due to cytoplasmic mixing and in certain chicken cells perhaps due to the change in cell-cell contacts. In both cases, dorsal axial specification requires neither specialized cellular pathways nor localized morphogens; yet a localized response is obtained.

Testing the model for dorsal axial specification and finding the common elements shared by vertebrates are important lines of future research. One path is to identify the oocyte UV target and to see how it is involved in dorsal axial development. Another path is to follow Vg1 protein and to see whether it acts as a dorsal mesodermal inducer. Once the key events of dorsal axial specification are identified in frogs, a search can be made for the same events in fish, birds, and mammals. The results of this search should provide the underlying basis for the conservation of the vertebrate body plan.

LITERATURE REFERENCES

Chung HM, Malacinski GM (1980) Establishment of the dorso/ventral polarity of the amphibian embryo: Use of ultraviolet irradiation and egg rotation as probes. Dev Biol 80:120-133

Cooke J, Symes K, Smith EJ (1989) Potentiation by the lithium ion of morphogenetic responses to a Xenopus inducing factor. Development 105:549-558

Dale L, Matthews G, Tabe L., Colman A (1989) Developmental expression of the protein product of Vg1, a localized maternal mRNA in the frog Xenopus laevis. EMBO J 8:1057-1065

Elinson RP (1983) Cytoplasmic phases in the first cell cycle of the activated frog egg. Dev Biol 100:440-451

Elinson RP (1985) Changes in levels of polymeric tubulin associated with activation and dorso-ventral polarization of the frog egg. Dev Biol 109:224-233

Elinson RP (1989) Microtubules and specification of the dorsoventral axis in frog embryos. BioEssays 11:(in press)

Elinson RP, Kao KR (1989) The location of dorsal information in frog early development. Diff Growth Dev 31:(in press)

Elinson RP, Pasceri P (1989) Two UV-sensitive targets in dorsoanterior specification of frog embryos. Development 106:511-518

Elinson RP, Rowning B (1988) A transient array of parallel microtubules: potential tracks for a cortical/cytoplasmic rotation that forms the grey crescent of frog eggs. Dev Biol 128:185-197

Eyal-Giladi H, Fabian BC (1980) Axis determination in uterine chick blastodiscs under changing spatial positions during the sensitive period for polarity. Dev Biol 77:228-232

Holwill S, Heasman J, Crawley CR, Wylie CC (1987) Axis and germ line deficiencies caused by u.v. irradiation of Xenopus oocytes cultured in vitro. Development 100:735-743

Imoh H, Okamoto M, Eguchi G (1983) Accumulation of annulate lamellae in the subcortical layer during progesterone-induced oocyte maturation in Xenopus laevis. Dev Growth Diff25:1-10

Kao KR, Elinson RP (1989) Dorsalization of mesoderm induction by lithium. Dev Biol 132:81-90

Kessel RG (1983) The structure and function of annulate lamellae: Porous cytoplasmic and intranuclear membranes. Int Rev Cytol 82:181-303

Kimelman D, Kirschner M (1987) Synergistic induction of mesoderm by FGF and TGF-ß and the identification of an mRNA coding for FGF in the early Xenopus embryo. Cell 51:869-877

Klymkowsky MW, Maynell LA (1989) MPF-induced breakdown of cytokeratin filament organization in the maturing Xenopus oocyte depends upon translation of maternal mRNAs. Dev Biol

134:479-485
Kochav S, Eyal-Giladi H (1971) Bilateral symmetry in chick embryo determination by gravity. Science 171:1027-1029
Manes ME, Elinson RP (1980) Ultraviolet light inhibits grey crescent formation on the frog egg. Roux's Arch Dev Biol 189:73-76
Manes ME, Elinson RP, Barbieri FD (1978) Formation of the amphibian egg grey crescent: Effects of colchicine and cytochalasin B. Roux's Arch Dev Biol 185:99-104
Melton DA (1987) Translocation of a localized maternal mRNA to the vegetal pole of Xenopus oocytes. Nature 328:80-82.
Pondel MD, King ML (1988) Localized maternal mRNA related to transforming growth factor ß mRNA is concentrated in a cytokeratin-enriched fraction from Xenopus oocytes. Proc Natl Acad Sci USA 85:7612-7616
Rosa F, Roberts AB, Danielpour D, Dart LL, Sporn MB, Dawid IB (1988) Mesoderm induction in amphibians: the role of TGF-ß2-like factors. Science 239:783-785.
Scharf SR, Gerhart JC (1980) Determination of the dorso-ventral axis in eggs of Xenopus laevis: Complete rescue of UV-impaired eggs by oblique orientation before first cleavage. Dev Biol 79:181-198
Slack JMW, Isaacs HV, Darlington BG (1988) Inductive effects of fibroblast growth factor and lithium ion on Xenopus blastula ectoderm. Development 103:581-590
Tannahill D, Melton DA (1989) Localized synthesis of the Vg1 protein during early Xenopus development. Development 106:775-785
Vincent JP, Gerhart JC (1987) Subcortical rotation in Xenopus eggs: an early step in embryonic axis specification. Dev Biol 123:526-539
Vincent JP, Oster GF, Gerhart JC (1986) Kinematics of gray crescent formation in Xenopus eggs: the displacement of subcortical cytoplasm relative to the egg surface. Dev Biol 113:484-500
Vincent JP, Scharf SR, Gerhart JC (1987) Subcortical rotation in Xenopus eggs: a preliminary study of its mechanochemical basis. Cell Motil Cytoskel 8:143-154
Weeks DL, Melton DA (1987) A maternal mRNA localized to the vegetal hemisphere in Xenopus eggs codes for a growth factor related to TGF-ß. Cell 51:861-868

THE CYTOSKELETON AND POLYSPERMY IN SEA URCHIN EGGS

L. SANTELLA
STAZIONE ZOOLOGICA "A. Dohrn" di Napoli
VILLA COMUNALE
80121 NAPOLI
ITALY

INTRODUCTION

At fertilization the sea urchin egg undergoes a dramatic structural reorganization of its surface. Within the first few seconds there is a depolarization of the egg plasma membrane and this is followed by a transient increase in intracellular Ca^{++}. The elevated Ca induces the exocytosis of cortical granules and the resulting fertilization membrane helps make the egg refractory to further insemination.

At about the same time, an efflux of protons from the egg causes an alkalinization of the cytoplasm (Epel 1978). During this period, the numerous short microvilli that cover the surface of the unfertilized egg elongate (Eddy and Shapiro 1976; Shroeder 1978; Shroeder 1979) and the elongated microvilli are known to contain core bundles of actin filaments. It was suggested that the increase in cytoplasmic pH which accompanies fertilization might cause the polymerization of cortical actin (Hamaguchi and Mabuchi 1988; Spudich and Spudich 1979; Wang and Taylor 1979), which in turn may cause extension of the microvilli.

SURFACE ACTIN POLYMERIZES IN A WAVE

The first event of sperm-egg interaction is a step like conductance event (Dale and de Santis 1981; Longo et al., 1986). The period between this initial interaction of the fertilizing spermatozoon with the egg and the initiation of the calcium wave that leads to egg activation is called the latent period. What happens during the latent period and its significance is not known.

There are no further electrical changes until the initiation of the cortical reaction when a larger conductance event occurs (Dale and de Santis 1981; Longo et al., 1986)

NATO ASI Series, Vol. H 45
Mechanism of Fertilization
Edited by B. Dale

Progression of the cortical reaction may be interrupted by a heat shock and such partially fertilized eggs are viable and may be re-fertilized (Allen and Hagström 1955; Hagström and Runnström 1959; Dale et al., 1989). At the scanning electron microscope, partially fertilized eggs show a fertilization membrane with a smooth surface, limited radially by small blebs of membrane and merging into a micropapillar zone that appears similar to that of the virgin egg. At the transmission electron microscope, outside the zone of activation on the "unactivated" part of the egg towards the antipode, the number and position of cortical granules resembles that of the virgin egg surface. By using the fluorescence staining method of Shroeder and Stricker (1983), we identified in the case of partially fertilized eggs, that the polymerized cortical actin cytoskeleton was in most cases restricted to the area of cortical granule exocytosis, suggesting that polymerization of actin also occurs in a wave starting from the point of sperm fusion and progressing to the antipode (Hamaguchi and Mabuchi, 1988).

Partially fertilized eggs may be refertilized and upon reinsemination they give rise to step depolarizations indicating that spermatozoa may interact and possibly fuse with the "unactivated surface". Although this suggests that the egg surface retains its capacity to fuse with spermatozoa, it is not clear why several spermatozoa are able to interact with the unactivated surface.

CYTOSKELETAL ELEMENTS REGULATE SPERM ENTRY

The sea urchin egg is a good model system for studying the cytoskeletal changes in the cell surface that occur following fertilization and this re-organization of cortical actin has been temporally correlated with the fertilization dependent intracellular Ca^{++} transient and cytoplasmic alkalinization (Begg DA and Rebhun LI, 1979; Carron and Longo 1982).

Many groups studying unfertilized sea urchin eggs have been unable to demonstrate the existence of cortical actin filaments (Spudich and Spudich, 1979; Tilney and Jaffe, 1980; Begg et al., 1982), while more recent work (Sardet, 1984; Kline and Shatten 1986; Hamaguchi and Mabuchi, 1988; Henson and Begg, 1988) has demonstrated the presence of polymerized actin in the unfertilized egg cortex. Henson and Begg (1988) using ultrastructural tecniques and fluorescent phallotoxin staining have shown that F-actin is present in the egg cortex and that it is concentrated within the microvilli.

By using the fluorescence stain rhodamine phalloidin (RdPh), we were not able to detect any significant fluorescence in the virgin sea urchin egg.

Since it is known that temperature affects actin organization we investigated the effect of temperature on the cortical actin system in sea urchin eggs (Santella and Monroy 1989). Eggs kept for 1 hour at 3 °C show a structural reorganization of actin elements, with the fluorescence distributed in rods in the cortex often grouped to give a granular appearance. Furthermore we observed that cold shocked eggs when fertilized at room temperature were polyspermic. This correlation between polyspermy and changes in the re-organization of cytoskeletal components was reversible in a time dependent fashion. That is, when cold shocked eggs were maintained for about 15' minutes at room temperature before staining with rhodamine phalloidin (RhPh), the fluorescence disappeared and the egg surface assumed the image of a control unfertilized egg.
The results showed that cold promotes polyspermy in sea urchin eggs and that this effect is completely reversible following a 7-20 minutes recovery period at room temperature.

CONCLUDING REMARKS

In the past, the notion has been held that eggs are endowed with membrane located mechanisms to repel supernumerary spermatozoa. Several seconds after fertilization eggs from most animals undergo a complete structural reorganization of their surface, called the cortical reaction, and while it is widely accepted that the cortical reaction will prevent the entry of spermatozoa, there is controversy over the existence of a second mechanism hypothesized by Rothschild and Swann (1952) and called the fast partial block.

The present report does not directly address the question of the existence of a fast block. Although work from our laboratory is still preliminary , the correlation between polyspermy and changes in the organization of cortical actin suggests that components of the cytoskeleton may limit or regulate entry of spermatozoa. This leads to an alternative hypothesis as to how sea urchin eggs prevent polyspermy.
It appears from our morphological studies that cold treated eggs are similar to oocytes with long microvilli extending into the jelly layer. The cortical granules, that in a control virgin egg are just beneath the plasma membrane, are seen to be deeper in the cytoplasm, as indeed in oocytes (Dale and Santella, 1985). It is well known that sea urchin oocytes do not possess mechanisms to prevent polyspermy. Oocytes are not capable of a cortical reaction and there is no evidence to suggest they have a fast partial block. An often ignored fact is that polyspermy in sea urchin eggs whether naturally occuring or induced by chemical or physical means, involves the entry of only two, or at maximum three spermatozoa, which usually enter at opposite poles.

This may mean that there are a limited number of sites on the egg surface through which spermatozoa can enter. (Dale, 1985; Dale, 1987). During cytoplasmic maturation there may be a decrease in number of sperm penetration sites, leading to a mature egg with a pre-determined preferential entry site (Boveri, 1901; Runnström, 1961; Dale, 1985). The transition from a polyspermic to a monospermic condition may be then correlated with the physiological state of cortical actin.

REFERENCES

Allen R, Hagström B (1955) Interruption of the cortical reaction by heat. Exp Cell Res 9: 157-167

Begg DA, Rebhurn LI (1979) pH regulates the polymerization of actin in the sea urchin egg cortex. J Cell Biol 83: 241-248

Begg DA, Rebhun LI, Hyatt H (1982) Structural organization of actin in the sea urchin egg cortex: microvillar elongation in tha absence of actin filament bundle formation. J Cell Biol 93: 24-32

Boveri T (1901) Die Polaritat von Ovocyte, Ei und Larvae des Strongylocentrotus lividus. Zool Jb (Morph) 14: 630-651

Carron CP, Longo FJ (1982) Relation of cytoplasmic alkalinization to microvillar elongation and microfilament formation in the sea urchin egg. Dev Biol 89: 128-137

Cline CA, Shatten G (1986) Microfilaments during sea urchin fertilization: Fluorescence detection with rhodaminyl phalloidin. Gamete Res 14: 277-291

Dale B (1985) Sperm receptivity in sea urchin oocytes and eggs. J Exp Biol 118: 85-97

Dale B (1987) Mechanism of fertilization Nature Lond 325: 762-763

Dale B, De Santis A (1981) Maturation and fertilization of the sea urchin oocyte: an electrophysiological study. Develop Biol 85: 474-484

Dale B, Santella L (1985) Sperm-oocyte interaction in the sea urchin. J Cell Sci 74: 153-167

Dale B, Hagström B, Santella L (1989) Partially fertilized sea urchin eggs: an electrophysiological and morphological study. Develop Growth & Differ 31: (2) 165-170

Eddy EM, Shapiro BM (1976) Changes in the topography of the sea urchin after fertilization. J Cell Biol 71: 35-48

Epel D (1978) Mechanism of activation of sperm and egg during fertilization of sea urchin gametes. Curr Top Dev Biol 12: 185-246

Hagström B, Runnström J (1959) Re-fertilization of partially fertilized sea urchin eggs. Protoplasma 87: 281-290

Hamaguchi Y, Mabuchi I (1988) Accumulation of fluorescent labeled actin in the cortical layer in sea urchin eggs after fertilization. Cell Motility & Cytoskeleton 9: 153-163

Henson JH, Begg DA (1988) Filamentous actin organization in the unfertilized sea urchin egg cortex. Dev Biol 127: 338-348

Longo F, Lynn J, McCulloh D, Chambers E (1986) Correlative ultrastructure and electrophysiological study of sperm egg interactions of the sea urchin Lytechinus variegatus. Develop Biol 118: 155-166

Rothschild L, Swann M (1952) The fertilization reaction in the sea urchin. The block to polyspermy. J Exp Biol 29: 469-483

Runnström J (1961) The mechanism of protection of the eggs against polyspermy. Experiments on the sea urchin Paracentrotus lividus. Ark Zool 13: 565-571

Santella L, Monroy A (1989) Cold shock induces actin reorganization and polyspermy in sea urchin eggs. J Exp Zool 252: 183-189

Sardet C (1984) The ultrastructure of the sea urchin eggs cortex isolated before and after fertilization Dev Biol 105: 196-210

Shroeder T (1978) Microvilli on sea urchin eggs: A second burst of elongation. Dev Biol 64: 324-346

Shroeder T (1979) Surface area change at fertilization: resorption of the mosaic membrane. Dev Biol 70: 306-326

Shroeder T, Stricker S (1983) Morphological changes during maturation of starfish oocytes: Surface ultrastructure and cortical actin. Develop Biol 98: 373-384

Spudich A, Spudich J (1979) Actin in triton-treated cortical preparations of unfertilized and fertilized sea urchin eggs. J Cell Biol 82: 212-226

Tilney LG, Jaffe LA (1980) Actin, microvilli and the fertilization cone of sea urchin eggs. J Cell Biol 87: 771-782

Wang Y, Taylor D (1979) Distribution of fluorescent labeled actin in living sea urchin eggs during early development. J Cell Biol 82: 672-679

CELL DIFFERENTIATION IN THE MOUSE PREIMPLANTATION EMBRYO

Tom P. Fleming
Department of Biology
Medical & Biological Sciences Building
University of Southampton
Southampton SO9 3TU
United Kingdom

NATO ASI Series, Vol. H 45
Mechanism of Fertilization
Edited by B. Dale

INTRODUCTION

Early embryonic development in eutherian mammals proceeds from fertilisation and the reinitiation of the cell cycle to a developmental programme in which cell differentiation and tissue segregation occur concurrently in a coordinated process culminating in the formation of a blastocyst. This programme has been studied most extensively in the mouse (Fig. 1), where the origin, character and fate of the two phenotypes of the blastocyst have been analysed in some detail. These comprise the outer trophectoderm epithelium composed of polarised cells (apical membranes facing outwards) linked by tight junctions and cell-cell adhesion systems, and the inner cell mass (ICM), an internal, eccentrically-placed cluster of relatively undifferentiated, non-epithelial cells from which will derive the entire foetus. The trophectoderm engages in vectorial fluid transport, driven by NaK-ATPase localised at basolateral membranes, to generate the blastocoele (reviewed in Wiley, 1988) that is maintained by the tight junction permeability barrier and lies adjacent to the ICM. After trophectoderm-uterine attachment and blastocyst implantation, the trophectoderm gives rise to most of the placental lineages of the developing conceptus, following a proliferative stimulus induced by the ICM (reviewed in Gardner, 1983). Blastocyst morphogenesis and foetal survival during pregnancy therefore depend upon the earliest developmental events regulating trophectoderm delamination and the segregation of the ICM. Firstly, this paper will summarise briefly our current understanding of these processes since they have been reviewed in detail elsewhere (Johnson & Maro, 1986; Johnson et al, 1986a; Fleming & Johnson, 1988; Pratt, 1989) and secondly will consider more thoroughly one feature of epithelial development, tight junction formation, that has recently been examined using antibody probes. Such an approach is instructive both for understanding how different epithelial features integrate during differentiation and how down-regulation of epithelial marker proteins might be achieved in the ICM lineage.

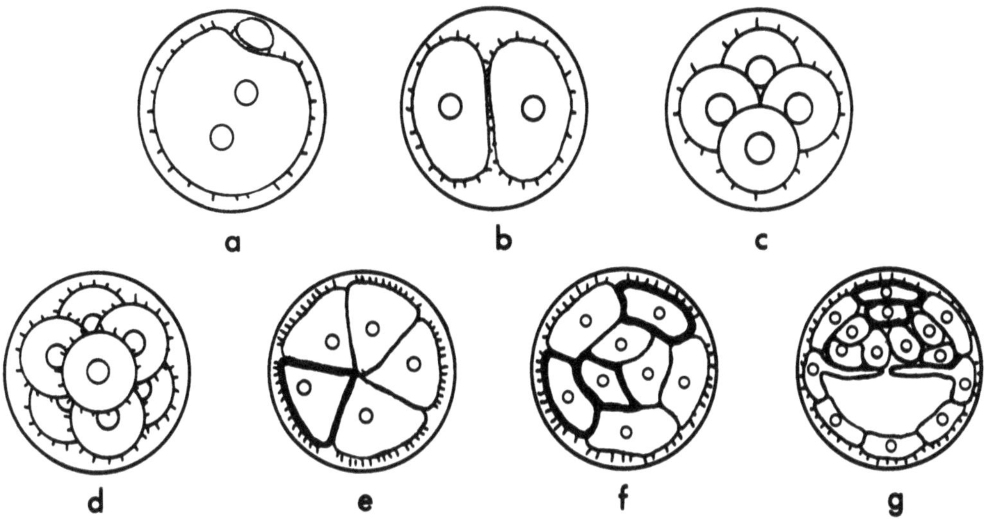

Figure 1. Diagram of early post-fertilization development in the mouse. Embryos are shown surrounded by the zona pellucida. Cell cycles are approx 12h duration except the zygote (18-22h) and 2-cell stage (18h). (a) Two pronuclear zygote, 12-18h after sperm entry. (b) 2-cell stage. (c) 4-cell stage. (d) Early 8-cell stage. (e) Later 8-cell stage following compaction; blastomeres are polarised with microvilli restricted to the outer apical membrane while cell adhesion occurs at contact sites. (f) 16-cell morula containing outer polar and inner nonpolar blastomeres. (g) Early blastocyst (32- to 64-cell stage) containing outer trophectoderm and inner cell mass; trophectoderm processes separate the ICM from the blastocoel. Bold cell outlines in e-g illustrate the role of differentiative divisions in generating the distinct phenotypes of the blastocyst.

TROPHECTODERM DIFFERENTIATION

Early embryonic cells only gradually acquire their differentiative features and it is not until the 8-cell stage at compaction (when blastomere outlines become indistinct due to intercellular flattening, Fig. 1e) that the process of trophectoderm biogenesis begins. Up until this stage, blastomeres are fully totipotent (Kelly, 1977) and display a relatively symmetrical, nonpolar organisation of cellular substructure, except at immediate points of cell contact where local changes in cortical morphology occur but which are lost upon cell repositioning (reviewed in Johnson et al, 1986a). At compaction, surface and cytoplasmic components become radically reorganised so that each blastomere acquires a stably polarised phenotype with the apicobasal axis of polarity running from the outer, contact-free membrane to the median site of cell apposition; for the whole embryo, cell polarity is radially arranged. The polarised state of 8-cell blastomeres is typically epithelial: the apical cell surface is microvillous and non-adhesive while basolateral membranes have few microvilli and progressively sequester, by homophilic cell-cell association, the calcium-dependent cell adhesion glycoprotein uvomorulin (E-cadherin), responsible for cell flattening at this stage (Ducibella et al, 1977; Handyside, 1980; Hyafil et al, 1980; Reeve & Ziomek, 1981; Johnson et al, 1986b; Vestweber et al, 1987). This polarisation of the cytocortex (membrane plus underlying skeleton) is also evident in the distribution of myosin, spectrin and vinculin (reviewed in Lehtonen et al, 1988), as well as in the emergence of gap (Ducibella et al, 1975; Lo & Gilula, 1979; Goodall & Johnson, 1984) and focal tight junctions (Ducibella & Anderson, 1975; Ducibella et al, 1975; Magnuson et al, 1975; Pratt 1985; Fleming et al, 1989; see later). In the cytoplasm, cytoskeletal elements (actin filaments, microtubules, pericentriolar material) and endocytic organelles also redistribute into a polarised state (Reeve, 1981; Johnson & Maro, 1984; Fleming & Pickering, 1985; Maro et al, 1985; Houliston et al, 1987; Houliston & Maro, 1989).

The expression of cell polarity appears to be temporally regulated at the post-translational level by the lifting of an inhibitory influence provided by an undefined compaction-restraining factor (Levy et al, 1986). Once this putative 'permissive' state is reached, the global restructuring of blastomeres is organised spatially in response to cell contact patterns (Ziomek & Johnson, 1980; Johnson & Ziomek, 1981a). A number of studies, using different approaches, have attempted to unravel the interrelationships between different events at compaction, thereby to identify a causal sequence leading to phenotypic change (see reviews cited in Introduction). These studies have indicated that (i) cytoplasmic reorganisation occurs in response to the asymmetry established in the cytocortex, as assayed by microvillous distribution (Johnson & Maro, 1985; Fleming et al, 1986; Wiley & Obasaju, 1988); (ii) microvillous cytocortical asymmetry can occur by trivial means due to the 'permissive' state of 8-cell blastomeres, but, under normal conditions, its expression is accelerated and its axis oriented by uvomorulin cell-cell adhesion (Johnson et al, 1986b); (iii) the instructive uvomorulin contact signal is likely to be transmitted by a non-cytoskeletal route, such as involving ion currents (Nuccitelli & Wiley, 1985) or second messenger molecules (Bloom, 1989) and leads to a stably polarised cytocortex in which the apical domain displays a polymerised organisation (eg, microvillous) and the basolateral contact area is relatively depolymerised (eg, non-microvillous); (iv) the stability of cytocortical polarity spans mitosis (Johnson & Ziomek, 1981b), thereby enabling later epithelial maturation to occur upon the 8-cell template (see below), but the structural basis for this stability is unknown and is unlikely to reside specifically in the apical pole of microvilli itself (Johnson et al, 1988).

The trophectoderm biogenesis programme, although initiated at compaction, does not result in a definitive epithelium until the late 32-cell stage (nascent blastocyst), two cell cycles and approximately 24 hours later. During this period the epithelial phenotype is elaborated only in outside cells; this maturation process contributes to trophectoderm functioning in at least two important respects. First, it provides the appropriate cytoplasmic and cytocortical organisation for acquiring and maintaining membrane domain polarity that is in turn necessary for vectorial fluid transport. Restriction of NaK-ATPase to basolateral membranes occurs at the time of cavitation (approx late 32-cell stage; Watson & Kidder, 1988) and is preceded during the 16-cell morula stage by the lateral spreading of tight junctions from focal sites, established at compaction, to a zonular ring around each polarised cell (Ducibella et al, 1975; Magnuson et al, 1977; see later). The physical segregation of membrane domains coincides with the cytoskeleton-mediated redistribution of mitochondria to basolateral sites where they appear to support localised NaK-ATPase activity (reviewed in Wiley, 1988). Other maturation changes occur in the endocytic system including the appearance of mature lysosomes and endocytic processing routes (Fleming & Pickering, 1985; Maro et al, 1985; Fleming & Goodall, 1986) that appear to provide a basis for maintaining membrane domain specificity, despite the potential randomising effect of transcytosis (Fleming, 1986).

A second aspect of the maturation process concerns an increase in the structural rigidity of the emerging epithelium necessary for withstanding the stresses imposed by blastocoele expansion. Thus, extensive assembly of cytokeratin filaments occurs in polar 16-cell blastomeres (Chisholm & Houliston, 1987) while desmosome formation is apparent from cavitation (Jackson et al, 1980), preceded by the synthesis and surface assembly of desmosomal molecular components (A Elsmore, T Fleming, D Garrod, in preparation). In addition, the epithelial adherens junction becomes evident in the late morula (Vestweber et al, 1987).

LINEAGE SEGREGATION IN THE EARLY EMBRYO

Different cell lineage marking techniques applied to cleaving mouse embryos have shown that the ICM lineage originates at the 8- to 16- and the 16- to 32-cell division cycles when an internal population of non-polar 16- and 32-cell blastomeres are generated, enclosed completely by the epithelial-like cells of the trophectoderm lineage (Balakier & Pedersen, 1982; Ziomek & Johnson, 1982; Pedersen et al, 1986; Fleming, 1987). Inside cells at these stages are totipotent and, like 8-cell blastomeres at compaction, will gradually express trophectodermal features (cell polarity, lineage-specific proteins) if manipulated to an outside position for a sufficient period of time (Ziomek & Johnson, 1981; Ziomek et al, 1982; Johnson & Ziomek, 1983). However, in the intact embryo totipotency expression is unusual and inside cells tend to give rise exclusively to ICM cells of the early blastocyst (Fleming, 1987).

The allocation of cells to the embryo interior does not result from migratory activity but rather from the differentiative division of parental cells that have already initiated differentiation towards trophectoderm (Fig. 1; Johnson & Ziomek, 1981b). In these cells, the cleavage plane is orthogonal to the apicobasal axis of polarity such that putative ICM progeny inherit non-microvillous, adhesive basolateral membrane and associated cytoplasm while the entire microvillous pole, together with the stabilising focus for maintaining cell polarity (see earlier), is segregated into sister cells that retain and subsequently extend their epithelial characteristics and remain on the embryo surface. The proportion of polarised 8-cell blastomeres dividing differentiatively is variable (usually 2-7, mean 5; Fleming, 1987), the remainder, in which the cleavage plane lies parallel to the axis of polarity, generating phenotypically identical progeny (conservative division), both of which occupy an outside position following a shared inheritance of the apical cytocortex (Fig. 1e-g; Johnson & Ziomek, 1981b).

The orientation of the cleavage plane in 8-cell blastomeres is influenced indirectly by cell interactions that take place, not at the time of imminent cell division, but earlier, at compaction, when the relative extent of cell flattening defines the relative size of the microvillous pole; cells with a smaller polar area tend to divide differentiatively more frequently (Pickering et al, 1988). This relationship between compaction and division orientation also accounts for a well-founded observation that earlier-dividing cells contribute disproportionately to the ICM (Kelly et al, 1978; Surani & Barton, 1984). Recent evidence suggests that older cells will flatten earlier, thereby occupying a deeper position at compaction and generating a smaller polar area (Garbutt et al, 1987).

After division, the appropriate spatial segregation of phenotypes within the 16-cell morula is a reflection of their differing properties. The differential adhesive character of polar cells ensures that their correct orientation and position is maintained with the non-adhesive apical membrane facing outwards and uniformly adhesive nonpolar cells remaining internalised (Ziomek & Johnson, 1981, 1982; Surani & Handyside, 1983). Cellular interactions play a more direct role in lineage allocation at division in the 16-cell morula than they do in the 8-cell embryo. The proportion of polar 16-cell blastomeres dividing differentiatively is regulated and compensates for the variable number of inside cells generated at the fourth cleavage division, so that trophectoderm and ICM cell population sizes in the nascent blastocyst become normalised (Fleming, 1987). At this division cycle, cell shape directly influences cleavage orientation and can be modified both by cell interactions (Johnson & Ziomek, 1983) and the relative number of inside and outside cells present. Cell diversification can therefore be divided into two phases, the first (8- to 16-cell transition) concerned with establishing *qualitative* differences between the two lineages, the second (16- to 32-cell transition) concerned with the *quantitative* regulation of lineage size.

TIGHT JUNCTION FORMATION AND EARLY DEVELOPMENTAL MECHANISMS

The preceding sections illustrate the intimate relationship existing between the dual processes of trophectoderm differentiation and tissue segregation in the early embryo. We are now in a position to assess this coordinated developmental programme at the level of individual proteins and to consider how, for example, cell phenotype and cell position might modify specific protein expression patterns. Such an approach would be particularly informative for proteins that are likely to have a key role in epithelial maturation and, in addition, display tissue specificity. In this context, tight junctions are structures characteristic of epithelia, playing a fundamental role in trophectoderm transport activity and are not present in the early ICM. Recently, protein components of the tight junction have been isolated; one of these, ZO-1, is a 225kD phosphorylated peripheral membrane protein localised cytoplasmically to the integral (and, as yet undefined) component responsible for the branching strand image revealed in freeze-fracture (Stevenson et al, 1986, 1988: Anderson et al, 1987).

ZO-1 monoclonal antibody applied to early mouse embryos identifies a 225kD polypeptide in immunoblots and labels the apical border of cell-cell contact in trophectoderm and polarised morula cells as an uninterrupted line (Fig. 2a; Fleming et al, 1989). ZO-1 is first detectable biochemically from the 4- to 8-cell stage and immunocytochemically as focal points of reactivity at the apical contact zone in compacting 8-cell embryos (Fig. 2b) and cell couplets derived from them (2/8 pairs, Fig. 2c, d). With time, the focal staining pattern extends laterally until the continuous linear image is formed; this temporal maturation correlates closely with the mode of appearance of tight junctions as revealed by ultrastuctural analysis.

The regulation of ZO-1 assembly *de novo* at the 8-cell stage has been studied to identify the significance, if any, of tight junction formation in the complex cellular changes that occur at compaction (Fleming et al, 1989). Since the cytocortex has been implicated a causal role in these changes (see earlier) and the tight junction is a specialised domain of the cytocortex circumscribing its polar axis, it is plausible that the tight junction band may represent a primary structural organisation in the overall polarisation process (for a wider perspective, see Izquierdo, 1977). The assembly results are summarised diagrammatically in Fig. 3. ZO-1 association at the cell membrane is delayed and its distribution randomised when the uvomorulin cell-cell adhesion system has been specifically neutralised; this pattern of perturbation resembles the effect on the generation of microvillous polarity which is also delayed and oriented randomly under these conditions (Johnson et al,

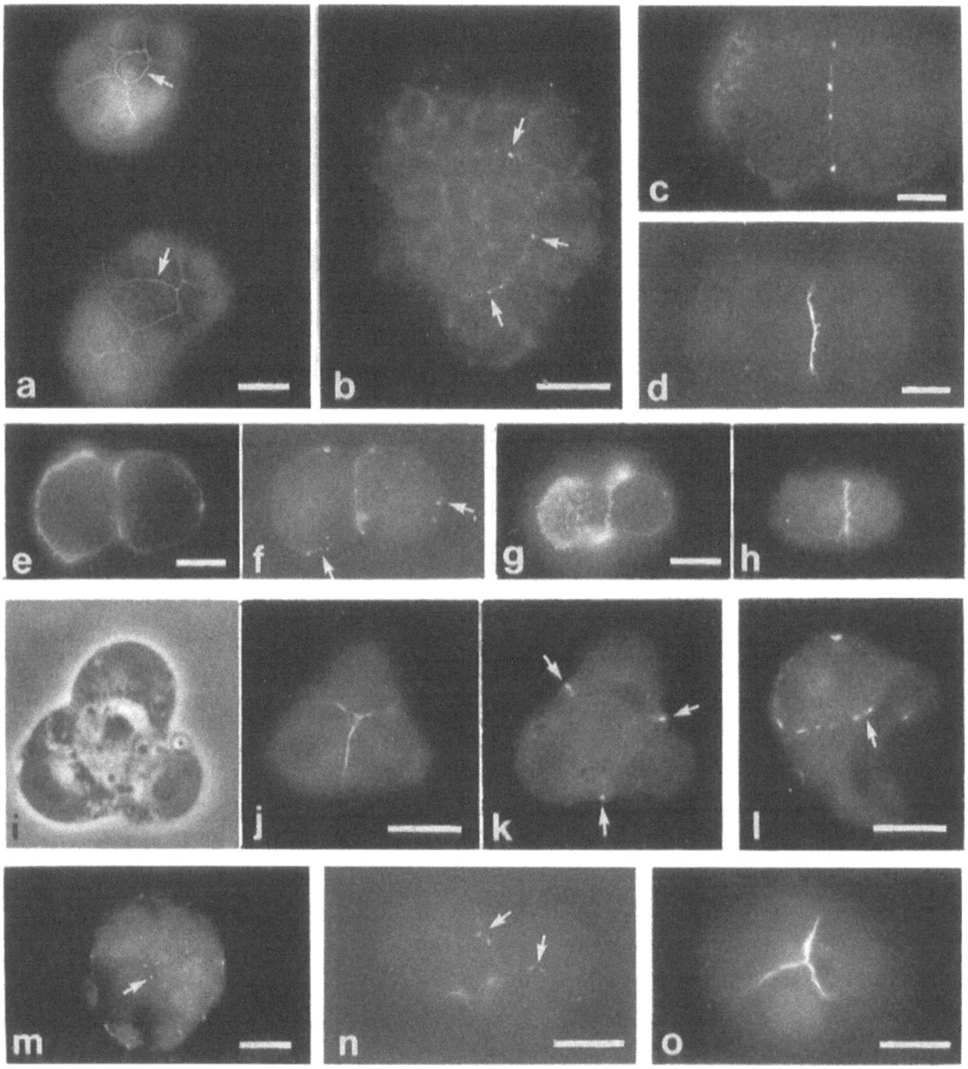

Figure 2. Immunofluorescent staining of ZO-1 in mouse embryos and derivative cell clusters. (a) Blastocysts showing beltlike ZO-1 labelling at the apicolateral cell borders in the trophectoderm. (b) Compacted 8-cell embryo viewed tangentially; discontinuous ZO-1 sites demarcate the apical cell border of the blastomere in the centre of the field. (c) 2/8 couplet 8h post-division from 1/4 blastomere; a series of ZO-1 spots are present at the periphery of the contact zone. (d) 2/8 couplet 10h post-division; a continuous linear ZO-1 pattern is present at the contact zone. (e, f) 2/16 polar:nonpolar couplet 1h post-division from 1/8 cell; the left cell is polar as shown by con A staining of microvilli (e) but both cells show randomly placed ZO-1 spots at the cell surface (f). (g, h) 2/16 polar:nonpolar couplet 5h post-division with left cell polar by con A staining (g); ZO-1 is localised to the periphery of the contact zone between cells (h). (i-k) 4/32 cluster derived from 1/8 cell and

containing three outer polar and one inner nonpolar cells as shown by phase contrast (i); ZO-1 labelling is restricted to the contact zone between outside cells only as shown in tangential (j) and mid-sectional (k) planes. (l) 4/32 cluster showing the rare condition of spot-like ZO-1 sites at the surface of an inside cell. (m) Intact blastocyst showing ZO-1 spots on the surface of certain ICM cells. (n) ICM isolated from early blastocyst and fixed immediately; ZO-1 is evident at the contact zone between some cells. (o) ICM cultured for 6h before fixation; a complete linear ZO-1 assembly pattern is present between outer cells, resembling the intact blastocyst (see a). Bar lines = 10 um except (a, m) = 25 um.

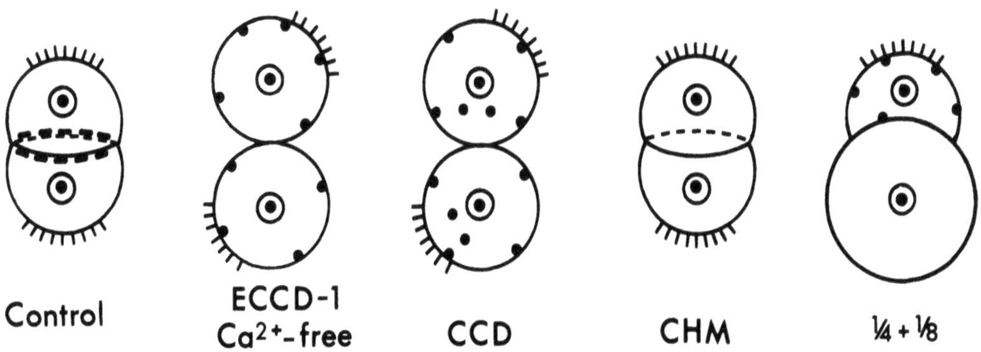

Figure 3. Diagram of typical ZO-1 staining patterns obtained in late 2/8 couplets (and in 1/8 cells combined with 1/4 cells) 9h post-division following various manipulations. ZO-1 sites are shown as solid dots (or discontinuous lines in control) at the cell surface or in the cytoplasm. In control pairs, cell flattening occurs and poles form opposite the contact point and ZO-1 localises at the periphery of the contact zone. Pairs incubated in ECCD-1 or Ca^{++}-free medium from division fail to flatten, generate poles along random axes and assemble ZO-1 at random surface sites. Cytochalasin D (CCD) treatment results in a similar condition except that cytoplasmic ZO-1 sites may also be present. In the presence of cycloheximide (CHM) from division, flattening and surface polarity still occur but ZO-1 assembly is abolished. In 1/8 cells coupled with 1/4 cells from division, flattening and polarity occur normally but ZO-1 assembly is at random surface sites.

1986), indicating that both features of the cytocortex may derive from the uvomorulin signal along a common pathway. However, two experimental situations suggest otherwise. First, protein synthesis inhibition throughout the fourth cell cycle blocks entirely the surface assembly of ZO-1 but fails to prevent the expression of cell adhesion and microvillous polarisation (Fig. 3). Second, in asynchronous aggregates of 1/4 and 1/8 cells (the former lacking ZO-1 expression), the 1/8 cell adheres to the 1/4 cell and develops a normal microvillous pole opposite the contact zone, but fails to assemble ZO-1 exclusively in this region (Fig. 3). These results suggest that ZO-1 expression is translationally controlled and assembly of the protein at the contact zone relies upon a mutual capacity for ZO-1 assembly in the companion cell, presumably reflecting specific cell-cell molecular interactions (Fleming et al, 1989). Thus, by analysing compaction events at the level of protein biosynthesis and assembly, further evidence of 'hierarchical' relationships become evident, in this case suggesting that tight junction formation occurs secondarily to the establishment of a stably polarised cytocortex that is characterised by apical microvilli and basolateral uvomorulin adhesion molecules.

The mode of expression of ZO-1 during cleavage has also been examined in relation to the molecular processes underlying tissue diversification (T. Fleming, in preparation). Although, during blastocyst formation, definitive tight junctions are exclusive to the trophectoderm lineage, the mechanisms that lead to differential protein expression (as exemplified by tight junction construction) are far from clear. Earlier studies have shown that tissue-specific polypeptide synthesis precedes commitment, is evident in the blastocyst (Van Blerkom et al, 1976), and is first detectable in the morula (Handyside & Johnson, 1978). However, tissue-specificity of protein pools could, in addition, arise from non-equivalent inheritance of preexisting components during cytokinesis (cf. the cytocortical pole). ZO-1 surface distribution randomises during mitosis in 8-cell blastomeres such that in polar:nonpolar 2/16 derivatives of 1/8 cells, the protein is inherited by both daughters. In these constructs, the differential adhesive character of the polar cell results in the gradual envelopment of the nonpolar cell (Johnson & Ziomek, 1983); during this period, ZO-1 progressively localises at the border of the advancing contact zone between both cells (Fig. 2e-h). Thus, the capacity for inside nonpolar 16-cell blastomeres to inherit ZO-1 and to make putative tight junctional contacts with outer polar cells in such couplets argues against differentiative division having a primary role in the segregation of this protein. Similar findings have been obtained for cytokeratin filaments in an analysis of their distribution during differentiative division at the 16-cell stage (Chisholm & Houliston, 1987). Although it is unknown whether these two epithelial components are typical, it is reasonable to propose at this time that the primary importance of inheritance in lineage diversification is the allocation of the stabilised focus of cytocortical asymmetry to the trophectoderm lineage (see earlier) rather than a global segregation of protein species.

How then is tissue-specific protein expression regulated? During the morula and early blastocyst stages, cell position has been shown to be largely responsible for maintaining the differentiation of inside cells towards an ICM fate, both morphologically (Ziomek & Johnson, 1982; Ziomek et al, 1982; Fleming et al, 1984) and biosynthetically (Johnson, 1979). This period coincides with ZO-1 expression in most intact embryos being restricted to outside cell contact points only (eg, Fig. 2a). A similar restriction in ZO-1 distribution is also usually evident in small clusters of 16- and 32-cells (derived in culture from single 4- and 8-cell blastomeres respectively) in which the identification of cell position in unsectioned material is more readily apparent than in whole embryos (Fig. 2i-k). This distribution pattern implies that in nonpolar cells ZO-1 levels resulting from earlier inheritance are gradually run down, providing that their internal position is maintained. It is noteworthy that in a small minority (approx 15%) of intact embryos (Fig. 2m) and isolated cell clusters (Fig. 2l), one or more internal cells display focal ZO-1 sites at the cell surface, a condition entirely consistent with their recent origin from the polar lineage by differentiative division.

The down regulation of ZO-1 in the ICM lineage presumably results from natural turnover in the absence of appreciable biosynthesis. The level at which biosynthetic regulation occurs has been examined by monitoring the regeneraton of trophectoderm features during culture of isolated totipotent ICMs, derived by

immunosurgery from nascent blastocysts (32- to 64-cell stage). In these specimens, surface assembly of ZO-1 at the outermost region of contact between outside cells, occurs rapidly. Within an hour of isolation, spot-like ZO-1 sites demarcate the cell border (Fig. 2n) and within 6 hours a complete zonular network is established (Fig. 2o). The rapid expression and spatial patterning of ZO-1 contrasts with the more protracted time course required for reexposed inside cells to display overt polarisation and commitment to phenotypic transformation (Ziomek & Johnson, 1981; Johnson & Ziomek, 1983; Fleming et al, 1984; A Perry & T Fleming, unpublished). The timing and extent of ZO-1 assembly in isolated ICMs is unchanged in the presence of the transcriptional inhibitor alpha-amanitin, administered prior to immunosurgery. However, cycloheximide treatment significantly retards, but does not completely inhibit, ZO-1 reexpression. These results suggest that ZO-1 appearance in isolated ICMs is not a consequence of new gene expression but is primarily a response to translational activity. Thus, down regulation of
ZO-1 within the ICM lineage *in situ* appears to involve a cessation or reduction in the synthesis of the protein, whilst appropriate mRNA is preserved, possibly by stabilisation or modification processes.

In an earlier study it was shown that during the period of ICM totipotency, the capacity of isolated ICMs to reestablish completely the trophectoderm protein synthetic profile was in fact dependent upon transqiptional activity (Johnson, 1979). In the presence of alpha-amanitin, only those polypeptides that are generated during the initial period following isolation were synthesised. It would appear, therefore, that the ICM lineage engages in differential biosynthetic regulation. Why might this be so? The earliest detectable morphological response in ICMs to isolation is the spreading of outside cells over core cells and the formation of a compacted cellular mass (Fleming et al, 1984; Perry & Fleming unpublished). These cell adhesion events, now shown to include ZO-1 assembly, can be interpreted as an immediate post-transcriptional response to reestablish positional differences between cells that are crucial for maintaining a viable ICM lineage. Once enacted, the process of trophectoderm regeneration, as a specific epithelial type, can proceed at a rate compatible with altered gene expression. The reservation, within the ICM cytoplasm, of templates encoding basic epithelial features is also reasonable in terms of the next phase in its developmental programme, that of delamination of primary endoderm at the interface with the blastocoel.

FUTURE PERSPECTIVES

The use of epithelial marker proteins to unravel early mammalian developmental mechanisms has led to a greater understanding of differentiation and diversification processes. This approach now needs to be extended, at more refined levels of molecular and biochemical analysis, to define with greater precision how cell position and phenotype influence biosynthetic and assembly events. In the absence of a comprehensive range of developmental mutants, a wider use of techniques to neutralise expression of specific proteins (eg, antibody microinjection) will also help identify protein function and interactions with other components during differentiation (Emerson, 1989).

ACKNOWLEDGEMENTS

Part of the work described here was financed by grants from the Medical Research Council to Drs MH Johnson and PR Braude and from The Wellcome Trust to TPF.

REFERENCES

Anderson JM, Stevenson BR, Jesaitis LA, Goodenough DA, Mooseker MS (1987) Characterization of ZO-1, a protein component of the tight junction from mouse liver and Madin-Darby canine kidney cells. J Cell Biol 106:1141-1149

Balakier H, Pedersen RA (1982) Allocation of cells to inner cell mass and trophectoderm lineages in preimplantation mouse embryos. Dev Biol 90:352-362

Bloom TL (1989) The effects of phorbol ester on mouse blastomeres: a role for protein kinase C in compaction? Development 106:159-171

Chisholm JC, Houliston E (1987) Cytokeratin filament assembly in the preimplantation mouse embryo. Development 101:565-582

Ducibella T, Albertini DF, Anderson E, Biggers JD (1975) The preimplantation mammalian embryo:characterization of intercellular junctions and their appearance during development. Dev Biol 45:231-250

Ducibella T, Anderson E (1975) Cell shape and membrane changes in the eight-cell mouse embryo: prerequisites for morphogenesis of the blastocyst. Dev Biol 47:45-58

Ducibella T, Ukena T, Karnovsky M, Anderson E (1977) Changes in cell surface and cortical cytoplasmic organization during early embryogenesis in the preimplantation mouse embryo. J Cell Biol 74:153-167

Emerson JA (1988) Disruption of the cytokeratin filament network in the preimplantation mouse embryo. Development 104:219-234

Fleming TP (1986) Endocytosis and epithelial biogenesis in the mouse early embryo. BioEssays 4:105-109

Fleming TP (1987) A quantitative analysis of cell allocation to trophectoderm and inner cell mass in the mouse blastocyst. Dev Biol 119:520-531

Fleming TP, Cannon P, Pickering SJ (1986) The cytoskeleton, endocytosis and cell polarity in the mouse preimplantation embryo. Dev Biol 113:406-419

Fleming TP, Goodall H (1986) Endocytic traffic in trophectoderm and polarised blastomeres of the mouse preimplantation embryo. Anat Rec 216:490-503

Fleming TP, Johnson MH (1988) From egg to epithelium. Ann Rev Cell Biol 4:459-485

Fleming TP, McConnell J, Johnson MH, Stevenson BR (1989) Development of tight junctions de novo in the mouse early embryo: control of assembly of the tight junction-specific protein, ZO-1. J Cell Biol 108:1407-1418

Fleming TP, Pickering SJ (1985) Maturation and polarization of the endocytotic system in outside blastomeres during mouse preimplantation development. J Embryol exp Morph 89:175-208

Fleming TP, Warren PD, Chisholm JC, Johnson MH (1984) Trophectodermal processes regulate the expression of totipotency within the inner cell mass of the mouse expanding blastocyst. J Embryo exp Morph 84:63-90

Garbutt CL, Johnson MH, George MA (1987) When and how does cell division order influence cell allocation to the inner cell mass of the mouse blastocyst? Development 100:325-332

Gardner RL (1983) Origin and differentiation of extraembryonic tissues in the mouse. Int Rev Exp Path 24:63-133

Goodall H, Johnson, MH (1984) The nature of intercellular coupling within the preimplantation mouse embryo. J Embryol exp Morph 79:53-76
Handyside AH (1980) Distribution of antibody- and lectin-binding sites on dissociated blastomeres of mouse morulae: evidence for polarization at compaction. J Embryol exp Morph 60:99-116
Handyside AH, Johnson MH (1978) Temporal and spatial patterns of the synthesis of tissue-specific polypeptides in the preimplantation mouse embryo. J Embryol exp Morph 44:191-199
Houliston E, Maro B (1989) Posttranslational modification of distinct microtubule subpopulations during cell polarization and differentiation in the mouse preimplantation embryo. J Cell Biol 108:543-551
Houliston E, Pickering SJ, Maro B (1987) Redistribution of microtubules and pericentriolar material during the development of polarity in mouse blastomeres. J Cell Biol 104:1299-1308
Hyafil F, Morello D, Babinet C, Jacob F (1980) A cell surface glycoprotein involved in the compaction of embryonal carcinoma cells and cleavage stage embryos. Cell 21:927-934
Izquierdo L (1977) Cleavage and differentiation. In 'Development in Mammals' vol 2:99-118, Johnson MH (ed) North-Holland Amsterdam
Jackson BW, Grund C, Schmid E, Burki K, Franke WW, Illmensee K (1980) Formation of cytoskeletal elements during mouse embryogenesis. Differentiation 17:161-179
Johnson MH (1979) Molecular differentiation of inside cells and inner cell masses isolated from the preimplantation mouse embryo. J Embryol exp Morph 53:335-344
Johnson MH, Chisholm JC, Fleming TP, Houliston E (1986a) A role for cytoplasmic determinants in the development of the mouse early embryo? J Embryo exp Morph 97(Suppl):97-121
Johnson MH, Maro B (1984) The distribution of cytoplasmic actin in mouse 8-cell blastomeres. J Embryol exp Morph 82:97-117
Johnson MH, Maro B (1985) A dissection of the mechanisms generating and stabilising polarity in mouse 8- and 16-cell blastomeres: the role of cytoskeletal elements. J Embryol exp Morph 90:311-334
Johnson MH, Maro B (1986) Time and space in the mouse early embryo: a cell biological approach to cell diversification. In 'Experimental Approaches to Mammalian Embryonic Development' Rossant J, Pedersen RA (eds) pp35-66 Cambridge University Press New York
Johnson MH, Maro B, Takeichi M (1986b) The role of cell adhesion in the synchronisation and orientation of polarization in 8-cell mouse blastomeres. J Embryol exp Morph 93:239-255
Johnson MH, Pickering SJ, Dhiman A, Radcliffe GS, Maro B (1988) Cytocortical organisation during natural and prolonged mitosis of mouse 8-cell blastomeres. Development 102:143-158
Johnson MH, Ziomek CA (1981a) Induction of polarity in mouse 8-cell blastomeres: specificity, geometry and stability. J Cell Biol 91:303-308
Johnson MH, Ziomek CA (1981b) The foundation of two distinct cell lineages within the mouse morula. Cell 24:71-80
Johnson MH, Ziomek CA (1983) Cell interactions influence the fate of mouse blastomeres undergoing the transition from the 16- to the32-cell stage. Dev Biol 95:211-218

Kelly SJ (1977) Studies of the developmental potential of 4- and 8-cell stage mouse blastomeres. J Exp Zool 200:365-376
Kelly SJ, Mulnard JG, Graham CF (1978) Cell division and cell allocationin early mouse development. J Embryol exp Morph 48:37-51
Lehtonen E, Ordonez G, Reima I (1988) Cytoskeleton in preimplantation mouse development. Cell Differentiation 24:165-178
Levy JB, Johnson MH, Goodall H, Maro B (1986) Control of the timing of compaction: a major developmental transition in mouse early development. J Embryol exp Morph 95:213-237
Lo CW, Gilula NB (1979) Gap junctional communication in the preimplantation mouse embryo. Cell 18:399-409
Magnuson T, Demsey A, Stackpole CW (1977) Characterization of intercellular junctions in the preimplantation mammalian embryo by freeze-fracture and thin-section electron microscopy. Dev Biol 61:252-261
Maro B, Johnson MH, Pickering SJ, Louvard D (1985) Changes in the distribution of membranous organelles during mouse early development. J Embryol exp Morph 90:287-309
Nuccitelli R, Wiley L (1985) Polarity of isolated blastomeres from mouse morulae: detection of transcellular ion currents. Dev Biol 109:452-463
Pedersen RA, Wu K, Balakier H (1986) Origin of the inner cell mass in mouse embryos: cell lineage analysis by microinjection. Dev Biol 117:581-595
Pickering SJ, Maro B, Johnson MH, Skepper JN (1988) The influence of cell contact on the division of mouse 8-cell blastomeres. Development 103;353-363
Pratt HPM (1985) Membrane organization in the preimplantation mouse embryo. J Embryol exp Morph 90:101-121
Pratt HPM (1989) Marking time and making space: chronology and topography in the early mouse embryo. Int Rev Cytol in press
Reeve WJD (1981) Cytoplasmic polarity develops at compaction in rat and mouse embryos. J Embryol exp Morph 62:351-367
Reeve WJD, Ziomek CA (1981) distribution of microvilli on dissociated blastomeres from mouse embryos: evidence for surface polarization at compaction. J Embryol exp Morph 62:339-350
Stevenson BR, Anderson JM, Bullivant S (1988) The epithelial tight junction: structure, function and preliminary biochemical characterization. Mol Cell Biochem 83:129-145
Stevenson BR, Siliciano JD, Mooseker MMS, Goodenough DA (1986) Identification of ZO-1:a high molecular weight polypeptide associated with the tight junction (*zonula occludens*) in a variety of epithelia. J Cell Biol 103:755-766
Surani MAH, Barton SC (1984) Spatial distribution of blastomeres is dependent on cell division order and interactions in mouse morulae. Dev Biol 102:335-343
Surani MAH, Handyside AH (1983) Reassortment of cells according to position in mouse morulae. J Exp Zool 225:505-511
Van Blerkom J, Barton SC, Johnson MH (1976) Molecular differentiation in the preimplantation mouse embryo. Nature 259:319-321
Vestweber D, Gossler A, Boller K, Kemler R (1987) Expression and distribution of cell adhesion molecule uvomorulin in mouse preimplantation embryos. Dev Biol 124:451-456

Watson AJ, Kidder CM (1988) Immunofluorescent assessment of the timing of appearance and cellular distribution of Na/K-ATPase during mouse embryogenesis. Dev Biol 126:80-90
Wiley LM (1988) Trophectoderm: the first epithelium to develop in the mammalian embryo. Scanning Microscopy 2:417-426
Wiley LM, Obasaju MF (1988) Induction of cytoplasmic polarity in heterokaryons of mouse 4-cell-stage blastomeres fused with 8-cell- and 16-cell-stage blastomeres. Dev Biol 130:276-284
Ziomek CA, Johnson MH (1980) Cell surface interaction induces polarization of mouse 8-cell blastomeres at compaction. Cell 21:935-942
Ziomek CA, Johnson MH (1981) Properties of polar and apolar cells from the 16-cell mouse morula. Roux Arch Dev Biol 190:287-296
Ziomek CA, Johnson MH (1982) The roles of phenotype and position in guiding the fate of 16-cell mouse blastomeres. Dev Biol 91:440-447
Ziomek CA, Johnson MH, Handyside AH (1982) The developmental potential of mouse 16-cell blastomeres. J Exp Zool 221:345-355

FERTILIZATION AND SEX DETERMINATION IN THE RHIZOCEPHALA (CIRRIPEDIA, CRUSTACEA)

R. Yanagimachi
Department of Anatomy and Reproductive Biology
University of Hawaii
Honolulu, Hawaii 96822
U.S.A.

Introduction

In the summer of 1985, I visited Dr. Alberto Monroy at the Statione Zoologica di Napoli after attending an international meeting (on mammalian reproduction) held on the island of Capri. A few days after I arrived in Napoli, I called Dr. Monroy from a friend's laboratory at the Universita di Napoli. He, despite my unexpected call and his busy schedule, kindly invited me to come to his laboratory that day. After discussing topics of fertilization for an hour or so, I told him that the Stazione Zoologica di Napoli was a "Mecca" for the study of Rhizocephala. The Rhizocephala are parasitic cirripeds which are not so familiar to most modern biologists. Some readers may recall the parasitic castration of female crabs by a rhizocephalan, Sacculina. Dr. Monroy knew this species very well, but he did not know that the study of Rhizocephala began in Naploli in 1787 by F.Cavolini. Geoffrey Smith of the University of Oxford spent three years in the Statione Zoologica and published a landmark monograph on the biology of Rhizocephala (Smith, 1906). Before I began studying mammalian fertilization, I spent several years studying the life-cycle and sexual organization of the Rhizocephala. As a graduate student at Hokkaido University, Japan in Japan, I dreamed of spending a few months or years in Napoli to study the same species of Rhizocephala Geoffrey Smith used for his study. When I briefly talked to Dr. Monroy about what I found in the Rhizocephala, he was very much interested and wanted to hear more. As I had no pictures or diagrams with me, I asked Dr. Monroy if he could find volume 29 of the "Fauna und Flora des Golfes von Neapel." This book, written by Smith, has many illustrations of the Rhizocephala. Dr. Monroy went to the Library, but could not find the book. Since I knew that I would be coming back to Italy the following year, I promised to give a seminar during my next visit to the Satatione Zoologica. In return, Dr Monroy promised to find the book by that time. Unfortunately, these promises were never kept due to Dr. Monroy's sudden death in 1986. The following is what I wanted to present at the seminar.

NATO ASI Series, Vol. H 45
Mechanism of Fertilization
Edited by B. Dale

Rhizocephalans Reproduce By Cross-Fertilization, Not By Self-Fertilization

Rhizocephalans are a group of parasitic cirripeds found on other crustaceans like crabs, hermit-crabs or shrimps (Fig. 1A and 1B). Well known examples of non-partasitic cirripeds are barnacles which are commonly found on rocks in the intertidal zone (Fig. 1C) and stalked (gooseneck) barnacles which are found on the surface of driftwood (Fig. 1D). These non-parasitic cirripeds are hermaphroditic. A mature animal has both a functional ovary and testes. Although self-fertilization may occur in these animals, cross-fertilization is generally the rule. They live gregariously and wherever adjacent individuals are available, they copulate using a long tubular penis (Barnes, 1980).

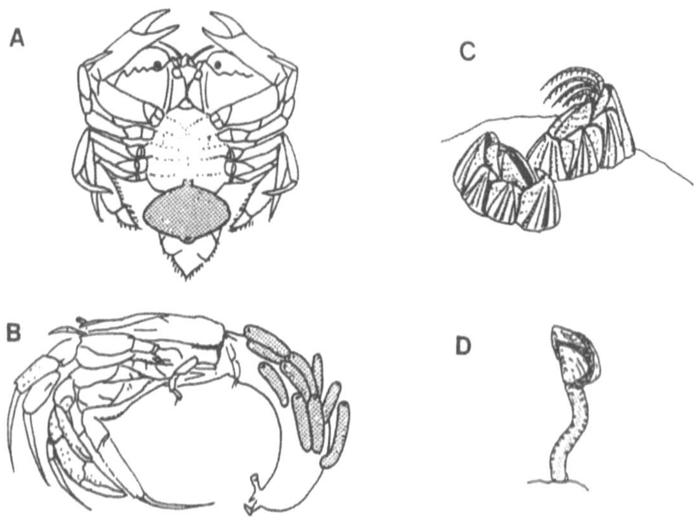

Fig. 1. Cirripeds in various forms. (A), *Sacculina carcini* parasitic to a crab. (B), *Peltogasterella gracilis* (= *P. socialis*) parasitic to a hermit-crab. (C), barnacles on rock in the intertidal zone. (D), gooseneck barnacle on a driftwood.

Before I began to study the Rhizocephala, people thought that all members of this subclass, with a few exceptions, were hermaphrodites which reproduced by self-fertilization. In fact, an adult rhizocephalan possesses both functional ovary and "testis." A mature individual can produce several broods of larvae even if it is separated from other members of the species. Professor Atsuhiko Ichikawa, the major Professor of my graduate study, wanted me to investigate the parasitic castration of male crabs by Rhizocephala. I can now confess that I was not much interested in the project. I was more interested in the Rhizocephala itself. By reading several books and papers (including the book written by by Smith in 1906), I found that rhizocephalans have a very interesting life-cycle (Fig. 2). Eggs are fertilized and develop in the mantle cavity of an adult animal (A). Hatched neuplius larvae (B)leave the mantle cavity and swim the ocean as a plankton without taking in any food (they have no digestive tract). In several days, they transform into cypris larvae (C). A cypris larva that has attached to the skin (carapace) of the host injects a mass of undifferentiated cells into the host (D). The injected cell mass (E) migrates toward the intestine of the host where it begins to form a root-like structure (F) around the intestine. From this root, sac-like structures develop and one or multiple "sacs" emerge from the host immediately after molting of the host. The "sac" (generally called externa) with a root left in the host is the juvenile form (G) of an adult rhizocephalan (A). Histologically, the externa is nothing but an giant ovary surrounded by a muscular capsule called the mantle. A pair of "testes" is found beneath the ovary. It is no wonder that all previous investigators

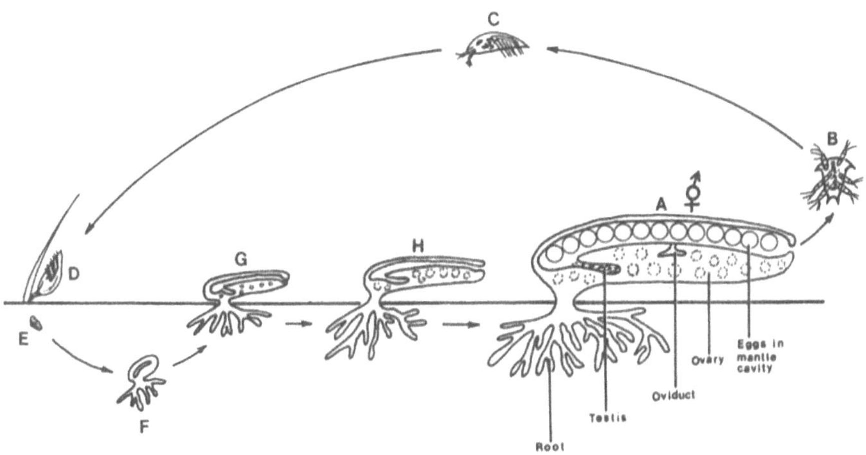

Fig. 2. The concept of the life-cycle of the Rhizocephala, before 1940. The horizontal line represents the integument of the crab host.

thought that rhizocephalans were hermaphroditic animals. Many investigators had noticed the presence of empty shells of cyrpis larvae around the opening of the mantle cavity of young externae which had recently emerged from the host. No one saw the larvae alive, so people thought that they were ordinary larvae that had attached there accidentally and then died (Smith, 1906). It was Reinhard (1942) who discovered that these cypris larvae expell their cellular contents into the mantle cavity of juvenile externa and these cells then migrate into the testis. Reinhard imagined that the spermatozoa of cypris origin mingle with the spermatozoa of the testis and effect fertilization of the first batch of eggs. Thus,

Reinhard maintained the view that rhizocephalans are hermaphrodites and that the cypris larvae in question are complemental males.

I was fortunate to be on the West coast of the Island of Hokkaido, where three species of the Rhizocephala (particularly, Peltogasterella gracilis = P. socialis) were abundant. With an ample supply of specimens, I soon found that the so-called testis was not really a testis, but a "pouch or receptacle" which accommodated the male. When a cypris larva (Fig. 3C) attaches to a juvenile female (G) of Peltogasterella, the larva (D') injects a mass of undiffentiated cells (E') into the mantle cavity of the female. This cellular mass migrates, perhaps by an amoeboid movement, into the "receptacle" where it differentiates into spermatogenic cells. All of the spermatozoa produced in the pouch ("testis") of the adult animal (A) originate from the cyris larva (male) attached to the mantle opening. If only one cypris male succeeds in injecting the cells, only one

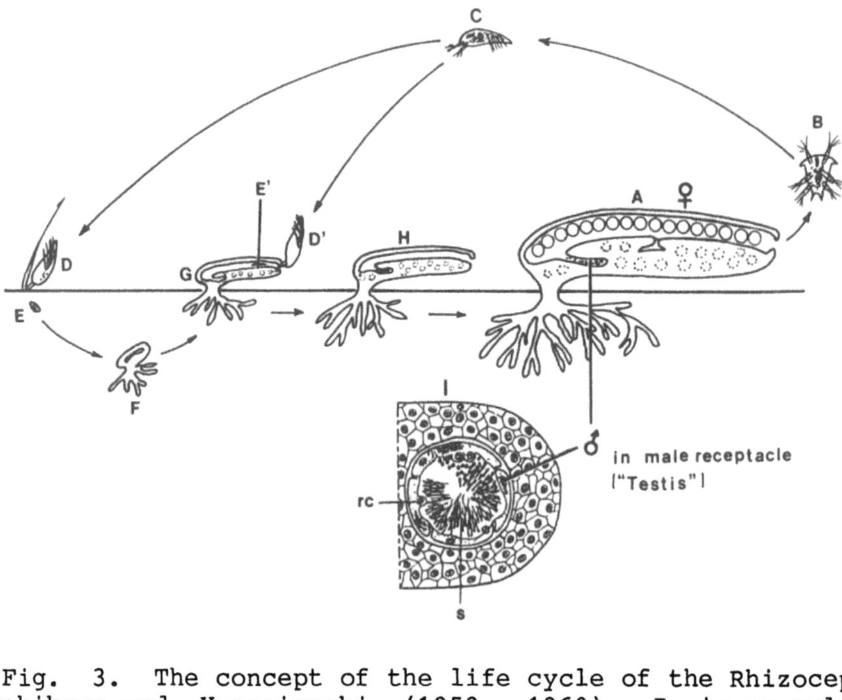

Fig. 3. The concept of the life cycle of the Rhizocephala by Ichikawa and Yanagimachi (1958, 1960). I is an enlarged cross-section of one of cyrpis male receptacles: rc, reserved cypris cells lining the inner wall of the receptacle; s, spermatozoa produced by cypris cells.

"testis" develops to produce spermatozoa. If two males succeed in injection the cells, both the right and left "testes" develop . Thus, the Rhizocephala are not hermaphroditic animals. The adult form we commonly see is a female which is hyper-parasitized by male(s). Males are so simplified that they consist of nothing but spermatogenic cells (I). I first discovered this in Peltogasterella (Ichikawa and Yanagimachi, 1958) and then in Peltogaster and Sacculina (Ichikawa and Yanagimachi, 1960). I did all of the work, but I put Professor Ichikawa's name as senior author in the first two papers because without his interest and encouragement, I would have never started the study of Rhizocephala.

The important point of above discovery is that rhizocephalans are not hermaphrodites. They are diecious animals reproducing by cross-fertilization. There are few species of the Rhizocephala which were thought to be females without males (e.g.,Sylon, Tompsonia and Mycetomorpha). It is now clear that they are all diecious animals. Modified males (cells injected by cypris males) reside within females as hyper-parasites (Reinhard and Evans, 1951; Yanagimachi and Fujimaki, 1967; Lutzen, 1981). Complemental males which were found by Smith (1906) in the mantle cavity of Duplorbis (cf. text-fig.24 of his book) are, in my opinion, true (not complemental) males residing in the mantle cavity of the female.

It is interesting that the juvenile externa (female) does not mature sexually unless it receives cells from cypris males (Ichikawa and Yanagimachi, 1958, 1960; Hoeg and Ritchie, 1985). It is quite possible that the cypris male cells secrete some substance which induces maturation of externa (female). Of the thousands of mature externae of Peltogasterella which I examined, only two did not have cypris male cells in their male receptacles ("testes") (Yanagimachi, unpublished observation). I witnessed their ovulation (discharge of eggs into mantle cavity), but none of the eggs underwent cleavage. These eggs became degenerate and were expelled from the mantle cavity. These two externae (females) died and dropped off the hosts, without ever producing live embryos. I imagine that these two externae (females) recieved cypris male cells and started to grow, but the male cells were either degenerate or expelled from the receptacles (testes) soon after externae (females) started to grow. It is well known that mature females of Cancricepon elegans (a parasitic epicardean isopod) and Bonellia viridis (an echiuran) always carry dwarf males (cf. Barnes, 1980). It is tempting to speculate that these female, like rhizicephalan females, cannot mature sexually unless they are parasitized by males.

In most species of the Rhizocephala, only one externa appears on one host at one time. After repeated release of several or more broods of nauplius larvae, the externa dies and drops from the host, and then a new juvenile externa emerges from the host. Perhaps, the old and new externae originate from the same root. In other words, one female rhizocephala has a chance to "mate" with many different males during its life time. In some species of the Rhizocephala, many externae (as many as one hundred) emerge from the host at a time. These are called gregarious species. Peltogasterella gracilis and Thompsonia japonica, which I used in my study, and Sacculina gregaria are examples. Definitely in Thompsonia (Potts, 1915) and most probably in other gregarious species, all externae originate from a single cypris female attached to the host. The first rizocephalan female that settles into a host seems to prevent infestation of the host by other females of the same and other species, although multiple infestation of a host by more than two rhizocephalan females may occur (cf. Rainbow et al., 1979).

Sex Is Determined During Oogenesis

It was Veillet (1943) who first noticed the presence of two types of cypris larvae in Lernaeodiscus galathea. They were considerably different in size. Veillet speculated that the larger ones develop into hermaphrodites (now we know these are females) and the smaller ones function as larval (cypris) males. According to him, the large cypri develop from large eggs and the small ones from smaller eggs. In other words, Veillet suggested that the sexes of Lernaeodiscus were pre-determined in the egg stage. Following this lead, I measured the size of eggs of Peltogasterella gracilis immediately after ovulation, and indeed, I found large (about 160 µm in diameter) and small eggs (about 145 µm in diameter). Interestingly, large eggs were produced by some females (externae) and smaller eggs by other females. In other words, each female produced either only large or only small eggs. All females (up to about 30) on a single host produced either only large or only small eggs (there were some exceptions to this rule, which will be discussed later). Large eggs developed into large cypris larvae (about 265 µm in body length) and small ones into small cypris larvae (about 239 µm in body length). Examination of settlement behavior of these larvae revealed that large ones were males (which attach to juvenile females) and smaller ones were females (which attach to the host) (Yanagimachi, 1961a). Hoeg (1984) and Walker (1985) confirmed that Sacculina carcini also produce large male and small female larvae. Thus, the sex of rhizocephalans seems to be determined before eggs are fertilized (Yanagimachi, 1961a).

What Determines the Sex of Egg?

As stated above, some of Peltogasterella females (externae) produced only large eggs (prospective males), while others produced only small eggs (prospective females). Examination of chromosomes of large and small eggs during meiosis (Yanagimachi, 1961a) revealed that all of the large eggs each receive 15 chromosomes, whereas small eggs receive either 15 chromosomes or 15 plus one extra chromosome. This extra chromosome will be called tentatively the F chromosome. None of large eggs receive the F chromosome. Spermatogonia have 30 chrosomes and all spermatides (spermatozoa) have 15 chromosomes. No F chromosome is present in spermatogonia or spermatocytes (Yanagimachi, unpublished data). Figure 4 is a diagram of the life cycle of Peltogasterella gracilis which was drawn based on these and previously described findings (cf. Yanagimachi, 1960, 1961a,b). Some (about half) of the adult females (externae) have 30 (2n) chromosomes. These females (Fig. 4A) produce large eggs which develop into large larvae with 30 (2n) chromosomes. They are males which attach to the juvenile females and inject their cellular contents into male-cell receptacles of the juvenile female, which had been previously called testes. The cells that entered the receptacle transform into spermatozoa. Each spermatozoon has 15 (n) chromosomes. Other adult females (A') have 31 (2n + F) chromosomes. They produce small eggs, half of which receiving 15 (n) chromosomes and the rest 16 (n + F) chromosomes. When fertilized by spermatozoa, half of the small fertilized eggs have 30 (2n) chromosomes and the rest 31 (2n + F) chromosomes. Although these two kinds of eggs are different in chromosome constitution, they are morphologically identical and both develop into small larvae. They are female larvae which attach to the hosts and inject their cellular contents into the hosts. A small larva with 30 (2n) chromosomes develops into a male-producing female (A) and the one with 31 (2n + F) chromosomes become a female-producing female (A').

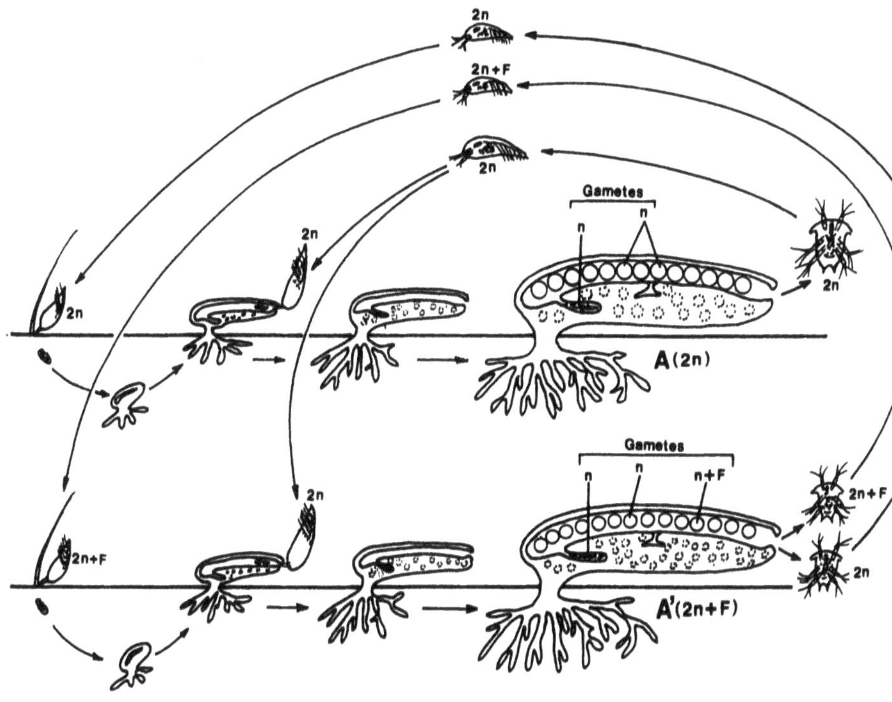

Fig. 4. A proposed mechanism of sex determination in Peltogasterella gracilis (Yanagimachi, 1960, 1961a,b).

Thus, it is the egg cytoplasm, not the chromosomes, that is directly responsible for sex determination in this animal. It is conceivable that some substance synthesized within the ooplasm under the action of F chromosome directs the oocytes to develop into small eggs which are destined to develop into female larvae. The oocytes which have matured without influence from the F chromosome become large eggs which develop into male larvae.

The Unanswered Question

During the course of my study on Peltogasterella gracilis (Yanagimachi, 1961a), I found three "unusual" externae (females) on two different hermit-crabs. Each of these externae produced both large and small eggs which developed into large and small cypris larvae, respectively. Chromosome analyses these eggs suggested that one externa on a hermit-crab had 31 (2n + F) chromsomes, and other two on another hermit-crab had 30 (2n) chromosomes. Thus, these two externae were typical female- and male-producing externae (females), and therefore they should have produced either only small eggs or only large eggs (cf. Fig. 4). Why, did these externae produce both small and large eggs?

Richie and Hoeg (1981) measured the size of larvae released from 21 mature externae of Lernaeodiscus porcellanea. Thirty-three per cent of these externae produced either only large or only small larave. The remaining externae, on the other hand, produced both large and small larvae. The ratio of large and small larvae varied from externa to externa, but usually one of them dominated the other (e.g., <5% large : >95% small). According to Ritchie and Hoeg (1981) who maintained Lernaeodiscus (on host crabs) in their laboratory for about one year, most of the externae kept producing either large or small larvae, as in Peltogasterella. There was one externa that produced only large (male) larvae during first month of captivity (May, 1974). It suddently started to produce a mixture of large and small (female) larvae the next month (June, 1974), and then the following month this externa started to produce only small (female) larvae (July, 1974). After 8 months (from May, 1975), it again started to produce only large (male) larvae. Ritchie and Hoeg found another externa like this. According to Hoeg (1984), many of mature externae of Sacculina carcini produced both large and small larvae simultaneously. Not all the larvae were tested for their sexes by settlement experiments, but at least three Sacculina externae produced both functional male and female larvae (Hoeg, 1984). Walker (1985,1988) found that the majority of larvae produced by mature externae of Sacculina carcini during the winter (November through April) were males. The proportion of female larvae increased in spring (May), and during summer (July and August) the majority of larvae were females. In the fall (September and October), the proportion of female larvae decreased again. Since this rhizocephalan lives in Northen Altantic, low sea water temperature and short days seem to be favourable for the production of male larvae, and higher temperature and long days for female larvae. However, Walker (1987) could not prove this experimentally. Even in the captivity with constant light and constant temperature, the ratio of male/female larvae produced by most of mature externae changed during 8 months of May thorough December. This lead Walker to infer that either the age of the mature externa or the hormone(s) from the host crab governs the sex of the larvae. I may have overlooked the seasonal change in the sex ratio of Peltogasterella larvae, because I carried out my study only during fall months (September through early November) (Yanagimachi, 1961). If I had studied throughout the year, I may have found seasonal changes in the sex ratio of larvae, as in Sacculina.

As stated already, I found two populations of mature Peltogasterella externae, one having 30 chromosomes (2n) and the other having 31 chromosomes (2n + F) (Yanagimachi, 1961a). The former produced large (male) larvae, and the latter produced small (female) larvae. I imagined that genes on F chromosome direct ovarian oocytes to develop into small (female) eggs. Obviously, this hypothesis can not explain why some externae of Peltogasterlla could produce both female and male larvae (Yanagimachi, 1961). One may wonder if chromosomes like F chromosome in Peltogasterella exist in other rhizocephalans. Clearly, further study is needed to elucidate the mechanism of sex determination in the Rhizocephala.

References

Barns RD (1980) Invertebrate Zoology (4th ed). Saunders College, Piladephia.

Hoeg JT (1984) Size and settling behavior in male and female cypris larvae of the parasitic banacle Sacculina carcini Thompson (Crustacea: Cirripedia: Rhizocephala). J Exp Mar Biol Ecol 76:145-156.

Hoeg JT, Rirchie LE (1985) Male cypris settlement and its effects on juvenile development in Lernaeodiscus porcellanea Muller (Crustacea: Cirripeidia: Rhizocephala). J Exp Mar Biol Ecol 87:1-11.

Ichikawa A, Yanagimachi R (1958) Studies on the sexual organization of the Rhizocephala. I. The nature of the "testis" of Peltogasterella socialis Kruger. Annot Zool Japon 31:82-96.

Ichikawa A, Yanagimachi R (1960) Studies on the Sexual organization of the Rhizocephala. II. The reproductive function of the larval (cypris) males of Peltogaster and Sacculina. Annot Zool Japon 33:42-56.

Lutzen J (1981) Observations on the rhizocephalan barnacle Sylon hippolytes M Sars parasitic on the prawn Spirontocaris lillikeborg (Danielssen). J Exp Mar Biol Ecol 50:231-254.

Lutzen J (1984) Growth, reproduction, and life span in Sacculina carcini Thompson (Cirripedia: Rhizocephala) in the isefjord, Denmark. Sarisia 69:91-106.

Potts FA (1915) On the rhizocephalan genus Thompsonia and its relation to the evolution of the group. Pap Dept Mar Biol Carnegie Wash Inst 8:1-32.

Rainbow PS, Ford MP, Hepplewhite I (1979) Absence of gregarious settling behavior by female cyrpis larvae of British parasitic rhizocephalan barnacles. J Mar Biol Assoc UK 59:591-596.

Reinhard EG (1942) The reproductive role of the complemental males of Peltogaster. J Morph 70:389-402.

Reinhard EG, Evans JT (1951) The spermatogenic nature of the "mantle bodies" in the aberrant rhizocephalids, Mycetomorpha. J Morph 89:59-69.

Ritchie LE, Hoeg JT (1981) The life history of Lernaeodiscus porcellanea (Cirripedia: Rhizocephala) and co-evolution with its procellanid host. J Crust Biol 1:334-347.

Smith G (1906) Rhizocephala. Fauna Flora Golfes Neapel 29:1-123. Verlag von R. Friedlander & Sohn, Berlin.

Veillet A (1943) Note sur le dismorphisme des larves de Lernaeodiscus galathea Normann et Scott et sur la nature des "males larvaires" des Rhizpcephales. Bull Inst Oceanograh Monaco 841:1-4.

Walker G (1985) The cypris larvae of Sacculina carcini Thompson (Crustacea: Cirripedia:Rhizocephala). J Exp Mar Biol Ecol 93:131-145.

Walker G (1987) Further studies concerning the sex ratio of the larvae of the parasitic barnacle, Sacculina carnicini Thompson. J Exp Mar Biol Ecol 106:151-163

Walker G (1988) Observations on the larval development of Sacculina carcini (Crustacea: Cirripedia: Rhizocephala). J Mar Biol Assoc U.K. 68:377-390.

Yanagimachi R (1960) The life cycle of Peltogasterella gracilis (Rhizocephala, Cirripedia). Bull Mar Biol Stat Asamushi Tohoku Univ 10:109-110.

Yanagimachi R (1961a) Studies on the sexual organization of the Rhizocephala. III. The mode of sex determination of Peltogasterella. Biol Bull 120:272-283.

Yanagimachi R (1961b) The life cycle of Peltogasterella (Cirripedia,Rhizocehala). Crustaceana 2:183-186.

Yanagimachi R, Fujimaki N (1967) Studies on the sexual organization of the Rhizocephala. IV. On the nature of the "testis" of Thompsonia. Annot Zool Japon 40:98-104.

NATO ASI Series H

Vol. 1: **Biology and Molecular Biology of Plant-Pathogen Interactions.** Edited by J. A. Bailey. 415 pages. 1986.

Vol. 2: **Glial-Neuronal Communication in Development and Regeneration.** Edited by H. H. Althaus and W. Seifert. 865 pages. 1987.

Vol. 3: **Nicotinic Acetylcholine Receptor: Structure and Function.** Edited by A. Maelicke. 489 pages. 1986.

Vol. 4: **Recognition in Microbe-Plant Symbiotic and Pathogenic Interactions.** Edited by B. Lugtenberg. 449 pages. 1986.

Vol. 5: **Mesenchymal-Epithelial Interactions in Neural Development.** Edited by J. R. Wolff, J. Sievers, and M. Berry. 428 pages. 1987.

Vol. 6: **Molecular Mechanisms of Desensitization to Signal Molecules.** Edited by T. M. Konijn, P. J. M. Van Haastert, H. Van der Starre, H. Van der Wel, and M. D. Houslay. 336 pages. 1987.

Vol. 7: **Gangliosides and Modulation of Neuronal Functions.** Edited by H. Rahmann. 647 pages. 1987.

Vol. 8: **Molecular and Cellular Aspects of Erythropoietin and Erythropoiesis.** Edited by I. N. Rich. 460 pages. 1987.

Vol. 9: **Modification of Cell to Cell Signals During Normal and Pathological Aging.** Edited by S. Govoni and F. Battaini. 297 pages. 1987.

Vol. 10: **Plant Hormone Receptors.** Edited by D. Klämbt. 319 pages. 1987.

Vol. 11: **Host-Parasite Cellular and Molecular Interactions in Protozoal Infections.** Edited by K.-P. Chang and D. Snary. 425 pages. 1987.

Vol. 12: **The Cell Surface in Signal Transduction.** Edited by E. Wagner, H. Greppin, and B. Millet. 243 pages. 1987.

Vol. 13: **Toxicology of Pesticides: Experimental, Clinical and Regulatory Perspectives.** Edited by L. G. Costa, C. L. Galli, and S. D. Murphy. 320 pages. 1987.

Vol. 14: **Genetics of Translation. New Approaches.** Edited by M. F. Tuite, M. Picard, and M. Bolotin-Fukuhara. 524 pages. 1988.

Vol. 15: **Photosensitisation. Molecular, Cellular and Medical Aspects.** Edited by G. Moreno, R. H. Pottier, and T. G. Truscott. 521 pages. 1988.

Vol. 16: **Membrane Biogenesis.** Edited by J. A. F. Op den Kamp. 477 pages. 1988.

Vol. 17: **Cell to Cell Signals in Plant, Animal and Microbial Symbiosis.** Edited by S. Scannerini, D. Smith, P. Bonfante-Fasolo, and V. Gianinazzi-Pearson. 414 pages. 1988.

Vol. 18: **Plant Cell Biotechnology.** Edited by M. S. S. Pais, F. Mavituna, and J. M. Novais. 500 pages. 1988.

Vol. 19: **Modulation of Synaptic Transmission and Plasticity in Nervous Systems.** Edited by G. Hertting and H.-C. Spatz. 457 pages. 1988.

Vol. 20: **Amino Acid Availability and Brain Function in Health and Disease.** Edited by G. Huether. 487 pages. 1988.

NATO ASI Series H

Vol. 21: **Cellular and Molecular Basis of Synaptic Transmission.**
Edited by H. Zimmermann. 547 pages. 1988.

Vol. 22: **Neural Development and Regeneration. Cellular and Molecular Aspects.**
Edited by A. Gorio, J. R. Perez-Polo, J. de Vellis, and B. Haber. 711 pages. 1988.

Vol. 23: **The Semiotics of Cellular Communication in the Immune System.**
Edited by E. E. Sercarz, F. Celada, N. A. Mitchison, and T. Tada. 326 pages. 1988.

Vol. 24: **Bacteria, Complement and the Phagocytic Cell.**
Edited by F. C. Cabello und C. Pruzzo. 372 pages. 1988.

Vol. 25: **Nicotinic Acetylcholine Receptors in the Nervous System.**
Edited by F. Clementi, C. Gotti, and E. Sher. 424 pages. 1988.

Vol. 26: **Cell to Cell Signals in Mammalian Development.**
Edited by S. W. de Laat, J. G. Bluemink, and C. L. Mummery. 322 pages. 1989.

Vol. 27: **Phytotoxins and Plant Pathogenesis.**
Edited by A. Graniti, R. D. Durbin, and A. Ballio. 508 pages. 1989.

Vol. 28: **Vascular Wilt Diseases of Plants. Basic Studies and Control.**
Edited by E. C. Tjamos and C. H. Beckman. 590 pages. 1989.

Vol. 29: **Receptors, Membrane Transport and Signal Transduction.**
Edited by A. E. Evangelopoulos, J. P. Changeux, L. Packer, T. G. Sotiroudis, and K. W. A. Wirtz. 387 pages. 1989.

Vol. 30: **Effects of Mineral Dusts on Cells.**
Edited by B. T. Mossman and R. O. Bégin. 470 pages. 1989.

Vol. 31: **Neurobiology of the Inner Retina.**
Edited by R. Weiler and N. N. Osborne. 529 pages. 1989.

Vol. 32: **Molecular Biology of Neuroreceptors and Ion Channels.**
Edited by A. Maelicke. 675 pages. 1989.

Vol. 33: **Regulatory Mechanisms of Neuron to Vessel Communication in Brain.**
Edited by F. Battaini, S. Govoni, M. S. Magnoni, and M. Trabucchi. 416 pages. 1989.

Vol. 34: **Vectors as Tools for the Study of Normal and Abnormal Growth and Differentiation.**
Edited by H. Lother, R. Dernick, and W. Ostertag. 477 pages. 1989.

Vol. 35: **Cell Separation in Plants: Physiology, Biochemistry and Molecular Biology.**
Edited by D. J. Osborne and M. B. Jackson. 449 pages. 1989.

Vol. 36: **Signal Molecules in Plants and Plant-Microbe Interactions.**
Edited by B. J. J. Lugtenberg. 425 pages. 1989.

Vol. 37: **Tin-Based Antitumour Drugs.** Edited by M. Gielen. 226 pages. 1990.

Vol. 38: **The Molecular Biology of Autoimmune Disease.**
Edited by A. G. Demaine, J-P. Banga, and A. M. McGregor. 404 pages. 1990.

Vol. 39: **Chemosensory Information Processing.** Edited by D. Schild. 403 pages. 1990.

Vol. 40: **Dynamics and Biogenesis of Membranes.**
Edited by J. A. F. Op den Kamp. 367 pages. 1990.

Vol. 41: **Recognition and Response in Plant-Virus Interactions.**
Edited by R. S. S. Fraser. 467 pages. 1990.

NATO ASI Series H

Vol. 42: **Biomechanics of Active Movement and Deformation of Cells.**
Edited by N. Akkaş. 524 pages. 1990.

Vol. 43: **Cellular and Molecular Biology of Myelination.**
Edited by G. Jeserich, H. H. Althaus, and T. V. Waehneldt. 565 pages. 1990.

Vol. 44: **Activation and Desensitization of Transducing Pathways.**
Edited by T. M. Konijn, M. D. Houslay, and P. J. M. Van Haastert. 336 pages. 1990.

Vol. 45: **Mechanism of Fertilization: Plants to Humans.**
Edited by B. Dale. 710 pages. 1990.

MIX
Papier aus verantwortungsvollen Quellen
Paper from responsible sources
FSC® C105338

If you have any concerns about our products,
you can contact us on
ProductSafety@springernature.com

In case Publisher is established outside the EU,
the EU authorized representative is:
Springer Nature Customer Service Center GmbH
Europaplatz 3, 69115 Heidelberg, Germany

Printed by Libri Plureos GmbH
in Hamburg, Germany